Partial differential equations

Partial differential equations

J. WLOKA

University of Kiel

TRANSLATED BY C.B AND M.J. THOMAS

CAMBRIDGE
UNIVERSITY PRESS

Published by the Press Syndicate of the University of Cambridge
The Pitt Building, Trumpington Street, Cambridge CB2 1RP
40 West 20th Street, New York, NY 10011-4211, USA
10 Stamford Road, Oakleigh, Victoria 3166, Australia

Originally published in German as *Partielle Differentialgleichungen*
by B.G. Teubner, Stuttgart, 1982 and © B.G. Teubner, Stuttgart 1982

First published in English by Cambridge University Press 1987
as *Partial differential equations*

Reprinted 1992

British Library cataloguing in publication data

Wloka, J.
Partial differential equations

1. Differential equations, Partial
I. Title II. Partielle Differentialgleichungen. *English*
515.3′53 QA374

Library of Congress cataloguing in publication data

Wloka, Joseph.
Partial differential equations.
Translation of: Partielle Differentialgleichungen.
Includes index.
1. Differential equations, Partial. I. Title.
QA374.W5613 1987 515.3′53 86-18817

ISBN 0 521 25914 2 hardback
ISBN 0 521 27759 0 paperback

Transferred to digital printing 2002

TM

Contents

Preface ix

I Sobolev spaces **1**

§1 Notation, basic properties, distributions **1**

1.1 Notation 1

1.2 Partition of unity 4

1.3 Regularisation of functions 8

1.4 Distributions 10

1.5 The support of a distribution 12

1.6 Differentiation and multiplication 14

1.7 Distributions with compact support 18

1.8 Convolution 20

1.9 The Fourier transformation 25

§2 Geometric assumptions for the domain Ω **35**

2.1 Segment and cone properties 36

2.2 The $N^{k,\kappa}$-property of Ω 38

2.3 (k,κ)-diffeomorphisms and (k,κ)-smooth Ωs 47

2.4 Normal transformations 52

2.5 Differentiable manifolds 58

§3 Definitions and density properties for the Sobolev–Slobodeckii spaces $W_2^l(\Omega)$ **61**

3.1 Definition of the Sobolev–Slobodeckii spaces $W_2^l(\Omega)$ 61

3.2 Density properties 64

§4 The transformation theorem and Sobolev spaces on differentiable manifolds **74**

4.1 The transformation theorem 75

4.2 Sobolev spaces on differentiable manifolds, and on the frontier $\partial\Omega$ of a (k,κ)-smooth region 87

§5 Definition of Sobolev spaces by the Fourier transformation and extension theorems **90**

5.1 Sobolev spaces and the Fourier transformation 91

5.2 Extension theorems 95

§6 Continuous embeddings and Sobolev's lemma **105**

§7 Compact embeddings **112**
§8 The trace operator **120**
§9 Weak sequential compactness and approximation of derivatives by difference quotients **133**

II Elliptic differential operators 139

§10 Linear differential operators **139**
§11 The Lopatinskiĭ–Šapiro condition and examples **148**
11.1 The Lopatinskiĭ–Šapiro condition 148
11.2 Examples 157
§12 Fredholm operators **165**
12.1 The Riesz–Schauder spectral theorem (compact operators) 165
12.2 Fredholm operators 168
12.3 A priori *estimates, the Weyl lemma and smoothable operators* 180
§13 The main theorem and some theorems on the index of elliptic boundary value problems **186**
13.1 The main theorems for elliptic boundary value problems 186
13.2 The index and spectrum of elliptic boundary value problems 209
§14 Green's formulae **213**
14.1 Normal boundary value operators and Dirichelet systems 214
14.2 The first Green formula 219
14.3 Adjoint boundary value operators and boundary value spaces 222
14.4 The second Green formula 231
14.5 The antidual operator L′ and the adjoint boundary value problem 235
§15 The adjoint boundary value problem and the connection with the image space of the original operator **239**
§16 Examples **252**

III Strongly elliptic differential operators and the method of variations 261

§17 Gelfand triples, the Lax–Milgram theorem, *V*-elliptic and *V*-coercive operators **261**
17.1 Gelfand triples 261
17.2 Representations for functionals on Sobolev spaces 268
17.3 The Lax–Milgram theorem 271
17.4 V-elliptic and V-coercive forms, solution theorems 273
17.5 The Green operator 275
17.6 The concepts V-elliptic and V-coercive for differential operators 279
§18 Agmon's condition **280**
§19 Agmon's theorem: conditions for the *V*-coercion of strongly elliptic differential operators **290**
19.1 The theorems of Gårding and Agmon 290
19.2 Examples, including the Dirichlet problem for strongly elliptic differential operators 302

§20 Regularity of the solutions of strongly elliptic equations **307**
§21 The solution theorem for strongly elliptic equations and examples **336**
§22 The Schauder fixed point theorem and a non-linear problem **361**
§23 Elliptic boundary value problems for unbounded regions **370**

IV Parabolic differential operators 376

§24 The Bochner integral **376**
24.1 Pettis' theorem 376
24.2 The Bochner integral 384
§25 Distributions with values in a Hilbert space *H* and the space *W*(0, *T*) 390
§26 The existence and uniqueness of the solution of a parabolic differential equation **395**
§27 The regularity of solutions of the parabolic differential equation **403**
27.1 An abstract regularity theorem 404
27.2 Differentiability with respect to t 411
27.3 Differentiability with respect to x, respectively t 414
§28 Examples **423**

V Hyperbolic differential operators 434

§29 Existence and uniqueness of the solution **434**
§30 Regularity of the solutions of the hyperbolic differential equation **442**
30.1 An abstract regularity theorem 442
30.2 Differentiability with respect to t 445
30.3 Differentiability with respect to x 447
§31 Examples **452**

VI Difference processes for the calculation of the solution of the partial differential equation 462

§32 Functional analytic concepts for difference processes **462**
§33 Difference processes for elliptic differential equations and for the wave equation **481**
33.1 Some important inequalities 481
33.2 Construction of a difference process for the Dirichlet problem 484
33.3 A difference process for the wave equation in several space variables 488
§34 Evolution equations **496**
34.1 The time-independent case 498
34.2 The time-dependent case 503
34.3 Stability behaviour of the perturbed process 505
34.4 Several step processes 507

References 511
Function and distribution spaces 515
Index 516

For Brigitte

Preface

Boundary value problems are the subject of this book. All boundary value conditions for elliptic differential operators are given, using the Lopatinskiĭ–Šapiro condition (= covering condition), which lead to the normal solvability of a boundary value problem. The variational method is also presented in detail, and questions of its connections with general elliptic theory considered. Those parabolic and hyperbolic equations for which the right-hand side (derivatives with respect to x) is an elliptic differential operator are considered, and the knowledge about elliptic operators is used in order to obtain insight into the solvability and regularity properties of the solution for mixed problems.

I have chosen a form of the Lopatinskiĭ–Šapiro condition which allows us to test immediately, whether or not given boundary value conditions satisfy it. It appears that all classical boundary value problems satisfy it, the examples are worked through individually.

In order not to overexpand the compass of the book, and to maintain its introductory character, I have not considered pseudo-differential operators; all the same I have proved the main theorem for elliptic boundary value problems by means of pseudo-differential operators – without calling them such.

Before the discussion of differential equations there is an introductory chapter on distributions and Sobolev spaces; here I have proceeded in an elementary way, working with the Fourier transformation, and not using interpolation theorems. This is possible without further assumptions as long as one remains inside L^2-theory. I have not considered the L^p-theory; this comes into its own for non-linear equations, see for example Lions [3], while in the linear case it does not bring any essentially new insights.

I have looked very precisely into the differentiability properties of the frontier $\partial\Omega$; in the end we want to solve the Dirichlet problem on the square as well as on the circle, and the proofs are not essentially simpler for the C^{∞}-theory.

Among the many procedures for the practical solution of a partial differential equation I have singled out the difference process – its appeal lies in its simplicity: derivatives are replaced by difference quotients and so we obtain a system of linear equations, which in the main we can solve by the usual methods. In the last chapter I wish less to lead into modern methods, than to transmit the feeling to the reader that it is actually possible numerically to solve partial differential equations.

The reader ought to be familiar with the language of functional analysis, at about the level of the books of Heuser [1], or Wloka [1]. He may find the basic theorems of functional analysis relevant for analysis, such as the Hahn–Banach, Banach–Steinhaus and Riesz theorems, the open mapping theorem..., collected and proved on pp. 12–27 in L.H. Loomis, *An introduction to abstract harmonic analysis*, New York (1953). In separate sections I have thoroughly considered less familiar material such as, for example, the theory of Fredholm operators, Gelfand triples, abstract Green solution operators, the Schauder fixed point theorem and the Bochner integral. In this way I hope to spare the reader a time-consuming hunt through the literature. Another aim, which I have included in these functional analytic sections, is wherever possible to replace hard analysis by soft analysis, and so restricted the difficult estimation machinery to an unavoidable minimum. In this way I also believe that I have given the reader a better view of the connections and possible generalisations.

I owe a very heavy debt of gratitude to R. Mennicken, G. Bauer and B. Sagraloff from Regensburg, to my students J. Benner, R. Janssen, R. Rath, and to Miss Inga Haecks. They have read the complete manuscript critically and carefully, and have provided many important remarks and improvements. I am particularly grateful to M. König for assistance in the laborious proof reading of the German edition.

I thank Professor Dr G. Köthe for encouraging me to write this book, and the publishers both for their assistance in the preparation of the manuscript and cooperation in its presentation.

Kiel 1980 J. Wloka

Note added in the English translation

I have made some corrections and slight changes, especially in §§ 14 and 15. To English-speaking readers, instead of the books of Heuser [1] and Wloka [1] (see above) I would recommend those of Schechter [1] or Taylor [1] as an introduction to the language of functional analysis.

Kiel 1986 J. Wloka

I

Sobolev spaces

§1 Notation, basic properties, distributions

In order to build up the theory of Sobolev spaces in a simple manner we need a generalised notion of differentiation; this we find in L. Schwartz' theory of distributions, and give here a short introduction to the theory of distributions, in so far as it is applicable to Sobolev spaces. In order to spare the reader additional labour we have been careful to give all proofs. Our development is concentrated on the following points: partition of unity, an aid which we shall use at many points in this book; the generalised notion of differentiation, for which we also present theorems illuminating the connection with classical differentiation; the regularisation (convolution) of functions and distributions, which play a great role in approximation and density problems; Fourier transformation, which we study in the spaces $\mathscr{S}, \mathscr{S}'$ and $L_2(\mathbb{R}^r)$, obtaining an important analytic tool, that will render us good service in many questions.

For similar introductions we refer to the books of Hörmander [1] and Rudin [3]; for a broad initiation into distribution theory the original book of L. Schwartz [1] is still to be recommended.

1.1 Notation

Let $\mathbb{R}$ be the set of real numbers, $\mathbb{C}$ the set of complex numbers; we denote by $\mathbb{R}^r$ and $\mathbb{C}^r$ the real and complex r-dimensional spaces equipped with the Euclidean norm

$$|x| = [|x_1|^2 + \cdots + |x_r|^2]^{1/2}, \quad x = (x_1, \ldots, x_r) \in \mathbb{R}^r \text{ (respectively } \mathbb{C}^r).$$

We use L. Schwartz' notation for derivatives, products, faculties, etc. Let

$s=(s_1,\ldots,s_r)$ be a multi-index, $s_i\in\mathbb{N}=\{0,1,\ldots\}$, $i=1,2,\ldots,r$, we write

$$\mathrm{D}^s=\frac{\partial^{s_1+\cdots+s_r}}{\partial x_1^{s_1}\cdots\partial x_r^{s_r}},\quad |s|=s_1+\cdots+s_r,\quad \Delta=\sum_{j=1}^r\frac{\partial^2}{\partial x_j^2},$$

$$x^s=x_1^{s_1}x_2^{s_2}\cdots x_r^{s_r},\quad 0^0=0,$$

$$s!=s_1!\cdots s_r!,\quad \binom{m}{s}=\binom{m_1}{s_1}\cdots\binom{m_r}{s_r}.$$

Let Ω be an open set in $\mathbb{R}^r$. The elements of $C^l(\Omega)$ are all (complex valued) functions $\varphi(x)$, $x\in\Omega$, which on Ω possess continuous and bounded derivatives $\mathrm{D}^s\varphi(x)$, $|s|\leqslant l$ (up to order l). We define the norm $\|\varphi\|_{C^l}$ on C^l by

$$\|\varphi\|_{C^l}=\sup_{\substack{|s|\leqslant l\\ x\in\Omega}}|\mathrm{D}^s\varphi(x)|.$$

Convergence in the space $C^l(\Omega)$ means uniform convergence in Ω not only of the sequence of functions itself, but also of the sequence of its sth partial derivatives ($|s|\leqslant l$). By $C^l(\bar{\Omega})$ we understand the proper subspace of $C^l(\Omega)$, consisting of all functions $\varphi\in C^l(\Omega)$ which, together with their derivatives up to order l, are also continuous on the frontier of Ω (that is, continuous on $\bar{\Omega}$). In the case that Ω is in addition bounded, we can use the norm

$$\|\varphi\|_{C^l(\bar{\Omega})}=\max_{\substack{|s|\leqslant l\\ x\in\bar{\Omega}}}|\mathrm{D}^s\varphi(x)|.$$

We denote by $\mathscr{E}^l(\Omega)$ the set of all functions which together with their derivatives up to order l are continuous on Ω. (Here no boundedness is asked for.) For $\mathscr{E}^\infty(\Omega)$ we also write $\mathscr{E}(\Omega)$ or $C^\infty(\Omega)$ – later we shall introduce on $\mathscr{E}(\Omega)$ the topology of an F-space.

We say that a function φ is λ-Hölder continuous on Ω; if

$$\frac{|\varphi(x)-\varphi(y)|}{|x-y|^\lambda}\leqslant C<\infty$$

holds for all $x,y\in\Omega$; here $0<\lambda\leqslant 1$ and $x\neq y$. (The λ-Hölder continuous functions on Ω for $\lambda>1$ are constant.) In the case $\lambda=1$ we also talk of Lipschitz continuous functions. We define the space $C^{l,\lambda}(\Omega)$ to be the collection of all l-fold continuous, differentiable, bounded functions φ on Ω (also let all derivatives $\mathrm{D}^s\varphi$, $|s|\leqslant l$ be bounded on Ω) for which the lth derivatives are λ-Hölder continuous. As norm in $C^{l,\lambda}(\Omega)$ we take the expression

$$\|\varphi\|_{l,\lambda}=\sup_{\substack{|s|\leqslant l\\ x\in\Omega}}|\mathrm{D}^s\varphi(x)|+\sup_{\substack{|s|=l\\ y,x\in\Omega\\ x\neq y}}\frac{|\mathrm{D}^s\varphi(x)-\mathrm{D}^s\varphi(y)|}{|x-y|^\lambda}.$$

All the spaces introduced are complete, and hence are Banach or Fréchet spaces; the simple proofs are left to the reader, see also Wloka [1], pp. 21 and 55. For the sake of completeness we define

$$C^{0,0}(\Omega) := C(\Omega), \quad C^{l,0}(\Omega) := C^l(\Omega).$$

We wish to define the $L_p(\Omega)$ spaces. First suppose that $1 \leqslant p < \infty$. The space $L_p(\Omega)$ consists of all Lebesgue measurable functions defined on $\Omega \subset \mathbb{R}^r$, whose pth power is integrable with respect to the Lebesgue measure $\mathrm{d}x = \mathrm{d}x_1 \cdots \mathrm{d}x_r = \mathrm{d}\mu$, that is,

$$\int_\Omega |\varphi(x)|^p \,\mathrm{d}x < \infty \text{ holds.}$$

As norm in $L_p(\Omega)$ we take the expression

$$\|\varphi\|_p = \left[\int_\Omega |\varphi(x)|^p \,\mathrm{d}x\right]^{1/p}.$$

Strictly speaking, in $L_p(\Omega)$ we have to deal not with single functions but with classes of functions, which differ from each other only on sets of measure zero. Thus if $\|\varphi\|_p = 0$, it follows that $\varphi(x) = 0$ almost everywhere on Ω, but not that $\varphi(x)$ is identically zero. We come now to the definition of the space $L_\infty(\Omega)$, $p = \infty$. This space consists of all functions φ defined on Ω which are Lebesgue measurable and bounded almost everywhere (φ is called bounded almost everywhere if the inequality $|\varphi(x)| \leqslant M < \infty$ holds outside a set $\{x : x \in A\}$ of measure zero). By means of

$$\|\varphi\|_\infty = \operatorname*{supess}_{x\in\Omega} |\varphi(x)| =: \inf_{\substack{A\subset\Omega\\ \mu(A)=0}} \sup_{x\in\Omega\setminus A} |\varphi(x)|,$$

we introduce a norm on $L_\infty(\Omega)$. We can show that the spaces $L_p(\Omega)$, $1 \leqslant p \leqslant \infty$ are complete, see for example Wloka [1], p. 52.

The set $L_1^{\mathrm{loc}}(\Omega)$ consists of all Lebesgue measurable functions on Ω, which are integrable on each compact subset $K \subset \Omega$, that is,

$$\int_K |\varphi(x)| \,\mathrm{d}x < \infty \quad \text{for all } K \subset\subset \Omega.$$

It is easy to show that all the C-spaces and L_p-spaces introduced here are contained in $L_1^{\mathrm{loc}}(\Omega)$.

The space $L_2(\Omega)$ is particularly important in what follows; it is a separable Hilbert space with the inner product

$$(\varphi, \psi) = \int_\Omega \varphi(x) \cdot \overline{\psi(x)} \,\mathrm{d}x; \quad \varphi; \psi \in L_2(\Omega).$$

For embedding theory we need a compactness criterion for L_p-spaces, this goes back to Kolmogorov; for the proof, see for example Wloka [1], p. 201.

Kolmogorov compactness criterion. Let M be a subset of the space $L_p(\Omega)$, $1 \leqslant p < \infty$. M is relatively compact if and only if the following three conditions are satisfied:

1. M is bounded in $L_p(\Omega)$, that is, $\sup_{\varphi \in M} \| \varphi \|_p = \sup_{\varphi \in M} [\int_\Omega |\varphi(x)|^p \, dx]^{1/p} < \infty$.
2. $\lim_{h \to 0} \int_\Omega |\varphi(x+h) - \varphi(x)|^p \, dx = 0$ holds uniformly for $\varphi \in M$.
3. $\lim_{\alpha \uparrow \beta} \int_{\{|x| > \alpha\} \cap \Omega} |\varphi(x)|^p = 0$ holds uniformly for $\varphi \in M$.

If the domain Ω is bounded, then condition 3 is superfluous.

1.2 Partition of unity

The partition of unity is known to be an important tool; we shall often make use of it in what follows. In order to introduce it we need some definitions. We say that $\{\Omega_i, i \in I\}$ – the indexing set I is arbitrary – is an open cover of $\mathbb{R}^r$, if the sets Ω_i are open and

$$\bigcup_{i \in I} \Omega_i = \mathbb{R}^r \text{ holds.} \tag{1}$$

We call a cover $\{U_j, j \in J\}$, a refinement of $\{\Omega_i, i \in I\}$, provided that for each j we can find some $i(j)$ with

$$U_j \subset \Omega_{i(j)}. \tag{2}$$

We say that a cover $\{\Omega_i, i \in I\}$ is locally finite, if for each point $x \in \mathbb{R}^r$ we can find a ball $B(x, \rho)$ for which $\Omega_i \cap B(x, \rho) \neq \emptyset$ at most finitely often. We want to show that $\mathbb{R}^r$ is paracompact, which means that for each open cover $\{\Omega_i, i \in I\}$ we can find a locally finite refinement $\{U_j, j \in J\}$. However, we prove much more:

Theorem 1.1 *Let $\{\Omega_i, i \in I\}$ be an open cover of $\mathbb{R}^r$, then there exists a locally finite open refinement $\{U_j, j \in \mathbb{N}\}$ with countable indexing set $J = \mathbb{N}$, and with $\bar{U}_j$ compact, $j \in \mathbb{N}$.*

Proof. Let B_n be the ball $B(0, n)$, $n = 1, 2, \ldots$; we cover $\mathbb{R}^r$ by the compact annuli

$$\bar{B}_1, \bar{B}_2 \backslash B_1, \ldots, \bar{B}_{n+1} \backslash B_n, \ldots.$$

Let $x \in \bar{B}_1$. Because of (1), x lies in some Ω_{i_x}, and we may choose $U_x = B(x, \rho)$ in such a way that

$$\bar{U}_x \subset \Omega_{i_x} \quad \text{and} \quad \rho < \tfrac{1}{2}. \tag{3}$$

The U_x form an open cover of the compact set $\bar{B}_1$, so already finitely many $U_1,\ldots,U_m$ cover the ball $\bar{B}_1$. In this way we have defined the refinement $U=\{U_j\}$ for $j=1,2,\ldots,m$. We next consider the annulus $\bar{B}_2\backslash B_1$ and proceed.

We choose V_x with

$$\bar{V}_x \subset \Omega_{i_x}, \quad V_x = B(x,\rho), \rho < \tfrac{1}{2}, x\in\bar{B}_2\backslash B_1. \tag{3'}$$

Since $\bar{B}_2\backslash B_1$ is compact we can select a finite subcover $\{V_1,\ldots,V_l\}$ from the cover $\{V_x\}$. We discard those Vs which already appear among $U_1,\ldots,U_m$ and call the remainder $U_{m+1},\ldots,U_k$. $\{U_1,\ldots,U_k\}$ covers $\bar{B}_2$. Continuing in this way we find the open, countable cover $\{U_j, j\in\mathbb{N}\}$ of $\mathbb{R}^r$. Because of (3) $\{U_j\}$ is a refinement of $\{\Omega_i\}$, and because $\rho<\frac{1}{2}$ the sets $\bar{U}_j$ are compact. We show that $\{U_j\}$ is locally finite. Let $x\in\bar{B}_n\backslash B_{n-1}$ and consider the ball $B(x,\frac{1}{4})$. Because $\rho<\frac{1}{2}$ only the sets U_j which cover $\bar{B}_n\backslash B_{n-1}$ and the neighbouring annuli $\bar{B}_{n+1}\backslash B_n$ and $\bar{B}_{n-1}\backslash B_{n-2}$ can actually meet the ball $B(x,\frac{1}{4})$; and there are – by construction – only finitely many such U_js. ■

Remark 1.1 Theorem 1.1 also holds for each locally compact manifold, which satisfies the second axiom of countability (that is, $M=\bigcup_{n=1}^{\infty}K_n, K_n$ compact).

For the sake of completeness we present here the simple proof of Warner [1], p. 9.

Proof. We take the K_n in the second axiom of countability to be compact, $K_n\subset K_{n+1}$, $\bigcup_{n=1}^{\infty}K_n=M$, and prove first the existence of a sequence of open sets G_n with $\bar{G}_n$ compact, $\bar{G}_n\subset G_{n+1}$, $\bigcup_{n=1}^{\infty}G_n=M$.

Let $x\in K_1$, because of local compactness we may choose a neighbourhood U_x with $\bar{U}_x$ compact and such that $K_1\subset\bigcup_{x\in K_1}U_x$. Therefore since K_1 is compact

$$K_1\subset\bigcup_{i=1}^{m}U_{x_i}.$$

We set $G_1:=\bigcup_{i=1}^{m}U_{x_i}$ and define the other $G_n, n\geqslant 2$, inductively. If G_{n-1} is already defined, we consider the compact set $K_n\cup\bar{G}_{n-1}$ and we find as before locally compact neighbourhoods $U^n_{x_i}$ with

$$K_n\cup\bar{G}_{n-1}\subset\bigcup_{i=1}^{m_n}U^n_{x_i}.$$

We set $G_n:=\bigcup_{i=1}^{m_n}U^n_{x_i}$, and can then easily verify the desired properties of G_n.

Now for the assertions of Theorem 1.1. Let $\{\Omega_i, i\in I\}$ be an arbitrary open cover of M. For $n \geqslant 3$ the set $\bar{G}_n \backslash G_{n-1}$ is compact and contained in the open set $G_{n+1} \backslash \bar{G}_{n-2}$. We consider the open cover

$$\{\Omega_i \cap (G_{n+1} \backslash \bar{G}_{n-2}), i \in I\} \text{ of } \bar{G}_n \backslash G_{n-1}$$

and select from it the finite cover

$$\bar{G}_n \backslash G_{n-1} \subset \bigcup_{j=1}^{m_n} U_j^n; \quad U_j^n = \Omega_j \cap (G_{n+1} \backslash \bar{G}_{n-2}).$$

Similarly we select from the open cover $\{\Omega_i \cap G_3, i \in I\}$ of $\bar{G}_2$ the finite cover

$$\bar{G}_2 \subset \bigcup_{j=1}^{m} U_j, \quad U_j = \Omega_j \cap G_3.$$

It is easily seen that the U's have the desired properties in Theorem 1.1, thus, for example, local finiteness: let $x \in M$, then there exists some $n \geqslant 3$ with $x \in G_n \backslash G_{n-1}$ (for $n = 2$ and $x \in G_2$), hence also $x \in \bar{G}_n \backslash G_{n-1}$. Since

$$x \in \bar{G}_n \backslash G_{n-1} \subset G_{n+1} \backslash \bar{G}_{n-2}$$

holds, and $G_{n+1} \backslash \bar{G}_{n-2}$ is open, we can find some neighbourhood U_x with

$$x \in U_x \subset G_{n+1} \backslash \bar{G}_{n-2}.$$

This U_x can only meet those U's that either lie in $G_{n+1} \backslash \bar{G}_{n-2}$ or in its neighbours, that is, in $G_{n+2} \backslash \bar{G}_{n-1}$, respectively $G_n \backslash \bar{G}_{n-3}$ (set $G_0 = \varnothing$). These are, however, by construction only finitely many in number, so that we have proved local finiteness (the proof for $n = 2$ is similar). ■

Corollary 1.1 *Let $\Omega \subset \mathbb{R}^r$ be open and let $\{\Omega_i, i \in I\}$ be an open cover of Ω. Then the assertions of Theorem* 1.1 *hold.*

It suffices to adjoin an additional set, for example $\Omega_0 = \mathbb{R}^r$, to the cover $\{\Omega_i\}$ in order to obtain a cover of $\mathbb{R}^r$, and be able to apply Theorem 1.1. We may also take balls $B(x_j, a_j)$ for the refinement $\{U_j\}$.

A continuous function $\varphi(x)$, $x \in \mathbb{R}^r$, defined on $\mathbb{R}^r$ is called finite if it vanishes outside a bounded subset. The support of φ, $\operatorname{supp} \varphi$, is the closure of all points x with $\varphi(x) \neq 0$, in symbols

$$\operatorname{supp} \varphi := \overline{\{x : \varphi(x) \neq 0\}}.$$

Following L. Schwartz we denote the set of all finite, infinitely differentiable functions with supports in Ω (Ω open) by the symbol $\mathscr{D}(\Omega)$ (or $C_0^\infty(\Omega)$), and call the elements of $\mathscr{D}(\Omega)$ fundamental functions. For $\mathscr{D}(\mathbb{R}^r)$ we simply write $\mathscr{D}$.

Example Let $g(t), t \in \mathbb{R}^1$ denote the function

$$g(t) = \begin{cases} 0 & \text{if } t \leqslant 0. \\ e^{-1/t} & \text{if } t > 0, \end{cases}$$

which belongs to C^∞ but not to $\mathscr{D}$. With the help of $g(t)$ we can easily construct a fundamental function $h \in \mathscr{D}$:

$$h(x) = g(1 - |x|^2) = \begin{cases} 0 & \text{if } |x| \geqslant 1, \\ \exp(-1/1 - |x|^2) & \text{if } |x| < 1, \end{cases}$$

where $|x|^2 = x_1^2 + \cdots + x_r^2$. We have

$$\operatorname{supp} h = \overline{B(0, 1)}.$$

For $a > 0$, $x_0 \in \mathbb{R}^r$, we set

$$\varphi_{x_0, a}(x) = h\left(\frac{x - x_0}{a}\right) \in \mathscr{D}(\mathbb{R}^r), \tag{4}$$

so that we have

$$\operatorname{supp} \varphi_{x_0, a} = \overline{B(x_0, a)}, \quad \text{and} \quad \varphi_{x_0, a}(x) > 0 \quad \text{for } x \in B(x_0, a). \tag{5}$$

By a *partition of unity* $\{\alpha_j, j \in J\}$ on Ω we understand a class of functions $\alpha_j \in \mathscr{D}(\Omega)$ with the following properties:

(a) the collection of supports $\{\operatorname{supp} \alpha_j : j \in J\}$ is locally finite,
(b) $0 \leqslant \alpha_j(x) \leqslant 1$, and $\sum_{j \in J} \alpha_j(x) \equiv 1$ for $x \in \Omega$.

Because of the local finiteness of the supports, the series appearing in (b) has only finitely many non-zero members for each $x \in \Omega$. We say that the partition of unity $\{\alpha_j, j \in J\}$ is subordinate to the cover $\{\Omega_i, i \in I\}$, if for each $j \in J$ there exists an $i(j) \in I$ with

$$\operatorname{supp} \alpha_j \subset \Omega_{i(j)}.$$

Theorem 1.2 (Partition of unity) *Let $\{\Omega_i, i \in I\}$ be an open cover of the open set $\Omega \subset \mathbb{R}^r$. There exists a partition of unity $\{\alpha_j, j \in \mathbb{N}\}$ subordinate to the cover $\{\Omega_i, i \in I\}$ with countable indexing set $J = \mathbb{N}$.*

Proof. By Theorem 1.1, Corollary 1.1, there exists a countable, locally finite refinement $\{U_j, j \in \mathbb{N}\}$ with $U_j = B(x_j, a_j)$ and $\bar{U}_j \subset \Omega_{i(j)}$. We take the functions (4), $\varphi_{x_j, a_j} \in \mathscr{D}(\Omega)$, and by (5) have $\operatorname{supp} \varphi_{x_j, a_j} = \bar{U}_j$. In the series $\psi(x) = \sum_j \varphi_{x_j, a_j}(x)$, because of local finiteness, there are only finitely many members differing from zero on $B(x, \rho)$. Hence $\psi \in C^\infty(\Omega)$, and by the covering property together with (5), $\psi(x) > 0$ for all $x \in \Omega$. We define $\alpha_j(x) = \varphi_{x_j, a_j}(x)/\psi(x)$ and see that (a) and (b) are satisfied. ■

Complement We can also arrange that $\sum_j \beta_j^2(x) = 1$, $\beta_j \in \mathscr{D}(\Omega)$. We simply put

$$\beta_j(x) = \frac{\varphi_{x_j,a_j}(x)}{[\sum_j(\varphi_{x_j,a_j}(x))^2]^{1/2}}.$$

Corollary 1.2 *Let A be closed in $\mathbb{R}^r$ and G open with $A \subset G$. There exists an infinitely differentiable function α with the properties:*

(a) $0 \leqslant \alpha(x) \leqslant 1$ *for all* $x \in \mathbb{R}^r$,
(b) $\alpha(x) = 1$ *for* $x \in A$,
(c) $\operatorname{supp} \alpha \subset G$, *and in particular* $\alpha(x) = 0$ *on the complement of* G, $x \in CG$.

Proof. We consider the cover $\{G, CA\}$ of $\mathbb{R}^r$. Let $\alpha_j, j \in \mathbb{N}$ be a subordinate partition of unity (Theorem 1.2), that is, the support of α_j lies either in G or in CA. We define α to be the sum of all those α_j, whose support lies in G. In short

$$M := \{j \in \mathbb{N} : \operatorname{supp} \alpha_j \subset G\}, \quad \alpha(x) = \sum_{j \in M} \alpha_j(x),$$

(a) is then clear. (c) follows from

$$\operatorname{supp} \alpha = \overline{\bigcup_{j \in M} \operatorname{supp} \alpha_j} = \bigcup_{j \in M} \operatorname{supp} \alpha_j \subset G, \tag{6}$$

where the equation in the middle holds for locally finite families. Now for (b) we rearrange

$$1 = \sum_j \alpha_j(x) = \sum_{j \in M} \alpha_j(x) + \sum{}' \alpha_j(x) = \alpha(x) + \sum{}' \alpha_j(x), \tag{7}$$

where $\sum' \alpha_j(x)$ is the sum of those α_j whose supports lie in CA, but not in G. If $x \in A$, then $\sum' \alpha_j(x) = 0$, so that given (7) we have proved (b). ■

1.3 Regularisation of functions

We return once more to the function $h \in \mathscr{D}(\mathbb{R}^r)$ – see (4). We have

$$\int_{\mathbb{R}^r} h(x)\,dx = C > 0,$$

and we set $\tilde{h}(x) = (1/C)h(x)$, so that we obtain

$$\int_{\mathbb{R}^r} \tilde{h}(x)\,dx = \int_{|x| \leqslant 1} \tilde{h}(x)\,dx = 1.$$

The function $h_\varepsilon(x) = (1/\varepsilon^r)\tilde{h}(x/\varepsilon)$ also has the property

$$\int_{\mathbb{R}^r} h_\varepsilon(x)\,dx = 1,$$

and is a fundamental function which vanishes for $|x| \geqslant \varepsilon$. The function $h_\varepsilon(x)$

plays an important part in so-called *regularisation*. We describe the *convolution integral*

$$\varphi_\varepsilon(x) = (\varphi * h_\varepsilon)(x) = \int_{\mathbf{R}^r} \varphi(y)\cdot h_\varepsilon(x-y)\,\mathrm{d}y = \int_{\mathbf{R}^r} h_\varepsilon(y)(x-y)\,\mathrm{d}y \tag{8}$$

as the regularisation of φ.

Theorem 1.3 *Let φ be integrable (that is, $\varphi \in L_1$) and let φ vanish outside a compact subset K of Ω. We have the following assertions:*

(a) *The support of $\varphi_\varepsilon = \varphi * h_\varepsilon$ is contained in $K_\varepsilon = \{x : d(x, K) \leqslant \varepsilon\}$, and is again compact.*
(b) *If $\varepsilon < d(K, C\Omega)$, then $\varphi_\varepsilon \in \mathscr{D}(\Omega)$.*
(c) *If $\varphi \in L_p(\Omega)$, $1 \leqslant p < \infty$, then $\lim_{\varepsilon \to 0} \| \varphi_\varepsilon - \varphi \|_p = 0$.*
(d) *If φ is continuous, then $\varphi_\varepsilon \to \varphi$ uniformly as $\varepsilon \to 0$.*

Proof. Formula (8) written as $\varphi_\varepsilon(x) = \int_K \varphi(y) h_\varepsilon(x-y)\,\mathrm{d}y$ shows that, because φ is integrable, the Lebesgue theorems on the interchange of integral and limit are applicable, and we obtain

$$\varphi_\varepsilon(x) \in C^\infty. \tag{9}$$

Let $\varphi_\varepsilon(x) \neq 0$; by formula (8) it must be that $x - y \in K$ and $|y| \leqslant \varepsilon$. If we write $K_\varepsilon := \{x : d(x, K) \leqslant \varepsilon\}$, we deduce that x must belong to K_ε, that is the support of φ_ε is a closed subset of K_ε, and we have proved (a), (b) follows immediately from (a) and (9). Suppose now that $\varphi \in L_p(\Omega)$, $1 < p < \infty$; the Hölder inequality (Schwarz inequality in the case $p = 2$) applied to (8) gives $(1 = 1/p + 1/q)$

$$\begin{aligned} \| \varphi_\varepsilon \|_p^p &\leqslant \int_\Omega \left[\int_{\mathbf{R}^r} |\varphi(y)| h_\varepsilon^{1/p + 1/q}(x-y)\,\mathrm{d}y \right]^p \mathrm{d}x \\ &\leqslant \int_\Omega \left[\int_{\mathbf{R}^r} |\varphi(y)|^p h_\varepsilon^{p/p}(x-y)\,\mathrm{d}y \right]^{p/p} \left[\int_{\mathbf{R}^r} h_\varepsilon^{q/q}(x-y)\,\mathrm{d}y \right]^{p/q} \mathrm{d}x \\ &= \int_{\mathbf{R}^r} |\varphi(y)|^p \left(\int_\Omega h_\varepsilon(x-y)\,\mathrm{d}x \right) \mathrm{d}y = \| \varphi \|_p^p \quad \text{for } \varepsilon < d(K, C\Omega), \end{aligned} \tag{10}$$

that is, φ_ε belongs to $L_p(\Omega)$. The Hölder inequality applied in the same way as in (10) gives

$$\begin{aligned} \| \varphi_\varepsilon - \varphi \|_p^p &= \int_\Omega |h_\varepsilon * \varphi(x) - \varphi(x)|^p\,\mathrm{d}x \leqslant \int_\Omega \left[\int_{\mathbf{R}^r} h_\varepsilon(x-y) |\varphi(y) - \varphi(x)|\,\mathrm{d}y \right]^p \mathrm{d}x \\ &\leqslant \int_\Omega \int_{|y| \leqslant \varepsilon} h_\varepsilon(y) |\varphi(x-y) - \varphi(x)|^p\,\mathrm{d}y\,\mathrm{d}x \\ &\leqslant \sup_{y \leqslant \varepsilon} \int_\Omega |\varphi(x-y) - \varphi(x)|^p\,\mathrm{d}x. \end{aligned}$$

Condition 2 (mean continuity) of the Kolmogorov compactness criterion (a single element $\{\varphi\}$ is compact!) conclude the proof of (c). The reader may carry out the appropriate alterations to the proof in the case $p = 1$ for himself.

Now to (d). Let φ be assumed continuous. Since $\int h_\varepsilon(y)\,\mathrm{d}y = 1$ we can write:

$$\varphi_\varepsilon(x) - \varphi(x) = \int_{y \leqslant \varepsilon} (\varphi(x-y) - \varphi(x))h_\varepsilon(y)\,\mathrm{d}y$$

and the uniform continuity of φ implies the uniform convergence of φ_ε to φ as ε tends to zero. ■

1.4 Distributions

Next we define distributions in the sense of L. Schwartz on Ω. We take $\mathscr{D}(\Omega)$ as the fundamental or test space. We easily see that $\mathscr{D}(\Omega)$ is a linear space, and we can consider linear functionals on $\mathscr{D}(\Omega)$, that is linear maps $T:\mathscr{D}(\Omega) \to \mathbb{C}$. We define:

Definition 1.1 *We call a linear functional $T:\mathscr{D}(\Omega) \to \mathbb{C}$ a* distribution, *if for each compact set $K \subset\subset \Omega$ we can find constants C and k, so that the estimate*

$$|T(\varphi)| \leqslant C \sum_{|s| \leqslant k} \sup_K |\mathrm{D}^s\varphi| \quad \textit{holds for all } \varphi \in \mathscr{D}(K). \tag{11}$$

Here the notation $K \subset\subset \Omega$ means that K is compact and $K \subset \Omega$. $\mathscr{D}(K)$ consists of all $\varphi \in \mathscr{D}(\Omega)$ with $\operatorname{supp}\varphi \subseteq K$. We denote the space of all distributions by $\mathscr{D}'(\Omega)$. In the case that we can choose the constant k independent of K, we say that the distribution T is of finite order. The smallest of these ks we call the order of T, and we denote the space of all distributions of finite order by $\mathscr{D}'_F(\Omega)$.

Example 1.1 Let $f \in L_1^{\mathrm{loc}}(\Omega)$. Then

$$T_f(\varphi) := \int f(x)\varphi(x)\,\mathrm{d}x, \quad \varphi \in \mathscr{D}(\Omega), \tag{12}$$

is a distribution of order 0.

Example 1.2 Let m be a Radon measure, that is, $m \in (C_0^0(\Omega))'$ (see Bourbaki [2]). Then

$$T_m(\varphi) := \int \varphi(x)\,\mathrm{d}m, \quad \varphi \in \mathscr{D}(\Omega), \tag{13}$$

is again a distribution of order 0.

Addendum. *The space $C_0^0(\Omega)$ is the space of all continuous functions on Ω with compact support,* $\operatorname{supp}\varphi \subset\subset \Omega$, *equipped with the inductive limit topology*

$\operatorname{ind}_{K \subset\subset \Omega} C_0(K)$, *where* $C_0(K)$ *carries the* sup norm:

$$C_0(K) = \left\{ \underset{\operatorname{supp}\varphi \subset K}{\varphi \text{ continuous}} : \sup_{x \in K} |\varphi(x)| < \infty \right\}.$$

We label the continuous linear functionals on $C_0^0(\Omega)$, *that is,* $m \in (C_0^0(\Omega))'$, *as Radon measures on* Ω.

Example 1.3 We define the Dirac delta distributions by

$$\delta^{(s)}(\varphi) := (-1)^{|s|} D^s \varphi(0), \quad \varphi \in \mathscr{D}(\Omega).$$

$\delta^{(s)}$ has order $|s|$.

Condition (11) in the definition can easily be verified for all three examples. In future we want to identify the distribution T_f (12) with the function f and T_m (13) with the measure m. In order to make this identification we need the map $f \mapsto T_f$ (respectively $m \mapsto T_m$) to be injective. By Theorem 1.3(d) $\mathscr{D}(\Omega)$ is a dense subset of $C_0^0(\Omega)$ and from the equation (see (13)) $T_m(\varphi) = \int \varphi(x)\,dm = 0$ for $\varphi \in \mathscr{D}(\Omega)$ it follows that $m = 0$. Hence $m \mapsto T_m$ is injective. In order to see that $f \mapsto T_f$ is injective for $f \in L_1^{\mathrm{loc}}(\Omega)$, we associate to the function f the absolutely continuous (Radon) measure $m = f \cdot dx$, and injectivity follows from what has just been proved.

We introduce sequential convergence on $\mathscr{D}(\Omega)$. We can also introduce a locally convex topology on $\mathscr{D}(\Omega)$, which gives (11) as continuity condition, either as an inductive limit topology, see for example L. Schwartz [1], or by a direct assignment of seminorms, see for example Hörmander [1]. We do not wish to do this, because nowhere do we use the locally convex topology.

Definition 1.2 *We say that the sequence* $\varphi_n \in \mathscr{D}(\Omega)$ *converges to* φ_0 *in* $\mathscr{D}(\Omega)$, *if the following two conditions are satisfied*:

1. $D^s \varphi_n$ *converges uniformly to* $D^s \varphi_0$ *for each multi-index* s, *and*
2. *There exists a compact subset* $K \subset\subset \Omega$ *with* $\operatorname{supp} \varphi_n \subset K$ *for* $n = 0, 1, \ldots$.

Theorem 1.4 *A linear functional* T *on* $\mathscr{D}(\Omega)$ *is a distribution if and only if* T *is sequentially continuous.*

Proof. If condition (11) is fulfilled, then T is sequentially continuous in the sense of the definition of convergence above. If (11) is not fulfilled, then there exists some $K_0 \subset\subset \Omega$, so that for each $C = k = n$ we can find some $\varphi_n \in \mathscr{D}(K_0)$ with $T(\varphi_n) = 1$ and $\sup_{K_0} |D^s \varphi_n| \leqslant 1/n$ for $|s| \leqslant n$, ((11) is homogeneous!) This sequence φ_n has the properties:

$$\varphi_n \to 0 \text{ in } \mathscr{D}(\Omega) \quad \text{and} \quad T(\varphi_n) \to 1 \neq 0,$$

a contradiction. Hence T cannot be sequentially continuous on $\mathscr{D}(\Omega)$. Clearly $\mathscr{D}'(\Omega)$ is again a linear space. We define $(\alpha_1 T_1 + \alpha_2 T_2)(\varphi) := \alpha_1 T_1(\varphi) + \alpha_2 T_2(\varphi)$ for $\alpha_1, \alpha_2 \in \mathbb{C}$. We equip $\mathscr{D}'(\Omega)$ with the weak topology, that is, as seminorms we take

$$p_\varphi(T) := |T(\varphi)|,$$

where φ is a chosen, fixed element in $\mathscr{D}(\Omega)$. A sequence of distributions T_n converges (weakly) to the distribution T, in short $T_n \to T$ in $\mathscr{D}'(\Omega)$, if for each $\varphi \in \mathscr{D}(\Omega)$ it is true that $T_n(\varphi) \to T(\varphi)$ (in the numerical sense).

With this definition of convergence we show that the embedding

$$L_2(\Omega) \subsetneq \mathscr{D}'(\Omega)$$

is continuous: since $f_n \to f$ in $L_2(\Omega)$ implies the weak convergence $f_n \rightharpoonup f$ in $L_2(\Omega)$, hence

$$\int_\Omega f_n \varphi \, \mathrm{d}x \to \int_\Omega f \varphi \, \mathrm{d}x \quad \text{for } \varphi \in \mathscr{D}(\Omega).$$

Using (12) this implies that

$$T_{f_n} \to T_f \text{ in } \mathscr{D}'(\Omega).$$

We shall also equip the other spaces of distributions $\mathscr{E}'(\Omega)$, $\mathscr{S}'(\Omega)$ – see later – with the weak topology.

Let Ω_0 be an open subset of Ω. We define the restriction of the distribution $T \in \mathscr{D}'(\Omega)$ by the restriction of T to the domain of definition $\mathscr{D}(\Omega_0)$. We denote the restriction $T|_{\mathscr{D}(\Omega_0)}$ by $T|_{\Omega_0}$. Two distributions T_1 and T_2 are called locally equal at $x \in \Omega$, if in some neighbourhood U of x the restrictions on U are equal, that is,

$$T_1|_U = T_2|_U,$$

in other words, if $T_1(\varphi) = T_2(\varphi)$ for all $\varphi \in \mathscr{D}(U)$. ■

1.5 The support of a distribution

The local behaviour of a distribution already determines its global behaviour; this is the content of L. Schwartz' 'Principe du recollement des morceaux' [1].

Theorem 1.5 *Let $\{\Omega_i, i \in I\}$ be an open cover of Ω and T_i, $i \in I$, be distributions from $\mathscr{D}'(\Omega_i)$; which agrees on $\Omega_i \cap \Omega_j$, that is,*

$$T_i|_{\Omega_i \cap \Omega_j} = T_j|_{\Omega_i \cap \Omega_j}. \tag{14}$$

Then there exists a unique distribution $T \in \mathscr{D}'(\Omega)$, which agrees with T_i on Ω_i, that *is,*

$$T(\varphi) = T_i(\varphi) \quad \text{for } \varphi \in \mathscr{D}(\Omega_i), i \in I. \tag{15}$$

Proof. Let $\{\alpha_k\}$ be a partition of unity subordinate to the cover $\{\Omega_i\}$, see Theorem 1.2, where $\operatorname{supp} \alpha_k \subset \Omega_{i(k)}$. We have

$$\varphi = \sum_k \alpha_k \cdot \varphi \quad \text{for } \varphi \in \mathscr{D}(\Omega),$$

we set

$$T(\varphi) := \sum_k T_{i(k)}(\alpha_k \varphi), \tag{16}$$

and using Theorem 1.4 must prove the sequential continuity of T. Let $\varphi_n \to 0$ in $\mathscr{D}(\Omega)$, then there exists a compact subset $K \subset\subset \Omega$ with

$$\operatorname{supp} \varphi_n \subset K \quad \text{for } n = 1, 2, \ldots.$$

Because of the local finiteness of $\{\alpha_k\}$ K is met by at most finitely many of the $\operatorname{supp} \alpha_k$ (K is compact!) and (16) takes the form

$$T(\varphi_n) = \sum_k^{\text{finite}} T_{i(k)}(\alpha_k \varphi_n), \quad \operatorname{supp} \varphi_n \subset K. \tag{17}$$

We easily see that $\varphi_n \to 0$ in $\mathscr{D}(\Omega)$ implies that $\alpha_k \varphi_n \to 0$ in $\mathscr{D}(\Omega_{i(k)})$, from which the sequential continuity in (17) is immediately apparent. We demonstrate (15). Let $\varphi \in \mathscr{D}(\Omega_i)$, then $\operatorname{supp} \alpha_k \varphi$ is contained in $\Omega_{i(k)} \cap \Omega_i$ and in the light of (14), we have

$$T_i(\varphi) = T_i\left(\sum_i \alpha_k \varphi\right) = \sum_k T_i(\alpha_k \varphi) = \sum_k T_{i(k)}(\alpha_k \varphi) = T(\varphi),$$

which is (15).

The global distribution T is already uniquely determined by (15). Let T' be another distribution which satisfies (15), and let

$$T'(\varphi_0) \neq T(\varphi_0) \quad \text{for some } \varphi_0 \in \mathscr{D}(\Omega). \tag{18}$$

Let $\operatorname{supp} \varphi_0 = K_0$, K_0 meets at most finitely many of the $\operatorname{supp} \alpha_k$, and since $\operatorname{supp} \alpha_k \varphi_0 \subset \Omega_{i(k)}$ we have

$$T'(\varphi_0) = T'\left(\sum_k \alpha_k \varphi_0\right) = \sum_k T'(\alpha_k \varphi_0) = \sum_k T_{i(k)}(\alpha_k \varphi_0) = T(\varphi_0)$$

contradicting (18). ■

Theorem 1.5 has as a corollary that a distribution which is locally zero is also globally zero. Given Theorem 1.5 we can define the support of a distribution $T \in \mathscr{D}'(\Omega)$.

Definition 1.3 *The support of T,* supp T, *is the complement of the set of points of Ω at which T is locally equal to zero.*

The set supp T is closed in Ω, since by definition its complement is open. We also have

$$T(\varphi)=0 \quad \text{for } \varphi\in\mathscr{D}(\Omega) \text{ with } \operatorname{supp} T\cap\operatorname{supp}\varphi=\varnothing, \tag{19}$$

and by Theorem 1.5 $\mathbf{C}$ (supp T) is the largest open set on which $T=0$. Thus the present definition agrees with the usual definition for continuous functions.

1.6 Differentiation and multiplication

Definition 1.4 *Let s be a multi-index and $T\in\mathscr{D}'(\Omega)$ a distribution. We define the derivative* D^sT *by*

$$(\mathrm{D}^sT)(\varphi):=(-1)^{|s|}T(\mathrm{D}^s\varphi).$$

D^sT is again a distribution, since from (11) we obtain the estimate

$$|\mathrm{D}^sT(\varphi)|\leqslant C\sum_{|\alpha|\leqslant k+s}\sup_K|\mathrm{D}^\alpha\varphi| \quad \text{for } \varphi\in\mathscr{D}(K).$$

Differentiation is continuous in the distributional sense.

Theorem 1.6

(a) *If $T_n\to T$ in $\mathscr{D}'(\Omega)$ then $\mathrm{D}^sT_n\to\mathrm{D}^sT$ in $\mathscr{D}'(\Omega)$.*

(b) *Let τ_h be the translation operator (see below) then for $(0,\ldots,0,h_j,0,\ldots,0)\to 0$, $j=1,\ldots,r$, we have*

$$\frac{1}{h_j}(\tau_{h_j}T-T)\to\frac{\partial T}{\partial x_j} \text{ in } \mathscr{D}'.$$

Proof. The proof of (a) is clear. To prove (b), we define $\tau_hT(\varphi):=T(\varphi(x-h))$ and see that τ_hT is again a distribution from $\mathscr{D}'$. We have

$$\frac{1}{h_j}(\tau_{h_j}T(\varphi)-T(\varphi))=T\left(\frac{\varphi(x-h_j)-\varphi(x)}{h_j}\right),$$

and, as we easily see (Definition 1.2), the difference quotient $(\varphi(x-h_j)-\varphi(x))/h_j$ converges in $\mathscr{D}$ to $-\partial\varphi/\partial x_j$ as $h_j\to 0$. Therefore

$$T\left(\frac{\varphi(x-h_j)-\varphi(x)}{h_j}\right)\to T\left(-\frac{\partial\varphi}{\partial x_j}\right)=\frac{\partial T}{\partial x_j}(\varphi),$$

with which we have proved (b). ∎

Theorem *Distributional derivatives commute:*

$$D^{\alpha}D^{\beta}T = D^{\beta}D^{\alpha}T. \tag{20}$$

Proof. $(D^{\alpha}D^{\beta}T)(\varphi) = (-1)^{|\alpha|}D^{\beta}T(D^{\alpha}\varphi) = (-1)^{|\alpha|+|\beta|}T(D^{\beta}D^{\alpha}\varphi)$
$= (-1)^{|\alpha|+|\beta|}T(D^{\alpha}D^{\beta}\varphi) = \cdots = (D^{\beta}D^{\alpha}T)(\varphi).$ ■

Example 1.4 Let δ be the Dirac delta distribution, then we have

$$D^{s}\delta = \delta^{(s)}$$

(see Example 1.3).

Example 1.5 Let $H(x)$, $x\in\mathbb{R}^1$ be the Heaviside function

$$H(x) = \begin{cases} 0 & \text{for } x \leqslant 0, \\ 1 & \text{for } x > 0. \end{cases}$$

We have $H' = \delta$.

Definition 1.5 *Let $T\in\mathscr{D}'(\Omega)$ and $a\in C^{\infty}(\Omega)$. We define the product aT by*

$$(aT)(\varphi) := T(a\varphi).$$

aT is again a distribution, since by using (11) and applying the Leibniz product rule we have

$$|(aT)(\varphi)| = |T(a\varphi)| \leqslant C \sum_{|s|\leqslant k} \sup_K |D^{s}(a\varphi)| \leqslant C' \sum_{|s|\leqslant k} \sup_K |D^{s}\varphi| \quad \text{for } \varphi\in\mathscr{D}(K).$$

Supp aT is contained in supp $a \cap$ supp T and from $T_n \to T$ in $\mathscr{D}'(\Omega)$ it follows that $aT_n \to aT$.

Warning *The product of two distributions $T_1, T_2\in\mathscr{D}(\Omega)$ can only be satisfactorily defined in special cases.*

We see that the new Definition 1.5 agrees with the usual pointwise definition of multiplication for functions and measures.

The Leibniz product rule holds:

$$D^{\alpha}(aT) = \sum_{\beta\leqslant\alpha} \binom{\alpha}{\beta} D^{\beta}a \cdot D^{\alpha-\beta}T \quad \text{for } a\in C^{\infty}, T\in\mathscr{D}'. \tag{21}$$

Proof. We have

$$-T\left(a\frac{\partial}{\partial x_i}\varphi\right) = T\left(\left(\frac{\partial a}{\partial x_i}\right)\varphi\right) - T\left(\frac{\partial}{\partial x_i}(a\varphi)\right),$$

that is,

$$\frac{\partial}{\partial x_i}(aT)=\frac{\partial a}{\partial x_i}\cdot T+a\cdot\frac{\partial T}{\partial x_i}$$

and (21) follows by induction.

For later use we write the Leibniz product rule (21) in Hörmander's form. Let $P(D)=\sum_{|s|\leqslant m}b_s\mathrm{D}^s$ be a linear differential operator with constant coefficients b_s, then

$$P(D)(a\cdot T)=\sum_{a}\frac{1}{\alpha!}(\mathrm{D}_a^\alpha)(P^{(\alpha)}(\mathrm{D})T)\quad \text{for } a\in C^\infty,\ T\in\mathscr{D}', \tag{21'}$$

where

$$P^{(\alpha)}(x)=\mathrm{D}_x^\alpha P(x)=\frac{\partial^{|\alpha|}}{\partial x_1^{\alpha_1}\cdots\partial x_r^{\alpha_r}}\sum_{|s|\leqslant m}b_s x_1^{s_1}\cdots x_r^{s_r}.$$

Next we give examples in which classical differentiation – we denote it by brackets [ ] – coincides with distributional differentiation.

Theorem 1.7

(a) Let $f\in\mathscr{E}^1(\Omega)$. *Then the classical partial derivatives agree with the distributional derivatives, that is, we have*

$$T_{[\partial f/\partial x_i]}=\frac{\partial T_f}{\partial x_i}\quad \textit{for } i=1,\ldots,r. \tag{22}$$

(b) *Let f and g be continuous on Ω, $f,g\in\mathscr{E}^0(\Omega)$, and suppose that in the distributional sense $\partial g/\partial x_i=f$ for some $i=1,\ldots,r$ (more precisely $\partial T_g/\partial x_i=T_f$), then the classical derivative $[\partial g/\partial x_i]$ also exists, and for all $x\in\Omega$, we have*

$$\left[\frac{\partial g}{\partial x_i}\right](x)=f(x).$$

Proof. (a) Under the assumptions of 1 the formula for partial integration is valid

$$\int_\Omega\frac{\partial f}{\partial x_i}\varphi\,\mathrm{d}x=-\int_\Omega f\frac{\partial\varphi}{\partial x_i}\mathrm{d}x\quad \text{for } \varphi\in\mathscr{D}(\Omega). \tag{23}$$

However, (23) is (22) written differently.

(b) Let $\chi\in\mathscr{D}(\Omega)$, then the functions χg and $\partial(\chi g)/\partial x_i=(\partial\chi/\partial x_i)g+\chi(\partial g/\partial x_i)$ are continuous and have compact support. Therefore it suffices to prove the theorem for continuous functions with compact support in Ω. With the notation of Theorem 1.3 we have $g_\varepsilon, f_\varepsilon \in C_0^\infty$ and $[\partial g_\varepsilon/\partial x_i] = f_\varepsilon$, because

$$\left[\frac{\partial g_\varepsilon}{\partial x_i}\right] = \int g(y)\left[\frac{\partial}{\partial x_i} h_\varepsilon(x-y)\right]\mathrm{d}y = -\int g(y)\left[\frac{\partial}{\partial y_i} h_\varepsilon(x-y)\right]\mathrm{d}y$$
$$= -T_g\left(\frac{\partial}{\partial y_i} h_\varepsilon(x-y)\right)_y = \left(\frac{\partial}{\partial y_i} T_g\right)(h_\varepsilon(x-y))_y$$
$$= T_f(h_\varepsilon(x-y))_y = \int f(y) h_\varepsilon(x-y)\,\mathrm{d}y = f_\varepsilon.$$

By Theorem 1.3 as $\varepsilon \to 0$ the convergence of both g_ε to g and $[\partial g_\varepsilon/\partial x_i]$ to f is uniform. Therefore by results in classical analysis (Rudin [1]), $[\partial g/\partial x_i]$ exists and must equal f. ■

Theorem 1.8 *Let $f \in C^{0,1}(\Omega)$. Then the classical partial derivatives $[\partial f/\partial x_i]$, $i = 1, \ldots, r$ exist almost everywhere on Ω. They are measurable and bounded,*

$$\left[\frac{\partial f}{\partial x_i}\right] \in L_\infty(\Omega), \tag{24}$$

and they agree with the distributional derivatives

$$\left[\frac{\partial f}{\partial x_i}\right] = \frac{\partial f}{\partial x_i}, \quad i = 1, \ldots, r.$$

By induction we immediately obtain: *let $f \in C^{l,1}(\Omega)$, then for $|s| \leqslant l+1$ $[\mathrm{D}^s f]$ exists almost everywhere, and*

$$[\mathrm{D}^s f] = \mathrm{D}^s f \quad \textit{for } |s| \leqslant l+1.$$

Proof. Since f belongs to $C^{0,1}(\Omega)$ we have

$$|f(x) - f(y)| \leqslant C|x-y|, \tag{25}$$

and $f(x) = f(x_1, \ldots, x_r)$ is absolutely continuous in each variable x_i, if the other variables $x_1, \ldots, x_{i-1}, x_{i+1}, \ldots, x_r$ are fixed. We apply Lebesgue's theorem (see Natanson [1], p. 274), find that the partial derivatives $[\partial f/\partial x_i]$ exist almost everywhere, and satisfy

$$\int_a^{x_i} \left[\frac{\partial f}{\partial x_i}\right] \mathrm{d}x_i = f(x_1, \ldots, x_i, \ldots, x_r) - f(x_1, \ldots, a, \ldots, x_r). \tag{26}$$

Here almost everywhere means for almost all x_i, while the other variables are held fixed. By the definition of the classical derivative,

$$\lim_{h_i \to 0} \frac{f(x+h_i) - f(x)}{h_i} = \left[\frac{\partial f}{\partial x_i}\right](x),$$

in the sense of pointwise convergence almost everywhere. By results from measure theory (see for example Natanson [1]) the measurability of $f(x)$ implies that of $[\partial f/\partial x_i]$. The inequality (25) gives the boundedness of derivatives. Since $f \in C^{0,1}(\Omega)$ and (24) holds, we have

$$f, \left[\frac{\partial f}{\partial x_i}\right] \in L_1^{\mathrm{loc}}(\Omega), \quad i = 1, \ldots, r,$$

and we can apply Fubini's theorem to the integrals appearing below. For $\varphi \in \mathscr{D}(\Omega)$ we get

$$\begin{aligned} \frac{\partial f}{\partial x_i}(\varphi) &= -T_f\left(\frac{\partial \varphi}{\partial x_i}\right) = -\int_\Omega f(x) \frac{\partial \varphi(x)}{\partial x_i} \mathrm{d}x \\ &= -\int_{\mathbf{R}^{r-1}} \left(\int_{\mathbf{R}} f(x) \frac{\partial \varphi}{\partial x_i} \mathrm{d}x_i \right) \mathrm{d}x_1 \cdots \mathrm{d}x_{i-1} \, \mathrm{d}x_{i+1} \cdots \mathrm{d}x_r \\ &= \int_{\mathbf{R}^{r-1}} \left(\int_{\mathbf{R}} \left[\frac{\partial f}{\partial x_i}\right] \varphi(x) \, \mathrm{d}x_i \right) \mathrm{d}x_1 \cdots \mathrm{d}x_{i-1} \, \mathrm{d}x_{i+1} \cdots \mathrm{d}x_r \\ &= \int_\Omega \left[\frac{\partial f}{\partial x_i}\right] \varphi(x) \, \mathrm{d}x, \end{aligned} \tag{27}$$

which gives the required agreement between classical and distributional derivation. Here the partial integration in the middle of (27) can be carried out because of (26). ∎

1.7 Distributions with compact support

Next we come to distributions with compact support. We give $C^\infty(\Omega) = \mathscr{E}(\Omega)$ the structure of an (F)-space by taking the expressions

$$p_{k,K}(\varphi) = \sum_{|s| \leq k} \sup_K |\mathrm{D}^s \varphi|, \quad \varphi \in \mathscr{E}(\Omega)$$

as a system of seminorms on $\mathscr{E}(\Omega)$. Here $k = 0, 1, \ldots,$ and K is a compact subset of Ω. Since countably many K_n already exhaust Ω, that is,

$$\Omega = \bigcup_{n=1}^{\infty} K_n, \quad \text{with } K_n \subset K_{n+1} \text{ compact}, \tag{28}$$

the space $\{\mathscr{E}(\Omega), p_{k,K}\}$ is metrisable and we easily see that it is complete. Hence $\mathscr{E}(\Omega)$ is an (F)-space. We recall that a sequence $\varphi_n \in \mathscr{E}(\Omega)$ converges to $\varphi \in \mathscr{E}(\Omega)$, in short $\varphi_n \to \varphi$ in $\mathscr{E}(\Omega)$, if for each pair k, K

$$p_{k,K}(\varphi_n - \varphi) = \sum_{|s| \leq k} \sup_K |\mathrm{D}^s \varphi_n - \mathrm{D}^s \varphi| \to 0.$$

Theorem 1.9 *The test space $\mathscr{D}(\Omega)$ is dense in $\mathscr{E}(\Omega)$.*

Proof. Let $\varphi\in\mathscr{E}(\Omega)$, by Corollary 1.2 there exists some $\chi_n\in\mathscr{D}(\Omega)$ with $\chi_n(x)=1$ for $x\in K_n$, and we have

$$p_{k,K}(\varphi-\chi_m\cdot\varphi)=0 \quad \text{for } m\geqslant n.$$

Since $\chi_n\varphi\in\mathscr{D}(\Omega)$, we have proved the density assertion. ∎

A linear functional $l(\varphi)$ on $\mathscr{E}(\Omega)$ is continuous, if and only if we can find k, K and a constant $C<\infty$, with

$$|l(\varphi)|\leqslant Cp_{k,K}(\varphi)=C\sum_{|s|\leqslant k}\sup_K|D^s\varphi|, \quad \varphi\in\mathscr{E}(\Omega). \tag{29}$$

We denote the dual space of all continuous, linear functionals on $\mathscr{E}(\Omega)$ by $\mathscr{E}'(\Omega)$, and we give $\mathscr{E}'(\Omega)$ the weak topology, that is, we take as seminorms

$$p_\varphi(l)=|l(\varphi)|, \quad \varphi \text{ a fixed element in } \mathscr{E}(\Omega).$$

Theorem 1.10 *Let $T\in\mathscr{D}(\Omega)$ be a distribution with compact support, $\operatorname{supp} T\subset\subset\Omega$. Then T has finite order and $T(\varphi)$ may be uniquely extended to a continuous linear functional on $\mathscr{E}(\Omega)$. Conversely each continuous, linear functional on $\mathscr{E}(\Omega)$ can be regarded as a distribution with compact support, and the inclusion*

$$\mathscr{E}'(\Omega)\hookrightarrow\mathscr{D}'(\Omega)$$

is continuous and injective.

Proof. Let $\operatorname{supp} T$ be compact, then by Corollary 1.2 we can find some $\chi\in\mathscr{D}(\Omega)$ with $\chi=1$ on $\operatorname{supp} T$. Let $K_0=\operatorname{supp}\chi$. By (19) ($\operatorname{supp} T\cap \operatorname{supp}(1-\chi)=\varnothing$) we have

$$T(\varphi)=T(\chi\varphi)+T((1-\chi)\varphi)=T(\chi\varphi) \quad \text{for } \varphi\in\mathscr{D}(\Omega), \tag{30}$$

and by (11) together with (30)

$$|T\varphi|=|T\chi(\varphi)|\leqslant C\sum_{|s|\leqslant k_0}\sup_{K_0}|D^s(\chi\varphi)|\leqslant C_0\sum_{|s|\leqslant k_0}\sup_{K_0}|D^s(\varphi)| \tag{31}$$

for all $\varphi\in\mathscr{D}(\Omega)$. Here we have used the Leibniz rule, and the fact that for all $\varphi\in\mathscr{D}(\Omega)$ $\operatorname{supp}\chi\varphi$ is contained in K_0. (31) shows that T is of finite order $\leqslant k_0$.

We define the extension of T to $\mathscr{E}(\Omega)$ by (30)

$$\tilde{T}(\varphi):=T(\chi\varphi), \quad \varphi\in\mathscr{E}(\Omega).$$

The estimate (31) also holds for all $\varphi\in\mathscr{E}(\Omega)$, for the inclusion $\operatorname{supp}\chi\varphi\subset K_0$ holds for all $\varphi\in\mathscr{E}(\Omega)$.

However, (31) is (29), that is, $\tilde{T}(\varphi)$ is a continuous linear functional on $\mathscr{E}(\Omega)$. The uniqueness of the extension $\tilde{T}(\varphi)$ follows from the density of $\mathscr{D}(\Omega)$ in $\mathscr{E}(\Omega)$. Conversely suppose that T is a continuous linear functional on $\mathscr{E}(\Omega)$. Then following (29) there exists a compact subset K (and numbers C, k) with

$$|T(\varphi)| \leqslant C \sum_{|s| \leqslant k} \sup_K |D^s\varphi| \quad \text{for all } \varphi \in \mathscr{E}(\Omega). \tag{32}$$

Given (32) = (11) the restriction of T to $\mathscr{D}(\Omega)$ is a distribution from $\mathscr{D}'(\Omega)$; from (32) we also know that if $K \cap \operatorname{supp} \varphi = \varnothing$, then $T(\varphi) = 0$ or $\operatorname{supp} T \subset K$. In this way we have characterised $\mathscr{E}'(\Omega)$ as a subspace of $\mathscr{D}'(\Omega)$. The injectivity of $\mathscr{E}' \hookrightarrow \mathscr{D}'$ follows from Theorem 1.9, while continuity is easy to see.

1.8 Convolution

We introduce some notation: the *translation operator* τ_x,

$$(\tau_x u)(y) := u(y - x) \quad \text{and} \quad \check{u}(y) := u(-y)$$

where in both cases u is a function on $\mathbb{R}^r$.

Let A and B be two subsets contained in $\mathbb{R}^r$, we write

$$A + B := \{x + y : x \in A, y \in B\}.$$

If A is closed and B is compact, then $A + B$ is closed. If both sets A and B are compact, then $A + B$ is also compact. For the sake of brevity we shall write $\mathscr{D}$ and $\mathscr{D}'$ instead of $\mathscr{D}(\mathbb{R}^r)$ and $\mathscr{D}'(\mathbb{R}^r)$, similarly for $\mathscr{E}$ and $\mathscr{E}'$. For two continuous functions f and g, of which one has compact support, we define the *convolution* of f and g by

$$(f * g)(x) = \int f(x - y) g(y)\, dy = \int f(y) g(x - y)\, dy = (g * f)(x),$$

which can also be written as

$$(f * g)(x) = \int_{\mathbb{R}^r} f(y)(\tau_x \check{g})(y)\, dy.$$

This leads to the following:

Definition 1.6 *Let $T \in \mathscr{D}'$ and $\varphi \in \mathscr{D}$. We define*

$$(T * \varphi)(x) := T_y(\varphi(x - y)) := T(\tau_x \check{\varphi}). \tag{33}$$

Here the subscript y in T_y means that T is regarded as a linear form on functions in the variable y, while x is held fixed. We note that $(T * \varphi)(0) = T(\check{\varphi})$ holds, which implies that if $T_1 * \varphi = T_2 * \varphi$ for all $\varphi \in \mathscr{D}$, then T_1 must equal T_2.

Theorem 1.11 *Let* $T \in \mathscr{D}'$ *and* $\varphi \in \mathscr{D}$. *Then* $T * \varphi \in C^\infty = \mathscr{E}$ *and* $\operatorname{supp}(T * \varphi) \subset \operatorname{supp} T + \operatorname{supp} \varphi$. *The derivatives of the convolution are given by*

$$D^s(T * \varphi) = (D^s T) * \varphi = T * (D^s \varphi). \tag{34}$$

Proof. For $\varphi \in \mathscr{D}$, $\tau_\alpha : \mathbb{R}^r \to \mathscr{D}$, $x \mapsto \tau_x \varphi$ is (sequentially) continuous (see Definition 1.2), $\check{}$ is a continuous map $\mathscr{D} \to \mathscr{D}$, and T is a continuous map $\mathscr{D} \to \mathbb{C}$ (Theorem 1.4). Hence by (33) $(T * \varphi)(x)$ is a continuous function of x. By (19), $(T * \varphi)(x) = 0$ unless the support of T meets $\operatorname{supp}_y \varphi(x - y)$, that is, unless there exists some $y \in \operatorname{supp} T$ with $x - y \in \operatorname{supp} \varphi$. However, in this case $x \in \operatorname{supp} \varphi + \operatorname{supp} T$, and we have proved the statement about supports. Let h_i be the vector $(0, \ldots, 0, h_i, 0, \ldots, 0)$ in $\mathbb{R}^r$, and form the difference quotient

$$\frac{1}{h_i}[(T * \varphi)(x + h_i) - (T * \varphi)(x)] = T_y\left(\frac{1}{h_i}[\varphi(x + h_i - y) - \varphi(x - y)]\right).$$

As we see from Definition 1.2, the difference quotient $(1/h_i)[\varphi(x + h_i - y) - \varphi(x - y)]$ converges as $h_i \to 0$ to $(\partial \varphi / \partial x_i)(x - y)$ in $\mathscr{D}$ (here we consider $(\partial \varphi / \partial x_i)(x - y)$ as a function of y for some fixed x). Since T is continuous on $\mathscr{D}$ (Theorem 1.4) we have shown that $[(\partial / \partial x_i)(T * \varphi)](x) = (T * [\partial \varphi / \partial x_i])(x)$, and the equation $D^s(T * \varphi) = T * (D^s \varphi)$ follows by induction. At the same time we have also shown that $T * \varphi \in C^\infty$. The remainder of (34) follows from the formula

$$\tau_x((D^s \varphi)\check{}\,) = (-1)^{|s|} D^s(\tau_x \check{\varphi}). \qquad \blacksquare$$

Theorem 1.12 *Let* $T \in \mathscr{D}'$ *and* $\varphi, \psi \in \mathscr{D}$. *Then the associative rule holds:*

$$(T * \varphi) * \psi = T * (\varphi * \psi).$$

Proof. We form the Riemann sum for the integral

$$\varphi * \psi = \int \varphi(x - y) \psi(y) \, dy, \varepsilon > 0,$$

$$f_\varepsilon(x) = \varepsilon^r \sum_g \varphi(x - g\varepsilon) \psi(g\varepsilon)$$

where g runs through all the integral lattice points of $\mathbb{R}^r$. We have

$$\operatorname{supp} f_\varepsilon \subset \operatorname{supp} \varphi + \operatorname{supp} \psi \subset\subset \mathbb{R}^r, \tag{35}$$

and for each s

$$D^s f_\varepsilon(x) = \varepsilon^r \sum_g D^s \varphi(x - g\varepsilon) \psi(g\varepsilon)$$

converges uniformly to

$$((D^s \varphi) * \psi)(x) = D^s(\varphi * \psi)(x) \text{ as } \varepsilon \to 0.$$

Together with (35) (supp f_ε is compact) this shows that the convergence $f_\varepsilon \to \varphi * \psi$ takes place in $\mathscr{D}$. Hence we have

$$T*(\varphi*\psi)(x) = \lim_{\varepsilon\to 0} T*f_\varepsilon(x) = \lim_{\varepsilon\to 0} \varepsilon^r \sum_g (T*\varphi)(x-g\varepsilon)\psi(g\varepsilon)$$
$$= ((T*\varphi)*\psi)(x) \qquad \blacksquare$$

The regularisation theorem 1.3 may be extended to distributions.

Theorem 1.13 *Let* $T\in\mathscr{D}'$. *Then* $T*h_\varepsilon\in C^\infty$, $\operatorname{supp}(T*h_\varepsilon)\subset \operatorname{supp} T + \overline{B(0,\varepsilon)} = (\operatorname{supp} T)_\varepsilon$ *and* $T*h_\varepsilon \to T$ *in* $\mathscr{D}'$ *as* $\varepsilon\to 0$.

Proof. Given Theorem 1.11 we need only prove the last assertion. We have $T(\psi) = (T*\check{\psi})(0)$, and have only to prove

$$((T*h_\varepsilon)*\check{\psi})(0) \to (T*\check{\psi})(0) \text{ as } \varepsilon\to 0. \tag{36}$$

By Theorem 1.12 we have

$$((T*h_\varepsilon)*\check{\psi})(0) = (T*(h_\varepsilon*\check{\psi}))(0),$$

and Theorem 1.3 together with (34) shows that the convergence $h_\varepsilon*\check{\psi}\to\check{\psi}$ takes place in $\mathscr{D}$, from which (36) follows. ■

Theorem 1.14 *Let* $T\in\mathscr{D}'$. *Convolution* $T*\mathscr{D}\to\mathscr{E}$ *is continuous (here sequentially continuous) and commutes with the translation operator* τ_h

$$T*(\tau_h\varphi) = \tau_h(T*\varphi) \quad \varphi\in\mathscr{D}. \tag{37}$$

Proof. Using (34), continuity follows from the definition (33), as does (37). ■

The converse theorem is also valid.

Theorem 1.15 *Let* U *be a linear, sequentially continuous map* $\mathscr{D}\to\mathscr{E}$ *which commutes with the translation operator* τ_h:

$$\tau_h(U\varphi) = U(\tau_h\varphi). \tag{38}$$

Then there exists a unique distribution $T\in\mathscr{D}'$ *with*

$$U\varphi = T*\varphi, \quad \varphi\in\mathscr{D}.$$

Proof. By the assumption of sequential continuity the linear functional $(U\varphi)(0)$ is a distribution. Let $T(\varphi) := (U\check{\varphi})(0)$, then

$$(U\varphi)(0) = T(\check{\varphi}) = (T*\varphi)(0).$$

We replace φ by $\tau_{-h}\varphi$, use (37), (38) and obtain

$$(U\varphi)(h) = \tau_{-h}(U\varphi(0)) = U(\tau_{-h}\varphi)(0) = (T*\tau_{-h}\varphi)(0)$$
$$= \tau_{-h}(T*\varphi)(0) = (T*\varphi)(h).$$

We now prove uniqueness. From $T*\varphi = 0$ for all $\varphi\in\mathscr{D}$ it follows that $T(\check{\varphi}) = (T*\varphi)(0) = 0$ for all $\varphi\in\mathscr{D}$, that is, $T=0$. ■

Let T be a distribution with compact support. By Theorem 1.10 T may be considered to be a continuous linear functional on $\mathscr{E}$, and the convolution $T*\varphi$, $\varphi\in\mathscr{E}$, can be defined by the same formula (33) as before:

$$(T*\varphi)(x) := T(\tau_h\check{\varphi}).$$

Theorem 1.16 *Let T be a distribution with compact support, $T\in\mathscr{E}'$, and $\varphi\in\mathscr{E}$. Then*

(a) $\tau_x(T*\varphi) = T*(\tau_x\varphi)$,
(b) $T*\varphi\in\mathscr{E}$ and $D^s(T*\varphi) = (D^sT)*\varphi = T*(D^s\varphi)$,
(c) *the map $T*\mathscr{E}\to\mathscr{E}$ is continuous,*
(d) $\operatorname{supp}(T*\varphi)\subset\operatorname{supp}T + \operatorname{supp}\varphi$.

If in addition $\psi\in\mathscr{D}$, then

(e) $T*\psi\in\mathscr{D}$,
(f) $T*(\varphi*\psi) = (T*\varphi)*\psi = (T*\psi)*\varphi$,
(g) *As $\varepsilon\to 0$, $T*h_\varepsilon\to T$ in $\mathscr{E}'$.*

The proof is analogous to the proofs of the theorems given above, and is left to the reader.

Definition 1.7 *Let $S, T\in\mathscr{D}'$, and suppose that at least one of these distributions has compact support. We define the map $U:\mathscr{D}\to\mathscr{E}$ by*

$$U\varphi = S*(T*\varphi), \quad \varphi\in\mathscr{D}.$$

The map U is well defined and the images $U\varphi$ all belong to $\mathscr{E}$.

1. If T has compact support, then by Theorem 1.16(e) $T*\varphi\in\mathscr{D}$ and $S*(T*\varphi)\in\mathscr{E}$
2. If the support of S is compact, then $T*\varphi\in\mathscr{E}$ and by Theorem 1.16(b) $S*(T*\varphi)\in\mathscr{E}$ also. By Theorem 1.16(a) (respectively Theorem 1.14) U commutes with the translation operator τ_x:

$$\tau_x(U\varphi) = \tau_x(S*(T*\varphi)) = S*(T*\tau_x\varphi) = U(\tau_x\varphi),$$

and Theorems 1.14 and 1.16(c) show that $U:\mathcal{D}\to\mathcal{E}$ is continuous. Hence the assumptions of Theorem 1.15 are satisfied, and there exists a distribution V with

$$V*\varphi = U\varphi = S*(T*\varphi). \tag{39}$$

We say that this distribution V represents the *convolution* $S*T$

$$V =: S*T, \tag{40}$$

and in the light of (39) we have

$$(S*T)*\varphi = S*(T*\varphi), \tag{41}$$

that is, the associative rule is satisfied.

We see immediately that if T is a function from $\mathcal{D}$ (respectively from $\mathcal{E}$), then the new definition of convolution (39), (40) agrees with the earlier one.

Example 1.6 $T*\delta = T$, $T\in\mathcal{D}'$.

Theorem 1.17 *Let* $S, T\in\mathcal{D}'$, *and let at least one of these distributions have compact support. We have:*

(a) $T*S = S*T$ (commutativity rule),
(b) $\operatorname{supp}(T*S)\subset\operatorname{supp} T+\operatorname{supp} S$,
(c) $D^s T = D^s\delta * T$ *and* $D^s(T*S) = (D^s T)*S = T*(D^s S)$,
(d) $S*(T*V) = (S*T)*V$ (associative rule),

where we assume that $S, T, V\in\mathcal{D}'$ *and that at least two of these distributions have compact support.*

Proof. (a) We observe that the two distributions $T_1, T_2\in\mathcal{D}'$ are equal if and only if for all $\varphi,\psi\in\mathcal{D}$ we have $T_1*(\varphi*\psi) = T_2*(\varphi*\psi)$. Indeed by Theorem 1.12 we have $(T_1*\varphi)*\psi = (T_2*\varphi)*\psi$ for all $\psi\in\mathcal{D}$, from which it follows that $T_1*\varphi = T_2*\varphi$ for all $\varphi\in\mathcal{D}$, that is, $T_1 = T_2$.

We use Theorem 1.12 (respectively Theorem 1.16(f)) together with the fact that the convolution of two functions is commutative, and have

$$\begin{aligned}(T*S)*(\varphi*\psi) &= T*(S*(\varphi*\psi)) = T*((S*\varphi)*\psi) = T*(\psi*(S*\varphi))\\ &= (T*\psi)*(S*\varphi).\end{aligned}$$

Similarly

$$(S*T)*(\varphi*\psi) = (S*T)*(\psi*\varphi) = (S*\varphi)*(T*\psi) = (T*\psi)*(S*\varphi).$$

Therefore $(T*S)*(\varphi*\psi) = (S*T)*(\varphi*\psi)$ for all $\varphi,\psi\in\mathcal{D}$ and the previous remark concludes the proof of (a).

(b) We regularise and by the property of the definition (41) have

$$(T*S)*h_\varepsilon = T*(S*h_\varepsilon).$$

By Theorem 1.11 (respectively Theorem 1.16 (d)) the support of $(T*S)*h_\varepsilon$ is contained in $\operatorname{supp} T + \operatorname{supp}(S*h_\varepsilon) \subset \operatorname{supp} T + \operatorname{supp} S + \overline{B(0,\varepsilon)}$. In the limit as $\varepsilon \to 0$ we obtain

$$\operatorname{supp}(T*S) \subset \operatorname{supp} T + \operatorname{supp} S.$$

(d) The assumption that at least two of the distributions under consideration have compact support, together with (b), imply that the convolutions $S*(T*V)$ and $(S*T)*V$ are defined. If $\varphi \in \mathscr{D}$ the property of the definition (41) implies that

$$(S*(T*V))*\varphi = S*((T*V)*\varphi) = S*(T*(V*\varphi)).$$

If the support of V is compact, then

$$((S*T)*V)*\varphi = (S*T)*(V*\varphi) = S*(T*(V*\varphi)),$$

since, in this case, by Theorem 1.16 (e) $V*\varphi \in \mathscr{D}$, and we have proved (d). If the support of V is not compact, then the support of S must be compact, and (a) together with the part of the associativity rule already proved, gives

$$S*(T*V) = S*(V*T) = (V*T)*S = V*(T*S) = V*(S*T) = (S*T)*V.$$

(c) A double application of (34) gives

$$(\mathrm{D}^s T)*\varphi = T*\mathrm{D}^s\varphi = T*\mathrm{D}^s\varphi*\delta = T*\mathrm{D}^s\delta*\varphi \quad \text{for all } \varphi \in \mathscr{D},$$

from which it follows that $\mathrm{D}^s T = \mathrm{D}^s\delta * T$.

If we apply the last formula together with (a) and (d) we have

$$\begin{aligned}\mathrm{D}^s(T*S) &= \mathrm{D}^s\delta*(T*S) = (\mathrm{D}^s\delta*T)*S = (\mathrm{D}^s T)*S = \mathrm{D}^s(S*T)\\ &= (\mathrm{D}^s S)*T = T*\mathrm{D}^s S.\end{aligned}$$

■

1.9 The Fourier transformation

For a function $f \in L_1(\mathbb{R}^r)$ its Fourier transformation is defined by

$$\hat{f}(\xi) := \int_{\mathbb{R}^r} \mathrm{e}^{-\mathrm{i}(x,\xi)} \cdot f(x)\mathrm{d}x =: \mathscr{F}f(\xi). \tag{42}$$

Here $(x_1\xi) = x_1\xi_1 + \cdots + x_r\xi_r$. If $\hat{f}$ is also integrable, we can express f in terms of $\hat{f}$ with the help of the Fourier inversion formula

$$f(x) = \frac{1}{(2\pi)^r}\int_{\mathbb{R}^r} \mathrm{e}^{\mathrm{i}(x,\xi)}\hat{f}(\xi)\mathrm{d}\xi = \mathscr{F}^{-1}\hat{f}(x). \tag{43}$$

Our aim is to extend Fourier transformation to spaces of distributions.

Definition 1.8 *$\mathcal{S} = \mathcal{S}(\mathbb{R}^r)$ denotes the space of all $\varphi \in C^\infty(\mathbb{R}^r)$ which have the property that*

$$\sup_{|s| \leqslant k} \sup_x |(1 + |x|^2)^m D^s \varphi(x)| < \infty \quad \text{for all } k \text{ and } m. \tag{44}$$

We give $\mathcal{S}$ the structure of an (F)-space by defining seminorms by means of the expressions

$$p_{k,m}(\varphi) = \sup_{|s| \leqslant k} \sup_x |(1 + |x|^2)^m D^s \varphi(x)|, \quad \varphi \in \mathcal{S}. \tag{44'}$$

Example 1.7 $\mathcal{D} \subset \mathcal{S}$, $e^{-|x|^2}$ belongs to $\mathcal{S}$ but not to $\mathcal{D}$.

We consider the dual space $\mathcal{S}'$ to $\mathcal{S}$. The elements S of $\mathcal{S}'$ (the continuous linear functionals on $\mathcal{S}$) are called *tempered distributions*. Thus S is a tempered distribution if and only if we can find C, k, m with

$$|S(\varphi)| \leqslant C p_{k,m}(\varphi) = C \sup_{\substack{x \\ |s| \leqslant k}} |(1 + |x|^2)^m D^s \varphi(x)| \quad \text{for } \varphi \in \mathcal{S}.$$

Example 1.8 Let f be measurable and let $|f(x)|/(1 + |x|^2)^n \leqslant C < \infty$ for some value of n; f is then a tempered distribution. Indeed we have

$$S_f(\varphi) := \int f(x) \cdot \varphi(x) dx = \int \frac{f(x)(1 + |x|^2)^{n+r}}{(1 + |x|^2)^{n+r}} \varphi(x) dx,$$

and can estimate.

$$\begin{aligned} |S_f(\varphi)| &\leqslant \int \frac{dx}{(1 + |x|^2)^r} \cdot \sup_x \frac{|f(x)|}{(1 + |x|^2)^n} \cdot \sup_x |(1 + |x|^2)^{n+r} \varphi(x)| \\ &= C' p_{0,n+r}(\varphi) \quad \text{for all } \varphi \in \mathcal{S}. \end{aligned}$$

Theorem 1.18

(a) *$\mathcal{D}$ is dense in $\mathcal{S}$.*
(b) *Each tempered distribution $S \in \mathcal{S}'$ can be extended to a distribution in $\mathcal{D}'$, and the induced inclusion $\mathcal{S}' \subset \mathcal{D}'$ is an injection.*

Proof. (b) If we fix a compact subset K, then, by looking at the seminorms (44), the sequential convergence induced on $\mathcal{D}(K)$ by $\mathcal{S}$ is identical with the usual as given by Definition 1.2, since each $(1 + |x|^2)^m$ is bounded on K. Let $S \in \mathcal{S}'$; then the restriction of S to $\mathcal{D}$ is continuous: $\mathcal{D} \to \mathbb{C}$, and represents a distribution from $\mathcal{D}'$. By (a) $\mathcal{D}$ is dense in $\mathcal{S}$, which implies that $\mathcal{S}' \subset \mathcal{D}'$ is injective.

(a) We take some $\psi \in \mathcal{D}$, such that $\psi = 1$ on the unit ball in $\mathbb{R}^r$ (Corollary 1.2).

Let $\varphi \in \mathscr{S}$, we set

$$\varphi_\varepsilon(x) = \varphi(x) \cdot \psi(\varepsilon x), \quad \varepsilon > 0$$

and have $\varphi_\varepsilon \in \mathscr{D}$. We assert that as $\varepsilon \to 0$

$$\varphi_\varepsilon \to \varphi \quad \text{in} \quad \mathscr{S}. \tag{45}$$

We have

$$D^s(\varphi - \varphi_\varepsilon) = \sum_{\beta \leqslant s} \binom{s}{\beta} D^\beta \varphi(x) \cdot \varepsilon^{|s-\beta|} D^{s-\beta}(1 - \psi(\varepsilon x)) \tag{46}$$

We form $\sup_{x,|s| \leqslant k} |(1 + |x|^2)^m D^s(\varphi - \varphi_\varepsilon)|$ and apply (46).

Since $\varphi \in \mathscr{S}$, the terms $(1 + |x|^2)^m D^s \varphi(x)$ are not only all bounded, but also converge to 0 as $|x| \to \infty$. This, the εs in (46), and the fact that $1 - \psi(\varepsilon x) = 0$ for $|x| \leqslant 1/\varepsilon$ ensure convergence in (45). ■

Theorem 1.19

(a) *$\mathscr{S}$ is dense in $\mathscr{E}$.*

(b) *Each distribution $T \in \mathscr{E}'$ with compact support may be extended to a tempered distribution, and the inclusion $\mathscr{E}' \subset \mathscr{S}'$ so obtained is injective.*

(c) *The inclusions $\mathscr{E}' \subset \mathscr{S}' \subset \mathscr{D}'$ are continuous (in the sense of weak convergence).*

Proof. (a) We have $\mathscr{D} \subset \mathscr{S} \subset \mathscr{E}$ and by Theorem 1.9 $\mathscr{D}$ is dense in $\mathscr{E}$.

(b) Let $T \in \mathscr{E}'$, we restrict T to $\mathscr{S}$. Since $\varphi_i \to \varphi$ in $\mathscr{S}$ implies that $\varphi_i \to \varphi$ in $\mathscr{E}$ ($(1 + |x|^2)^m$ is bounded on each compact subset K) $T|_{\mathscr{S}}$ is also continuous on $\mathscr{S}$, that is, $T|_{\mathscr{S}} \in \mathscr{S}'$. Because of (a) the inclusion of $\mathscr{E}'$ in $\mathscr{S}'$ obtained in this way is injective.

(c) is trivial. ■

Before the next theorem we make some remarks. The seminorms (44) which define the space $\mathscr{S}$ force each $P(x)\varphi$ and $Q(D)\varphi$ to belong to $\mathscr{S}$ whenever φ does, where P and Q are arbitrary polynomials with constant coefficients. We also have

$$\mathscr{S} \subset L_1(\mathbb{R}^r),$$

for, since $\sup_x[(1 + |x|^2)^r \varphi(x)] \leqslant C < \infty$ ($\varphi \in \mathscr{S}$), it follows that

$$|\varphi(x)| \leqslant \frac{C}{(1 + |x|^2)^r},$$

or

$$\|\varphi\|_{L_1} = \int |\varphi(x)| \mathrm{d}x \leqslant C \int \frac{\mathrm{d}x}{(1 + |x|^2)^r} = c' < \infty.$$

We now come to the Fourier transformation.

Theorem 1.20 *The Fourier transformation* (42) *is a continuous map* $\mathcal{F}:\mathcal{S}\to\mathcal{S}$ *and the following formulae hold*

$$\mathcal{F}\left(\frac{\partial\varphi}{\partial x_j}\right)=\mathrm{i}\xi_j\mathcal{F}(\varphi) \quad and \quad \mathcal{F}(x_j\varphi)=\mathrm{i}\frac{\partial}{\partial\xi_j}(\mathcal{F}\varphi),\ \ j=1,\dots,r. \tag{47}$$

Proof. Let $\varphi\in\mathcal{S}$, by the preceding remark φ and $\partial\varphi/\partial x_j$ belong to L_1, and partial integration gives the first formula in (47). To prove the second formula, we interchange $\mathrm{i}\partial/\partial\xi_j$ and the integral sign, which is permissible since given $x_j\cdot\varphi\in L_1$ the new integral $\int e^{-\mathrm{i}(x,\xi)}x_j\cdot\varphi(x)\,dx=\mathcal{F}(x_j\varphi)(\xi)$ is uniformly convergent. Finally for continuity, definition (42) shows that for $\varphi\in L_1$ the function $\mathcal{F}\varphi$ is bounded. Let $\varphi\in\mathcal{S}$, then $(1-\Delta)^m(x^\alpha\varphi(x))\in\mathcal{S}\subset L_1$, where $\Delta=(\partial/\partial x_1)^2+\cdots+(\partial/\partial x_r)^2$ is the Laplace operator. If we apply (47) we see that $(1+|\xi|^2)^m\cdot \mathrm{i}^{|\alpha|}D^\alpha\mathcal{F}\varphi$ is bounded, so that $\mathcal{F}\varphi\in\mathcal{S}$. Furthermore, for each m and α, the following estimate holds:

$$\begin{aligned}\sup_\xi|(1+|\xi|^2)^m D^\alpha\mathcal{F}\varphi(\xi)|&=\sup_\xi\left|\int e^{-\mathrm{i}(\xi,x)}(1-\Delta)^m(x^\alpha\varphi(x))\,dx\right|\\ &\leqslant\int\frac{(1+|x|^2)^r|(1-\Delta)^m(x^\alpha\varphi(x))|\,dx}{(1+|x|^2)^r}\\ &\leqslant\int\frac{dx}{(1+|x|^2)^r}Cp_{2m,r+|\alpha|}(\varphi)\\ &=C'p_{2m,r+|\alpha|}(\varphi),\end{aligned}$$

and hence we have proved continuity. ■

Theorem 1.21 *The Fourier inversion formula* (43) *holds in* $\mathcal{S}$.

Proof. For the proof we need the function $\psi(x)=e^{-|x|^2/2}\in\mathcal{S}$, for which

$$\hat{\psi}(y)=(\mathcal{F}\psi)(y)=(2\pi)^{r/2}e^{-|y|^2/2} \quad \text{and} \quad \int\hat{\psi}(y)\,dy=(2\pi)^r.$$

Also for each $\varphi\in\mathcal{F}$

$$\widehat{\varphi(\varepsilon x)}(y)=\frac{1}{\varepsilon^r}\hat{\varphi}\left(\frac{y}{\varepsilon}\right) \tag{48}$$

holds. Let $\varphi\in\mathcal{S}$, we must work out the repeated integral

$$\int e^{\mathrm{i}(x,\xi)}\int\varphi(y)e^{-\mathrm{i}(y,\xi)}\,dy\,d\xi. \tag{49}$$

Since the function under the integral sign does not belong to $L_1(\mathbb{R}^r_y)\times$

$L_1(\mathbb{R}^r_\xi)$ we cannot apply Fubini's theorem. We get round this by introducing the function $\psi(\xi)$ as an additional factor in the integral (49). Because $\psi(\xi)\in \mathscr{S}\subset L_1$ the function under the integral sign is now absolutely integrable,

$$|e^{i(x,\xi)}\varphi(y)\psi(\xi)e^{-i(y,\xi)}| = |\varphi(y)\psi(\xi)|\in L_1\times L_1,$$

and we can freely interchange the integrals:

$$\int \hat{\varphi}(\xi)\psi(\xi)e^{i(x,\xi)}d\xi = \int \hat{\psi}(y-x)\varphi(y)dy = \int \hat{\psi}(y)\varphi(x+y)\,dy. \tag{50}$$

In (50) replace $\psi(\xi)$ by $\psi(\varepsilon\xi)$ and use formula (48):

$$\int \hat{\varphi}(\xi)\psi(\varepsilon\xi)e^{i(x,\xi)}d\xi = \int \hat{\psi}(y)\varphi(x+\varepsilon y)\,dy.$$

The functions $\hat{\varphi}$ and $\hat{\psi}$ are in $\mathscr{S}\subset L_1$, φ and ψ are bounded and continuous, hence we can allow $\varepsilon\to 0$ in the limit under the integral sign, and obtain

$$\psi(0)\int \hat{\varphi}e^{i(x,\xi)}d\xi = \varphi(x)\int \hat{\psi}\,dy,$$

that is,

$$\int \hat{\varphi}e^{i(x,\xi)}d\xi = (2\pi)^r\varphi(x),$$

the inversion formula. ∎

Corollary 1.3 *The Fourier transformation is a linear, topological isomorphism*

$$\mathscr{F}:\mathscr{S}\leftrightarrow\mathscr{S}.$$

We collect together into a theorem further basic properties of the Fourier transformation:

Theorem 1.22 *If* $\varphi,\psi\in\mathscr{S}$, *then*

$$\int \hat{\varphi}\cdot\psi\,dx = \int \varphi\cdot\hat{\psi}\,dx, \tag{51}$$

$$\int \omega\cdot\bar{\psi}\,dx = \frac{1}{(2\pi)^r}\int \hat{\varphi}\cdot\bar{\hat{\psi}}\,dx \quad \text{(Parseval equation)}, \tag{52}$$

$$\widehat{\varphi*\psi} = \hat{\varphi}\cdot\hat{\psi}, \tag{53}$$

$$\widehat{\varphi\cdot\psi} = \frac{1}{(2\pi)^r}\hat{\varphi}*\hat{\psi}. \tag{54}$$

Proof. Equation (50) is of course true for each φ and ψ from $\mathscr{S}$. If we take

$x=0$, we obtain (51). In order to prove (52), we put $\chi=(1/(2\pi)^r)\hat{\psi}$, for which the Fourier inversion formula (43) (see Theorem 1.21) gives

$$\hat{\hat{\chi}}(\xi)=\frac{1}{(2\pi)^r}\int\hat{\psi}(x)e^{i(x,\xi)}\,dx=\psi(\xi),$$

and we obtain (52) by substituting χ for ψ in (51). We now prove (53):

$$\begin{aligned}\widehat{\varphi*\psi}&=\int e^{-i(x,\xi)}\left(\int\varphi(y)\psi(x-y)dy\right)dx=\int\varphi(y)\left(\int e^{-i(x,\xi)}\psi(x-y)dx\right)dy\\&=\int\varphi(y)\left(\int e^{-i(t+y,\xi)}\psi(t)dt\right)dy\\&=\int\varphi(y)e^{-i(y,\xi)}\left(\int e^{-i(t,\xi)}\psi(t)dt\right)dy=\hat{\varphi}\cdot\hat{\psi}.\end{aligned}$$

The second equality holds since the function under the integral sign belongs to $L_1\times L_1$, allowing us to apply Fubini's theorem, and so freely interchange the integral signs. (54) is obtained from (53) by means of the Fourier inversion formula (43) (Theorem 1.21). ■

We carry over the Fourier transformation to $\mathscr{S}'$ by means of duality. Indeed we have obtained all the operations defined on distributions by means of duality.

Definition 1.9 *Let $T\in\mathscr{S}'$. We define the Fourier transform $\hat{T}$ by*

$$\hat{T}(\varphi):=T(\hat{\varphi}),\quad \varphi\in\mathscr{S}.$$

Theorem 1.20 shows that $\hat{T}\in\mathscr{S}'$, and formula (51) shows that, if $T\in L_1$, the new definition agrees with the usual definition of the Fourier transformation. We may also write the Fourier inversion formula (43) (Theorem 1.21) as $\hat{\hat{\varphi}}=(2\pi)^r\check{\varphi}$ where, as already indicated, $\check{\varphi}$ equals $\varphi(-x)$.

Having set $\check{T}(\varphi):=T(\check{\varphi})$, we may formulate:

Theorem 1.23 *The Fourier inversion formula holds for each $T\in\mathscr{S}'$, that is,*

$$\hat{\hat{T}}=(2\pi)^r\check{T}.$$

The Fourier transformation is an isomorphism $\mathscr{F}:\mathscr{S}'\leftrightarrow\mathscr{S}'$ which, if we take the weak topology on $\mathscr{S}'$, is continuous in both directions. Furthermore the formulae (47) may be carried over immediately to $T\in\mathscr{S}'$.

Proof. We have $\hat{\hat{T}}(\varphi)=T(\hat{\hat{\varphi}})=(2\pi)^rT(\check{\varphi})=(2\pi)^r\check{T}(\varphi)$. The remaining assertions of the theorem are clear. ■

We come now to the important Parseval theorem, namely that the Fourier transformation, applied to $L_2(\mathbb{R}^r)$ defines an isometry. We show first that L_2 admits an embedding in $\mathscr{S}'$. Let $f\in L_2(\mathbb{R}^r)$, applying the Schwarz inequality we obtain

$$\begin{aligned}|T_f(\varphi)| = \left|\int f\varphi\,\mathrm{d}x\right| &\leqslant \left[\int |f|^2\,\mathrm{d}x\right]^{1/2}\left[\int |\varphi|^2\,\mathrm{d}x\right]^{1/2}\\ &\leqslant c\left[\int \frac{|\varphi|^2(1+|x|^2)^{2r}}{(1+|x|^2)^{2r}}\,\mathrm{d}x\right]^{1/2}\\ &\leqslant c\left[\int \frac{\mathrm{d}x}{(1+|x|^2)^{2r}}\right]^{1/2}\sup|\varphi(x)(1+|x|^2)^r|\\ &= c'p_{0,r}(\varphi),\end{aligned}$$

that is, $T_f\in\mathscr{S}'$. This means that each $f\in L_2$ represents an element in $\mathscr{S}'$. Since $\mathscr{S}$ is dense in L_2, the inclusion $L_2\to\mathscr{S}'$ is injective. We also have that if $T\in\mathscr{S}'$ with $|T(\varphi)|\leqslant c\|\varphi\|_{L_2}$ then $T=T_f$ for some $f\in L_2$. ■

Theorem 1.24 *Let $f\in L_2(\mathbb{R}^r)$, then $\hat{f}\in L_2(\mathbb{R}^r)$ and we have Parseval's equation*

$$\int |\hat{f}|^2\,\mathrm{d}x = (2\pi)^r\int |f|^2\,\mathrm{d}x. \tag{55}$$

By application of the parallelogram equation to (55) *we obtain the full Parseval equation*

$$\int \hat{f}\cdot\bar{\hat{g}}\,\mathrm{d}x = (2\pi)^r\int f\cdot\bar{g}\,\mathrm{d}x,\quad f,g\in L_2(\mathbb{R}^r).$$

Proof. Definition 1.9 and formula (52) give

$$|\hat{f}(\varphi)| = |f(\hat{\varphi})| \leqslant \|f\|_2\|\hat{\varphi}\|_2 = (2\pi)^{r/2}\|f\|_2\|\varphi\|_2,\quad \varphi\in\mathscr{D}.$$

Since $\mathscr{D}$ is dense in $L_2(\mathbb{R}^r)$ (apply Theorem 1.3 or look at Theorem 3.4), we can extend the last estimate to L_2

$$|\hat{f}(\varphi)| \leqslant (2\pi)^{r/2}\|f\|_2\|\varphi\|_2,\quad \varphi\in L_2$$

and apply the Riesz theorem. Denoting the Riesz element again by $\hat{f}$ we have

$$\hat{f}(\varphi) = \int \hat{f}\varphi\,\mathrm{d}x,$$

with

$$\|\hat{f}\|_2 \leqslant (2\pi)^{r/2}\|f\|_2,$$

see the equation above. Iterating the last inequality twice yields

$$(2\pi)^r \|f\|_2 = \|\hat{\hat{f}}\|_2 \leqslant (2\pi)^{r/2} \|\hat{f}\|_2 \leqslant (2\pi)^r \|f\|_2,$$

which is (55). ∎

Theorem 1.25 *Let T be a distribution with compact support, $T \in \mathscr{E}'$. Then its Fourier transform is the function*

$$\hat{T}(\xi) = T_x(\mathrm{e}^{-\mathrm{i}(x,\xi)}). \tag{56}$$

The right-hand side of (56) is also defined for complex values $\xi \in \mathbb{C}^r$, and represents an entire, analytic function, the so-called Laplace–Fourier transform of T.

Proof. The proof is trivial for functions with compact support. In order to prove the general case, we regularise. By Theorem 1.16 (g), as $\varepsilon \to 0$ we have $T * h_\varepsilon \to T$ in the weak topology on $\mathscr{E}'$, hence also in $\mathscr{S}'$. By Theorem 1.23, as $\varepsilon \to 0$, $T * h_\varepsilon$ converges to $\hat{T}$ in $\mathscr{S}'$.

On the other hand we have

$$\widehat{T * h_\varepsilon} = (T * H_\varepsilon)(\mathrm{e}^{-\mathrm{i}(x,\xi)}) = T(\tilde{h}_\varepsilon * \mathrm{e}^{-\mathrm{i}(x,\xi)}) = \hat{\tilde{h}}(\varepsilon\xi)\, T(\mathrm{e}^{-\mathrm{i}(x,\xi)}),$$

and $\hat{\tilde{h}}(\varepsilon\xi)$ tends to $\hat{\tilde{h}}(0) = \int \tilde{h}(x)\mathrm{d}x = 1$, and from what we already know we obtain

$$T(\mathrm{e}^{-\mathrm{i}(x,\xi)}) = \hat{T} \quad \text{for } \xi \in \mathbb{R}^r.$$

If $\xi \in \mathbb{C}^r$, then on each compact subset of $\mathbb{C}^r$ the entire, analytic functions $(T * h_\varepsilon)$ $(\mathrm{e}^{-\mathrm{i}(x,\xi)})$ converge uniformly towards $T(\mathrm{e}^{-\mathrm{i}(x,\xi)})$ as $\varepsilon \to 0$. By Weierstrass' theorem $T(\mathrm{e}^{-\mathrm{i}(x,\xi)})$ is again an entire, analytic function of $\xi \in \mathbb{C}^r$. ∎

Theorem 1.26 *Let $T_1 \in \mathscr{E}'$ and $T_2 \in \mathscr{S}'$. Then $T_1 * T_2$ belongs to $\mathscr{S}'$, and the Fourier transform of $T_1 * T_2$ equals $\hat{T}_1 \cdot \hat{T}_2$:*

$$\widehat{T_1 * T_2} = \hat{T}_1 \cdot \hat{T}_2. \tag{57}$$

By Theorem 1.25 $\hat{T}_1 \in C^\infty$ and the product in (57) is well-defined.

Proof. Let $\varphi \in \mathscr{D}$, by definition of convolution we have

$$(T_1 * T_2)(\varphi) = T_1 * (T_2 * \check{\varphi})(0) = \check{T}_1(T_2 * \check{\varphi}).$$

Since $T_1 \in \mathscr{E}'$, there exists a compact subset $K \subset \mathbb{R}^r$ and constants C, k with

$$|(T_1 * T_2)(\varphi)| = |\check{T}_1(T_2 * \check{\varphi})| \leqslant C \sum_{|s| \leqslant k} \sup_K |\mathrm{D}^s(T_2 * \check{\varphi})|$$

$$= C \sum_{|s| \leqslant k} \sup_K |(\mathrm{D}^s T_2) * \check{\varphi}| = C \sum_{|s| \leqslant k} \sup_K |\mathrm{D}^s T_2(\varphi(x-y))_y|.$$

Since all $D^s T_2$ (finitely many!) are in $\mathscr{S}'$, we can further estimate

$$|((T_1 * T_2)\varphi)| \leqslant C' \sup_{x\in K} p_{k',m'}(\varphi(x-y))_y \leqslant C'' p_{k',m'}(\varphi),$$

where $p_{k',m'}$ is some seminorm (44) on $\mathscr{S}$. By Theorem 1.18 (a) $\mathscr{D}$ is dense in $\mathscr{S}$, and by a density argument we have

$$|(T_1 * T_2)(\varphi)| \leqslant C'' p_{k',m'}(\varphi), \quad \varphi \in \mathscr{S},$$

that is, $T_1 * T_2 \in \mathscr{S}'$.

In order to prove (57), we suppose first that $T_1 \in \mathscr{D}$. Let $\psi \in \mathscr{S}$ and $\hat{\psi} \in \mathscr{S}$, then we have

$$(T_1 * T_2)(\hat{\psi}) = T_2(\check{T}_1 * \hat{\psi}) = \hat{T}_2(\hat{T}_1 \psi) = \hat{T}_1 \cdot \hat{T}_2(\psi).$$

Here we have used (54) and Theorem 1.21:

$$\widehat{\hat{T}_1 \cdot \psi} = \check{T}_1 * \hat{\psi}.$$

Since $\mathscr{D}$ is dense in $\mathscr{S}$, we can drop the assumption that $\hat{\psi} \in \mathscr{D}$, and have that $(T_1 * T_2)(\hat{\psi}) = \hat{T}_1 \cdot \hat{T}_2(\psi)$ for all $\psi \in \mathscr{S}$, that is, we have proved (57) in the special case that $T_1 \in \mathscr{D}$, $T_2 \in \mathscr{S}'$. Suppose now that in general $T_1 \in \mathscr{E}'$ and $T_2 \in \mathscr{S}'$; let $\varphi \in \mathscr{D}$, we have $(\varphi * T_1) * T_2 = \varphi * (T_1 * T_2)$. Applied to this equation the Fourier transformation gives

$$(\widehat{\varphi * T_1}) \cdot \hat{T}_2 = \hat{\varphi} \cdot (\widehat{T_1 * T_2}), \quad \text{since } \varphi \in \mathscr{D} \text{ and } \varphi * T_1 \in \mathscr{D}.$$

In this way we show that $\widehat{\varphi * T_1} = \hat{\varphi} \cdot \hat{T}_1$, and taking everything together

$$\hat{\varphi} \cdot \hat{T}_1 \cdot \hat{T}_2 = \hat{\varphi} \cdot (\widehat{T_1 * T_2}). \tag{58}$$

It is always possible to choose $\varphi \in \mathscr{D}$, so that for an arbitrarily preassigned $\xi_0 \in \mathbb{R}^r$, $\hat{\varphi}(\xi_0) \neq 0$. Therefore from (58) we conclude that

$$\widehat{T_1 * T_2} = \hat{T}_1 \cdot \hat{T}_2.$$ ■

Exercises

1.1 Let T be a distribution with supp $T \subset \{0\}$. Show that there is a representation $T = \sum_{|\alpha| \leqslant m} d_\alpha D^\alpha \delta$, where m is finite and the d_α are constants.

1.2 Let E be an arbitrary closed subset of $\mathbb{R}^r$. Show that there exists a function $f \in \mathscr{E}(\mathbb{R}^r)$ with $f(x) = 0$ for $x \in E$ and $f(x) > 0$ otherwise.

1.3 Let $\Omega = (0, \infty)$, and define the distribution T by

$$T(\varphi) := \sum_{m=1}^{\infty} D^m \varphi\left(\frac{1}{m}\right), \quad \varphi \in \mathscr{D}((0,\infty)).$$

Show that the order of T is infinite, and that T is not extendable to a distribution from $\mathscr{D}'(\mathbb{R})$.

1.4 For which functions $f\in C^{\infty}$ is it true that $f\cdot\delta'=0$? The same question for $f\cdot\delta''=0$?

1.5 Show that the sequences

$$\tfrac{1}{2}ne^{-n|x|},\quad \frac{2}{\pi}\frac{n}{e^{nx}+e^{-nx}},\quad \frac{n}{\pi(1+n^2x^2)}$$

in $\mathscr{D}'(\mathbb{R})$ converge to δ.

1.6 If $\varphi,\psi\in\mathscr{S}$, show $\varphi*\psi$ also belongs to $\mathscr{S}$.

1.7 If $T\in\mathscr{D}'(\mathbb{R})$ and P is a polynomial of degree k in x, show that $T*P$ is also a polynomial of degree at most k.

1.8 If $S,T\in\mathscr{E}'(\mathbb{R})$ and $k\in\mathbb{N}$, show that $x^k(S*T)=\sum_{v=0}^{k}\binom{k}{v}(x^vS*x^{k-v}T)$.

1.9 Calculate the Fourier transforms of the following functions

$$x^ke^{-ax}H(x),\quad a>0, k\in\mathbb{N};\quad e^{-ax^2},\quad a>0;\quad (a^2+x^2)^{-1}, a\neq 0.$$

1.10 Let $f\in L_1(\mathbb{R}^r)$, show that for $|x|\to\infty$ the Fourier transform $\hat{f}(x)\to 0$.

1.11 Let $f\in L_1(\mathbb{R}^r)$, show that the Fourier transform $\hat{f}$ is a continuous function.

1.12 Show that $e^x\notin\mathscr{S}'$, $e^x\cos e^x\in\mathscr{S}'$.

1.13 *Periodic distributions.* Let T^r be an r-dimensional torus; we may interpret functions defined on T^r as periodic: to each function φ on T^r we may associate a function $\tilde{\varphi}$ of period 2π in each variable by

$$\tilde{\varphi}(x_1,\dots,x_r)=\varphi(e^{ix_1},\dots,e^{ix_r}).$$

Let $\mathscr{D}(T^r)=\mathscr{E}(T^r)$ the set of all C^{∞}-functions on T^r given the $\mathscr{E}$-topology (see §1.7), then $\mathscr{D}(T^r)$ is an (F)-space. We define the Fourier coefficients c_k of $\varphi\in\mathscr{D}(T^r)$ by

$$c_k=\int_{T^r}\varphi(x)e^{-i(x,k)}\,dx,\quad k=(k_1,\dots,k_r)\in\mathbb{Z}^r,$$

where dx is the Haar measure on T^r. We have

$$\varphi(x)=\sum_k c_ke^{i(x,k)}\quad\text{(convergence in } \mathscr{D}(T^r))$$

and the topology determined by the seminorms

$$P_N(\varphi)=\left[\sum_k(1+|k|^2)^N|c_k|^2\right]^{1/2}$$

is equivalent to the $\mathscr{E}$-topology on $\mathscr{D}(T^r)$. Let $\mathscr{D}'(T^r)$ be the dual space; its elements are called periodic distributions. For $T\in\mathscr{D}'(T^r)$ we can again calculate the Fourier coefficients by means of

$$c_k:=T(e^{i(k,x)}),\quad k\in\mathbb{Z}^r$$

and we have

$$T = \sum_k c_k e^{i(x,k)}$$

by weak convergence in $\mathscr{D}'(T^r)$, T belongs to $\mathscr{D}'(T^r)$ if and only if we can find N and C with

$$|c_k| \leqslant C(1 + |k|)^N \quad \text{for all } k \in \mathbb{Z}^r.$$

The convolution $T * S$ of two periodic distributions is most simply defined by means of the Fourier coefficients: $c_k(T) \cdot c_k(S)$; that is, $(T * S)\,(e^{i(k,x)}) := c_k(T) \cdot c_k(S)$ and the results for convolution in §1.5 remain valid with distinctly simpler proofs.

1.14 Show that the distribution $1/|x|^{r-2}$ is an elementary solution for the Laplace operator $-\Delta$, that is,

$$-\Delta \frac{1}{|x|^{r-2}} = \frac{(r-2)2\pi^{r/2}}{(r/2)} \delta(x) = N\delta(x), \quad r \geqslant 3,$$

or

$$-\Delta_x \frac{1}{|x-y|^{r-2}} = N\delta(x-y), \quad r \geqslant 3.$$

Similarly for $r = 2$

$$-\Delta \ln \frac{1}{|x|} = 2\pi\delta(x)$$

1.15 For the biharmonic operator Δ^2 and $r = 2$ show that

$$\Delta^2 \frac{1}{8\pi} |x|^2 \ln |x| = \delta.$$

Further elementary solutions are given in L. Schwartz [1].

§2 Geometric assumptions for the domain Ω

For the embedding theorems in the theory of Sobolev spaces we need various geometric assumptions for the domain Ω. We begin with segment and cone properties, and consider two notions of regularity for the frontier $\partial\Omega$ of Ω: $N^{k,\kappa}$ and $C^{k,\kappa}$-regularity. We exhibit various connections between the notions introduced. In questions of integration it is important to know when $\partial\Omega$ has measure zero. Theorems 2.7, 2.8 and 2.9 give sufficient conditions for this. We need normal and admissible coordinate transformations for Green's formulae and elliptic differential operators; we introduced these transformations in Definitions 2.8 and 2.9, and ensure their existence by Theorems 2.11 and 2.12. The remainder of this section is devoted to differentiable manifolds and submanifolds.

2.1 Segment and cone properties

For most theorems we need only assume that Ω is an open subset of $\mathbb{R}^r$ bounded or unbounded. For some theorems we must assume that Ω is a region in $\mathbb{R}^r$, that is, an open and connected set. We denote the frontier (or boundary) of Ω by $\partial\Omega = \bar{\Omega} \cap C\Omega$.

Definition 2.1 *We say that Ω has the segment property, if for each $x \in \partial\Omega$ we can find a neighbourhood U_x and a vector $y_x \neq 0$, such that for each $z \in \bar{\Omega} \cap U_x$ the point $z + ty_x$ belongs to Ω for all $0 < t < 1$ (see Fig. 2.1).*

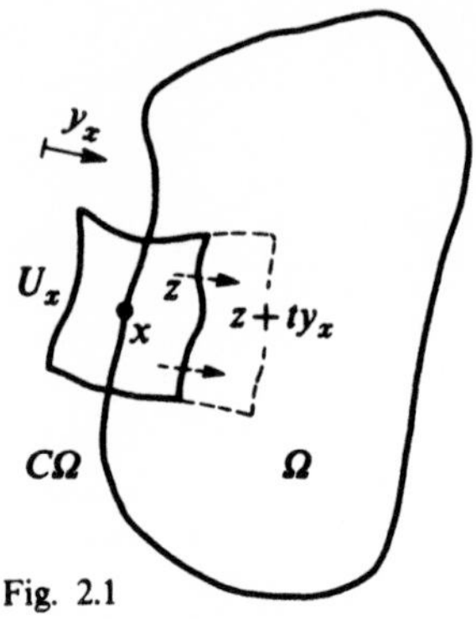

Fig. 2.1

Let $B(x, \rho) = \{y : |y - x| < \rho\}$ be an open ball with x as centre. By a cone $C(x, \rho, \Sigma)$ with vertex at x we mean the set

$$C(x, \rho, \Sigma) := B(x, \rho) \cap \{\lambda(y - x) : y \in \Sigma, \lambda > 0\},$$

where Σ is an open non-empty set on the sphere $S(x, \rho) = \{y : |y - x| = \rho\}$ bounding $B(x, \rho)$. Cones are open sets that do not contain the vertex x. Without changing the definition we can also choose Σ to lie on the unit sphere $S(x, 1)$ (see Fig. 2.2).

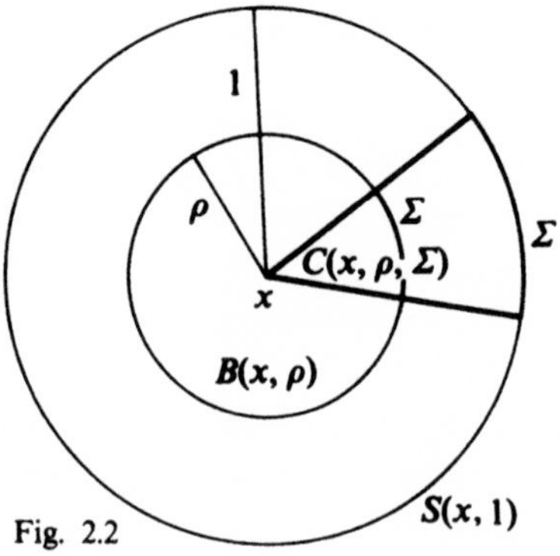

Fig. 2.2

Definition 2.2 *We say that Ω has the cone property, if for each $x \in \bar{\Omega}$ we can find a cone $C(x, \rho, \Sigma)$ in Ω, with vertex at x:*

$$C(x, \rho, \Sigma) \subset \Omega,$$

which is congruent to some fixed given cone C_0. We will call C_0 the standard cone. Here congruent means that translations and rotations are permitted (see *Fig.* 2.3).

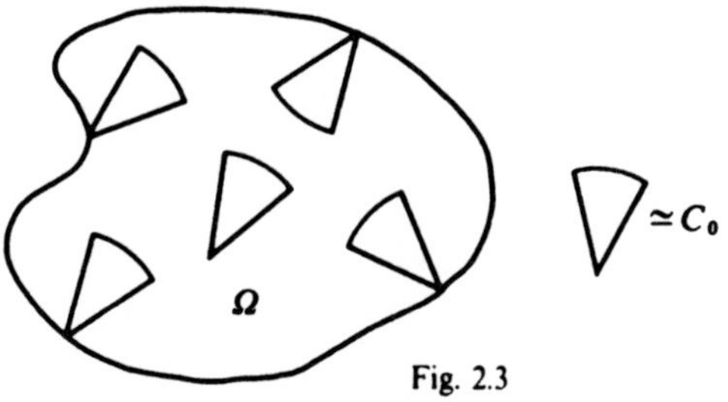

Fig. 2.3

Remark 2.1 We can call the cone property given in Definition 2.2 *closed* ($x \in \bar{\Omega}$), in contrast to *open* ($x \in \Omega$), where we only demand that for $x \in \Omega$ there exists a cone $C(x, \rho, \Sigma) \subset \Omega$, where $C(x, \rho, \Sigma)$ is congruent to some fixed cone C_0. However, by Lemma 2.2 the closed and open cone properties are equivalent, and we do not need to distinguish between them.

Definition 2.3 *We say that Ω has the uniform cone property, if for each $x \in \bar{\Omega}$ we can find a neighbourhood U_x and a cone C^x, such that C^x is congruent to a fixed given cone C_0 with the property*

$$\textit{for each } z \in \bar{\Omega} \cap U_x \textit{ the subset } z + C^x \textit{ is contained in } \Omega. \qquad (1)$$

The uniform cone property thus implies that each translation by $z, z \in U_x \cap \bar{\Omega}$ leaves the cone C^x inside Ω (see Fig. 2.4).

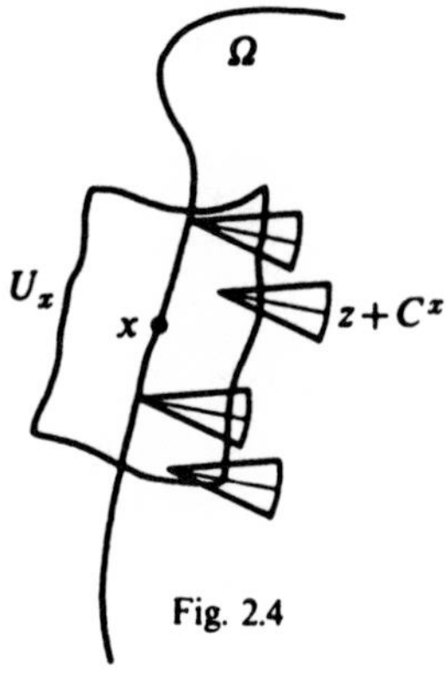

Fig. 2.4

Remark 2.2 Here also we can distinguish between the closed and open uniform cone properties, that is weaken (1) to say

$$\text{for each } z \in \Omega \cap U_x \text{ the subset } z + C^x \subset \Omega. \tag{2}$$

A simple continuity argument gives the equivalence: let $z \in \partial\Omega \cap U_x$, and let the cone $z + C^x$ possess an interior point x_0 from $C\Omega$. Then there exists a small neighbourhood V of z, $V \subset U_x$, with the property that all cones $y + C^x$, $y \in V$, also have the point $x_0 \in C\Omega$ as interior point. However, since z was a boundary point, V always contains points y from Ω, and we arrive at a contradiction to (2).

We see immediately that the uniform cone property implies the cone property. The uniform cone property also implies the segment property; in order to see this it suffices to take as y_x in Definition 2.1 an arbitrary vector $0 \neq y_x \in C^x$ originating at the vertex of C^x.

2.2 The $N^{k,\kappa}$-property of Ω

We formulate the $N^{k,\kappa}$-property, where $k = 0, 1, \dots$ and $\kappa \in [0, 1]$. Again we give a local definition:

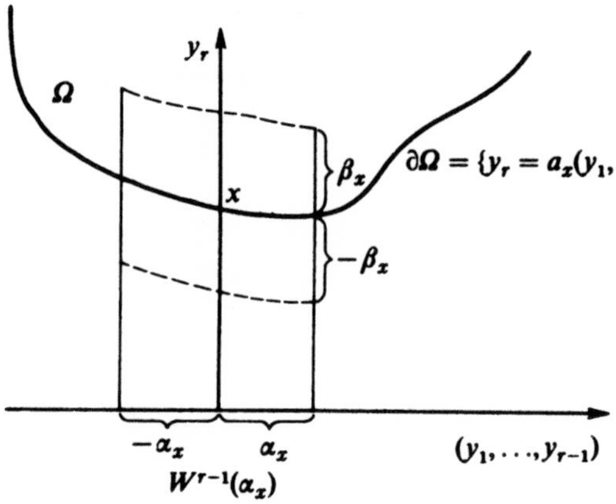

Fig. 2.5

Definition 2.4 (see Fig. 2.5) *We say that Ω has the $N^{k,\kappa}$-property, if for each $x \in \partial\Omega$ we can find a neighbourhood U_x, an orthogonal coordinate transformation $A_x: \mathbb{R}^r \to \mathbb{R}^r$, two numbers $\alpha_x, \beta_x > 0$ and a function $a_x: \mathbb{R}^{r-1} \to \mathbb{R}$ with the following properties.*

Let $y_1, \dots, y_r$ be the new Cartesian (orthogonal) coordinates given by the *transformation A_x, and let $W^{r-1}(\alpha_x)$ be the cube*

$$\{(y_1,\ldots,y_{r-1}):|y_i|<\alpha_x, i=1,\ldots,r-1\} \text{ in } \mathbb{R}^{r-1}.$$

We require that:

1. $a_x \in C^{k,\kappa}(W^{r-1}(\alpha_x))$.
2. *The portion $U_x \cap \partial\Omega$ of the frontier may be described in the new coordinates by the function $y_r = a_x(y_1,\ldots,y_{r-1})$, that is,*
$$U_x \cap \partial\Omega = \{(y_1,\ldots,y_r):(y_1,\ldots,y_{r-1}) \in W^{r-1}(\alpha_x); y_r = a_x(y_1,\ldots,y_{r-1})\}.$$
3. *If we displace the portion $U_x \cap \partial\Omega$ of the frontier by β_x, then we remain inside Ω, that is,*
$$U_x \cap \Omega = \{(y_1,\ldots,y_r):|y_i|<\alpha_x, \quad i=1,\ldots,r-1,$$
$$a_x(y_1,\ldots,y_{r-1}) < y_r < a_x(y_1,\ldots,y_{r-1}) + \beta_x\} = V_x^+.$$
4. *If we displace the portion $U_x \cap \partial\Omega$ of the frontier in the reverse direction by $-\beta_x$, then we immediately find ourselves outside Ω:*
$$U_x \cap \mathbf{C}\bar{\Omega} = \{(y_1,\ldots,y_r):|y_i|<\alpha_x, \quad i=1,\ldots,r-1,$$
$$a_x(y_1,\ldots,y_{r-1}) - \beta_x < y_r < a_x(y_1,\ldots,y_{r-1})\} = V_x^-.$$

Requirements 3 and 4 imply that locally Ω lies on one side of the frontier $\partial\Omega$.

Convention *If $k=0$, then we only consider the classes $N^{0,1}$ and $N^{0,0}$, or in other words: if $k=0$, only the choice $\kappa=1$ or $\kappa=0$ is permitted.*

Example Cubes in $\mathbb{R}^r$ and arbitrary polyhedral regions belong to $N^{0,1}$.

Theorem 2.1 *Let Ω be bounded and have the property $N^{0,1}$. Then Ω also has the uniform cone property.*

Since by our convention $N^{k,\kappa} \subset N^{0,1}$ for all $(k,\kappa) \neq 0$, all bounded $N^{k,\kappa}$-regions also have the uniform cone property for $(k,\kappa) \neq 0$.

Proof. We first consider the frontier $\partial\Omega$ of Ω. Let $x \in \partial\Omega$; we choose U_x, $\alpha_x > 0$, $\beta_x > 0$ etc. according to Definition 2.4, and for the sake of simplicity omit the subscript x for the moment. We consider $U \cap \Omega$, which has property 3, we halve the value of α and label the region thus obtained as $U' \cap \Omega$. Let δ be the distance of the lower bounding surface from the upper bounding surface (see Definition 2.4.3):

$$\delta := \operatorname{dist}(a(y_1,\ldots,y_{r-1}), a(y_1,\ldots,y_{r-1}) + \beta) \leqslant \beta.$$

The distance δ must be positive, for a point common to both bounding surfaces must have the coordinates $(y_1, \ldots, y_{r-1})$, but above $(y_1, \ldots, y_{r-1})$ both surfaces are already separated by β. Let $y_0 \in \partial\Omega \cap U' \cap \Omega$. By Definition 2.4.2 y_0 has coordinates $(y_1^0, \ldots, y_{r-1}^0, a(y_1^0, \ldots, y_{r-1}^0))$. We wish to construct a cone parallel to the y_r-axis and with vertex at y_0, which lies in $U \cap \Omega$ (see Fig. 2.6). We first construct a cone K determining data (v, ρ), which stands perpendicular to a hyperplane. We obtain the desired cone $C(y_0, \rho, \Sigma)$ as the intersection of the cone $K(v, \rho)$ with the ball $B(y_0, \rho)$

$$C(y_0, \rho, \Sigma) := K(v, \rho) \cap B(y_0, \rho). \tag{3}$$

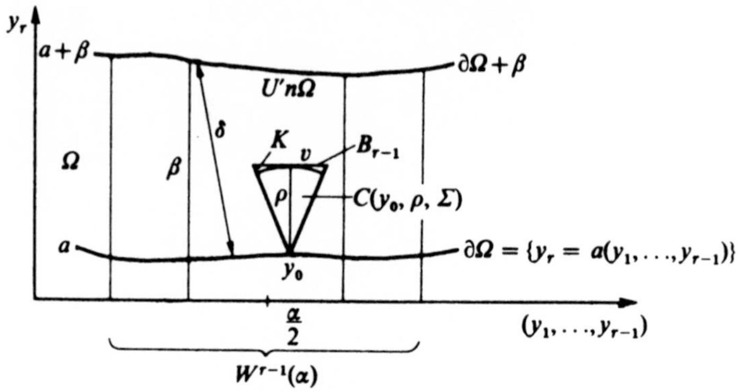

Fig. 2.6

As the principal axis of K we take the line $(y_1^0, \ldots, y_{r-1}^0)$ parallel to the axis y_r which passes through the vertex y_0. Perpendicular to the line $(y_1^0, \ldots, y_{r-1}^0)$ at distance ρ from y_0 we pass an $(r-1)$-hyperplane and mark out on it an $(r-1)$-ball B_r with radius v and centred at the intersection point of the line $(y_1^0, \ldots, y_{r-1}^0)$ and the hyperplane. We obtain the cone $K(v, \rho)$ by joining the points of B_{r-1} to y_0. We submit the determining data (v, ρ) of the cone $K(v, \rho)$ to the conditions

$$0 < \rho < \delta, \quad 0 < v \left\langle \frac{\alpha}{2}, \quad \frac{\rho}{v} \right\rangle k, \tag{4}$$

where k is the Lipschitz constant for the function $a(y') = a(y_1, \ldots, y_{r-1})$. By Definition 2.4.1 we have $a \in C^{0,1}(W^{r-1}(\alpha))$, which implies that there exists some $0 < k < \infty$ such that for all $y_1', y_2' \in W^{r-1}(\alpha)$,

$$|a(y_1') - a(y_2')| \leqslant k |y_1' - y_2'|_{r-1}. \tag{5}$$

Conditions (4) ensure that the cone $C(y_0, \rho, \Sigma)$ given by (3) lies in $U \cap \Omega$. We check this. Since $y_0 \in a$ and $\rho < \delta$, the cone $C(y_0, \rho, \Sigma)$ cannot meet the upper surface $a+\beta$; since $v<\alpha/2$, $C(y_0,\rho,\Sigma)$ lies above the $(r-1)$-cube $W(\alpha)$, and $\rho/v>k$ implies that because of (5) the only intersection point of $C(y_0,\rho,\Sigma)$ with the lower boundary surface a is the point y_0. Taken together all this shows that

$$C(y_0,\rho,\Sigma)\subset U\cap\Omega.$$

Conditions (4) can always be satisfied; this follows trivially from $\delta>0$, $\alpha>0$ and $0<k<\infty$. Since conditions (4) are independent of y_0, all cones $C(y_0,\rho,\Sigma)$ $y_0\in\partial\Omega\cap\bar{U}'$ are equivalent under parallelism or translation. We can therefore write $C(y_0,\rho,\Sigma)=y_0+C(0,\rho,\Sigma)$, where the cone $C(0,\rho,\Sigma)$ does not depend on y_0. Let U'' be the neighbourhood from Definition 2.4, corresponding to the numbers $\alpha/2$, $\beta/2$. If in addition we require that $\rho<\beta/2$, then for $z\in U''\cap\bar{\Omega}$ the cones $z+C(0,\rho,\Sigma)$ all lie in Ω, thus we have shown that the uniform cone condition is satisfied by Ω near the frontier.

The proof concludes with a compactness argument. Cover $\bar{\Omega}$, which is compact, with finitely many sets $U_1,\ldots,U_m$ with the properties:

(a) Let $U_i=U''$ be a cover of the frontier ($U_i\cap\partial\Omega\neq\emptyset$) as above, then the translation equivalent cones $z+C(0,\rho_i,\Sigma_i)=z+C^i$, $z\in\bar{\Omega}\cap U_i$ all lie in Ω.

(b) Let U_i be an 'interior' cone, that is, $U_i\subset\Omega$.

Here we can set $U_i=B(a_i,r_i/2)$, where $B(a_i,r_i)\subset\Omega$; as cone C^i we take some subcone of $B(0,r_i/2)$. The property $z+C^i\subset\Omega$ for $z\in U_i$ holds because certainly $B(a_i,r_i)\subset\Omega$. For the standard cone, which determines the uniform cone property in Definition 2.3, we take the smallest of the cones C^i, $i=1,\ldots,m$ (congruences are admitted). ■

Theorem 2.2 *If $\Omega\in N^{0,0}$ the segment property holds.*

This follows immediately from Definition 2.4.3.

Basically, the most important theorems for the Sobolev spaces $W_2^l(\Omega)$ are proved by imposing the cone condition as an assumption on Ω. The following theorems show this; in part they go back to Gagliardo [1]. We present Theorem 2.3 (Gagliardo) and 2.4 for the sake of completeness; we use them nowhere in what follows.

Lemma 2.1 *Let Ω be open in $\mathbb{R}^r$, and let Ω have the cone property. Then there exist finitely many cones $C_j(0,\rho,\Sigma_j)$, $j=1,\ldots,m$, congruent to some fixed cone $C(0,\rho,\Sigma)$, for which we have*

$$\Omega=\bigcup_{j=1}^{m}\Omega_j,\quad \Omega_j=\bigcup_{x\in A_j}(x+C_j),\quad A_j\subset\Omega. \tag{6}$$

Proof. Let C_0 be the standard cone of Definition 2.2. By a covering argument for the sphere in $B(0,\rho)$ we can find finitely many cones $C_j = C_j(0,\rho,\Sigma_j)$, $j=1,\dots,m$, with the property that each cone C congruent to C_0 with vertex at the origin contains at least one of the cones C_j. Precisely we ask that $\bar{C}_j\setminus\{0\}\subset C$. For each $x\in\Omega$ we can therefore find some j, $1\leqslant j\leqslant m$, with

$$x+C_j\subset C(x,\dots)\subset\Omega \quad\text{and}\quad x+C_j\subset\Omega. \tag{7}$$

Since Ω is open and $\overline{x+C_j}$ is compact, it follows that $y+C_j\subset\Omega$ also for all y sufficiently close to x. This implies that for each $x\in\Omega$ we can find some j and some $y\in\Omega$ with

$$x\in y+C_j\subset\Omega. \tag{8}$$

If we define A_j as $A_j=\{x\in\Omega : x+C_j\subset\Omega\}$, then because of (8) we have proved the existence of the decomposition (6). ■

Lemma 2.2 *Let Ω be open in $\mathbb{R}^r$ and have the open cone property. Then Ω also has the closed cone property* (*see Definition* 2.2).

Proof. Let $z\in\partial\Omega$, and let C_j, $j=1,\dots,m$ be the cones from Lemma 2.1. It is enough to show that for at least one C_j it is true that $z+C_j\subset\Omega$. We suppose the opposite, then each of the cones $z+C_j$, $j=1,\dots,m$ contains at least one point from $\mathbf{C}\Omega$ as an interior point. As in Remark 2.2 we can construct a small neighbourhood V of z with the property that all cones $y+C_j$, $j=1,\dots,m$, $y\in V$, have points in common with $\mathbf{C}\Omega$. However, since $V\cap\Omega\neq\emptyset$, and by (7) for each $y\in\Omega$ there is a j with $x+C_j\subset\Omega$, we obtain a contradiction. ■

Theorem 2.3 *Let Ω be open and bounded in $\mathbb{R}^r$ and have the cone property. Then there exist finitely many parallelepipeds P_j, $j=1,\dots,m$, which for each $\rho>0$ give the following decomposition*

$$\Omega=\bigcup_{i=1}^{n}\Omega_i,\quad \Omega_i=\bigcup_{x\in A_i}(x+P_i),\quad A_i\subset\Omega, \tag{9}$$

where the diameter of $A_i\leqslant\rho$, and the P_is belong to the already determined parallelepipeds P_j, $j=1,\dots,m$. If ρ is sufficiently small, then in addition we have that

$$\Omega_i\in N^{0,1}\quad \textit{for } i=1,\dots,n. \tag{10}$$

Proof. (see Fig. 2.7). Since we can fit a parallelepiped P_j (all P_j congruent), inside each cone C_j, with vertex at 0, we can repeat the proof of Lemma 2.1

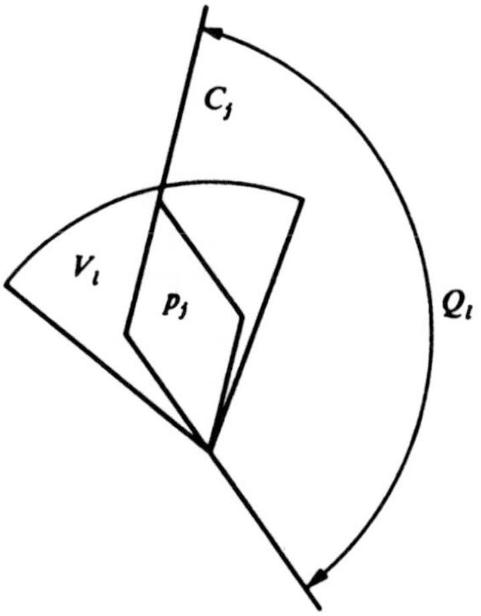

Fig. 2.7

writing P_j instead of C_j throughout, obtaining

$$\Omega = \bigcup_{j=1}^{m} \Omega_j, \quad \Omega_j = \bigcup_{x \in A_j} (x + P_j), \quad A_j \subset \Omega, \quad j = 1, \ldots, m.$$

If the diameters $d(A_j) \leqslant \rho$, $j = 1, \ldots, m$, we have finished, otherwise we take the finite decomposition

$$A_j = \bigcup_{k=1}^{N} A_{jk}, \quad \text{where } d(A_{jk}) \leqslant \rho. \tag{11}$$

Such a decomposition is always possible, since $A_j \subset \Omega$ and Ω is bounded. If we put $\Omega_{jk} = \bigcup_{x \in A_{jk}} (x + P_j)$ and change the numbering, we obtain (9).

Now for the Lipschitz property (10). Let P be congruent to P_j, $j = 1, \ldots, m$, we put

$$\delta := \text{dist}\,(\text{midpoint of } P, \partial P), \quad \sigma := \delta/2, \quad \text{and take } 0 < \rho < \delta. \tag{12}$$

We consider $\Omega_i = \bigcup_{x \in A_i} (x + P_i)$, $d(A_i) \leqslant \rho$, and for the sake of simplicity omit the subscript i, thus

$$\Omega = \bigcup_{x \in A} (x + P), \quad d(A) \leqslant \rho.$$

Let v_l be a corner of P, and let

$$Q_l = \{y = v_l + \lambda(x - v)_l : x \in P, \lambda > 0\},$$

the pyramid generated by P with vertex at v_l. We have $P = \bigcap Q_l$, where the intersection is taken over all 2^r corners of P. Let

$$\Omega_{(l)} := \bigcup_{x \in A} (x + Q_l),$$

and let B be an arbitrary ball with radius $\sigma = \delta/2$. Let x be a fixed point of A, because of (12) the ball B cannot meet two opposite side faces of $x + P$. Therefore there exists a corner v_l of P with the property that $x + v_l$ lies on all side faces of $x + P$ meeting B, in the event that these actually exist. Hence $B \cap (x + P) = B \cap (x + Q_l)$. Next suppose $x, y \in A$, and let us assume that B meets two relatively opposite side faces of $x + P$ and $y + P$, that is, there exist points a and b on opposite side faces of P with

$$x + a \in B \quad \text{and} \quad y + b \in B.$$

Then we would have

$$\rho \geqslant \operatorname{dist}(x, y) = \operatorname{dist}(x + b, y + b) \geqslant \operatorname{dist}(x + b, x + a) - \operatorname{dist}(x + a, y + b)$$
$$\geqslant 2\delta - 2\delta = \delta,$$

contradicting (12). For $\rho < \delta$ B cannot meet relatively opposite side faces of $x + P$ and $y + P$; here x, y are arbitrary. Hence

$$B \cap (x + P) = B \cap (x + Q_l),$$

where l is independent of $x \in A$, from which it follows that

$$B \cap \Omega = B \cap \Omega_{(l)}. \tag{13}$$

We take Cartesian coordinates $(\xi', \xi_r) = (\xi_1, \ldots, \xi_{r-1}, \xi_r)$ in the ball B, so that the ξ_r-axis runs parallel to the vector from the midpoint of P to the corner v_l. We can characterise the region $(x + Q_l) \cap B$ in B by the inequality $\xi_r > a_x(\xi')$, where the function $a_x(\xi')$ satisfies a Lipschitz condition with constant k independent of x (see Fig. 2.8). Therefore given (13) we can characterise

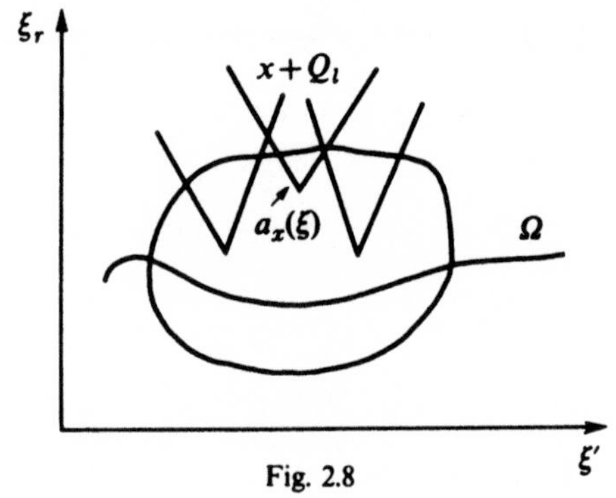

Fig. 2.8

$B \cap \Omega_{(l)}$ and also $B \cap \Omega$ by the inequality $\xi_r > a(\xi')$, where the function $a(\xi') = \inf_{x\in A} a_x(\xi')$ is again Lipschitz (see below). Since B can be a neighbourhood of an arbitrary point z of $\partial\Omega$, we see that we can satisfy the conditions of Definition 2.4, thus proving (10). It remains to prove the Lipschitz nature of $a(\xi')$. Let $\varepsilon > 0$ be arbitrarily preassigned; from the definition of the infimum we have

$$\begin{aligned} |a(\xi'_1) - a(\xi'_2)| &= \left|\inf_{x\in A} a_x(\xi'_1) - \inf_{x\in A} a_x(\xi'_2)\right| \\ &= |a_{x_1}(\xi'_1) - \delta_1 - (a_{x_2}(\xi'_2) - \delta_2)| \\ &\leqslant |a_{x_1}(\xi'_1) - a_{x_2}(\xi'_2)| + \delta_1 + \delta_2, \end{aligned}$$

where $0 \leqslant \delta_1 + \delta_2 < \varepsilon$. We join the points ξ'_1 and ξ'_2 by a two-dimensional plane in the ξ_r-direction and argue in this plane. Since a_{x_1} and a_{x_2} are the bounding surfaces of translation-equivalent pyramids, they must meet at least one point ξ'_0, therefore $a_{x_1}(\xi'_0) = a_{x_2}(\xi'_0)$. At this point we first consider the possibility that ξ'_0 lies between ξ'_1 and ξ'_2. We then have (a_{x_1} and a_{x_2} have the same Lipschitz contant k):

$$\begin{aligned} |a_{x_1}(\xi'_1) - a_{x_2}(\xi'_2)| + \delta_1 + \delta_2 &\leqslant |a_{x_1}(\xi'_1) - a_{x_1}(\xi'_0)| + |a_{x_2}(\xi'_0) - a_{x_2}(\xi'_2)| + \varepsilon \\ &\leqslant k|\xi'_1 - \xi'_0|_{r-1} + k|\xi'_0 - \xi'_2|_{r-1} + \varepsilon \\ &= k|\xi'_1 - \xi'_2|_{r-1} + \varepsilon, \end{aligned}$$

or, since ε was arbitrary,

$$|a(\xi'_1) - a(\xi'_2)| \leqslant k|\xi'_1 - \xi'_2|_{r-1}.$$

If ξ'_0 lies to the left of ξ'_i (and similarly if ξ'_0 lies to the right of ξ'_2), then $a_{x_2}(\xi'_1)$ must lie beneath $a_{x_1}(\xi'_1)$, see Fig. 2.9, and we then take $a(\xi'_1) =$

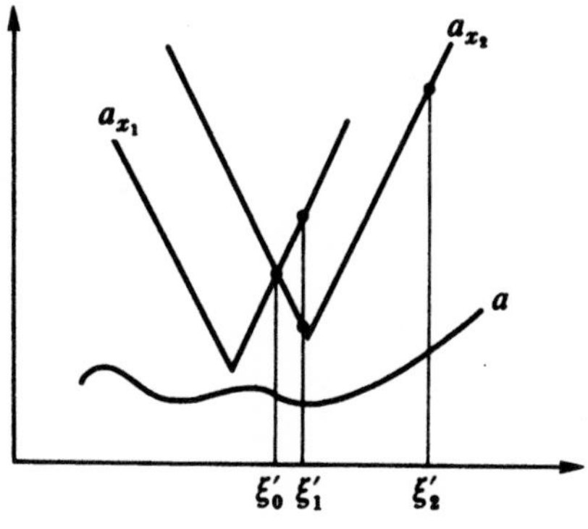

Fig. 2.9

$a_{x_2}(\xi'_1) - \delta'_1$. Since $a(\xi'_1) = \inf_{x \in A} a_x(\xi'_1)$, we must then have $0 \leqslant \delta'_1 \leqslant \delta_1$, and we obtain

$$|a(\xi'_1) - a(\xi'_2)| = |a_{x_2}(\xi'_1) - \delta'_1 - (a_{x_2}(\xi'_2) - \delta_2)|$$
$$\leqslant |a_{x_2}(\xi'_1) - a_{x_2}(\xi'_2)| + \delta'_1 + \delta_2 \leqslant k|\xi'_1 - \xi'_2|_{r-1} + \varepsilon.$$

With this, $\varepsilon > 0$ being arbitrary, we have proved that $a(\xi')$ is Lipschitz continuous. ■

With minor changes Theorem 2.3 can be carried over to unbounded domains.

Theorem 2.4 *Let Ω be an arbitrary open set in $\mathbb{R}^r$, bounded or unbounded, with the cone property. Then there exist finitely many parallelepipeds $P_j, j = 1, \ldots, m$, which for each $\rho > 0$ induce the following decomposition*

$$\Omega = \bigcup_{i=1}^{\infty} \Omega_i, \quad \Omega_i = \bigcup_{x \in A_i} (x + P_i), \quad A_i \subset \Omega, d(A_i) \leqslant \rho,$$

where the families $\{\Omega_i\}$ and $\{A_i\}$ are locally finite, and the P_is are taken from among the already determined parallelepipeds $P_j, j = 1, \ldots, m$. If ρ is sufficiently small, then in addition we have

$$\Omega_i \in N^{0,1} \quad \text{for all } i = 1, 2, \ldots.$$

We only wish to consider those points in the proof of Theorem 2.3 that need to be changed in order to arrive at a proof of Theorem 2.4.

1. The P_js and their total number m depend only on the standard cone C_0 and not on the size of the region Ω. We decompose the A_js into disjoint unions $A_j = \bigcup_{k=1}^{\infty} A_{jk}$ (see (11)), with $d(A_{jk}) \leqslant \rho$, where the families $\{A_{jk}\}_{k=1}^{\infty}$ are locally finite; this is always possible. In this way the families

$$\Omega_{jk} = \bigcup_{x \in A_{jk}} (x + P_j), \quad k = 1, 2, \ldots$$

are also locally finite. Finitely many locally finite families, $j = 1, \ldots, m$, together form another locally finite family (change the numbering!)

$$\{A_i\} = \{A_{jk}\}, \quad \{\Omega_i\} = \{\Omega_{jk}\}.$$

2. To prove that $\Omega_{jk} = \Omega_i \in N^{0,1}$ we need the constants (12). Since these only depend on $P_j, j = 1, \ldots, m$, that is, on C_0, we can take over this part of the proof unchanged, and obtain

$$\Omega_i \in N^{0,1}.$$

2.3 (k, κ)-diffeomorphisms and (k, κ)-smooth Ωs

We introduce the concept of a k-diffeomorphism, and more generally that of a (k, κ)-diffeomorphism.

Definition 2.5 *Let Φ be a 1–1 transformation of the domain $\Omega \subset \mathbb{R}^r$ onto the domain $\Omega' \in \mathbb{R}^r$. We say that Φ is a (k, κ)-diffeomorphism if the following conditions are satisfied:*

1. *The coordinate functions of the map Φ,*

$$\begin{aligned} y_1 &= \varphi_1(x_1, \ldots, x_r) \\ &\vdots \\ y_r &= \varphi_r(x_1, \ldots, x_r) \end{aligned}$$

   belong to the class $C^{k,\kappa}(\bar{\Omega})$.

2. *The coordinate functions of the inverse map Φ^{-1},*

$$\begin{aligned} x_1 &= \psi_1(y_1, \ldots, y_r) \\ &\vdots \\ x_r &= \psi_r(y_1, \ldots, y_r) \end{aligned}$$

   belong to the class $C^{k,\kappa}(\bar{\Omega}')$.

3. *If $k \geqslant 1$, we require that the Jacobian determinant of Φ satisfies*

$$0 < c \leqslant \left| \det \frac{\partial \Phi}{\partial x}(x) \right| \leqslant C \quad \textit{for all } x \in \bar{\Omega},$$

   where c, C are constants independent of x. We abbreviate 'Φ is a (k, κ)-diffeomorphism' by $\Phi \in C^{k,\kappa}$.

Remark 2.3 If the domain Ω is bounded, $\bar{\Omega}$ compact, $k \geqslant 1$, then condition 3 is a consequence of 1 and 2, since

$$\det\left(\frac{\partial \Phi}{\partial x}\right) \cdot \det\left(\frac{\partial \Phi^{-1}}{\partial y}\right) = 1$$

implies that $\det(\partial \Phi / \partial x)$ can never vanish.

Remark 2.4 For $k \geqslant 1$ condition 2 follows from conditions 1 and 3:

Let $k = 1$, by the inverse function theorem we have $\Phi^{-1} \in C^1$, and the Jacobian matrix satisfies

$$\frac{\partial \Phi}{\partial x} \cdot \frac{\partial \Phi^{-1}}{\partial y} = I, \tag{14}$$

that is, the derivatives of the inverse map Φ^{-1} may be linearly expressed in terms of the derivatives of Φ, divided by $\det(\partial\Phi/\partial x)$. Since $0 < c \leqslant |\det(\partial\Phi/\partial x)|$, if $\partial\Phi/\partial x \in C^{0,k}$, so does $\partial\Phi^{-1}/\partial y$ (see the proof of Theorem 2.6 where this assertion is worked out in detail), that is, $\Phi^{-1} \in C^{1,\kappa}$. For $k > 1$ our claim follows by the differentiation of (14) and induction.

Definition 2.6 *We also call a $(k, 0)$-diffeomorphism, $k \geqslant 1$, a k-diffeomorphism.*

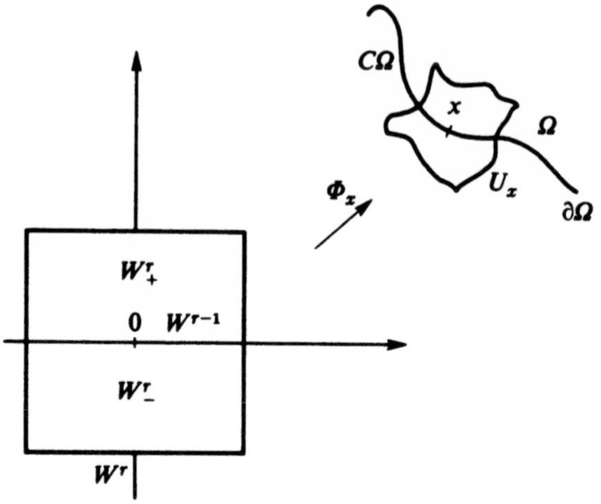

Fig. 2.10

Definition 2.7 (see Fig. 2.10). *We say that the region Ω is (k, κ)-smooth, if for each $x \in \partial\Omega$ we can find a neighbourhood U_x with the following properties:*

1. *U_x is (k, κ)-diffeomorphic to the unit cube $W^r = \{x: -1 < x_i < 1, i = 1, \ldots, r\}$. If Φ_x is the 1–1 transformation $U_x \leftrightarrow W^r$ we require further that:*
2. *$U_x \cap \partial\Omega$ is mapped 1–1 onto the central plane $x_r = 0$ of the cube W^r, $U_x \cap \partial\Omega \leftrightarrow W^{r-1}$ $(x_r = 0.)$*
3. *$U_x \cap \Omega \leftrightarrow W^r_+ = \{0 < x_r < 1\} \cap W^r$.*
4. *$U_x \cap C\Omega \leftrightarrow W^r_- = \{-1 < x_r < 0\} \cap W^r$.*

Remark *We have taken the standard domain for a map to be a cube; in many other books it is usual to take a ball. However, since all considerations are local, the difference is of no importance.*

For Ω bounded and (k, κ)-*smooth we may also say* (k, κ)-regular, and in this case we can cover $\partial\Omega$, which is compact, with finitely many U_x with the properties prescribed in Definition 2.7. We abbreviate 'Ω is (k, κ)-smooth' by $\Omega \in C^{k,\kappa}$, or $\partial\Omega \in C^{k,\kappa}$.

Theorem 2.5 *If the domain Ω has the $N^{k,\kappa}$-property of Definition 2.4, then it is (k,κ)-smooth.*

Proof. With the notation of Definition 2.4 we may write

$$W_x = W^{r-1}(\alpha_x) \times (-\beta_x, \beta_x), \quad \Phi_x : U_x \to W_x,$$

where

$$\Phi_x = \begin{cases} z_1 = y_1 \\ z_2 = y_2 \\ \cdot \\ \cdots \\ \cdot \\ z_{r-1} = y_{r-1} \\ z_r = y_r - a(y_1, \dots, y_{r-1}) \end{cases}, \quad \Phi_x^{-1} = \begin{cases} y_1 = z_1 \\ y_2 = z_2 \\ \cdot \\ \cdots \\ \cdot \\ y_{r-1} = z_{r-1} \\ y_r = z_r + a(z_1, \dots, z_{r-1}), \end{cases} \tag{15}$$

and by means of stretching the axes we can easily map the cube W_x into the unit cube. (Stretching of axes and orthogonal transformations do not affect (k,κ)-smoothness.) Here the Jacobian matrices have the form

$$\frac{\partial z}{\partial y} = \begin{bmatrix} 1 & & \cdots & & 0 \\ 0 & 1 & \cdots & & 0 \\ & & \cdots & & \\ 0 & & \cdots & 1 & 0 \\ -\dfrac{\partial a}{\partial y_1} & & \cdots & -\dfrac{\partial a}{\partial y_{r-1}} & 1 \end{bmatrix}, \quad \frac{\partial y}{\partial z} = \begin{bmatrix} 1 & & \cdots & & 0 \\ 0 & 1 & \cdots & & 0 \\ & & \cdots & & \\ 0 & & \cdots & 1 & 0 \\ \dfrac{\partial a}{\partial z_1} & & \cdots & \dfrac{\partial a}{\partial z_{r-1}} & 1 \end{bmatrix},$$

and $(\partial z/\partial y) = 1$.

From the special form of the maps (15) we deduce that (k,κ)-smoothness occurs if and only if the function $a(y_1, \dots, y_{r-1})$ belongs to $C^{k,\kappa}(W^{r-1}(\alpha_x))$. ∎

Theorem 2.6 *Let $k \geqslant 1$. The $N^{k,\kappa}$-property is equivalent to (k,κ)-smoothness.*

Proof. We have just seen that the domains Ω that have the $N^{k,\kappa}$-property are (k,κ)-smooth. We now wish to prove the converse. Let $x_0 \in \partial\Omega$, let U be a neighbourhood of x_0 with the properties given in Definition 2.7, and let $\Phi : W \to U$, $y \mapsto x$ be the (k,κ)-transformation. Locally we must represent the 'surface' $\partial\Omega$ in Cartesian coordinates by means of a function a (see Definition 2.4). By Definition 2.7.2 locally $\partial\Omega$ is given in the W-coordinates y by the equation

$$y_r = 0 \quad \text{or} \quad y_r(x_1, \dots, x_r) = 0, \tag{16}$$

where the W-coordinates y do not need to be Cartesian (the x-coordinates are). Since by Definition 2.5.3 the Jacobian determinant $\det(\partial y/\partial x) \neq 0$, by expansion, we see at least one of the derivatives $\partial y_1/\partial x_1 \cdots \partial y_r/\partial x_r$ must differ from zero; suppose that

$$\frac{\partial y_r}{\partial x_r} \neq 0 \quad \text{on } U, \tag{17}$$

(choosing U sufficiently smaller). By the implicit function theorem we can solve (16) for x_r in U,

$$x_r = a(x_1, \ldots, x_{r-1}), \tag{18}$$

where the xs are the Cartesian coordinates for U. Here the $x_1, \ldots, x_{r-1}$ may be allowed to vary over an $(r-1)$-dimensional cube $W^{r-1}(\alpha)$. Let $V_{-\varepsilon} = \{x \in U : d(x, CU) > \varepsilon\}$, where $\varepsilon > 0$ is chosen small enough that the initial point $x_0 \in V_{-\varepsilon}$. We restrict (18) to $V_{-\varepsilon}$ and label the surface so obtained as $\partial\Omega_{-\varepsilon}$. We push $\partial\Omega_{-\varepsilon}$ along the x_r-axis upwards to $+\varepsilon/2$ and downwards to $-\varepsilon/2$; because of the choice of ε we remain inside U. If we put $\beta = \varepsilon/2$ we obtain

$$\tilde{V} := \{x : (x_1, \ldots, x_{r-1}) \in W^{r-1}(\alpha), a(x_1, \ldots, x_{r-1}) - \beta < x_r < a(x_1, \ldots, x_{r-1}) + \beta\}$$

as a neighbourhood of $x_0 \in \partial\Omega$ with the properties required in Definition 2.4. We check property 2.4.1; the remaining properties are clear, since $\tilde{V} \subset U$. Let $k = 1$. Because of (16) we have

$$\frac{\partial a}{\partial x_i}(x_1, \ldots, x_{r-1}) = -\frac{\dfrac{\partial y_r}{\partial x_i}(x_1, \ldots, x_{r-1}, a(x_1, \ldots, x_{r-1}))}{\dfrac{\partial y_r}{\partial x_r}(x_1, \ldots, x_{r-1}, a(x_1, \ldots, x_{r-1}))}, \quad i = 1, \ldots, r-1. \tag{19}$$

Since by assumption the function $y_r \in C^{1,\varkappa}$ and (17) is valid in the form $|\partial y_r/\partial x_r| \geq c > 0'$, for $i = 1, \ldots, r-1$, we can estimate

$$\begin{aligned}
\left|\frac{\partial a}{\partial x_i}(x) - \frac{\partial a}{\partial x_i}(z)\right| &= \left|\frac{\partial y_r}{\partial x_i}(x)\cdot\left(\frac{\partial y_r}{\partial x_r}(x)\right)^{-1} - \frac{\partial y_r}{\partial x_i}(z)\cdot\left(\frac{\partial y_r}{\partial x_r}(x)\right)^{-1}\right. \\
&\quad \left. + \frac{\partial y_r}{\partial x_i}(z)\left(\frac{\partial y_r}{\partial x_r}(x)\right)^{-1} - \frac{\partial y_r}{\partial x_i}(z)\cdot\left(\frac{\partial y_r}{\partial x_r}(z)\right)^{-1}\right| \\
&\leq \left|\frac{\partial y_r}{\partial x_r}\right|^{-1}\cdot\left|\frac{\partial y_r}{\partial x_i}(x) - \frac{\partial y_r}{\partial x_i}(z)\right| + \left|\frac{\partial y_r}{\partial x_r}(x)\right|^{-1} \\
&\quad \cdot\left|\frac{\partial y_r}{\partial x_r}(z)\right|^{-1}\cdot\left|\frac{\partial y_r}{\partial x_i}(z)\right|\cdot\left|\frac{\partial y_r}{\partial x_r}(x) - \frac{\partial y_r}{\partial x_r}(z)\right| \\
&\leq \frac{k}{c}|x-z|^{\varkappa} + \frac{k}{c^2}|x-z|^{\varkappa} = \left(\frac{k}{c} + \frac{k}{c^2}\right)|x-z|^{\varkappa},
\end{aligned}$$

that is, $\partial a/\partial x_i \in C^{0,\kappa}$ for $i=1,\dots,r-1$, that is, $a\in C^{1,\kappa}$. For $k>1$ the proof uses differentiation of the formula (19) and induction. ■

Maps Φ in $C^{0,1}$ have an important property which, in general, continuous functions do not have.

Lemma 2.3 *Let $\Phi:\Omega\to\Omega'$ belong to $C^{0,1}$ and let $A\subset\Omega$ with $\mu_r(A)=0$ (μ_r denotes Lebesgue measure in $\mathbb{R}^r$). Then $\mu_r(\Phi(A))=0$.*

Proof. From $|\Phi(x)-\Phi(x')|\leqslant K|x-x'|$, $x,x'\in\Omega$, where $K<\infty$ is the Lipschitz constant, it follows that the image of a ball of radius ρ is contained in a ball of radius $K\rho$. $\mu_r(A)=0$ means that for each $\varepsilon>0$ we can find a covering of A by countably many balls $B(a_n,\rho_n)$ with

$$A\subset\bigcup_{n=1}^{\infty}B(a_n,\rho_n)\quad\text{and}\quad\sum_{n=1}^{\infty}\mu_r(B(a_n,\rho_n))<\varepsilon.$$

The Lipschitz property implies that $\Phi(A)$ is covered by the balls $B(\Phi(a_n),K\rho_n)$, where $\mu_r(B(\Phi(a_n),K\rho_n))=K^r\mu_r(B(a_n,\rho_n))$ and we have

$$\Phi(A)\subset\bigcup_{n=1}^{\infty}B(\Phi(a_n),K\rho_n),$$

$$\sum_{n=1}^{\infty}\mu_rB(\Phi(a_n),K\rho_n)=K^r\sum_{n=1}^{\infty}\mu_r(B(a_n,\rho_n))<K^r\varepsilon,$$

that is, $\Phi(A)$ also has measure zero. ■

Theorem 2.7 *Let Ω be (k,κ)-smooth with $k+\kappa\geqslant 1$. Then the frontier $\partial\Omega$ has Lebesgue measure zero, that is, $\mu_r(\partial\Omega)=0$.*

Proof. We cover $\partial\Omega$ with neighbourhoods U_i having the properties listed in Definition 2.7. By Theorem 1.1 the open cover $\{U_i\}$ of $\partial\Omega$ can be chosen to be countable, so $i=1,2,\dots$. Since $\mu_r(W^{r-1})=0$, by Lemma 2.3 we have $\mu_r(\partial\Omega\cap U_i)=0$ for $i=1,2,\dots$, from which, because

$$\partial\Omega=\bigcup_{i=1}^{\infty}(\partial\Omega\cap U_i),$$

our theorem follows. ■

Theorem 2.8 *If $\Omega\in N^{k,\kappa}$ with $k+\kappa\geqslant 0$, then the frontier $\partial\Omega$ has Lebesgue measure zero, $\mu_r(\partial\Omega)=0$.*

For $k+\kappa\geqslant 1$ this follows immediately from Theorems 2.5 and 2.7. For $k+\kappa\geqslant 0$ there is another simpler proof straight from Definition 2.4.2: continuous functions $a(x_1,\dots,x_{r-1})$ are known to be Riemann integrable, the portion of the boundary $U_x \cap \partial\Omega$ can therefore be sandwiched arbitrarily closely between an upper and lower sum, which implies that $\mu_r(U_x \cap \partial\Omega) = 0$. The choice of a countable cover by Theorem 1.1 concludes the proof.

Theorem 2.9 *If the open set Ω has the cone property (see Definition 2.2), then* $\mu_r(\partial\Omega) = 0$.

Proof. Let $\{U_j\}$ be an open countable cover of $\partial\Omega$ by bounded sets U_j, $j = 1, 2, \ldots$. By Theorem 2.4 we have

$$\partial\Omega \cap U_j \subset \bigcup_{i=1}^{n} \partial\Omega_{ij} \quad \text{with } \Omega_{ij} \in N^{0,1},$$

from which, given Theorem 2.8, our assertion follows. ■

Let Ω belong to the class $C^{k,\kappa}$, then we can consider the local transformations $y = \Phi$ (see Definition 2.7) as local coordinates y in the neighbourhood of $\partial\Omega$, and say that the local coordinates y belong to $C^{k,\kappa}$.

2.4 Normal transformations

For later use we need special coordinate transformations which leave the normal direction unchanged. Let $\Omega \in N^{k,\kappa}$, $k + \kappa \geqslant 1$, then for almost all $x_0 \in \partial\Omega$, with the help of Theorem 1.8, we can define the normal $\mathbf{n}$ to $\partial\Omega$ through x_0 (Ω lies in $\mathbb{R}^r$). For example we may write

$$\mathbf{n} := \frac{1}{K}\left(\frac{\partial a}{\partial x_1}(x_0'), \ldots, \frac{\partial a}{\partial x_{r-1}}(x_0'), 1\right),$$

where $x_0 = (x_0', a(x_0')) \in \partial\Omega$,

$$K := \left[1 + \sum_{i=1}^{r-1}\left(\frac{\partial a}{\partial x_i}\right)^2\right]^{1/2},$$

where $x_r = a(x_1, \ldots, x_{r-1})$ is the (local) equation of $\partial\Omega$ in the Cartesian coordinates $(x_1, \ldots, x_r)$, see Fig. 2.11. For $k \geqslant 1$, by Theorem 2.6 for $\Omega \in C^{k,\kappa}$, the normal $\mathbf{n}$ is defined everywhere.

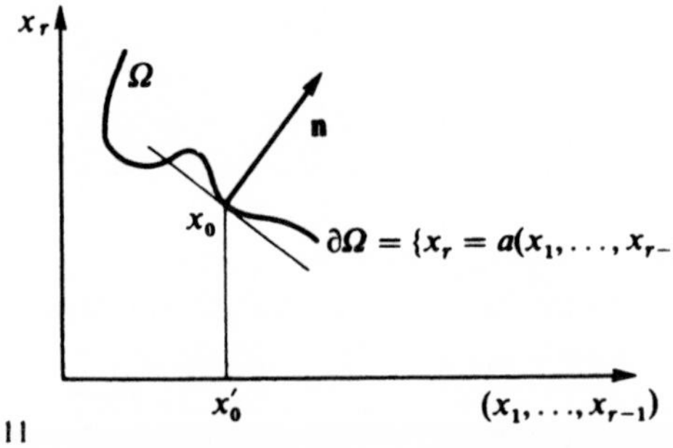

Fig. 2.11

We first define normal transformations (respectively normal coordinates). Let $\Phi: U \to U'$, $x \mapsto y$ be a $C^{k,\kappa}$-transformation, $k \geq 1$. We fix the point x and thereby $y = \Phi(x)$, and consider the vector space T_x of all directional derivatives at x; T_x is spanned by $\partial/\partial x_1, \ldots, \partial/\partial x_r$. The transformation Φ induces a linear isomorphism between the vector spaces T_x and T_y, given by the Jacobian matrix

$$\frac{\partial \Phi^{-1}}{\partial x} : T_x \to T_y$$

or as formulae for the basis

$$\frac{\partial}{\partial y_j} = \sum_{i=1}^{r} \frac{\partial x_i}{\partial y_i} \frac{\partial}{\partial x_i}, \quad j = 1, \ldots, r \tag{20}$$

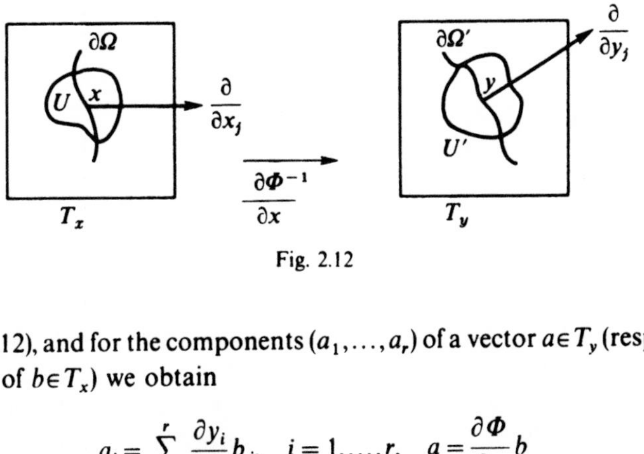

Fig. 2.12

(see Fig. 2.12), and for the components $(a_1, \ldots, a_r)$ of a vector $a \in T_y$ (respectively $(b_1, \ldots, b_r)$ of $b \in T_x$) we obtain

$$a_i = \sum_{j=1}^{r} \frac{\partial y_i}{\partial x_j} b_j, \quad i = 1, \ldots, r, \quad a = \frac{\partial \Phi}{\partial x} b \tag{21}$$

(see, for example, Warner [1], p. 16 etc., or Holmann & Rummler [1], §9). If $x \in \partial\Omega$ and Φ takes $\partial\Omega$ into $\partial\Omega'$, then we can distinguish the normal vector $\mathbf{n}$ in T_x ($\mathbf{n}$ normal to $\partial\Omega$ at x), and similarly $\mathbf{n}' \in T_y$ ($\mathbf{n}'$ normal to $\partial\Omega'$ at y); neither $\mathbf{n}$ nor $\mathbf{n}'$ need have unit length.

We make the following:

Definition 2.8 *We say that the transformation Φ is normal at the point $x_0 \in \partial\Omega$, if $\mathbf{n}_0$ is mapped to $\mathbf{n}'_0$ under $\partial\Phi^{-1}/\partial x: T_{x_0} \to T_{y_0}$ – in short, the normal direction at x_0 is invariant under the transformation Φ (caution: the direction of the normal remains invariant, not necessarily the length of the vector $\mathbf{n}$). If Φ is normal for all points $x \in \partial\Omega \cap U$, we speak of a normal transformation, without specific mention of points. If the coordinate transformation $y = \Phi$:*

$W^r \to U$, $W^{r-1} \to \partial\Omega \cap U$ (see Definition 2.7) is normal, we say that the coordinates y are normal; pictorially this means that through each $x \in \partial\Omega \cap U$ the coordinate line y_r runs in the direction of the normal to $\partial\Omega$. In order to fix the direction of $\mathbf{n}$ once and for all, we always orient the normal towards the interior of Ω.

We have a simple criterion for normal transformations at a point $x_0 \in \partial\Omega$.

Theorem 2.10 *Let $\Phi: U \to U'$, $x \mapsto y$ be a coordinate transformation in $C^{k,\kappa}$, $k \geqslant 1$, which preserves the direction of the inward normal at the point $x_0 \in \partial\Omega \cap U$, $\Omega \in C^{k,\kappa}$, $k \geqslant 1$. We orient the coordinate axis x_r in the direction of the inward normal through $x_0 \in \partial\Omega$ and y_r in the direction of the inward normal through $y_0 \in \partial\Omega$. A necessary and sufficient condition for the map Φ to be normal at $x_0 \in \partial\Omega$ is that the Jacobian matrix of Φ at x_0 takes the form*

$$\frac{\partial y}{\partial x}(x_0) = \begin{bmatrix} \dfrac{\partial y_1}{\partial x_1} & \cdots & \dfrac{\partial y_1}{\partial x_{r-1}} & 0 \\ & \cdots & & 0 \\ \dfrac{\partial y_{r-1}}{\partial x_1} & \cdots & \dfrac{\partial y_{r-1}}{\partial x_{r-1}} & 0 \\ \dfrac{\partial y_r}{\partial x_1} & \cdots & \dfrac{\partial y_r}{\partial x_{r-1}} & \dfrac{\partial y_r}{\partial x_r} \end{bmatrix}, \quad \text{with } \frac{\partial y_r}{\partial x_r} > 0. \tag{22}$$

Proof. The orientation of the axes x_r and y_r implies that $\mathbf{n} = (0, 0, \ldots, 1)$ and $\mathbf{n}' = (0, 0, \ldots, 1)$. Let Φ be normal at x_0; we put $b = \mathbf{n} = (0, 0, \ldots, 1)$ in (21) and must have $a = \sigma \cdot \mathbf{n}' = (0, 0, \ldots, \sigma)$, $\sigma > 0$, which is only possible if $\partial y_i/\partial x_r = 0$ for $i = 1, \ldots, r-1$ and $\partial y_r/\partial x_r = \sigma > 0$. This proves the necessity of (22). Suppose conversely that (22) is satisfied. We put $b = \mathbf{n} = (0, 0, \ldots, 1)$ in (21) obtaining $a_i = 0$ for $i = 1, \ldots, r-1$ and $a_r = (\partial y_r/\partial x_r)(x_0) > 0$. Therefore $a = (\partial y_r/\partial x_r) \cdot \mathbf{n}'$, hence we have shown the normality of Φ at $x_0 \in \partial\Omega$. ■

Definition 2.9 *We call a coordinate transformation $\Phi \in C^{k,\kappa}$, $k \geqslant 1$, admissible at the point $x_0 \in \partial\Omega$, if the Jacobian matrix at the point x_0 has the form*

$$\frac{\partial y}{\partial x}(x_0) = \begin{bmatrix} \dfrac{\partial y_1}{\partial x_1} & \cdots & \dfrac{\partial y_1}{\partial x_{r-1}} & 0 \\ & \cdots & & \\ \dfrac{\partial y_{r-1}}{\partial x_1} & \cdots & \dfrac{\partial y_{r-1}}{\partial x_{r-1}} & 0 \\ 0 & \cdots & 0 & \dfrac{\partial y_r}{\partial x_r} \end{bmatrix}, \quad \text{with } \frac{\partial y_r}{\partial x_r} = \sigma > 0. \tag{23}$$

Here again we have oriented the x_r and y_r coordinate axes in the direction of the appropriate normals. We can also characterise admissible transformations geometrically. Let $T_{\partial\Omega}$ be the tangent hyperplane at the point x_0 (respectively $T_{\partial\Omega'}$); we decompose: $T_{x_0} = T_{\partial\Omega} \oplus \mathbf{n}_{x_0}$, $T_{y_0} = T_{\partial\Omega'} \oplus \mathbf{n}'_{y_0}$ then an admissible transformation is characterised by the fact that it takes $\mathbf{n}_{x_0}$ to $\mathbf{n}'_{y_0}$ and $T_{\partial\Omega}$ into $T_{\partial\Omega'}$. While the first characteristic property follows from Theorem 2.10, we use formula (21) again to exhibit the second. By $a \in T_{\partial\Omega'}$ (respectively $b \in T_{\partial\Omega}$) we mean that $a_r = 0$ (respectively $b_r = 0$). Thus from $b \in T_{\partial\Omega}$ it follows that $a \in T_{\partial\Omega'}$ if and only if

$$0 = a_r = \sum_{j=1}^{r-1} \frac{\partial y_r}{\partial x_j} b_j.$$

If we allow b to vary in $T_{\partial\Omega}$, we obtain $\partial y_r/\partial x_1 = 0, \ldots, \partial y_r/\partial x_{r-1} = 0$ as characteristic conditions for the invariance of $T_{\partial\Omega}$ under $\partial\Phi^{-1}/\partial x$.

For each point $x_0 \in \partial\Omega$ we can introduce admissible coordinates; this is the assertion of the next theorem.

Theorem 2.11 *Let Ω be (k,κ)-smooth and $k \geqslant 1$. Then each point $x_0 \in \partial\Omega$ possesses a neighbourhood U_{x_0} and a coordinate transformation $\Phi_{x_0}: U_{x_0} \to W^r$ with the properties:*

(a) $\Phi_{x_0} \in C^{k,\kappa}$.
(b) Φ_{x_0} *is admissible for* $x_0 \in \partial\Omega$.
(c) *The transformation induces* 1–1 *correspondences (see Definition* 2.7)

$$U_{x_0} \leftrightarrow W^r, \quad U_{x_0} \cap \Omega \leftrightarrow W^r_+, \quad U_{x_0} \cap \mathbf{C}\bar{\Omega} \leftrightarrow W^r_-, \quad U_{x_0} \cap \partial\Omega \leftrightarrow W^{r-1}.$$

Property (c) *implies amongst other things that* $(x_1, \ldots, x_{r-1})$ *are coordinates for* $\partial\Omega$.

Proof. Since $k \geqslant 1$ by Theorem 2.6 Ω has the $N^{k,\kappa}$-property, and we can find some U_{x_0} with

$$\partial\Omega \cap U_{x_0} = \{x_r = a(x_1, \ldots, x_{r-1})\}, \quad a \in C^{k,\kappa},$$

where we can additionally require that at the point x_0 x_r runs in the direction of the inward normal to $\partial\Omega$. For the transformation $\Phi_{x_0}: x \mapsto y$ we take

$$\begin{aligned} y_1 &= x_1 \\ &\vdots \\ y_{r-1} &= x_{r-1} \\ y_r &= x_r - a(x_1, \ldots, x_{r-1}), \end{aligned} \tag{24}$$

and can read off properties (a) to (c) directly from (24); for (b) we have, for example,

$$\frac{\partial y}{\partial x}(x_0)=\begin{bmatrix} 1 & \cdots & 0 & 0 \\ & \cdots & & \\ 0 & \cdots & 1 & 0 \\ -\dfrac{\partial a}{\partial x_1} & \cdots & -\dfrac{\partial a}{\partial x_{r-1}} & 1 \end{bmatrix}=\begin{bmatrix} 1 & \cdots & 0 & 0 \\ & \cdots & & \\ 0 & \cdots & 1 & 0 \\ 0 & \cdots & 0 & 1 \end{bmatrix}. \tag{25}$$

Because x_r runs in the direction of the normal to $\partial\Omega$ at x_0, and the coordinates $(x_1,\ldots,x_r)$ are mutually perpendicular, we have

$$\frac{\partial a}{\partial x_1}(x_0')=0, \quad \text{therefore} \quad \frac{\partial y}{\partial x_i}(x_0)=-\frac{\partial a}{\partial x_i}(x_0')=0 \quad \text{for } i=1,\ldots,r-1.$$

Here $x_0=(x_0',a(x_0'))$ and we have established the second equation in (25). ■

Theorem 2.12 *Let Ω be (k,κ)-smooth and $k\geqslant 2$. Then each point $x_0\in\partial\Omega$ possesses a neighbourhood U_{x_0} with admissible coordinates $y=(y_1,\ldots,y_r)$, with $y\in C^{k-1,\kappa}$. More precisely there exists an admissible coordinate transformation $y=\Phi_{x_0}: W^r\to U_{x_0}$, which is a $(k-1,\kappa)$-diffeomorphism and which maps W^r_+ to $\Omega\cap U_{x_0}$, W^{r-1} to $\partial\Omega\cap U_{x_0}$. Moreover for each point $x\in\partial\Omega\cap U_{x_0}$, Φ_{x_0} is admissible and $(y_1,\ldots,y_{r-1})$ are the local coordinates of $\partial\Omega$.*

Complement *The reader may easily see that with the help of Theorem 1.8 the condition $k\geqslant 2$ may be weakened to $k+\kappa\geqslant 2$.*

Proof. By Theorem 2.6 we may assume that Ω has the $N^{k,\kappa}$-property, that is, for each $x_0\in\partial\Omega$ we can find a neighbourhood U_{x_0} and Cartesian coordinates z with

$$\partial\Omega\cap U_0=\{z_r=a(z_1,\ldots,z_{r-1})\},\quad a\in C^{k,\kappa}.$$

The inward normal $\mathbf{n}$ to $\partial\Omega$ is then given by

$$\mathbf{n}=(n_1,\ldots,n_r)=\frac{1}{K}\left(\frac{\partial a}{\partial z_1},\ldots,\frac{\partial a}{\partial z_{r-1}},1\right) \text{ with } K=\left[1+\sum_{i=1}^{r-1}\left(\frac{\partial a}{\partial z_i}\right)^2\right]^{1/2}. \tag{26}$$

We take different coordinates (x',t) and put

$$\begin{aligned} z_1 &= x_1+tn_1(x'),\\ &\vdots\\ z_{r-1} &= x_{r-1}+tn_{r-1}(x'), \quad (x',t)=(x_1,\ldots,x_{r-1},t)\\ z_r &= a(x')+tn_r(x'). \end{aligned} \tag{27}$$

Geometrically this means: we go along the normal $\mathbf{n}$ from the point $P(z)$ to

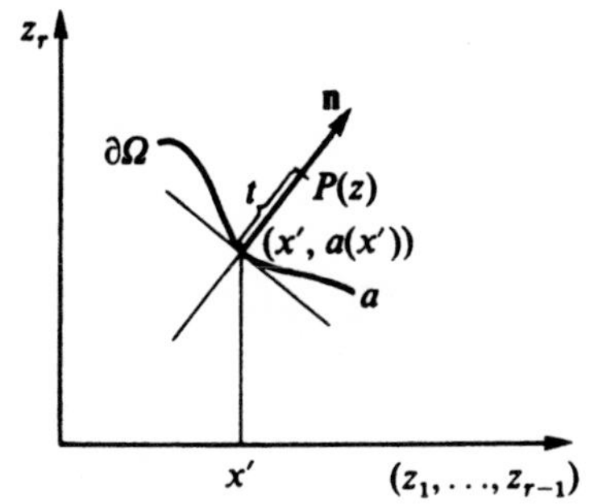

Fig. 2.13

the intersection point $(x', a(x'))$ with $\partial\Omega$. In this way we go t steps. As the new coordinates of the point $P(z)$ we take (x', t); see Fig. 2.13. In order to show that the mapping $z \mapsto (x', t)$ is 1–1 we calculate the Jacobian determinant at x_0 $(t = 0)$:

$$\det \frac{\partial z}{\partial(x', t)}(x_0)$$

$$= \begin{bmatrix} 1 & \cdots 0 & 0 & n_1 \\ & \cdots & & \\ 0 & \cdots & 1 & n_{r-1} \\ \dfrac{\partial a}{\partial x_1} & \cdots & \dfrac{\partial c}{\partial x_{r-1}} & n_r \end{bmatrix} = \frac{1}{K} \begin{bmatrix} 1 & 0 & 0 & \dfrac{\partial a}{\partial x_1} \\ & \cdots & & \\ 0 & \cdots & 1 & \dfrac{\partial a}{\partial x_{r-1}} \\ \dfrac{\partial a}{\partial x_1} & \cdots & \dfrac{\partial a}{\partial x_{r-1}} & 1 \end{bmatrix}$$

$$= \frac{1}{K}\left[1 + \sum_{i=1}^{r-1} \left(\frac{\partial a}{\partial x_i}\right)^2\right] = K. \tag{28}$$

Since $K \neq 0$ there exists a small neighbourhood V_{x_0} of x_0, which we again label U_{x_0}, where $\det(\partial z/\partial(x', t)) \neq 0$, and where we can apply the implicit function theorem. This ensures the 1–1 nature of $(x', t) \leftrightarrow z$, and we denote the transformation $(x', t) \mapsto z$ by Φ_{x_0}. Since the normal is given by the derivatives (26), the coordinate transformation Φ_{x_0} (27) belongs to $C^{k-1,\kappa}$. In order to see that Φ_{x_0} is admissible for each $x \in \partial\Omega \cap U_{x_0}$, we must take z_r in the direction of the normal through x while t is already normal by construction; see Definition 2.9. However, the point $(z_1, \ldots, z_{r-1})$, respectively $(x_1, \ldots, x_{r-1})$ (Cartesian coordinates!), is then tangential to $\partial\Omega$ and for $x \in \partial\Omega \cap U_{x_0}$

$$\frac{\partial a}{\partial x_i} = 0 \quad (i = 1, \ldots, r-1) \text{ holds.}$$

If the last expression is substituted in the second matrix of (28) we obtain the admissibility of Φ_{x_0} at $x \in \partial\Omega \cap U_{x_0}$. We see that the formula (28) holds not only for x_0 but for all $x \in \partial\Omega \cap U_{x_0}$, since $\partial\Omega \cap U_{x_0}$ is characterised by $t = 0$. The remaining properties of Φ_{x_0} can be easily checked. ■

2.5 Differentiable manifolds

For the sake of completeness we give the definition of a differentiable manifold. Let M be a topological space which satisfies the second axiom of countability, that is, $M = \bigcup_{n=1}^{\infty} K_n$, K_n compact. Let $U_i, i \in I$, be a system of open subsets of M. To each subset U_i let there exist a bijective map α_i onto some region $\alpha_i(U_i)$ of $\mathbb{R}^r$. We call the pair (U_i, α_i) a *chart* or *choice of coordinates* for M. Let $\kappa = 0$ or 1; the collection of all charts $a = \{(U_i, \alpha_i) : i \in I\}$ is called a $C^{k,\kappa}$-*atlas* or *system of coordinates* if the following conditions are satisfied:

1. $\bigcup_{i \in I} U_i = M$.
2. The maps $\alpha_i \circ \alpha_j^{-1} : \alpha_j(U_i \cap U_j) \to \alpha_i(U_i \cap U_j)$ (in $\mathbb{R}^r$) belong to $C^{k,\kappa}$, $\kappa = 0$ or 1.

Two coordinate systems a and b are called equivalent, if $a \cup b$ is again a coordinate system, which is equivalent to saying that for arbitrary charts $(U_i, \alpha_i) \in a$ and $(V_j, \beta_j) \in b$ the compositions

$$\alpha_i \circ \beta_j^{-1} : \beta_j(U_i \cap V_j) \to \alpha_i(U_i \cap V_j) \quad \text{and} \quad \beta_j \circ \alpha_i^{-1} : \alpha_i(U_i \cap V_j) \to \beta_j(U_i \cap V_j)$$

also belong to $C^{k,\kappa}$. Since for $\kappa = 0$ or 1 the class $C^{k,\kappa}$ is invariant with respect to composition of functions, we do have a genuine equivalence relation.

Definition 2.10 *If $\kappa = 0$ or 1 the equivalence classes under this relation are called $C^{k,\kappa}$-structures on the manifold M, or in short $C^{k,\kappa}$-manifold. A $C^{k,\kappa}$-structure is already determined by a single coordinate system a.*

Definitiom 2.11 *Let M be a $C^{k,\kappa}$-manifold and $a = \{(U_i, \alpha_1), i \in I\}$ an atlas. We say that the function $f : M \to \mathbb{C}$ belongs to the class $C^{l,\lambda}$, $l + \lambda \leqslant k + \kappa$ if all the functions (with domain of definition in $\mathbb{R}^r$)*

$$f \circ \alpha_i^{-1} : \alpha_i(U_i) \to \mathbb{C}, \quad i \in I,$$

belong to the class $C^{l,\lambda}(\alpha_i(U_i))$.

We see immediately that this definition is independent of the coordinate system. In a similar way we can define when a map $f : M \to N$ between two $C^{k,\kappa}$-manifolds belongs to class $C^{l,\lambda}$, $l + \lambda \leqslant k + \kappa$: the map $\beta_j \circ f \circ \alpha_i^{-1}$ must belong to class $C^{l,\lambda}$, where $\{U_i, \alpha_i\}$ and $\{V_j, \beta_j\}$ are coordinate systems for M and N respectively.

As for $\mathbb{R}^r$ the partition of unity is an important tool for the study of the differentiable manifold M.

Theorem 2.13 (partition of unity) *Let $\{\Omega_i \in i \in I\}$ be an open cover of the $C^{k,\kappa}$-manifold M. There exists a partition of unity $\{\varphi_j : j \in \mathbb{N}\}$ subordinate to the cover $\{\Omega_i : i \in I\}$ with countable indexing set $J = \mathbb{N}$ and with $\varphi_j \in C_0^{k,\kappa}(M)$.*

Proof. We take the coordinate neighbourhoods U_x, $x \in M$, choose $W_x \neq \emptyset$ open with $\bar{W}_x \subset\subset U_x$, and consider the refinement $\{W_x \cap \Omega_i : i \in I, x \in M\}$. By Remark 1.1 we can find a locally finite refinement $\{V_j, j \in \mathbb{N}\}$ with countable indexing set $J = \mathbb{N}$, for which

$$\bar{V}_j \subset \bar{W}_{x_j} \cap \bar{\Omega}_{i_j} \subset \bar{W}_{x_j} \subset\subset U_{x_j},$$

that is, $\bar{V}_j$ compact $\subset\subset U_{x_j}$, $j \in \mathbb{N}$, and such that $U_j := U_{x_j}$ is a coordinate neighbourhood of $x_j \in M$.

Let α_j be the coordinate map associated to U_j; we have

$$\alpha_j(\bar{V}_j) \text{ compact} \subset\subset \alpha_j(U_j) \quad \text{open in } \mathbb{R}^r.$$

By Corollary 1.2 there exists a function ψ_j, $0 \leqslant \psi_j \in \mathscr{D}(\alpha_j(U_j))$, with compact support, and with the property

$$\psi_j = \begin{cases} 1 & \text{on} \quad \alpha_j(\bar{V}_j), \\ 0 & \text{on} \quad C\alpha_j(U_j), \\ \text{between 0 and 1 otherwise.} \end{cases}$$

We lift the function ψ_j to the manifold M, that is, we form $\psi_j \circ \alpha_j$ and have

$$\psi_j \circ \alpha_j \in C_0^{k,\kappa}(M).$$

We obtain the desired partition of unity $\{\varphi_j, j \in \mathbb{N}\}$ by means of

$$\varphi_j := \frac{\psi_j \circ \alpha_j}{\sum_j \psi_j \circ \alpha_j}. \qquad \blacksquare$$

We next explain what we mean by a submanifold N of M.

Definition 2.12 *Let M be a manifold with a $C^{k,\kappa}$-structure a. We say that $N \subset M$ is a submanifold of (M, a) if for each $p \in N$ we can find a chart $(U, \alpha) \in a$, $p \in U$, $\alpha = (x_1, \ldots, x_r)$ with*

$$N \cap U = \{z \in U : x_1(z) = 0, \ldots, x_m(z) = 0\}, \tag{29}$$

where $0 \leqslant m \leqslant r$. Here the number $d = r - m$ is called the dimension of the submanifold.

For other equivalent definitions of a submanifold (particularly for $k \geq 1$) see Warner [1], Holmann & Rummler [1].

Theorem 2.14 *The submanifold N carries a canonical $C^{k,\kappa}$-manifold structure, induced from (M, α). It is therefore again a $C^{k,\kappa}$-manifold.*

Proof. We must construct an atlas a' on N. From the atlas a on M we take all those charts (U, α) with $U \cap N \neq \varnothing$, satisfying (29). We must set $U' = U \cap N$, $\alpha' = (x_{m+1}, \ldots, x_r)$, and the desired atlas is the collection of all such pairs (U', α'). We easily see that because of (29) α' is a bijection between $U \cap N$ and an open subset of $\mathbb{R}^{r-m}$. We must check that conditions 1 and 2 of the definition of an atlas hold for a'. Condition 1 is trivial; for condition 2 we specify the exact form of the mapping $\alpha'_i \circ \alpha'^{-1}_j$. Let

$$\alpha'_i = (x_{m+1}, \ldots, x_r), \quad \alpha'_j = (y_{m+1}, \ldots, y_r),$$

then we have

$$\alpha'_i \circ \alpha'^{-1}_j = (x_{m+1}(0, \ldots, 0, y_{m+1}, \ldots, y_r), \ldots, x_r(0, \ldots, 0, y_{m+1}, \ldots, y_r)),$$

and this map, as the restriction of a $C^{k,\kappa}$-map, again belongs to the class $C^{k,\kappa}$. ■

Theorem 2.15 *Let Ω be (k, κ)-smooth. Then the frontier $\partial\Omega$ is an $(r-1)$-dimensional $C^{k,\kappa}$-submanifold of $\mathbb{R}^r$, for $\kappa = 0, 1$, and for $0 < \kappa < 1$ a $C^{k,0}$-submanifold.*

Proof. We take the neighbourhoods U_x and the transformations $\Phi_x, x \in \partial\Omega$ in Definition 2.7.

$$M := \bigcup_x U_x \tag{30}$$

is open in $\mathbb{R}^r$, therefore has the structure of a C^∞-manifold, and the identity map $I: M \to M$ is C^∞. The (U_x, Φ_x) form an atlas on M; the atlas condition 1 is (30), and condition 2:

$$\Phi_x \circ \Phi_{x'}^{-1} = \Phi_x \circ I \circ \Phi_{x'}^{-1} : W^{r-1} \to W^{r-1} \in C^{k,\kappa}, \kappa = 0, 1$$
$$(\text{respectively } C^{k,0}, 0 < \kappa < 1)$$

is also satisfied.

Condition 2 in the definition shows that $\partial\Omega$ is a submanifold of M; we can apply Theorem 2.14 to complete the proof.

Exercises

We introduce two definitions. The region $\Omega \subset \mathbb{R}^r$ is called *star-shaped* with respect to a point x_0 if each ray leaving this point has exactly one intersection point with $\partial\Omega$. A region Ω is called *star-shaped* with respect to a ball contained in Ω if it is *star-shaped* with respect to each point of the ball.

2.1 Show that convex regions are star-shaped.

2.2 Let Ω be bounded and let the frontier $\partial\Omega$ have an explicit representation in the form $r = r(\omega)$ in polar coordinates. Show that Ω is *star-shaped* with respect to a ball with centre at the origin if and only if the function $r(\omega)$ is Lipschitz continuous, $r \in C^{0,1}$.

2.3 If Ω is bounded and star-shaped with respect to some ball, then show that $\Omega \in N^{0,1}$.

2.4 Let Ω be bounded and satisfy the cone condition. Prove Ω is the union of a finite number of regions, which are *star-shaped* with respect to a ball.

2.5 Prove that convex bounded regions satisfy the cone condition.

2.6 Does there exist a region Ω in $\mathbb{R}^2$, which satisfies the cone condition but not the uniform cone condition?

§3 Definitions and density properties for the Sobolev–Slobodeckiĭ spaces $W_2^l(\Omega)$

Here we introduce the definitions of the W_2^l-spaces; for l non-integral we adhere to the recipe of Slobodeckiĭ [1], by which we avoid interpolation theory, see for example Triebel [2], and can construct the theory of the W_2^l-spaces in an elementary manner. We also show in this section that the various classes of C^∞-functions are dense in the W_2^l-spaces; for many estimates and identities, as these appear for differential operators, this leads to a great saving of labour. We only need to prove them for C^∞-functions; the usual density arguments then show the validity of the estimates, identities, etc. for all functions from the appropriate W_2^l-spaces.

3.1 Definition of the Sobolev–Slobodeckiĭ spaces $W_2^l(\Omega)$

In this book we restrict ourselves to the Hilbert spaces $W_2^l(\Omega)$, $l \in \mathbb{R}_+$. For the general case of the W_p^l-spaces, $1 \leqslant p \leqslant \infty$, we refer to the books of Adams [1], Nečas [1], or Triebel [2].

Definition 3.1 *Suppose first that l is integral, that is, $l = 0, 1, \ldots$. We define the Sobolev space $W_2^l(\Omega)$ to be the set of all functions $\varphi \in L_2(\Omega)$ for which the distributional derivatives $D^s\varphi$ are again elements of $L_2(\Omega)$ for $|s| \leqslant l$:*

$$W_2^l(\Omega) := \{\varphi \in L_2(\Omega) : D^s\varphi \in L_2(\Omega) \text{ for } |s| \leqslant l\}.$$

We introduce a scalar product on $W_2^l(\Omega)$ by means of

$$(\varphi, \psi)_l := \sum_{|s| \leqslant l} \int_\Omega D^s\varphi(x)\overline{D^s\psi(x)}\,dx \tag{1}$$

where, as usual, functions which are equal almost everywhere are identified.

Now suppose that l is not integral, therefore $l = [l] + \lambda$, $0 < \lambda < 1$. As scalar product we take

$$(\varphi, \psi)_l := \sum_{|s| \leqslant [l]} \int_\Omega D^s\varphi(x) \cdot \overline{D^s\psi(x)}\, dx$$

$$+ \sum_{|s| \leqslant [l]} \iint_{\Omega \times \Omega} \frac{(D^s\varphi(x) - D^s\varphi(y))\overline{(D^s\psi(x) - D^s\psi(y))}\, dy\, dx.}{|x - y|^{r + 2\lambda}} \tag{1'}$$

We wish to abbreviate the norm defined by (1′) *as*

$$\|\varphi\|_l^2 := \|\varphi\|_{[l]}^2 + \sum_{|s| \leqslant [l]} I_\lambda(D^s\varphi),$$

where we have written

$$I_\lambda(\varphi) := \iint_{\Omega \times \Omega} \frac{|\varphi(x) - \varphi(y)|^2}{|x - y|^{r + 2\lambda}}\, dx\, dy.$$

For $l = [l] + \lambda$ not integral we define:

$$W_2^l(\Omega) = \{\varphi \in L_2(\Omega) : D^s\varphi \in L_2(\Omega) \text{ for } |s| \leqslant [l] \text{ and } I_\lambda(D^s\varphi) < \infty\}.$$

Hence $W_2^l(\Omega)$ consists of functions φ from $W_2^{[l]}(\Omega)$ with the additional property that $I_\lambda(D^s\varphi) < \infty$ for $|s| \leqslant [l]$.

Theorem 3.1 *The Sobolev space $W_2^l(\Omega)$ is a separable Hilbert space, hence $W_2^l(\Omega)$ has a countable basis.*

Proof. Suppose first that l is integral. We easily see that $W_2^l(\Omega)$ is a pre-Hilbert space. In particular the Schwarz inequality, which holds in $L_2(\Omega)$, implies that (1) is defined for all $\varphi, \psi \in W_2^l(\Omega)$. We check completeness. For a Cauchy sequence φ_n from $W_2^l(\Omega)$ the sequences $D^s(\varphi_n)$ are also Cauchy sequences in $L_2^l(\Omega)$ for each $|s| \leqslant l$ because of the norm on $W_2^l(\Omega)$ given by (1)

$$\|\varphi\|_l = \left(\sum_{|s| \leqslant l} \int_\Omega |D^s\varphi(x)|^2\, dx \right)^{1/2}. \tag{2}$$

Because of the completeness of the space $L_2(\Omega)$ there exist functions φ^s for $|s| \leqslant l$ with

$$D^s\varphi_n \to \varphi^s \text{ in } L_2(\Omega). \tag{3}$$

It remains to show that $D^s\varphi^0 = \varphi^s$. On the one hand from (3) it follows that

$$\int_\Omega D^s\varphi_n(x)\psi(x)\, dx \to \int_\Omega \varphi^s(x)\psi(x)\, dx \quad \text{for } \psi \in \mathscr{D}(\Omega), \tag{4}$$

and on the other hand, by use of distributional derivation we obtain, see Theorem 1.6,

$$\int_\Omega D^s\varphi_n\cdot\psi\,dx=(-1)^{|s|}\int_\Omega \varphi_n\cdot D^s\psi\,dx\to(-1)^{|s|}\int_\Omega \varphi^0\cdot D^s\psi\,dx$$

$$=\int_\Omega D^s\varphi^0\cdot\psi\,dx \quad \text{for } \psi\in\mathscr{D}(\Omega). \tag{5}$$

Thus (4) and (5) yield

$$\int_\Omega \varphi^s\cdot\psi\,dx=\int_\Omega D^s\varphi^0\cdot\psi\,dx \quad \text{for all } \psi\in\mathscr{D}(\Omega),$$

or

$$D^s\varphi^0=\varphi^s$$

in the distributional sense. In order to prove completeness for l not integral, we have in addition to show that the integrals $I_\lambda(D^s\varphi_n-D^s\varphi_0)$ converge to zero for $|s|\leqslant[l]$. We have just shown that $D^s\varphi_n\to D^s\varphi_0$ in $L^2(\Omega)$. By a measure theoretic result of Riesz (see for example Natanson [1], p. 106, or Rudin [2], p. 70) there exists a subsequence m_n with the property that

$$D^s\varphi_{m_n}(x)\to D^s\varphi^0(x)$$

almost everywhere. From this it follows that the sequence

$$\frac{|D^s\varphi_{m_n}(x)-D^s\varphi_{m_n}(y)|^2}{|x-y|^{r+2\lambda}}\to\frac{|D^s\varphi^0(x)-D^s\varphi^0(y)|^2}{|x-y|^{r+2\lambda}}$$

converges almost everywhere in $\Omega\times\Omega$. Fatou's lemma (see Natanson [1], p. 155 or Halmos [1]) implies that, because of the Cauchy property $I_\lambda(\varphi_n-\varphi_m)\leqslant\varepsilon$,

$$I_\lambda(D^s\varphi^0)\leqslant\sup_{m_n} I_\lambda(D^s\varphi_{m_n})\leqslant M<\infty,$$

which means that we have obtained

$$\varphi^0\in W_2^l(\Omega)=W_2^{[l]+\lambda}(\Omega).$$

Fatou's lemma applied once more shows that

$$I_\lambda(D^s\varphi_{m_n}-D^s\varphi^0)\to 0 \quad \text{for } n\to\infty,$$

and with the triangle inequality

$$I_\lambda(D^s\varphi_n-D^s\varphi^0)\to 0 \quad \text{for } n\to\infty, |s|\leqslant[l].$$

This, together with the convergence $\|\varphi_n-\varphi^0\|_{[l]}\to 0$ proved already, gives the completeness of $W_2^l(\Omega)$. By means of the (isometric) map

$$W_2^l(\Omega)\ni\varphi\mapsto\left\{D^s\varphi, |s|\leqslant[l]; \frac{D^s\varphi(x)-D^s\varphi(y)}{|x-y|^{r/2+\lambda}}, |s|\leqslant[l]\right\}$$
$$\in \mathop{\times}_{|s|\leqslant[l]} L_2(\Omega)\times \mathop{\times}_{|s|\leqslant[l]} L_2(\Omega\times\Omega),$$

we can regard $W_2^l(\Omega)$ as a closed subspace of the Cartesian product

$$\mathop{\times}_{|s|\leqslant[l]} L_2(\Omega)\times \mathop{\times}_{|s|\leqslant[l]} L_2(\Omega\times\Omega).$$

Since

$$\mathop{\times}_{|s|\leqslant[l]} L_2(\Omega)\times \mathop{\times}_{|s|\leqslant[l]} L_2(\Omega\times\Omega)$$

is separable, so also is $W_2^l(\Omega)$; this follows from the lemma below.

Lemma 3.1 *Let H be a separable Hilbert space and U a subspace. Then U is again separable.*

Proof. Let $\{h_n\}$ be a dense sequence in H. We consider countably many balls $B(h_n, 1/m)$ and choose in each an element $u_{m,n}\in U$, if such exists. In this way we obtain countably many $u_{n,m}$, forming a dense subset in U. Let u be an arbitrary point in U. Then by assumption there exists a subsequence h_n' of $\{h_n\}$ with $h_n'\to u$. For each m we choose the index m_n large enough that $u\in B(h'_{n_m}, 1/m)$. Then since $B(h'_{n_m}, 1/m)\cap U$ is non-empty, there exists some $u_{n_{m,m}}\in U$ and we have $u_{n_{m,m}}\to u$. ∎

3.2 Density properties

Theorem 3.2 *Functions in $W_2^l(\Omega)$ with bounded support are dense in $W_2^l(\Omega)$.*

Remark By the definition of supp φ, see Definition 1.3, this is relatively closed with respect to Ω. It can, however extend to the frontier $\partial\Omega$, and a function $\varphi\in W_2^l(\Omega)$, $l\geqslant 1$, can have a sharp discontinuity on the frontier and not be extendable by 0. The situation when supp φ is strictly contained in Ω is completely different; then the function is extendable by 0, see Theorems 3.3 and 3.7.

For the proof of Theorem 3.2 we need the following:

Lemma 3.2 *Let $a\in C^{[l]+1}(\mathbb{R}^r)$. Then for all $\varphi\in W_2^l(\Omega)$*

$$a\cdot\varphi\in W_2^l(\Omega) \quad \textit{and} \quad \|a\varphi\|_l\leqslant C\cdot\|a\|_{C^{[l]+1}}\|\varphi\|_l,$$

where the constant C is independent of a and φ.

Proof. For integral values $l = 0, 1, \dots$, Lemma 3.2 follows from the Leibniz product rule, and we can make do with the assumption that $a \in C^l(\Omega)$. Let $l = [l] + \lambda$, $0 < \lambda < 1$. For $|s| \leqslant [l]$ the mean value theorem of differential calculus gives

$$|D^s a(x) - D^s a(y)| \leqslant \max |D^{s+1} a| \cdot |x - y| \leqslant \|a\|_{C^{[l]+1}} \cdot |x - y|,$$

from which it follows (see Proposition 4.2 below) that

$$|D^s a(x) - D^s a(y)| \leqslant 4 \|a\|_{C^{[l]+1}} \frac{|x - y|}{1 + |x - y|}, \quad |s| \leqslant [l]. \qquad (*)$$

We have still to estimate the integrals $I_\lambda(D^\alpha(a \cdot \varphi))$ for $|\alpha| \leqslant [l]$. The Leibniz product rule reduces this to estimating the integrals $I_\lambda(D^s a \cdot D^t \varphi)$, where $|s + t| \leqslant [l]$. By addition and subtraction of $D^s a(x) \cdot D^t \varphi(y)$ we have

$$I_\lambda(D^s a \cdot D^t \varphi) = \iint_{\Omega \times \Omega} \frac{|D^s a(x) \cdot D^t \varphi(x) - D^s a(y) \cdot D^t \varphi(y)|^2}{|x - y|^{r + 2\lambda}} \mathrm{d}x \, \mathrm{d}y$$

$$\leqslant \iint_{\Omega \times \Omega} |D^s a(x)|^2 \frac{|D^t \varphi(x) - D^t \varphi(y)|^2}{|x - y|^{r + 2\lambda}} \mathrm{d}x \, \mathrm{d}y$$

$$+ \iint_{\Omega \times \Omega} |D^t \varphi(y)|^2 \frac{|D^s a(x) - D^s a(y)|}{|x - y|^{r + 2\lambda}} \mathrm{d}x \, \mathrm{d}y.$$

We can immediately estimate the first integral

$$I_1 \leqslant \|a\|^2_{C^{[l]}} I_\lambda(D^t \varphi) \leqslant \|a\|^2_{C^{[l]+1}} \|\varphi\|^2_l,$$

while for the second, by application of Fubini's theorem, we have

$$I_2 = \int_\Omega |D^t \varphi(y)|^2 \left(\int_\Omega \frac{|D^s a(x) - D^s a(y)|^2}{|x - y|^{r + 2\lambda}} \mathrm{d}y \right) \mathrm{d}y.$$

we apply the inequality $(*)$ to the inner integral and estimate

$$\int_\Omega \frac{|D^s a(x) - D^s a(y)|^2}{|x - y|^{r + 2\lambda}} \mathrm{d}x \leqslant 4^2 \|a\|^2_{C^{[l+1]}} \int_\Omega \frac{\mathrm{d}x}{(1 + |x - y|)^2 |x - y|^{r + 2\lambda - 2}}$$

$$\leqslant C \|a\|^2_{C^{[l]+1}}.$$

The last follows, since the integral

$$\int_{\mathbb{R}^r} \frac{\mathrm{d}z}{(1 + |z|^2)|z|^{r + 2\lambda - 2}}$$

is convergent. Therefore we have

$$I_2 \leqslant \|D^t \varphi\|^2_0 \cdot C \cdot \|a\|^2_{C^{[l]+1}},$$

and putting everything together

$$\|a\cdot\varphi\|_l^2=\|a\varphi\|_{[l]}^2+\sum_{|\alpha|\leqslant[l]}I_\lambda(D^\alpha(a\cdot\varphi))$$

$$=\sum_{|\alpha|\leqslant[l]}\|D^\alpha(a\cdot\varphi)\|_0^2+\sum_{|\alpha|\leqslant[l]}I_\lambda(D^\alpha(a\cdot\varphi))$$

$$\leqslant C\sum_{|\alpha|\leqslant[l]}\sum_{s+t=\alpha}\binom{s}{\alpha}^2(\|D^s aD^t\varphi\|_0^2+I_\lambda(D^s aD^t\varphi))\leqslant C\|a\|_{C^{[l]+1}}^2\|\varphi\|_l^2.$$

■

Now for the proof of Theorem 3.2:

Let f be a fixed function in $\mathscr{D}(\mathbb{R}^r)$ with the properties:

1. $f(x)=1$ for $|x|\leqslant 1$.
2. $f(x)=0$ for $|x|\geqslant 2$.
3. $|D^s f(x)|\leqslant M$ for all x and $|s|\leqslant[l]+1$ (f exists by Corollary 1.2).

We put $f_\varepsilon(x)=f(\varepsilon x)$ for $\varepsilon>0$ and have $f_\varepsilon(x)=1$ for $|x|\leqslant 1/\varepsilon$, $|D^s f(x)|\leqslant M\varepsilon^{|s|}\leqslant M$ for $0<\varepsilon\leqslant 1$, $|s|\leqslant[l]+1$. By Lemma 3.2 if $\varphi\in W_2^l(\Omega)$ the function $\varphi_\varepsilon=f_\varepsilon\cdot\varphi$ again belongs to $W_2^l(\Omega)$ and has bounded support in Ω. If we set $\Omega^\varepsilon=\{x\in\Omega:|x|>1/\varepsilon\}$, then by virtue of Lemma 3.2 we have for $0<\varepsilon\leqslant 1$

$$\|\varphi-\varphi_\varepsilon\|_{l,\Omega}=\|\varphi-\varphi_\varepsilon\|_{l,\Omega^\varepsilon}\leqslant\|\varphi\|_{l,\Omega^\varepsilon}+\|\varphi_\varepsilon\|_{l,\Omega^\varepsilon}$$
$$\leqslant\|\varphi\|_{l,\Omega^\varepsilon}+cM\|\varphi\|_{l,\Omega^\varepsilon}=(1+cM)\|\varphi\|_{l,\Omega^\varepsilon}.$$

The right-hand side goes to 0 as $\varepsilon\to 0$ ($\Omega^\varepsilon\to\varnothing$), and hence we have proved our theorem. We now want to define the space $\mathring{W}_2^l(\Omega)$.

Definition 3.2 *We denote the closed hull of $\mathscr{D}(\Omega)$ in $W_2^l(\Omega)$ as the space $\mathring{W}_2^l(\Omega)$, that is,*

$$\mathring{W}_2^l(\Omega):=\overline{\mathscr{D}(\Omega)}^{W_2^l}. \tag{6}$$

By Lemma 3.1 $\mathring{W}_2^l(\Omega)$ is again a separable Hilbert space.

In order to obtain further density properties for the spaces $\mathring{W}_2^l(\Omega)$ and $W_2^l(\Omega)$ we need a lemma:

Lemma 3.3 *Let $\varphi\in W_2^l(\Omega)$ and let φ have compact support in $\Omega'\subset\subset\Omega$. Then for the regularisation $\varphi_\varepsilon=h_\varepsilon*\varphi$ (see Definition (8) in §1), we have*

(a) $\lim_{\varepsilon\to 0}\varphi_\varepsilon=\varphi$ *in* $W_2^l(\Omega)$.

(b) $\varphi_\varepsilon\in\mathscr{D}(\Omega)$ *for ε sufficiently small, that is, expressed precisely,*

$$\operatorname{supp}\varphi_\varepsilon\subset\Omega'+\varepsilon\subset\subset\Omega\quad\textit{for }\varepsilon\subseteq\varepsilon_0=\operatorname{dist}(\Omega',C\Omega).$$

Proof. Let $l=0,1,\ldots$. For $l=0$ we have the statement of Theorem 1.3(c).

Because of Theorem 1.11 we have

$$\mathrm{D}^s\varphi_\varepsilon = h_\varepsilon * \mathrm{D}^s\varphi = (\mathrm{D}^s\varphi)_\varepsilon \quad \text{for } |s| \leqslant l,$$

and another application of Theorem 1.3(c) gives

$$\mathrm{D}^s\varphi_\varepsilon \to \mathrm{D}^s\varphi \quad \text{in } L_2(\Omega), \quad |s| \leqslant l.$$

Since the norm in $W_2^l(\Omega)$, $l = 0, 1, \ldots$ is defined by $\|\varphi\|_l^2 = \sum_{|s| \leqslant l} \|\mathrm{D}^s\varphi\|_0^2$, we have proved our lemma for $l = 0, 1, \ldots$.

Suppose now that l is not integral, therefore $l = [l] + \lambda$, $0 < \lambda < 1$. We have only still to show the convergence of the integrals $I_\lambda(\mathrm{D}^s\varphi_\varepsilon - \mathrm{D}^s\varphi)$ to zero, as $\varepsilon \to 0$ for $|s| \leqslant [l]$. We have

$$\begin{aligned} &\{(\mathrm{D}^s\varphi_\varepsilon(x) - \mathrm{D}^s\varphi(x)) - (\mathrm{D}^s\varphi_\varepsilon(y) - \mathrm{D}^s\varphi(y))\} \\ &= \int_{|z| \leqslant \varepsilon} h_\varepsilon(z) \cdot \{(\mathrm{D}^s\varphi(x+z) - \mathrm{D}^s\varphi(y+z)) - (\mathrm{D}^s\varphi(x) - \mathrm{D}^s\varphi(y))\}\,\mathrm{d}z, \end{aligned}$$

which by manipulations of the kind carried out in the proof of Theorem 1.3 gives

$$\begin{aligned} &I_\lambda(\mathrm{D}^s\varphi_\varepsilon - \mathrm{D}^s\varphi) \\ &= \iint_{\Omega\times\Omega} \frac{|\{(\mathrm{D}^s\varphi_\varepsilon(x) - \mathrm{D}^s\varphi(x)) - (\mathrm{D}^s\varphi_\varepsilon(y) - \mathrm{D}^s\varphi(y))\}|^2}{|x-y|^{r+2\lambda}}\,\mathrm{d}x\,\mathrm{d}y \\ &\leqslant \sup_{|z| \leqslant \varepsilon} \iint_{\Omega\times\Omega} \frac{|(\mathrm{D}^s\varphi(x+z) - \mathrm{D}^s\varphi(y+z)) - (\mathrm{D}^s\varphi(x) - \mathrm{D}^s\varphi(y))|^2}{|x-y|^{r+2\lambda}}\,\mathrm{d}x\,\mathrm{d}y. \end{aligned} \tag{7}$$

By assumption the function $|\mathrm{D}^s\varphi(x) - \mathrm{D}^s\varphi(y)|/|x-y|^{r/2+\lambda}$ belongs to $L^2(\Omega\times\Omega)$, so it is therefore mean continuous (§1.1, Kolmogorov compactness criterion). This implies that we can take the $\sup_{|z| \leqslant \varepsilon} \iint_{\Omega\times\Omega} \cdots$ in (7) to be as small as we like, thus we have proved our lemma. ■

Theorem 3.3 *The functions $\varphi \in W_2^l(\Omega)$, for which the support is compact and contained in Ω, form a dense subset of $\mathring{W}_2^l(\Omega)$, that is,*

$$\mathring{W}_2^l(\Omega) = \overline{\{\varphi : \varphi \in W_2^l(\Omega), \operatorname{supp}\varphi \subset\subset \Omega\}}^{W_2^l}. \tag{8}$$

Proof. Let $\varphi \in W_2^l(\Omega)$ and let φ have compact support, $\operatorname{supp}\varphi =: K \subset\subset \Omega$. Since by Definition 3.2 $\mathscr{D}(\Omega)$ is dense in $\mathring{W}_2^l(\Omega)$, we have proved the theorem if we can approximate φ by functions from $\mathscr{D}(\Omega)$. We apply Lemma 3.3, with $0 < \varepsilon < \varepsilon_0 = \operatorname{dist}(K, \mathrm{C}\Omega)$, then $\varphi_\varepsilon \in \mathscr{D}(\Omega)$ and $\varphi_\varepsilon \to \varphi$ in $W_2^l(\Omega)$. ■

Theorem 3.4 *We have $\mathring{W}_0^0(\Omega) = W_2^0(\Omega) = L_2(\Omega)$, or $\mathscr{D}(\Omega)$ is dense in $L_2(\Omega)$.*

Proof. Let $\varphi \in L_2(\Omega)$, and let K_n be a sequence of open, bounded sets with

$\bar{K}_n \subset K_{n+1} \subset \cdots \subset \Omega$ and $\bigcup_n K_n = \Omega$, for which

$$\int_{K_n} |\varphi(x)|^2 \, dx \to \int_{\Omega} |\varphi(x)|^2 \, dx \quad \text{as } n \to \infty. \tag{9}$$

Let χ_n be the characteristic function of K_n, $\chi_n \cdot \varphi \in L_2(\Omega)$ and has compact support $\operatorname{supp}(\chi_n \cdot \varphi) \subset \bar{K}_n \subset \Omega$. Therefore by Theorem 3.3 $\chi_n \cdot \varphi \in \mathring{W}_2^0(\Omega)$, that is, $\chi_n \cdot \varphi$ can be approximated in $L_2(\Omega)$ by functions in $\mathscr{D}(\Omega)$. So if we can show that $\chi_n \cdot \varphi \to \varphi$ in $L_2(\Omega)$ we have proved everything. But, given (9),

$$\| \varphi - \chi_n \cdot \varphi \|^2 = \int_{\Omega \setminus K_n} |\varphi(x)|^2 \, dx \to 0, \text{ since } \int_{\Omega} |\varphi(x)|^2 \, dx < \infty. \qquad \blacksquare$$

If $\Omega = \mathbb{R}^r$ from Theorems 3.2 and 3.3 we deduce:

Corollary 3.1 *For all values of l, $W_2^l(\mathbb{R}^r) = \mathring{W}_2^l(\mathbb{R}^r)$.*

Proof. In the case $\Omega = \mathbb{R}^r$ 'supp φ bounded' implies that supp φ is already compact and strictly contained in $\mathbb{R}^r$. We see therefore that the same subset, namely $\{\varphi : \operatorname{supp} \varphi \text{ is compact}\}$ is dense in both $W_2^l(\mathbb{R}^r)$ and $\mathring{W}_2^l(\mathbb{R}^r)$. This proves equality. $\blacksquare$

Theorem 3.4 and Corollary 3.1 describe almost all general situations in which $\mathscr{D}(\Omega)$ is dense in $W_2^l(\Omega)$. If $\emptyset \neq \Omega$ is open and bounded, then for $l \geqslant 1$, $\mathring{W}_2^l(\Omega)$ is a proper subspace of $W_2^l(\Omega)$, see Corollary 7.1 below.

It is interesting that the next theorem holds in general, without special assumptions on the frontier $\partial\Omega$ of Ω. Here we give the proof of Floret [1], see also Meyers & Serrin [1].

Theorem 3.5 *$W_2^l(\Omega) \cap \mathscr{E}(\Omega)$ is dense in $W_2^l(\Omega)$.*

Proof. Let $\{\Omega_i : i \in \mathbb{N}\}$ be a countable, locally finite cover of Ω by bounded open sets with $\bar{\Omega}_i \subset \Omega$; by Theorem 1.1 such always exists. Let $\{\alpha_j\}$ be a partition of unity subordinate to $\{\Omega_i\}$. If $\varphi \in W_2^l(\Omega)$, then $\varphi = \sum_i \varphi \alpha_i$ on Ω and $\varphi \cdot \alpha_i \in \mathring{W}_2^l(\Omega_i)$ (by Theorem 3.3). Because of Definition 3.2 there exist sequences $\varphi_i^{(n)}$, $n \in \mathbb{N}$, in $\mathscr{D}(\Omega_i)$ with

$$\varphi_i^{(n)} \xrightarrow[n \to \infty]{} \varphi \cdot \alpha_i \quad \text{in } \mathring{W}_2^l(\Omega_i) \text{ or } \mathring{W}_2^l(\Omega).$$

Without loss of generality we can take

$$\| \varphi_i^{(n)} - \varphi \cdot \alpha_i \|_l \leqslant \frac{1}{n \cdot 2^i}.$$

Since the cover $\{\Omega_i\}$ is locally finite, we have

$$\varphi^{(n)} = \sum_{i=1}^{\infty} \varphi_i^{(n)} \in \mathscr{E}(\Omega).$$

We now show that the sequence $\varphi^{(n)} - \varphi$ is a null sequence in $W_2^l(\Omega)$ (so that also $\varphi^{(n)} = (\varphi^{(n)} - \varphi) + \varphi \in W_2^l(\Omega)$).

$$(\varphi^{(n)} - \varphi) = \left(\sum_i \varphi_i^{(n)}\right) - \left(\sum_i \varphi \cdot \alpha_i\right) = \sum_i (\varphi_i^{(n)} - \varphi\alpha_i).$$

Because

$$\left\| \sum_{i=N+1}^{N+p} (\varphi_i^{(n)} - \varphi\alpha_i) \right\| \leqslant \sum_{i=N+1}^{N+p} \| \varphi_i^{(n)} - \varphi\alpha_i \|_l \leqslant \sum_{i=N+1}^{N+p} \frac{1}{n2^i} \leqslant \frac{1}{n2^N},$$

the series $\sum_i (\varphi_i^{(n)} - \varphi\alpha_i)$ converges for each $n \in \mathbb{N}$ in $W_2^l(\Omega)$. From the estimate

$$\left\| \sum_{i=1}^{N} (\varphi_i^{(n)} - \varphi\alpha_i) \right\|_l \leqslant \sum_{i=1}^{N} \| \varphi_i^{(n)} - \varphi\alpha_i \|_l \leqslant \sum_{i=1}^{N} \frac{1}{n2^i} \leqslant \frac{1}{n},$$

it then follows (because of the continuity of the norm) that as $N \to \infty$

$$\left\| \sum_{i=1}^{\infty} (\varphi_i^{(n)} - \varphi\alpha_i) \right\|_l = \| (\varphi^{(n)} - \varphi) \|_l \leqslant \frac{1}{n}.$$

Hence as $n \to \infty$, it follows that $\varphi^{(n)} \to \varphi$ in $W_2^l(\Omega)$. ∎

Theorem 3.5 implies that the set of C^∞-functions that satisfy the defining conditions for $W_2^l(\Omega)$ with classical continuous derivatives, forms a dense subset of $W_2^l(\Omega)$. In this way we obtain a new approach to the spaces $W_2^l(\Omega)$ which avoids the concept of weak derivation.

Until now we have imposed no restrictions on the region Ω. We begin now with the segment property, see Definition 2.1.

Theorem 3.6 *Let Ω have the segment property. Then the restrictions to Ω of functions from $\mathscr{D}(\mathbb{R}^r)$ form a dense subset of $W_2^l(\Omega)$:*

$$\overline{\mathscr{D}(\mathbb{R}^r)|_\Omega}^{W_2^l} = W_2^l(\Omega).$$

From Theorem 3.6 it immediately follows that all the spaces $C_0^k(\bar{\Omega})$, $k \geqslant l$ are dense in $W_2^l(\Omega)$.

Proof. For the first step, we localise. Let $\varphi \in W_2^l(\Omega)$. Because of Theorem 3.2 we may assume that $\operatorname{supp}_\Omega \varphi = K$ is bounded. The set $F = \bar{K} \backslash (\bigcup_{x \in \partial\Omega} U_x)$ is compact and contained in Ω; here we make the choice of the neighbourhoods U_x according to the segment property of Definition 2.1. There exists an open set U_0 with $F \subset\subset U_0 \subset \bar{U}_0 \subset\subset \Omega$, U_0 open. Since $\bar{K}$ is compact, we can find finitely many U_x (we label them $U_1, \ldots, U_k$) with the properties of Definition 2.1, and such that $\bar{K} \subset U_0 \cup U_1 \cup \cdots \cup U_k$. We can also shrink the U_j so that

$$\tilde{U}_j \subset\subset U_j, \quad j = i, \ldots, k, \quad \tilde{U}_0 = U_0,$$

and so that $\bar{K} \subset \tilde{U}_0 \cup \tilde{U}_1 \cup \cdots \cup \tilde{U}_k$ still holds. Let $\{\alpha_j\}$ be a partition of unity subordinate to the cover $\{\tilde{U}_j\}$. We set $u_j = \alpha_j \cdot \varphi$ and have $\operatorname{supp} u_j \subset \tilde{U}_j \cap \Omega$. If for each j we can find some $\varphi_j \in \mathscr{D}(U_j)$ with

$$\| u_j - \varphi_j \|_{l, U_j \cap \Omega} \leqslant \frac{\varepsilon}{k+1}, \quad j = 0, \ldots, k, \tag{10}$$

then, were we to put $\psi := \sum_{j=0}^{k} \varphi_j$, we would have proved everything, because $\psi \in \mathscr{D}(\mathbb{R}^r)$ and we have

$$\| \varphi - \psi \|_{l, \Omega} = \left\| \sum_{j=0}^{k} (\alpha_j \varphi - \varphi_j) \right\|_{j, \Omega} \leqslant \sum_{j=0}^{k} \| u_j - \varphi_j \|_{l, \Omega}$$

$$= \sum_{j=0}^{k} \| u_j - \varphi_j \|_{l, U_j \cap \Omega} \leqslant \frac{(k+1)\varepsilon}{k+1} = \varepsilon.$$

For the second step, we approximate. We demonstrate (10) for a single value of j. Let $j = 0$, since $\operatorname{supp} u_0 \subset\subset U_0 \subset \Omega$; by Lemma 3.3 we can always find some $\varphi_0 \in \mathscr{D}(U_0)$ for which (10) holds with $j = 0$. We now take some j with $1 \leqslant j \leqslant k$.

The basic idea of the proof which follows is this: the function u_j can only have a singularity on the frontier $\partial\Omega$. By the segment property we push the singularity away from the frontier into the exterior; it now lies in Γ_t, see Fig. 3.1. Next we can smooth the function u_j near the frontier $\partial\Omega$ and extend

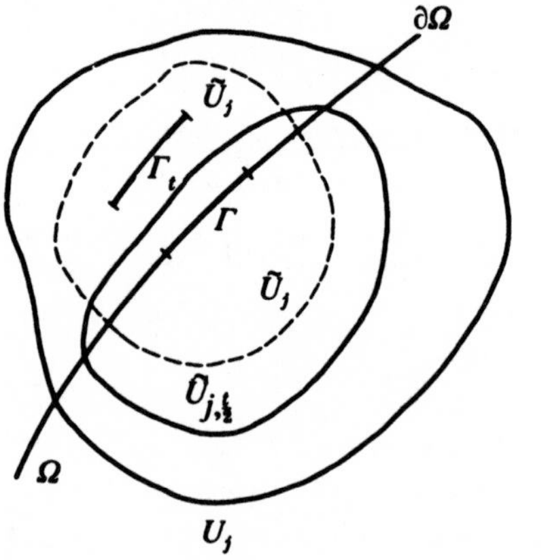

Fig. 3.1

by zero to a function in $\mathscr{D}(\mathbb{R}^r)$. All the steps described are precise up to some ε, which is the approximation we wish to justify.

Now for the details. We extend u_j, $1 \leqslant j \leqslant k$ by means of 0. For the extended function we have $u_j \in W_2^l(U_j \setminus \Gamma)$, where $\Gamma := \bar{\tilde{U}}_j \cap \partial\Omega$ (u_j may be singular on Γ). Let y_j be the vector distinct from zero in Definition 2.1 contained in the neighbourhood U_j. We define $\Gamma_t = \Gamma - ty_j$, where we choose t in such a way that

$$0 < t < \min\left(1, \operatorname{dist}\frac{(\tilde{U}_j, CU_j)}{|y_j|}\right).$$

Then $\Gamma_t \subset U_j$ and $\Gamma_t \cap \bar{\Omega} = \varnothing$; indeed were Γ_t and $\bar{\Omega}$ to have a point z in common, then $z + ty$ would belong to Ω by the segment property. However, since $z + ty_j \in \Gamma_t + ty_j = \Gamma \subset \partial\Omega$, and $\partial\Omega$ does not lie in Ω, we would have a contradiction. For later use we note

$$\operatorname{dist}(\Gamma_t, \bar{\Omega}) = \delta(t) > 0, \text{ since } \Gamma_t \text{ is compact.} \tag{11}$$

We set $u_{j,t}(x) = u_j(x + ty_j)$ and have $u_{j,t} \in W_2^l(U_j \setminus \Gamma_t)$, where again $u_{j,t}$ may be singular on Γ_t. Translation is continuous in W_2^l – this follows from continuity in the mean for functions from $L_2(\Omega)$, respectively $L_2(\Omega \times \Omega)$ (see the proof of Lemma 3.3), and from the interchangeability of the differential operator D^s with the translation operator. Put another way, after restriction to $U_j \cap \Omega$ we have

$$\| u_{j,t} - u_j \|_{l, U_j \cap \Omega} \leqslant \varepsilon/2(k+1) \quad \text{for } 0 < t < \eta. \tag{12}$$

Also t can be chosen so small that the portion of the support of $u_{j,t}$ lying inside Ω is already contained in $\tilde{U}_j \cap \Omega$:

$$\operatorname{supp} u_{j,t} \cap \Omega \subset \tilde{U}_j \cap \Omega \quad \text{for } 0 < t < \eta. \tag{13}$$

Let $\tilde{U}_{j,t/2} = \{x : \operatorname{dist}(\tilde{U}_j \cap \Omega, x) < \delta(t)/2\}$ with $\delta(t)$ from (11). Here let t be so small that

$$\tilde{U}_{j,t/2} \subset\subset U_j \quad \text{for } 0 < t < \eta.$$

We choose a function $\alpha_{j,t} \in C_0^\infty$ with

$$\alpha_{j,t} = \begin{cases} 1 & \text{on } \overline{\tilde{U}_j \cap \Omega}, \\ 0 & \text{on } C\tilde{U}_{j,t/2}, \\ \text{between 0 and 1 otherwise.} \end{cases}$$

The function $u_{j,t} \cdot \alpha_{j,t}$ has compact support in U_j, which by construction lies away from the singularity Γ_t. Therefore $u_{j,t} \cdot \alpha_{j,t} \in W_2^l(U_j)$, and again we apply

Lemma 3.3, obtaining a function $\varphi_j \in \mathscr{D}(U_j)$ with

$$\| u_{j,t} \cdot \alpha_{j,t} - \varphi_j \|_{l,U_j} \leqslant \frac{\varepsilon}{2(k+1)}. \tag{14}$$

The properties of the supports of $u_{j,t}$ (see (13)) and $\alpha_{j,t}$ imply that on $U_j \cap \Omega$ $u_{j,t} \cdot \alpha_{j,t} = u_{j,t}$ holds. Thus from (12) and (14) we obtain

$$\| u_j - \varphi_j \|_{l,U_j \cap \Omega} \leqslant \frac{\varepsilon}{k+1},$$

which is (10) and concludes the proof of Theorem 3.6. ∎

In order to give a further useful characterisation of the space $\mathring{W}_2^l(\Omega)$ we need a simple extension theorem.

Lemma 3.4 *Let $l = 0, 1, \ldots$. Functions from $\mathring{W}_0^l(\Omega)$ can be extended by 0 to all of $\mathbb{R}^r$ in a norm preserving manner, that is,*

$$\| F\varphi \|_{l,\mathbb{R}^r} = \| \varphi \|_{l,\Omega}, \quad \varphi \in \mathring{W}_2^l(\Omega), \quad \operatorname{supp} F\varphi \subset \bar{\Omega}. \tag{15}$$

Proof. Let $\varphi \in \mathring{W}_2^l(\Omega)$, then by Definition 3.2 there exists a sequence $\varphi_n \in \mathscr{D}(\Omega)$ with $\varphi_n \to \varphi$ in $\mathring{W}_2^l(\Omega)$. We extend the function φ_n by 0 and obtain $\varphi_n = F\varphi_n \in \mathscr{D}(\mathbb{R}^r)$,

$$\| F\varphi_n \|_{l,\mathbb{R}^r} = \| \varphi_n \|_{l,\Omega}. \tag{16}$$

The functions $F\varphi_n$ form a Cauchy sequence in $W_2^l(\Omega)$ ((16)!), let their limit value in $W_2^l(\mathbb{R}^r)$ be $F\varphi$. Then we have $F\varphi|_\Omega = \varphi$ and also (15), as we easily see. We now prove the support property. Since we have extended by 0 we have $\int_{C\Omega} |\varphi_n|^2 \, dx = 0$, and also $\int_{C\Omega} |F\varphi|^2 \, dx = 0$, which implies that $\operatorname{supp} F\varphi \subset \bar{\Omega}$. ∎

Remark Lemma 3.4 and the following Theorem 3.7 also hold for all $l \neq m + \frac{1}{2}$, where $m = 0, 1, \ldots$, see Lions & Magenes [1].

Theorem 3.7 *Let $l = 0, 1, \ldots$, and let the region Ω have the segment property of Definition 2.1. The space $\mathring{W}_2^l(\Omega)$ consists of the restrictions to Ω of all functions from $W_2^l(\mathbb{R}^r)$, which are supported in $\bar{\Omega}$:*

$$\mathring{W}_2^l(\Omega) = \{ \varphi \in W_2^l(\mathbb{R}^r) : \operatorname{supp} \varphi \subset \bar{\Omega} \}. \tag{17}$$

Proof. Let $\mathring{\mathscr{W}}_2^l(\Omega)$ be the space on the right in (17); that $\mathring{W}_2^l(\Omega) \subset \mathring{\mathscr{W}}_2^l(\Omega)$ follows immediately from Lemma 3.4. In order to prove the reverse inclusion we push the functions inward by means of the segment property (in the proof of Theorem 3.6 we pushed them outwards). We carry out the proof in two steps.

First, suppose in addition that Ω is bounded, and let $\{U_0, U_1, \dots, U_m\}$ be a cover of $\bar{\Omega}$, where $U_0 \subset\subset \Omega$ and the $U_1, \dots, U_m$ have the displacement property of Definition 2.1. Let $\{\alpha_i : i = 1, \dots, m\}$ be a partition of unity subordinate to the cover $\{U_0, \dots, U_m\}$; let $\varphi \in W_2^l(\mathbb{R}^r)$ with $\operatorname{supp} \varphi \subset \bar{\Omega}$. Since $\operatorname{supp}(\alpha_0 \cdot \varphi) \subset\subset \Omega$, by Lemma 3.3 there exists a sequence

$$\varphi_{n,0} \in \mathscr{D}(\Omega) \quad \text{with } \varphi_{n,0} \to \alpha_0 \cdot \varphi \text{ in } \mathring{W}_2^l(\Omega).$$

We next consider the functions $\varphi_i := \alpha_i \cdot \varphi$, for $i = 1, \dots, m$. Because of the segment property the displaced functions

$$\varphi_{i,t}(x) := \varphi_i(x - ty_i), \quad 0 < t < 1, i = 1, \dots, m$$

have their support in Ω. Here $y_i \neq 0$ is the vector from Definition 2.1. Since the translation operator is continuous, we can find some t_n with

$$\| \varphi_i - \varphi_{i,t_n} \|_{l,\Omega} \leqslant \frac{1}{n}, \quad i = 1, \dots, m,$$

and by Lemma 3.3 ($\operatorname{supp} \varphi_{i,t_n} \subset\subset \Omega$) there exists some $\varphi_{i,n} \in \mathscr{D}(\Omega)$ with

$$\| \varphi_{i,t_n} - \varphi_{i,n} \|_{l,\Omega} \leqslant \frac{1}{n}, \quad i = 1, \dots, m.$$

Therefore we have

$$\tilde{\varphi}_n := \sum_{i=0}^{m} \varphi_{i,n} \in \mathscr{D}(\Omega), \quad \text{and} \quad \tilde{\varphi}_n \to \varphi \text{ in } W_2^l(\Omega),$$

hence we have proved that $\varphi \in \mathring{W}_2^l(\Omega)$.

For the second step, let Ω be unbounded. Let $\varphi \in W_2^l(\mathbb{R}^r)$ with $\operatorname{supp} \varphi \subset \bar{\Omega}$. As in the proof of Theorem 3.2, for each ε we can find a function $\psi \in W_2^l(\mathbb{R}^r)$ with

$$\| \varphi - \psi \|_{l,\mathbb{R}^r} \leqslant \frac{\varepsilon}{2}, \tag{18}$$

$\operatorname{supp} \psi = K$ compact and $K \subset \bar{\Omega}$.

We now proceed as in the proof of Theorem 3.6; we cover $K \subset U_0 \cup U_1 \cup \dots \cup U_k$, where $U_0 \subset\subset \Omega$ and the $U_1, \dots, U_k$ are relatively compact with the displacement property of Definition 2.1. Let $\{\alpha_j\}$ be a partition of unity subordinate to the cover $\{U_j\}$, $j = 0, 1, \dots, k$. We consider the functions $u_j = \alpha_j \cdot \psi$, $j = 0, \dots, k$. For $j = 0$ we have $\operatorname{supp} u_0 \subset\subset U_0 \subset\subset \Omega$, we can apply Lemma 3.3 and find a function $\psi_0 \in \mathscr{D}(\Omega)$ with

$$\| u_0 - \psi_0 \|_{l,\Omega} \leqslant \frac{\varepsilon}{2(k+1)}. \tag{19}$$

Suppose next that $1 \leqslant j \leqslant k$. We push the support of u_j into $U_j \cap \Omega$, that is, we consider $u_{j,t}(x_0) := u_j(x - ty_j)$, $0 < t$ sufficiently small, and as in the first step have

$$\|u_j - u_{j,t}\|_{l, U_j \cap \Omega} \leqslant \frac{\varepsilon}{4(k+1)}. \tag{20}$$

Lemma 3.3 applied again gives some $\psi_j \in \mathscr{D}(\Omega)$ with

$$\|u_{j,t} - \psi_j\|_{l, U_j \cap \Omega} \leqslant \frac{\varepsilon}{4(k+1)}. \tag{21}$$

(20) and (21) give

$$\|u_j - \psi_j\|_{l, U_j \cap \Omega} \leqslant \frac{\varepsilon}{2(k+1)}. \tag{22}$$

We set $\chi = \sum_{j=0}^{k} \psi_j$, have $\chi \in \mathscr{D}(\Omega)$, and using (19) and (22) deduce that

$$\|\psi - \chi\|_{l,\Omega} = \left\| \sum_{j=0}^{k} (\alpha_j \psi - \psi_j) \right\|_{l,\Omega} \leqslant \sum_{j=0}^{k} \|u_j - \psi_j\|_{l,\Omega}$$

$$\leqslant \sum_{j=0}^{k} \|u_j - \psi_j\|_{l,\Omega \cap U_j} \leqslant \frac{\varepsilon}{2}.$$

This, together with (18) implies that we have found a function $\chi \in \mathscr{D}(\Omega)$ with

$$\|\varphi - \chi\|_{l,\Omega} \leqslant \varepsilon \quad \text{or} \quad \varphi \in \mathring{W}_2^l(\Omega). \qquad ■$$

Exercises

3.1 Starting with the spaces $L_p(\Omega)$, $1 \leqslant p \leqslant \infty$, define the spaces $W_p^l(\Omega)$. Which statements of §3 continue to hold in the spaces $W_p^l(\Omega)$?

3.2 Give examples of regions Ω in $\mathbb{R}^2$ for which the approximation theorem (3.6) is not valid.

§4 The transformation theorem and Sobolev spaces on differentiable manifolds

For boundary value problems we need W_2^l-spaces on differentiable manifolds, for example on $\partial\Omega$, and introduce them in this section. For this we need to know the behaviour of the W_2^l-spaces under change of coordinates; Sobolev spaces on manifolds should be independent, up to equivalence, of specially chosen coordinates. We prove Theorem 4.1, the transformation theorem; with this result to hand it is easy to define the space $W_2^l(M)$ for a compact differentiable manifold M, see Definition 4.4. Further on in §4 we examine the first simple properties of these spaces $W_2^l(M)$, such as equivalent definitions, density properties, etc.

4.1 The transformation theorem

In order to prove the invariance of Sobolev–Slobodeckiĭ spaces under coordinate transformations we must acquaint ourselves with multipliers of W-spaces. We study these in what follows.

We need some new function spaces for intermediate constructions.

Definition 4.1 *Let $l\in\mathbb{N}$. We say that $\varphi\in A^l(\Omega)$ if, almost everywhere in Ω, φ possesses bounded and measurable derivatives up to order l. We put*

$$\|\varphi\|_{A^l}=\max_{|\alpha|\leqslant l}\operatorname*{sup\,ess}_{x\in\Omega}|D^\alpha\varphi(x)|.$$

Now let $l\in\mathbb{R}_+$, $l=[l]+\lambda$, where $0<\lambda<1$. We say that $\varphi\in A^l(\Omega)$ if $\varphi\in A^{[l]}(\Omega)$ and the functions

$$q_\lambda^2(D^\alpha\varphi)(x)=\int_\Omega\frac{|D^\alpha\varphi(x)-D^\alpha\varphi(y)|^2}{|x-y|^{r+2\lambda}}\,dy,\quad |\alpha|\leqslant[l],$$

are bounded almost everywhere in Ω. We put

$$\|\varphi\|_{A^l}=\max\left\{\|\varphi\|_{A^{[l]}},\ \max_{|\alpha|\leqslant[l]}\operatorname*{sup\,ess}_{x\in\Omega}q_\lambda(D^\alpha\varphi)(x)\right\}.$$

We now consider a variant of Hölder space, obtained by incorporating the mean value theorem of the differential calculus into the definition.

Definition 4.2 *Let $k\in\mathbb{N}$ and $0\leqslant\kappa\leqslant 1$. We say that the function φ belongs to $\tilde{C}^{k,\kappa}(\Omega)$ if $\varphi\in C^k(\Omega)$ and the numbers*

$$\sup_{x\in\Omega}|D^\alpha\varphi(x)|,\quad |\alpha|\leqslant k; \tag{1}$$

$$\sup_{x,y\in\Omega}\frac{|D^\alpha\varphi(x)-D^\alpha\varphi(y)|}{|x-y|},\quad |\alpha|\leqslant k-1;\quad \sup_{x,y}\frac{|D^\beta\varphi(x)-D^\beta\varphi(y)|}{|x-y|^\kappa},\quad |\beta|=k;$$

are all finite. As norm in $\tilde{C}^{k,\kappa}(\Omega)$ we take the maximum among the numbers in (1) and denote this by $\|\varphi\|_{\widetilde{k,\kappa}}$.

We see immediately that the spaces $\tilde{C}^{k,\kappa}(\Omega)$ form a scale, that is,

$$\tilde{C}^{k_1,\kappa_1}(\Omega)\subsetneq\tilde{C}^{k_2,\kappa_2}(\Omega)\quad\text{if } k_1+\kappa_1\leqslant k_2+\kappa_2, k_1\geqslant k_2.$$

The definition has been so chosen that we have

$$\tilde{C}^{k+1,0}(\Omega)\subsetneq\tilde{C}^{k,1}(\Omega),$$

something which need not necessarily occur for Hölder spaces.

Trivially we have $\tilde{C}^{k,\kappa}(\Omega) \subsetneq C^{k,\kappa}(\Omega)$, and we now give a simple theorem, which for one region included in another induces the reverse inclusion.

Proposition 4.1 *Let the open set Ω be strictly contained in Ω'. This means $\Omega \subset \Omega'$ and* $\operatorname{dist}(\Omega, C\Omega') = d_0 > 0$. *Then each function φ which belongs to $C^{k,\kappa}(\Omega')$ also belongs to $\tilde{C}^{k,\kappa}(\Omega)$; thus:*

$$C^{k,\kappa}(\Omega')|_{\Omega} \subset \tilde{C}^{k,\kappa}(\Omega). \tag{2}$$

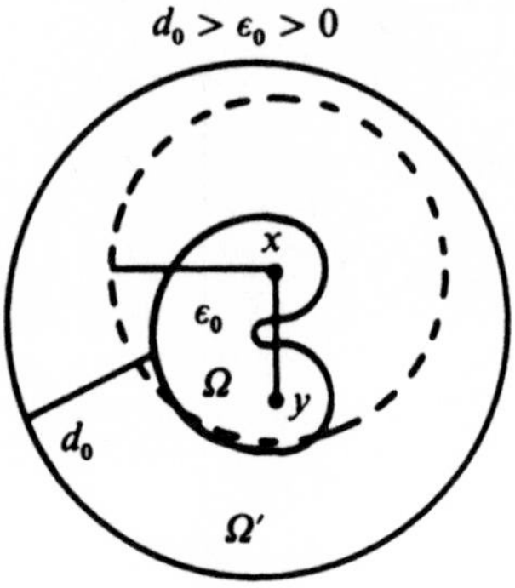

Fig. 4.1

Proof (see Fig. 4.1). Let $d_0 > \varepsilon_0 > 0$; for $x, y \in \Omega$ with $|x - y| < \varepsilon_0$ we have $[x, y] \subset \Omega'$ (distance assumption) and the mean value theorem of differential calculus gives

$$|D^\alpha\varphi(x) - D^\alpha\varphi(y)| \leqslant |D^{\alpha+1}\varphi||\alpha - y| \quad \text{for } |\alpha| \leqslant k - 1,$$

for $x, y \in \Omega$ with $|x - y| \geqslant \varepsilon_0$ we have

$$|D^\alpha\varphi(x) - D^\alpha\varphi(y)| \leqslant 2|D^\alpha\varphi| \leqslant 2|D^\alpha\varphi|\frac{|x - y|}{\varepsilon_0}, \quad |\alpha| \leqslant k - 1,$$

which together with the (k, κ)-norm gives the estimate

$$\|\varphi\|_{\widetilde{k,\kappa};\Omega} \leqslant C\|\varphi\|_{k,\kappa;\Omega'},$$

hence we have proved (2). ■

For convex sets Ω by the same method of proof we obtain

$$\tilde{C}^{k,\kappa}(\Omega) \simeq C^{k,\kappa}(\Omega).$$

Furthermore for bounded, sufficiently regular regions we can prove the equivalence of the Hölder space $C^{k,\kappa}(\Omega)$ and $\tilde{C}^{k,\kappa}(\Omega)$ by means of compactness properties (Ehrling's lemma, see Theorem 7.3 below and Wloka [1]). In a similar way to Definition 2.5 we can define a $\tilde{C}^{k,\kappa}$-diffeomorphism.

Definition 4.3 *Let $\Phi:\Omega\to\Omega'$ be a 1–1 transformation. We say that $\Phi\in\tilde{C}^{k,\kappa}$ if the following hold:*

1. $\Phi\in\tilde{C}^{k,\kappa}(\Omega)$,
2. $\Phi^{-1}\in\tilde{C}^{k,\kappa}(\Omega')$.

Of course 1 and 2 are to be understood componentwise.

3. *If $k\geqslant 1$ we require that the Jacobian determinant Φ satisfy the inequality*

$$0<c\leqslant\left|\det\frac{\partial\Phi}{\partial x}(x)\right|\leqslant C<\infty,\quad x\in\Omega,$$

where c and C are constants independent of x.

Proposition 4.2 *If $\varphi\in\tilde{C}^{0,\kappa}(\Omega)$, $0\leqslant\kappa\leqslant 1$, then*

$$|\varphi(x)-\varphi(y)|\leqslant 4\|\varphi\|_{\tilde{C}^{0,\kappa}}\cdot\frac{|x-y|^\kappa}{1+|x-y|^\kappa}.$$

Proof. By definition of $\tilde{C}^{0,\kappa}$ we have

$$|\varphi(x)|\leqslant\|\varphi\|_{\tilde{C}^{0,\kappa}},\tag{3}$$

and

$$|\varphi(x)-\varphi(y)|\leqslant\|\varphi\|_{\tilde{C}^{0,\kappa}}|x-y|^\kappa.\tag{4}$$

Let $|x-y|\geqslant 1$, then (3) implies that

$$|\varphi(x)-\varphi(x)|\leqslant 2\|\varphi\|_{\tilde{C}^{0,\kappa}}=2\|\varphi\|_{\tilde{C}^{0,\kappa}}\cdot\frac{|x-y|^\kappa}{1+|x-y|^\kappa}\left[1+\frac{1}{|x-y|^\kappa}\right]$$

$$\leqslant 4\|\varphi\|_{\tilde{C}^{0,\kappa}}\cdot\frac{|x-y|^\kappa}{1+|x-y|^\kappa}.$$

Now let $|x-y|<1$, then with (4) we have

$$|\varphi(x)-\varphi(y)|\leqslant\|\varphi\|_{\tilde{C}^{0,\kappa}}|x-y|^\kappa\frac{1+|x-y|^\kappa}{1+|x-y|^\kappa}\leqslant 2\|\varphi\|_{\tilde{C}^{0,\kappa}}\frac{|x-y|^\kappa}{1+|x-y|^\kappa}.\quad\blacksquare$$

Proposition 4.3 *We have a continuous embedding*

$$\tilde{C}^{k,\kappa}(\Omega)\hookrightarrow A^l(\Omega),$$

for $l\leqslant k+\kappa$ if l is integral and for $l<k+\kappa$ otherwise.

Proof. First let l be integral, $k = l$ and $\kappa = 0$. The existence of the embedding $\tilde{C}^{l,0}(\Omega) \hookrightarrow A^l(\Omega)$ follows immediately from the definition of the norms. For the case $k = l - 1$, $\kappa = 1$ we note that (1) gives the inequality

$$|D^\alpha \varphi(x) - D^\alpha \varphi(y)| \leqslant \|\varphi\|_{\tilde{C}^{l-1,1}} \cdot |x - y| \quad \text{for } |\alpha| = l - 1.$$

Thus the functions $D^\alpha \varphi$, $|\alpha| = l - 1$ are absolutely continuous, the partial derivatives $D^{\alpha+1}\varphi$ exist almost everywhere and are bounded almost everywhere by $\|\varphi\|_{\tilde{C}^{l-1,1}}$. This shows the existence of the embedding $\tilde{C}^{l-1,1}(\Omega) \hookrightarrow A^l(\Omega)$. The general embedding for l integral is a consequence of the scale property of the $\tilde{C}$-spaces

$$\tilde{C}^{k,\kappa}(\Omega) \hookrightarrow \tilde{C}^{l-1,1}(\Omega) \hookrightarrow A^l(\Omega), \quad l \leqslant k + \kappa.$$

Second, suppose now that l is not integral, thus $l = [l] + \lambda$, $0 < \lambda < 1$. We first show $\tilde{C}^{0,\kappa}(\Omega) \hookrightarrow A^\lambda(\Omega)$ for $0 < \lambda < \kappa \leqslant 1$. From Proposition 4.2, after integration we obtain

$$\begin{aligned} q_\lambda^2(\varphi)(x) &= \int_\Omega \frac{|\varphi(x) - \varphi(y)|^2}{|x - y|^{r+2\lambda}} \, dy \\ &\leqslant 16 \|\varphi\|_{\tilde{C}^{0,\kappa}}^2 \cdot \int_\Omega \frac{dy}{(1 + |x - y|^\kappa)^2 |x - y|^{r+2(\lambda-\kappa)}} \\ &\leqslant 16 \|\varphi\|_{\tilde{C}^{0,\kappa}}^2 \int_{\mathbb{R}^r} \frac{dz}{(1 + |z|^\kappa)^2 |z|^{r+2(\lambda-\kappa)}} \leqslant C^2 \|\varphi\|_{\tilde{C}^{0,\kappa}}^2. \end{aligned}$$

since the integral in the second line converges. The inequality proved shows that

$$\|\varphi\|_{A^\lambda} \leqslant C \|\varphi\|_{\tilde{C}^{0,\kappa}}.$$

The continuity of the embedding $\tilde{C}^{[l],\kappa}(\Omega) \hookrightarrow A^{[l]+\lambda}(\Omega)$ follows from those already established:

$$\begin{aligned} &\|\varphi\|_{A^{[l]}} \leqslant \|\varphi\|_{\tilde{C}^{[l],0}} \leqslant \|\varphi\|_{\tilde{C}^{[l],\kappa}}, \\ &\|D^\alpha \varphi\|_{A^\lambda} \leqslant c \|D^\alpha \varphi\|_{\tilde{C}^{0,\kappa}} \quad \text{for } |\alpha| = [l], \end{aligned}$$

thus $\|\varphi\|_{A^{[l]+\lambda}} \leqslant c \|\varphi\|_{\tilde{C}^{0,\kappa}}$. The general case $l < [l] + \lambda < k + \kappa$ again follows from the scale property.

Proposition 4.4 *If the functions φ, ψ belong to $A^l(\Omega)$, then their product also belongs to $A^l(\Omega)$ and we have*

$$\|\varphi \cdot \psi\|_{A^l} \leqslant K \|\varphi\|_{A^l} \cdot \|\psi\|_{A^l} \tag{5}$$

where the constant K depends only on l.

Proof. For integral values of l (5) is clear. Suppose therefore that $l = [l] + \lambda$ with $0 < \lambda < 1$. It follows from

$$(\Phi(x)\Psi(x) - \Phi(y)\Psi(y)) = \Phi(x)(\Psi(x) - \Psi(y)) + \Psi(y)(\Phi(x) - \Phi(y)),$$

that

$$|\Phi(x)\Psi(x) - \Phi(y)\Psi(y)|^2 \leqslant 2|\Phi(x)|^2|\Psi(x) - \Psi(y)|^2 + 2|\Psi(y)|^2|\Phi(x) - (y)|^2,$$

and from the Leibniz product rule we obtain

$$|D^\alpha(\varphi\cdot\psi)(x) - D^\alpha(\varphi\cdot\psi)(y)|^2 \leqslant K \sum_{|\beta|\leqslant|\alpha|} [|D^\beta\varphi(y)|^2|D^{\alpha-\beta}\psi(x) - D^{\alpha-\beta}\psi(y)|^2 + |D^{\alpha-\beta}\psi(x)|^2|D^\beta\varphi(x) - D^\beta\varphi(y)|^2]. \tag{6}$$

If $|\alpha| \leqslant [l]$, and we integrate, then

$$\begin{aligned} q_\lambda^2(D^\alpha(\varphi\cdot\psi))(x) &= \int_\Omega \frac{|D^\alpha(\varphi\cdot\psi)(x) - D^\alpha(\varphi\cdot\psi)(y)|^2}{|x-y|^{r+2\lambda}}\,dy \\ &\leqslant K\left\{\|\varphi\|_{A^l}^2 \cdot \sum_{|\beta|\leqslant|\alpha|}\int_\Omega \frac{|D^\beta\psi(x) - D^\beta\psi(y)|^2}{|x-y|^{r+2\lambda}}\,dy \right.\\ &\quad \left. + \|\psi\|_{A^l}^2 \cdot \sum_{|\beta|\leqslant|\alpha|}\int_\Omega \frac{|D^\beta\varphi(x) - D^\beta\varphi(y)|^2}{|x-y|^{r+2\lambda}}\,dy\right\}, \end{aligned}$$

which together with (5) for $[l]$ concludes the proof. ■

Now for the multiplier property of A^l:

Proposition 4.5 *Let $\varphi \in W_2^l(\Omega)$ and $\psi \in A^l(\Omega)$. Then $\varphi\cdot\psi \in W_2^l(\Omega)$ and*

$$\|\varphi\cdot\psi\|_{W_2^l} \leqslant K\|\varphi\|_{W_2^l}\|\psi\|_{A^l} \tag{7}$$

where K depends only on l.

Proof. Once more if l is integral, (7) follows simply from the Leibniz product rule. Suppose therefore that $l = [l] + \lambda$ with $0 < \lambda < 1$. Then by integration it follows from (6) that

$$\begin{aligned} &\iint_{\Omega\times\Omega} \frac{|D^\alpha(\varphi\cdot\psi)(x) - D^\alpha(\varphi\cdot\psi)(y)|^2}{|x-y|^{r+2\lambda}}\,dx\,dy \\ &\leqslant K\sum_{|\beta|\leqslant|\alpha|}\left\{\iint_{\Omega\times\Omega} \frac{|D^\beta\varphi(y)|^2|D^{\alpha-\beta}\psi(x) - D^{\alpha-\beta}\psi(y)|^2}{|x-y|^{r+2\lambda}}\,dx\,dy \right. \qquad (8)\\ &\quad \left. + \iint_{\Omega\times\Omega} \frac{|D^{\alpha-\beta}\psi(x)|^2|D^\beta\varphi(x) - D^\beta\varphi(y)|^2}{|x-y|^{r+2\lambda}}\,dx\,dy\right\}. \end{aligned}$$

We apply Fubini's theorem and estimate

$$\iint_{\Omega\times\Omega}\frac{|D^\beta\varphi(y)|^2|D^{\alpha-\beta}\psi(x)-D^{\alpha-\beta}\psi(y)|^2}{|x-y|^{r+2\lambda}}\,dx\,dy$$

$$=\int_\Omega|D^\beta\varphi(y)|^2\left(\int_\Omega\frac{|D^{\alpha-\beta}\psi(x)-D^{\alpha-\beta}\psi(y)|^2}{|x-y|^{r+2\lambda}}\,dx\right)dy \tag{9}$$

$$\leqslant\|\psi\|^2_{\Lambda^l}\int_\Omega|D^\beta\varphi(y)|^2\,dy\leqslant\|\psi\|^2_{\Lambda^l}\|\varphi\|^2_{W^l_2},$$

the last because $|\beta|\leqslant|\alpha|\leqslant[l]$.

The second integral in (8) is even simpler to estimate. Since $|\alpha-\beta|\leqslant[l]$ we have

$$\iint_{\Omega\times\Omega}\frac{|D^{\alpha-\beta}\psi(x)|^2|D^\beta\varphi(x)-D^\beta\varphi(y)|^2}{|x-y|^{r+2\lambda}}\,dx\,dy$$

$$\leqslant\|\psi\|^2_{\Lambda^l}\iint_{\Omega\times\Omega}\frac{|D^\beta\varphi(x)-D^\beta\varphi(y)|^2}{|x-y|^{r+2\lambda}}\,dx\,dy\leqslant\|\psi\|^2_{\Lambda^l}\|\varphi\|^2_{W^l_2}. \tag{10}$$

Substitution of the estimates (9) and (10) in (8) gives (7). ■

We wish to define Sobolev spaces on differentiable manifolds, for example on $\partial\Omega$. They should be independent of any specially chosen coordinates. To this end we must know how Sobolev spaces behave under coordinate transformations. It happens that a far-reaching equivalence theorem, Theorem 4.1, holds. We will use this theorem again and again, and therefore give the full, distinctly long proof for all $l\in\mathbb{R}_+$. Sections of the proof can be found in the literature, see for example Fichera [1], Adams [1], Slobodeckiĭ [1], Nečas [1].

Theorem 4.1 (transformation theorem) *Let Ω, Ω' be two regions in $\mathbb{R}^r$, and let Φ be a $\tilde{C}^{k,\kappa}$-diffeomorphism between them, $\Phi:\Omega\to\Omega'$, $\Phi^{-1}:\Omega'\to\Omega$. Here, if $k=0$ we only consider the case $\kappa=1$. The pullback operators* (12) *(see below)*

$$^*\Phi:W^l_2(\Omega')\to W^l_2(\Omega),\quad {}^*\Phi^{-1}:W^l_2(\Omega)\to W^l_2(\Omega')$$
$$^*\Phi:\mathring{W}^l_2(\Omega')\to\mathring{W}^l_2(\Omega),\quad {}^*\Phi^{-1}:\mathring{W}^l_2(\Omega)\to\mathring{W}^l_2(\Omega')$$

are continuous for $l\leqslant k+\kappa$, if $k+\kappa$ is integral, and for $l<k+\kappa$ otherwise. Hence the Sobolev spaces

$$W^l_2(\Omega)\simeq W^l_2(\Omega')\quad\text{and}\quad\mathring{W}^l_2(\Omega)\simeq\mathring{W}^l_2(\Omega') \tag{11}$$

are equivalent.

Remark 4.1 With somewhat stronger assumptions on the diffeomorphism

Φ, for example by requiring that $\Phi \in A^{k+\kappa}(\Omega)$ and $\Phi^{-1} \in A^{k+\kappa}(\Omega')$, we can also obtain the equivalences (11) for $l = k + \kappa$, $0 < \kappa < 1$, for this see Slobodeckiĭ [1].

We break the proof of Theorem 4.1 into several steps:

1. $k \geqslant 1$, $\kappa = 0$, $l \leqslant k$, l integral,
2. $k \geqslant 1$, $\kappa = 0$, $l < k$, l non-integral,
3. $k \geqslant 1$, $0 < \kappa < 1$, $l \leqslant k + \kappa$, l integral,
4. $k \geqslant 1$, $0 < \kappa < 1$, $l < k + \kappa$, l non-integral,
5. $k = 0$, $\kappa = 1$, $l < 1$,
6. $k = 0$, $\kappa = 1$, $l = 1$,
7. $k \geqslant 1$, $\kappa = 1$, $l \leqslant k + 1$, l integral,
8. $k \geqslant 1$, $\kappa = 1$, $l < k + 1$, l non-integral.

In all steps of the proof we use the fact that C^∞-functions are dense in $W_2^l(\Omega)$, see Theorem 3.5. Theorem 4.1 is concerned with the continuity of the pullback operators $*\Phi, *\Phi^{-1}$. These are defined by

$$*\Phi\varphi := \varphi(\Phi(x)), \quad \varphi \in W_2^l(\Omega'); *\Phi^{-1}\varphi := \varphi(\Phi^{-1}(y)), \quad \varphi \in W_2^l(\Omega), \tag{12}$$
$$*\Phi : W_2^l(\Omega') \to W_2^l(\Omega) \quad \text{and} \quad *\Phi^{-1} : W_2^l(\Omega) \to W_2^l(\Omega')$$

both are linear, and we have

$$*\Phi \circ *\Phi^{-1} = \mathrm{Id}_{W_2^l(\Omega)}, *\Phi_0^{-1} \circ \Phi = \mathrm{Id}_{W_2^l(\Omega')}. \tag{$*$}$$

The proof of Theorem 4.1 consists in first showing that the image sets of $*\Phi$ and $*\Phi^{-1}$ are contained respectively in $W_2^l(\Omega)$ and $W_2^l(\Omega')$ (that they actually equal $W_2^l(\Omega)$ and $W_2^l(\Omega')$ follows from ($*$)), and second to prove their continuity. Since $*\Phi$ and $*\Phi^{-1}$ play the same role, it suffices to consider one of them, for example $*\Phi$. Indeed it is enough to prove the estimate

$$\| *\Phi\varphi \|_{l,\Omega} \leqslant c \| \varphi \|_{l,\Omega'} \quad \text{for } \varphi \in C^\infty \cap W_2^l(\Omega'); \tag{13}$$

the usual density argument, Theorem 3.5, then gives

$$\| *\Phi\varphi \|_{l,\Omega} \leqslant c \| \varphi \|_{l,\Omega'} \quad \text{for all } \varphi \in W_2^l(\Omega').$$

From this both the assertion for the image set of $*\Phi$ and the continuity follow.

We prove (13) in the eight individual cases, each case involves a lemma; we have chosen the presumptions for the individual lemmata to be as general as possible.

Lemma 4.1 *Let $\Phi : \Omega \to \Omega'$ be a $C^{k,0}$-diffeomorphism, with $k \geqslant 1$, $\kappa = 0$. Let l be integral and $l \leqslant k$. Then the spaces $W_2^l(\Omega) \simeq W_2^l(\Omega')$ are equivalent. As announced we show only that $*\Phi : W_2^l(\Omega') \to W_2^l(\Omega)$ is continuous.*

Proof. We must prove (13). Let

$$\Phi:\begin{array}{l}\Omega\to\Omega'\\ x\mapsto y,\end{array}$$

We also write

$$\Phi(x)=y(x)=(y_1(x),\ldots,y_r(x)).$$

Let $\varphi\in C^\infty\cap W_2^l(\Omega')$. The rule for the differentiation of composed functions gives for $j=1,\ldots,r$

$$\frac{\partial^*\Phi\varphi}{\partial x_j}=\frac{\partial\varphi(\Phi(x))}{\partial x_j}=\frac{\partial\varphi(y(x))}{\partial x_j}=\sum_{i=1}^{r}\frac{\partial\varphi}{\partial y_i}(\Phi(x))\frac{\partial y_i}{\partial x_j}(x),$$

or in general

$$\mathrm{D}_x^\alpha(^*\Phi\varphi)=\mathrm{D}_x^\alpha\varphi(y(x))=\sum_{|\beta|\leqslant|\alpha|}M_{\alpha\beta}(x)\cdot{}^*\Phi(\mathrm{D}_y^\beta\varphi).$$

Here $M_{\alpha\beta}$ is a polynomial (of degree $\leqslant|\beta|$) in the derivatives (up to order $|\alpha|$) of the components of $\Phi=(y_1(x),\ldots,y_r(x))$, and $M_{00}\equiv1$. We have

$$\|{}^*\Phi\varphi\|_{l,\Omega}^2=\sum_{|\alpha|\leqslant l}\int_\Omega|\mathrm{D}_x^\alpha(^*\Phi\varphi)|^2\,\mathrm{d}x$$
$$\leqslant\sum_{|\alpha|\leqslant l}\sum_{|\beta|\leqslant|\alpha|}c_1\int_\Omega M_{\alpha\beta}^2(x)|{}^*\Phi(\mathrm{D}_y^\beta\varphi)|^2\,\mathrm{d}x.$$

Since the derivatives of Φ are bounded (see Definition 4.3), so is $M_{\alpha\beta}(x)$ and after a change of variables we obtain

$$\leqslant c_2\sum_{|\beta|\leqslant l}\int_\Omega|{}^*\Phi(\mathrm{D}_y^\beta\varphi)|^2\,\mathrm{d}x=c_2\sum_{|\beta|\leqslant l}\int_{\Omega'}|\mathrm{D}_y^\beta\varphi|^2\cdot\left|\det\frac{\partial x}{\partial y}\right|\mathrm{d}y$$
$$\leqslant c_3\sum_{|\beta|\leqslant l}\int_{\Omega'}|\mathrm{D}_y^\beta\varphi|^2\,\mathrm{d}y=c_3\|\varphi\|_{l,\Omega'}^2.$$ ■

Lemma 4.2 *Let $\Phi:\Omega\to\Omega'$ be a $\tilde{C}^{k,0}$-diffeomorphism, $k\geqslant1$, $l=[l]+\lambda$ with $0<\lambda<1$ and $[l]\leqslant k-1$. Then the map ${}^*\Phi:W_2^l(\Omega')\to W_2^l(\Omega)$ is continuous, respectively the spaces $W_2^l(\Omega)\simeq W_2^l(\Omega')$ are equivalent.*

Proof. Because $l=[l]+\lambda$, W_2^l carries the Slobodeckiĭ norm

$$\|\varphi\|_{[l]+\lambda}^2=\|\varphi\|_{[l]}^2+\sum_{|\alpha|\leqslant[l]}\iint_{\Omega\times\Omega}\frac{|\mathrm{D}^\alpha\varphi(x)-\mathrm{D}^\alpha\varphi(x')|^2}{|x-x'|^{r+2\lambda}}\,\mathrm{d}x\,\mathrm{d}x'$$
$$=\|\varphi\|_{[l]}^2+\sum_{|\alpha|\leqslant[l]}I_\lambda\mathrm{D}^\alpha\varphi.$$

Once more we must prove (13). To this end we first estimate the part $\|{}^*\Phi\varphi\|^2_{[l]}$ of the norm by the method of Lemma 4.1.

Now for $I_\lambda(D^\alpha_x({}^*\Phi\varphi))$. We again apply the rule for the differentiation of the composition of two functions, obtaining

$$I_\lambda(D^\alpha_x{}^*\Phi\varphi) \leqslant c_1 \sum_{|\beta|\leqslant|\alpha|} I_\lambda(M_{\alpha\beta}\cdot{}^*\Phi D^\beta_y\varphi). \tag{14}$$

As we know, $M_{\alpha\beta}$ is a polynomial in the derivatives up to order $|\alpha| \leqslant [l] \leqslant k-1$ of the components of Φ. Since $\Phi\in\tilde{C}^{k,0}$, $M_{\alpha\beta}\in\tilde{C}^{1,0}$, and by Proposition 4.3 $M_{\alpha\beta}\in A^1$. I_λ forms part of the Slobodeckiĭ norm for $W^{0+\lambda}_2$ and we can refine the estimate (14)

$$c_1\sum I_\lambda(M_{\alpha\beta}\cdot{}^*\Phi D^\beta_y\varphi) \leqslant c_2 \sum_{|\beta|\leqslant[l]} \|{}^*\Phi D^\beta_y\varphi\|^2_\lambda$$

$$= c_2\left\{\sum_{|\beta|\leqslant[l]} \|{}^*\Phi D^\beta_y\varphi\|^2_0 + \sum_{|\beta|\leqslant[l]} I_\lambda({}^*\Phi D^\beta_\varphi)\right\}. \tag{15}$$

As in Lemma 4.1 on the first sum we need only to introduce the change of coordinates Φ in order to estimate it by

$$c_3 \sum_{|\beta|\leqslant[l]} \|D^\beta_y\varphi\|^2_{0,\Omega'} \leqslant c_3\|\varphi\|^2_{[l],\Omega'}. \tag{16}$$

Now for the second sum in (15). We have

$$I_\lambda({}^*\Phi D^\beta_y\varphi) = \iint_{\Omega\times\Omega} \frac{|{}^*\Phi D^\beta_y\varphi(x) - {}^*\Phi D^\beta_y\varphi(x')|^2}{|x-x'|^{r+2\lambda}}\,dx\,dx'. \tag{17}$$

Since $\Phi = y(x)\in\tilde{C}^{k,0}$, the mean value inequality applies; we write it, putting $x(y) = \Phi^{-1}(y)$ in the form

$$|y-y'| \leqslant C|x(y)-x(y')|. \tag{18}$$

By change of variable and use of (18) we can estimate the integral (17) as

$$I_\lambda({}^*\Phi D^\beta_y\varphi) \leqslant c' \iint_{\Omega\times\Omega'} \frac{|D^\beta_y\varphi(y) - D^\beta_y\varphi(y')|^2}{|y-y'|^{r+2\lambda}} \left|\det\frac{\partial x}{\partial y}\right|^2 dy\,dy'$$

$$= c' I_\lambda(D^\beta_y\varphi)_{\Omega'} \leqslant c'\|\varphi\|^2_{[l]+\lambda,\Omega'}. \tag{19}$$

(19) and (16), applied to the estimate (15) (respectively (14)), give the desired conclusion and we have proved Lemma 4.2 ■

Lemma 4.3 *In case* 3 *we must have* $l \leqslant k$, *and we proceed as in Lemma* 4.1. *It is enough to assume* $\Phi\in C^{k,0}$.

Lemma 4.4 *Let* $\Phi:\Omega\to\Omega'$ *be a* $\tilde{C}^{k,\kappa}$*-diffeomorphism with* $k \geqslant 1$, $0<\kappa<1$, *let*

l be non-integral and $l < k + \kappa$. The map ${}^*\Phi: W_2^l(\Omega') \to W_2^l(\Omega)$ is continuous, respectively the spaces $W_2^l(\Omega) \simeq W_2^l(\Omega')$ are equivalent.

Proof. We choose some number κ, with $l \leqslant k + \kappa_1 < k + \kappa$, $0 < \kappa_1 < \kappa < 1$. Since $\Phi \in \tilde{C}^{k,\kappa}$ Proposition 4.3 implies that

$$\Phi \in A^{k+\kappa_1}. \tag{20}$$

We have only to consider the case $l = k + \kappa$ (the cases $l = k$ and $l < k$ are already covered by Lemmas 4.1 and 4.2). We have

$$\| {}^*\Phi\varphi \|_{k+\kappa_1}^2 = \| {}^*\Phi\varphi \|_k^2 + \sum_{|\alpha| \leqslant k} I_{\kappa_1}(\mathrm{D}_x^\alpha {}^*\Phi\varphi).$$

Again, we estimate the part $\| {}^*\Phi\varphi \|_k^2$ of the norm by means of Lemma 4.1. For the second part of the norm we proceed as in Lemma 4.2

$$I_{\kappa_1}(\mathrm{D}_x^\alpha {}^*\Phi\varphi) \leqslant c \sum_{|\beta| \leqslant |\alpha|} I_{\kappa_1}(M_{\alpha\beta} \cdot {}^*\Phi \mathrm{D}_y^\beta \varphi),$$

where this time, because of (20) and Proposition 4.4 we have $M_{\alpha\beta} \in A^{\kappa_1}$. We recall that elements from A^{κ_1} are multipliers on $W_2^{\kappa_1}$, and so the final part of the proof runs word for word as in Lemma 4.2, up to the change $M_{\alpha\beta} \in A^{\kappa_1}$. ■

We come now to case 5: $\Phi \in \tilde{C}^{0,1}$ and $l < 1$. The difficulty which confronts us is that we no longer have the transformation of variables formula at our disposal, $k = 0$!

Detailed results from measure and differentiation theory help us further at this point. $\Phi \in \tilde{C}^{0,1}$ implies that Φ and Φ^{-1} are Lipschitz continuous, that is, they satisfy the inequalities

$$\begin{aligned} |\Phi(x) - \Phi(x')| &\leqslant K|x - x'|, \qquad x, x' \in \Omega, \\ |\Phi^{-1}(y) - \Phi^{-1}(y')| &\leqslant K'|y - y'|, \quad y, y' \in \Omega'. \end{aligned} \tag{21}$$

Lipschitz continuous functions are absolutely continuous and take measurable sets to measurable sets (see Natanson [1], p. 276), that is, if M is measurable, then $\Phi(M)$ is again measurable. From (21) it further follows that the image of a ball of radius R is contained in a ball of radius $K \cdot R$. Therefore we obtain

$$\mu(\Phi(M)) \leqslant K_1 \mu(M) \quad \text{and} \quad \mu(M) \leqslant K_2 \mu(\Phi(M)), \tag{22}$$

where μ denotes Lebesgue measure in $\mathbb{R}^r$. The inequalities (22) take over the role of the change of variable formula.

Lemma 4.5 *Let $\Phi: \Omega \to \Omega'$ be a $C^{0,1}$-'diffeomorphism' and let $0 \leqslant l < 1$. Then* *the map* $^*\Phi: W_2^l(\Omega') \to W_2^l(\Omega)$ *is continuous; respectively the spaces* $W_2^l(\Omega) \simeq W_2^l(\Omega')$ *are equivalent.*

Proof. First we prove the case $l=0$, that is, $W_2^0 = L_2$. We take a lattice in $\mathbb{R}^r$ generated by hyperplanes parallel to the axes separated by distance d. Let W_i be the cubes of this lattice, by means of which we exhaust Ω:

$$\Omega = \lim_{\substack{d\to 0\\ m\to\infty}} \bigcup_{i=1}^{m} W_i \quad \text{with } W_i \subset \Omega.$$

We have by the properties of the Lebesgue integral (see Natanson [1])

$$\| {}^*\Phi\varphi \|_0^2 = \int_\Omega |\varphi(\Phi(x))|^2 \, dx = \lim_{\substack{d\to 0\\ m\to\infty}} \sum_{i=1}^{m} d^r \inf_{x\in W_i} |\varphi\Phi(x)|^2, \tag{23}$$

and because of (22)

$$d^r = \mu(W_i) \leqslant K_2 \mu(\Phi W_i).$$

This allows us further to estimate (23)

$$\| {}^*\Phi\varphi \|_0^2 \leqslant K_2 \lim_{\substack{d\to 0\\ m\to\infty}} \sum_{i=1}^{m} \mu(\Phi(W_i)) \inf_{x\in W_i} |\varphi(\Phi(x))|^2 \leqslant K_2 \int_{\Omega'} |\varphi(y)|^2 \, dy.$$

The last inequality holds because the $\{\Phi(W_i)\}$ exhaust the region $\Phi(\Omega) = \Omega'$. Suppose next that $0 < l = \lambda < 1$. We have

$$\| {}^*\Phi\varphi \|_l^2 = \| {}^*\Phi\varphi \|_0^2 + \iint_{\Omega\times\Omega} \frac{|\varphi(\Phi(x)) - \varphi(\Phi(x'))|^2}{|x-x'|^{r+2l}} \, dx \, dx' \tag{24}$$

where we can already estimate $\| {}^*\Phi\varphi \|_{0,\Omega}^2$ by means of $K_2 \| \varphi \|_{0,\Omega'}^2$. Finally for $I_l({}^*\Phi\varphi)$ in (24). We use the first inequality (21) obtaining

$$\begin{aligned} I_l({}^*\Phi\varphi) &\leqslant K^{r+2l} \iint_{\Omega\times\Omega} \frac{|\varphi(\Phi(x)) - \varphi(\Phi(x'))|^2}{|\Phi(x) - \Phi(x')|^{r+2l}} \, dx \, dx' \\ &\leqslant K^{r+2} \cdot K_2^2 \cdot I_l(\varphi) \end{aligned}$$

where the integral obtained is estimated by the method of (23). ■

We come now to case 6.

Lemma 4.6 *Let* $\Phi: \Omega \to \Omega'$ *be a* $C^{0,1}$*-'diffeomorphism' and let* $l = 1$. *Then the map* $^*\Phi: W_2^l(\Omega') \to W_2^l(\Omega)$ *is continuous, respectively the spaces* $W_2^l(\Omega) \simeq W_2^l(\Omega')$ *are equivalent.*

Proof. We have

$$\|{}^*\Phi\varphi\|_1^2 = \|{}^*\Phi\varphi\|_0^2 + \sum_{i=1}^{r}\left\|\frac{\partial\varphi(\Phi(x))}{\partial x_i}\right\|_0^2,$$

and we can estimate $\|{}^*\Phi\varphi\|_0^2$ by means of Lemma 4.5. It remains to estimate $\|\partial\varphi(\Phi(x))/\partial x_i\|_0^2$. We return to (21). The components of $\Phi(x) = (y_1(x), \ldots, y_r(x))$ are absolutely continuous, hence they possess partial derivatives almost everywhere, which are measurable and locally integrable (see for example Natanson [1], 274, or Saks [1]) and by (21) these partial derivatives are bounded by K. For $\varphi \in C^\infty$ almost everywhere we have

$$\frac{\partial\varphi(y(x))}{\partial x_i} = \sum_{j=1}^{r}\frac{\partial\varphi(y(x))}{\partial y_i}\cdot\frac{\partial y_i(x)}{\partial x_i} \tag{25}$$

By Theorem 1.8 (25) is also valid in the distributional sense, and the expression $\|\partial\varphi(y(x))/\partial x_i\|_0^2$ has meaning. We can estimate

$$\left\|\frac{\partial\varphi(y(x))}{\partial x_i}\right\|_0^2 \leqslant c\sum_{j=1}^{r}\left\|\frac{\partial\varphi(y(x))}{\partial y_j}\frac{\partial y_j}{\partial x_i}\right\|_0^2 \leqslant cK^2\sum_{j=1}^{r}\left\|\frac{\partial\varphi(y(x))}{\partial y_j}\right\|_0^2$$
$$\leqslant cK^2K_2\sum_{j=1}^{r}\left\|\frac{\partial\varphi}{\partial y_j}\right\|_{0,\Omega'}^2,$$

where the last inequality follows from Lemma 4.5. ■

For the proof of cases 7 and 8 we use the information already accumulated.

Lemma 4.7 *Let $\Phi:\Omega\to\Omega'$ be a $C^{k,1}$-diffeomorphism, l integral and $l \leqslant k+1$. Then the map ${}^*\Phi: W_2^l(\Omega') \to W_2^l(\Omega)$ is continuous, respectively the spaces $W_2^l(\Omega) \simeq W_2^l(\Omega')$ are equivalent.*

Proof. If $l < k+1$, Lemma 4.1 applies. If $l = k+1$ we write

$$\|{}^*\Phi\varphi\|_{k+1}^2 = \|{}^*\Phi\varphi\|_k^2 + \sum_{|\alpha|=k+l}\|D_x^\alpha{}^*\Phi\varphi\|_0^2,$$

where $\|{}^*\Phi\varphi\|_k$ may again be estimated by Lemma 4.1. For $D_x^\alpha{}^*\Phi\varphi$ we have

$$D_x^\alpha{}^*\Phi\varphi = \sum_{|\beta|\leqslant|\alpha|} M_{\alpha\beta}(x)\cdot{}^*\Phi(D_y^\beta\varphi),$$

where the highest derivatives in $M_{\alpha\beta}(x)$, namely with $|\alpha| = k+1$, exist and are bounded as in Lemma 4.6 by measure theoretic considerations. The method of Lemma 4.6 concludes the proof. ■

Lemma 4.8 *Let $\Phi:\Omega\to\Omega'$ be a $\tilde{C}^{k,1}$-diffeomorphism and let $l < k+1$. Then the* *map* $^*\Phi: W_2(\Omega') \to W_2^l(\Omega)$ *is continuous, respectively the spaces* $W_2^l(\Omega) \simeq W_2^l(\Omega')$ *are equivalent.*

Proof. We choose some κ with

$$l < k + \kappa < k + 1, \quad 0 \leqslant \kappa \leqslant 1. \tag{26}$$

A $\tilde{C}^{k,1}$-diffeomorphism is also a $\tilde{C}^{k,\kappa}$-diffeomorphism and given (26) we may apply Lemma 4.4.

The equivalence of the $\mathring{W}$-spaces in (11) follows from the remark that a continuous map Φ takes compact sets (supp φ, for example) into compact sets. ∎

With this we have completed the proof of the transformation theorem.

4.2 Sobolev spaces on differentiable manifolds, and on the frontier $\partial\Omega$ of a (k, κ)-smooth region

Let $\kappa = 0$ or 1. We consider a compact $C^{k,\kappa}$-manifold M, see Definition 2.10. Let (U_i, α_i) be an admissible $C^{k,\kappa}$-atlas for M, and β_i a subordinate partition of unity, see Theorem 2.13. Since M is compact, we may take the indexing set $i \in I$ to be finite. By assumption the maps $\alpha_i \circ \alpha_j^{-1}$ are (k, κ)-diffeomorphisms, possibly by shrinking the U_i we can always ensure that they are $\tilde{C}^{k,\kappa}$-diffeomorphisms, see Proposition 4.1. We now define the spaces $W_2^l(M)$.

Definition 4.4 *Let* $l \leqslant k + \kappa$. *We say that the function* $\varphi: M \to \mathbb{C}$ *belongs to* $W_2^l(M)$, *if all functions*

$$(\varphi \cdot \beta_i) \circ \alpha_i^{-1} : \alpha_i(U_i) \to \mathbb{C}, \quad i \in I, \tag{27}$$

belong to $\mathring{W}_2^l(\alpha_i(U_i))$, *respectively* $\mathring{W}_2^l(\mathbb{R}^r)$ *(the support of* (27) *is compact and contained in* $\alpha_i(U_i)$*). We make* $W_2^l(M)$ *into a Hilbert space by defining the scalar product by*

$$(\varphi, \psi)_l = \sum_i ((\varphi \cdot \beta_i) \circ \alpha_i^{-1}, \quad (\psi \cdot \beta_i) \circ \alpha_i^{-1})_{W_2^l(\mathbb{R}^r)}. \tag{28}$$

As we see from Definition 4.4 the space $W_2^l(M)$ and the scalar product depend on the choice of atlas $a = (U_i, \alpha_i)$ and partition of unity $\{\beta_i\}$. We show next with the help of the transformation theorem that this dependence is not essential.

Theorem 4.2 *Let* M *be a compact* $C^{k,\kappa}$*-manifold, where* $\kappa = 0$ *or* 1, *and let* $a_1 = (U_i^1, \alpha_{i,1})$ *and* $a_2 = (U_j^2, \alpha_{j,2})$ *be two equivalent coordinate systems, and* $\{\beta_i^1\}\{\beta_j^2\}$ *two corresponding subordinate partitions of unity. Then the spaces* $W_2^{l,a_1}(M)$ *and* $M_2^{l,a_2}(M)$ *are equivalent. Here as in Definition* 4.4, $l \leqslant k + \kappa$.

Proof. We have

$$\|\varphi\|^2_{l,a_1} = \sum_i \|(\varphi\beta^1_i)\circ\alpha^{-1}_{i,1}\|^2_{W^l_2(\alpha_i(U^1_i))}$$

$$\leqslant c\sum_i\sum_j \|(\varphi\beta^1_i\cdot\beta^2_j)\circ\alpha^{-1}_{i,1}\|_{W^l_2(\alpha_{i,1}(U^1_i\cap U^2_j))}.$$

However, because of the assumed equivalence of a_1 and a_2 the maps

$$\alpha_{j,2}\circ\alpha^{-1}_{i,1}:\alpha_{i,1}(U^1_i\cap U^2_j)\to\alpha_{j,2}(U^1_i\cap U^2_j)$$

are $\tilde{C}^{k,\kappa}$-diffeomorphisms. We can apply the transformation theorem 4.1 and estimate further

$$\leqslant c_2\sum_i\sum_j \|(\varphi\cdot\beta^1_i\cdot\beta^2_j)\circ\alpha^{-1}_{j,2}\|^2_{W^l_2(\alpha_{j,2}(U^1_i\cap U^2_j))}.$$

The functions $\beta^1_i\circ\alpha^{-1}_{j,2}$ belong to $\tilde{C}^{k,\kappa}$, and by Proposition 4.3 also to A^l. As such they are multipliers on W^l_2 and we have

$$\leqslant c_3\sum_j \|(\varphi\beta^2_j)\circ\alpha^{-1}_{j,2}\|^2_{W^l_2(\alpha_{j,2}(U^1_i\cap U^2_j))}$$

$$\leqslant c_3\sum_j \|(\varphi\beta^2_j)\circ\alpha^{-1}_{j,2}\|_{W^l_2(\alpha_{j,2}(U^2_j))}$$

$$= c_3\|\varphi\|^2_{l,a_2}.$$

In this way we did prove that the embedding $W^{l,a_2}_2(M)\subsetneq W^{l,a_1}_2(M)$ is continuous. Interchanging the roles of a_1 and a_2 we obtain the desired equivalence. ∎

In the light of the previous theorem we may use the shorter notation $W^{l,a}_2(M)$.

Theorem 4.3 *Functions from* $C^{k,\kappa}(M)=C^{k,\kappa}_0(M)$ *are dense in* $W^l_2(M)$. *Here once more, as in Definition* 4.4, *let* $l\leqslant k+\kappa$.

Proof. Before the proof we make the following remark: if the support of $\varphi\in W^l_2(M)$ is contained in a coordinate neighbourhood (U_{i_0},α_{i_0}), that is, $\operatorname{supp}\varphi\subset U_{i_0}$, then

$$\|\varphi\|_{l,M} = \|\varphi\alpha^{-1}_{i_0}\|_{W^l_2(\alpha_{i_0}(U_{i_0}))}. \tag{29}$$

The proof of the remark follows from the fact that we can choose a partition of unity with $\beta_{i_0}\equiv 1$ on $\operatorname{supp}\varphi$, hence $\varphi=\beta_{i_0}\cdot\varphi$, which substituted in (28) gives (29).

Let $\varphi=\sum_i\varphi_i$ with $\operatorname{supp}\varphi_i\subset U_i$. Such a decomposition always exists and is obtained by means of a suitable partition of unity. By (29) $\varphi_i\circ\alpha_i^{-1}\in$

$\mathring{W}_2^l(\alpha_i(U_i))$, and there exists some $\psi_{i,n} \in \mathscr{D}(\alpha_i(U_i))$ with

$$\| \varphi_i \circ \alpha_i^{-1} - \psi_{i,n} \| \leqslant \frac{1}{mn}, \tag{30}$$

where m is the number of charts in the atlas $\{U_i, \alpha_i\}$. We put $\psi_n = \sum_i \psi_{i,n} \circ \alpha_i$ and by (30) have

$$\| \varphi - \psi_n \|_{l,M} = \| \sum_i (\varphi_i - \psi_{i,n} \circ \alpha_i) \|_{l,M} \leqslant \sum_i \| \varphi_i \circ \alpha_i^{-1} - \psi_{i,n} \|_l \leqslant \frac{1}{n}.$$

Since $\psi_n \in C^{k,\kappa}(M)$ we have proved our theorem. ■

We can now define the space $W_2^l(\partial\Omega)$ for Ω bounded and (k, κ)-smooth, where $\kappa = 0,1$. We have only to observe that by Theorem 2.15 $\partial\Omega$ is a compact $C^{k,\kappa}$-manifold; the Sobolev space $W_2^l(\partial\Omega)$ is well defined by Definition 4.4, and by Theorem 4.2, up to equivalence, independent of the coordinates chosen. If Ω is bounded and has the $N^{k,\kappa}$-property (see Definition 2.4) we can give the norm on $W_2^l(\partial\Omega)$ by means of a coordinate invariant surface integral. Thus, let $\varphi \in W_2^l(\partial\Omega)$, we again decompose φ as a sum $\sum_i \varphi_i$, $\operatorname{sup} \varphi_i \subset U_i$, where the coordinate neighbourhoods U_i have the properties of Definition 2.4.

We define:

$$\| \varphi_i \|_l^2 = \sum_{|\alpha| \leqslant [l]} \int_{\partial\Omega} |D^\alpha \varphi_i|^2 \mathrm{d}\sigma + \sum_{|\alpha| \leqslant [l]} \iint_{\partial\Omega \times \partial\Omega} \frac{|D^\alpha \varphi_i(x) - D^\alpha \varphi_i(y)|}{|x-y|^{r-1+2\lambda}} \mathrm{d}\sigma\, \mathrm{d}\sigma$$

and $\| \varphi \|_l^2 = \sum_i \| \varphi_i \|_l^2$. Here $l = [l] + \lambda$, $\partial\Omega \cap U_i$ is given by the equation $x_r = a_i(x_1, \ldots, x_{r-1})$, and the derivatives D^α are taken with respect to the variables $x_1, \ldots, x_{r-1}$. The surface element $\mathrm{d}\sigma$ can be written in terms of coordinates $x_1, \ldots, x_{r-1}$,

$$\mathrm{d}\sigma = \left[1 + \sum_{j=1}^{r-1} \left(\frac{\partial a_i}{\partial x_j} \right)^2 \right]^{1/2} \mathrm{d}x_1 \cdots \mathrm{d}x_{r-1},$$

and since

$$1 \leqslant \left[1 + \sum_{j=1}^{r-1} \left(\frac{\partial a_i}{\partial x_j} \right)^2 \right]^{1/2} \leqslant C < \infty,$$

noting that $a_i(x_1, \ldots, x_{r-1})$ belongs at least to the class $C^{0,1}$, that is, almost everywhere the derivatives $\partial a_i / \partial x_j$ are bounded, the surface norm (31) is equivalent to the manifold norm (28).

If Ω is bounded and $k \geqslant 1$ we can also introduce a coordinate invariant surface norm on $W_2^l(\partial\Omega)$ by means of (31) ($\partial\Omega$ is (k, κ)-smooth). Indeed by Theorem 2.6, if $k \geqslant 1$, the properties (k, κ)-smooth and $N^{k,\kappa}$ are the same.

In general, if the compact $C^{k,\kappa}$-manifold carries a Riemannian structure

$\{g_{ij}\}$, instead of using (28) we may define the norm on $W_2^l(M)$ by means of a coordinate invariant volume integral. The volume element is $dv = \sqrt{(\det g_{ij})}\, dx$, for $k \geqslant 1$ we have $0 < c \leqslant \det g_{ij} \leqslant C < \infty$, and the new norm is again equivalent to the earlier one given by (28).

Suppose now that $0 < \kappa < 1$ and that $\partial\Omega$ is the boundary of a $C^{k,\kappa}$-smooth, bounded region Ω. In this case, using Definition 4.4, it is again possible to define the space $W_2^l(\partial\Omega)$ for $l < k + \kappa$; to do this we take $\alpha_i = \alpha_x = \Phi_x|_{\partial\Omega}$, where the Φ_x are the transformations from Definition 2.7. However, it is now not possible to prove Theorem 4.2, that is, the independence from coordinate transformations. But, and this is important, Theorem 4.3 remains true. We also remark that in the proof of the trace theorems (where we use the spaces $W_2^l(\partial\Omega)$), Theorem 4.2 is not needed.

Exercises

4.1 Show that $C^{0,1}(\Omega)$ consists of precisely those bounded continuous functions whose distributional derivative belong to $L^\infty(\Omega)$ (are bounded).

4.2 Show that the embedding of $C^{0,1}(a,b)$ in $C^0(a,b)$ is compact. What can one say about the embedding

$$C^{k,\kappa}(a,b) \subset C^{l,\lambda}(a,b)$$

for $k + \kappa > l + \lambda$?

4.3 Is it true that $C^{0,1}(a,b) \neq C^{1,0}(a,b)$?

4.4 Let T^r be the r-dimensional torus. We consider the space $W_2^l(T^r)$ (see also Exercise 1.13) consisting of periodic functions. Show that an equivalent norm on $W_2(T^r)$ is given by

$$[\sum_k (1 + |k|^2)^l |c_k|^2]^{1/2}, \tag{1}$$

where the c_k are the Fourier coefficients.

4.5 Prove the density theorem 4.3 on the torus directly by means of the norm (1) and the results of Exercise 1.13.

§5 The definition of Sobolev spaces by means of the Fourier transformation and extension theorems

We again turn our attention to the Fourier transformation, see §1.9, and develop this as an important tool for Sobolev spaces. We exhibit first an isomorphism theorem (Theorem 5.2), which characterises the behaviour of $W_2^l(\mathbb{R}^r)$ under Fourier transformation, and finally show how we can define the Sobolev spaces by means of the Fourier transformation. We then use the Fourier transformation to prove the extension theorem of Calderon–

Zygmund for integral values of l: suppose that Ω is bounded and satisfies the uniform cone condition, then functions from $W_2^l(\Omega)$ can be extended to functions from $W_2(\mathbb{R}^r)$, and the extension operator

$$F: W_2^l(\Omega) \to W_2^l(\mathbb{R}^r)$$

can be chosen to be linear and continuous. For non-integral values of l we prove the extension theorem of Hestenes, without the Fourier transformation; however, here we must impose stronger conditions on Ω, for example (k, κ)-smoothness and $l < k + \kappa$.

The applications of the extension theorems are numerous; we will become acquainted with them in later sections.

5.1 Sobolev spaces and the Fourier transformation

At this point we apply the Fourier transformation, see Definition 1.9, in order to study the Sobolev–Slobodeckiĭ spaces. It turns out that we can define the spaces H^l for all real values $l \in \mathbb{R}$.

Definition 5.1 *We define the space H^l by means of the Fourier transformation $\hat{\varphi} = \mathscr{F}\varphi$ as*

$$H^l = \left\{ \varphi \in \mathscr{S}' : \|\varphi\|_{H^l}^2 = \int_{\mathbb{R}^r} |\hat{\varphi}|^2 (1 + |\xi|^2)^l \, d\xi < \infty \right\}. \tag{1}$$

We see immediately that H^l is a Hilbert space.

Theorem 5.1 *$\mathscr{D}$ is dense in each $H^l, l \in \mathbb{R}$.*

Proof. First we consider the weighted Hilbert space L_2^l with the norm $\|\varphi\|_{L_2^l}^2 = \int_{\mathbb{R}^r} |\varphi|^2 (1 + |\xi|^2)^l \, d\xi$ and show that $\mathscr{S}$ and $\mathscr{D}$ are dense in L_2^l (because $\mathscr{D} \subset \mathscr{S}$ it is enough to show this for $\mathscr{D}$). Let $\psi \in L_2^l$, $\varepsilon > 0$, $\eta(\xi) := (1 + |\xi|^2)^{l/2}$ then by Theorem 3.4 there exists $\varphi \in \mathscr{D}$ with $\|\varphi - \eta \cdot \psi\|_{L_2} < \varepsilon$. Therefore φ/η belongs to $\mathscr{D}$ and

$$\begin{aligned} \|\varphi/\eta - \psi\|_{L_2^l}^2 &= \int_{\mathbb{R}^r} \left| \frac{\varphi(\xi)}{(1 + |\xi|^2)^{l/2}} - \psi(\xi) \right|^2 (1 + |\xi|^2)^l \, d\xi \\ &= \int_{\mathbb{R}^r} |\varphi(\xi) - (1 + |\xi|^2)^{l/2} \psi(\xi)|^2 \, d\xi \\ &= \|\varphi - \eta \cdot \psi\|_{L_2}^2 < \varepsilon^2, \end{aligned} \tag{2}$$

hence we have shown $\mathscr{D}$ to be dense in L_2^l.

Now to the proof of our theorem. We have $\varphi \in H^l$ if and only if $\hat{\varphi} \in L_2^l$, and must therefore show that $\mathscr{Z} := \mathscr{F}\mathscr{D}$ is dense in L_2^l. By Theorem 1.18

$\mathscr{D}$ is dense in $\mathscr{S}$, and by Corollary 1.3 the Fourier transformation is an isomorphism $\mathscr{F}:\mathscr{S}\leftrightarrow\mathscr{S}$, hence $\mathscr{Z}$ is dense in $\mathscr{S}$. Convergence in $\mathscr{S}$ implies convergence in L_2^l, this follows from $(1+|\xi|^2)^l\leqslant(1+|\xi|^2)^m$, m a natural number $m\geqslant l$, and from the form of the seminorms in $\mathscr{S}$ (see §1, (44)). Therefore we have $\mathscr{Z}$ L_2^l-dense in $\mathscr{S}$, and $\mathscr{S}$ dense in L_2^l (see above); hence we have shown that $\mathscr{D}$ is dense in H^l. ■

In the case of the Sobolev–Slobodeckiĭ spaces, that is, for $l\geqslant 0$, we show that we obtain no new class of spaces in Definition 5.1.

Theorem 5.2 *For $l\geqslant 0$ the Fourier transformation is a (linear, topological) isomorphism between the spaces*

$$\mathscr{F}:W_2^l(\mathbb{R}^r)\to L_2^l. \tag{3}$$

From this it follows, with the notation of Definition 5.1, that the norms $\|\varphi\|_{H^l}$ and $\|\varphi\|_{W_2^l}$ are equivalent on $W_2(\mathbb{R}^r)$, that is, there exist constants c, $c'>0$ with $c\|\varphi\|_{H^l}\leqslant\|\varphi\|_{W_2^l}\leqslant c'\|\varphi\|_{H^l}$, or $W_2^l(\mathbb{R}^r)\simeq H^l$. We call $\|\varphi\|_{H^l}$ the Fourier norm of $\varphi\in W_2^l(\mathbb{R}^r)$, $l\geqslant 0$.

Proof. First suppose that l is integral, $l\in\mathbb{N}$; the Parseval equation (Theorem 1.24) together with §1, (47) gives

$$\|\varphi\|_{W_2^l}^2=\sum_{|s|\leqslant l}\int_{\mathbb{R}^r}|\mathrm{D}^s\varphi|^2\,\mathrm{d}x=\frac{1}{(2\pi)^r}\int_{\mathbb{R}^r}\Big(\sum_{|s|\leqslant l}|\xi^{2s}|\Big)|\hat{\varphi}|^2\,\mathrm{d}\xi,$$

and the elementary inequalities

$$\frac{1}{2^{2l}}(1+|\xi|^2)^l\leqslant\sum_{|s|\leqslant l}|\xi^{2s}|\leqslant(1+|\xi|^2)^l \tag{4}$$

which ends the first part of the proof.

Suppose now that $l>0$ and not integral, that is, $l=[l]+\lambda$, $0<\lambda<1$. We prove the equality

$$\iint_{\mathbb{R}^r\times\mathbb{R}^r}\frac{|\varphi(x+z)-\varphi(x)|^2}{|z|^{r+2\lambda}}\,\mathrm{d}x\,\mathrm{d}x=\frac{c_{r,\lambda}}{(2\pi)^r}\int|\xi|^{2\lambda}|\hat{\varphi}(\xi)|^2\,\mathrm{d}\xi, \tag{5}$$

where the constants $c_{r,\lambda}>0$ do not depend on φ. We have (see Theorem 1.23)

$$\varphi(x+z)-\varphi(x)=\frac{1}{(2\pi)^r}\int\mathrm{e}^{\mathrm{i}\langle x,\xi\rangle}(\mathrm{e}^{\mathrm{i}\langle z,\xi\rangle}-1)\hat{\varphi}(\xi)\,\mathrm{d}\xi,$$

and by the Parseval equation (Theorem 1.24)

$$\int|\varphi(x+z)-\varphi(x)|^2\,\mathrm{d}x=\frac{1}{(2\pi)^r}\int|\hat{\varphi}(\xi)|^2|\mathrm{e}^{\mathrm{i}\langle z,\xi\rangle}-1|^2\,\mathrm{d}\xi,$$

which gives

$$\iint_{\mathbb{R}^r\times\mathbb{R}^r}\frac{|\varphi(x+z)-\varphi(x)|^2}{|z|^{r+2\lambda}}\,dx\,dz=\frac{1}{(2\pi)^r}\int|\xi|^{2\lambda}|\hat{\varphi}(\xi)|^2\left(\int\frac{|e^{i\langle z,\xi\rangle}-1|^2}{|\xi|^{2\lambda}|z|^{r+2\lambda}}\,dz\right)d\xi.$$

We label the inner integral $c(\xi)$, and see immediately that $c(\xi)$ is spherically symmetric, hence $c(\xi)=c(|\xi|)$. The variable change $z'=z\cdot|\xi|$ gives $0<c(\xi)=c(1)=c_{r,\lambda}$, with which we have proved (5). Let $l=[l]+\lambda$, by (5) we have

$$\begin{aligned}\|\varphi\|^2_{W_2^l}&=\sum_{|s|\leqslant[l]}(\|D^s\varphi\|_0^2+I_\lambda(D^s\varphi))\\&=\frac{1}{(2\pi)^r}\sum_{|s|\leqslant[l]}\int(|\xi^{2s}|+c_{r,\lambda}|\xi^{2s}|\cdot|\xi|^{2\lambda})|\hat{\varphi}|^2\,d\xi\\&=\frac{1}{(2\pi)^r}\int\left(\sum_{|s|\leqslant[l]}|\xi^{2s}|\right)(1+c_{r,\lambda}|\xi|^{2\lambda})|\hat{\varphi}|^2\,d\xi,\end{aligned}$$

and the inequalities

$$c'(1+|\xi|^2)^\lambda\leqslant(1+c_{r,\lambda}|\xi|^{2\lambda})\leqslant c(1+|\xi|^2)^\lambda,$$

together with (4) give the proof of (3). ■

Let Ω be open in $\mathbb{R}^r$, we can now – again for all real l – define the spaces $H^l(\Omega)$. However, the equivalence $H^l(\Omega)\simeq W_2^l(\Omega)$, $l\geqslant 0$, holds only under special assumptions on the frontier $\partial\Omega$ of Ω, see later.

Definition 5.2 *The space $H^l(\Omega)$ consists of the restrictions to Ω of all distributions from H^l; the norm on $H^l(\Omega)$ is given by*

$$\|\varphi\|_{H^l(\Omega)}=\inf\|\varphi^c\|_{H^l},\tag{6}$$

where the infimum is taken over all distributions $\varphi^c\in H^l$, whose restriction to Ω gives the element φ, that is,

$$R_\Omega\varphi^c=\varphi\in H^l(\Omega).$$

It is easy to see that (6) has the properties of a norm.

We wish to prove the completeness of $H^l(\Omega)$. To this end we remark that if φ,ψ are two elements from $H^l(\Omega)$ and the extension φ^c is preassigned, then by (6) there exists a unique extension ψ^c with

$$\|\varphi^c-\psi^c\|_{H^l}\leqslant 2\|\varphi-\psi\|_{H^l(\Omega)}.\tag{7}$$

First we take a Cauchy sequence φ_n in $H^l(\Omega)$ with $\sum_{n=1}^\infty\|\varphi_n-\varphi_{n+1}\|_{H^l(\Omega)}<\infty$. Choose an arbitrary extension φ_1^c of φ_1, then by (7) there exists some extension φ_2^c of φ_2 with

$$\|\varphi_1^c-\varphi_2^c\|_{H^l}\leqslant 2\|\varphi_1-\varphi_2\|_{H^l(\Omega)},$$

and in general some extension φ_{n+1}^c of φ_{n+1} with

$$\|\varphi_n^c-\varphi_{n+1}^c\|_{H^l}\leqslant 2\|\varphi_n-\varphi_{n+1}\|_{H^l(\Omega)}.$$

From $\sum_{n=1}^{\infty}\|\varphi_n^c-\varphi_{n+1}^c\|_{H^l}<\infty$ it follows, however, that φ_n^c is a Cauchy sequence in H^l, and in this space has the limit φ_0^c. Because

$$R_\Omega\varphi_0^c=:\varphi_0 \quad\text{and}\quad \|\varphi_n-\varphi_0\|_{H^l(\Omega)}\leqslant\|\varphi_n^c-\varphi_0^c\|_{H^l},$$

it is also true that $\varphi_n\to\varphi_0$ in $H^l(\Omega)$.

If φ_n is an arbitrary Cauchy sequence in $H^l(\Omega)$, then there is a subsequence φ_{n_k} with $\sum_{k=1}^{\infty}\|\varphi_{n_k}-\varphi_{n_{k+1}}\|_{H^l(\Omega)}<\infty$. This subsequence has a limit φ_0 which is also the limit of the whole sequence.

By Theorem 5.1 $\mathscr{D}$ is dense in each H^l, and from this follows:

Lemma 5.1 *The restrictions to Ω of functions from $\mathscr{D}(\mathbb{R}^r)$ are dense in $H^l(\Omega)$.*

Proof. Let $\varphi\in H^l(\Omega)$ and let φ^c be an extension from H^l. Then there exists some $\psi_\varepsilon\in\mathscr{D}$ with $\|\varphi^c-\psi_\varepsilon\|_{H^l}\leqslant\varepsilon$ and we have

$$\|\varphi-R_\Omega\psi_\varepsilon\|_{H^l(\Omega)}\leqslant\|\varphi^c-\psi_\varepsilon\|_{H^l}\leqslant\varepsilon.$$

We can also define the space $\mathring{H}^l(\Omega)$ to be the closure of the function space $\mathscr{D}(\Omega)$ in the topology of H^l. Since all functions from $\mathscr{D}(\Omega)$ are extendable by zero we obtain nothing new for $l\geqslant 0$, so for $l\geqslant 0$ we do have $\mathring{H}^l(\Omega)\simeq\mathring{W}_2^l(\Omega)$, set theoretically and topologically.

Things are different for the spaces $H^l(\Omega), l\geqslant 0$; they can be genuine subspaces of $W_2^l(\Omega)$, see Volevič & Panejach [1]. We will, however, prove Theorem 5.3 that, if there exists a continuous extension operator F_Ω: $W_2^l(\Omega)\to W_2^l(\mathbb{R}^r)$, then the equivalence $H^l(\Omega)\simeq W_2^l(\Omega)$, $l\geqslant 0$ holds. ■

First we want to make the concepts precise. Elements from $W_2^l(\Omega)$, respectively $H^l(\Omega)$, are distributions, therefore the restriction operator $R_{\Omega'}^{\Omega}$ to smaller set $\Omega'\subset\Omega$ is always defined. We see immediately the continuity of

$$R_{\Omega'}^{\Omega}: W_2^l(\Omega)\to W_2^l(\Omega'), \quad\text{respectively}\quad R_{\Omega'}^{\Omega}: H^l(\Omega)\to H^l(\Omega').$$

In one case this is a matter of integrals in the norm and we have

$$\|R_{\Omega'}^{\Omega}\varphi\|_{W_2^l(\Omega')}\leqslant\|\varphi\|_{W_2^l(\Omega)}$$

and in the other case we work with inf norms, for which

$$\|R_{\Omega'}^{\Omega}\varphi\|_{H^l(\Omega')}\leqslant\|\varphi\|_{H^l(\Omega)} \quad\text{for}\quad \Omega'\subset\Omega.$$

Definition 5.3 *By an extension operator $F_{\Omega'}^{\Omega}$ from a smaller open set Ω' to a* *larger Ω, $\Omega' \subset \Omega$, we understand a continuous, linear operator*

$$F^{\Omega}_{\Omega'}: W^l_2(\Omega') \to W^l_2(\Omega), \quad l \geqslant 0 \tag{8}$$

with the property

$$R^{\Omega}_{\Omega'} \circ F^{\Omega}_{\Omega'} = I_{W^l_2(\Omega')}, \quad \Omega' \subset \Omega. \tag{9}$$

There are similar definitions for the families of spaces H^l and $\mathring{W}^l_2$. If $\Omega = \mathbb{R}^r$ we simply write $F^{\mathbb{R}^r}_{\Omega'} =: F_{\Omega'}$.

We proved in Lemma 3.4 that the spaces $\mathring{W}^l_2(\Omega')$, $l = 0, 1, \ldots$, always admit a continuous extension operator to an arbitrary, larger, open set indeed we can always extend functions from $\mathring{W}^l_2(\Omega')$ by zero.

Theorem 5.3 *For the space $W^l_2(\Omega)$, $l \geqslant 0$, let there exist a continuous extension operator $F_\Omega: W^l_2(\Omega) \to W^l_2(\mathbb{R}^r)$. Then*

$$W^l_2(\Omega) \simeq H^l(\Omega), \tag{10}$$

both set theoretically and topologically.

Proof. Definition 5.2 implies that $H^l(\Omega)$ consists of those functions $\varphi \in W^l_2(\Omega)$ that have at least one extension $\varphi^c \in W^l_2(\mathbb{R}^r)$. Since by assumption we may put $\varphi^c = F_\Omega \varphi$, we have proved set theoretic equality in (10). From (6) and from the continuity of F_Ω it follows that

$$\|\varphi\|_{H^l(\Omega)} \leqslant \|F_\Omega \varphi\|_{W^l_2(\mathbb{R}^r)} \leqslant c\|\varphi\|_{W^l_2(\Omega)},$$

while the continuity of the restriction operator $R := R^{\mathbb{R}^r}_\Omega$ gives

$$\|\varphi\|_{W^l_2(\Omega)} \leqslant \|R\varphi^c\|_{W^l_2(\Omega)} \leqslant \|\varphi^c\|_{W^l_2(\mathbb{R}^r)}.$$

So, after taking the infimum we obtain

$$\|\varphi\|_{W^l_2(\Omega)} \leqslant \|\varphi\|_{H^l(\Omega)},$$

with which we have proved the equivalence (10). ■

5.2 Extension theorems

We want now to give concrete extension theorems for the spaces $W^l_2(\Omega)$, $l \geqslant 0$. First we present the extension theorem of Calderon–Zygmund [1]; it holds under very general conditions on Ω, it being sufficient that Ω is bounded and fulfils the uniform cone condition. By Theorem 2.1 this is also the case for $\Omega \in N^{0,1}$.

Theorem 5.4 (Calderon–Zygmund) *Let Ω be bounded and fulfil the uniform cone condition (see Definition 2.3). Then for $l = 0, 1, \ldots$ there exists a continuous,* linear extension operator

$$F_\Omega: W_2^l(\Omega) \to W_2^l(\mathbb{R}^r).$$

Proof. The case $l = 0$ is trivial, and therefore we take $l = 1, 2, \ldots$. From the uniform cone condition it follows that the segment property holds; therefore by Theorem 3.6 the $C_0^\infty(\mathbb{R}^r)|_\Omega$-functions are dense in $W_2^l(\Omega)$, and we can assume sufficient smoothness for the elements of $W_2^l(\Omega)$. We fix a point $x \in \Omega$, let U_x be the 'cone neighbourhood' from Definition 2.3, that is, $z \in \bar{\Omega} \cap U_x$ implies $z + C^{(x)} \subset \Omega$. We may assume that U_x is bounded. For the sake of brevity we omit the subscript x and write $U := U_x, K := \Omega \cap U, C := C^{(x)}$. Let $C = C(0, \rho, \Sigma)$, where we choose the defining set Σ for the cone C to lie on the unit sphere $S(0, 1)$; see the text following Definition 2.1 and Fig. 2.2.

Let u be a function from $C^\infty(\bar{K})$ with support in $\bar{\Omega} \cap U = K \cup (\partial\Omega \cap U)$. Let $\sigma \in \Sigma$ then for $x \in \bar{K}$ we have:

$$\int_0^\infty t^{l-1} \frac{\mathrm{d}^l}{\mathrm{d}t^l}[\mathrm{e}^{-t} u(x + t\sigma)]\,\mathrm{d}t = (-1)^l (l-1)!\, u(x) \tag{11}$$

and

$$\frac{\mathrm{d}^l}{\mathrm{d}t^l}[\mathrm{e}^{-t} u(x + t\sigma)] = \mathrm{e}^{-t}(-1)^l \sum_{|\alpha| \leqslant l} \mathrm{D}^\alpha u(x + t\sigma)\sigma^\alpha \frac{|\alpha|!}{\alpha!}\binom{l}{|\alpha|}(-1)^{|\alpha|}. \tag{12}$$

Let $\nu(\sigma)$ be an infinitely differentiable function defined on the sphere $S(0, 1)$, whose support is contained in Σ, and which has the properties:

$$\nu(\sigma) \geqslant 0, \qquad \int_S \nu(\sigma)\,\mathrm{d}\sigma = 1.$$

If we multiply both sides of (11) by $\nu(\sigma)$ and integrate over S, (11) and (12) give

$$u(x) = \frac{1}{(l-1)!}\int_S \nu(\sigma)\,\mathrm{d}\sigma \int_0^\infty t^{l-1}\mathrm{e}^{-t} \sum_{|\alpha| \leqslant l} (-1)^{|\alpha|}\frac{|\alpha|!}{\alpha!}\binom{l}{|\alpha|}\sigma^\alpha \mathrm{D}^\alpha u(x + \sigma t)\,\mathrm{d}t,$$

or after substituting $y = x + t\sigma$, and with $\mu(x) := \nu(-x)$:

$$u(x) = \frac{1}{(l-1)!}\int_{\mathbb{R}^r}\left[\sum_{|\alpha| \leqslant l}\frac{|\alpha|!}{\alpha!}\binom{l}{|\alpha|}\frac{(x-y)^\alpha}{|x-y|^{|\alpha|+r-l}}\mathrm{D}^\alpha u(y)\right]$$

$$\cdot \mathrm{e}^{-|x-y|}\mu\left(\frac{x-y}{|x-y|}\right)\mathrm{d}y, \quad x \in \bar{K}. \tag{13}$$

We wish to make this formula, which goes back to Sobolev [1], the basis of the extension operator. Here we must observe that because of the support property of μ and the cone condition, only points y from $\bar{\Omega} \cap U$ make a contribution to the integral. First we prove a lemma.

Lemma 5.2 *We denote by W the functions u from $W_2^l(\Omega) \cap C(\bar{K})$ which have support in $K \cup (\partial\Omega \cap U) = \bar{\Omega} \cap U$. Then there exists a continuous extension operator*

$$F: W \to W_2^l(\mathbb{R}^r) \quad \textit{with } R_K \circ Fu = u.$$

Proof. Let $u \in W$, for $|\alpha| \leqslant l$ we set

$$f_\alpha(x) = \begin{cases} D^\alpha u(x), & x \in K, \\ 0 \quad , & x \notin K, \end{cases}$$

and define for $x \in \mathbb{R}^r$

$$v(x) := Fu$$

$$= \frac{1}{(l-1)!} \int_{\mathbb{R}^r} \left[\sum_{|\alpha| \leqslant l} \frac{|\alpha|!}{\alpha!} \binom{l}{|\alpha|} \frac{(x-y)^\alpha}{|x-y|^{|\alpha|+r-l}} f_\alpha(y) \right] e^{-|x-y|} \mu\left(\frac{x-y}{|x-y|} \right) dy$$

$$= \sum_{|\alpha| \leqslant l} \int_{\mathbb{R}^r} I_\alpha(x-y) f_\alpha(y) dy. \tag{14}$$

For $x \in K$ we have immediately from (13) that $v(x) = u(x)$ or $R_K \circ Fu = u$, that is, F is an extension operator. For later use (see Addendum 5.1) we note that $v(x) = u(x)$ for points $x \in \partial\Omega \cap \bar{K}$ also. Indeed by Theorem 2.9 $\partial\Omega$ has Lebesgue measure zero and altering the definition of f_α appropriately on the set of measure zero $\partial\Omega \cap \bar{K}$ makes no change in the integral (14), so we can apply (13). In order to see the continuity of F we use the Fourier transformation and Theorem 5.2. Since it is clear that $\int_{\mathbb{R}^r} |I_\alpha(x)| dx < \infty$ for $|\alpha| \leqslant l$, that is $I_\alpha \in l_1 \subset \mathscr{S}'$, and $f_\alpha \in \mathscr{E}'$ if $|\alpha| \leqslant l$, we can apply the convolution theorem 1.26, and by Fourier transformation obtain from (14) that

$$\hat{v}(\xi) = \sum_{|\alpha| \leqslant l} \hat{I}_\alpha(\xi) \cdot \hat{f}_\alpha(\xi). \tag{15}$$

If for $|\alpha| \leqslant l$ we can prove

$$|\hat{I}_\alpha(\xi)| \leqslant c_1 (1 + |\xi|^2)^{-l/2}, \tag{16}$$

we will have proved everything, since (15) squared and multiplied by $(1 + |\xi|^2)^l$ gives, with due attention paid to (16) and to Theorem 5.2,

$$\| v \|_{W_2^l(\mathbb{R}^r)} \leqslant C \| u \|_{W_2^l(K)},$$

which is the continuity of F.

Now for (16). We consider the integral

$$\hat{J}_\alpha(\xi) = \frac{\alpha!}{|\alpha|!}\binom{l}{|\alpha|}^{-1}\hat{I}_\alpha(\xi) = \frac{1}{(l-1)!}\int_{\mathbb{R}^r}\frac{x^\alpha e^{-|x|}}{|x|^{|\alpha|+r-1}}\mu\left(\frac{x}{|x|}\right)e^{-i(x,\xi)}\,dx$$

$$= \frac{1}{(l-1)!}\int_S \sigma^\alpha\mu(\sigma)\int_0^\infty t^{l-1}e^{-t(1+i(\sigma,\xi))}\,dt\,d\sigma = \int_S \frac{\sigma^2\mu(\sigma)}{[1+i(\sigma,\xi)]^l}\,d\sigma,$$

by partial integration.

For $|\xi| \leqslant 1$ we clearly have

$$|\hat{J}_\alpha(\xi)| \leqslant c_2. \tag{17}$$

Therefore it is enough to consider the case $|\xi| > 1$.

Let $\eta = \xi/|\xi|$. Without loss of generality we may assume that $\operatorname{supp}\mu$ is so small that $\mu(\sigma) \neq 0$ implies that $(\sigma, \sigma_0) > \frac{1}{2}$, where $\sigma_0 = (0, 0, \ldots, -1)$. We can then find some $\varepsilon > 0$ with the properties:

1. If $|\eta_r| \leqslant \varepsilon$ then $\eta \notin \operatorname{supp}\mu$.
2. If $|\eta_r| > \varepsilon$ and $\sigma \in \operatorname{supp}\mu$, then $|(\sigma, \eta)| \geqslant c_3$. Here $\eta = (\eta_1, \ldots, \eta_r)$.

First we consider the possibility 1: $|\eta_r| \leqslant \varepsilon$. We introduce a new coordinate system, where the axes are determined by the vectors $\sigma^1, \ldots, \sigma^r$ (see Fig. 5.1),

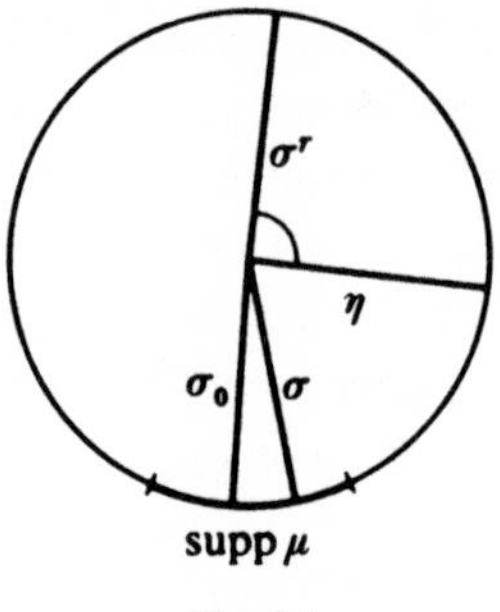

Fig. 5.1

and where

$$-(\sigma^r, \sigma_0) = (1 - \eta_r^2)^{1/2}, \quad \sigma^1 = \eta.$$

The choice of the coordinate system has the effect that if $\sigma \in \operatorname{supp}\mu$ then $(\sigma, \sigma^r) \neq 0$. We label the new coordinates $\tau_1, \ldots, \tau_r$ and put $\tau = (\tau', \tau_r)$. After this coordinate transformation $\hat{J}_\alpha(\xi)$ becomes, noting $\sqrt{(1 - \tau_1^2 - \cdots - \tau_{r-1}^2)} = \tau_r \neq 0$,

$$\hat{J}_\alpha(\xi) = \int_{|\tau|=1}\frac{\mu(\tau)\,d\tau}{[1 + i\tau_1|\xi|]^l} = \int_{|\tau'|<1}\frac{\lambda(\tau')\,d\tau'}{(1 + i\tau_1|\xi|)^l},$$

where λ is infinitely differentiable in $|\tau'|<1$, and $\operatorname{supp}\lambda\subset\{|\tau'|<1\}$. After integration by parts we obtain

$$\int_{|\tau'|<1}\frac{\lambda(\tau')\,d\tau'}{(1+i\tau_1|\xi|)^l}=\frac{1}{(l-1)!}\frac{1}{(i|\xi|)^{l-1}}\int_{|\tau'|<1}\frac{\left(\dfrac{\partial^{l-1}\lambda}{\partial\tau_1^{l-1}}\right)d\tau'}{1+i\tau_1|\xi|}$$

$$=\frac{1}{(l-1)!}\frac{1}{(i|\xi|)^{l-1}}\int_{|\tau'|<1}\frac{\delta(1-i\tau_1|\xi|)}{1+\tau_1^2|\xi|^2}d\tau' \tag{18}$$

(where we have put $\partial^{l-1}\lambda/\partial\tau^{l-1}=\delta$). However, now we have

$$\int_{|\tau'|<1}\frac{\delta(\tau')|\xi|\tau_1\,d\tau'}{1+\tau_1^2|\xi|^2}=\int_{|\tau'|<1}\frac{\delta(0,\tau_2,\ldots,\tau_{r-1})|\xi|\tau_1}{1+\tau_1^2|\xi|^2}d\tau'$$

$$+\int_{|\tau'|<1}\frac{\delta(\tau')-\delta(0,\tau_2,\ldots,\tau_{r-1})}{1+\tau_1^2|\xi|^2}|\xi|\tau_1\,d\tau', \tag{19}$$

and the first integral on the right-hand side of (19) equals zero, while for the second, because $|\delta(\tau')-\delta(0,\tau_2,\ldots,\tau_{r-1})|\leqslant c_5|\tau_1|$, we have the following estimate

$$\left|\int_{|\tau'|<1}\frac{\delta(\tau')-\delta(0,\tau_2,\ldots,\tau_{r-1})}{1+\tau_1^2|\xi|^2}|\xi|\tau_1\,d\tau'\right|\leqslant\frac{c_6}{|\xi|}. \tag{20}$$

We have further

$$\left|\int_{|\tau'|<1}\frac{\delta\,d\tau'}{1+\tau_1^2|\xi|^2}\right|\leqslant c_7\int_{-1}^{1}\frac{d\tau_1}{1+|\xi|^2\tau_1^2}\leqslant\frac{c_8}{|\xi|}. \tag{21}$$

(20) and (21) give the estimate (16) for (18). For the second remaining possibility $|\eta_r|>\varepsilon$, $|(\sigma,\eta)|>c_3$, for $\sigma\in\operatorname{supp}\mu$, and we see estimate (16) without further argument. With this we have proved Lemma 5.2. ■

The proof of Theorem 5.4 now follows by means of a partition of unity. We cover $\bar{\Omega}$ by 'cone neighbourhoods' U_x, since $\bar{\Omega}$ is compact we need only finitely many, $\bar{\Omega}\subset\bigcup_{k=1}^m U_{x_k}$. We take a partition of unity $\{\varphi_k\}$ subordinate to the cover $\{U_{x_k}\}$, and define

$$F\psi=\sum_{k=1}^m F_k(\varphi_k\psi),\quad \psi\in W_2^l(\Omega),$$

where the F_k are the extension operators constructed in Lemma 5.2. Continuity (8) and property (9) are then seen without difficulty. ■

Addendum 5.1 *From the formulae* (11)–(14) *we see immediately that the extension operator F_Ω has the property*

$$F_\Omega \varphi|_{\partial\Omega} = \varphi|_{\partial\Omega} \quad \textit{for all } \varphi \in C^{l-1,1}(\bar{\Omega}).$$

Proof. By Theorem 1.8 $\varphi \in C^{l-1,1}$ possesses everywhere continuous derivatives up to order $l-1$, and bounded derivatives of order l almost everywhere; furthermore the distributional derivatives agree almost everywhere with the classical derivatives for $|s| \leqslant l$. Sets of measure zero play no role in the integral formulae (11)–(14) and we read the restriction property $F\varphi|_{\partial\Omega}$ directly from (13) and (14). ■

Corollary 5.1 *Let Ω be bounded and satisfy the uniform cone condition. Denote by Ω_ε the ε-neighbourhood of Ω*

$$\Omega_\varepsilon = \{x \in \mathbb{R}^r : d(x, \Omega) < \varepsilon\}.$$

There exists a continuous linear extension operator

$$F^\varepsilon_\Omega : W^l_2(\Omega) \to \mathring{W}^l_2(\Omega_\varepsilon). \tag{22}$$

Proof. Let $\varphi_\varepsilon \in \mathscr{D}$ with $\varphi_\varepsilon = 1$ on Ω and $\varphi_\varepsilon = 0$ on $\mathbf{C}\Omega_\varepsilon$, let F_Ω be the extension operator from Theorem 5.4, put

$$F^\varepsilon_\Omega \varphi := \varphi_\varepsilon \cdot (F_\Omega \varphi) \quad \text{for } \varphi \in W^l_2(\Omega),$$

and we have proved the corollary. ■

We now prove two extension theorems for which the method of proof goes back to Hestenes [1]. In Theorem 5.6 the assumptions on Ω are certainly stronger than in Theorem 5.4; in contrast to Theorem 5.4, however, it holds for all $l \in \mathbb{R}_+$.

Theorem 5.5 *Let $\mathbb{R}^r_+ = \{x \in \mathbb{R}^r : x_r > 0\}$. For each $l \geqslant 0$ there exists a continuous linear extension operator*

$$F : W^l_2(\mathbb{R}^r_+) \to W^l_2(\mathbb{R}^r).$$

F has the property that if supp φ *is bounded, then* supp $F\varphi$ *is also bounded. From* (24) *we see immediately that we obtain the support of $F\varphi$ from that of* supp φ *by stretching and by reflection in the hyperplane $x_r = 0$; $\varphi \in W^l_2(\mathbb{R}^r_+)$. Furthermore, let $0 < L < \infty$ be given. Then for each $l \leqslant L$ the extension operator of Theorem 5.5, $F_l : W^l_2(\mathbb{R}^r_r) \to W^l_2(\mathbb{R}^r)$ can be chosen to be independent of l.*

Remark Let $\varphi \in W^l_2(W^r_+)$ and suppose that the support of φ meets the frontier ∂W^r_+ of W^r_+ only in W^{r-1}. Then we conclude from (24) and the following proof that

$$\operatorname{supp} F\varphi \subset\subset (L+1)\cdot W^r = \{x\in\mathbb{R}^r:|x_1|\leqslant L+1,\ldots,|x_r|\leqslant L+1\},$$
$$F\varphi\in \mathring{W}^l_2((L+1)W^r),$$

and we have

$$\|F\varphi\|_{W^l_2}((L+1)W^r)\leqslant c\|\varphi\|_{W^l_2(W^r_+)}.$$

Proof. In proving the theorem we prove simultaneously independence from l, and, without loss of generality, suppose that L is a natural number. Since $\mathbb{R}^r_+$ has the segment property (Definition 2.1) it is enough in the proof to consider sufficiently smooth functions with bounded support, by Theorem 3.6 these are dense in $W^l_2(\mathbb{R}^r_+)$. Suppose first that l is integral, thus $l=0,1,\ldots$. We consider the system of equations

$$\sum_{m=0}^{L}\left(-\frac{1}{m+1}\right)^q\beta_m=1,\quad q=0,\ldots,L. \tag{23}$$

Since the (Vandermonde) determinant of this system is different from zero, it is uniquely solvable for β_m. Let

$$\varphi\in C^l_0(\bar{\mathbb{R}}^r_+)\ (\operatorname{supp}\varphi \text{ bounded})\quad\text{and}\quad \mathbb{R}^r_-=\{x\in\mathbb{R}^r:x_r<0\}.$$

We put

$$\tilde{\varphi}(x):=\sum_{m=0}^{L}\beta_m\varphi\left(x',\frac{-x_r}{m+1}\right)\quad\text{for } x=(x^1,x_r)\in\mathbb{R}^r_-, \tag{24}$$

and define the extension operator by

$$(F\varphi)(x):=\begin{cases}\varphi(x), & x\in\bar{\mathbb{R}}^r_+,\\ \tilde{\varphi}(x), & x\in\bar{\mathbb{R}}^r_-,\end{cases} \tag{25}$$

it is independent of l for $l\leqslant L$.

We show first that if $\varphi\in C^l_0(\bar{\mathbb{R}}^r_+)$ then $F\varphi\in C^l_0(\mathbb{R}^r)$. For this we need only consider the limiting values of $\tilde{\varphi}$ and the derivatives on the hyperplane $x_r=0$. Let p_1 be an index and p_2 a multiindex with $p_1+|p_2|\leqslant l$, using (23) and (24) we have

$$\lim_{x_r=0}\frac{\partial^{p_1}}{\partial x_x^{p_1}}D^{p_2}_{x'}\tilde{\varphi}=\lim_{x_r=0}\sum_{m=0}^{L}\beta_m\frac{\partial^{p_1}}{\partial x_r^{p_1}}D^{p_2}_{x'}\varphi\left(x',\frac{-x_r}{m+1}\right)$$
$$=\sum_{m=0}^{L}\beta_m\left(-\frac{1}{m+1}\right)^{p_1}\frac{\partial^{p_1}}{\partial x_x^{p_1}}D^{p_2}_{x'}\varphi\bigg|_{x_r=0}=\frac{\partial^{p_1}}{\partial x_r^{p_1}}D^{p_2}_{x'}\varphi\bigg|_{x_r=0},$$

and so we have proved that $F\varphi\in C^l_0(\mathbb{R}^r)$. The hyperplane $x_r=0$ has $\mathbb{R}^r$-

measure 0 and, because of (24), for $\varphi \in C_0^l(\mathbb{R}_+^r)$ we have

$$\|F\varphi\|_{W_2^l(\mathbb{R}^2)}^2 = \sum_{|s| \leqslant l} \|D^s(F\varphi)\|_0^2 = \sum_{|s| \leqslant l} \left(\int_{\mathbb{R}_+^r} |D^s\varphi|^2 \, dx + \int_{\mathbb{R}^r} |D^s\tilde{\varphi}| dx \right)$$
$$\leqslant C \sum_{|s| \leqslant l} \int_{\mathbb{R}_+^r} |D^s\varphi|^2 \, dx = C\|\varphi\|_{W_2^l(\mathbb{R}_+^r)}^2,$$

and so (by the usual density considerations) we have proved the continuity of $F: W_2^l(\mathbb{R}_+^r) \to W_2^l(\mathbb{R}^r)$. Property (9) of Definition 5.3 is an immediate consequence of the definition (25) of $F\varphi$.

Suppose now that $l = [l] + \lambda$ with $0 < \lambda < 1$. It remains only to prove the estimate

$$I_\lambda(D^s(F\varphi)) = \iint_{\mathbb{R}^r \times \mathbb{R}^r} \frac{|D^s(F\varphi)(x) - D^s(F\varphi)(y)|^2}{|x-y|^{r+2\lambda}} \, dx \, dy$$
$$\leqslant C \iint_{\mathbb{R}_+^r \times \mathbb{R}_+^4} \frac{|D^s\varphi(x) - D^s\varphi(y)|^2}{|x-y|^{r+2\lambda}} \, dx \, dy = CI_\lambda(D^s\varphi), \qquad (26)$$

for $|s| \leqslant [l]$, $\varphi \in C_0^{[l]}(\mathbb{R}_+^r)$ (supp φ bounded), $F\varphi$ defined by (25). Let I_1 be the first integral in (26); we have

$$I_1 = \iint_{\mathbb{R}_+^r + \mathbb{R}_+^r} \frac{|D^sF\varphi(x) - D^sF\varphi(y)|^2}{|x-y|^{r+2\lambda}} \, dx \, dy$$
$$+ \iint_{\mathbb{R}_-^r + \mathbb{R}_-^r} \frac{|D^sF\varphi(x) - D^sF\varphi(y)|^2}{|x-y|^{r+2\lambda}} \, dx \, dy$$
$$+ 2\iint_{\mathbb{R}_+^r + \mathbb{R}_-^r} \frac{|D^sF\varphi(x) - D^sF\varphi(y)|^2}{|x-y|^{r+2\lambda}} \, dx \, dy = I_2 + I_3 + 2I_4, \qquad (27)$$

since, because $\varphi \in C_0^{[l]}(\mathbb{R}_+^r)$, $F\varphi \in C_0^{[l]}(\mathbb{R}^r)$, we are allowed to neglect sets of measure zero in integration (classical derivatives agree with distributional derivatives by Theorem 1.7).

By (24) we have

$$D^s\tilde{\varphi}(x) = \sum_{m=0}^{L} \beta_m \left(\frac{-1}{m+1} \right)^{p_1} D_x^s\varphi\left(x', -\frac{x_r}{m+1} \right), \qquad (28)$$

which substituted in the integral I_3 gives

$$I_3 \leqslant c_2 \iint_{\mathbb{R}_+^r \times \mathbb{R}_+^r} \frac{|D^s\varphi(x) - D^s\varphi(y)|^2}{|x-y|^{r+2\lambda}} \, dx \, dy = c_1 I_1 = c_1 I_\lambda(D^s\varphi). \qquad (29)$$

We now estimate the integral I_4. After transformation of the variable in x

(23) and (28) give

$$I_4 = \iint_{\mathbb{R}^r_+ \times \mathbb{R}^r_-} \left| \sum_{m=0}^{L} \beta_m \left(\frac{-1}{m+1}\right)^{p_2} \left[\mathrm{D}^s_x \varphi\left(x', -\frac{x_r}{m+1}\right) - \mathrm{D}^s_y \varphi(y) \right] \right|^2 \frac{\mathrm{d}x\,\mathrm{d}y}{|x-y|^{r+2\lambda}}$$

$$\leqslant \iint_{\mathbb{R}^r_+ \times \mathbb{R}^r_+} \sum_{m=0}^{L} \frac{|\mathrm{D}^s\varphi(x) - \mathrm{D}^s\varphi(y)|^2}{|x^{(m)} - y|^{r+2\lambda}} \mathrm{d}x\,\mathrm{d}y, \tag{30}$$

where we have put $x^{(m)} = (x', -(m+1)x_r)$.

For $x, y \in \mathbb{R}^r_+$ we have, however,

$$\begin{aligned} |x^{(m)} - y| &= [|x'-y'|^2 + |(m+1)x_r + y_r|^2]^{1/2} \\ &= [|x'-y'|^2 + (m+1)^2 x_r^2 \pm 2(m+1)x_r y_r + y_r^2]^{1/2} \\ &\geqslant [|x'-y'|^2 + x_r^3 - 2x_r y_r + y_r^2]^{1/2} = |x-y|, \end{aligned}$$

and in (30) we may estimate further

$$I_4 \leqslant C_3 I_3 = C_3 I_\lambda(\mathrm{D}^s\varphi). \tag{31}$$

(27), (29), (30) and (31) give the desired estimate (26). ■

With the help of the transformation theorem 4.1, Theorem 5.5 gives:

Theorem 5.6 *Let Ω be bounded and (k, κ)-smooth. Then for $0 \leqslant l < k + \kappa$ (if $k + \kappa$ is integral, we may also take $l = k + \kappa$) there exists a continuous linear extension operator $F_\Omega: W^l_2(\Omega) \to \mathring{W}^l_2(\tilde{\Omega})$, where $\tilde{\Omega} \supset \bar{\Omega}$ and F_Ω is independent of l in the range $l \leqslant k + \kappa = L$.*

Proof. We cover $\bar{\Omega}$ with small neighbourhoods U_j, $j = 0, \dots, m$, which have the properties of the (k, κ)-smoothness definition 2.7, and choose a partition of unity $\{\alpha_j\}$ *subordinate to the cover* $\{U_j\}$. Let $\Phi_j: U_j \to W^r$ be the (k, κ)-transformation given by 4.1,† and $^*\Phi_j: W^l_2(W^r) \to W^l_2(U_j)$ the associated pull-back operator. We put

$$F_\Omega \varphi := \sum_{j=0}^{m} {}^*\Phi_j \circ F \circ {}^*\Phi_j^{-1}(\alpha_j \cdot \varphi), \quad \varphi \in W^l_2(\Omega); \tag{32}$$

† By perhaps shrinking U_j (see Proposition 4.1) we can always suppose that the Φ_j are $\tilde{C}^{k,\kappa}$-diffeomorphisms, so that Theorem 4.1 is applicable.

shrinking by a factor $1/(L+1)$ we may suppose that

$$\operatorname{supp} {}^*\Phi_j^{-1}(\alpha_j\varphi) \subset W^r_+/(L+1) = x: \begin{Bmatrix} 0 < x_1 < 1/(L+1), \ldots, \\ \vdots \\ 0 < x_r < 1/(L+1), \ldots, \end{Bmatrix}.$$

The support of ${}^*\Phi_j^{-1}(\alpha_j\varphi)$ only meets the frontier of $W^r_+/(L+1)$ in $W^{r-1}/(L+1)$. By Theorem 5.5 and the following remark, for the extension $F \circ \Phi_j^{-1}(\alpha_j\varphi)$ we have $\operatorname{supp} F \circ {}^*\Phi_j^{-1}(\alpha_j\varphi) \subset\subset W^r$, and

$$\| F \circ {}^*\Phi_j^{-1}(\alpha_j\varphi) \|_{W_2^l(W^r)} \leqslant c \| {}^*\Phi_j^{-1}(\alpha_j\varphi) \|_{W_2^l(W^r_+)}. \tag{33}$$

from which we obtain the continuity of

$$\varphi \mapsto F \circ {}^*\Phi_j^{-1}(\alpha_j\varphi): W_2^l(\Omega) \to \mathring{W}_2^l(W^r).$$

Another application of Theorem 4.1 gives the continuity of

$${}^*\Phi_j \circ F \circ {}^*\Phi_j^{-1}(\alpha_j\varphi): W_2^l(\Omega) \to \mathring{W}_2^l(U_j).$$

Therefore the map (32)

$$F_\Omega: W_2^l(\Omega) \to \mathring{W}_2^l\left(\bigcup_{j=0}^m U_j\right) = \mathring{W}_2^l(\tilde{\Omega})$$

is continuous, where we set $\tilde{\Omega} = \bigcup_{j=0}^m U_j$.

We easily check property (9) using the definition (32). Independence from l follows from (32) using Theorem 5.5; so F is independent of l. ■

Analogously to Corollary 5.1 we prove the corollary:

Corollary 5.2 *Let Ω be bounded and (k, κ)-smooth. Then if $0 \leqslant l < k + \kappa$ there exists for small $\varepsilon > 0$ a continuous linear extension operator (independent of l)*

$$F^\varepsilon_\Omega: W_2^l(\Omega) \to \mathring{W}_2^l(\Omega_\varepsilon).$$

$\tilde{\Omega} \supset \bar{\Omega}$ implies that $\operatorname{dist}(\bar{\Omega}, C\tilde{\Omega}) > 0$ *and we put* $0 < \varepsilon < \operatorname{dist}$.

The analogue of Addendum 5.1 is also valid.

Addendum 5.2 *The extension operator from Theorems 5.5 and 5.4 (or from Corollary 5.2) has the property*

$$F_\Omega \varphi|_{\partial\Omega} = \varphi|_{\partial\Omega}, \tag{34}$$

for all $\varphi \in C^{[l]-1,1}(\bar{\Omega})$ (respectively $\varphi \in C^0(\bar{\Omega})$ in the case $[l] = 0$).

For the proof we must examine the extension definition (24) more closely. For $\varphi \in C^{[l]-1,1}(\bar{\mathbb{R}}^r_+)$ by means of (25) and obvious estimates we deduce that

$F\varphi \in C^{[l]-1,1}(\mathbb{R}^r)$. Again we apply Theorem 1.8 and obtain the equality of the distributional derivatives $D^s(F\varphi)$ with the classical derivatives (almost everywhere) for $|s| \leqslant [l]$. This shows that the entire proof of Theorem 5.4 can be carried out for $\varphi \in C^{[l]-1,1}(\bar{\mathbb{R}}^r_+)$, and the definition (24) shows the validity of (34) directly. In order to prove (34) for F_Ω or for F^ε_Ω we remark that the transformation theorem 4.1 leaves the relation (34) invariant; the fact that $\varphi \in C^{[l]-1,1}$ also remains unchanged.

For later use in §13 we need an almost trivial extension theorem.

Theorem 5.7 *Let $f \in C^k(B(0,\rho))$ and let $|f(x)-c| \leqslant \varepsilon$ on $B(0,\rho)$, where c is a constant. Then f may be extended from $B(0,\rho/2)$ to $\mathbb{R}^r$, and for the extension $\tilde{f}$ we have*

$$\tilde{f} \in C^k(\mathbb{R}^r) \text{ and } |\tilde{f}(x)-c| \leqslant \varepsilon \quad \text{for all } x \in \mathbb{R}^r.$$

Remark The same extension theorem is also valid for vector valued functions $f(x)$, and the same method of proof also gives an extension $\tilde{f} \in C^k(\mathbb{R}^r_+)$ from $B_+(0,\rho/2)$ to $\mathbb{R}^r_+$ with $|\tilde{f}(x)-c| \leqslant \varepsilon$ for $x \in \mathbb{R}^r_+$ under the assumption that $f \in C^k(B_+(0,\rho))$ and $|f(x)-c| \leqslant \varepsilon$ for $x \in B_+(0,\rho)$.

Proof. Let $\alpha \in \mathscr{D}(\mathbb{R}^r)$ with $\alpha = 1$ on $B(0,\rho/2)$, $\alpha = 0$ on $\mathbf{C}B(0,\rho)$, and $0 \leqslant \alpha \leqslant 1$ otherwise, see Corollary 1.2. In abbreviated form we put $\tilde{f} = \alpha(f-c)+c$, and have $|\tilde{f}-c| = \alpha|f-c| \leqslant \alpha\varepsilon \leqslant \varepsilon$. The other properties of f are easily deduced. ■

Exercises

5.1 Let T^r be the r-dimensional torus. We define $W_2^{-l}(T^r)$ to be the dual space to $W_2^l(Tr)$, hence $f(\cdot)$ is a continuous linear functional on $W_2^l(T^r)$, and we define the Fourier coefficients of f by means of

$$c_k := f(e^{-i(k,x)}), \quad k \in \mathbb{Z}^r.$$

Then $f \in W_2^{-l}(T^r)$ if and only if

$$\sum_k (1+|k|^2)^{-l}|c_k|^2 < \infty \quad \text{and} \quad p(f) = \left[\sum_k (1+|k|^2)^{-1}|c_k|^2\right]^{1/2}$$

is an equivalent norm on $W_2^{-l}(T^r)$.

5.2 $\mathscr{D}(T^r)$ is dense in $W_2^{-l}(T^r)$.

5.3 We have $f = \sum_k c_k e^{i(k,x)}$, convergence being understood in $W_2^{-l}(T^r)$.

§6 Continuous embeddings and Sobolev's lemma

The Sobolev spaces form a scale, that is, we have the continuous embeddings

$$W_2^{l_1}(\Omega) \subsetneq W_2^{l_2}(\Omega) \quad \text{for } l_1 \geqslant l_2 \text{ (Theorem 6.1).}$$

If l is sufficiently large then $W_2^l(\Omega)$ consists only of continuous functions, indeed continuous differentiable functions. This is the content of the so-called Sobolev embedding theorem

$$W_2^l(\Omega) \hookrightarrow C^{l_1}(\bar{\Omega}) \quad \text{for } l - l_1 > r/2.$$

We present two proofs of the Sobolev embedding theorem, the first for integral values of l, under very general assumptions on the region Ω, and the second using the Fourier transformation and the extension theorems, which also covers non-integral values of l, but under stronger assumptions on the region Ω.

We begin with the scaling of the W- and H-spaces.

Theorem 6.1 *The following embeddings are continuous for $l_2 > l_1$:*

(a) $H^{k_2} \hookrightarrow H^{l_1}$ *(see §5), $l_1, l_2 \in \mathbb{R}$ arbitrary.*
(b) $H^{l_2}(\Omega) \hookrightarrow H^{l_1}(\Omega)$, *$l_1, l_2 \in \mathbb{R}$ arbitrary, Ω arbitrary.*
(c) $\mathring{H}^{l_1}(\Omega) \hookrightarrow \mathring{H}^{l_1}(\Omega)$, *$l_1, l_2 \in \mathbb{R}$ arbitrary, Ω arbitrary.*
(d) $\mathring{W}_2^{l_2}(\Omega) \hookrightarrow \mathring{W}_2^{l_2}(\Omega)$, *$l_1, l_2 \in \mathbb{R}_+$ arbitrary, Ω arbitrary.*
(e) $W_2^{l_2}(\Omega) \hookrightarrow W_2^{l_1}(\Omega)$, *$l_1, l_2 \in \mathbb{N}$, Ω arbitrary.*
(f) $W_2^{l_2}(\Omega) \hookrightarrow W_2^{l_1}(\Omega)$, *$l_1, l_2 \in \mathbb{R}_+$ arbitrary, and Ω has the extension property, that is, there exist continuous extension operators*

$$F_\Omega^{l_i}: W_2^{l_i}(\Omega) \to W_2^{l_i}(\mathbb{R}^r), \quad i = 1, 2 \text{ (see §5).}$$

(g) *Let M be a compact $C^{k,\kappa}$-manifold and $l_1 < l_2 \leqslant k + \kappa$, $\kappa = 0, 1$. Then $W_2^{l_2}(M) \hookrightarrow W_2^{l_1}(M)$ for $l_1, l_2 \in \mathbb{R}_+$.*

Proof. We have $L_2^{l_2} \hookrightarrow L_2^{l_1}$ – see §5 – from which, by applying the Fourier transformation, (a) follows. From (a), by taking the infimum for the norm (see Definition 5.2) we obtain (b). Statement (c) follows from (b) by restriction to $\mathscr{D}(\Omega)$, and since $\mathring{W}_2^l(\Omega) \simeq \mathring{H}^l(\Omega)$ for $l \in \mathbb{R}_+$, we have proved (d) also. (e) is trivial; we have only to write down the norms. (f): because of the assumed extension property, by Theorem 5.3, we have $W_2^{l_i}(\Omega) \simeq H^{l_i}(\Omega)$, $i = 1, 2$, so that (b) implies statement (f). (g) follows from (d), see the definition (4.4) of the norms on $W_2(M)$. ■

We now prove Sobolev's lemma for the spaces $W_2^l(\Omega)$, $l \in \mathbb{R}_+$; for this we have to impose geometric conditions on the region Ω. Here we require that Ω has the segment property (Definition 2.1) and the cone property (Definition 2.2). Both conditions follow, for example, from the uniform cone condition (Definition 2.3) or, if Ω is bounded, from condition $N^{0,1}$, see Theorem 2.1. If Ω is bounded it is enough to assume no more than the cone condition, see

Gagliardo's theorem (2.3), and to vary the proof of Sobolev's lemma (6.2) slightly (we work not with a single Ω but with several Ω_i).

First we prove Sobolev's lemma for integral $l \in \mathbb{N}$, therefore $l = 1, 2, \ldots$. As just remarked, in this case very weak assumptions on the region are sufficient, see Theorem 6.2 and the related results.

Theorem 6.2 (Sobolev's lemma) *Let the region Ω have the segment property, satisfy the cone condition and let $l > r/2$, $l \in \mathbb{N}$. Then the elements of $W_2^l(\Omega)$ are bounded continuous functions on $\bar{\Omega}$, and the embedding*

$$W_2^l(\Omega) \subsetneq C(\bar{\Omega}) \tag{1}$$

is continuous.

Remark The statement that elements from $W_2^l(\Omega)$ are bounded continuous functions is of course to be understood as: each element from $W_2^l(\Omega)$ represents an equivalence class of functions which agree up to sets of measure zero, and inside each class we can find a bounded, continuous function.

Proof. Since Ω has the segment property, the functions φ from $W_2^l(\Omega) \cap C^\infty(\bar{\Omega})$ are dense in $W_2^l(\Omega)$ by Theorem 3.6, and for these functions first we prove the inequality

$$\sup_{x \in \Omega} |\varphi(x)| \leqslant C \|\varphi\|_l. \tag{2}$$

Let $x \in \bar{\Omega}$ and $C_x \subset \Omega$ be the cone from the cone condition. This cone is a subset of the ball $B(x, R)$

$$C_x \subset B(x, R), \tag{3}$$

and from the surface of this ball it cuts out the defining subset Σ with surface measure $\operatorname{mes} \Sigma > 0$, see the remarks associated with the cone construction preceding Definition 2.2.

Let $e_R(y) = f_{2/R}(y - x)$, where f_ε, $\varepsilon = 2/R$, is the function from the proof of Theorem 3.2. We have

$$e_R(y) = \begin{cases} 1 & \text{for } |y - x| < R/2, \\ 0 & \text{for } |y - x| \geqslant R, \end{cases}$$

and

$$|D^s e_R(y)| \leqslant \frac{M}{R^{|s|}}, \quad |s| \leqslant l, l \text{ fixed}. \tag{4}$$

Let $\varphi \in W_2^l(\Omega) \cap C^\infty(\bar{\Omega})$ and $x \in \bar{\Omega}$. We consider, as above, the balls $B(x, R)$ and $B(x, R/2)$ about x, and integrate along the ray $y = x + \mathbf{n}\rho$ through x, where **n** is a unit vector. We have, see the definition of $e_R(y)$,

$$\varphi(x) = -e_R(x+\mathbf{n}\rho)\varphi(x+\mathbf{n}\rho)|_{\rho=0}^{\rho=R} = -\int_0^R \frac{\partial(e_R\varphi)}{\partial\rho}\,\mathrm{d}\rho,$$

where in what follows we omit the argument of $e_R\varphi$. By partial integration we further obtain

$$\varphi(x) = \cdots = \frac{(-1)^l}{(l-1)!}\int_0^R \rho^{l-1}\frac{\partial^l(e_R\varphi)}{\partial\rho^l}\,\mathrm{d}\rho.$$

If we integrate over Σ on the surface of the ball $\overline{B(x,R)}$, we obtain, after application of the Schwarz inequality

$$\begin{aligned}|\mathrm{mes}\,\Sigma\cdot\varphi(x)| &= \frac{1}{(l-1)!}\left|\int_{C_x}\frac{\rho^{l-r}\cdot\partial^l(e_R\varphi)}{\partial\rho^l}\,\mathrm{d}y\right| \\ &\leqslant \frac{1}{(l-1)!}\left[\int_{C_x}\rho^{2(l-r)}\,\mathrm{d}y\right]^{1/2}\left[\int_{C_x}\left|\frac{\partial^l(e_R\varphi)}{\partial\rho^l}\right|^2\mathrm{d}y\right]^{1/2} \qquad (5)\\ &\leqslant \frac{1}{(l-1)!}\left[\int_{C_x}\rho^{2(l-r)}\,\mathrm{d}y\right]^{1/2}\cdot\|e_R\varphi\|_{l,\Omega}\end{aligned}$$

where the last inequality follows from the fact that $\bar{C}_x \subset \bar{\Omega}$ (cone condition). By Theorem 2.9 the frontier $\partial\Omega$ of Ω has measure zero, therefore in (5) we can replace $\|e_R\varphi\|_{l,\bar{\Omega}}$ by $\|e_R\varphi\|_{l,\Omega}$. Since the cone $\bar{C}_x$ is contained in the ball $\overline{B(x,R)}$ by (3) we have further

$$\begin{aligned}\int_{\bar{C}_x}\rho^{2(l-r)}\,\mathrm{d}y &= \int_{\bar{C}_x}|y-x|^{2(l-r)}\,\mathrm{d}y \leqslant \int_{\overline{B(x,R)}}|y-x|^{2(l-r)}\,\mathrm{d}y \\ &= \int_{\overline{B(0,R)}}|y|^{2(l-r)}\,\mathrm{d}y = \frac{2\pi^{r/2}R^{2(l-r)+r}}{\Gamma\left(\frac{r}{2}\right)\cdot(2(l-r)+r)} =: BR^{2l-r}. \qquad (6)\end{aligned}$$

If we apply the Leibniz product rule to $\|e_R\varphi\|_l$ in (5) and refer back to (4) we obtain

$$\|e_R\varphi\|_{l,\Omega} \leqslant C\max_{|s|\leqslant l}\frac{1}{R^{|s|}}\cdot\|\varphi\|_{l,\Omega}. \qquad (7)$$

We substitute (6) and (7) in (5) and have

$$|\mathrm{mes}\,\Sigma\cdot\varphi(x)| \leqslant C'\cdot R^{l-(r/2)}\cdot\max_{|s|\leqslant l}\frac{1}{R^{|s|}}\|\varphi\|_l,$$

which is (2). Since $W_2^l(\Omega)\cap C^\infty(\bar{\Omega})$ is dense in $W_2^l(\Omega)$ by continuous extension we deduce from (2) that

$$\|\varphi\|_{C(\Omega)} \leqslant c\|\varphi\|_{l,\Omega} \quad \text{for all } \varphi \in W_2^l(\Omega), \tag{8}$$

which is (1). ■

By (8) the continuous extension property also takes place in the sup norm, and therefore the limit functions obtained by the extension process are again bounded, continuous functions.

Corollary 6.1 *Let the region Ω have the segment property, satisfy the cone condition, and let $l_1 - l_2 > r/2$. Then the elements from $W_2^{l_1}(\Omega)$ are l_2-fold continuous differentiable functions, and the embedding*

$$W_2^{l_1}(\Omega) \subsetneq C^{l_2}(\bar{\Omega}) \tag{9}$$

is continuous.

Proof. If $\varphi \in W_2^{l_1}(\Omega)$, then $D^s\varphi \in W_2^{l_1-|s|}(\Omega)$ for $|s| \leqslant l_1$, and by Sobolev's lemma $D^s\varphi$ is continuous on $\bar{\Omega}$ for $|s| \leqslant l_2$. Furthermore by (8)

$$\sup_{x\in\bar{\Omega}} |D^s\varphi(x)| \leqslant C\|D^s\varphi\|_{l_1-|s|} \leqslant C\cdot\|\varphi\|_{l_1},$$

for all $|s| \leqslant l_2$, that is, the inclusion (9) is continuous. ■

Sobolev's lemma holds for the space $\mathring{W}_2^l(\Omega)$ without any restrictions on the region Ω.

Corollary 6.2 *Let $\Omega \subset \mathbb{R}^r$ be an arbitrary open set, and let $l_1 - l_2 > r/2$. Then elements from $\mathring{W}_2^{l_1}(\Omega)$ are l_2-fold continuous differentiable functions, and the inclusion*

$$\mathring{W}_2^{l_1}(\Omega) \subsetneq \mathring{C}^{l_2}(\bar{\Omega})$$

is continuous.

Here $\mathring{C}^{l_2}(\bar{\Omega})$ is the space of all continuous functions with continuous bounded derivatives up to order l_2, which together with their derivatives up to order l_2 take the value zero on the frontier $\partial\Omega$, that is, $\mathring{C}^{l_2}(\bar{\Omega}) = \overline{\mathscr{D}(\Omega)}^{C^{l_2}}$.

Proof. The set $\Omega = \mathbb{R}^r$ satisfies all cone conditions and by Corollary 6.1 we have

$$\|\varphi\|_{C^{l_2}(\mathbb{R}^r)} \leqslant c\|\varphi\|_{l_1} \quad \text{for } \varphi \in W_2^{l_1}(\mathbb{R}^r) = \mathring{W}_2^{l_1}(\mathbb{R}^r) \tag{10}$$

Therefore (10) is also valid for all functions $\varphi \in \mathscr{D}(\Omega)$ and we can now write (10) as

$$\|\varphi\|_{C^{l_2}(\Omega)} \leqslant C\|\varphi\|_{l_1}, \quad \varphi \in \mathscr{D}(\Omega). \tag{11}$$

By continuous extension it follows from (11) that

$$\|\varphi\|_{\mathring{C}^{l_2}(\Omega)} \leqslant C \|\varphi\|_{W_2^{l_1}(\Omega)}, \varphi \in \mathring{W}_2^{l_1}(\Omega)$$

which is the assertion of the corollary.

We want now to prove Sobolev's lemma for all $l \in \mathbb{R}_+$.

Theorem 6.3 *Let $l > r/2$. Then functions from $W_2^l(\mathbb{R}^r)$ are continuous, perhaps after change on a set of measure zero, belong to $\mathring{C}(\mathbb{R}^r)$, and the inclusion*

$$W_2^l(\mathbb{R}^r) \subsetneq \mathring{C}(\mathbb{R}^r) \tag{12}$$

is continuous. Here we have put $\mathring{C}(\mathbb{R}^r) = \overline{\mathscr{D}(\mathbb{R}^r)}^{C(\mathbb{R}^r)}$, similarly for the spaces $\mathring{C}^k(\mathbb{R}^r)$, $\mathring{C}^k(\bar{\Omega})$, etc.

Proof. As in the previous results it is enough to prove (12) for a dense subset in W_2^l. By Theorem 5.1 $\mathscr{D}$ is dense in $W_2^l(\mathbb{R}^r) \simeq H^l$, and we have now to prove the estimate

$$\sup_x |\varphi(x)| \leqslant c \|\varphi\|_l \quad \text{for } \varphi \in \mathscr{D}(\mathbb{R}^r). \tag{13}$$

By Theorem 5.2 $W_2^l(\mathbb{R}^r) \simeq H^l$ for $l \in \mathbb{R}_+$ and we can use the Fourier transformation for the proof of (13). By Theorem 1.21, if $\varphi \in \mathscr{D}$, $x \in \mathbb{R}^r$,

$$\varphi(x) = \frac{1}{(2\pi)^r} \int \hat{\varphi}(\xi) e^{i\langle x, \xi \rangle} \, d\xi,$$

and the Schwarz inequality yields

$$\begin{aligned} |\varphi(x)| &\leqslant \frac{1}{(2\pi)^r} \int |\hat{\varphi}(\xi)| (1 + |\xi|^2)^{l/2} \frac{1}{(1 + |\xi|^2)^{l/2}} \, d\xi \\ &\leqslant \frac{1}{(2\pi)^r} \|\varphi\|_{H^l} \left[\int \frac{d\xi}{(1 + |\xi|^2)^l} \right]^{1/2} \leqslant c \|\varphi\|_{W_2^l}. \end{aligned} \tag{14}$$

Since we assume that $l > r/2$ the integral $\int [1/(1 + |\xi|^2)^l] \, d\xi$ converges. With (14) we have proved (13), from which (12) follows. ■

With little effort we can prove the converse of Theorem 6.3: the condition $l > r/2$ is necessary for the validity of (12). Indeed from (12) it follows (again we use Theorem 5.2: $W_2^l \simeq H^l$) that

$$\|\varphi(0)\| \leqslant c \|\varphi\|_{H^l},$$

that is, the functional $\delta: \varphi \mapsto \varphi(0)$ is continuous, $\delta: H^l \to \mathbb{C}$, that is, $\delta \in (H^l)'$. However, it is known that $(H^l)' = H^{-l} = \mathscr{F}(L_2^{-l})$, see Volevič & Panejach [1], and once more we use the spaces L_2^l introduced in the proof of Theorem 5.1, obtaining $\mathscr{F}\delta = 1 \in L_2^{-l}$ or $(1+|\xi|^2)^{-l/2} \in L_2$. This is the case if and only if $l > r/2$.

By induction we obtain (see Corollary 6.1):

Corollary 6.3 *Let $k \in \mathbb{N}$, $l \in \mathbb{R}_+$ and $l-k > r/2$. Then the elements from $W_2^l(\mathbb{R}^r)$ are k-fold continuous differentiable functions and the inclusion*

$$W_2^l(\mathbb{R}^r) \subsetneq \mathring{C}^k(\mathbb{R}^r)$$

is continuous.

By restriction to an arbitrary open subset we obtain (here we use once more $\mathring{H}^l(\Omega) \simeq \mathring{W}_2^l(\Omega) = \overline{\mathscr{D}(\Omega)}$):

Corollary 6.4 *Let $k \in \mathbb{N}$, $l \in \mathbb{R}_+$ and $l-k > r/2$. Then the elements from $\mathring{W}_2^l(\mathbb{R}^r)$ are k-fold continuous differentiable functions and the inclusion*

$$\mathring{W}_2^l(\Omega) \subsetneq \mathring{C}^k(\bar{\Omega})$$

is continuous.

In conclusion we use an extension theorem to prove Sobolev's lemma for all $l \in \mathbb{R}_+$.

Theorem 6.4 *Let $k \in \mathbb{N}$, $l \in \mathbb{R}_+$ and $l-k > r/2$. Let Ω be bounded and (m,μ)-smooth, where $m+\mu > l$. Then elements from $W_2^l(\Omega)$ are k-fold continuous differentiable functions on $\bar{\Omega}$ and the inclusion*

$$W_2^l(\Omega) \subsetneq C^k(\bar{\Omega})$$

is continuous.

We obtain the proof from the diagram

$$W_2^l(\Omega) \underset{F_{\Omega}^{\Omega_\varepsilon}}{\longrightarrow} \mathring{W}_2^l(\Omega_\varepsilon) \subsetneq \mathring{C}^k(\bar{\Omega}_\varepsilon) \underset{R_{\Omega}^{\Omega_\varepsilon}}{\longrightarrow} C^k(\bar{\Omega}),$$

where $F_{\Omega^\varepsilon}^{\Omega}$ is the continuous extension operator from Corollary 5.2, and E^{Ω^ε} is the restriction operator from $\bar{\Omega}_\varepsilon$ to $\bar{\Omega}$ $(\bar{\Omega}_\varepsilon \supset \bar{\Omega})$.

If l is integral we can use Corollary 5.1 (the extension theorem of Calderon–Zygmund) instead of Corollary 5.2, and obtain:

Theorem 6.2 *Let $l, k \in \mathbb{N}$ and $l-k > r/2$. Let Ω be bounded and satisfy the uniform cone condition, then $W_2^l(\Omega) \subsetneq C^k(\bar{\Omega})$.*

It is possible to omit the condition $k \in \mathbb{N}$ everywhere. We have then only to prove Corollary 6.3 in the form

$$W_2^l(\mathbb{R}^r) \subsetneq \mathring{C}^{k,\kappa}(\mathbb{R}^r),$$

for $l-k-\kappa > r/2$, which was done by Slobodeckiĭ [1].

Theorem 6.4 may be translated immediately to compact $C^{m,\mu}$-manifolds.

Theorem 6.5 *Let $k \in \mathbb{N}$, $l \in \mathbb{R}_+$ and $l-k > r/2$; let M be a compact $C^{m,\mu}$-manifold, where $m+\mu \geq l$, and $\mu = 0, 1$. Then the functions from $W_2^l(M)$ are k-fold continuously differentiable, and the inclusion*

$$W_2^l(M) \subsetneq C^k(M) \tag{15}$$

is continuous.

Proof. Let $\{U_j\}$ be a (finite) atlas with the properties of Definition 2.10, let $j=1,\ldots,n$, $\Phi_j: U_j \to W^r$, the $C^{m,\mu}$-charts, and let $\{\alpha_j\}$ be an associated partition of unity. In order to obtain the embedding I of (15), we write

$$I\varphi = \sum_{j=1}^{n} I(\alpha_j \varphi), \quad \varphi \in W_2^l(M),$$

and factorise, $j=1,\ldots,n$

$$I(\alpha_j \cdot): W_2^l(M) \underset{\alpha_j}{\longrightarrow} W_2^l(U_j) \underset{*\Phi_j^{-1}}{\longrightarrow} W_2^l(W^r) \subsetneq \mathring{C}^k(\bar{W}^r) \underset{*\Phi_j}{\longrightarrow} \mathring{C}^k(\bar{U}_j) \subsetneq C^k(M).$$

By Theorems 4.1 and 6.4 each arrow in (16) is continuous, hence we have proved the continuity of (15). It is also possible to prove Theorem 6.5 on the frontier $\partial\Omega$ of a $C^{m,\mu}$-smooth bounded region Ω in $\mathbb{R}^r$, for $0<\mu<1$. ■

Exercise

6.1 Prove the various versions of Sobolev's lemma for the spaces $W_2^l(T^r)$ using the Fourier series norm

$$p(f) = \left[\sum_k (1+|k|^2)^l |c_k|^2\right]^{1/2}.$$

§7 Compact embeddings

We wish to study compact embeddings between Sobolev spaces, and proceed as in §6, that is, first we prove embedding theorems for $l \in \mathbb{N}$ under the most general hypotheses on Ω, and then consider the general case $l \in \mathbb{R}_+$. For the $\mathring{W}$-spaces we can again manage without geometric hypotheses on the frontier of Ω; Ω must only be bounded; we cannot, however, essentially weaken the boundedness hypothesis.

For partial differential equations compact embeddings play a decisive role, we need them in the proof of the main theorem §13; they make the Green solution operator compact, see §17.

Theorem 7.1 *Let $\Omega \subset \mathbb{R}^r$ be open and bounded, and let $l_2 < l_1$; $l_1, l_2 \in \mathbb{N}$. Then the embedding $\mathring{W}_2^{l_1}(\Omega) \hookrightarrow \mathring{W}_2^{l_2}(\Omega)$ is compact.*

Proof. We must show that the unit ball of $\mathring{W}_2^{l_1}(\Omega)$ is relatively compact in $\mathring{W}_2^{l_2}(\Omega)$. To this end we use the Kolmogorov compactness criterion from §1.1. Let $\varphi \in \mathscr{D}(\Omega)$ with $\|\varphi\|_{l_1} \leqslant 1$ and $|s| \leqslant l_1 - 1$. We extend the function φ by zero to all of $\mathbb{R}^r$. Then for $[x, x+t]$ we have

$$\mathrm{D}^s\varphi(x+t) - \mathrm{D}^s\varphi(x) = \int_0^1 \frac{\partial}{\partial \tau} \mathrm{D}^s\varphi(x+\tau t)\,\mathrm{d}\tau, \tag{1}$$

Therefore

$$R := \int_\Omega |\mathrm{D}^s\varphi(x+t) - \mathrm{D}^s\varphi(x)|^2\,\mathrm{d}x = \int_\Omega \left| \int_0^1 t\cdot \operatorname{grad} \mathrm{D}^s\varphi(x+\tau t)\,\mathrm{d}\tau \right|^2 \mathrm{d}x.$$

We estimate by means of the Schwarz inequality and obtain

$$\begin{aligned} R &\leqslant \int_\Omega \left[\int_0^t |t|\,|\operatorname{grad} \mathrm{D}^s\varphi(x+\tau t)|\,\mathrm{d}\tau \right]^2 \mathrm{d}x \\ &\leqslant |t|^2 \int_0^1 \left(\int_\Omega |\operatorname{grad} \mathrm{D}^s\varphi(x+\tau t)|^2\,\mathrm{d}x \right) \mathrm{d}\tau \\ &\leqslant |t|^2 \int_0^1 \left(\int_\Omega \sum_{|s| \leqslant l_1} |\mathrm{D}^s\varphi(x+\tau t)|^2\,\mathrm{d}x \right) \mathrm{d}\tau \\ &\leqslant |t|^2 \int_0^1 \left[\sum_{|s| \leqslant l_1} \int_\Omega |\mathrm{D}^s\varphi(y)|^2\,\mathrm{d}y \right] \mathrm{d}\tau = |t|^2 \|\varphi\|_{l_1}^2 \leqslant |t|^2. \end{aligned}$$

Because by Definition 3.2 $\mathscr{D}(\Omega)$ is dense in $\mathring{W}_2^{l_1}(\Omega)$, we obtain that the estimate

$$\int_\Omega |\mathrm{D}^s\varphi(x+t) - \mathrm{D}^s\varphi(x)|^2\,\mathrm{d}x \leqslant |t|^2, \tag{2}$$

is valid for all elements of the set

$$M := \{\mathrm{D}^s\varphi : \varphi \in \mathring{W}_2^{l_1}(\Omega), \quad \|\varphi\|_{l_1} \leqslant 1, |s| \leqslant l_1 - 1\}.$$

Because $\|\mathrm{D}^s\varphi\|_0 \leqslant \|\varphi\|_{l_1} \leqslant 1$ the set M is bounded in $L_2(\Omega)$, this together with (2) shows that the conditions of Kolmogorov's Theorem in §1 are satisfied, and hence that M is relatively compact in $L_2(\Omega)$. A choice of subsequence concludes the proof: let $\{\varphi_n\}$ be a sequence in $\mathring{W}_2^{l_1}(\Omega)$ with $\|\varphi_n\|_{l_1} \leqslant 1$. Since each element of the sequences $\{D^s\varphi_n\}$, $0 \leqslant |s| \leqslant l_2$ belongs to the relatively compact set M, it is possible to find a subsequence $\{\varphi_{n_k}\}$, such that each of the sequences $\{D^s\varphi_{n_k}\}$, $0 \leqslant |s| \leqslant l_2$ converges in $L_2(\Omega)$, which is equivalent to convergence in $\mathring{W}_2^{l_2}(\Omega)$. ■

The extension theorems proved in §5 make it possible to prove compact embedding results also for the spaces $W_2^l(\Omega)$, without the little circle above.

Theorem 7.2 *Let Ω be bounded and satisfy the uniform cone condition (or $\Omega \in N^{0,1}$), and let $l_2 < l_1, l_1, l_2, \in \mathbb{N}$. Then the embedding $W_2^{l_1}(\Omega) \subsetneq W_2^{l_2}(\Omega)$ is compact.*

Proof. We consider the diagram

$$W_2^{l_1}(\Omega) \xrightarrow[F_\Omega^\varepsilon]{} \mathring{W}_2^{l_1}(\Omega_\varepsilon) \subsetneq \mathring{W}_2^{l_2}(\Omega_\varepsilon) \xrightarrow[R_\Omega^{\Omega_\varepsilon}]{} W_2^{l_2}(\Omega).$$

By Corollary 5.1 the extension operator F_Ω^ε is continuous; Ω_ε is bounded, by Theorem 7.1 the embedding $\subsetneq$ is compact, and the restriction operator $R_\Omega^{\Omega_\varepsilon}(\Omega_\varepsilon \supset \Omega)$ is trivially continuous. The composite map, which is $W_2^{l_1}(\Omega) \subsetneq W_2^{l_2}(\Omega)$ is therefore compact. ■

We now want to prove Ehrling's lemma. First the abstract version:

Theorem 7.3 *Let X_1, X_2, X_3 be normed spaces, $A: X_1 \to X_2$ compact, $T: X_2 \to X_3$ a continuous injection. Then for each $\varepsilon > 0$ there exists a constant $c(\varepsilon)$ with*

$$\|AX\|_2 \leqslant \varepsilon \|x\|_1 + c(\varepsilon)\|TAx\|_3 \quad \textit{for all } x \in X_1. \tag{3}$$

Proof. We assume that (3) fails for some $\varepsilon_0 > 0$. Then for each $n \in \mathbb{N}$ there exists some $x_n \in X_1$ (without loss of generality let $\|x_n\|_1 = 1$) with

$$\|Ax_n\|_2 > \varepsilon_0 + n\|TAx_n\|_3. \tag{4}$$

Since A is continuous we have $\|Ax_n\|_2 \leqslant \|A\|$ for all $n \in \mathbb{N}$. Therefore we have

$$\|TAx_n\|_3 < \frac{\|A\|}{n} - \frac{\varepsilon_0}{n},$$

or

$$TAx_n \to 0 \quad \text{as } n \to \infty. \tag{5}$$

On the other hand by assumption $\{Ax_n\}$ is relatively compact. Hence there exists a subsequence, we again denote it by $\{Ax_n\}$, which converges to some element y. Since T is continuous we have $TAx_n \to Ty$ as $n \to \infty$, which in conjunction with (5) gives $Ty = 0$. T was assumed to be an injection, so y must equal zero. Therefore we have

$$Ax_n \to 0, \tag{6}$$

and from (4) we obtain $\|Ax_n\|_2 > \varepsilon_0$, contradicting (6). ■

Theorem 7.4 (Ehrling's lemma). *Let Ω be bounded (and satisfy the uniform cone condition). Then for each $\varepsilon > 0$ there exists a constant $c(\varepsilon)$, so that for all $\varphi \in \mathring{W}_2^l(\Omega)$ (for all $\varphi \in W_2^l(\Omega)$) we have*

$$\|\varphi\|_{l-k} \leqslant \varepsilon\|\varphi\|_l + c(\varepsilon)\|\varphi\|_0, \quad l \geqslant k \geqslant 1.$$

The proof follows from Theorems 7.1, 7.2 and 7.3.

We can use Ehrling's lemma to introduce new equivalent norms on the space $\mathring{W}_2^l(\Omega)$ (respectively $W_2^l(\Omega)$). We want to formulate the appropriate theorem for $\mathring{W}_2^l(\Omega)$, it holds also for $W_2^l(\Omega)$ under additional assumptions, for example when Ω satisfies the uniform cone condition, $l \in \mathbb{N}$.

Theorem 7.5 *Let Ω be bounded and $l \in \mathbb{N}$. The norms defined by the equations*

$$\|\varphi\|_l^2 = \sum_{|s| \leqslant l} \|D^s\varphi\|_0^2$$

and

$$\|\varphi\|_{l,1}^2 := \|\varphi\|_0^2 + \sum_{|s| = l} \|D^s\varphi\|_0^2$$

on $\mathring{W}_2^l(\Omega)$ are equivalent, $l = 1, 2, \ldots$.

Proof. It is clear that $\|\varphi\|_{l,1} \leqslant \|\varphi\|_l$ for all $\varphi \in \mathring{W}_2^l(\Omega)$. By Theorem 7.4 for $\varepsilon = \frac{1}{2}$ there exists some $c = c(\frac{1}{2})$ with

$$\|\varphi\|_{l-1} \leqslant \tfrac{1}{2}\|\varphi\|_l + c\|\varphi\|_0,$$

respectively

$$\begin{aligned}\|\varphi\|_{l-1}^2 &\leqslant 2(\tfrac{1}{4}\|\varphi\|_l^2 + c^2\|\varphi\|_0^2)\\ &= \tfrac{1}{2}\|\varphi\|_{l-1}^2 + \frac{1}{2}\sum_{|s|=1}\|D^s\varphi\|_0^2 + 2c^2\|\varphi\|_0^2,\end{aligned}$$

respectively

$$\|\varphi\|_{l-1}^2 \leqslant 4c^2\|\varphi\|_0^2 + \sum_{|s|=l}\|D^s\varphi\|_0^2.$$

Therefore we obtain

$$\begin{aligned}\|\varphi\|_l^2 = \|\varphi\|_{l-1}^2 + \sum_{|s|=l}\|D^s\varphi\|_0^2 &\leqslant 4c^2\|\varphi\|_0^2 + 2\sum_{|s|=l}\|D^s\varphi\|_0^2\\ &\leqslant \max(2, 4\cdot c^2)\|\varphi\|_{l,1}^2 \quad \text{for all } \varphi \in \mathring{W}_2^l(\Omega).\end{aligned}$$ ■

As a corollary to Theorem 7.5 we note that each other norm $\| \quad \|_l'$ on $\mathring{W}_2^l(\Omega)$ with

$$c_1 \|\varphi\|_{l,1} \leqslant \|\varphi\|_l' \leqslant c_2 \|\varphi\|_l, \quad \varphi \in \mathring{W}_2^l(\Omega),$$

is also equivalent to the $\| \quad \|_l$-norm on $\mathring{W}_2^l(\Omega)$. ■

Theorem 7.6 (first Poincaré inequality) *Let Ω be bounded and $l = 1, 2, \ldots$. Then there exists a constant c dependent only on the diameter of Ω, such that for all $\varphi \in \mathring{W}_2^l(\Omega)$*

$$\|\varphi\|_l^2 \leqslant c \sum_{|s|=l} \int_\Omega |D^s\varphi(x)|^2 \, dx, \tag{7}$$

that is, the square root of the right-hand side of (7) *is an equivalent norm on* $\mathring{W}_2^l(\Omega)$.

Proof. We prove (7) in the case $l = 1$. The general case then follows by induction.

By integration by parts we obtain the following estimate for $\varphi \in \mathscr{D}(\Omega)$:

$$\|\varphi\|_0^2 = \frac{1}{r} \sum_{i=1}^r \int_\Omega |\varphi(x)|^2 \cdot 1 \cdot dx = -\frac{1}{r} \sum_{i=1}^r \int_\Omega \frac{\partial(\varphi(x) \cdot \overline{\varphi(x)})}{\partial x_i} x_i \, dx$$

$$= -\frac{1}{r} \sum_{i=1}^r \int_\Omega \frac{\partial \varphi}{\partial x_i} \cdot \overline{\varphi(x)} \cdot x_i \cdot dx - \frac{1}{r} \sum_{i=1}^r \int_\Omega \varphi(x) \frac{\partial \bar{\varphi}}{\partial x_i} x_i \, dx.$$

Let $\Omega \subset \{x : |x_i| \leqslant d, \ i = 1, \ldots, r\}$, we apply the Schwarz inequality:

$$\|\varphi\|_0^2 \leqslant \frac{2}{r} \sum_{i=1}^r \int_\Omega \left| \frac{\partial \varphi}{\partial x_i} \right| \cdot |\varphi| \cdot |x_i| dx \leqslant \frac{2d}{r} \|\varphi\|_0 \sum_{i=1}^r \left(\int_\Omega \left| \frac{\partial \varphi}{\partial x_i} \right|^2 dx \right)^{1/2}.$$

After manipulation this gives

$$\|\varphi\|_0 \leqslant \frac{2d}{\sqrt{r}} \left[\sum_{i=1}^r \int_\Omega \left| \frac{\partial \varphi}{\partial x_i} \right|^2 dx \right]^{1/2}.$$

From this finally follows

$$\|\varphi\|_1^2 = \|\varphi\|_0^2 + \sum_{|s|=1} \|D^s\varphi\|_0^2 \leqslant \left(\frac{4d^2}{r} + 1 \right) \sum_{|s|=1} \|D^s\varphi\|_0^2,$$

which is (7) for $\varphi \in \mathscr{D}(\Omega)$. The usual density inference concludes the proof. ■

Remark The first Poincaré inequality also holds in regions which are bounded in only one direction. For example suppose $|x_1| \leqslant d$, then we have

$$\|\varphi\|_0^2 = \int_\Omega |\varphi(x)|^2 \cdot 1 \cdot \mathrm{d}x = -\int_\Omega \frac{\partial(\varphi\bar{\varphi})}{\partial x_1} \cdot x_1 \cdot \mathrm{d}x \leqslant 2d \cdot \|\varphi\|_0 \cdot \left[\int_\Omega \left|\frac{\partial \varphi}{\partial x_1}\right|^2 \mathrm{d}x\right]^{1/2},$$

from which the further proof, that is, (7), follows.

Corollary 7.1 *Let Ω be bounded and $l = 1, 2, \ldots$, then $\mathring{W}_2^l(\Omega) \neq W_2^l(\Omega)$.*

Proof. By (7) the function $\varphi \equiv 1$ does not belong to $\mathring{W}_2^l(\Omega)$. ■

Theorem 7.7 (second Poincaré inequality) *Let Ω be bounded and satisfy the uniform cone condition, or $\Omega \in N^{0,1}$. Then for all $\varphi \in W_2^l(\Omega)$, $l = 1, 2, \ldots$ we have the inequality*

$$\|\varphi\|_l^2 \leqslant c\left[\sum_{|s|=l} \int_\Omega |\mathrm{D}^s\varphi|^2 \mathrm{d}x + \sum_{|s|<l} \left|\int_\Omega \mathrm{D}^s\varphi \, \mathrm{d}x\right|^2\right]. \tag{8}$$

Proof. We assume that (8) is not correct, then there exists a sequence $\varphi_n \in W_2^l(\Omega)$ with

$$\|\varphi_n\|^l = 1, \tag{9}$$

and

$$1 = \|\varphi_n\|_l^2 > n\left[\sum_{|s|=l} \|\mathrm{D}^s\varphi_n\|_0^2 + \sum_{|s|<l} \left|\int_\Omega \mathrm{D}^s\varphi_n \mathrm{d}x\right|^2\right],$$

from which it follows that

$$\mathrm{D}^s\varphi_n \to 0 \text{ in } L^2(\Omega) \quad \text{for } |s| = l. \tag{10}$$

By Theorem 7.2 $\{\varphi_n\}$ is relatively compact in $W_2^{l-1}(\Omega)$, therefore there exists a subsequence, for the sake of simplicity we again denote if by $\{\varphi_n\}$, with $\varphi_n \to \varphi$ in $W_2^{l-1}(\Omega)$, which together with (10) gives (use Cauchy sequences!) $\varphi_n \to \varphi$ in $W_2^l(\Omega)$. Therefore by (10) $\mathrm{D}^s\varphi \equiv 0$ for $|s| = l$ and φ is a polynomial of degree $\leqslant l-1$. Apply (9) once more, which gives

$$\lim_{n+\infty} \int_\Omega \mathrm{D}^s\varphi_n \mathrm{d}x = \int_\Omega \mathrm{D}^s\varphi \, \mathrm{d}x = 0$$

for $|s| < l$. The polynomial φ must therefore vanish identically, $\varphi \equiv 0$, contradicting (9), $1 = \lim \|\varphi_n\|_l = \|\varphi\|_l$. ■

We next give compact embeddings for $l \in \mathbb{R}_+$, that is, for the Slobodeckiĭ spaces.

Theorem 7.8 *Let $\Omega \subset \mathbb{R}^r$ be open and bounded, and let $l_2 < l_1$; $l_1, l_2 \in \mathbb{R}_+$. Then the embedding $\mathring{W}_2^{l_1}(\Omega) \subsetneq \mathring{W}_2^{l_2}(\Omega)$ is compact.*

Proof. We need a lemma.

Lemma 7.1 *The following inequality holds for all* $\xi, \eta \in \mathbb{R}^r$:

$$(1+|\xi|^2)^l \leqslant 2^l(1+|\xi-\eta|^2)^l(1+|\eta|^2)^l. \tag{11}$$

Indeed for $x, y \in \mathbb{R}^r$ we have

$$1+|x+y|^2 \leqslant 1+|x|^2+2|x||y|+|y|^2 \leqslant 2(1+|x|^2)(1+|y|^2).$$

Substituting $x=\xi-\eta$, $y=\eta$ gives

$$(1+|\xi|^2) \leqslant 2(1+|\xi-\eta|^2)(1+|\eta|^2),$$

which on taking powers leads to (11).

For the proof of Theorem 7.8 we again use the Fourier transformation, by Theorem 5.2 for $l \in \mathbb{R}_+$ we have

$$\mathring{W}^l_2(\Omega) \simeq \mathring{H}^l(\Omega).$$

We must show that each sequence $\varphi_n \in \mathring{W}^{l_1}_n(\Omega)$ with $\|\varphi_n\|_{l_1} \leqslant 1$ possesses a convergent subsequence in $\mathring{W}^{l_2}_2(\Omega)$, again denoted by φ_n. Let $\alpha \in \mathscr{D}(\mathbb{R}^r)$ (with compact support) be so chosen that $\alpha = 1$ in $\bar{\Omega}$. Because $\operatorname{supp}\varphi_n \subset \bar{\Omega}$ we have $\varphi_n = \alpha \cdot \varphi_n$, which after Fourier transformation, Theorem 1.22, gives

$$\hat{\varphi}_n(\xi) = \frac{1}{(2\pi)}r \int \hat{\alpha}(\xi-\eta)\hat{\varphi}_n(\eta)\,\mathrm{d}\eta. \tag{12}$$

We multiply by $(1+|\xi|^2)^{l_1/2}$ and obtain

$$\hat{\varphi}_n(\xi)(1+|\xi|^2)^{l_1/2} = \frac{1}{(2\pi)^r}\int \hat{\alpha}(\xi-\eta)\hat{\varphi}_n(\eta)(1+|\xi|^2)^{l_1/2}\,\mathrm{d}\eta.$$

We use (11), estimate by means of the Schwarz inequality, and calculate in the Fourier norm §5(3)

$$|\hat{\varphi}_n(\xi)|(1+|\xi|^2)^{l_1/2} \leqslant \frac{2^{l_1/2}}{(2\pi)^r}\int |\hat{\alpha}(\xi-\eta)|(1+|\xi-\eta|^2)^{l_1/2}|\hat{\varphi}_n(\eta)|(1+|\eta|^2)^{l_1/2}\,\mathrm{d}\eta$$

$$\leqslant \frac{2^{l_1/2}}{(2\pi)^r}\|\alpha\|_{l_1}\cdot\|\varphi_n\|_{l_1} \leqslant \frac{2^{l_1/2}}{(2\pi)^r}\|\alpha\|_{l_1}$$

since by assumption $\|\varphi_n\|_{l_1} \leqslant 1$. By Theorem 1.23 the functions $\hat{\varphi}_n, \hat{\alpha}$ are highly differentiable, indeed even entire analytic, and from (12) we obtain in the same way for arbitrary s

$$|\mathrm{D}^s\hat{\varphi}_n(\xi)|(1+|\xi|^2)^{l_1/2} \leqslant \frac{2^{l_1/2}}{(2\pi)^r}\|\mathscr{F}^{-1}(\mathrm{D}^s\hat{\alpha})\|_{l_1}. \tag{13}$$

Let K be an arbitrary compact subset of $\mathbb{R}^r_\xi$, then from (13) it follows that

$$\sup_{\xi\in K}|D^s\hat{\varphi}_n(\xi)| \leqslant c \sup_{\xi\in K}(1+|\xi|^2)^{-l_1/2},$$

that is, the sequence $\hat{\varphi}_n(\xi)$ is uniformly bounded and uniformly continuous on each compact subset K. Therefore we can find a subsequence $\hat{\varphi}_{m_n}(\xi)$, which converges uniformly on K. By exhausting $\mathbb{R}^r_\xi$ by compact subsets K_m ($\mathbb{R}^r_\xi = \bigcup_{m=1}^\infty K_m$, $K_{m+1} \supset K_m$), and by a diagonal procedure we obtain a subsequence $\hat{\varphi}_{n_n}(\xi)$ which converges on $\mathbb{R}^r_\xi$, and uniformly on each K. For the sake of brevity we again call this subsequence $\hat{\varphi}_n(\xi)$. Let $\varepsilon > 0$, and choose K so that

$$\frac{(1+|\xi|^2)^{l_2}}{(1+|\xi|^2)^{l_1}} \leqslant \varepsilon^2 \quad \text{for } \xi\in\mathbb{R}^r\backslash K, \tag{14}$$

here we have used the assumption that $l_1 > l_2$. (14) leads to the estimate

$$\begin{aligned}
&\|\varphi_n - \varphi_m\|^2_{l_2}\\
&= \int_K |\hat{\varphi}_n - \hat{\varphi}_m|^2(1+|\xi|^2)^{l_2}\,d\xi + \int_{\mathbb{R}^r\backslash K} |\hat{\varphi}_n - \hat{\varphi}_m|^2(1+|\xi|^2)^{l_2}\,d\xi\\
&= \int_K |\hat{\varphi}_n - \hat{\varphi}_m|(1+|\xi|^2)^{l_2}\,d\xi + \int_{\mathbb{R}^r\backslash K} |\varphi_n - \varphi_m|^2(1+|\xi|^2)^{l_1}\cdot\frac{(1+|\xi|^2)^{l_2}}{(1+|\xi|^2)^{l_1}}\,d\xi\\
&\leqslant \max_K |\hat{\varphi}_n - \hat{\varphi}_m|^2\cdot M + \varepsilon^2\|\varphi_n - \varphi_m\|^2_{l_1} \leqslant M\cdot\max_K|\hat{\varphi}_n - \hat{\varphi}_m|^2 + 2^2\cdot\varepsilon^2,
\end{aligned}$$

which because of the uniform convergence of $\hat{\varphi}_n$ on K shows that φ_n is a Cauchy sequence in $\mathring{W}^{l_2}_2(\Omega)$. ■

Theorem 7.9 *Let $\Omega \subset \mathbb{R}^r$ be bounded and (k,κ)-smooth. Let $l_2 < l_1 \leqslant k+\kappa$; l_1, $l_2 \in \mathbb{R}_+$. Then the embedding*

$$W^{l_1}_2(\Omega) \hookrightarrow W^{l_2}_2(\Omega) \tag{15}$$

is compact.

Proof. We consider the diagram

$$W^{l_1}_2(\Omega) \xrightarrow[F^\varepsilon_\Omega]{} \mathring{W}^{l_1}_2(\Omega_\varepsilon) \hookrightarrow \mathring{W}^{l_2}_2(\Omega_\varepsilon) \xrightarrow[R^{\Omega_\varepsilon}_\Omega]{} W^{l_2}_2(\Omega). \tag{16}$$

By Corollary 5.2 F^ε_Ω is continuous, by Theorem 7.8 the injection $\hookrightarrow$ is compact and the restriction map $R^{\Omega_\varepsilon}_\Omega$ ($\Omega_\varepsilon \supset \Omega$) is trivially continuous, therefore the composition in (16), which is (15), is compact. ■

In conclusion we consider Sobolev spaces on manifolds.

Theorem 7.10 *Let M be a compact $C^{k,\kappa}$ manifold and $l_2 < l_1 \leqslant k+\kappa; l_1, l_2 \in \mathbb{R}_+$, $\kappa = 0, 1$. The embedding*

$$W_2^{l_1}(M) \subsetneq W_2^{l_2}(M) \tag{17}$$

is compact.

Proof. Let $\{U_j\}$ be a (finite) atlas with the properties of Definition 2.10, let $\Phi_j: U_j \to W^r$ be the $C^{k,\kappa}$-transformations, let $\{\alpha_j\}$ be an associated partition of unity. We decompose the embedding I, (17), into

$$I\Phi = \sum_{j=1}^{m} I(\alpha_j \varphi)$$

and factorise $I(\alpha_j \cdot)$:

$$W_2^{l_1}(M) \xrightarrow[\alpha_j \cdot]{} \mathring{W}_2^{l_1}(U_j) \xrightarrow[{}^*\Phi_j^{-1}]{} \mathring{W}_2^{l_1}(W^r) \underset{\text{compact}}{\subsetneq} \mathring{W}_2^{l_2}(W^r) \xrightarrow[{}^*\Phi_j]{} \mathring{W}_2^{l_2}(U_j) \subsetneq \mathring{W}_2^{l_2}(M)$$

Since $I(\alpha_j \cdot)$ is compact so is I. ■

Theorem 7.10 also remains true for the compact frontier $\partial\Omega$ of a $C^{k,\kappa}$-domain, where $0 < \kappa < 1$.

Given Theorems 7.8, 7.9 and 7.10 we can prove, as above, Ehrling's Theorems, and provide other equivalent norms on the W_2^l-spaces. In order to quote it later, we formulate Ehrling's theorem.

Theorem 7.11 *Let Ω be open and bounded (respectively M compact), let $0 \leqslant l_2 < l_1; l_1, l_2 \in \mathbb{R}_+$. Then for each $\varepsilon > 0$ there exists a constant $c(\varepsilon)$, such that for all $\varphi \in \mathring{W}_2^{l_1}(\Omega)$ (respectively for all $\varphi \in W_2^{l_1}(\Omega)$ under the assumptions of Theorem 7.9 or for all $\varphi \in W_2^{l_1}(M)$ under the assumptions of Theorem 7.10) we have*

$$\|\varphi\|_{l_2} \leqslant \varepsilon \|\varphi\|_{l_1} + c(\varepsilon)\|\varphi\|_0.$$

Exercises

7.1 Show that the embedding $W_2^1(\mathbb{R}) \subset L_2(\mathbb{R})$ is not compact.

7.2 Prove all the theorems of the last section for Fourier series, that is, for the spaces $W_2^l(T^r)$, where T^r is a torus.

§8 The trace operator

Let V be a submanifold of $\bar{\Omega}$, V can also be the frontier $\partial\Omega$ of Ω. We wish to define a trace operator T; it should associate a function given on $\bar{\Omega}$ with a function given on V, and in such a way that for 'sufficiently smooth' functions $\varphi \in C^m(\bar{\Omega})$ we have

$$T\varphi = \varphi|_V = \text{'}\varphi \text{ restricted to } V\text{'}. \tag{1}$$

We wish also to solve the converse problem (so-called inverse theorems): to extend a function given on V to a function defined on Ω. The importance of the trace operator and of inverse theorems for boundary value problems is clear.

We begin with $\Omega = \mathbb{R}^r$, $V = \mathbb{R}^{r-1}$, we only need the case that $\dim V = r-1$, that is, V is a hypersurface; let $\mathbb{R}^{r-1}$ be given by the coordinates $(x_1, \ldots, x_{r-1}, 0) = (x', 0)$.

We consider the general case $l \in \mathbb{R}_+$.

Theorem 8.1 *Let $l > \frac{1}{2}$. There exists a continuous linear map*

$$T_0 : W_2^l(\mathbb{R}^r) \to W_2^{l-1/2}(\mathbb{R}^{r-1}),$$

called the trace operator, *with the property* (1), *that is, for $\varphi \in \mathscr{D}(\mathbb{R}^r)$*

$$T_0\varphi = \varphi(x', 0) \quad \textit{for } (x', 0) = (x_1, \ldots, x_{r-1}, 0) \in \mathbb{R}^{r-1}. \tag{2}$$

Proof. We must prove the inequality

$$\| T_0\varphi \|_{W_2^{l-1/2}(\mathbb{R}^{r-1})} \leqslant c \| \varphi \|_{W_2^l(\mathbb{R}^r)} \quad \text{for } \varphi \in \mathscr{D}(\mathbb{R}^r), \tag{3}$$

$\mathscr{D}$ is dense in $W_2^l(\mathbb{R}^r)$. We use the Fourier transformation and the Fourier norm of §5(3). We define T_0 for $\varphi \in \mathscr{D}(\mathbb{R}^r)$ by (2)

$$T_0\varphi := \varphi(x', 0) =: g(x'), \quad x' \in \mathbb{R}^{r-1},$$

and obtain the general definition of the trace operator T_0, as usual, by continuous extension. For $\varphi \in \mathscr{D}(\mathbb{R}^r)$ we have

$$\varphi(x', 0) = \frac{1}{2\pi} \int_{-\infty}^{+\infty} \hat{\varphi}^{\xi_r}(x', \xi_r) e^{i\langle \xi_r, 0 \rangle} \, d\xi_r,$$

from which by Fourier transformation in $(x' \in) \mathbb{R}^{r-1}$ it follows that

$$\hat{g}(\xi') = \frac{1}{2\pi} \int_{-\infty}^{+\infty} \hat{\varphi}(\xi) \, d\xi_r. \tag{4}$$

We estimate, using (4) and Theorem 5.2,

$$\begin{aligned}
\| g \|_{l-1/2}^2 &\leqslant c_1 \int_{\mathbb{R}^{r-1}} |\hat{g}(\xi')|^2 (1 + |\xi'|^2)^{l-1/2} \, d\xi' \\
&= \frac{c_1}{4\pi^2} \int_{\mathbb{R}^{r-1}} \left| \int_{-\infty}^{+\infty} \hat{\varphi}(\xi) \, d\xi_r \right|^2 (1 + |\xi'|^2)^{l-1/2} \, d\xi' \\
&\leqslant c_2 \int_{\mathbb{R}^{r-1}} \left[(1 + |\xi'|^2)^{l-1/2} \int_{-\infty}^{\infty} |\hat{\varphi}(\xi)|^2 (1 + |\xi|^2)^l \, d\xi_r \right. \\
&\quad \times \left. \int_{-\infty}^{\infty} (1 + |\xi|^2)^{-l} \, d\xi_r \right] d\xi'.
\end{aligned}$$

If $l > \frac{1}{2}$ we have

$$\int_{-\infty}^{\infty} (1+|\xi|^2)^{-l} \mathrm{d}\xi_r = \pi(1+|\xi'|^2)^{-l+1/2}$$

$$= c_2\pi \int_{\mathbb{R}} \int_{-\infty}^{\infty} |\hat{\varphi}(\xi)|^2 (1+|\xi|^2)^l \mathrm{d}\xi_r \mathrm{d}\xi' \leqslant c_3 \|\varphi\|_l^2, \quad (5)$$

with which we have proved (3). ∎

Addendum 8.1 *The trace formula* (2)

$$T_0\varphi = \varphi(x', 0), \quad (x', 0) \in \mathbb{R}^{r-1},$$

is also valid for all $\varphi \in C_0^l(\mathbb{R}^r)$, *if* l *is integral, and for* $\varphi \in C_0^{[l]+1}(\mathbb{R}^r)$ *otherwise.* C_0 *means that the functions have compact support.*

Proof. We have, by examining norms and estimating integrals

$$C_0^l(\Omega) \subsetneq W_2^l(\Omega), \quad l = 0, 1, 2, \ldots$$

and

$$\tilde{C}_0^{[l],\lambda_1}(\Omega) \subsetneq W_2^{[l]+\lambda}(\Omega) \quad \text{for } \lambda_1 > \lambda_2.$$

Here the symbol $\subsetneq$ also denotes the sequential continuity of the inclusions, provided that by the convergence $\varphi_n \to \varphi$ in $C_0^l(\Omega)$ (respectively $\tilde{C}_0^{[l],\lambda_1}(\Omega)$) one understands that all φ_n, φ have their support in a fixed compact subset $K \subset\subset \Omega$ and that $\mathrm{D}^s\varphi_n \to \mathrm{D}^s\varphi$, $|s| \leqslant l$ uniformly on K (respectively in the norm of $\tilde{C}^{[l],\lambda_1}(K)$). For $\tilde{C}^{[l],\lambda_1}$ see Definition 4.2.

Suppose now that $\varphi \in C_0^{[l]+1}(\mathbb{R}^r)$, $\operatorname{supp}\varphi = K \subset\subset \mathbb{R}^r$. We regularise $\varphi_\varepsilon = \varphi * h_\varepsilon$; by Theorems 1.3 and 1.11 we have for $0 < \varepsilon < 1$ that $\operatorname{supp}\varphi_\varepsilon \subset K_\varepsilon \subset K_1 \subset\subset \mathbb{R}^r$, $\varphi_\varepsilon \in \mathscr{D}$ and

$$\varphi_\varepsilon \to \varphi \text{ for } \varepsilon \to 0 \text{ in } C_0^{[l]+1}(\mathbb{R}^r);$$

because

$$C_0^{[l]+1}(\mathbb{R}^r) \subsetneq \tilde{C}_0^{[l],\lambda_1}(\mathbb{R}^r) \subsetneq W_2^{[l]+\lambda=l}(\mathbb{R}^r),$$

φ_ε converges to φ also in $W_2^l(\mathbb{R}^r)$.

On the other hand we can, since it is a matter of uniform convergence, restrict to the set $\mathbb{R}^{r-1}$, obtaining $\varphi_\varepsilon(x', 0) \to \varphi(x', 0)$ for $\varepsilon \to 0$ in $C_0^{[l]+1}(\mathbb{R}^{r-1})$, and, as above, the inclusions

$$C_0^{[l]+1}(\mathbb{R}^{r-1}) \subsetneq \cdots \subsetneq W_2^{l-1/2}(\mathbb{R}^{r-1})$$

show that $\varphi_\varepsilon(x', 0)$ converges to $\varphi(x', 0)$ in $W_2^{l-1/2}(\mathbb{R}^{r-1})$. Continuity of the trace operator T_0 gives

$$\begin{array}{ccc} T_0\varphi_\varepsilon \to T_0\varphi & & \\ \| & & \text{in } W_2^{l-1/2}(\mathbb{R}^{r-1}), \\ \varphi_\varepsilon(x',0) \to \varphi(x',0) & & \end{array}$$

which is $T_0\varphi = \varphi(x',0)$ for $\varphi \in C_0^{[l]+1}(\mathbb{R}^r)$. ■

For integral l the proof is easier.

Addendum 8.2 *Let $\varphi \in W_2^l(\mathbb{R}^r)$ have compact support. Then $T_0\varphi \in W_2^{l-1/2}(\mathbb{R}^{r-1})$ also has compact support, and*

$$\operatorname{supp}(T_0\varphi) \subset \operatorname{supp}\varphi \cap \mathbb{R}^{r-1}.$$

Proof. We need a simple lemma on distributions.

Lemma 8.1 *Let the sequence of distribution $T_n \in \mathscr{D}'$ converge in $\mathscr{D}'$ to T, $T_n \to T$ in $\mathscr{D}'$, and suppose that* $\operatorname{supp} T_n \subset K$ *for almost all T_n, where K is closed. Then it is also true that* $\operatorname{supp} T \subset K$.

For the proof we take $\psi \in \mathscr{D}$ with $\operatorname{supp}\psi \subset \mathrm{C}K$, so that $T_n(\psi) \to 0$ or $\to T(\psi)$ as $n \to \infty$. Hence $T(\psi) = 0$, which implies that $\operatorname{supp} T \subset K$.

Now for the proof of Addendum 8.2. Let $\varphi \in W_2^l(\mathbb{R}^r)$ with $\operatorname{supp}\varphi = K$ compact in $\mathbb{R}^r$. We regularise $\varphi_\varepsilon = \varphi * h_\varepsilon$ and by Lemma 3.3 have that $\varphi_\varepsilon \to \varphi$ in $W_2^l(\mathbb{R}^r)$ and $\operatorname{supp}\varphi_\varepsilon \subset K_{\varepsilon'}$ for $\varepsilon < \varepsilon'$. Theorem 8.1 and (2) give $T_0\varphi_\varepsilon \to T_0\varphi$ in $W_2^{l-1/2}(\mathbb{R}^{r-1})$ and $\operatorname{supp}(T_0\varphi_\varepsilon) \subset K_\varepsilon \cap \mathbb{R}^{r-1}$ for $\varepsilon < \varepsilon'$. Therefore by Lemma 8.1 $\operatorname{supp}(T_0\varphi) \subset K_\varepsilon \cap \mathbb{R}^{r-1}$ for all $\varepsilon' > 0$ and after taking the intersection

$$\operatorname{supp}(T_0\varphi) \subset \bigcap_{\varepsilon' > 0} K_\varepsilon \cap \mathbb{R}^{r-1} = K \cap \mathbb{R}^{r-1}.$$ ■

We can combine Theorem 8.1 and Addendum 8.2 as:

Theorem 8.2 *Let $l > \frac{1}{2}$ and let W^r be the unit cube in $\mathbb{R}^r$. The trace operator acts linearly and continuously*

$$T_0 : \mathring{W}_2^l(W^r) \to \mathring{W}_2^{l-1/2}(W^{r-1}),$$

and (2) *holds for all $\varphi \in C_0^l(W^r)$, respectively all $\varphi \in C_0^{[l]+1}(W^r)$.*

We wish to solve the inverse problem.

Theorem 8.3 *Let $l > \frac{1}{2}$. There exists a continuous linear extension operator*

$$Z_0 : W_2^{l-1/2}(\mathbb{R}^{r-1}) \to W_2^l(\mathbb{R}^r)$$

with the property

$$T_0 \circ Z_0 \varphi = \varphi \quad \textit{for all } \varphi \in W_2^{l-1/2}(\mathbb{R}^{r-1}). \tag{6}$$

With this the trace operator T_0 from Theorem 8.1 is onto.

Proof. Since by Theorem 5.1 $\mathscr{D}(\mathbb{R}^{r-1})$ is dense in $W_2^{l-1/2}(\mathbb{R}^{r-1})$, we need only define $Z_0\varphi$ for $\varphi \in \mathscr{D}(\mathbb{R}^{r-1})$ and show (6) and continuity for these φs. The usual density argument concludes the proof.

Suppose therefore that $\varphi(x') \in \mathscr{D}(\mathbb{R}^{r-1})$, $x' \in \mathbb{R}^{r-1}$. We put

$$Z_0\varphi := u(x) := \frac{1}{(2\pi)^{r-1}c_l} \int_{\mathbb{R}^r} e^{i(x,\xi)} \frac{(1+|\xi'|^2)^{l-1/2}}{(1+|\xi|^2)^l} \hat{\varphi}(\xi')\, d\xi, x \in \mathbb{R}^r. \tag{7}$$

Formula (5) and the fact that $\varphi(\xi')$ is a strongly decreasing function show that the integral on the right-hand side of (7) exists; $u(x)$ is therefore a well-defined function. For $x = (x', 0) \in \mathbb{R}^{r-1}$ we have because of (5)

$$\begin{aligned} u(x', 0) &= \frac{1}{(2\pi)^{r-1}c_l} \int_{\mathbb{R}^{r-1}} e^{i(x',\xi')}(1+|\xi'|^2)^{l-1/2}\hat{\varphi}(\xi') \int_{-\infty}^{\infty} \frac{d\xi_r}{(1+|\xi|^2)^l}\, d\xi' \\ &= \frac{1}{(2\pi)^{r-1}} \int_{\mathbb{R}^{r-1}} e^{i(x',\xi')}\hat{\varphi}(\xi')\, d\xi' = \varphi(x'), \end{aligned}$$

with which we have checked (6).

Next the continuity condition (see (7) and Theorem 5.2)

$$\begin{aligned} \|u\|_l^2 &\leqslant c_1 \int_{\mathbb{R}^r} |\hat{u}(\xi)|^2(1+|\xi|^2)^l\, d\xi \\ &= c_2 \int_{\mathbb{R}^{r-1}} (1+|\xi'|^2)^{l-1}|\hat{\varphi}(\xi')|^2 \int_{-\infty}^{\infty} \frac{d\xi r}{(1+|\xi|^2)^l}\, d\xi' \qquad (8) \\ &= c_3 \int_{\mathbb{R}^{r-1}} (1+|\xi'|^2)^{l-1/2}|\hat{\varphi}(\xi')|^2\, d\xi' \leqslant c_4 \|\varphi\|_{l-1/2}^2. \end{aligned}$$

■

We again wish to localise (see Theorem 8.2). Let $\varepsilon > 0$ and let $\alpha_\varepsilon \in \mathscr{D}(W^r + B(0,\varepsilon))$ be such that $\alpha_\varepsilon(x) = 1$ on W^r. We set $W_\varepsilon^r = W^r + B(0,\varepsilon)$. The operator

$$\alpha_\varepsilon \cdot Z_0 : \mathring{W}_2^{l-1/2}(W^{r-1}) \subsetneq W_2^{l-1/2}(\mathbb{R}^{r-1}) \underset{Z_0}{\rightarrow} W_2^l(\mathbb{R}^r) \underset{\alpha_\varepsilon \cdot}{\longrightarrow} \mathring{W}_0^l(W_\varepsilon^r)$$

is continuous and as one sees the following holds:

Theorem 8.4 *Let $l > \frac{1}{2}$. The extension operator*

$$Z_0^\varepsilon = \alpha_\varepsilon \cdot Z_0 : \mathring{W}_2^{l-1/2}(W^{r-1}) \rightarrow \mathring{W}_2^l(W_\varepsilon^r)$$

is linear and continuous, and has the property

$$T_0 \circ Z_0^{\varepsilon}\varphi|_{W^{r-1}} = \varphi \quad \textit{for all } \varphi \in \mathring{W}_2^{l-1/2}(W^{r-1}).$$

Here we also use the fact following from (2) that T_0 is multiplicative for $\alpha \in \mathscr{D}(\mathbb{R}^r)$

$$T_0(\alpha\varphi) = \alpha|_{\mathbb{R}^{r-1}} \cdot T_0\varphi.$$

We wish to take into the trace operator derivatives in the normal direction, here $\partial/\partial x_r$. Let $\varphi \in W_2^l(\mathbb{R}^r)$, then $\partial^k\varphi/\partial x_r^k \in W_2^{l-k}(\mathbb{R}^r)$ for $k = 0, 1, \ldots, m$; $l \geq m$. For $l - m > \frac{1}{2}$, $m \in \mathbb{N}$ we define

$$T_m\varphi = \left(T_0\varphi, T_0\frac{\partial\varphi}{\partial x_r}, \ldots, T_0\frac{\partial^m\varphi}{\partial x_r^m} \right), \tag{9}$$

and have for $\varphi \in \mathscr{D}(\mathbb{R}^r)$ (respectively $\varphi \in C_0^{[l]+m+1}$)

$$T_m\varphi = \left(\varphi(x', 0), \frac{\partial\varphi}{\partial x_r}(x', 0), \ldots, \frac{\partial^m\varphi}{\partial x_r^m}(x', 0) \right), \quad x' \in \mathbb{R}^{r-1}.$$

The following simple corollary to Theorem 8.1 holds

Theorem 3.5 *The trace operator T_m defined by (9) acts continuously and linearly*

$$T_m : W_2^l(\mathbb{R}^r) \to \mathop{\times}_{k=0}^{m} W_2^{l-k-1/2}(\mathbb{R}^{r-1}),$$

where $l - m > \frac{1}{2}$, $m \in \mathbb{N}$.

The trace operator T_m defined by (9) again possesses a continuous right inverse Z_m.

Theorem 8.6 *There exists a continuous, linear map*

$$Z_m : \mathop{\times}_{k=0}^{m} W_2^{l-k-1/2}(\mathbb{R}^{r-1}) \to W_2^l(\mathbb{R}^r) \tag{10}$$

with the property

$$T_m \circ Z_m = I. \tag{11}$$

or putting $u = Z_m\varphi$, $\varphi = (\varphi_0, \varphi_1, \ldots, \varphi_m) \in \times_{k=0}^{m} \mathscr{D}(\mathbb{R}^{r-1})$

$$\frac{\partial^k u}{\partial x_r^k}(x', 0) = \varphi_k(x'), \quad x' \in \mathbb{R}^{r-1}, k = 0, 1, \ldots, m. \tag{11'}$$

Here again $l - m > \frac{1}{2}$, $m \in \mathbb{N}$.

Proof. We introduce the factor $x_r^k/k!$ into the defining formula for Z_0 and write $l-k$ instead of l; thus

$$v(x) := \tilde{Z}_k \varphi := \frac{x_r^k}{\pi k!(2\pi)^{r-1}} \int_{\mathbb{R}^r} e^{i(x,\xi)} \frac{(1+|\xi'|^2)^{l-k-1/2}}{(1+|\xi|^2)^{l-k}} \hat{\varphi}(\xi')\, d\xi.$$

We obtain a continuous linear map,

$$\tilde{Z}_k : W_2^{l-k-1/2}(\mathbb{R}^{r-1}) \to W_2^l(\mathbb{R}^r), \tag{12}$$

we see the continuity similarly to (8), and with the additional property

$$(\tilde{Z}_k\varphi)(x',0) = 0, \quad \frac{\partial \tilde{Z}_k\varphi}{\partial x_r}(x',0) = 0, \ldots, \quad \frac{\partial^k \tilde{Z}_k\varphi}{\partial x_r^k}(x',0) = \varphi(x'), \tag{13}$$

for $\varphi \in W_2^{l-k-1/2}(\mathbb{R}^{r-1})$, $k = 0, 1, \ldots, m$.

We define the operator Z_m by

$$Z_m\varphi := \tilde{Z}_0\varphi_0 + \tilde{Z}_1\left(\varphi_1 - \frac{\partial}{\partial x_r}\tilde{Z}_0\varphi_0\right)$$
$$+ \frac{1}{2!}\tilde{Z}_2\left(\varphi_2 - \frac{\partial^2}{\partial x_r^2}\tilde{Z}_0\varphi_0 - \frac{\partial^2}{\partial x_r^2}\tilde{Z}_1\left(\varphi_1 - \frac{\partial}{\partial x_r}\tilde{Z}_0\varphi_0\right)\right) + \cdots,$$

where $\varphi = (\varphi_0, \varphi_1, \ldots, \varphi_m)$. The continuity of Z_m, (10), follows immediately from the continuity of $\tilde{Z}_k$, (12), $k = 0, \ldots, m$, while property (11) is a consequence of (13). ∎

Remark 8.1 Of course we can localise Theorems 8.5 and 8.6, and state results similar to 8.2 and 8.4. We next want to prove the trace theorem for general bounded regions Ω. In doing this we make do in the first part of Theorem 8.7, existence of T_0, with every general assumptions on Ω; in the second part we must leave the normal direction invariant under transformation. This requires adding 1 to the smoothness assumption. For the smoothness assumption see the transformation theorem 4.1.

Theorem 8.7 (trace theorem)

(a) *Let Ω be a bounded (k,κ)-smooth region; let $\frac{1}{2} < l \leqslant k + \kappa$, for l integral, $k = l - 1$, $\kappa = 1$ is admissible. There exists a continuous linear trace operator*

$$T_0 : W_2^l(\Omega) \to W_2^{l-1/2}(\partial\Omega),$$

with the property

$$T_0\varphi = \varphi|_{\partial\Omega} \quad \textit{for } \varphi \in C^l(\bar{\Omega}), \tag{14}$$

respectively $\varphi \in C^{[l]+1}(\bar{\Omega})$ for l non-integral.

(b) *Again suppose that Ω is bounded and (k,κ)-smooth, let $l-m>\frac{1}{2}$, $m\in\mathbb{N}$, $l+1\leqslant k+\kappa$, again for integral $l, k=l, \kappa=1$ is admissible. Then there exists a continuous, linear trace operator*

$$T_m: W_2^l(\Omega)\to \mathop{\times}_{i=0}^{m} W_2^{l-i-1/2}(\partial\Omega)$$

with the property

$$T_m\varphi=\left(\varphi|_{\partial\Omega}, \frac{\partial\varphi}{\partial n}\bigg|_{\partial\Omega},\ldots,\frac{\partial^m\varphi}{\partial n^m}\bigg|_{\partial\Omega}\right) \quad \text{for } \varphi\in C^{l+m}(\bar{\Omega}), \tag{14'}$$

respectively $\varphi\in C^{[l]+m+1}(\bar{\Omega})$ for l non-integral. Here $\partial/\partial n$ is the derivation in the direction of the inward pointing normal.

Remark 8.2 Since by Theorem 3.6 functions from $C^\infty(\bar{\Omega})$ are dense in $W_2^l(\Omega)$, the trace operator is uniquely determined by (14) and the continuity requirement.

Proof of Theorem 8.7(a). We cover $\partial\Omega$ with neighbourhoods U_j, as they appear in the Definition 2.7 of (k,κ)-smoothness. Since $\partial\Omega$ is compact, finitely many $U_j, j=1,\ldots,n$ suffice. Let $\{\alpha_j\}$ be a partition of unity subordinate to the cover $\{U_j\}$, First we map U_j to the cube W^r by a (k,κ)-transformation Φ_j. By the transformation Theorem 4.1 we have $^*\Phi_j(\alpha_j\cdot\varphi)\in W_2^l(W_+^r)$. By the choice of the α_j the support of $^*\Phi_j(\alpha_j\cdot\varphi)$ can only meet the frontier of W_+^r in W^{r-1}; we therefore have

$$\|^*\Phi_j(\alpha_j\varphi)\|_{l,W_+^r}=\|^*\Phi_j(\alpha_j\varphi)\|_{l,\mathbb{R}_+^r},$$

and by Hestenes (see Theorem 5.5) we can extend; let F_j be the extension operator.

We obtain

$$F_j\circ{}^*\Phi_j(\alpha_j\varphi)\in \mathring{W}_2^l(W^r)$$

(see the remark about the support in Theorem 5.5). We apply the trace operator T_0 from Theorem 8.2; $T_0\circ F_j\circ{}^*\Phi_j(\alpha_j\varphi)$ now belongs to $W_2^{l-1/2}(W^{r-1})$. We transform back, this time restricting to W^{r-1} and obtain

$$^*\Phi_1^{-1}\circ T_0\circ F_j\circ{}^*\Phi_j(\alpha_j\varphi)\in \mathring{W}_2^{l-1/2}(\partial\Omega\cap U_j).$$

We define

$$T_0\varphi=\sum_{j=1}^{n}{}^*\Phi_1^{-1}\circ T_0\circ F_j\circ{}^*\Phi_j(\alpha_j\varphi), \tag{15}$$

and write out the defining composition for the individual summands

$$W_2^l(\Omega) \xrightarrow[\alpha_j]{} W_2^l(U_j^+) \xrightarrow[{*\Phi_j}]{} W_2^l(W_+^r) \xrightarrow[F_j]{} \mathring{W}_2^l(W^r)$$
$$\xrightarrow[T_0]{} \mathring{W}_0^{l-1/2}(W^{r-1}) \xrightarrow[{*\Phi_j^{-1}}]{} \mathring{W}_2^{l-1/2}(\partial\Omega \cap U_j) \subsetneq W_2^{l-1/2}(\partial\Omega). \quad (16)$$

By quoted results all the arrows in (16) are continuous and linear. T_0 defined by (15) therefore represents a continuous linear operator. Now for property (14). For this we have only to check the behaviour of functions from $C^l(\bar{\Omega})$ on $\partial\Omega$ under the transformations (16); α_j multiplied by the values $\varphi|_{\partial\Omega}$ gives $(\alpha\varphi)|_{\partial\Omega}$; C^l-functions go over under $*\Phi_j$ to C^l-functions; the values $(\alpha_j\varphi)|_{\partial\Omega}$ and $*\Phi_j(\alpha_j\varphi)$ correspond to each other; by Addendum 5.2 F_j leaves the values of $C^l = C^{l,0} \subset C^{l-1,1}$-functions on W^{r-1} unchanged, T_0 leaves them unchanged (Property (2), Addendum 8.1); $*\Phi_j^{-1}$ associates $*\Phi_j(\alpha_j\varphi)$ with the corresponding value $*\Phi_j^{-1} \circ *\Phi_j(\alpha_j\varphi)|_{\partial\Omega} = (\alpha_j\varphi)|_{\partial\Omega}$. Putting all this together we have (15)

$$T_0\varphi|_{\partial\Omega} = \sum_{j=1}^{n} \alpha_j\varphi|_{\partial\Omega} = \varphi|_{\partial\Omega} \quad \text{for } \varphi \in C^l(\bar{\Omega}),$$

which is (14), and so we have proved (a). For non-integral l, that is, for $\varphi \in C^{[l]+1}(\bar{\Omega})$, (14) may be similarly established.

(b) Again suppose that $\{U_j\}$, $j = 1, \ldots, n$ is a covering of $\partial\Omega$ and $\{\alpha_j\}$, $j = 1, \ldots, n$, a subordinate partition of unity. Since $m \geq 1$ and we wish to describe the trace in the normal direction, we must so choose the transformations $\Phi_j: W^r \leftrightarrow U_j$, that they leave the normal direction invariant, this is possible by Theorem 2.12. By Theorem 2.12 Φ_j belongs to $C^{k-1,\kappa}$ if Ω is (k, κ)-smooth; our stronger hypothesis gives

$$l > \tfrac{1}{2} + m \geq \tfrac{1}{2} + 1, \quad l \leq (k-1) + \kappa,$$

which is

$$\tfrac{1}{2} + 1 < (k+1) + \kappa \quad \text{or} \quad k \geq 2,$$

hence the assumptions of Theorem 2.12 may be satisfied. The conditions of the transformation theorem 4.1 may also be satisfied for $\Phi_j \in C^{k-1,\kappa}$ $k \geq 2$ because $l \leq (k-1) + \kappa$. The remainder of the proof of (b) develops as in (a), but instead of T_0 we use T_m: we define

$$T_m\varphi := \sum_{j=1}^{n} *\Phi_j^{-1} \circ T_m \circ F_j \circ *\Phi_j(\alpha_j\varphi)$$

by the scheme (see (16))

$$W_2^l(\Omega) \xrightarrow[\alpha_j]{} W_2^l(U_j) \xrightarrow[{*\Phi_j}]{} W_2^l(W_+^r) \xrightarrow[F_j]{} \mathring{W}_2^l(W^r) \xrightarrow[T_m]{} \bigtimes_{i=0}^{m} \mathring{W}_2^{l-i-1/2}(W^{r-1})$$
$$\xrightarrow[\times_{i=0}^{m} *\Phi_j^{-1}]{} \bigtimes_{i=0}^{m} \mathring{W}_2^{l-i-1/2}(U_j \cap \partial\Omega) \subsetneq \bigtimes_{i=0}^{m} W_2^{l-i-1/2}(\partial\Omega),$$

where T_m is the localised map (see Remark 8.1, Theorem 8.2) from Theorem 8.5. The proof of property (14) also proceeds as in (a), we use here the fact that Φ_j and Φ_j^{-1} leave the normal·direction invariant. ■

We now want to show that the trace operator T_m is onto. Because we can again weaken the assumptions for $m=0$, we distinguish between the two cases $m=0$ and $m \geqslant 1$.

Theorem 8.8 (inverse theorem)

(a) *Let Ω be a bounded (k, κ)-smooth region, let $\frac{1}{2} < l \leqslant k+\kappa$ (for integral l, $k = l-1$, $\kappa = 1$ is admissible). There exists a continuous, linear extension operator*

$$Z_0 : W_2^{l-1/2}(\partial\Omega) \to W_2^l(\Omega),$$

with the property

$$T_0 \circ Z_0 \varphi = \varphi \quad \text{for all } \varphi \in W_2^{l-1/2}(\partial\Omega). \tag{17}$$

(b) *Again let Ω be bounded and (k, κ)-smooth, let $l-m > \frac{1}{2}$, $m \in \mathbb{N}$, $l+1 < k+\kappa$ (again for integral $l, k = l, \kappa = 1$ is admissible). There exists a continuous, linear extension operator*

$$Z_m : \mathop{\times}_{i=0}^{m} W_2^{l-i-1/2}(\partial\Omega) \to W_2^l(\Omega),$$

with the property

$$T_m \circ Z_m = I \text{ on } \mathop{\times}_{i=0}^{m} W_2^{l-i-1/2}(\partial\Omega), \tag{17'}$$

where T_m is the trace operator from Theorem 8.7.

Proof. (a) We cover Ω in $\mathbb{R}^r$ with neighbourhoods $U_j, j = 1, \ldots, n$, so that the neighbourhoods $U_{j,\varepsilon}$ enlarged by $\varepsilon > 0$ have the properties from Definition 2.7. Moreover we choose U_0 open with $\bar{\Omega} \subset \bigcup_{j=0}^{n} U_j$. Let $V_j = U_j \cap \partial\Omega$, the V_j form an open cover in $\partial\Omega$ of the (k, κ)-submanifold $\partial\Omega$ (Theorem 2.15), let $\{\beta_j\}$ be a partition of unity in $\partial\Omega$ subordinate to the cover V_j. Let $\Phi_j : W_\varepsilon^r \leftrightarrow U_{j,\varepsilon}$ be the (k, κ)-transformations from Definition 2.7, where $\Phi_j : W^r \leftrightarrow U_j$ and $\Phi_j : W^{r-1} \leftrightarrow V_j$. We have the diagram

$$W_2^{l-1/2}(\partial\Omega) \underset{\beta_j}{\longrightarrow} \mathring{W}_2^{l-1/2}(V_j) \underset{{}^*\Phi_j}{\longrightarrow} \mathring{W}_2^{l-1/2}(W^{r-1}) \underset{Z_0^\varepsilon}{\longrightarrow} \mathring{W}_2^l(W_\varepsilon^r)$$

$$\underset{{}^*\Phi_j^{-1}}{\longrightarrow} \mathring{W}_2^l(U_{j,\varepsilon}) \subsetneq \mathring{W}_2^l\left(\bigcup_{j=1}^{n} U_{j,\varepsilon}\right) \underset{R_\Omega}{\longrightarrow} W_2^l(\Omega). \tag{18}$$

We explain this, ${}^*\Phi_j$ and ${}^*\Phi_j^{-1}$ are continuous by the transformation

Theorem 4.1, Z_0^ε is the map from Theorem 8.4, it is continuous. R_Ω is the restriction operator from $\bigcup_{j=1}^n U_{j,\varepsilon} \cup U_0$ on Ω (we have $\bigcup_{j=1}^n U_{j,\varepsilon} \cup U_0 \supset \bigcup_{j=0}^n U_j \supset \bar{\Omega} \supset \Omega$).

For $\varphi \in W_2^{l-1/2}(\partial\Omega)$ we define

$$Z_0\varphi := \sum_{j=1}^{n} R_\Omega \circ {}^*\Phi_j^{-1} \circ Z_0^\varepsilon \circ {}^*\Phi_j(\beta_j \cdot \varphi),$$

all the arrows in (18) are continuous and linear, so the mapping

$$Z_0 : W_2^{l-1/2}(\partial\Omega) \to W_2^l(\Omega).$$

is also linear and continuous. In order to check (17) (it is enough to consider $\varphi \in C^{k,\kappa}(\partial\Omega)$, since by Theorem 4.3, $C^{k,\kappa}(\partial\Omega)$ is dense in $W_2^{l-1/2}(\partial\Omega)$), we must follow the values of the functions through diagram (18) and use the equation $T_0^\varepsilon \circ Z_0 = I$ from Theorem 8.4. In comparison with Theorem 8.7 nothing new is required, so we leave the details to the reader.

(b) Let $m \geqslant 1$, since the trace operator T_m is defined in the normal direction, we must take transformations Φ_j which leave the normal direction invariant. This is possible by Theorem 2.12, and we have $\Phi_j \in C^{k-1,\kappa}$. Let Z_m^ε be the localised operator from Theorem 8.6, see Remark 8.1, we define

$$Z_m\varphi := \sum_{j=1}^{n} R_\Omega \circ {}^*\Phi_j^{-1} \circ Z_m^\varepsilon \circ \mathop{\times}_{i=0}^{m} {}^*\Phi_j(\beta_j \cdot \varphi),$$

and read the properties of Z_m from the diagram:

$$\mathop{\times}_{i=0}^{m} W_2^{l-i-1/2}(\partial\Omega) \xrightarrow[\beta_j]{} \mathop{\times}_{i=0}^{m} W_2^{l-i-1/2}(V_j) \xrightarrow[\times_{i=0}^{m} {}^*\Phi_j]{} \mathop{\times}_{i=0}^{m} \mathring{W}_2^{l-i-1/2}(W^{r-1})$$

$$\xrightarrow[Z_m^\varepsilon]{} \mathring{W}_2^l(W_\varepsilon^r) \xrightarrow[{}^*\Phi_j^{-1}]{} \mathring{W}_2^l(U_{j,\varepsilon}) \subsetneq \mathring{W}_2^l\left(\bigcup_{j=1}^{n} U_{j,\varepsilon} \cup U_0\right) \xrightarrow[R_\Omega]{} W_2^l(\Omega).$$

In the proof of (17), again by means of a diagram chase, we need the normal coordinates Φ_j in the transition $W^r \to U_j$, indeed with respect to these, the derivatives in the normal direction become normally directed derivatives and (11) from Theorem 8.6 becomes (17). ∎

By means of Theorem 8.8 we have characterised the image set of the trace operator T_m, we now want to find the kernel of T_m, here we take $l \in \mathbb{N}$ integral and $m = l$.

Theorem 8.9

(a) *Let Ω be a bounded* (0, 1)*-smooth region, and let* $T_0 : W_2^1(\Omega) \to W_2^{1/2}(\Omega)$

be the trace operator. Then

$$\ker T_0 = \mathring{W}_2^1(\Omega),$$

or put otherwise

$$T_0\varphi = 0 \quad \textit{if and only if} \quad \varphi \in \mathring{W}_2^1(\Omega).$$

(b) *Let Ω be a bounded, $(l+1,1)$-smooth region, $l \geqslant 1$, and let*

$$T_l: W_2^{l+1}(\Omega) \to \mathop{\times}_{k=0}^{l} W_2^{l-k+1/2}(\Omega),$$

be the trace operator from Theorem 8.7. Then

$$\ker T_l = \mathring{W}_2^{l+1}(\Omega),$$

or put otherwise

$$T_l\varphi = 0 \quad \textit{if and only if} \quad \varphi \in \mathring{W}_2^{l+1}(\Omega).$$

Proof. Before the proof we make two observations about the trace operator:
(a) The trace operator is multiplicative, that is,

$$T_0(\alpha\varphi) = \alpha|_{\partial\Omega} \cdot T_0\varphi, \alpha \in C^\infty(\bar{\Omega}), \tag{19}$$
$$T_l(\alpha\varphi) = \left(\alpha|_{\partial\Omega} \cdot T_0\varphi, \alpha|_{\partial\Omega} \cdot T_0 \frac{\partial\varphi}{\partial n} + \left.\frac{\partial\alpha}{\partial n}\right|_{\partial\Omega} \cdot T_0\varphi, \ldots\right),$$

which follows immediately from (14).

Let $\Phi: \Omega \to \Omega'$ be a $(0,1)$-transformation, respectively an $(l,1)$-normal transformation; the trace operator T_l then commutes with the pull-back operator $*\Phi$

$$*\Phi \circ T_l\varphi = T_l \circ *\Phi\varphi. \tag{20}$$

The trace formula (14) is also valid for all $\varphi \in C^{l,1}(\bar{\Omega})$ (see the proof of Addendum 8.1, we start there with $C^{l,1}(\bar{\Omega}) \subsetneq W_2^{l+1}(\Omega)$, where this last follows from Theorem 1.8). We need only prove (20) for $\varphi \in C^\infty(\bar{\Omega})$ ($C^\infty(\bar{\Omega})$ is dense in $W_2^{l+1}(\Omega)$ by Theorem 3.6), we then have $*\Phi\varphi \in C^{l,1}(\bar{\Omega}')$, and $*\Phi$ applied to (14) gives (20). Of course in the second case, $l \geqslant 1$, we need to assume that $*\Phi$ is a normal transformation. (19) and (20) also hold for general trace operators T_m under the hypotheses of Theorem 8.7.

Now for the proof of Theorem 8.9(a). We clearly have $\mathring{W}_2^1(\Omega) \subset \ker T_0$. Let $\varphi \in W_2^1(\Omega)$ with $T_0\varphi = 0$. We cover (see the proof of Theorem 8.7) $\partial\Omega$ with the U_j, choose a partition of unity α_j, transform by Φ_j and obtain

$$v := *\Phi_j(\alpha_j, \varphi) \in W_2^1(W'_+),$$

where because of (19) and (20)

$$T_0 v = 0. \tag{21}$$

we extend v to W^r by means of 0 and denote it by $F_0 v$. We show

$$\frac{\partial(F_0 v)}{\partial x_i} = F_0 \frac{\partial v}{\partial x_i} \quad \text{for } i = 1, \ldots, r. \tag{22}$$

Let $\psi \in \mathscr{D}(W^r)$, for $i = 1, \ldots, r-1$, we have

$$\int_{W^r} \frac{\partial \psi}{\partial x_i} F_0 v \,\mathrm{d}x = \int_{W^r_+} \frac{\partial \psi}{\partial x_i} v \,\mathrm{d}x = -\int_{W^r_+} \psi \frac{\partial v}{\partial x_i} \mathrm{d}x = -\int_{W^r} \psi F_0 \frac{\partial v}{\partial x_i} \mathrm{d}x \tag{23}$$

and because of (21)

$$\int_{W^r} \frac{\partial \psi}{\partial x_r} F_0 v \,\mathrm{d}x = \int_{W^r_+} \frac{\partial \psi}{\partial x_r} \cdot v \,\mathrm{d}x = \int_{W^{r-1}} \psi \cdot T_0 v \,\mathrm{d}x' - \int_{W^r_+} \psi \frac{\partial v}{\partial x_r} \mathrm{d}x$$
$$= -\int_{W^r} \psi F_0 \frac{\partial v}{\partial x_r} \mathrm{d}x.$$

with this we have proved (22) in the sense of distributions. From (22) it follows that $F_0 v \in \mathring{W}^1_2(W^r)$, while (21) and the definition of F_0 gives $\operatorname{supp} F_0 v \subset \bar{W}^r_+$. We transform backwards with Φ_j^{-1} and obtain

$$\tilde{\varphi} = \sum_{j=1}^{n} {}^*\Phi_j^{-1} \circ F_0 {}^*\Phi_j(\alpha_j \cdot \varphi) \in \mathring{W}^1_2(\mathbb{R}^r), \quad \operatorname{supp} \tilde{\varphi} \subset \bar{\Omega}, \tilde{\varphi}|_\Omega = \varphi,$$

from which by Theorem 3.7 it follows that $\varphi \in \mathring{W}^1_2(\Omega)$.

(b) Let $l \geqslant 1$. With the notation of (a) $T_l v = 0$ implies that $v(x', 0) = 0$, $\partial v(x', 0)/\partial x_r = 0$, etc. from which it follows that all derivatives $\mathrm{D}^\alpha v(x', 0)$ up to order $|\alpha| \leqslant l$ vanish on $W^{r-1} = \{x_r = 0\}$. Multiple partial integration as in (23) gives

$$\mathrm{D}^\alpha(F_0 v) = F_0 \mathrm{D}^\alpha v, \quad |\alpha| \leqslant l + 1,$$

and the remainder of the proof proceeds as in (a). ■

Exercises

8.1 Show that there is no continuous linear map $W^l_2(\mathbb{R}^r) \to W^l_2(\mathbb{R}^{r-1})$ which agrees on $\mathscr{D}(\mathbb{R}^r)$ with the restriction given by $\varphi \to \varphi(\cdot, 0)$.

8.2 Prove the trace theorem and the inverse theorem for Fourier series, that is, prove the continuity of

$$T_0 : W^l_2(T^r) \to W^{l-1/2}_2(T^{r-1}),$$
$$Z_0 : W^{l-1/2}_2(T^{r-1}) \to W^l_2(T^r),$$

where T_0 is defined on $\mathscr{D}(T^r)$ by $T_0 \varphi = \varphi(\cdot, 0)$. How does one define Z_0?

§9 Weak sequential compactness and approximation of derivatives by difference quotients

Regularity theorems for partial differential equations, see §20, are most simply proved by difference procedures, that is, one approximates the derivatives by difference quotients. For this we must know the behaviour of difference quotients in Sobolev spaces – §9 is devoted to this.

In Hilbert spaces closed balls are weakly sequentially compact; we start with the simple proof.

Theorem 9.1 *Let H be a Hilbert space. Let the sequence $h_n \in H$ be bounded, $\|h_n\| \leqslant K$, $n \in \mathbb{N}$. Then there exists a subsequence $h_{n'}$, which converges weakly to an element $h \in H$, and we have $\|h\| \leqslant K$.*

Proof. We set $S = [h_n : n \in \mathbb{N}]$, the linear hull of (h_n). Then the orthogonal decomposition theorem holds: $H = \bar{S} \oplus \bar{S}^{\perp}$. For $g = \bar{S}^{\perp}$ the scalar product $(g, h_n) = 0$ converges trivially. We prove the assertion for $g \in \bar{S}$. For this we apply the diagonal process. By the Schwarz inequality

$$|(h_n, h_1)| \leqslant K^2, \quad n \in \mathbb{N},$$

so there exists a subsequence h_{n_1} with

$$(h_{n_1}, h_1) \text{ convergent and with } |(h_{n_1}, h_1)| \leqslant K^2.$$

We form the scalar product (h_{n_1}, h_2) and in the same way obtain a subsequence h_{n_2} of h_{n_1} with

$$(h_{n_2}, h_2) \text{ convergent and } |(h_{n_2}, h_2)| \leqslant K^2.$$

In general we obtain a subsequence $H_{n_{k+1}}$ of h_{n_k} with

$$(h_{n_{k+1}}, h_{k+1}) \text{ convergent and } |(h_{n_{k+1}}, h_{k+1})| \leqslant K^2.$$

We choose the diagonal sequence h_{n_n} from the array below

$$\begin{aligned} h_{n_1} &= h_{1_1}, h_{2_1}, h_{3_1}, h_{4_1}, \ldots \\ h_{n_2} &= h_{1_2}, h_{2_2}, h_{3_2}, h_{4_2}, \ldots \\ h_{n_3} &= h_{1_3}, h_{2_3}, h_{3_3}, h_{4_3}, \ldots \\ h_{n_4} &= h_{1_4}, h_{2_4}, h_{3_4}, h_{4_4}, \ldots \end{aligned}$$

so that defining $h_{n'} := h_{n_n}$ $(h_{n'}, h_m)$ converges for all $m \in \mathbb{N}$.

Therefore $(h_{n'}, g)$ also converges for all $g \in S$. Suppose now that $g \in \bar{S}$ and that $\varepsilon > 0$ is arbitrarily preassigned. Then there exists $f \in S$ with

$\|f-g\| \leq \varepsilon/4K$, and there exists some $N_0(\varepsilon)\in\mathbb{N}$ with

$$|(h_{k'}-h_{m'},f)| \leq \frac{\varepsilon}{2} \quad \text{for all } k,\, m \geq N_0(\varepsilon).$$

For $k, m \geq N_0(\varepsilon)$ it then follows that

$$\begin{aligned}|(h_{k'}-h_{m'},g)| &\leq |(h_{k'}-h_{m'},f)| + |(h_{k'}-h_{m'},g-f)| \\ &\leq \frac{\varepsilon}{2} + \|h_{k'}-h_{m'}\|\cdot\|g-f\| \leq \frac{\varepsilon}{2} + 2K\frac{\varepsilon}{4K} = \varepsilon.\end{aligned}$$

Therefore the limit exists

$$F(g) = \lim_{k\to\infty}(h_{k'},g) \quad \text{for all } g\in H.$$

F is linear and we have the estimate

$$|F(g)| \leq \lim_{k\to\infty}(h_{k'},g)| \leq \lim_{k\to\infty}\|h_{k'}\|\,\|g\| \leq K\|g\|,$$

That is, F is continuous. By the Riesz representation theorem there exists $h\in H$ with

$$F(g) = (g,h) \quad \text{with } \|h\| = \|F\| \leq K, \qquad \blacksquare$$

As an abbreviation for the difference quotients of a function we introduce the following notation, $i = 1,\ldots,r$,

$$\Delta_h^i\varphi(x) = \frac{1}{h}[\varphi(x+he_i)-\varphi(x)] = \frac{1}{h}[\varphi(x_1,\ldots,x_i+h,\ldots)-\varphi(x_1,\ldots,x_i,\ldots)],$$

where $e_i = (0,\ldots,1,0,\ldots,0)$ and $h\in\mathbb{R}$.

The following theorems hold for Sobolev spaces, that is, for $l = 0,\ 1, 2,\ldots$.

Theorem 9.2 *Let $\Omega\subset\mathbb{R}^r$ be open and let $\varphi\in\mathring{W}_2^1(\Omega)$. Then*

$$\|\Delta_h^i\varphi\|_0 \leq \|\varphi\|_1 \quad \textit{and} \quad \left\|\Delta_h^i\varphi - \frac{\partial\varphi}{\partial x_i}\right\|_0 \to 0 \quad \textit{as } h\to 0.$$

Here – so as to have no difficulty with the definition of $\Delta_h^i\varphi$ – we have extended the function $\varphi\in\mathring{W}_2^1(\Omega)$ by zero to all of $\mathbb{R}^r$ (see Lemma 3.4). $\partial\varphi/\partial x_i$ is the distributional derivative.

Proof. Suppose first that $\varphi\in\mathscr{D}(\Omega)$ and $\delta > 0$ is chosen fixed – for the choice of

δ see (3). For all $|h| < \delta$ we have by the mean value theorem that

$$\Delta_h^i \varphi(x) - \frac{\partial}{\partial x_i}\varphi(x) = \int_0^1 \left[\frac{\partial}{\partial x_i}\varphi(x + he_i t) - \frac{\partial}{\partial x_i}\varphi(x)\right] dt$$

or

$$\Delta_h^i \varphi(x) = \int_0^1 \frac{\partial}{\partial x_i}\varphi(x + he_i t)\, dt.$$

By the Schwarz inequally we obtain

$$\begin{aligned}\left\|\Delta_h^i \varphi - \frac{\partial}{\partial x_i}\varphi\right\|_0^2 &= \int_\Omega \left|\int_0^1 \left[\frac{\partial}{\partial x_i}\varphi(x + he_i t) - \frac{\partial}{\partial x_i}\varphi(x)\right] dt\right|^2 dx \qquad (1)\\ &\leqslant \int_\Omega \left(\int_0^1 \left|\frac{\partial}{\partial x_i}\varphi(x + he_i t) - \frac{\partial}{\partial x_i}\varphi(x)\right|^2 dt\right) dx\\ &= \int_0^1 \int_\Omega \left|\frac{\partial}{\partial x_i}\varphi(x + he_i t) - \frac{\partial}{\partial x_i}\varphi(x)\right|^2 dx\, dt\end{aligned}$$

or

$$\|\Delta_h^i \varphi\|_0^2 = \int_\Omega \left|\int_0^1 \frac{\partial}{\partial x_i}\varphi(x + he_i t)\, dt\right|^2 dx \leqslant \int_0^1 \int_\Omega \left|\frac{\partial}{\partial x_i}\varphi(x + he_i t)\right|^2 dx\, dt.$$

Change of coordinates gives

$$\|\Delta_h^i \varphi\|_0^2 \leqslant \int_0^1 \left\|\frac{\partial}{\partial x_i}\varphi\right\|_0^2 dt = \left\|\frac{\partial}{\partial x_i}\varphi\right\|_0^2 \leqslant \|\varphi\|_1^2. \qquad (2)$$

A single function forms a compact set in $L^2(\Omega)$, therefore by the Kolmogorov theorem, §1 (continuity in the mean), we have

$$\int_\Omega \left|\frac{\partial}{\partial x_i}\varphi(x + he_i t) - \frac{\partial}{\partial x_i}\varphi(x)\right|^2 dx \leqslant \varepsilon^2 \quad \text{for } |th| \leqslant |h| \leqslant \delta_\varepsilon. \qquad (3)$$

Therefore by (1) we obtain

$$\left\|\Delta_h^i \varphi - \frac{\partial}{\partial x_i}\varphi\right\|_0 \leqslant \varepsilon \quad \text{for } |h| \leqslant \delta_\varepsilon.$$

Since $\mathscr{D}(\Omega)$ is dense in $\mathring{W}_2^1(\Omega)$, both (2) and the last inequality hold for all $\varphi \in \mathring{W}_2^1(\Omega)$, and so we have proved our theorem. ■

Induction on l and the application of Theorem 9.2 to $D^s\varphi, |s| \leqslant l-1$ give:

Theorem 9.3 *Let $\varphi \in \mathring{W}^l_2(\Omega)$ and $l = 1, 2, \ldots$. Then*

$$\| \Delta^i_h \varphi \|_{l-1} \leqslant \| \varphi \|_l \quad \text{and} \quad \left\| \Delta^i_h \varphi - \frac{\partial \varphi}{\partial x_i} \right\|_{l-1} \to 0 \quad \text{as } h \to 0.$$

We wish to extend Theorem 9.3 to functions from $W^l_2(\Omega)$. In order to be able to define $\Delta^i_h(\varphi)$ without restriction we must be able to extend functions from $W^l_2(\Omega)$. In particular this is the case if an extension theorem of the kind considered in §5 holds for Ω, for example if Ω is bounded and has the uniform cone property.

Theorem 9.4 *Let Ω admit a continuous extension operator*

$$F_\Omega : W^l_2(\Omega) \to \mathring{W}^l_2(\mathbb{R}^r) = W^l_2(\mathbb{R}^r),$$

for example suppose Ω is bounded and has the uniform cone property. For $l = 1, 2, \ldots, \varphi \in W^l_2(\Omega)$, it is then true that

$$\| \Delta^i_h(F_\Omega \varphi) \|_{l-1} \leqslant c \| \varphi \|_l \quad \text{and} \quad \left\| \Delta^i_h \left(F_\Omega \varphi - \frac{\partial \varphi}{\partial x_i} \right) \right\|_{l-1} \to 0 \quad \text{as } h \to 0.$$

Proof. For $F_\Omega \varphi$ the assertions of Theorem 9.3 hold for $\mathring{W}^l_2(\mathbb{R}^r) = W^l_2(\mathbb{R}^r)$. By restriction to Ω we obtain

$$\| \Delta^i_h(F_\Omega \varphi) \|_{\Omega, l-1} \leqslant \| \Delta^i_h(F_\Omega \varphi) \|_{\mathbb{R}^r, l-1} \leqslant \| F_\Omega \varphi \|_{\mathbb{R}^r, l} \leqslant c \| \varphi \|_{\Omega, l}$$

and

$$\left\| \Delta^i_h(F_\Omega \varphi) - \frac{\partial \varphi}{\partial x_i} \right\|_{\Omega, l-1} \leqslant \left\| \Delta^i_h(F_\Omega \varphi) - \frac{\partial (F_\Omega \varphi)}{\partial x_i} \right\|_{\mathbb{R}^r, l-1} \to 0. \qquad \blacksquare$$

We now prove an important existence theorem.

Theorem 9.5 *Let $\varphi \in L_2(\Omega)$ and $\delta > 0$. For each $\Omega' \subset \Omega$ with $\Omega'_\delta \subset \Omega$* let

$$\| \Delta^i_h \varphi \|_{0, \Omega'} \leqslant M \quad \text{for } |h| < \delta, \tag{4}$$

with M independent of δ, Ω' and h. Then the distributional derivative $\partial \varphi / \partial x_i$ belongs to $L_2(\Omega)$, and we have

$$\left\| \frac{\partial \varphi}{\partial x_i} \right\|_{0, \Omega} \leqslant M. \tag{5}$$

Remark (4) is, for example, satisfied, if φ possesses an extension $\tilde{\varphi}$ to $\mathbb{R}^r$ with $\| \Delta^i_h(\tilde{\varphi}) \|_{0, \Omega} \leqslant M$.

Proof. Since $\varphi \in L_2(\Omega)$, $\varphi \in \mathscr{D}'(\Omega)$ and by Theorem 1.6 $\Delta^i_h \varphi$ converges to

$\partial\varphi/\partial x_i$ as $h \to 0$ in the distributional sense, and we have only to show that $\partial\varphi/\partial x_i \in L_2(\Omega)$. We use weak convergence in $L_2(\Omega)$. By Theorem 9.1, (4) implies that as $h \to 0$ there exists a subsequence for which

$$\Delta_h^i \varphi \to w \text{ weakly in } L_2(\Omega'),$$

and $\| w \|_{0,\Omega'} \leqslant M$. Since trivially weak convergence in $L_2(\Omega')$ implies (weak) distributional convergence, we have $w = \partial\varphi/\partial x_i$ in $L_2(\Omega')$, and

$$\left\| \frac{\partial \varphi}{\partial x_i} \right\|_{0,\Omega'}^2 = \int_{\Omega'} \left| \frac{\partial \varphi}{\partial x_i} \right|^2 \mathrm{d}x \leqslant M^2, \tag{6}$$

with M independent of $\Omega' \subset \Omega$. Among other things (6) says that $\partial\varphi/\partial x_i \in L^2_{\text{loc}}(\Omega)$, from which it follows that $\partial\varphi/\partial x_i$ is measurable on Ω. We can therefore carry out the limiting process $\Omega' \to \Omega$ in (6) and obtain

$$\left\| \frac{\partial \varphi}{\partial x_i} \right\|_{0,\Omega}^2 = \int_{\Omega} \left| \frac{\partial \varphi}{\partial x_i} \right|^2 \mathrm{d}x \leqslant M^2. \tag{6'}$$

We can generalise: ■

Theorem 9.6 *Let $\varphi \in W_2^{l-1}(\Omega)$, $l = 1, 2, \ldots$ and $\delta > 0$. For each $\Omega' \subset \Omega$ with $\Omega'_\delta \subset \Omega$ let*

$$\| \Delta_h^i \varphi \|_{l-1,\Omega'} \leqslant M \quad \text{for } |h| < \delta, i = 1, \ldots, r,$$

with M independent of δ, Ω' and h. Then $\varphi \in W_2^l(\Omega)$.

Proof. The assumptions imply that $\mathrm{D}^s\varphi \in L^2(\Omega)$ for $|s| \leqslant l-1$, and $\| \Delta_h^i(\mathrm{D}^s\varphi) \|_{0,\Omega'} \leqslant M$ for $|s| \leqslant l-1$, $|h| < \delta$, $i = 1, \ldots, r$. By Theorem 9.5 we obtain

$$\frac{\partial}{\partial x_i} \mathrm{D}^s \varphi \in L^2(\Omega) \quad \text{for } |s| \leqslant l-1, i = 1, \ldots, r,$$

which is equivalent to $\varphi \in W_2^l(\Omega)$. ■

If we additionally require Ω to have the extension property (see Theorem 9.4) then using Theorem 9.4 we obtain:

Addendum 9.1 *We have in addition to the assertions of Theorem 9.6:*

$$\| \Delta_h^i(F_\Omega \varphi) \|_{l-1} \leqslant cM, c \text{ independent of } \varphi, \tag{7}$$

and

$$\left\| \Delta_h^i(F_\Omega \varphi) - \frac{\partial \varphi}{\partial x_i} \right\|_{l-1,\Omega} \to 0 \quad \text{as } h \to 0, i = 1, \ldots, r. \tag{8}$$

Proof. (7) follows from (5) and Theorem 9.4, while (8) represents the second statement of Theorem 9.4. ■

Exercises

9.1 Prove the theorems of this section for Fourier series directly, that is, for the spaces $W_2^l(T^r)$, where T^r is the torus.

9.2 Attempt to formulate optimal theorems for the torus T^r, for example by Theorem 9.3 the difference operator Δ_h^i acts continuously for $\lambda \geqslant 1$:

$$\Delta_h^i : W_2^l(T^r) \to W_2^{l-\lambda}(T^r).$$

How small is one allowed to make λ?

II

Elliptic differential operators

§10 Linear differential operators

In this section we consider some simple properties of linear differential operators. We define when a differential operator is elliptic, when strongly elliptic, and when properly elliptic. We show the invariance of these concepts under coordinate transformations, and prove simple theorems about the order of elliptic and strongly elliptic operators.

Let Ω be an open set in $\mathbb{R}^r$. We consider the linear partial differential operator A defined by

$$A(x, \mathrm{D})\varphi := \sum_{|s| \leq n} a_s(x)\mathrm{D}^s\varphi, \quad x \in \bar{\Omega}. \tag{1}$$

The number n is called the order of the differential operator (1), if there exists some s with $|s| = n$ and $a_s(x) \neq 0$ on Ω. Since $\mathrm{D}^s: W_2^{l+|s|}(\Omega) \to W_2^l(\Omega)$ is continuous (look at the norms), and multiplication by a function a: $W_2^l(\Omega) \to W_2^l(\Omega)$ is continuous, if $a \in C^l(\bar{\Omega})$ (respectively $a \in C^{[l]+1}(\bar{\Omega})$ for l not integral, for more precise multiplier classes A^l see §4), the operator

$$A: W_2^{l+n}(\Omega) \to W_2^l(\Omega) \text{ is continuous,} \tag{2}$$

if, for example, all $a_s(x) \in C^l(\bar{\Omega})$, $|s| \leq n$ (respectively $a \in C^{[l]+1}(\bar{\Omega})$, l non-integral). It is of course possible to further weaken the assumptions on the a_s, under which (2) is continuous.

Definition 10.1 *We call the operator*

$$A^*\varphi = \sum_{|s| \leq n} (-1)^{|s|}\mathrm{D}^s(\overline{a_s(x)}\,\varphi) \tag{3}$$

(formally) adjoint to (1); *we obtain it by taking adjoints in the sense of distribution*

theory, or in L^2. *For* $\varphi, \psi \in \mathscr{D}(\Omega)$ *we indeed have by partial integration that*

$$(A\varphi, \psi)_0 = \int_\Omega A\varphi(x)\overline{\psi(x)}\,\mathrm{d}x = \int_\Omega \varphi(x)\overline{A^*\varphi(x)}\,\mathrm{d}x = (\varphi, A^*\psi)_0.$$

In order that (3),

$$A^*: W_2^{l+n}(\Omega) \to W_2^l(\Omega) \text{ be continuous,}$$

we must, for example, require that $a_s \in C^{n+l}(\bar{\Omega})$ (respectively $a_s \in C^{n+[l]+1}(\bar{\Omega})$), for all $|s| \leqslant n$.

Definition 10.2 *As principal part* A^{H} *of the operator* A *we take the differential expression*

$$A^{\mathrm{H}}\varphi := \sum_{|s|=n} a_s(x)\mathrm{D}^s\varphi. \tag{4}$$

By means of the Leibniz product rule we obtain the principal part $(A^*)^{\mathrm{H}}$ of the operator A^* adjoint to A. It is

$$(A^*)^{\mathrm{H}}\varphi = (-1)^n \sum_{|s|=n} \overline{a_s(x)}\mathrm{D}^s\varphi.$$

Definition 10.3 *We now associate with the principal part* A^{H} *of the differential operator* A *the polynomial*

$$A^{\mathrm{H}}(x, \xi) := \sum_{|s|=n} a_s(x)\xi^s, \quad \xi \in \mathbb{R}^r, x \in \bar{\Omega},$$

called the principal part polynomial, with coefficients depending on $x \in \bar{\Omega}$. *Here, see §1,* ξ^s *means* $\xi_1^{s_1} \cdots \xi_r^{s_r}$.

Definition 10.4 A *is called elliptic at the point* $x \in \bar{\Omega}$, *if* $A^{\mathrm{H}}(x, \xi) \neq 0$ *for all* $0 \neq \xi \in \mathbb{R}^r$. A *is called elliptic on* $\bar{\Omega}$, *if* A *is elliptic at all points* $x \in \bar{\Omega}$. A *is called strongly elliptic at the point* $x \in \bar{\Omega}$, *if there exists a complex constant* γ *with*

$$\operatorname{Re}(\gamma \cdot A^{\mathrm{H}}(x, \xi)) \neq 0 \quad \textit{for all } 0 \neq \xi \in \mathbb{R}^r.$$

A *is called strongly elliptic in* $\bar{\Omega}$, *if* A *is strongly elliptic at all points* $x \in \bar{\Omega}$ *with the constant* γ *independent of* x.

An immediate consequence of these definitions is:

Theorem 10.1

(a) *If* A *is strongly elliptic at* $x \in \bar{\Omega}$, *then* A *is elliptic at* $x \in \bar{\Omega}$.
(b) *If* A *is elliptic at* $x \in \bar{\Omega}$, *then* A^* *is elliptic at* $x \in \bar{\Omega}$.
(c) *If* A *is strongly elliptic at* $x \in \bar{\Omega}$, *then* A^* *is strongly elliptic at* $x \in \bar{\Omega}$.

(d) *If A is elliptic at* $x\in\bar{\Omega}$, *then* $A\circ A^*$ *is strongly elliptic at* $x\in\bar{\Omega}$. *Furthermore the converses of* (b), (c) *and* (d) *all hold.*

In order to clarify the concepts just introduced we give some examples.

Example 10.1 If $r=1$, then the operator defined by (1) is an ordinary differential operator. It is

$$A\varphi=\sum_{j=0}^{n}a_j(x)\varphi^{(j)},$$
$$A^{\mathrm{H}}\varphi=a_n(x)\cdot\varphi^{(n)}\quad\text{and}\quad A^{\mathrm{H}}(x,\xi)=a_n(x)\cdot\xi_1^n\quad\text{with } \xi_1\in\mathbb{R}^1.$$

A is elliptic and strongly elliptic at a point x, if and only if $a_n(x)\neq 0$; A is strongly elliptic for all $x\in[a,b]$, if there exists constants $\gamma_1,\gamma_2\in\mathbb{R}$ with

$$\gamma_1\operatorname{Re}a_n(x)-\gamma_2\operatorname{Im}a_n(x)\neq 0.$$

Example 10.2 The Cauchy–Riemann operator, defined by

$$A\varphi=\frac{\partial\varphi}{\partial x}+\mathrm{i}\frac{\partial\varphi}{\partial y},$$

is elliptic in $\Omega=\mathbb{R}^2$, for it is $A^{\mathrm{H}}(x,\xi)=\xi_1+\mathrm{i}\xi_2\neq 0$ for all $\xi=(\xi_1,\xi_2)\neq 0$, $\xi\in\mathbb{R}^2$. However, at no point $x\in\mathbb{R}^2$ is it strongly elliptic, for $\operatorname{Re}(\gamma A^{\mathrm{H}}(\xi))=\xi_1\cdot\gamma_1-\xi_2\cdot\gamma_2=0$ does not only imply that $\xi_1=\xi_2=0$.

Example 10.3 The operator defined by

$$A\varphi=\sum_{j=1}^{r}a_j(x)\frac{\partial\varphi}{\partial x_j}+a_0(x)\varphi$$

has already been covered by Example 10.1 for $r=1$. Example 10.2 is a special case for $r=2$. For $r\geqslant 2$ A is elliptic at x if and only if

$$\sum_{j=1}^{r}\xi_j\operatorname{Re}a_j(x)\neq 0\quad\text{or}\quad\sum_{j=1}^{r}\xi_j\operatorname{Im}a_j(x)\neq 0,$$

for all $0\neq\xi\in\mathbb{R}^r$. This is clearly only possible for $r=2$. In this case A is elliptic if and only if

$$\operatorname{Re}a_1(x)\cdot\operatorname{Im}a_2(x)-\operatorname{Re}a_2(x)\cdot\operatorname{Im}a_1(x)\neq 0.$$

For $r=2$ A cannot be strongly elliptic, for

$$\operatorname{Re}(\gamma A^{\mathrm{H}}(x,\xi))=\sum_{j=1}^{2}(\operatorname{Re}a_j\cdot\gamma_1-\operatorname{Im}a_j\cdot\gamma_2)\xi_j=0,$$

does not only imply that $\xi_1=\xi_2=0$. Here we have written $\gamma=\gamma_1+\gamma_2\cdot\mathrm{i}$. For $r>2$ A is not elliptic, and hence also not strongly elliptic.

Example 10.4 The operator defined by

$$A\varphi = \frac{\partial \varphi}{\partial x} + \mathrm{i}\frac{\partial \varphi}{\partial y} + (\mathrm{i}x - y)\frac{\partial \varphi}{\partial t}$$

is not elliptic in the (x, y, t)-space, for if $\xi = (y, -x, 1) \neq 0$, $A^{\mathrm{H}}(x, y, t; \xi) = \xi_1 + \mathrm{i}\xi_2 + (\mathrm{i}x - y)\xi_3 = 0$. See also Example 10.3. In 1957 H. Lewy [1] showed that for this operator there exist functions $f \in C^\infty$ such that the linear differential equation $Au = f$ is neither solvable, nor local in the sense of distribution theory.

Example 10.5 In the case ord $A = n = 2$, $r \geqslant 2$, let

$$A\varphi = \sum_{j,k=1}^{r} a_{jk}(x)\frac{\partial^2 \varphi}{\partial x_j \partial x_k} + \sum_{j=1}^{r} a_j(x)\frac{\partial \varphi}{\partial x_j} + a_0(x)\varphi.$$

$A^{\mathrm{H}}(x, \xi) = \sum_{j,k=1}^{r} a_{jk}(x)\xi_j \cdot \xi_k$ is a quadratic form. If the coefficients $a_{jk}(x)$ are real, then A is strongly elliptic at x, if the quadratic form $A^{\mathrm{H}}(x, \xi)$ is positive or negative definite. In particular the Laplace operator Δ,

$$\Delta\varphi = \sum_{j=1}^{r} \frac{\partial^2 \varphi}{\partial x_j^2}, \quad \Delta(\xi) = \xi_1^2 + \cdots + \xi_1^2 = |\xi|^2,$$

and all its powers Δ^n are strongly elliptic.

Theorem 10.2 *Let*

$$A_1\varphi = \sum_{|s| \leqslant m} a_s(x)\mathrm{D}^s\varphi, \quad A_2\varphi = \sum_{|t| \leqslant n} b_1(x)\mathrm{D}^t\varphi,$$

and $A := A_1 \circ A_2$.

(a) *If A_1 and A_2 are elliptic at x, then so is A.*
(b) *If A_1 and A_2 are strongly elliptic at x, and the coefficients of one operator are real that is, either $a_s(x)$, $|s| = m$ or $b_t(x)$, $|t| = n$, are real, then A is strongly elliptic at x.*
(c) *If A_1 has real coefficients for $|s| = m$, then ellipticity and strong ellipticity are the same.*

Proof. The assertions follow immediately from the relations

$$A^{\mathrm{H}}(x, \xi) = A_1^{\mathrm{H}}(x, \xi) \cdot A_2^{\mathrm{H}}(x, \xi), \quad \text{respectively} \quad \mathrm{Re}(\gamma A^{\mathrm{H}}(x, \xi))$$
$$= \mathrm{Re}\, A_1^{\mathrm{H}}(x, \xi) \cdot \mathrm{Re}(\gamma A_2^{\mathrm{H}}(x, \xi))$$

where in the last equation the assumption for (b) is used. (c) is trivial. ∎

Theorem 10.3 *Let $A(x, \mathrm{D})$ be a linear differential operator on $\tilde{\Omega}$ defined by* (1). *Let $\Phi: \Omega \to \tilde{\Omega}$, $x \mapsto y(x)$ be a $C^{n,0}$-diffeomorphism. Then for the principal part of* *the Φ-transformed differential operator $\tilde{A}$*

$$\tilde{A}(y, \mathrm{D}_y)\varphi := \sum_{|s| \leqslant n} a_s(x(y))\mathrm{D}_x^s \varphi(x(y)),$$

we have the relation

$$\tilde{A}^{\mathrm{H}}(y, \eta) = A^{\mathrm{H}}\left(x(y), {}^{\mathrm{T}}\left\{ \frac{\partial y}{\partial x} \right\} \eta \right), \quad \eta \in \mathbb{R}^r. \tag{5}$$

Here ${}^{\mathrm{T}}\{\partial y/\partial x\}$ is the transposed Jacobi matrix of Φ. $\tilde{A}$ has the same order as A.

Proof. We write $A(x, \mathrm{D}_x)$ in the form

$$A\varphi = \sum_{i_1, \ldots, i_n = 1}^{r} a_{i_1, \ldots, i_n}(x) \frac{\partial^n \varphi}{\partial x_{i_1} \cdots \partial x_{i_n}} + \sum_{|s| < n} a_s(x)\mathrm{D}_x^s \varphi.$$

In order to calculate the principal part of the transformed operator $\tilde{A}$, we apply the chain rule repeatedly and pay attention only to those components which contribute to $\tilde{A}^{\mathrm{H}}$. We write $v(y) := \varphi(x(y))$, $\Phi^{-1}: x(y) \leftarrow\!\!\dashv y$ and have

$$\frac{\partial^n \varphi}{\partial x_{i_1} \cdots \partial x_{i_n}} = \sum_{j_1, \ldots, j_n = 1}^{r} \frac{\partial^n v}{\partial y_{j_1} \cdots \partial y_{j_n}} \cdot \frac{\partial y_{j_1}}{\partial x_{i_1}} \cdots \frac{\partial y_{j_n}}{\partial x_{i_n}} + \sum_{|s| < n} c_s(y)\mathrm{D}_y^s v.$$

Therefore

$$\begin{aligned} \tilde{A}^{\mathrm{H}}(y, \eta) &= \sum_{i_1, \ldots, i_n = 1}^{r} a_{i_1, \ldots, i_n}(x(y)) \sum_{j_1, \ldots, j_n = 1}^{r} \eta_{j_1} \frac{\partial y_{j_1}}{\partial x_{i_1}} \cdots \eta_{j_n} \frac{\partial y_{j_n}}{\partial x_{i_n}} \\ &= \sum_{i_1, \ldots, i_n = 1}^{r} a_{i_1, \ldots, i_n}(x(y)) \xi_{i_1} \cdots \xi_{i_n} = A^{\mathrm{H}}\left(x(y), {}^{\mathrm{T}}\left\{ \frac{\partial y}{\partial x} \right\} \eta \right), \end{aligned}$$

if we write

$$\xi_i = \sum_{j=1}^{r} \frac{\partial y_j}{\partial x_i} \eta_j, \quad \text{hence } \xi = {}^{\mathrm{T}}\left\{ \frac{\partial y}{\partial x} \right\} \eta.$$

The order of A is unchanged. Indeed if A has order n and $\tilde{A}$ has order $\tilde{n}$, then it clearly follows that $\tilde{n} \leqslant n$. Considering the inverse map $\Phi^{-1}: \tilde{\Omega} \to \Omega$ it follows analogously that $n \leqslant \tilde{n}$. ∎

Addendum 10.1 *Theorem* 10.3 *and the corollaries following from it are also valid for* $\operatorname{ord} A = n \geqslant 2$ *under the weaker assumption:* $\Phi \in C^{n-1,1}$, *see Theorem* 1.8.

Corollary 10.1 *Let A, given by* (1), *be elliptic (strongly elliptic), then the transformed operator $\tilde{A}$ is also elliptic (strongly elliptic).*

The proof follows immediately from (5), if one notes that the linear transformation given by the matrix ${}^{\mathrm{T}}\{\partial y/\partial x\}$ is non-singular, $\det\{\partial y/\partial x\} \neq 0$.

Definition 10.5 *Let $r \geqslant 2$. We call the elliptic operator $A(x, \mathrm{D})$ proper (at $x \in \bar{\Omega}$), if for each $0 \neq \xi' \in \mathbb{R}^{r-1}$ the polynomial $P(z) = A^{\mathrm{H}}(x, (\xi', z))$, $z \in \mathscr{C}$, has as many roots in the upper half plane* $\operatorname{Im} z > 0$ *as in the lower half plane* $\operatorname{Im} z < 0$, *counting multiplicities. We say also that in this case $P(z)$ is proper.*

Since A is elliptic, $P(z)$ can have no roots on the real axis, and we deduce that $\operatorname{ord} A = \operatorname{degree} P(z) = 2m$, that is, the order of a proper, elliptic differential operator is even. Since in $\mathbb{R}^r$ for $r \geqslant 2$ one can map two arbitrary, linearly independent vectors to two given linearly independent vectors by means of a non-singular linear transformation, for elliptic operators we have the equivalent definitions:

1. $P(z) = A^{\mathrm{H}}(x; (\xi', z)) = A^{\mathrm{H}}(x; (\xi', 0) + z \cdot (0, \ldots, 0, 1))$ is proper.
2. If $\xi_1, \xi_2 \in \mathbb{R}^r$ are linearly independent, the polynomial in the complex variable z, $A^{\mathrm{H}}(x; \xi_1 + z\xi_2)$ has m roots in $\operatorname{Im} z < 0$ and m roots in $\operatorname{Im} z > 0$.
3. The polynomial $A^{\mathrm{H}}(x; (1, \ldots, 0) + (0, \ldots, 1)z)$ has m roots in $\operatorname{Im} z < 0$ and m in $\operatorname{Im} z > 0$.

Remark 10.1 If the frontier $\partial\Omega$ of Ω is connected, $A(x, \mathrm{D})$ is elliptic in $\bar{\Omega}$, and the coefficients $a_s(x)$ are continuous in $\bar{\Omega}$, then it suffices for A to be proper elliptic on $\bar{\Omega}$, that the root condition 2 from Definition 10.5 hold for only one point $x_0 \in \partial\Omega$, and only one pair of linearly independent vectors $\xi_1, \xi_2 \in \mathbb{R}^r$.

Example 10.6 (Schechter [4]). In $\mathbb{R}^3$ the operator

$$A = \frac{\partial^4}{\partial x_1^4} + \frac{\partial^4}{\partial x_2^4} - \frac{\partial^4}{\partial x_3^4} + \mathrm{i}\left(\frac{\partial^2}{\partial x_1^2} + \frac{\partial^2}{\partial x_2^2}\right)\frac{\partial^2}{\partial x_3^2}$$

is properly elliptic but not strongly elliptic.

Example 10.7 Let $A(x, \mathrm{D})$ be an elliptical differential operator ($r \geqslant 2$) for which the coefficients $a_s(x)$ for $|s| = 2m = \operatorname{ord} A$ are real-valued. Then A is properly elliptic. Indeed the polynomial $P(z) = A^{\mathrm{H}}(x; (\xi', z))$ has real coefficients, no roots on the real axis, and, since then the roots appear in complex conjugate pairs, $P(z)$, respectively $A(x, \mathrm{D})$, is properly elliptic.

Corollary 10.2 *Let $A(x, \mathrm{D}_x)$ given by* (1) *be properly elliptic, let $\Phi: \Omega \to \tilde{\Omega}$ be a $C^{n,0}$-diffeomorphism, then the transformed operator $\tilde{A}(y, \mathrm{D}_y)$ is also properly elliptic.*

Proof. We apply formula (5) to Definition 10.5.2:

$$\tilde{A}^{\mathrm{H}}(y; \eta_1 + z\eta_2) = A^{\mathrm{H}}\left(x(y); {}^{\mathrm{T}}\left\{\frac{\partial y}{\partial x}\right\} \cdot (\eta_1 + z\eta_2)\right)$$

$$= A^{\mathrm{H}}\left(x(y); {}^{\mathrm{T}}\left\{\frac{\partial y}{\partial x}\right\}\eta_1 + z\cdot {}^{\mathrm{T}}\left\{\frac{\partial y}{\partial x}\right\}\eta_2 \right)$$

$$= A^{\mathrm{H}}(x(y); \eta'_1 + z\eta'_2)$$

and observe that because $\det\{\partial y/\partial x\} \neq 0$, if η_1, η_2 are linearly independent then so are the vectors η'_1, η'_2 after transformation by ${}^{\mathrm{T}}\{\partial y/\partial x\}$. ■

Theorem 10.4 *Let the differential operator* $A(x, \mathrm{D})$ *given by* (1) *be elliptic at* $x \in \bar{\Omega}$, *and let* $r \geqslant 3$. *Then* $A(x, \mathrm{D})$ *is properly elliptic, in particular the order of* $A(x, \mathrm{D})$ *is an even number* $n = 2m$.

Proof. We fix $0 \neq \xi' \in \mathbb{R}^{r-1}$. Because $r - 1 \geqslant 2$ there exists a continuous path $\eta(t): [0, 1] \to \mathbb{R}^{r-1}$, with $\eta(0) = \xi'$, $\eta(1) = -\xi'$ and $\eta(t) \neq 0$ for all $t \in [0, 1]$.

The number of zeros z with $\operatorname{Im} z > 0$ of the polynomial

$$A^{\mathrm{H}}(x, (\eta(t), z)) = 0, \quad t \in [0, 1],\ x \text{ fixed}, \tag{6}$$

is independent of t. Namely, since the zeros of the polynomial (6) depend continuously on the coefficients of the polynomial and hence on t, there would otherwise exist some $t_0 \in (0, 1)$ and some real z_0 with

$$A^{\mathrm{H}}(x; (\eta(t_0), z_0)) = 0, \quad \eta(t_0) \neq 0,$$

contradicting the ellipticity of the operator A. Hence the equations

$$A^{\mathrm{H}}(x; (\xi', z)) = 0 \quad \text{and} \quad A^{\mathrm{H}}(x; (-\xi', z)) = 0$$

both have the same number ($= m$) of zeros in $\operatorname{Im} z > 0$. Since A^{H} is homogeneous of degree n, we have

$$A^{\mathrm{H}}(x; (\xi', -z)) = (-1)^n A^{\mathrm{H}}(x; (-\xi', z)),$$

that is, the equation $A^{\mathrm{H}}(x, (\xi', z)) = 0$ possesses exactly m zeros in $\operatorname{Im} z < 0$. ■

Theorem 10.5 *Let the differential operator* $A(x, \mathrm{D})$ *given by* (1) *be strongly elliptic at* $x \in \bar{\Omega}$, *and let* $r \geqslant 2$. *Then* n, *the order of the operator* A, *is an even number*, $n = 2m$.

Proof. We have

$$\operatorname{Re}(\gamma A^{\mathrm{H}}(x, -\xi)) = (-1)^n \operatorname{Re}(\gamma A^{\mathrm{H}}(x, \xi)), \tag{7}$$

for $\xi \in \mathbb{R}^r$, γ some constant (see Definition 10.4) $x \in \bar{\Omega}$ fixed. Because $r \geqslant 2$ there exists for $\bar{\xi} \in \mathbb{R}^r$ a continuous path $\xi(t): [0, 1] \to \mathbb{R}^r$ with

$$\xi(0) = \bar{\xi}, \quad \xi(1) = -\bar{\xi} \quad \text{and} \quad \xi(t) \neq 0 \quad \text{for all } t \in [0, 1].$$

Since as a consequence of strong ellipticity $\operatorname{Re}(\gamma A^{\mathrm{H}}(x, \bar{\xi})) \neq 0$, for example

$\operatorname{Re}(\gamma A^{\mathrm{H}}(x;\bar{\xi})) > 0$, then by (7) in the case that n is odd,

$$\operatorname{Re}(\gamma A^{\mathrm{H}}(x;-\bar{\xi})) < 0.$$

Hence there would exist some $t_0 \in [0,1]$ with

$$\operatorname{Re}(\gamma A^{\mathrm{H}}(x;\xi(t_0))) = 0 \quad \text{and} \quad \xi(t_0) \neq 0,$$

contradicting the strong ellipticity of A. Therefore n must be even, $n = 2m$. ∎

Theorem 10.6 *Let the differential operator A given by (1) be elliptic, respectively strongly elliptic in Ω. Then for each compact subset $K \subset\subset \Omega$ there exists some number $c_0 > 0$, such that*

$$|A^{\mathrm{H}}(x;\xi)| \geqslant c_0 |\xi|^n,$$

respectively

$$|\operatorname{Re}(\gamma A^{\mathrm{H}}(x;\xi))| \geqslant c_0 |\xi|^n,$$

for all $x \in K$ and all $\xi \in \mathbb{R}^r$.

Proof. Let $\Sigma := \{\xi : \xi \in \mathbb{R}^r, |\xi| = 1\}$. If A is elliptic in Ω, then on the compact subset $K \times \Sigma \subset \mathbb{R}^{2r-1}$ the continuous positive function $|A^{\mathrm{H}}(x;\xi)|$ takes a positive minimum value $c_0 > 0$. Therefore for arbitrary $0 \neq \xi \in \mathbb{R}^r$ we have

$$\left| A^{\mathrm{H}}\left(x, \frac{\xi}{|\xi|}\right) \right| = \frac{1}{|\xi|^n} |A^{\mathrm{H}}(x,\xi)| \geqslant c_0 > 0,$$

for all $x \in K$. If A is strongly elliptic one considers the continuous function $|\operatorname{Re}(\gamma A^{\mathrm{H}}(x,\xi))|$ and proceeds analogously. ∎

Addendum 10.2 *Let $r \geqslant 2$. If the compact set $K \subset \Omega$ is connected, then $K \times \Sigma$ is also connected. Therefore* $\operatorname{Re}(\gamma A^{\mathrm{H}}(x,\xi))$ *has constant sign on $K \times \Sigma$, where γ is the constant appearing in the Definition* 10.4 *of strong ellipticity. Therefore either*

$$\operatorname{Re}(\gamma A^{\mathrm{H}}(x,\xi)) \geqslant c_0 |\xi|^n,$$

or

$$-\operatorname{Re}(\gamma A^{\mathrm{H}}(x;\xi)) \geqslant c_0 |\xi|^n,$$

for all $x \in K$ and all $\xi \in \mathbb{R}^r$.

Theorem 10.6 suggests the following definitions:

Definition 10.6 *We call the linear differential operator* $A(x, \mathrm{D})$ *defined by* (1) *uniformly elliptic on Ω, if there exists some constant $c_0 > 0$ with*

$$|A^{\mathrm{H}}(x;\xi)| \geqslant c_0 |\xi|^n,$$

for all $x \in \bar{\Omega}$ *and* $\xi \in \mathbb{R}^r$. $A(x, \mathrm{D})$ *is called strongly uniform elliptic, if there exist constants* γ *and* $c_0 > 0$ *with*

$$\mathrm{Re}\,(\gamma A^{\mathrm{H}}(x; \xi)) \geqslant c_0 |\xi|^n \tag{8}$$

for all $x \in \bar{\Omega}$ *and* $\xi \in \mathbb{R}^r$.

By Theorem 10.6 and Addendum 10.2, Definition 10.2 only provides something new for unbounded regions.

Definition 10.7 *We call the differential operator A defined by* (1) *(formally) self-adjoint, if* $A = A^*$ *(compare* (3)).

Theorem 10.7 *Let the differential operator A be (formally) self-adjoint and elliptic at* $x \in \bar{\Omega}$. *Then A is also strongly elliptic.*

Proof. For the coefficients of the principal part A^{H} of a formally self-adjoint operator we have

$$a_s(x) = (-1)^n \overline{a_s(x)}, \quad |s| = n, \tag{9}$$

this follows from (3) and (4). In the proof we distinguish between two cases:

(a) $\operatorname{ord} A = 2m$, then it follows from (9) that the coefficients of the principal part A^{H} are real, and A is strongly elliptic with $\gamma = 1$.

(b) $\operatorname{ord} A = 2m - 1$, then it follows from (9) that the coefficients of the principal part A^{H} are pure imaginary, and that A is strongly elliptic with $\gamma = -i$. ■

Remark Let $r \geqslant 2$, in the future we shall always multiply the coefficients of $A(x, \mathrm{D})$ by the constant γ in the inequality (8), and hence always characterise uniform strong ellipticity by the inequality

$$\mathrm{Re}\, A^{\mathrm{H}}(x; \xi) \geqslant c_0 |\xi|^n, \quad x \in \bar{\Omega}, \xi \in \mathbb{R}^r. \tag{10}$$

Exercises

10.1 Write the Laplace operator $\Delta = \partial^2/\partial x^2 + \partial^2/\partial y^2$ in polar coordinates (ρ, θ).

10.2 Let Ω be the open disc, $B = \{(x, y) \in \mathbb{R}^2 : x^2 + y^2 < 1\}$. We develop the functions u from $W_2^l(B)$ in Fourier series (polar coordinates)

$$u(\rho, \theta) = \sum_{k=-\infty}^{\infty} c_k(\rho) e^{ik\theta}. \tag{1}$$

Use Exercise 4.4 in order to give necessary and sufficient conditions on the Fourier coefficients $c_k(\rho)$ so that $u \in W_2^l(B)$.

10.3 Use the results of Exercise 10.2 in order to show that

$$T_0 : W_2^l(B) \to W_2^{l-1/2}(T^1) \text{ is continuous } l \geqslant 1;$$

(T_0 the trace operator $B \to \partial B = T^1$)

$$T_0\left(\frac{\partial}{\partial\rho}\right)^j : W_2^l(B) \to W_2^{l-j-1/2}(T^1) \text{ is continuous } l-j \geq 1.$$

10.4 Solve using equation (1) above the (partially homogeneous) Dirichlet problem

$$\begin{aligned} -\Delta u &= 0 && \text{on } B \\ u|_{\partial B} &= T_0 u = g, && \text{on } \partial B = T^1. \end{aligned} \tag{2}$$

Show that for $g \in W_2^{l-1/2}(T^1)$, $l \geq 1$ there exists a unique solution $u \in W_2^l(B)$ of (2) which depends linearly and continuously on g.

10.5 Solve the Dirichlet problem

$$\begin{aligned} -\Delta u &= f && \text{on } B, \\ u|_{\partial B} &= T_0 u = g && \text{on } \partial B, \end{aligned}$$

where $f \in W_2^{l-2}(B)$, $g \in W_2^{l-1/2}(T^1)$ and $u \in W_2^l(B)$.

§11 The Lopatinskiĭ–Šapiro condition and examples

In this section we consider the Lopatinskiĭ–Šapiro condition (L.Š. for short) for boundary value problems of an elliptic differential operator. Here we have formulated the L.Š. condition as an initial value problem for ordinary differential equations and not algebraically as a 'covering condition'. Indeed in examples it becomes apparent (see §11.2) that one can easily verify the L.Š. condition in the first form, but that the algebraic form is difficult to handle.

We give the L.Š. condition three equivalent formulations, the third, a theorem which goes back to Hörmander [1], finds central application in the proof of the main Theorem 13.1 for elliptic boundary value problems. One assertion of the main Theorem 13.1 is that the ellipticity of the differential operator plus the L.Š. condition are equivalent to the Fredholm property for a boundary value problem.

11.1 The Lopatinskiĭ–Šapiro (L.Š.) condition

Let Ω be an open, bounded set in $\mathbb{R}^r$ with a sufficiently smooth frontier $\partial\Omega$; in §13 we shall make the exact smoothness conditions precise. By a boundary value operator on $\partial\Omega$ we understand the expression

$$b(x, \mathrm{D})\varphi = \sum_{|s| \leq m} b_s(x) T_0(\mathrm{D}^s\varphi), \quad x \in \partial\Omega, \quad m = \operatorname{ord} b,$$

where T_0 is the trace operator from §8. In order that $b(x, \mathrm{D})$ be continuous, for example

$$b(x, \mathrm{D}) : W_2^{l+m+1/2}(\Omega) \to W_2^l(\partial\Omega) \text{ is continuous,}$$

it is sufficient to require of the coefficients $b_s(x)$ that they belong to $C^l(\partial\Omega)$, $|s| \leqslant m$, or that $b_s(x) \in C^{[l]+1}(\partial\Omega)$ if l is not integral. If we consider $b(x, \mathrm{D})$ as a map

$$b(x, \mathrm{D}): W_2^l(\Omega) \to W_2^{l-m-1/2}(\partial\Omega),$$

Then a typical requirement for continuity is

$$b_s(x) \in C^{[l-m-1/2]+1}(\partial\Omega).$$

Higher smoothness assumptions are needed for conversions thus, for example, we require that the coefficients $b_s(x)$ may be extended to $\bar{\Omega}$ and belong to $C^{m+l+1}(\bar{\Omega})$; in this case because of the multiplicative property of the trace operator (see the proof of Theorem 8.9) we have that

$$b(x, \mathrm{D})\varphi = \sum_{|s| \leqslant m} b_s(x) T_0(\mathrm{D}^s\varphi) = T_0\left(\sum_{|s| \leqslant m} b_s(x)\mathrm{D}^s\varphi \right).$$

We introduce the exact conditions on the coefficients $b_s(x)$ of $b(x, \mathrm{D})$ later in §13; in this section our studies are essentially restricted to one dimension only and for operators with constant coefficients.

Suppose we are given an elliptic differential operator $A(x, \mathrm{D})$ on $\bar{\Omega}$ of order $2m$, and m boundary value operators on $\partial\Omega$, $b_j(x, \mathrm{D}), j = 1, \ldots, m$, with corresponding orders m_j, where we suppose that $0 \leqslant m_j \leqslant 2m - 1$, $j = 1, \ldots, m$. In order that we can operate more easily with the Fourier transformation, from now on we will always equip the basic derivatives $\partial/\partial x_1, \ldots, \partial/\partial x_r$ with the factor $1/\mathrm{i}$. $(1/\mathrm{i})(\partial/\partial x_1), \ldots, (1/\mathrm{i})(\partial/\partial x_r)$ is then transformed by $\mathscr{F}$ (the Fourier transformation) to $(\xi_1, \ldots, \xi_r)$, see §1, (47).

We want to formulate the L.Š. condition at a boundary point $x_0 \in \partial\Omega$. Let $A^{\mathrm{H}}(x, \mathrm{D})$ be the principal part of $A(x, \mathrm{D})$, we fix a point $x_0 \in \partial\Omega$, take x_0 as origin of coordinates and choose the coordinate axis x_r in the direction of the inward pointing normal, the other coordinates perpendicular to x_r. The differential operator $A^{\mathrm{H}}(x_0, \mathrm{D})$ now has constant coefficients, we carry out the Fourier transformation $\mathscr{F}_{r-1}$ (see §1) on $(1/\mathrm{i})(\partial/\partial x_1), \ldots, (1/\mathrm{i})(\partial/\partial x_{r-1})$, which becomes $(\xi_1, \ldots, \xi_{r-1}) = \xi' \in T_{\partial\Omega}$, and we obtain with $x_r = t$,

$$\mathscr{F}_{r-1} A^{\mathrm{H}}(x_0, \mathrm{D}) = A^{\mathrm{H}}\left(x_0; \xi', \frac{1}{\mathrm{i}} \frac{\partial}{\partial x_r} \right) = A^{\mathrm{H}}\left(x_0; \xi', \frac{1}{\mathrm{i}} \frac{\mathrm{d}}{\mathrm{d}t} \right).$$

We fix $\xi' \neq 0$ and for $t \geqslant 0$ we consider the ordinary, linear differential equation with constant coefficients

$$A^{\mathrm{H}}\left(x_0; \xi', \frac{1}{\mathrm{i}} \frac{\mathrm{d}}{\mathrm{d}t} \right) v(t) = 0, \quad 0 \neq \xi' \in T_{\partial\Omega} \simeq \mathbb{R}^{r-1}, \tag{1}$$

where $T_{\partial\Omega}$ is the tangential hyperplane to $\partial\Omega$ at the point x_0. Clearly the solution space $\mathscr{M}$ of (1) decomposes as the direct sum

$$\mathscr{M} = \mathscr{M}^+ \oplus \mathscr{M}^- \oplus \mathscr{M}^0,$$

where $\mathscr{M}^+$ is the solution space for characteristic roots (that is, roots of $P(z) = A^{\mathrm{H}}(x_0, \xi', z)$) in the upper half plane $\operatorname{Im}\lambda > 0$. $\mathscr{M}^-$ is the solution space with characteristic values in the lower half plane $\operatorname{Im}\lambda < 0$. On the real axis, because of the assumed ellipticity of A^{H}, there exist no characteristic values, so that $\mathscr{M}^0 = (0)$. Here and in what follows we write the exponential solutions with an $\mathrm{i} = \sqrt{(-1)}$, hence $\mathrm{e}^{\mathrm{i}\lambda t}$, where λ is the characteristic root. Since $\mathscr{M}^0 = (0)$ we can equivalently characterise the solution space $\mathscr{M}^+$ as the space of stable solutions v, that is, $v(t) \to 0$ as $t \to \infty$, or as the space of the bounded solutions, or as the space of the tempered solutions v, that is, $v(t) = 0(t^m)$, or as the space of solutions in $W_2^{2m}(\mathbb{R}_+)$.

From the boundary value operators b_j, $j = 1, \ldots, m$, we now form the initial value conditions for $t = 0$

$$b_1^{\mathrm{H}}\left(x_0; \xi', \frac{1}{\mathrm{i}}\frac{\mathrm{d}}{dt}\right)v(0) = 0, \ldots, b_m^{\mathrm{H}}\left(x_0; \xi', \frac{1}{\mathrm{i}}\frac{\mathrm{d}}{dt}\right)v(0) = 0. \tag{2}$$

here we obtain the expressions $b_j^{\mathrm{H}}(x_0; \xi', (1/\mathrm{i})(\mathrm{d}/\mathrm{d}t))$, $j = 1, \ldots, m$, in the same way (that is, fixing $x_0 \in \partial\Omega$, applying the Fourier transformation $\mathscr{F}_{r-1}$, etc.) from the b_js as $A^{\mathrm{H}}(x_0; \xi', (1/\mathrm{i})(\mathrm{d}/\mathrm{d}t))$ from A. The L.Š. condition reads:

Condition 11.1 (Lopatinskiĭ–Šapiro – L.Š.) *The initial value problem* (1), (2) *admits in* $\mathscr{M}^+$ *only the trivial solution for all* $0 \neq \xi' \in T_{\partial\Omega} \simeq \mathbb{R}^{r-1}$; *or equivalently, see above,* (1), (2) *admits in* $W_2^{2m}(\mathbb{R}_+)$ *only the trivial solution.*

Remark In order to formulate the L.Š. condition we do not need the Fourier transformation we only need it later as a method of proof. What we must do is the following: let $T_{x_0} = T_{\partial\Omega} \oplus \mathbf{n}_{x_0}$ (see §2, Definition 2.9), we choose coordinates in such a way that

$$T_{\partial\Omega} = \left\{\frac{\partial}{\partial x_1}, \ldots, \frac{\partial}{\partial x_{r-1}}\right\}, \quad \mathbf{n}_{x_0} = \left\{\frac{\partial}{\partial x_r}\right\} = \left\{\frac{\mathrm{d}}{\mathrm{d}t}\right\},$$

and in the differential expressions replace

$$\frac{1}{\mathrm{i}}\frac{\partial}{\partial x_1}, \ldots, \frac{1}{\mathrm{i}}\frac{\partial}{\partial x_{r-1}} \text{ by the tangent vector } \xi' \in T_{\partial\Omega} \simeq \mathbb{R}^{r-1}.$$

We wish to illuminate the relevance of the L.Š. condition by a theorem going back to Hörmander [1]. For the sake of simplicity we take homogeneous differential polynomials with constant coefficients and

$$\Omega = \mathbb{R}^r_+, \quad \partial\Omega = \mathbb{R}^{r-1} = \{x : x \in \mathbb{R}^r, x_r = 0\}.$$

Theorem 11.1 *We consider the boundary value problem*

$$A(\mathrm{D})u = 0 \ \text{in} \ \bar{\mathbb{R}}^r_+, \quad b_j(\mathrm{D})u = 0 \ \text{in} \ \mathbb{R}^{r-1}, \quad j = 1, \ldots, m. \tag{3}$$

We assume the following regularity: if the solution u of (3) *belongs to some* $C^N(\bar{\mathbb{R}}^r_+)$, $N \in \mathbb{N}$, *then u must already belong to* $C^\infty(\bar{\mathbb{R}}^r_+)$. *The operators A and* $b_j, j = 1, \ldots, m$, *must then satisfy the L.Š. condition.*

Proof. By an exponential solution of (3) we understand an expression of the form

$$u(x) = \mathrm{e}^{\mathrm{i}(x', \xi')} v(x_r) = \mathrm{e}^{\mathrm{i}(x', \xi')} v(t), \tag{4}$$

where

$$x' = (x_1, \ldots, x_{r-1}), \ \xi' = (\xi_1, \ldots, \xi_{r-1}), \ x_r = t \text{ and } (x', \xi') = x_1 \cdot \xi_1 + \cdots + x_{r-1} \cdot \xi_{r-1}.$$

A solution is called tempered, if as $|x| \to \infty$ in $\bar{\mathbb{R}}^r_+$ we have $u(x) = O(|x|^M)$, $M \in \mathbb{N}$. We show first that (3) has no non-zero tempered exponential solution u for $\xi' \neq 0$.

We equip the spaces $C^N(\bar{\mathbb{R}}^r_+)$ and $C^\infty(\bar{\mathbb{R}}^r_+)$ with an (F)-space topology by means of the seminorms $\sup_{x \in K} \max_{|s| \leq N} |\mathrm{D}^s \varphi(x)|$, respectively $\sup_{x \in K} \max_{|s| \leq n} |\mathrm{D}^s \varphi(x)|$, where $K \subset\subset \bar{\mathbb{R}}^r_+$ and $n \in \mathbb{N}$. In order that we deal only with classical derivatives in $A(\mathrm{D})$ and $b_j(\mathrm{D})$ we can assume that $N \geq \operatorname{ord} A$, $\operatorname{ord} b_j$, $j = 1, \ldots, m$. Let L_N be the subspace of $C^N(\bar{\mathbb{R}}^r_+)$ of solutions of (3), L_∞ the corresponding subspace of $C^\infty(\bar{\mathbb{R}}^r_+)$. One sees that L_N, respectively L_∞, are closed, hence are again (F)-spaces. By the regularity assumption the natural map

$$L_\infty \to L_N$$

is surjective, and it is of course continuous. We can apply the open mapping theorem, and obtain the continuity of the inverse map

$$L_N \to L_\infty.$$

We put $n = N + 1$, then for each compact subset $K \subset\subset \bar{\mathbb{R}}^r_+$ we can find a constant C and some other compact subset $K' \subset\subset \bar{\mathbb{R}}^r_+$ with

$$\sup_{x \in K} \max_{|s| \leq N+1} |\mathrm{D}^s u(x)| \leq C \cdot \sup_{x \in K'} \max_{|s| \leq N} |\mathrm{D}^s u(x)| \quad \text{for all } u \in L_N. \tag{5}$$

Here we can choose both compact subsets K, K' as neighbourhoods of 0 in $\bar{\mathbb{R}}^r_+$. Let $u(x) \neq 0$ (see (4)) be a tempered exponential solution of (3) with

$\xi' \neq 0$. Then ξ' must be real, since otherwise the increase of u would be exponential. From (3) and (4) it follows that the function $v(t)$ satisfies the differential equation

$$A\left(\xi', \frac{1}{\mathrm{i}} \frac{\mathrm{d}}{\mathrm{d}t}\right) v(t) = 0, \quad t \geqslant 0, \tag{6}$$

with the initial conditions

$$b_1\left(\xi', \frac{1}{\mathrm{i}} \frac{\mathrm{d}}{\mathrm{d}t}\right) v(0) = 0, \ldots, b_m\left(\xi', \frac{1}{\mathrm{i}} \frac{\mathrm{d}}{\mathrm{d}t}\right) v(0) = 0. \tag{7}$$

The solution space $\mathscr{M}$ of (6) has the form

$$\mathscr{M} = \mathscr{M}^+ \oplus \mathscr{M}^0 \oplus \mathscr{M}^-,$$

where $\mathscr{M}^+$ and $\mathscr{M}^-$ have the same significance as above, and $\mathscr{M}^0$ contains all exponential, polynomial solutions with characteristic values on the real axis. The tempered solutions $v(t)$ of (6) lie in $\mathscr{M}^0 \oplus \mathscr{M}^+$.

Along with $u(x)$, $u_\lambda(x) = \mathrm{e}^{\mathrm{i}\lambda\langle x', \xi' \rangle} v(\lambda t)$ is also a tempered solution of (3), and the estimate (5) applied to $u_\lambda(x)$ implies that the right-hand side of (5) is $O(\lambda^{M+N})$ as $\lambda \to \infty$. Since $\xi' \neq 0$ at least one of the tangential derivatives, that is, one of the derivatives with respect to $(\partial/\partial x_1) \cdots (\partial/\partial x_{r-1})$ on the left-hand side of (5) must be $O(\lambda^{M+N+1})$ at least, which gives a contradiction. Therefore (3) can possess no tempered solution of the form (4) other than zero, or (6) with the initial conditions (7) admits only the zero solution in $\mathscr{M}^+ \oplus \mathscr{M}^0$. This implication also holds on restriction of the solution space to $\mathscr{M}^+$, hence we have deduced the L.Š. condition. ∎

From Theorem 11.1 we see that if we require regularity, in the sense of Weyl's lemma, then the boundary value problem must satisfy the L.Š. condition; here the differential operator does not even need to be elliptic. And for elliptic boundary value problems the requirement of Weyl's lemma is natural, since we expect a far-reaching regularity for elliptic problems.

We return to the case of an elliptic differential operator $A(x, D)$.

Theorem 11.2 *Let $A(x, \mathrm{D})$ be an elliptic differential operator of order $2m$, and $b_1(x, \mathrm{D}), \ldots, b_m(x, \mathrm{D})$ boundary operators. Let the L.Š. condition 11.1 be satisfied at $x_0 \in \partial\Omega$. Then the solution space of (1) $\mathscr{M} = \mathscr{M}^+ \oplus \mathscr{M}^-$ satisfies*

$$\dim \mathscr{M}^+ = \dim \mathscr{M}^- = m, \tag{8}$$

that is, the operator $A(x, \mathrm{D})$ is properly elliptic at $x_0 \in \partial\Omega$.

Proof. We examine the dependence on $0 \neq \xi' \in \mathbb{R}^{r-1}$:

$$\begin{aligned} &\mathcal{M}(\xi') = \mathcal{M}^{+}(\xi') \oplus \mathcal{M}^{-}(\xi'),^{\dagger} \\ &\dim \mathcal{M}(\xi') = \dim \mathcal{M}^{+}(\xi') + \dim \mathcal{M}^{-}(\xi'), \\ &\dim \mathcal{M}^{+}(\xi') = \text{Number of zeros in } \operatorname{Im} z > 0 \text{ of } P(z) = A^{\mathrm{H}}(\xi', z), \\ &\dim \mathcal{M}^{-}(\xi') = \text{Number of zeros in } \operatorname{Im} z < 0 \text{ of } P(z) = A^{\mathrm{H}}(\xi', z). \end{aligned} \tag{9}$$

Since A^{H} is homogeneous ($A^{\mathrm{H}}(-\xi', -z) = A^{\mathrm{H}}(\xi', z)$), we have

$$\mathcal{M}^{+}(-\xi') = \mathcal{M}^{-}(\xi), \quad \mathcal{M}^{-}(-\xi') = \mathcal{M}^{+}(\xi'). \tag{10}$$

The L.Š. condition 11.1 gives

$$\dim \mathcal{M}^{+}(\xi') \leqslant m \quad \text{for all } 0 \neq \xi' \in \mathbb{R}^{r-1}.$$

If we substitute $-\xi'$ and take account of (9), we obtain $\dim \mathcal{M}^{-}(\xi') = \dim \mathcal{M}^{+}(-\xi') \leqslant m$, and after addition

$$2m = \dim \mathcal{M}(\xi') = \dim \mathcal{M}^{+}(\xi') + \dim \mathcal{M}^{-}(\xi') \leqslant 2m,$$

which is only possible with equality everywhere. Therefore we have

$$\dim \mathcal{M}^{+} = \dim \mathcal{M}^{-} = m,$$

which because of (9) implies that the operator $A(x, \mathrm{D})$ is properly elliptic at $x_0 \in \partial\Omega$. ∎

Now we give the L.Š. condition in different equivalent versions, assuming here that the operator $A(x, \mathrm{D})$ is elliptic.

Condition 11.2 (L.Š.). *For all $0 \neq \xi' \in \mathbb{R}^{r-1}$ the initial value problem*

$$A^{\mathrm{H}}\left(x_0, \xi', \frac{1}{\mathrm{i}}\frac{\mathrm{d}}{\mathrm{d}t}\right) v(t) = 0 \quad t \geqslant 0 \tag{1}$$

$$b_1^{\mathrm{H}}\left(x_0, \xi', \frac{1}{\mathrm{i}}\frac{\mathrm{d}}{\mathrm{d}t}\right) v(0) = h_1, \ldots, b_m^{\mathrm{H}}\left(x_0, \xi', \frac{1}{\mathrm{i}}\frac{\mathrm{d}}{\mathrm{d}t}\right) v(0) = h_m, \tag{2'}$$

$$h = (h_1, \ldots, h_m) \text{ arbitrary from } \mathbb{R}^m \text{ respectively } \mathbb{C}^m,$$

is uniquely solvable in $\mathcal{M}^{+}$.†

The equivalence with L.Š. condition 11.1 follows immediately from Theorem 11.2, (8). For the third equivalent version of the L.Š. condition we

† We emphasise once more that A elliptic means that $\mathcal{M}^{0}(\xi') = (0)$ for all $0 \neq \xi' \in \mathbb{R}^{r-1}$.

need the assumption

$$0 \leqslant m_j = \operatorname{ord} b_j \leqslant 2m - 1. \tag{11}$$

Let L^{H} be the initial value operator

$$L^{\mathrm{H}} = \left(A^{\mathrm{H}}\left(x_0, \xi', \frac{1}{\mathrm{i}} \frac{\mathrm{d}}{\mathrm{d}t} \right), b_1^{\mathrm{H}}\left(x_0, \xi', \frac{1}{\mathrm{i}} \frac{\mathrm{d}}{\mathrm{d}t} \right)\Big|_{t=0}, \ldots, b_m^{\mathrm{H}}\left(x_0, \xi', \frac{1}{\mathrm{i}} \frac{\mathrm{d}}{\mathrm{d}t} \right)_{t=0} \right), \tag{12}$$

acting between the Banach spaces

$$L^{\mathrm{H}}: W_2^{2m+l}(\mathbb{R}_+) \to W_2^l(\mathbb{R}_+) \times \mathbb{C}^m. \tag{13}$$

Condition 11.3 (L.Š) *Let $l \geqslant 0$. For all $0 \neq \xi' \in \mathbb{R}^{r-1}$ the operator L^{H} (see (12)) is a topological and algebraic isomorphism between the spaces $W_2^{2m+l}(\mathbb{R}_+)$ and $W_2^l(\mathbb{R}_+) \times \mathbb{C}^m$. In particular we have the estimate*

$$\|\varphi\|_{2m+l} \leqslant C_{\xi'} \left(\left\| A^{\mathrm{H}}\left(x_0, \xi', \frac{1}{\mathrm{i}} \frac{\mathrm{d}}{\mathrm{d}t} \right) \varphi(t) \right\|_l + \sum_{j=1}^{m} \left| b_j\left(x_0, \xi', \frac{1}{\mathrm{i}} \frac{\mathrm{d}}{\mathrm{d}t} \right) \varphi(0) \right| \right)$$
$$\cdot \textit{for all } \varphi \in W_2^{2m+l}(\mathbb{R}_+), \tag{14}$$

and the constant $C_{\xi'}$ depends continuously on $\xi' \in \mathbb{R}^{r-1}\{0\}$.

Proof. That L.Š. Condition 11.3 implies L.Š. Condition 11.1 is immediately clear from the estimate (14), since $\mathscr{M}^+ \subset W_2^{2m+l}(\mathbb{R}_+)$.

Suppose conversely that the L.Š. Condition 11.2 is satisfied. We first show the continuity of the map L^{H} (13). For fixed $0 \neq \xi'$, $A^{\mathrm{H}}(\cdots(1/\mathrm{i})(\mathrm{d}/\mathrm{d}t))$ is a differential operator of order $2m$ with constant coefficients, which therefore acts continuously

$$A^{\mathrm{H}}: W_2^{2m+l}(\mathbb{R}_+) \to W_2^l(\mathbb{R}_+).$$

Now we deal with the boundary operators b_j^{H}. As Sobolev spaces, respectively C-spaces, over a point 0 we define

$$W_2^l(0) := C^{k,\kappa}(0) := \mathbb{C},$$

where we give $\mathbb{C}$ the usual topology. For b_j^{H} we have from (12) by Theorem 8.1, after a previous extension using Theorem 5.5:

$$b_j^{\mathrm{H}}: W_2^{2m+l}(\mathbb{R}_+) \to W_2^{2m+l-(2m-1)-1/2}(0) = W_2^{l+1/2}(0) = \mathbb{C} \text{ continuous}, \tag{15}$$

and thus we have proved the continuity of (13). We remark once more that in proving (15) we have made use of assumption (11). We now prove that (13) is injective and surjective, the inverse mapping theorem then gives the continuity of the inverse map

$$(L^{\mathrm{H}})^{-1}: W_2^l(\mathbb{R}_+) \times \mathbb{C}^m \to W_2^{2m+l}(\mathbb{R}_+),$$

and simultaneously the estimate (14).

L^{H} is injective on $W_2^{2m+l}(\mathbb{R}_+)$: this follows immediately from the L.Š. Condition 11.1, since $\mathcal{M} \cap W_2^{2m+l}(\mathbb{R}_+) = \mathcal{M}^+$.

L^{H} is surjective: let $(c_1, \ldots, c_m) \in \mathbb{C}^m$ and $f \in W_2^l(\mathbb{R}_+^1)$ be preassigned. By Theorem 5.5 we can extend f to the whole axis $\mathbb{R}^1$, where for the extension $\tilde{f}$ we have

$$\tilde{f} \in W_2^l(\mathbb{R}^1), \quad \|\tilde{f}\|_{l,\mathbb{R}^1} \leqslant c \|f\|_{l,\mathbb{R}_+^1}$$

with c independent of f. We apply the one-dimensional Fourier transformation (for the variable t), use the space $L_2^l(\mathbb{R})$ from Theorem 5.1 and have

$$\mathscr{F}\tilde{f} =: \hat{\tilde{f}}(\tau) \in L_2^l(\mathbb{R}).$$

We form $\hat{\psi}(\tau) := \hat{\tilde{f}}(\tau)/A^{\mathrm{H}}(\cdots, \tau)$ and assert that $\mathscr{F}^{-1}\hat{\psi} =: \psi(t)$ belongs to $W_2^{2m+l}(\mathbb{R})$. Because of the ellipticity of A, $(1 + |\tau|^2)^m / A^{\mathrm{H}}(\cdots, \tau)$ is bounded for $\tau \in \mathbb{R}$, from which it follows that $\hat{\psi}(\tau) \in L_2^{l+2m}(\mathbb{R})$, which is equivalent to $\psi(\tau) \in H^{2m+l}(\mathbb{R}) \simeq W_2^{2m+l}(\mathbb{R})$. One also sees by means of the Fourier transformation that $\psi(\tau)$ satisfies the equation

$$A^{\mathrm{H}}\left(\cdots, \frac{1}{\mathrm{i}}\frac{\mathrm{d}}{\mathrm{d}t}\right)\psi(t) = \tilde{f}(t) \quad \text{for all } t \in \mathbb{R}'.$$

Let $\psi_1(t)$ be the restriction of $\psi(t)$ to $\mathbb{R}_+$, we have $\psi_1 \in W_2^{2m+l}(\mathbb{R}_+)$. By (15) the boundary value operators $b_j^{\mathrm{H}}: W_2^{2m+l}(\mathbb{R}_+) \to \mathbb{C}$ act continuously, we can therefore calculate the boundary values

$$b_j^{\mathrm{H}}\left(\cdots, \frac{1}{\mathrm{i}}\frac{\mathrm{d}}{\mathrm{d}t}\right)\psi_1(t)|_{t=0} \quad \gamma_j, \quad j = 1, \ldots, m,$$

By Condition 11.2 we can find a solution $\psi_2(t)$ in $\mathcal{M}^+ \subset W_2^{2m+l}(\mathbb{R}_+)$ of

$$A^{\mathrm{H}}\left(\cdots, \frac{1}{\mathrm{i}}\frac{\mathrm{d}}{\mathrm{d}t}\right)\psi_2(t) = 0, \qquad t \in \mathbb{R}_+$$

and

$$b_j^{\mathrm{H}}\left(\cdots, \frac{1}{\mathrm{i}}\frac{\mathrm{d}}{\mathrm{d}t}\right)\psi_2(0) = c_j - \gamma_j, \quad j = 1, \ldots, m.$$

It is easily seen that $\varphi(t) := \psi_1(t) + \psi_2(t)$ solves the initial value problem

$$A^{\mathrm{H}}\left(\cdots, \frac{1}{\mathrm{i}}\frac{\mathrm{d}}{\mathrm{d}t}\right)\varphi(t) = f(t), \quad t > 0,$$

$$b_j\left(\cdots, \frac{1}{\mathrm{i}}\frac{\mathrm{d}}{\mathrm{d}t}\right)\varphi(0) = c_j, \qquad j = 1, \ldots, m,$$

hence we have proved the surjectivity of L^{H} (13). To conclude the proof we show that the constant $C_{\xi'}$ in (14) depends continuously on $\xi' \in \mathbb{R}^{r-1}\{0\}$. By this we understand that when $\tilde{\xi}'$ is close to ξ' then $C_{\tilde{\xi}'}$ can also be chosen close to $C_{\xi'}$, and such that for all $\varphi \in W_2^{2m+l}(\mathbb{R}_+)$

$$\|\varphi\|_{2m+l} \leqslant C_{\tilde{\xi}'}\left(\left\|A^{\mathrm{H}}\left(x_0, \tilde{\xi}', \frac{1}{\mathrm{i}}\frac{\mathrm{d}}{\mathrm{d}t}\right)\varphi\right\|_l + \sum_{j=1}^{m}\left|b_j\left(x_0, \tilde{\xi}', \frac{1}{\mathrm{i}}\frac{\mathrm{d}}{\mathrm{d}t}\right)\varphi(0)\right|\right). \tag{16}$$

We take $\tilde{\xi}'$ so close to ξ' that we have

$$\left\|A^{\mathrm{H}}\left(x_0, \xi', \frac{1}{\mathrm{i}}\frac{\mathrm{d}}{\mathrm{d}t}\right)\varphi - A^{\mathrm{H}}\left(x_0, \tilde{\xi}', \frac{1}{\mathrm{i}}\frac{\mathrm{d}}{\mathrm{d}t}\right)\varphi\right\|_l \leqslant \varepsilon\|\varphi\|_{2m+l},$$

$$\sum_{j=1}^{m}\left|b_j^{\mathrm{H}}\left(x_0, \xi', \frac{1}{\mathrm{i}}\frac{\mathrm{d}}{\mathrm{d}t}\right)\varphi(0) - b_j^{\mathrm{H}}\left(x_0, \tilde{\xi}', \frac{1}{\mathrm{i}}\frac{\mathrm{d}}{\mathrm{d}t}\right)\varphi(0)\right| \leqslant \varepsilon\|\varphi\|_{2m+l},$$

which because of the continuous dependence of the coefficients of the expressions A^{H}, b_j^{H} on $\xi' \in \mathbb{R}^{r-1}\{0\}$ is always possible; here $\varepsilon > 0$ is arbitrarily preassigned. Combining (14) and (17) (triangle inequality) gives

$$\|\varphi\|_{2m+l} \leqslant C_{\xi'}\left(\left\|A^{\mathrm{H}}\left(x_0, \tilde{\xi}', \frac{1}{\mathrm{i}}\frac{\mathrm{d}}{\mathrm{d}t}\right)\varphi(t)\right\|_l + \sum_{j=1}^{m}\left|b_j\left(x_0, \tilde{\xi}', \frac{1}{\mathrm{i}}\frac{\mathrm{d}}{\mathrm{d}t}\right)\varphi(0)\right|\right)$$
$$+ 2\varepsilon C_{\xi'}\|\varphi\|_{2m+l}.$$

If we impose on $\varepsilon > 0$ the restriction $1 - 2\varepsilon C_{\xi'} > 0$ and we substitute $C_{\tilde{\xi}'} := C_{\xi'}/(1 - 2\varepsilon C_{\xi'})$ then we see that (16) holds, so that $C_{\tilde{\xi}'}$ can be arbitrarily close to $C_{\xi'}$. ∎

To conclude this section we wish to show that the L.Š. condition does not depend on the choice of coordinates at $x_0 \in \partial\Omega$. More precisely, we wish to show that the L.Š. condition is invariant under admissible coordinate transformations, see Definition 2.9 (the normal directions and the tangent hyperplanes must correspond to one another at a point).

Theorem 11.3 *Let $\Phi: x \to y$ be an admissible coordinate transformation in some neighbourhood of $x_0 \in \partial\Omega$. If the L.Š. Condition 11.1 is satisfied for the x-coordinates then it is also satisfied for the y-coordinates.*

Proof. We write down (1) and (2) for the y-coordinates, where $y_0 = \Phi(x_0)$,

$$\tilde{A}^{\mathrm{H}}\left(y_0, \eta', \frac{1}{\mathrm{i}}\frac{\mathrm{d}}{\mathrm{d}\tau}\right)\tilde{v}(\tau) = 0, \quad \tau > 0, \tag{18}$$

$$\tilde{b}_j^{\mathrm{H}}\left(y_0, \eta', \frac{1}{\mathrm{i}}\frac{\mathrm{d}}{\mathrm{d}\tau}\right)\tilde{v}(0) = 0, \quad j = 1, \ldots, m. \tag{19}$$

By Definition 2.9 the Jacobi matrix $\partial y/\partial x$ of an admissible transformation decomposes as

$$\frac{\partial y}{\partial x} = \begin{pmatrix} & J & & \begin{matrix} 0 \\ \vdots \\ 0 \end{matrix} \\ 0 \cdots 0, & & & \sigma \end{pmatrix}, \quad \text{where } \det J \neq 0 \text{ and } \sigma > 0.$$

Because of Theorem 10.3 we have ($\eta' \in T_{\partial\Omega'}$, ${}^{\mathrm{T}}J\eta' \in T_{\partial\Omega}$)

$$\tilde{A}^{\mathrm{H}}\left(y_0, \eta', \frac{1}{\mathrm{i}}\frac{\mathrm{d}}{\mathrm{d}t}\right) = A^{\mathrm{H}}\left(x(y_0), {}^{\mathrm{T}}J\eta', \frac{\sigma}{\mathrm{i}}\frac{\mathrm{d}}{\mathrm{d}\tau}\right), \tag{20}$$

and

$$\tilde{b}_j^{\mathrm{H}}\left(y_0, \eta', \frac{1}{\mathrm{i}}\frac{\mathrm{d}}{\mathrm{d}\tau}\right) = b_j^{\mathrm{H}}\left(x(y_0), {}^{\mathrm{T}}J\eta', \frac{\sigma}{\mathrm{i}}\frac{\mathrm{d}}{\mathrm{d}\tau}\right), \quad j = 1, \ldots, m. \tag{21}$$

From (20) it is immediate that the transformed operator $\tilde{A}(y, \mathrm{D}_y)$ remains properly elliptic, and that also the solution spaces $\mathcal{M}^+$ and $\mathcal{M}^-$ remain invariant (after multiplication by $\sigma > 0$ the characteristic roots remain in the corresponding half planes $\operatorname{Im}\lambda > 0$, respectively $\operatorname{Im}\lambda < 0$). If we apply the formulae (20) and (21) to the initial value problem (18) and (19), we see that the function $\tilde{v}(\tau)$ satisfies the initial value problem

$$A^{\mathrm{H}}\left(x(y_0), {}^{\mathrm{T}}J\eta', \frac{\sigma}{\mathrm{i}}\frac{\mathrm{d}}{\mathrm{d}\tau}\right)\tilde{v}(\tau) = 0,$$

$$b_j^{\mathrm{H}}\left(x(y_0), {}^{\mathrm{T}}J\eta', \frac{\sigma}{\mathrm{i}}\frac{\mathrm{d}}{\mathrm{d}\tau}\right)\tilde{v}(0) = 0, \quad j = 1, \ldots, m,$$

or after the variable change

$$A^{\mathrm{H}}\left(x(y_0), {}^{\mathrm{T}}J\eta', \frac{1}{\mathrm{i}}\frac{\mathrm{d}}{\mathrm{d}t}\right)u(t) = 0 \tag{22}$$

$$b_j^{\mathrm{H}}\left(x(y_0), {}^{\mathrm{T}}J\eta', \frac{1}{\mathrm{i}}\frac{\mathrm{d}}{\mathrm{d}t}\right)u(0) = 0, \quad j = 1, \ldots, m, \tag{23}$$

where we have written $u(t) = \tilde{v}(\sigma t) = \tilde{v}(\tau)$. Because $\sigma > 0, u(t)$ is again in $\mathcal{M}^+$, if $\tilde{v}(\tau)$ was in $\mathcal{M}^+$. With $\eta' \neq 0$ it is also true that ${}^{\mathrm{T}}J\eta' \neq 0$ ($\det J \neq 0$) and Condition 11.1 applied to (22) and (23) gives $u = 0$, hence also $\tilde{v} = 0$, thus we have shown that Condition 11.1 also holds for the y-coordinates. ■

11.2 Examples

We work out the L.Š. conditions for different differential operators.

We begin with the Laplace operator

$$-\Delta = -\sum_{j=1}^{r}\frac{\partial^2}{\partial x_j^2}, \quad \text{ord}\,\Delta = 2, m = 1. \tag{24}$$

Since $m = 1$, we may put a single boundary condition b_1 on $\partial\Omega$ (ord $b_1 \leqslant 1$)

$$b_1(x, \mathrm{D}) = b_0(x) + \sum_{j=1}^{r-1} b_j(x)\frac{\partial}{\partial x_j} + b_r(x)\frac{\partial}{\partial x_r}. \tag{25}$$

For the sake of simplicity we so choose the local coordinates so that $\partial/\partial x_1, \ldots, \partial/\partial x_{r-1}$ are directed tangentially to $\partial\Omega$, and $\partial/\partial x_r$ is directed along the inner normal. Here the operator from (1) has the form

$$A^{\mathrm{H}}\left(x; \xi', \frac{1}{\mathrm{i}}\frac{\mathrm{d}}{\mathrm{d}t}\right) = |\xi'|^2 + \left(\frac{1}{\mathrm{i}}\frac{\mathrm{d}}{\mathrm{d}t}\right)^2 = |\xi'|^2 - \frac{\mathrm{d}^2}{\mathrm{d}t^2}, \quad \xi' \in \mathbb{R}^{r-1}. \tag{26}$$

In order to determine $\mathcal{M}^+$ we calculate the characteristic roots λ of (26)

$$|\xi'|^2 + \lambda^2 = 0.$$

We have $\lambda_{1,2} = \pm \mathrm{i}|\xi'|$, we take the root in the upper half plane, thus $\lambda_1 = \mathrm{i}|\xi'|$ and have, using our notational convention,

$$\mathcal{M}^+ = \{c_1 \mathrm{e}^{\mathrm{i}\lambda_1 t} = c_1 \mathrm{e}^{\mathrm{i}(\mathrm{i}|\xi'|)t} = c_1 \mathrm{e}^{-|\xi'|t} : c_1 \in \mathbb{C}\}.$$

For the L.Š. Condition 11.1 (2) we distinguish between two cases.

Case 1. ord $b_1(x, \mathrm{D}) = 0$, therefore $b_1(x, \mathrm{D}) = b_0(x)$. $b_1 v(0)$ for $v \in \mathcal{M}^+$ gives $=$ $b_0(x)c_1$, and the L.Š. Condition 11.1 is then satisfied if and only if for all $0 \neq \xi' \in \mathbb{R}^{r-1}$ from $b_0(x)c_1 = 0$ it follows that $c_1 = 0$, which happens if and only if

$$b_0(x) \neq 0. \tag{27}$$

In case 1, up to multiplication by $b_0(x) \neq 0$, we have to do the Dirichlet problem.

Case 2. ord $b_1(x, \mathrm{D}) = 1$.
Now

$$b_1^{\mathrm{H}}\left(x; \xi', \frac{1}{\mathrm{i}}\frac{\mathrm{d}}{\mathrm{d}t}\right) = \mathrm{i}\sum_{j=1}^{r-1} b_j(x)\xi'_j + \mathrm{i}b_r(x)\frac{1}{\mathrm{i}}\frac{\mathrm{d}}{\mathrm{d}t},$$

and for $v \in \mathcal{M}^+$ we have

$$b_1^{\mathrm{H}}\left(x, \xi', \frac{1}{\mathrm{i}}\frac{\mathrm{d}}{\mathrm{d}t}\right)v(0) = \left(\mathrm{i}\sum_{j=1}^{r-1} b_j(x)\xi'_j - b_r(x)|\xi'|\right)c_1.$$

In this case Condition 11.1 is equivalent to

$$\mathrm{i}\sum_{j=1}^{r-1} b_j(x)\xi_j - b_r(x)|\xi'| \neq 0 \quad \text{for all } 0 \neq \xi' \in \mathbb{R}^{r-1}. \tag{28}$$

The following classical boundary value problems for the Laplace operator Δ satisfy the L.Š. Condition 11.1, and therefore the main Theorem 13.1 is applicable to them.

Example 11.1 Dirichlet problem $u|_{\partial\Omega} = g_1$. Here $b_0 = 1$, $b_1 = \cdots = b_r = 0$, by (27) Condition 11.1 is satisfied.

Example 11.2 Neumann problem $\partial u/\partial n|_{\partial\Omega} = g_1$. Here we have $b_0 = 0, b_1 = \cdots = b_{r-1} = 0$, $b_r = 1$, by (28) Condition 11.1 is satisfied.

Example 11.3 The so-called third boundary value problem

$$b_0 u + b_r \frac{\partial u}{\partial n}\bigg|_{\partial\Omega} = g_1.$$

Since here $\operatorname{ord} b_1(x, \mathrm{D}) = 1$, we can again apply (28) with $b_1 = \cdots = b_{r-1} = 0$, and obtain that Condition 11.1 is equivalent to $b_r(x) \neq 0$.

Example 11.4 Boundary value problem with skew derivative. Here for $b_1(x, \mathrm{D})$ we can take the full expression (25), where we suppose that the coefficients of (25) are real. (28) shows that for $r > 2$ Condition 11.1 is equivalent to $b_r(x) \neq 0$. In this case the directional derivative $\partial/\partial\mu := \sum_{j=1}^{r} b_j(x)\partial/\partial x_j$ does not lie in the tangent plane to $\partial\Omega$, hence the name 'skew derivative'.

Example 11.5 Let $r = 2$, in this case Condition 11.1 for $\operatorname{ord} b_1(x, 0) = 1$ is equivalent to

$$\mathrm{i}b_1(x)\xi'_1 - b_2(x)|\xi'_1| \neq 0, \quad 0 \neq \xi'_1 \in \mathbb{R}^1,$$

that is, each preassigned, non-zero, vector field μ on the curve $\partial\Omega$, $\mu = (b_1, b_2) \neq 0$, b_1, b_2 real, gives an elliptic boundary value problem

$$-\Delta u = f \text{ on } \Omega, \quad \partial u/\partial\mu = g \quad \text{on } \partial\Omega.$$

Here the vector field μ may also be tangentially directed, $\mu = (b_1, 0)$, but with $b_1 \neq 0$.

For $r > 2$ the tangential boundary value operator

$$b_0(x) + \sum_{j=1}^{r+1} b_j(x)\frac{\partial}{\partial x_j}$$

(at least one $b_j \neq 0, j = 1, \ldots, r-1$ and $b_r = 0$) does not satisfy Condition 11.1, for the equation (see (28))

$$\sum_{j=1}^{r-1} b_j(x)\xi'_j = 0$$

may be satisfied for at least $0 \neq \xi' \in \mathbb{R}^{r-1}$. Therefore this boundary value problem is not elliptic in the sense of §13, and the main Theorem 13.1 is not applicable.

Furthermore for a general, properly elliptic second-order differential operator

$$A(x, \mathrm{D}) = -\sum_{j,k}^{r} a_{jk} \frac{\partial}{\partial x_j} \frac{\partial}{\partial x_k} + \sum_{j=1}^{r} a_j \frac{\partial}{\partial x_j} + a_0 \tag{29}$$

we can give all boundary conditions which satisfy Condition 11.1 explicitly – the reader may carry out the corresponding discussion for himself. Here we want to consider the Neumann problem and the so-called third boundary value problem.

We determine $\mathcal{M}^+$ for (29). The characteristic equation belonging to $A^{\mathrm{H}}(x; \xi', (1/\mathrm{i})(\mathrm{d}/\mathrm{d}t))$ is given by (29)

$$\sum_{k,j=1}^{r-1} a_{jk}\xi'_j\xi'_k + 2\lambda \sum_{j=1}^{r-1} a_{jr}\xi'_j + a_{rr}\lambda^2 = 0$$

and the characteristic roots are

$$\lambda_{1,2} = \frac{-\sum_{j=1}^{r-1} a_{jr}\xi'_j \pm \sqrt{\left[\left(\sum_{j=1}^{r-1} a_{jr}\xi'_j\right)^2 - a_{rr}\left(\sum_{k,j=1}^{r-1} a_{jk}\xi'_j\xi'_k\right)\right]}}{a_{rr}}. \tag{30}$$

The expression inside the radical must be different from zero:

$$\left(\sum_{j=1}^{r-1} a_{jr}\xi'_j\right)^2 - a_{rr}\left(\sum_{k,j=1}^{r-1} a_{jk}\xi'_j\xi'_k\right) \neq 0 \quad \text{for } 0 \neq \xi' \in \mathbb{R}^{r-1}, \tag{31}$$

otherwise the characteristic equation would have a double root. However, a properly elliptic form (29) of second order has by definition two distinct roots, one in the upper and one in the lower half plane.

Let λ_1 be the root in the upper half plane $\operatorname{Im}\lambda > 0$, we have

$$\mathcal{M}^+ = \{c_1 \mathrm{e}^{\mathrm{i}\lambda_1 t} : c_1 \in \mathbb{C}\}.$$

In order to formulate the Neumann problem for (29) we introduce the conormal derivative Nu of u

$$Nu := \sum_{j,k=1}^{r} a_{jk}(x) \frac{\partial u}{\partial x_j} \cos(\mathbf{n}, x_k), \quad x \in \partial\Omega,$$

here $\mathbf{n}$ is the unit vector in the direction of the inner normal to Ω. By the Neumann problem for (29) we understand the boundary value problem:

Example 11.6

$$Au = f \text{ in } \Omega, \quad Nu = g_1 \text{ on } \partial\Omega, \tag{32}$$

and by the third boundary value problem:

Example 11.7

$$Au = f \text{ in } \Omega, \quad b_0(x)u + Nu = g_1 \text{ on } \partial\Omega. \tag{33}$$

We wish to show that both boundary value problems satisfy Condition 11.1. Both times we have

$$b_1^{\mathrm{H}} = N\left(\xi', \frac{1}{\mathrm{i}}\frac{\mathrm{d}}{\mathrm{d}t}\right);$$

as local coordinates at $x \in \partial\Omega$ we again take $(x_1, \ldots, x_{r-1})$ tangential and x_r in the direction of the inner normal. N then takes the simple form

$$N = \sum_{j=1}^{r-1} a_{jr}\frac{\partial}{\partial x_j} + a_{rr}\frac{\partial}{\partial x_r},$$

($a_{rr} \neq 0$ because of the ellipticity of (29)) and

$$N\left(\xi', \frac{1}{\mathrm{i}}\frac{\mathrm{d}}{\mathrm{d}t}\right) = \mathrm{i}\sum_{j=1}^{r-1} a_{jr}\xi'_j + \mathrm{i}a_{rr}\frac{1}{\mathrm{i}}\frac{\mathrm{d}}{\mathrm{d}t}.$$

Let $v \in \mathcal{M}^+$, we calculate

$$N\left(\xi', \frac{1}{\mathrm{i}}\frac{\mathrm{d}}{\mathrm{d}t}\right)v(0) = \left(\mathrm{i}\sum_{j=1}^{r-1} a_{jr}\xi'_j\right)c_1 + a_{rr}c_1\mathrm{i}\lambda_1 = \mathrm{i}c_1\left(\sum_{j=1}^{r-1} a_{jr}\xi'_j + a_{rr}\lambda_1\right)$$

$$= \mathrm{i}c_1\sqrt{\left[-a_{rr}\left(\sum_{k,j=1}^{r-1} a_{jk}\xi'_j\xi'_k\right) + \left(\sum_{i=1}^{r-1} a_{jr}\xi'_j\right)^2\right]},$$

where the last follows by substituting (30) for λ_1. Because of (31) the radical expression differs from zero for $0 \neq \xi' \in \mathbb{R}^{r-1}$ and $N(\xi', (1/\mathrm{i})(\mathrm{d}/\mathrm{d}t))v(0) = 0$ gives $c_1 = 0$, that is, both boundary value problems (32) and (33) satisfy Condition 11.1.

Example 11.8 Let Δ^2 be the biharmonic operator, here $\operatorname{ord}\Delta^2 = 4$, $m = 2$, and we may put two boundary conditions $b_1(x, \mathrm{D})$, $b_2(x, \mathrm{D})$ on $\partial\Omega$ with $0 \leqslant \operatorname{ord} b_1$, $b_2 \leqslant 3$. In order to cover all boundary conditions, which satisfy Condition 11.1, we must distinguish between the ten cases, considered according to the orders of b_1 and b_2,

$$(\operatorname{ord} b_1, \operatorname{ord} b_2) = (0,0), (0,1), (1,1), (0,2), (1,2), (2,2), (0,3), (1,3), (2,3), (3,3).$$

Here we wish to discuss the first six cases (0, 0)–(2, 2), that is, boundary value operators of order up to 2, and to give examples for the remaining cases. The reader may work through the remaining four cases, with boundary value operators of order 3, as an exercise.

First $\mathcal{M}^+$. The characteristic equation for Δ^2 reads

$$(|\xi'|^2 + \lambda^2)^2 = 0,$$

and $\lambda_1 = \mathrm{i}|\xi'|$ is the double root, which lies in $\operatorname{Im}\lambda > 0$. $\mathcal{M}^+$ consists of all functions of the form

$$\mathcal{M}^+ = \{c_1 \mathrm{e}^{-|\xi'|t} + c_2 t\mathrm{e}^{-|\xi'|t} : c_1, c_2 \in \mathbb{C}\}$$

(linear ordinary differential equations!).

A general boundary value operator of order 2 has the form

$$b(x, \mathrm{D}) = b_0 + \sum_{j=1}^{r} b_r \frac{\partial}{\partial x_j} + \sum_{j,k=1}^{r} b_{jk} \frac{\partial}{\partial x_j} \cdot \frac{\partial}{\partial x_k}.$$

As local coordinates at $x \in \partial\Omega$ we again take $(x_1, \dots, x_{r-1})$ tangential and x_r normal and for the principal part we obtain

$$b^{\mathrm{H}}\left(x; \xi', \frac{1}{\mathrm{i}}\frac{\mathrm{d}}{\mathrm{d}t}\right)$$

$$= \begin{cases} b_0, & \text{if } \operatorname{ord} b_1 = 0, \\ \mathrm{i}\sum_{j=1}^{r-1} b_j \xi'_j + b_r \dfrac{\mathrm{d}}{\mathrm{d}t}, & \text{if } \operatorname{ord} b_1 = 1, \\ -\sum_{j,k=1}^{r-1} b_{jk}\xi'_j\xi'_k + 2\mathrm{i}\left(\sum_{j=1}^{r-1} b_{jr}\xi'_j\right)\dfrac{\mathrm{d}}{\mathrm{d}t} + b_{rr}\dfrac{\mathrm{d}^2}{\mathrm{d}t^2}, & \text{if } \operatorname{ord} b_1 = 2. \end{cases}$$

For $b^{\mathrm{H}}(x; \xi', (1/\mathrm{i})(\mathrm{d}/\mathrm{d}t))v(0)$, $v \in \mathcal{M}^+$ we obtain

$$b^{\mathrm{H}}\left(x; \xi', \frac{1}{\mathrm{i}}\frac{\mathrm{d}}{\mathrm{d}t}\right)v(0)$$

$$= \begin{cases} b_0 c_1, & \text{if } \operatorname{ord} b_1 = 0, \\ c_1 \mathrm{i}\sum_{j=1}^{r-1} b_j\xi'_j - c_1 b_r|\xi'| + c_2 b_r, & \text{if } \operatorname{ord} b_1 = 1, \\ -c_1\sum_{j,k}^{r-1} b_{jk}\xi'_j\xi'_k - 2\mathrm{i}c_1|\xi'|\left(\sum_{j=1}^{r-1} b_{jr}\xi'_j\right) + 2\mathrm{i}c_2\left(\sum_{j=1}^{r-1} b_{jr}\xi'_j\right) & \\ \quad + b_{rr}|\xi'|^2 c_1 - 2c_2 b_{rr}|\xi'|, & \text{if } \operatorname{ord} b_1 = 2. \end{cases}$$

Now we can easily give criteria for the validity of Condition 11.1.

Case (0,0). From $b_0^{(1)}c_1=0$, $b_0^{(2)}c_1=0$ it ought to follow that $c_1=c_2=0$, but this is not true and Condition 11.1 cannot be satisfied.

Case (0,1). From

$$c_1 b_0^{(1)}=0,\quad c_1\left[\mathrm{i}\sum_{j=1}^{r-1} b_j^{(2)}\xi_j'-b_r^{(2)}|\xi'|\right]+c_2 b_r^{(2)}=0,$$

it follows that $c_1=c_2=0$ if and only if the determinant of this system of equations is different from zero, that is,

$$b_0^{(1)}b_r^{(2)}\neq 0 \tag{34}$$

is equivalent to Condition 11.1. This case also contains the Dirichlet problem

$$u|_{\partial\Omega}=g_1,\quad \left.\frac{\partial u}{\partial n}\right|_{\partial\Omega}=g_2\quad \text{with } b_0^{(1)}=1,\quad b_r^{(2)}=1,$$

Case (1,1). Here we have $c_1=c_2=0$ follows from

$$c_1\left[\mathrm{i}\sum_{j=1}^{r-1} b_j^{(1)}\xi_j'-b_r^{(1)}|\xi'|\right]+c_2 b_r^{(1)}=0,$$

$$c_1\left[\mathrm{i}\sum_{j=1}^{r-1} b_j^{(2)}\xi_j'-b_r^{(2)}|\xi'|\right]+c_2 b_r^{(2)}=0,$$

if and only if

$$b_r^{(2)}\left[\mathrm{i}\sum_{j=1}^{r-1} b_j^{(1)}\xi_j'-b_r^{(1)}|\xi'|\right]-b_r^{(1)}\left[\mathrm{i}\sum_{j=1}^{r-1} b_j^{(2)}\xi_j'-b_r^{(2)}|\xi'|\right]\neq 0. \tag{35}$$

For $0\neq\xi'\in\mathbb{R}^{r-1}$ (35) is equivalent to Condition 11.1.

Case (0,2). From $c_1 b_0^{(1)}=0$,

$$c_1\left[-\sum_{j,k}^{r-1} b_{jk}^{(2)}\xi_j'\xi_k'-2\mathrm{i}|\xi'|\sum_{j=1}^{r-1} b_{jr}^{(2)}\xi_j'+b_{rr}^{2}|\xi'|^2\right]$$
$$+c_2\left[2\mathrm{i}\sum_{j=1}^{r-1} b_{jr}^{(2)}\xi_j'-2b_{rr}^{(2)}|\xi'|\right]=0,$$

it follows that $c_1=c_2=0$ if and only if

$$b_0^{(1)}\left[2\mathrm{i}\sum_{j=1}^{r-1} b_{jr}^{(2)}\xi_j'-2b_{rr}^{(2)}|\xi'|\right]\neq 0, \tag{36}$$

and for $0\neq\xi'\in\mathbb{R}^{r-1}$ (36) is equivalent to Condition 11.1.

Case (1,2). Condition 11.1 is satisfied if and only if for all $0 \neq \xi' \in \mathbb{R}^{r-1}$ we have

$$\left[\mathrm{i}\sum_{j=1}^{r-1} b_j^{(1)}\xi'_j - b_r^{(1)}|\xi'|\right]\left[2\mathrm{i}\sum_{j=1}^{r-1} b_{jr}^{(2)}\xi'_j - 2b_{rr}^{(2)}|\xi'|\right]$$
$$- b_r^{(1)}\left[-\sum_{j,k}^{r-1} b_{jk}^{(2)}\xi'_j\xi'_k - 2\mathrm{i}|\xi'|\sum_{j=1}^{r-1} b_{jr}^{(2)}\xi'_j + b_{rr}^{(2)}|\xi'|^2\right] \neq 0.$$

Case (2,2). Condition 11.1 is satisfied if and only if for all $0 \neq \xi' \in \mathbb{R}^{r-1}$ we have

$$\left[-\sum_{j,k}^{r-1} b_{jk}^{(1)}\xi'_j\xi'_k - 2\mathrm{i}|\xi'|\sum_{j=1}^{r-1} b_{jr}^{(1)}\xi'_j + b_{rr}^{(1)}|\xi'|^2\right]\left[2\mathrm{i}\sum_{j=1}^{r-1} b_{jr}^{(2)}\xi'_j - 2b_{rr}^{(2)}|\xi'|\right]$$
$$-\left[-\sum_{j,k}^{r-1} b_{jk}^{(1)}\xi'_j\xi'_k - 2\mathrm{i}|\xi'|\sum_{j=1}^{r-1} b_{jr}^{(2)}\xi'_j + b_{rr}^{(2)}|\xi'|^2\right]$$
$$\cdot\left[2\mathrm{i}\sum_{j=1}^{r-1} b_{jr}^{(1)}\xi'_j - 2b_{rr}^{(1)}|\xi'|\right] \neq 0.$$

Example 11.9 For the biharmonic operator Δ^2 we consider the following boundary value operators

$$b_1 = I, \quad b_2 = \frac{\partial}{\partial n}, \quad b_3 = \Delta =: M, \quad b_4 = -\frac{\partial\Delta}{\partial n} =: T. \tag{37}$$

We have

$$\mathcal{M}^+ = \{c_1 \mathrm{e}^{-|\xi'|t} + c_2 t\mathrm{e}^{-|\xi'|t} : c_1, c_2 \in \mathbb{C}\}$$

and for $v \in \mathcal{M}^+$

$$\begin{aligned}
b_1^{\mathrm{H}}\left(x;\xi',\frac{1}{\mathrm{i}}\frac{\mathrm{d}}{\mathrm{d}t}\right)v(0) &= Iv(0) = c_1,\\
b_2^{\mathrm{H}}\left(x;\xi',\frac{1}{\mathrm{i}}\frac{\mathrm{d}}{\mathrm{d}t}\right)v(0) &= \frac{\mathrm{d}}{\mathrm{d}t}v(0) = -|\xi'|c_1 + c_2,\\
b_3^{\mathrm{H}}\left(x;\xi',\frac{1}{\mathrm{i}}\frac{\mathrm{d}}{\mathrm{d}t}\right)v(0) &= \left(|\xi'|^2 - \left(\frac{\mathrm{d}}{\mathrm{d}t}\right)^2\right)v(0) = 2|\xi'|c_2,\\
b_4^{\mathrm{H}}\left(x;\xi',\frac{1}{\mathrm{i}}\frac{\mathrm{d}}{\mathrm{d}t}\right)v(0) &= -\left(|\xi'|^2\frac{\mathrm{d}}{\mathrm{d}t} - \frac{\mathrm{d}^3}{\mathrm{d}t^3}\right)v(0) = 2|\xi'|^2 c_2.
\end{aligned} \tag{38}$$

With the operators (37) we can form the following boundary value problems

$$\left(\Delta^2; I, \frac{\partial}{\partial n}\right), \quad (\Delta^2; I, \Delta), \quad \left(\Delta^2; \frac{\partial}{\partial n}, \Delta\right),$$

$$\left(\Delta^2; I, -\frac{\partial\Delta}{\partial n}\right), \quad \left(\Delta^2; \frac{\partial}{\partial n}, \frac{-\partial\Delta}{\partial n}\right), \quad \left(\Delta^2; \Delta, -\frac{\partial\Delta}{\partial n}\right). \tag{39}$$

From (38) it follows immediately that all the boundary value problems (39), apart from the last, satisfy the L.Š. Condition 11.1.

Exercises

11.1 For a second-order properly elliptic operator find all boundary conditions which satisfy the L.Š. condition.

§12 Fredholm operators

This section is devoted to the basic functional analytic properties of Fredholm operators. In contrast to the treatments in Schechter [1] or Heuser [1] we do not consider the most general case, but assume that all operators act linearly and continuously between Banach spaces. For our purposes, which concern elliptic differential operators, this is sufficient. In this section we present the functional analytic connection between Fredholm operators in a Schauder scheme and *a priori* estimates, and also the theory of so-called smoothable operators, which allow us to present Weyl's lemma in an abstract form.

12.1 The Riesz–Schauder spectral theorem (compact operators)

Let X be a Banach space. We denote its dual space by X'. For our purposes, differential operators in Hilbert space, it makes more sense to work with the antidual space X^*, the antidual space consists of all antilinear continuous functionals $f: X \to \mathbb{C}$. We write $f \in X^*$, $f(x) = \langle f, x\rangle$ and have

$$\langle f, \alpha x + \beta y\rangle = \bar{\alpha}\langle f, x\rangle + \bar{\beta}\langle f, y\rangle \quad \text{for } x, y, \in X, \quad \alpha, \beta \in \mathbb{C}.$$

The replacement of X' by X^* is not essential (we only take the conjugate in $\mathbb{C}$), it does, however, introduce a simplification for Hilbert spaces H, by Riesz' theorem we then have $H^* = H$ (isomorphic), and antidual operators A^* coincide with adjoint operators, see also formula (3) in §10.

Let X, Y be Banach spaces, $A: X \to Y$ a continuous, linear map. We define the antidual map $A^*: Y^* \to X^*$ by $\langle A^*y^*, x\rangle_X = \langle y^*, Ax\rangle_Y$. We have that $\|A^*\| = \|A\|$ and therefore A^* is a continuous linear map.

Since in this book we will always work with antidual spaces and antidual maps, we wish to reserve the star * for adjoint differential operators and adjoint boundary value problems, and otherwise (that is, for antidualities) simply write a prime sign ′.

For the convenience of the reader, we formulate the Riesz–Schauder spectral theorems for compact operators as completely as possible. The reader will find the demonstrations in Taylor [1], or in Wloka [1], p. 217. We will frequently return in this book to the Riesz–Schauder theorems.

Let X be a Banach space. We say that the complex number $\lambda \in \mathbb{C}$ is a regular value of the continuous operator $A: X \to X$, if the operator $A_\lambda = A - \lambda I$ (I equals the identity map $I: X \to X$) possesses a continuous inverse

$$(A - \lambda I)^{-1}: X \to X.$$

We call those complex numbers λ, which are not regular for A *spectral values* of A; we call the set of all spectral values the *spectrum* of A, denoted by Sp A. If the complex number λ has the property that the kernel, $\ker(A - \lambda I)$ of $A - \lambda I$ is different from $\{0\}$

$$\ker(A - \lambda I) \neq \{0\},$$

then we say that λ is an eigenvalue of A, of course the eigenvalues are spectral values of A. The vectors $x \neq 0$, which belong to $\ker(A - \lambda I)$, that is, those $x \neq 0$ with

$$Ax - \lambda x = 0$$

are called the eigenvectors of A corresponding to the eigenvalue λ. The eigenvectors together with 0 form the eigenspace $E(\lambda)$:

$$E(\lambda) = E(\lambda, A) := \ker(A - \lambda I).$$

Remark In the same way we can define the spectrum of a map $A: X \to Y$, if a second map, for example an embedding $J: X \to Y$ is given: we say that $\lambda \in \mathbb{C}$ is regular if the inverse

$$(A - \lambda J)^{-1}: Y \to X$$

exists and is continuous. Otherwise we say that λ is a spectral value. We will use this definition for differential operators.

Riesz–Schauder spectral theorem *Let $A: X \to X$ be a linear, compact operator. Then we have:*

(a) *The spectrum* Sp A *of A is contained in the disc $\{\lambda : |\lambda| \leqslant \|A\|\} \subset \mathbb{C}$, and consists of either a finite set, or of* 0 *and a (countable) sequence λ_n converging to* 0. *If X is infinite dimensional,* 0 *always belongs to the spectrum.*

(b) *Each number $\lambda_n \neq 0$ from the spectrum is an eigenvalue, and the eigenspace $E(\lambda_n)$ is finite dimensional.*

(c) *$A': X' \to X'$ is compact if and only if $A: X \to X$ is. If $\operatorname{Sp} A = \{\lambda_n, 0\}$ is the spectrum of A then $\{\bar{\lambda}_n, 0\} = \operatorname{Sp} A'$ is the spectrum of A', therefore*

$$\operatorname{Sp} A' = \overline{\operatorname{Sp} A},$$

and the eigenspaces $E(\lambda_n, A)$, $E(\bar{\lambda}_n, A')$ have the same (finite) dimension

$$\dim E(\lambda_n, A) = \dim E(\bar{\lambda}_n, A') < \infty,$$

(d) *Fredholm alternative. The equation*

$$(A - \lambda I)x = y, \quad \lambda \neq 0,$$

has at least one solution x if and only if $y \in \ker(A' - \bar{\lambda} I)^{\perp}$, or in other words, if and only if $A'f' = \bar{\lambda} f'$ implies that $\langle y, f' \rangle = 0$. The equation

$$(A' - \bar{\lambda} I)f = g, \quad \lambda \neq 0$$

has a solution if and only if $g \in \ker(A - \lambda I)^{\perp}$, or, if and only if $Ax = \lambda x$ implies that $\langle x, g \rangle = 0$.
For Hilbert spaces $H = X, (x, y)_H$ the inner or scalar product, and compact, self-adjoint operators $A = A'$, that is, $(Ax, y)_H = (x, Ay)_H$, $x, y \in H$, we can say more:

(e) *The spectrum $\operatorname{Sp} A$ of A lies on the real axis; if λ, μ are two distinct eigenvalues of A, then the eigenspaces $E(\lambda)$ and $E(\mu)$ are mutually orthogonal.*

(f) *There exists at least one eigenvalue λ with $\lambda = \|A\|$ or $\lambda = -\|A\|$.*

(g) *For each $x \in H$ we have the decomposition*

$$Ax = \sum_n \lambda_n (x, x_n) x_n, \quad x_n \in E(\lambda_n),$$

that is,

$$\overline{\operatorname{im} A} = \overline{A(H)} = \bigoplus_k E(\lambda_k),$$

and also

$$H = \ker A \oplus \overline{\operatorname{im} A} = E(0) \bigoplus_k E(\lambda_k).$$

As a consequence of (e), (f) *and* (g) *we obtain:*

(h) *If $\overline{\operatorname{im} A} = H$ and H is infinite dimensional, then there exist infinitely many distinct eigenvalues λ_n and the eigenvectors $\{x_n\}$, suitably normalised, form an orthonormal basis for H.*

Let $T = A - \lambda I$, $\lambda \neq 0$, $A: X \to X$ a compact operator. Then by (b)

$$\alpha(T) := \dim \ker T = \dim E(\lambda) < \infty,$$

equally whether $\lambda \neq 0$ is or is not an eigenvalue, and by (d) we have

$$\beta(T) := \dim \operatorname{coker} T := \dim (X/\operatorname{im} T) = \dim \ker T' = \alpha(T') < \infty,$$

while (c) gives

$$\alpha(T') = \dim \ker T' = \dim \ker T = \alpha(T).$$

Putting together

$$\beta(T) = \dim \operatorname{coker} T = \alpha(T) < \infty,$$

and

$$\operatorname{ind} T := \alpha(T) - \beta(T) = 0.$$

The defect numbers $\alpha(T)$ and $\beta(T)$ are important for the next definition.

12.2 Fredholm operators

Definition 12.1 *Let X and Y be two Banach spaces, we say that a continuous linear operator $T: X \to Y$, in short $T \in L(X, Y)$, is a Fredholm operator, in short $T \in F(X, Y)$ if T has the two properties*

$$\alpha(T) = \dim \ker T < \infty, \tag{I}$$

$$\beta(T) = \dim \operatorname{coker} T = \dim (Y/\operatorname{im} T) < \infty. \tag{II}$$

The following integer is called the index of T:

$$\operatorname{ind} T = \alpha(T) - \beta(T).$$

Following immediately from the definition of a Fredholm operator we have:

Theorem 12.1 *Let X_1, X_2, Y_1, and Y_2 be Banach spaces, and let $T \in F(X_1, X_2)$ be a Fredholm operator. If $U_1: Y_1 \to X_1$, $U_2: X_2 \to Y_2$ are top linear isomorphisms, then $\tilde{T} = U_2 \circ T \circ U_1$ is again a Fredholm operator $\in F(Y_1, Y_2)$, and*

$$\operatorname{ind} \tilde{T} = \operatorname{ind} T.$$

Therefore the index is a topological linear invariant.

Proof. Clearly $\ker \tilde{T} = U_1^{-1}(\ker T)$ and $\operatorname{im} \tilde{T} = U_2(\operatorname{im} T)$. The isomorphism property makes the defect numbers $\alpha(\tilde{T}) = \alpha(T)$ and $\beta(\tilde{T}) = \beta(T)$ equal, from which the assertion of our theorem follows. ■

Theorem 12.2 *Let X and Y be Banach spaces and $T \in L(X, Y)$. If $\beta(T) < \infty$, then the image* im T *is closed in Y.*

Proof. We factorise T in the following way

$$\begin{array}{ccc} X & \xrightarrow{T} & \operatorname{im} T \subset Y \\ & {\scriptstyle P}\searrow \quad \nearrow{\scriptstyle S} & \\ & X/\ker T & \end{array} \tag{1}$$

Here let $P: X \to X/\ker T$ be the canonical projection, and $S: X/\ker T \to \operatorname{im} T$ be the uniquely defined, continuous, linear, bijective map defined by $S\hat{x} := Tx (x \in \hat{x})$. Now let W be a complementary subspace for $\operatorname{im} T$ in Y. By hypothesis W is finite dimensional, hence a Banach space. We define the map

$$\bar{S}: (X/\ker T) \times W \to Y$$

by

$$\bar{S}(\hat{x}, w) := S\hat{x} + w.$$

The Cartesian product of two Banach spaces is again a Banach space, so that $(X/\ker T) \times W$ is a Banach space. The mapping $\bar{S}$ is continuous, linear and bijective, hence by the open mapping theorem $\bar{S}^{-1}$ is continuous. Since $(X/\ker T) \times \{0\}$ is closed in $(X/\ker T) \times W$, $\operatorname{im} T = \bar{S}((X/\ker T) \times \{0\})$ is closed in Y. ■

In the definition of a Fredholm operator T we can therefore incorporate the property

$$\operatorname{im} T \text{ is closed in } Y, \tag{III}$$

without altering the class of Fredholm operators.

To deduce further properties of Fredholm operators, we need further functional analytic concepts and results; we quote them without proof or reference. For the detailed arguments we refer to textbooks on functional analysis, for the most part we are concerned with well-known facts.

Let K be a set in X, the set of all elements $x' \in X'$, which are orthogonal to K, that is, $\langle x', x \rangle = 0$ for all $x \in K$, is called the orthogonal space $K^\perp$. (Dually one has $M^\perp \subset X$ if $M \subset X'$.) We have the following easy equalities for $T \in L(X, Y)$:

$$\ker T = (\operatorname{im} T)^\perp, \quad \ker T' = (\operatorname{im} T)^\perp, \tag{2}$$

and the theorems:

Let M be a subspace X, then there is a norm isomorphism

$$X'/M^\perp \simeq M', \tag{3}$$

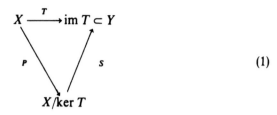

and if M is closed a norm isomorphism

$$(X/M)' \simeq M^{\perp}, \tag{4}$$

see for example Wloka [1], p. 99.

The following result goes back to Banach:

Theorem 12.3 (closed range theorem) *Let X, Y be Banach spaces and $T: X \to Y$ a continuous, linear map. Then the following assertions are equivalent*

(a) im T *is closed in* Y,
(b) im T' *is closed in* X',
(c) $\text{im}\, T = (\ker T')^{\perp}$,
(d) $\text{im}\, T' = (\ker T)^{\perp}$.

For the proof we refer to Wloka [1], p. 144, or Yosida [1].

Theorem 12.4 *If T is a Fredholm operator, $T \in F(X, Y)$, then T' is also a Fredholm operator, $T' \in F(Y', X')$ and conversely. For the defect numbers we have*

$$\alpha(T') = \beta(T) \quad \textit{and} \quad \beta(T') = \alpha(T), \tag{5}$$

from which it follows that $\text{ind}\, T' = -\text{ind}\, T$.

Proof. Let $T \in F(X, Y)$, write $M = \text{im}\, T$, so that by (2) and (4)

$$\ker T' = (\text{im}\, T)^{\perp} \simeq (Y/\text{im}\, T)' = (\text{coker}\, T)',$$

which is $\alpha(T') = \beta(T)$.

We now prove the second formula in (5). Since $T \in F(X, Y)$, im T is closed (Theorem 12.2) and we can apply Theorem 12.3, by which

$$\text{im}\, T' = (\ker T)^{\perp}.$$

In (3) $M = \ker T$ substituted, gives

$$(\ker T)' \simeq X'/(\ker T)^{\perp} = X'/\text{im}\, T' = \text{coker}\, T',$$

with which we have proved that $\beta(T') = \alpha(T)$, and simultaneously that $T' \in F(Y', X')$. The proof of the converse is similar, since all the dual formulae hold. ■

Definition 12.2 *Let $T \in L(X, Y)$. The operator $R_l \in L(X, Y)$, respectively the operator $R_r \in L(X, Y)$ is called the left regulariser, respectively the right regulariser, for T, if the operators*

$$k_1 := R_l T - I_X, \textit{ respectively } k_2 := R_r T - I_Y$$

are compact. Here I_X and I_Y denote the identity maps on the spaces X and Y.

Theorem 12.5 Let $T \in L(X, Y)$.

(a) *If T possesses a left regulariser R_l, then $\alpha(T) < \infty$ and* $\operatorname{im} T$ *is closed.*
(b) *If T possesses a right regulariser R_r, then $\beta(T) < \infty$ and* $\operatorname{im} T$ *is again closed (Theorem* 12.2).
(c) *If T possesses left and right regularisers R_l and R_r, then T is a Fredholm operator and conversely. In the last case one can take $R_r = R_l$.*

Proof.(a) From $\ker T \subset \ker(R_1 \circ T) = \ker(I_X + k_1)$ it follows that $\alpha(T) < \infty$, for $\ker(I_X + k_1)$ is the eigenspace of the compact operator k_1 associated with the eigenvalue $\lambda = -1$ (or regular value). It is finite dimensional by statement (b) of the Riesz–Schauder spectral theorem. In order to show that $\operatorname{im} T$ is closed we factorise T as in (1). We have $T = S \circ P$. Here $S:(X/\ker T) \to \operatorname{im} T$ is continuous, linear and bijective, and S^{-1} exists. It suffices to show that S^{-1} is continuous. We assume that there exists a sequence $\hat{x}_n \in X/\ker T$ with $\|\hat{x}_n\| = 1$ and

$$\|\hat{x}_n\| > n\|S\hat{x}_n\| \quad \text{for all } n \in \mathbb{N}.$$

Then $\|S\hat{x}_n\| \to 0$ as $n \to \infty$. If $x_n \in \hat{x}_n$ we have

$$R_1 \circ Tx_n - x_n = R_1 \circ S\hat{x}_n - x_n = k_1 x_n. \tag{6}$$

Without loss of generality let $\|x_n\| \leqslant 2$; since k_1 is compact there exists a subsequence x_{n_k} so that $k_1 x_{n_k}$ is convergent. Because $\|S\hat{x}_{n_k}\| \to 0$ it follows from (6) that $x_{n_k} \to x$, hence that $\hat{x}_{n_k} \to \hat{x}$ and further that $\|\hat{x}\| = 1$. On the other hand $\hat{x} = 0$, for if $x \neq 0$, one would obtain from the convergence of $S\hat{x}_{n_k}$ to 0 a contradiction to S being 1–1. Therefore we have obtained a contradiction, and there exists some constant $C < \infty$ with

$$\|\hat{x}\| \leqslant C\|S\hat{x}\| \quad \text{for all } x \in X/\ker T.$$

This implies the continuity of S^{-1}, and we have shown that $\operatorname{im} T$ is a closed set.

(b) By (2) we have

$$(\operatorname{im}(TR_r))^{\perp} = (\operatorname{im}(k_2 + I_Y))^{\perp} = \ker(k_2' + I_{Y'}). \tag{7}$$

Now $\operatorname{im}(TR_r) \subset \operatorname{im} T$ implies that $(\operatorname{im} T)^{\perp} \subset (\operatorname{im} TR_r)^{\perp}$, which with (4) and (7) gives

$$(\operatorname{coker} T)' = (Y/\operatorname{im} T)' \simeq (\operatorname{im} T)^{\perp} \subset (\operatorname{im} TR_r)^{\perp} = \ker(k_2' + I_{Y'}). \tag{8}$$

Now with k_2 the operator k_2' is also compact, and $\ker(k_2' + I_{Y'})$ is the eigenspace associated with the eigenvalue $\lambda = -1$ of the compact operator k_2

and again using statement (b) from the Riesz–Schauder spectral theorem we have

$$\beta(T) = \dim \operatorname{coker} T = \dim (\operatorname{coker} T)' \leqslant \dim \ker(k_2' + I_{Y'}) < \infty.$$

Here in (8) we have used that $\operatorname{im} T$ is closed.

$$\operatorname{im}(k_2 + I_Y) = \operatorname{im}(TR_r) \subset \operatorname{im} T. \tag{9}$$

By statement (d) of the Riesz–Schauder spectral theorem $\operatorname{im}(k_2 + I_Y)$ has finite codimension, by functional analysis and (9) $\operatorname{im} T$ also has finite codimension and is closed.

(c) The first part follows directly from (a) and (b). Suppose conversely that $T \in F(X, Y)$. By Theorem 12.2 $\operatorname{im} T$ is closed and we have

$$X = \ker T \oplus V, \quad Y = \operatorname{im} T \oplus W,$$

where by known results (see Wloka [1], p. 60, or Köthe [1], p. 159) all direct summands are closed, one summand is finite dimensional. We consider the commutative diagram

$$\begin{array}{ccc} \ker T \oplus V = X & \xrightarrow{\;T\;} & Y = \operatorname{im} T \oplus W \\ \downarrow & & \downarrow P \\ V & \xrightarrow{\;S\;} & \operatorname{im} T, \end{array}$$

where $S = T/V$ is the restriction of T to V, and P is the projection of Y onto $\operatorname{im} T$. V and $\operatorname{im} T$ are again Banach spaces, V closed in X, $\operatorname{im} T$ closed in Y, and S acts in a continuous bijective manner. By the open mapping theorem S^{-1} is also continuous. We put

$$R = S^{-1} \circ P,$$

and have

$$R \circ T - I_X = -\operatorname{proj}_{\ker T}, \quad T \circ R - Y_Y = -\operatorname{proj}_W.$$

If we write $k_1 = -\operatorname{proj}_{\ker T}$, $k_2 = -\operatorname{proj}_W$, we have proved our theorem, since with finite dimensional images the continuous operators k_1 and k_2 are compact. ∎

Remark 12.1 From the proof one sees that Theorem 12.5 remains true if on the remainder operators k_1, k_2 one imposes the stronger condition 'k_1 and k_2 have finite dimensional images' instead of 'compact'.

Theorem 12.6 *Let $T_1 \in F(X, Y)$, $T_2 \in F(Y, Z)$ be Fredholm operators, then the composite operator $T := T_2 \circ T_1 \in F(X, Z)$ is again a Fredholm operator, and we have* $\operatorname{ind}(T_2 \circ T_1) = \operatorname{ind} T_1 + \operatorname{ind} T_2$.

Proof. By Theorem 12.5 (c) there exist two continuous, linear operators $R_1 \in L(Y, X)$ and $R_2 \in L(Z, Y)$ such that the operators

$$k_{11} := R_1 T_1 - I_X, \quad k_{21} := T_1 R_1 - I_Y$$
$$k_{12} := R_2 T_2 - I_Y, \quad k_{22} := T_2 R_2 - I_Z$$

are compact. We put $R := R_1 \circ R_2$ and obtain

$$RT - I_X = k_{11} + R_1 k_{12} T_1 =: k_1,$$
$$TR - I_Z = k_{22} + T_2 k_{21} R_2 =: k_2,$$

where k_1 and k_2 are again compact. Therefore $T \in F(X, Z)$. In order to prove the index formula we proceed as follows. Let $\{\psi_1, \ldots, \psi_{\beta(T_1)}\}$ be a basis of $\ker T'$ and $\{\varphi_1, \ldots, \varphi_{\alpha(T_2)}\}$ a basis of $\ker T_2$. The equation $T_2 \circ T_1 x = 0$ is equivalent to $T_1 x = \sum_{k=1}^{\alpha(T_2)} c_k \varphi_k$; because $\operatorname{im} T_1 = (\ker T_1')^\perp$ (Theorem 12.3(c)) it must be that

$$\sum_{k=1}^{\alpha(T_2)} c_k \langle \psi_j, \varphi_k \rangle = 0 \quad \text{for } j = 1, \ldots, \beta(T_1).$$

Let $s = \operatorname{rank} \{\langle \psi_j, \varphi_k \rangle\}$, then $Tx = T_2 T_1 x = 0$ has exactly $\alpha(T) = \alpha(T_1) + (\alpha(T_2) - s)$ linearly independent solutions. If one proceeds similarly with the equation $T'z = T_1' T_2' z = 0$, one deduces that this equation has $\alpha(T') = \alpha(T_2') + (\alpha(T_1') - s)$ solutions. Therefore we have

$$\begin{aligned} \operatorname{ind} T &= \alpha(T) - \beta(T) = \alpha(T) - \alpha(T') \\ &= \alpha(T_1) + \alpha(T_2) - s - \alpha(T_2') - \alpha(T_1') + s = \operatorname{ind} T_1 + \operatorname{ind} T_2. \end{aligned}$$ ∎

Theorem 12.7 *If $T \in F(X, Y)$ is a Fredholm operator and if R is a left regulariser for T, then R is again a Fredholm operator, $R \in F(X, Y)$ and we have*

$$\operatorname{ind} R = -\operatorname{ind} T. \tag{10}$$

Proof (a) We have

$$\dim \ker R \circ T = \dim \ker (I_x + k) < \infty$$
$$\dim \ker R \leqslant \dim \ker R \circ T + \operatorname{codim} \operatorname{im} T \text{ (linear algebra!)}.$$

Since by assumption $\operatorname{codim} \operatorname{im} T = \dim \operatorname{coker} T < \infty$, it follows that $\dim \ker R < \infty$, hence we have shown that $\alpha(R) < \infty$.

(b) By assumption T is a right regulariser for R, by Theorem 12.5(b) $\beta(R)$ is therefore less than ∞.

By (a) and (b) R is Fredholm.

Now for the proof of the index formula (10). We have

$$R \circ T = I_X + k,$$

apply Theorem 12.6 and obtain

$$\operatorname{ind}(R\circ T)=\operatorname{ind}R+\operatorname{ind}T=\operatorname{ind}(I_X+k).$$

I_X+k is a Riesz–Schauder operator, therefore $\operatorname{ind}(I_X+k)=0$ (see the text following the spectral theorem) and therefore we have proved (10). ∎

Theorem 12.8 *If $T\in F(X,Y)$ is a Fredholm operator, and if $k\in K(X,Y)$ is compact, then $T+k\in F(X,Y)$ is again a Fredholm operator, and we have*

$$\operatorname{ind}(T+k)=\operatorname{ind}T.$$

Proof. We show that with T, $T+k$ is also both left and right regularisable, and this with the same regularisers. With

$$k_1=R_1\circ T-I_X \quad\text{and}\quad k_2=T\circ R_r-I_Y$$

the operators

$$k_1+R_1k=R_1(T+k)-I_X, \quad\text{and}\quad k_2+kR_r=(T+k)R_r-I_Y$$

are compact. By Theorem 12.5 $T+k\in F(X,Y)$. Finally given Theorem 12.7 we have

$$\operatorname{ind}(T+k)=-\operatorname{ind}R_1=\operatorname{ind}T.$$ ∎

We wish to illustrate the importance of the regularisers in yet another way. It turns out that by means of them one can reduce the equation $Tx=y$ to a Riesz–Schauder equation $Iu+Ku=v$, K-compact, indeed even to a finite dimensional system of equations $Iu+Au=v$, $\dim\operatorname{im}A<\infty$.

Let $T\in F(X,Y)$ be a Fredholm operator and R a left regulariser of T; we say that R (left) reduces if the equation $Tx=y$ is equivalent to the equation $RTx=Ry$, that is, each solution of the first equation is a solution of the second equation, and conversely. One sees immediately that in order for R to be (left) reduced it is necessary and sufficient that $\alpha(R)=0$. By Definition 12.2, see also Remark 12.1, we can also write the equation $RTx=Ry$ as $Ix+Ax=Ry$ with $\dim\operatorname{im}A<\infty$, after which the reduction is apparent. The following theorem, which dates back to Michlin [1], holds. This theorem is not numbered since we do not use either it or Vekua's theorem subsequently.

Theorem *Let $T\in F(X,Y)$ be a Fredholm operator with* $\operatorname{ind}T\geqslant 0$. *Then there exists a (left) reducing (left) regulariser R_0.*†

Proof. Let R be a left regulariser of T (this exists by Theorem 12.5); by

† The condition $\operatorname{ind}T\geqslant 0$ is also necessary for the existence of a reducing left regulariser.

Theorem 12.7 we have

$$\operatorname{ind} R = \alpha(R) - \beta(R) = -\operatorname{ind} T \leqslant 0, \text{ hence } \alpha(R) \leqslant \beta(R).$$

Let $\{\varphi_1, \dots, \varphi_{\alpha(R)}\}$ be a basis of $\ker R$, let the functionals $\{f_1, \dots, f_{\alpha(R)}\}$ be biorthogonal to this basis (see Wloka [1], p. 93), that is, $f_i(\varphi_j) = \delta_{i,j}$, $i, j = 1, \dots, \alpha(R)$, and let $\{\psi_1, \dots, \psi_{\beta(R)}\}$ be a basis of coker R, $\{g_1, \dots, g_{\beta(R)}\}$ biorthogonal to the ψs. We put

$$R_0 y = Ry + \sum_{i=1}^{\alpha(R)} f_i(y)\psi_i.$$

Since the sum in the definition of R_0 is finite dimensional, R_0 is again a left regulariser of T, and we have only to show that

$$\ker R_0 = (0).$$

Let $R_0 y_0 = R y_0 + \sum_{i=1}^{\alpha(R)} f_i(y_0)\cdot\psi_i = 0$, then because of the direct sum decomposition $X = \operatorname{im} R \oplus \operatorname{coker} R$ it follows that

$$R y_0 = 0 \quad \text{and} \quad \sum_{i=1}^{\alpha(R)} f_i(y_0)\cdot\psi_i = 0,$$

which is $y_0 \in \ker R$, and $f_i(y_0) = 0$ for $i = 1, \dots, \alpha(R)$. Since $y_0 \in \ker R$ we can write $y_0 = \sum_{i=1}^{\alpha(R)} c_i \cdot \varphi_i$, which on substitution in f_j gives:

$$0 = f_j(y_0) = \sum_{i=1}^{\alpha(R)} c_i f_j(\varphi) = c_j, \quad \text{or } y_0 = 0.$$

This means that $\ker R_0 = (0)$, completing our proof. ■

Remark Equations $Tx = y$ with $\operatorname{ind} T < 0$ cannot be left reduced; suppose that R is an arbitrary left regulariser, by Theorem 12.6 we have

$$\operatorname{ind} R = \alpha(R) - \beta(R) = -\operatorname{ind} T > 0,$$

therefore

$$\alpha(R) > \beta(R) \geqslant 0,$$

and we can never attain $\alpha(R) = 0$, or $\ker R = (0)$.

However, in this case we can right reduce and again succeed in finding a simpler equation. Let T be a Fredholm operator and R_r a right regulariser of T. We say that R_r is (right) reducing, if the equation $R_r z = x$ is solvable for all $x \in X$, in other words: $\operatorname{im} R_r = X$, or equivalently coker $R_r = (0)$. In this situation the equation $Tx = y$ is equivalent to $TR_r z = Iz + A_r z$ in the sense that both equations are simultaneously solvable or non-solvable. Here each solution x of $Tx = y$ may be represented by $x = R_r z$, where z is a solution of the reduced equation $Iz + A_r z = y$ and once more $\dim \operatorname{im} A_r < \infty$. We can prove a theorem which dates back to Vekua [1].

Theorem *Let* $T \in F(X, Y)$ *be a Fredholm operator with* $\operatorname{ind} T < 0$. *Then there exists a right reducing regulariser* R_r.

Proof. We take a right regulariser R of T, which exists by Theorem 12.5 and can also be assumed to be Fredholm. By Theorems 12.6 and 12.8 we have

$$0 = \operatorname{ind}(T \circ R) = \operatorname{ind} T + \operatorname{ind} R, \text{ therefore } \operatorname{ind} R = -\operatorname{ind} T > 0,$$

which is $\alpha(R) > \beta(R)$. With the relations above we write

$$R_r z = Rz + \sum_{i=1}^{\beta(R)} f_i(z)\psi_i,$$

and again obtain a right regulariser. In order to show that $\operatorname{im} R_r = X$, we apply Theorem 12.3 and show that $\ker R_r' = (0)$. Let $x' \in X'$; calculating with duality brackets gives the form of the dual operator

$$R_r' x' = R' x' + \sum_{i=1}^{\beta(r)} \langle x', \varphi_i \rangle f_i,$$

Let $R_r' x_0' = 0$, we form $\langle R_r' x_0', \varphi_j \rangle$ and have

$$\begin{aligned} 0 = \langle R_r' x_0', \varphi_j \rangle &= \langle x', R\varphi_j \rangle + \sum_{i=1}^{\beta(R)} \langle x_0', \psi_i \rangle \langle f_i, \varphi_j \rangle \\ &= \langle x_0', \psi_j \rangle, \quad j = 1, \ldots, \beta(R). \end{aligned}$$

We have $R\varphi_j = 0$, since by definition φ_j belongs to $\ker R$. Substituting $\langle x_0, \psi_j \rangle = 0$ in the definition of $R_r' x_0'$, we obtain $R' x_0' = 0$, that is,

$$x_0' \in \ker R' = (\operatorname{coker} R)',$$

and we can expand $x_0' = \sum_{i=1}^{\beta(r)} c_i g_i$. Substituting this in $\langle x_0', \psi_j \rangle = 0$ gives

$$0 = \langle x_0', \psi_j \rangle = \sum_{i=1}^{\beta(r)} c_i \langle g_i, \psi_j \rangle = c_j, \qquad \blacksquare$$

which is $x_0' = 0$ or $\ker R_r' = (0)$.

By Theorem 12.8 the index of a Fredholm operator is invariant under compact 'perturbations'. Before we study continuous perturbations of Fredholm operators we need some tools.

Definition 12.3 *We call a toplinear isomorphism* T *from* $L(X, X)$ *an* automorphism. *The automorphisms* $T \in L(X, X)$ *form a group* G, *the automorphism group of* X.

Lemma 12.1 *Let X be a Banach space and $T \in L(X, X)$. Then we have:*

(a)

$$\| T^n \| \leqslant \| T \|^n \quad \text{for } n = 0, 1, \ldots \tag{11}$$

(b) *If* $\| T \| < 1$, *then there exists* $(I - T)^{-1}$, *and it is*

$$(I - T)^{-1} = \sum_{k=0}^{\infty} T^k \in L(X, X). \tag{12}$$

(c) *The group G of automorphisms $T \in L(X, X)$ is an open set in $L(X, X)$. If $X = H$ is a Hilbert space, then G is also a connected set. For Banach spaces G is in general not connected, see for example Douady* [1].

Proof. (a) (11) holds for $n = 0$, since $\| T^0 \| = \| I \| = 1$. Since

$$\| T^n x \| = \| T(T^{n-1} x) \| \leqslant \| T \| \, \| T^{n-1} x \| \leqslant \| T \|^n \| x \|$$

(11) holds for all $n \in \mathbb{N}$.

(b) We put $T_n = \sum_{k=0}^{n} T^k$ and obtain

$$T_n(I - T) = (I - T)T_n = I - T^{n+1}. \tag{13}$$

Because $\| T \| < 1$ and because of (11) we obtain

$$\lim_{n \to \infty} \| T^n \| = 0 \quad \text{and} \quad \lim_{n \to \infty} (I - T^{n+1}) = I.$$

On the other hand (T_n) is convergent, since the sequence (T_n) is a Cauchy sequence in the complete space $L(X, X)$:

$$\| T_{n+p} - T_n \| = \left\| \sum_{k=n+1}^{n+p} T^k \right\| \leqslant \sum_{k=n+1}^{n+p} \| T^k \| \leqslant \sum_{k=n+1}^{n+p} \| T \|^k < \varepsilon.$$

From (13) we obtain (12) as $n \to \infty$.

(c) If $T_0 \in G$ and $\| T - T_0 \| < 1/\| T_0^{-1} \|$, then

$$\| I - T_0^{-1} T \| \leqslant \| T_0^{-1} \| \, \| T_0 - T \| < 1.$$

By (b) the operator $(I - (I - T_0^{-1} T))^{-1} = (T_0^{-1} T)^{-1} =: S$ exists. From this it follows that the operator T^{-1} exists and equals $S T_0^{-1}$, that is, $T \in G$ proving the openness of G in $L(X, X)$.

Suppose now that $X = H$ is a Hilbert space. If $T \in G$, $T' \circ T$ is a strictly positive operator ($(T'Tx, x) = (Tx, Tx) > 0$ for $x = 0$) and $T'T$ possesses a strictly positive square root A. We write $U = TA^{-1}$ and have $U'U = A^{-1} T' T A^{-1} = A^{-1} A^2 A^{-1} = I$, that is, U is unitary. The set of strictly positive operators is convex in G and we can link A and I by means of an interval

$$[A, I] = [\lambda A + (1 - \lambda) I : 0 \leqslant \lambda \leqslant 1] \subset G.$$

By multiplication by U we see that we can also link $T = UA$ and U by means of an interval in G. We now show that we can link U with I by means of a continuous path in G. Let $U = \int_{|\lambda|=1} e^{i\lambda}\, dE_\lambda$ be the spectral decomposition of U. The map $W:[0,1] \to L(H,H)$ defined by

$$W(t) = \int_{|\lambda|=1} e^{i\lambda t}\, dE_\lambda$$

is a continuous path in G (all $W(t)$s are unitary operators), with $W(0) = I$ and $W(1) = U$, since the integral converges in the operator norm. Let T_1, $T_2 \in G$. By the argument above, we can use continuous paths in G to link

$$T_1 - U_1 - I - U_2 - T_2,$$

that is, G is path-connected, and hence connected. ∎

We next study 'continuous perturbations' of Fredholm operators.

Theorem 12.9 *If* $T \in F(X, Y)$ *is a Fredholm operator with regulariser* R *and if* $A \in L(X, Y)$ *is such that* $\|A\|\,\|R\| < 1$, *then* $T + A \in F(X, Y)$ *is again a Fredholm operator and* $\operatorname{ind}(T + A) = \operatorname{ind} T$.

Remark 12.2 By Theorem 12.5(c) for any $T \in F(X, Y)$ there exist regularisers R, which are simultaneously right and left regularisers, we take one such for our proof.

Proof. Assuming $\|A\|\ \|R\| < 1$, Lemma 12.1(b) yields the existence and continuity of the operators $(I_X + RA)^{-1}$ and $(I_Y + AR)^{-1}$. It is easily shown that

$$R_1 = (I_X + RA)^{-1} R \text{ is a left regulariser}$$

and

$$R_r = R(I_Y + AR)^{-1} \text{ is a right regulariser}$$

for $T + A$, so that $T + A \in F(X, Y)$. Since the equations $R_1 y = 0$ and $Ry = 0$ are equivalent we have

$$\alpha(R_1) = \alpha(R). \tag{14}$$

Furthermore, because $\|A'\|\ \|R'\| = \|A\|\ \|R\| < 1$ we have the relation

$$\begin{aligned} R_1' &= R'(I_X + A'R')^{-1} = R'(I_X - A'R' + \cdots) \\ &= (I_X - R'A' + \cdots)R' = (I_X + R'A')^{-1} R' \end{aligned}$$

by which we see the equivalence of the equations $R_1' x = 0$ and $R'x = 0$. Therefore

$$\alpha(R_1') = \alpha(R'). \tag{15}$$

We apply Theorem 12.7, and using (5), (14) and (15) obtain

$$\operatorname{ind}(T+A) = -\operatorname{ind} R_1 = \beta(R_1) - \alpha(R_1) = \\ = \alpha(R_1') - \alpha(R_1) = \alpha(R') - \alpha(R) = -\operatorname{ind} R = \operatorname{ind} T. \qquad \blacksquare$$

Theorem 12.10 *$F(X, Y)$ is an open set in $L(X, Y)$, and the index is constant on each connected component of $F(X, Y)$.*

Proof. The first part follows directly from Theorem 12.9. If the operators T_1 and T_2 belong to the same connected component $M \subset F(X, Y)$, then there exists a continuous path in M which links T_1 and T_2. Since this path, as continuous image of the compact interval $[0, 1]$ is compact, one may apply Theorem 12.9 finitely often, and if follows that $\operatorname{ind} T_1 = \operatorname{ind} T_2$. ■

In Hilbert spaces, the converse of Theorem 12.10 holds. In order to prove it we need a preliminary lemma.

Lemma 12.2 *Let $X = H_1$, $Y = H_2$ be two Hilbert spaces, let $T \in F(H_1, H_2)$ be a Fredholm operator with* $\operatorname{ind} T \geqslant 0$, *and further let G_1 be a subspace of H_1 with* $\dim G_1 = \operatorname{ind} T$. *Then in the same component of $F(H_1, H_2)$ to which T belongs, there exists a Fredholm operator T_1 with* $\ker T_1 = G_1$ *and* $\operatorname{im} T_1 = H_2$.

Proof. By hypothesis $\dim \ker T - \dim(\operatorname{im} T)^{\perp} = \operatorname{ind} T \geqslant 0$, therefore $\dim \ker T \geqslant \dim(\operatorname{im} T)^{\perp}$. We can therefore find a subspace G of $\ker T$ and a topological isomorphism $\tilde{U}: G \to (\operatorname{im} T)^{\perp}$. By $U := \tilde{U} \circ \operatorname{proj}_G$ we obtain an operator $U \in L(H_1, H_2)$. The operators tU, $0 \leqslant t \leqslant 1$, have finite dimensional image sets and thus are compact. By Theorem 12.8 the operators $S_t = T + tU$ are in $F(H_1, H_2)$. Since the set $\{S_t : 0 \leqslant t \leqslant 1\}$ as a continuous image of the interval $[0, 1]$ is connected, the operator S_1 is in the same component as $S_0 = T$. $S_1 = T + U$ is surjective: $\operatorname{im} U = (\operatorname{im} T)^{\perp}$, $H_2 = \operatorname{im} T \oplus (\operatorname{im} T)^{\perp}$. Therefore by Theorem 12.10 we have $\dim \ker S_1 = \operatorname{ind} S_1 = \operatorname{ind} S_0 = \operatorname{ind} T = \dim G_1$, where the last equation is assumed. There then exists a regular map A of H_1 onto itself with $A(G_1) = \ker S_1$: let $\dim G_1 = n$, we have $H_1 = G_1 \oplus G_1^{\perp} = \ker S_1 \oplus \ker S_1^{\perp}$, we choose as first orthonormal-basis in $H_1 \{\varphi_1, \ldots, \varphi_n, \varphi_{n+1}, \ldots\}$, where $L[\varphi_1, \ldots, \varphi_n] = G_1$, as second orthonormal-basis $\{\psi_1, \ldots, \psi_n, \psi_{n+1}, \ldots\}$ with $L[\psi_1, \ldots, \psi_n] = \ker S_1$. We may define $A\varphi_i = \psi_i$, $i = 1, 2, \ldots$, and extend linearly and continuously. we write $T_1 := S_1 \circ A$. It is then true that $\ker T_1 = G_1$ and $\operatorname{im} T_1 = H_2$. Since by Lemma 12.1 (c) the group G of automorphisms is open and connected, there exists a continuous path in this automorphism group, $A(t)$, $0 \leqslant t \leqslant 1$, which links I_{H_1} and A. As automorphism the operators $A(t)$ are Fredholm operators. By Theorem 12.6 $S_1 \circ A(t)$ are again Fredholm operators and therefore elements of a continuous path linking S_1 and T_1 in $F(H_1, H_2)$. In all we have continuous paths connecting $T - S_1 - T_1$, and so T_1 belongs to the same component of $F(H_1, H_2)$ as T. ■

Theorem 12.11 *Let $S, T \in F(H_1, H_2)$ be Fredholm operators, H_1, H_2 Hilbert spaces and* $\operatorname{ind} S = \operatorname{ind} T$. *Then S and T belong to the same component of $F(H_1, H_2)$.*

Proof. Since $T \to T'$ determines a topological isomorphism from $F(H_1, H_2)$ to $F(H_2', H_1')$ (Theorem 12.4) with $\operatorname{ind} T' = -\operatorname{ind} T$, it is enough to consider the case $\operatorname{ind} S = \operatorname{ind} T \geqslant 0$. By Lemma 12.2 we may choose operators S_1 and T_1 in the components containing S and T respectively, to satisfy $\ker S_1 = \ker T_1 = G_1$ (G_1 is a subspace of H_1 with $\dim G_1 = \operatorname{ind} S = \operatorname{ind} T$) and $\operatorname{im} S_1 = \operatorname{im} T_1 = H_2$. By the open mapping theorem, S_1 and T_1 map the space $G_1^\perp$ topologically and isomorphically onto H_2. Therefore $A := T_1^{-1} \circ S$ is an automorphism of $G_1^\perp$. Let $A(t)$, $0 \leqslant t \leqslant 1$, be a continuous path in the automorphism group of $G_1^\perp$, which links A with $I_{G_1^\perp}$ (Lemma 12.1(c)). Because $A(t) \in F(G_1^\perp, G_1^\perp)$ and $P := \operatorname{proj}_{G_1^\perp} \in F(H_1, G_1^\perp)$ it follows by Theorem 12.6 that $T_1 \circ A(t) \circ P$ is a continuous path in $F(H_1, H_2)$ which links $T_1 AP = S_1$ and $T_1 \circ I_{G_1^\perp} \circ P = T_1 P = T_1$. Therefore S_1 and T_1, and also S and T belong to the same connected component. ■

12.3 *A priori* estimates, Weyl's lemma and smoothable operators

Theorem 12.12 (*a priori* estimate) *Let X, Y, Z be three Banach spaces. Let $X \subsetneq Y$ be a compact embedding, let $T: X \to Z$ be a linear continuous map. Then the following conditions are equivalent:*

(a) $\alpha(T) < \infty$ *and* $\operatorname{im} T$ *is closed in Z.*
(b) *There exists a constant $c > 0$, so that for all $x \in X$ we have*

$$\|x\|_X \leqslant c(\|x\|_Y + \|Tx\|_Z). \tag{16}$$

Remarks 12.3

1. We shall call the diagram

$$\begin{array}{ccc} Y & & \\ \text{compact} \uparrow & & \\ X & \xrightarrow[T]{} & Z \end{array},$$

presupposed above, a Schauder scheme.
2. (16) is called an *a priori* estimate; or Schauder's or Friedrichs' estimate.

3. One obtains the implication $1 \Rightarrow 2$ without compactness of the embedding $X \subsetneq Y$.

Proof. Let $\alpha(T) < \infty$ and $\operatorname{im} T$ closed in Z. Since $\ker T = X_0$ is finite dimensional, we can decompose into a direct sum $X = X_0 \oplus X_1$, where X_1 is closed. The restriction of T to X_1 is continuous, bijective, $T: X_1 \to \operatorname{im} T$. By the open mapping theorem the inverse $T|_{X_1}^{-1}$ is also continuous and we have the estimate

$$\|x_1\|_X \leqslant c_1 \|Tx_1\|_Z \quad \text{for all } x_1 \in X_1. \tag{17}$$

Because $X \subsetneq Y$ we can consider $\ker T = X_0$ as a subspace of Y, since $\ker T$ is finite dimensional, all norms on $\ker T$ are equivalent and $\ker T$ is again closed in Y. We may write $Y = \ker T \oplus Y_1$, where the projection

$$P^y: Y \to \underset{\text{in } Y}{\ker T} \to \underset{\text{in } X}{\ker T}$$

is continuous, that is, we have

$$\|x_0\|_X = \|P^Y y\|_X \leqslant c_2 \|y\|_Y \quad \text{for all } y \in Y \quad \text{and} \quad P^Y y = x_0,$$

or restricted to X

$$\|x_0\|_X = \|P^Y x\|_X \leqslant c_2 \|x\|_Y \quad \text{for all } x \in X. \tag{18}$$

From (17) and (18) with $x = x_0 + x_1$, $x_0 \in X_0$, $x_1 \in X_1$, $Tx_1 = Tx$, we have

$$\begin{aligned}\|x\|_X &\leqslant \|x_0\|_X + \|x_1\|_X \leqslant c_2 \|x\|_Y + c_1 \|Tx_1\|_Z \\ &\leqslant c(\|x\|_Y + \|Tx\|_Z) \quad \text{for all } x \in X,\end{aligned}$$

which is the *a priori* estimate (16).

Conversely if (16) holds, then on the one hand $\|x\|_X \leqslant c\|x\|_Y$ for all $x \in \ker T$; on the other hand the continuity of the embedding $X \subsetneq Y$ gives the existence of a number $c_4 > 0$ with $\|x\|_Y \leqslant c_4 \|x\|_X$ for all $x \in X$. The norms $\|\circ\|_X$ and $\|\circ\|_Y$ are therefore equivalent on $\ker T$. Since because of the compact embedding $X \subsetneq Y$ the unit ball is compact in $\ker T$, we have $\dim \ker T = \alpha(Z) < \infty$, and thus we have proved the first part of 1. Now for the second statement.

We again factorise $X = X_0 \oplus X_1$, $X_0 = \ker T$ and consider the restriction of T to X_1; $S := T|_{X_1}$, $S: X_1 \to \operatorname{im} T$ is injective and we show

$$\|x_1\|_X \leqslant c \|Sx_1\|_Z \quad \text{for all } x_1 \in X_1. \tag{19}$$

Supposing the converse to hold, there exists a sequence

$$x_n \in X_1 \quad \text{with} \quad \|x_n\|_X = 1 \quad \text{and} \quad 1 = \|x_n\|_X > n \|Sx_n\|_Z.$$

Since the embedding $X \subsetneq Y$ is compact, we can find a subsequence x_{k_n} with $x_{k_n} \to x$ in Y, because of (16) and $Sx_n \to 0$ in Z, x_{k_n} is a Cauchy sequence in X and we also have $x_{k_n} \to \tilde{x}$ in X, or in X_1, since X_1 is closed. By the continuity of S we have

$$Sx_{k_n} \to S\tilde{x} = 0 \quad \text{or} \quad \tilde{x} = 0 \quad \text{contradicting} \quad \|x_n\| = 1.$$

Therefore we have proved (19); (19) implies the continuity of S^{-1}: im $T \to X_1$. Relative to the continuous map S^{-1}, im T, as preimage of the closed set X_1, is closed. ■

By Theorem 12.12 it follows that Fredholm operators in a Schauder scheme satisfy an *a priori* estimate, and conversely, if T and T' satisfy *a priori* estimates then T is a Fredholm operator, see Theorem 12.4.

For applications to elliptic differential operators we need a partially stronger notion of regularisability than that considered in Definition 12.2.

Definition 12.4 *Let X_1, X_2, Y_1, Y_2, be four Banach spaces, let $X_2 \subsetneq X_1$, $Y_2 \subsetneq Y_1$ and let $T: X_1 \to Y_1$ be linear and continuous.*

$$\begin{array}{ccc} X_1 & \xrightarrow{T} & Y_1 \\ \uparrow & & \uparrow \\ X_2 & & Y_2 \end{array} \tag{20}$$

We say that T is smoothable, if there exist operators $R_1, R_r \in L(Y_1, X_1)$ with

$$g_X := R_1 \circ T - I_{X_1} \in L(X_1, X_2)$$
$$g_Y := T \circ R_r - I_{Y_1} \in L(Y_1, Y_2)$$

and with the property

$$R_1|_{Y_2}, R_r|_{Y_2} \in L(Y_2, X_2),$$

We call the operators g_X, g_Y smoothing operators, and the operator R_1 (respectively R_r) a smoothing left (respectively right) regulariser.

Weyl's lemma holds for smoothable operators.

Lemma 12.3 *Let T be smoothable, then from $Tx = y$, $x \in X_1$, $y \in Y_2$ it follows that x is already an element of X_2.*

Proof. We have $g_X x = R_1 \circ Tx - I_{X_1} x$ or $x = R_1 x - g_X x \in X_2$ by the defining properties. ■

Lemma 12.3 implies that ker T is already contained in X_2.

Remark 12.4 As we see from the proof, Weyl's lemma holds under much more general assumptions, we need only the representation formulae

$$g_X x = R_1 \circ Tx - I_{X_1} x \quad \text{and} \quad R_1|_{Y_2} \in L(Y_2, X_2).$$

Theorem 12.13 *Assume given the scheme* (20) *and let* T *be smoothable. If the embeddings* $X_2 \hookrightarrow X_1$ *and* $Y_2 \hookrightarrow Y_1$ *are compact, then* T *is a Fredholm operator* $T \in F(X_1, Y_1)$.

Proof. We extend the smoothing operators

$$g_X : X_1 \to X_2 \hookrightarrow X_1; \quad g_Y : Y_1 \to Y_2 \hookrightarrow Y_1,$$

obtain compact operators and apply Theorem 12.5(c). ∎

For elliptic operators the Fredholm operators act in scales. It is important that the defect numbers α, β and the index do not depend on the place in a scale. We can prove this property abstractly; it is a consequence of the so-called lemma of Gothberg & Kreĭn [1].

Lemma 12.4 *Let* N *be a finite dimensional subspace of the Banach space* X *with the direct sum decomposition* $X = N \oplus R$. *If* D *is a dense subspace in* X *we have*:

(a) *The subspace* $D \cap R$ *is dense in* R.
(b) *There exists a subspace* $\tilde{N}$ *with* $\dim \tilde{N} = \dim N$, $\tilde{N} \subset D$ *and the direct sum decomposition* $X = \tilde{N} \oplus R$.

Proof. Let $\{e_1, \ldots, e_n\}$ be a basis for N, let $\{f_1, \ldots, f_n\}$ be continuous linear functionals from X' with

$$\langle f_j, c_k \rangle = \delta_{j,k}, \, k = 1, \ldots, n \quad \text{and} \quad \langle f_j, x \rangle = 0 \quad \text{for } x \in R, j = 1, \ldots, n,$$

(see for example Wloka [1], p. 93). We choose elements $\tilde{c}_k \in D$, $k = 1, \ldots, n$ 'near to' e_k, so that

$$\det\{\langle f_j, \tilde{e}_k \rangle\} \neq 0. \tag{21}$$

This is certainly possible since D is dense in X, and both the f_js and the determinant are continuous. Let $y \in R$ be arbitrary. There exists a sequence $z_m \in D$ with $z_m \to y$ as $m \to \infty$. We form the sequence $\tilde{z}_m$,

$$\tilde{z}_m = z_m + \sum_{k=1}^{n} \alpha_{m,k} \tilde{e}_k \tag{22}$$

with coefficients $\alpha_{m,k}$ yet to be determined. We have $\tilde{z}_m \in D$. We choose the coefficients $\alpha_{m,k}$ so that $\tilde{z}_m \in D \cap R$. This is the case if and only if

$$\langle f_j, \tilde{z}_m \rangle = 0 \quad \text{for} \quad j = 1, \ldots, n; \; m = 1, 2, \ldots .$$

respectively

$$\sum_{k=1}^{n} \alpha_{m,k} \langle f_j, \tilde{e}_k \rangle = -\langle f_j, z_m \rangle \quad \text{for} \quad j = 1, \ldots, n, \; m = 1, 2, \ldots . \tag{23}$$

Because of (21), (23) is uniquely solvable for each m, and the $\alpha_{m,k}$ found in this way are linear combinations of the expressions $\langle f_j, z_m \rangle$ with coefficients independent of m. Because

$$\lim_{m \to \infty} \langle f_j, z_m \rangle = \langle f_j, y \rangle = 0, \quad j = 1, \ldots, n,$$

we have

$$\lim_{m \to \infty} \alpha_{m,k} = 0 \quad \text{for } k = 1, \ldots, n.$$

Therefore (22) implies that

$$\lim_{m \to \infty} \tilde{z}_m = \lim_{m \to \infty} z_m = y,$$

and so we have proved (a).

The elements $\tilde{e}_k$, $k = 1, \ldots, n$ are linearly independent. From

$$\sum_{k=1}^{n} c_k \tilde{e}_k = 0$$

it indeed follows that

$$\sum_{k=1}^{n} c_k \langle f_j, \tilde{e}_k \rangle = 0, \quad j = 1, \ldots, n,$$

and hence recalling (21) $c_1 = \cdots = c_n = 0$. Let

$$\tilde{N} = L[\tilde{e}_1, \ldots, \tilde{e}_n].$$

Then $\dim \tilde{N} = n = \dim N$, and by construction $\tilde{N} \subset D$. Let $x \subset \tilde{N} \cap R$. Then on the one hand $x = \sum_{k=1}^{n} \gamma_k \tilde{e}_k$, and on the other $\langle f_j, x \rangle = 0$ for $j = 1, \ldots, n$. It follows that

$$\sum_{k=1}^{n} \gamma_k \langle f_j, \tilde{e}_k \rangle = 0, \quad j = 1, \ldots, n,$$

from which, because of (21), $\gamma_1 = \cdots = \gamma_n = 0$, and therefore $x = 0$. Thus the sum $\tilde{N} \oplus R$ is direct, and for dimensional reasons we must have $X = \tilde{N} \oplus R$. ■

Theorem 12.14 *Suppose given the scheme* (20):

$$\begin{array}{ccc} X_1 & \xrightarrow{T} & Y_1 \\ \uparrow\!\!\rfloor & & \uparrow\!\!\rfloor \\ X_2 & & Y_2 \end{array}$$

and that the following additional conditions are satisfied:

(a) $T_{X_2} \in L(X_2, Y_2)$,
(b) *T is left smoothable, that is, there exists some $R_1 \in L(Y_1, X_1)$ with $gx = R_1 \circ Tx - I_{X_1} x \in X_2$ for $x \in X_1$ and $R_1|_{Y_2} \in L(Y_2, X_2)$, see Remark* 12.4.
(c) *The operator $T: X_1 \to Y_1$ is Fredholm.*
(d) *Y_2 is dense in Y_1.*

Then the defect numbers of the operators $T|_{X_1}$ and $T|_{X_2}$ are equal, that is, we *have*

$$\alpha(T|_{X_1}) = \alpha(T|_{X_2}), \quad \beta(T|_{X_1}) = \beta(T|_{X_2}),$$

from which it follows that $T|_{X_2}$ is again Fredholm, and that

$$\operatorname{ind}(T|_{X_1}) = \operatorname{ind}(T|_{X_2}).$$

Proof. The equation for α follows as in Weyl's lemma, 12.3, see Remark 12.4, from $x = R_1 \circ Tx - gx$. For β we have, by definition,

$$Y_1 = T(X_1) \oplus Q; \quad \dim Q = \beta(T|_{X_1}).$$

Since Y_2 is dense in Y_1 we can apply Lemma 12.4 and take $Q \subset Y_2$. We then have

$$Y_2 = Y_1 \cap Y_2 = Y_2 \cap T(X_1) \oplus Q. \tag{24}$$

We show that $Y_2 \cap T(X_1) = T(X_2)$. Let $y \in Y_2 \cap T(X_1)$, then $y = Tx$, $y \in Y_2$ and $x = R_1 y - gx$ shows that $x \in X_2$, with which we have shown one inclusion; the other follows immediately from (a). Therefore we can write (24) as

$$Y_2 = T(X_2) \oplus Q,$$

and we have $\beta(T|_{X_2}) = \dim Q = \beta(T|_{X_1})$. ∎

Exercises

12.1 Let $T \in F(X, X)$ be a Fredholm operator and $K \in K(X, X)$ be compact. Examine the spectrum of $T + \lambda K$, $\lambda \in \mathbb{C}$.

12.2 Let $T \in F(X, X)$ be a Fredholm operator with $\operatorname{ind} T = 0$. Show that there is a representation $T = U + K$ where U is an isomorphism and K is a compact operator.

12.3 Let l^2 be the space of complex sequences $x=(x_1,x_2,\ldots)$ with $\sum_{n=1}^{\infty}|x_n|^2<\infty$. Show that the shift operators

$$T^+:x\mapsto(0,x_1,x_2,\ldots),\quad T^-:x\mapsto(x_2,x_3,\ldots)$$

are Fredholm operators and find the indices.

12.4 For two Fredholm operators $T_1:X_1\to X_1, T_2:X_2\to X_2$ consider the direct sum

$$T_1\oplus T_2:X_1\oplus X_2\to X_1\oplus X_2.$$

Show $T_1\oplus T_2$ is a Fredholm operator with $\operatorname{ind}(T_1\oplus T_2)=\operatorname{ind}T_1+\operatorname{ind}T_2$.

12.5 Let $T\in F(X,X)$ be a Fredholm operator. Show that there exists some $\rho>0$ with $\alpha(T+\lambda I)$ and $\beta(T+\lambda I)$ constant for $0<|\lambda|<\rho$.

12.6 Prove that a compact operator $K\in K(X,Y)$ is a Fredholm operator if and only if X and Y are finite dimensional.

12.7 Let $K\in L(X,X)$ and suppose that there is a natural number $m>1$ such that K^m is compact. Show that all statements of the spectral theorem in §12.1 hold for K. Hint: Use the identity

$$\lambda^m I-K^m=(\lambda I-K)(\lambda^{m-1}I+\lambda^{m-2}K+\cdots+\lambda K^{m-2}+K^{m-1}).$$

§13 The main theorem and some theorems on the index of elliptic boundary value problems

We prove the main theorem, that is, we state the equivalence between the ellipticity of a boundary value problem, the *a priori* estimate, the Fredholm property and smoothability. Here we associate Weyl's lemma with the concept of smoothability. In conclusion we show that the index depends only on the principal parts $A^{\mathrm{H}}, b_1^{\mathrm{H}},\ldots,b_m^{\mathrm{H}}$, and calculate the index of the Dirichlet problem, it equals zero. Further indices are calculated in §16. We consider also the spectral value problem

$$Au-\lambda u=f,\quad b_1u=g_1,\ldots,b_mu=g_m,\quad \lambda\in\mathbb{C},$$

and prove Theorem 13.4: if $L=(A,b_1,\ldots,b_m)$ is elliptic, with $\operatorname{ind}L=0$, and there exists at least one $\lambda_0\in\mathbb{C}$, which is not an eigenvalue of L, then our spectral value problem may be posed in the framework of Riesz–Schauder theory.

13.1 The main theorem for elliptic boundary value problems

We formulate the differentiability assumptions for this section. Let $r\geqslant 2$; the case $r=1$ is well-known, see for example Naĭmark [1], and therefore we do not consider it.

1. Let the region $\Omega \subset \mathbb{R}^r$ be bounded and belong to the class $C^{2m+k;\kappa}$ $k+\kappa \geqslant 1$, $m \geqslant 1$ (see Definition 2.7).

2. Let $A(x, \mathrm{D})$ be a linear differential operator of order $2m$ with coefficients dependent on x, which belong to $C^{k+1}(\bar{\Omega})$.

3. Suppose given m boundary value operators $b_j(x, \mathrm{D})$, $j = 1, \ldots, m$; where the b_js are linear differential operators of order m_j, $0 \leqslant m_j \leqslant 2m-1$, with coefficients dependent on $x \in \partial\Omega$, which may be extended to functions belonging to $C^{2m-m_j+k+1}(\bar{\Omega})$. We interpret the boundary operators $b_j(x, \mathrm{D})$ with the help of the trace operator T_0, see §11:

$$b_j(x, \mathrm{D})\varphi = \sum_{|s| \leqslant m_j} b_{j,s}(x) T_0(\mathrm{D}^s \varphi) = T_0\left(\sum_{|s| \leqslant m_j} b_{j,s}(x) \mathrm{D}^s \varphi \right).$$

We recall the definitions:

We say that $A(x, \mathrm{D})$ is elliptic on $\bar{\Omega}$, if:

I. $A^{\mathrm{H}}(x, \xi) \neq 0$ for all $0 \neq \xi \in \mathbb{R}^r$, $x \in \bar{\Omega}$ (see §10). By Theorem 10.6 $A(x, \mathrm{D})$ is then uniformly elliptic on $\bar{\Omega}$, that is, we have

$$|A^{\mathrm{H}}(x, \xi)| \geqslant c_0 |\xi|^{2m} \quad \text{for all } \xi \in \mathbb{R}^r, x \in \bar{\Omega}.$$

We consider the boundary value problem

$$\begin{aligned} &A(x, \mathrm{D})u(x) = f(x) \quad \text{on} \quad \Omega, \\ &b_1(x, \mathrm{D})u(x) = g_1(x), \ldots, b_m(x, \mathrm{D})u(x) = g_m(x) \quad \text{on} \quad \partial\Omega, \end{aligned} \tag{1}$$

and say that it is elliptic, if in addition to I (ellipticity of A) for each boundary point $x \in \partial\Omega$ we have

II. The L.Š. Condition 11.1 is satisfied for the principal parts $A^{\mathrm{H}}(x, \mathrm{D})$, $b_1^{\mathrm{H}}(x, \mathrm{D}), \ldots, b_m^{\mathrm{H}}(x, \mathrm{D})$.

By Theorem 11.2 the operator $A(x, \mathrm{D})$ is then properly elliptic for $x \in \partial\Omega$, that is,

III. For all $0 \neq \xi' \in \mathbb{R}^{r-1}$ the polynomial $P(z) = A^{\mathrm{H}}(x, (\xi', z)) = 0$ has in the upper half plane $\operatorname{Im} z > 0$ exactly m zeros (multiplicities counted).

Assuming 1, 2 and 3 we find that the differential operators

$$\begin{aligned} &A(x, \mathrm{D}): W_2^{2m+l}(\Omega) \to W_2^l(\Omega) \\ &b_j(x, \mathrm{D}): W_2^{2m+l}(\Omega) \to W_2^{2m+l-m_j-1/2}(\partial\Omega), \quad j = 1, \ldots, m, \end{aligned}$$

are continuous for $0 \leqslant l \leqslant k+\kappa$ ($l = k+\kappa$ is allowed if l is integral), see §10 and §11. Here $m_j = \operatorname{ord} b_j(x, \mathrm{D})$. Of course by introducing the exact multiplier classes for the spaces W_2^l we might further weaken the assumptions on the coefficients of $A(x, \mathrm{D})$ and $b_j(x, \mathrm{D})$. However, in doing this we would not gain much and would lose in transparency.

For short we write

$$H^l := H^l(\Omega, \partial\Omega) := W_2^l(\Omega) \times_{j=1}^{m} W_2^{2m+l-m_j-1/2}(\partial\Omega),$$
$$L := L(x, \mathrm{D}) := (A(x, \mathrm{D}), b_1(x, \mathrm{D}), \ldots, b_m(x, \mathrm{D}))$$

and consider the continuous diagram

$$\begin{array}{ccc} W_2^{2m+l-1}(\Omega) & \xrightarrow{L} & H^{l-1}(\Omega, \partial\Omega) \\ \uparrow & & \uparrow \\ W_2^{2m+l}(\Omega) & \xrightarrow{L} & H^l(\Omega, \partial\Omega) \end{array} \tag{2}$$

for $1 \leqslant l \leqslant k + \kappa$. Strictly speaking we must give L the subscript l, since, however, for the restriction $|_l$ we have $L_{l-1}|_l = L|_l$, we can save ourselves this subscript.

We formulate the main theorem for elliptic boundary value problems.

Theorem 13.1 *Suppose that the differentiability assumptions* 1–3 *are satisfied. Then for* $1 \leqslant l \leqslant k + \kappa$ *the following four statements are equivalent*:

(a) *The boundary value problem* (1) *is elliptic, that is, A is elliptic for all* $x \in \bar{\Omega}$ *and* $(A, b_1, \ldots, b_m)$ *satisfies the L.Š. Condition* 11.1 *for all* $x \in \partial\Omega$.
(b) *The operator L in* (2) *is smoothable, see Definition* 12.4.
(c) *The operator* $L: W_2^{2m+l-1}(\Omega) \to H^{l-1}$ *is Fredholm.*
(d) *For all* $\varphi \in W_2^{2m+l-1}(\Omega)$ *we have the a priori estimate*

$$\|\varphi\|_{2m+l-1} \leqslant c(\|L\varphi\|_{H^{l-1}} + \|\varphi\|_{2m+l-2})$$

or expanded

$$\|\varphi\|_{2m+l-1} \leqslant c\left\{\|A\varphi\|_{l-1} + \sum_{j=1}^{m} \|b_j\varphi\|_{2m+l-1-m_j-1/2,\partial\Omega} + \|\varphi\|_{2m+l-2}\right\}.$$

Before we pass to the proof of the main theorem, we wish to draw some consequences. Suppose therefore that one of the equivalent statements (a)–(d) of the main Theorem 13.1 is satisfied.

Corollary 13.1 *The global Weyl's lemma holds: Let* $Lu = F$, $u \in W_2^{2m+l-1}(\Omega)$ $F \in H^l$, *then u is already in* $W_2^{2m+l}(\Omega)$, *that is, if* $F \in H^{k+\kappa}$, *then* $u \in W_2^{2m+k+\kappa}(\Omega)$.

This follows immediately from Lemma 12.3, if we use statement (b) of the main theorem. We obtain the classical form of Weyl's lemma, if we allow l to take all values and apply Sobolev's lemma, Theorem 6.4:

$$\textit{If } Lu = F \textit{ and } F \in C^\infty \textit{ it follows that } u \in C^\infty.$$

Corollary 13.2 *The defect numbers and the index of the operator L are independent of l.*

Given statement (b) this follows from Theorem 12.14 if H^l is dense in H^{l-1}. The latter follows, however, from Theorems 3.6 and 4.3.

Now for the proof of the main theorem, we start with the simple implications (b)⇒(c)⇒(d)⇒(a).

By Theorem 12.13 we have the implication (b)⇒(c), if we know that the vertical inclusions in (2) are compact. However, this compactness is guaranteed by Theorems 7.9 and 7.10.

For the implication (c)⇒(d) we consider the Schauder scheme

$$\begin{array}{l} W_2^{2m+l-2} \\ \uparrow \\ W_2^{2m+l-1} \xrightarrow{L} H^{l-1} \end{array}$$

and apply Theorem 12.12.

Now the proof that statement (a) follows from the estimate (d). Since we work here with pointwise admissible transformations, Theorem 2.11, and start out from the *a priori* estimate (d) for $l-1=0$, that is, $l=1$, we succeed with minimal assumptions

$$\Omega \in C^{2m-1,1}, \quad a_s(x) \in C^0(\bar{\Omega}), \quad b_{j,s}(x) \in C^{2m-m_j}(\bar{\Omega}).$$

We fix a point $x_0 \in \bar{\Omega}$, let U be a sufficiently small neighbourhood of x_0. Since Ω belongs to the class $C^{2m-1,1}$, by Theorem 2.11 we can find an admissible transformation $\Phi \in C^{2m-1,1}$, which takes $U \cap \Omega$ to the half cube W^r_+ and takes $U \cap \partial\Omega$ to $\{x_r = 0\} = W^{r-1}$, $\begin{pmatrix} U \to W^r \\ x_0 \mapsto 0 \end{pmatrix}$.

By the transformation theorem 4.1 the *a priori* estimate from (d) becomes

$$\|\varphi\|_{2m,W^r_+} \leq c\left\{\|\tilde{A}\varphi\|_{0,W^r_+} + \sum_{j=1}^m \|\tilde{b}_j\varphi\|_{2m-m_j-1/2,W^{r-1}} + \|\varphi\|_{2m-1,W^r_+}\right\}, \quad (3)$$

for all $\varphi \in W_2^{2m}(W^r_+)$, whose supports do not meet $\partial W^r_+ \backslash W^{r-1}$. $\tilde{A}, \tilde{b}_j$ are the transformed differential operators. We shrink the unit cube $W^r = W^r(1)$ onto the cube $W^r(\delta)$ centred at 0 with edge length 2δ, and use corresponding notation for $W^r_+(\delta)$ and $W^{r-1}(\delta)$. Of course the inequality (3) holds for all $(\delta \leqslant 1)$,

$$\varphi \in W_2^{2m}(W^r_+(\delta)).$$

We will fix the magnitude of δ later, see the text after formula (7). If x_0 lies not on the frontier but in the interior of Ω, we do not need to transform. We decompose

$$\begin{aligned} \tilde{A}(x, \mathrm{D}) &= \tilde{A}_0(\mathrm{D}) + A_1(x, \mathrm{D}) + A_2(x, \mathrm{D}), \\ \tilde{b}_j(x, \mathrm{D}) &= \tilde{b}_{j,0}(\mathrm{D}) + b_{j,1}(x, \mathrm{D}) + b_{j,2}(x, \mathrm{D}), \quad j = 1, \dots, m, \end{aligned}$$

where

$$\begin{aligned} &\tilde{A}_0(\mathrm{D}) = \tilde{A}^{\mathrm{H}}(0, \mathrm{D}), \quad A_1(x, \mathrm{D}) = \tilde{A}^{\mathrm{H}}(x, \mathrm{D}) - \tilde{A}^{\mathrm{H}}(0, \mathrm{D}), \quad A_2 = \tilde{A} - \tilde{A}_0 - A_1 \\ &\tilde{b}_{j,0}(\mathrm{D}) = \tilde{b}_j^{\mathrm{H}}(0, \mathrm{D}), \quad b_{j,1}(x, \mathrm{D}) = \tilde{b}_j^{\mathrm{H}}(x, \mathrm{D}) - \tilde{b}_j^{\mathrm{H}}(0, \mathrm{D}), \quad b_{j,2} = \tilde{b}_j - \tilde{b}_{j,0} - b_{j,1}. \end{aligned}$$

We estimate

$$\| A_1 \varphi \|_{0, W^r_+(\delta)} \leqslant c_1(\delta) \| \varphi \|_{2m, W^r_+(\delta)} \tag{4}$$

$$\begin{aligned} \| b_{j,1} \varphi \|_{2m - m_j - 1/2, W^{r-1}(\delta)} &= \| T_0(b_{j,1} \varphi) \|_{2m - m_j - 1/2, W^{r-1}(\delta)} \leqslant c \| b_{j,1} \varphi \|_{2m - m_j, W^r_+(\delta)} \\ &\leqslant c_2(\delta) \| \varphi \|_{2m, W^r_+(\delta)} + c_3 \| \varphi \|_{2m-1, W^r_+(\delta)}, \end{aligned} \tag{5}$$

here

$$c_1(\delta) = \max_{|s| = 2m} \max_{x \in W^r_+(\delta)} |\tilde{a}_s(x) - \tilde{a}_s(0)|, c_2(\delta) = \max_{|s| = m_j} \max_{x \in W^r_+(\delta)} |\tilde{b}_{j,s}(x) - \tilde{b}_{j,s}(0)|$$

and c_3 is a constant.

In deducing (5) we have used the trace theorem 8.1 and the Leibniz product rule, and have therefore also used the differentiability assumptions for the coefficients of $A(x, \mathrm{D})$ and $b_j(x, \mathrm{D})$. We note further that as $\delta \to 0$, $c_1(\delta)$, $c_2(\delta) \to 0$. We also have

$$\begin{aligned} &\| A_2 \varphi \|_0 \leqslant c_4 \| \varphi \|_{2m-1}, \\ &\| b_{j,2} \varphi \|_{2m - m_j - 1/2, W^{r-1}(\delta)} = \| T_0(b_{j,2} \varphi) \|_{2m - m_j - 1/2, W^{r-1}(\delta)} \\ &\qquad \leqslant c \| b_{j,2} \varphi \|_{2m - m_j, W^r_+(\delta)} \leqslant c_5 \| \varphi \|_{2m-1}, \end{aligned} \tag{6}$$

since the order of A_2 is less than $2m$, and the order of $b_{j,2}$ is smaller than m_j.

If we substitute (4), (5) and (6) in (3) we obtain

$$\begin{aligned} \| \varphi \|_{2m, W^r_+(\delta)} &\leqslant c(\delta) \| \varphi \|_{2m, W^r_+(\delta)} \\ &+ c_6 \left\{ \| \tilde{A}_0(\mathrm{D}) \varphi \|_{0, W^r_+(\delta)} + \sum_{j=1}^{m} \| \tilde{b}_{j,0}(\mathrm{D}) \varphi \|_{2m - m_j - 1/2, W^{r-1}(\delta)} + \| \varphi \|_{0, W^r_+(\delta)} \right\}, \end{aligned} \tag{7}$$

we can give the last norm $\| \varphi \|_0$ the subscript 0, because $2m - 1 > 0$. If we choose the edge length 2δ of $W^r_+(\delta)$ so small that $c(\delta) < \frac{1}{2}$ and $\delta < 1$, then (7) becomes

$$\|\varphi\|_{2m,W^r_+(\delta)}$$
$$\leqslant c\left\{\|\tilde{A}_0(\mathrm{D})\varphi\|_{0,W^r_+(\delta)} + \sum_{j=1}^{m}\|\tilde{B}_{j,0}(\mathrm{D})\varphi\|_{2m-m_j-\frac{1}{2},W^{r-1}(\delta)} + \|\varphi\|_{0,W^r_+(\delta)}\right\}. \quad (8)$$

We show that the ellipticity of $\tilde{A}_0(\mathrm{D})$ follows from (8). Because of invariance under transformation, Corollary 10.1, we will then also have proved the ellipticity of $A(x_0, \mathrm{D})$ at $x_0 \in \bar{\Omega}$.

Assume the opposite, $\tilde{A}_0(\xi) = 0$ for some $0 \neq \xi \in \mathbb{R}^r$, and take $0 \leqslant \chi(x)$ in $\mathscr{D}(W^r_+(\delta))$. We write

$$u_\lambda(x) := \frac{e^{i\lambda(\xi,x)}\chi(x)}{\lambda^{2m}} = \chi(x)v_\lambda(x).$$

Because of the support property of $\chi(x)$ we have $\tilde{B}_{j,0}(\mathrm{D})u_\lambda(x) = 0$ for $j = 1, \ldots, m$ on the frontier W^{r-1}; for large λ we have

$$0 \neq O(1) = \|u_\lambda\|_{2m}, \quad \|u_\lambda\|_0 < O(\lambda^{-1}). \quad (9)$$

In order to estimate $\tilde{A}_0 u_\lambda$ we apply the Leibniz product rule in the form

$$\tilde{A}_0(\mathrm{D})u_\lambda = \chi(x)\tilde{A}_0(\mathrm{D})\frac{e^{i\lambda(\xi,x)}}{\lambda^{2m}} + \sum_{|\alpha|>0} \mathrm{D}^\alpha\chi \cdot \tilde{A}_0^{(\alpha)}(\mathrm{D})v_\lambda.$$

Since by hypothesis $\chi(x)\tilde{A}_0(\mathrm{D})v_\lambda = \chi(x)\tilde{A}_0(\xi)\cdots = 0$ we have

$$\|\tilde{A}_0 u_\lambda\|_0 < O(\lambda^{-1}), \quad (10)$$

substituting (9) and (10) in (8) gives a contradiction.

We show next that from (8) we can deduce the L.Š. Condition 11.1 for the operators $\tilde{A}_0(\mathrm{D})$, $\tilde{B}_{j,0}(\mathrm{D})$, $j = 1, \ldots, m$. Let $w(t) \neq 0$ be a solution from $\mathscr{M}^+$ of

$$\begin{aligned}&\tilde{A}_0\left(\xi', \frac{1}{i}\frac{d}{dt}\right)w(t) = 0, \quad t > 0,\\ &\tilde{B}_{1,0}\left(\xi', \frac{1}{i}\frac{d}{dt}\right)w(0) = 0, \ldots, \tilde{B}_{m,0}\left(\xi', \frac{1}{i}\frac{d}{dt}\right)w(0) = 0,\end{aligned} \quad (11)$$

for a specific $0 \neq \xi' \in \mathbb{R}^{r-1}$. On grounds of homogeneity $w(\lambda t)$, $\lambda > 0$ satisfies the equations

$$\begin{aligned}&\tilde{A}_0\left(\lambda\xi', \frac{1}{i}\frac{d}{dt}\right)w(\lambda t) = 0,\\ &\tilde{B}_{j,0}\left(\lambda\xi', \frac{1}{i}\frac{d}{dt}\right)w(\lambda t)|_{t=0} = 0, \quad j = 1, \ldots, m,\end{aligned} \quad (12)$$

We put

$$v_\lambda(x) := \frac{e^{i\lambda(x',\xi')} w(\lambda x_r)}{\lambda^{2m}}, \quad x = (x', x_r), \quad x_r = t,$$

and see that because of (12) $v_\lambda(x)$ satisfies the equations

$$\tilde{A}_0(D) v_\lambda(x) = 0, \quad \tilde{B}_{j,0}(D) v_\lambda(x', 0) = 0, \quad j = 1, \ldots, m. \tag{13}$$

$w(t)$ from $\mathcal{M}^+$ is a linear combination of expressions of the form $t^\kappa \exp\{i\lambda_l t\}$, where the characteristic values λ_l lie in $\operatorname{Im}\lambda > 0$; thus $w(t)$ and all its derivatives are square integrable on $\mathbb{R}^1_+$. We again take a function $0 \leqslant \psi(x) \leqslant \mathcal{D}(W^r(\delta))$, which equals 1 in a neighbourhood $W^r(\varepsilon)$ of the origin $0 \in \mathbb{R}^r$, and we put

$$u_\lambda(x) := \psi(x) \cdot v_\lambda(x).$$

We have $u_\lambda(x) \in W_2^{2m}(W^r_+(\delta))$, where $\operatorname{supp} u_\lambda \cap \partial W^r_+ \setminus W^{r-1} = \varnothing$ and therefore the estimate (18) holds. We have (again apply (13) and the Leibniz rule)

$$\| u_\lambda \|_0 \leqslant O(\lambda^{-2m}), \quad \| \tilde{A}_0 u_\lambda \|_0 \leqslant O(\lambda^{-1}). \tag{14}$$

It is also true (Leibniz rule, trace theorem 8.1 and (13)) that

$$\begin{aligned} \| \tilde{B}_{j,0}(D) u_\lambda \|_{2m - m_j - 1/2, W^{r-1}} &= \| T_0(\tilde{B}_{j,0}(D) u_\lambda) \|_{2m - m_j - 1/2, W^{r-1}} \\ &\leqslant c \| \tilde{B}_{j,0}(D) u_\lambda \|_{2m - m_j, W^r_+} = \Big\| \psi(x) \tilde{B}_{j,0}(D) v_\lambda(x) + \sum_{|\alpha| > 0} D^\alpha \psi \cdot \tilde{B}^{(\alpha)}_{j,0}(D) v_\lambda \Big\|_{2m - m_j, W^r_+} \\ &= \Big\| \sum_{|\alpha| > 0} D^\alpha \psi \cdot \tilde{B}^{(\alpha)}_{j,0}(D) v_\lambda \Big\|_{2m - m_j, W^r_+} \\ &\leqslant c \| v_\lambda \|_{2m - m_j + m_j - 1, W^r_+} = c \| v_\lambda \|_{2m - 1, W^r_+} \leqslant O(\lambda^{-1}). \end{aligned} \tag{15}$$

Hence for the right-hand side of (8) we have the estimate $\leqslant O(\lambda^{-1})$.

We now show that for the left-hand side of (8) we obtain

$$\| u_\lambda \|_{2m} \cdot \lambda^{1/2} \geqslant c > 0 \quad \text{for large } \lambda, \tag{16}$$

where c is independent of λ. We use the defining property of $\psi(x)$, namely that $\psi(x) = 1$ on $W^r(\varepsilon)$, and take in $\| u_\lambda \|^2_{2m}$ a derivative of order $2m$, for example $\partial^{2m}/\partial t^{2m}$. Then we have

$$\begin{aligned} \| u_\lambda \|^2_{2m} &\geqslant \int_{W^r_+(\varepsilon)} \left| \frac{\partial^{2m}}{\partial t^{2m}} \left(\frac{\psi(x) e^{i\lambda(x',\xi')} w(\lambda t)}{\lambda^{2m}} \right) \right|^2 dx' \, dt \\ &= \int_{W^{r-1}(\varepsilon)} dx' \cdot \int_0^\varepsilon | w^{(2m)}(\lambda t) |^2 \, dt \\ &= \int_{W^{r-1}(\varepsilon)} dx' \cdot \int_0^{\lambda\varepsilon} | w^{(2m)}(\tau) |^2 \frac{d\tau}{\lambda}. \end{aligned}$$

Since the limit of $\int_0^{\lambda\varepsilon}|w^{(2m)}(\tau)|^2\,\mathrm{d}\tau$ equals $\int_0^{\infty}|w^{(2m)}(\tau)|^2\,\mathrm{d}\tau > 0$, and is independent of λ, we have proved (16). The estimate (16), however, contradicts $\|u_\lambda\|_{2m} \leqslant O(\lambda^{-1})$, with which we have checked the L.Š. Condition 11.1 for the operators $\tilde{A}_0(\mathrm{D})$, $\tilde{b}_{j,0}(\mathrm{D})$, $j = 1,\dots,m$. By Theorem 11.3, Condition 11.1 is invariant under admissible coordinate transformations, and therefore Condition 11.1 holds for the original operators $A(x_0, \mathrm{D})$, $b_j(x_0, \mathrm{D})$, $j = 1,\dots,m$ at the point $x_0 \in \partial\Omega$, and we have shown the implication (d)$\Rightarrow$(a). We recall that by Theorem 11.2 conditions I and II imply condition III.

We now come to the long part of the proof, namely the implication (a)$\Rightarrow$(b). In order to make the proof of this implication transparent we first present some lemmata. First Lemma 13.1, the *a priori* estimate for homogeneous differential operators in the half space $\mathbb{R}^r_+$.

Lemma 13.1 *Let $\mathbb{R}^r_+$ be the half space $x_r > 0$ bounded by the hyperplane $\mathbb{R}^{r-1}$ $(x_r = 0)$. Let $A(\mathrm{D})$, $b_j(\mathrm{D})$, $j = 1,\dots,m$ be homogeneous differential operators with constant coefficients of orders, $\operatorname{ord} A = 2m$, $0 \leqslant \operatorname{ord} b_j = m_j \leqslant 2m-1$, $j = 1,\dots,m$. Let $A(\mathrm{D})$ be elliptic and let the L.Š. Condition 11.3 be satisfied. Then for all $\varphi \in W^{2m+l}(\mathbb{R}^r_+)$, $l \geqslant 0$ we have the estimate*

$$\|\varphi\|_{2m+l,\mathbb{R}^r_+} \leqslant c\left\{\|A\varphi\|_{l,\mathbb{R}^r_+} + \sum_{j=1}^{m}\|b_j\varphi\|_{2m+l-m_j-1/2,\mathbb{R}^{r-1}}\right\}. \tag{17}$$

Proof. Let $l = [l] + \lambda$. We use the estimate (14) from Condition 11.3 and write it out explicitly

$$\begin{aligned}
&\sum_{i=0}^{2m+[l]}\left(\int_0^\infty\left|\frac{\mathrm{d}^i h(\tau)}{\mathrm{d}\tau^i}\right|^2\mathrm{d}\tau + \int_0^\infty\int_0^\infty\frac{\left|\dfrac{\mathrm{d}^i h(\tau)}{\mathrm{d}\tau'} - \dfrac{\mathrm{d}^i h(\tau')}{\mathrm{d}\tau'^i}\right|^2}{|\tau-\tau'|^{1+2\lambda}}\mathrm{d}\tau\,\mathrm{d}\tau'\right)\\
&\leqslant C(\xi')\left[\sum_{i=0}^{[l]}\left(\int_0^\infty\left|\frac{\mathrm{d}^i}{\mathrm{d}\tau^i}A\left(\xi',\frac{1}{\mathrm{i}}\frac{\mathrm{d}}{\mathrm{d}\tau}\right)h(\tau)\right|^2\mathrm{d}\tau\right.\right.\\
&\quad\left.+\int_0^\infty\int_0^\infty\frac{\left|\dfrac{\mathrm{d}^i}{\mathrm{d}\tau^i}A\left(\xi',\dfrac{1}{\mathrm{i}}\dfrac{\mathrm{d}}{\mathrm{d}\tau}\right)h(\tau) - \dfrac{\mathrm{d}^i}{\mathrm{d}\tau'^i}A\left(\xi',\dfrac{1}{\mathrm{i}}\dfrac{\mathrm{d}}{\mathrm{d}\tau'}\right)h(\tau')\right|^2}{|\tau-\tau'|^{1+2\lambda}}\mathrm{d}\tau\,\mathrm{d}\tau'\right)\\
&\quad\left.+\sum_{j=1}^{m}\left|b_j\left(\xi',\frac{1}{\mathrm{i}}\frac{\mathrm{d}}{\mathrm{d}\tau}\right)h(0)\right|^2\right],\qquad h\in W_2^{2m+l}(\mathbb{R}^1_+),
\end{aligned}$$

where $C(\xi')$ depends continuously on $0 \neq \xi' \in \mathbb{R}^{r-1}$, but is independent of h. Here because of Theorem 3.6 h can be taken from $C^\infty[0,\infty) \cap W^{2m+l}(\mathbb{R}^1_+)$, and h can also depend on some parameter, for example, $\xi' : h'(\xi', \tau)$. Since the unit sphere in $\mathbb{R}^{r-1}$ is compact, and $C(\xi')$ depends continuously on ξ' we have the inequality

$$\sum_{i=0}^{2m+[l]}\left(\int_0^\infty\left|\frac{d^i h(\xi',\tau)}{d\tau^i}\right|^2 d\tau+\int_0^\infty\int_0^\infty\frac{\left|\dfrac{d^i h(\xi',\tau)}{d\tau^i}-\dfrac{d^i h(\xi',\tau')}{d\tau'^i}\right|^2}{|\tau-\tau'|^{1+2\lambda}}\,d\tau\,d\tau'\right)$$

$$\leqslant C\left[\sum_{i=0}^{[l]}\left(\int_0^\infty\left|\frac{d^i}{d\tau^i}A\left(\frac{\xi'}{|\xi'|},\frac{1}{i}\frac{d}{d\tau}\right)h(\xi',\tau)\right|^2 d\tau\right.\right.$$

$$\left.+\int_0^\infty\int_0^\infty\frac{\left|\dfrac{d^i}{d\tau^i}A\left(\dfrac{\xi'}{|\xi'|},\dfrac{1}{i}\dfrac{d}{d\tau}\right)h(\xi',\tau)-\dfrac{d^i}{d\tau^i}A\left(\dfrac{\xi'}{|\xi'|},\dfrac{1}{i}\dfrac{d}{d\tau'}\right)h(\xi',\tau')\right|^2}{|\tau-\tau'|^{1+2\lambda}}\,d\tau\,d\tau'\right)$$

$$\left.+\sum_{j=1}^{m}\left|b_j\left(\frac{\xi'}{|\xi'|},\frac{1}{i}\frac{d}{d\tau}\right)h(\xi',0)\right|^2\right], \tag{18}$$

for all $0\neq\xi'\in\mathbb{R}^{r-1}$ and $h(\xi',\tau)\in C^\infty[0,\infty)\cap W^{2m+l}(\mathbb{R}^1_+)$, where now $C=\max_{|\xi'|=1}C(\xi')$ is independent of ξ'. Together with $v(\xi',\tau)$ the function $v(\xi',\tau/|\xi'|)$ belongs to $C^\infty[0,\infty)\cap W^{2m+l}(\mathbb{R}^1_+)$ (ξ' fixed), and $h(\xi',\tau)=v(\xi',\tau/|\xi'|)$ substituted in (18) gives

$$\sum_{i=0}^{2m+(l)}\left(\int_0^\infty\left|\frac{d^i v\left(\xi',\dfrac{\tau}{|\xi'|}\right)}{d\tau^i}\right|^2 d\tau+\int_0^\infty\int_0^\infty\frac{\left|\dfrac{d^i v\left(\xi',\dfrac{\tau}{|\xi'|}\right)}{d\tau^i}-\dfrac{d^i v\left(\xi',\dfrac{\tau'}{|\xi'|}\right)}{d\tau'^i}\right|^2}{|\tau-\tau'|^{1+2\lambda}}\,d\tau\,d\tau'\right)$$

$$\leqslant C\left[\sum_{i=0}^{[l]}\left(\int_0^\infty\left|\frac{d^i}{dt^i}A\left(\frac{\xi'}{|\xi'|},\frac{1}{i}\frac{d}{dt}\right)v\left(\xi',\frac{\tau}{|\xi'|}\right)\right|^2 d\tau\right.\right.$$

$$\left.+\int_0^\infty\int_0^\infty\frac{\left|\dfrac{d^i}{d\tau^i}A\left(\dfrac{\xi'}{|\xi'|},\dfrac{1}{i}\dfrac{d}{d\tau}\right)v\left(\xi',\dfrac{\tau}{\xi'}\right)-\dfrac{d^i}{d\tau'^i}A\left(\dfrac{\xi'}{|\xi'|},\dfrac{1}{i}\dfrac{d}{d\tau'}\right)v\left(\xi',\dfrac{\tau}{|\xi'|}\right)\right|^2}{|\tau-\tau'|^{1+2\lambda}}\,d\tau\,d\tau'\right)$$

$$\left.+\sum_{j=1}^{m}\left|b_j\left(\frac{\xi'}{|\xi'|},\frac{1}{i}\frac{d}{d\tau}\right)v(\xi',0)\right|^2\right],$$

The substitution $t=\tau/|\xi'|$ gives, because of the homogeneity of the expressions $A, b_j, j=1,\ldots,m$,

$$\sum_{i=0}^{2m+[l]}\left(|\xi'|^{2(2m-i)}\int_0^\infty\left|\frac{d^i v(\xi',t)}{dt^i}\right|^2 dt+|\xi'|^{2(2m-i-\lambda)}\int_0^\infty\int_0^\infty\frac{\left|\dfrac{d^i v(\xi',t)}{dt^i}-\dfrac{d^i v(\xi',t')}{dt'^i}\right|^2}{|t-t'|^{l+2\lambda}}\,dt\,dt'\right)$$

$$\leqslant C\left[\sum_{i=0}^{[l]}\left(|\xi'|^{-2i}\int_0^\infty\left|\frac{d^i}{dt^i}A\left(\xi',\frac{1}{i}\frac{d}{dt}\right)v(\xi',t)\right|^2 dt\right.\right.$$

$$+|\xi'|^{2i-2\lambda}\int_0^\infty\int_0^\infty \frac{\left|\frac{d^i}{dt^i}A\left(\xi',\frac{1}{i}\frac{d}{dt}\right)v(\xi',t)-\frac{d^i}{dt'^i}A\left(\xi',\frac{1}{i}\frac{d}{dt'}\right)v(\xi',t')\right|^2}{|t-t'|^{1+2\lambda}}\,dt\,dt'\Bigg)$$

$$+\sum_{j=1}^{m}|\xi'|^{2(2m-m_j)-1}\left|b_j\left(\xi',\frac{1}{i}\frac{d}{dt}\right)v(\xi',0)\right|^2\Bigg]. \tag{19}$$

We multiply (19) by $|\xi'|^{2l}$, extend t from $\mathbb{R}^1_+$ to $\mathbb{R}^1$, take the Fourier transformation with respect to t, apply the inverse Fourier transformation for all variables (ξ', t), and restrict t from $\mathbb{R}^1$ to $\mathbb{R}^1_+$. This gives the required estimate (17). ■

For step III we require a further lemma.

Lemma 13.2 *Let B be a continuous operator from $W_2^l(\mathbb{R}^r)\to W_2^{l+N}(\mathbb{R}^r)$ (or better from $H^l\to H^{l+N}$, see §5) for which an estimate of type (40) (see later) holds:*

$$\|B\varphi\|_{l+N}\leqslant K_1\|\varphi\|_l+K_2\|\varphi\|_{l-1},\quad l \text{ arbitrary} \in\mathbb{R}^1.$$

Then for each $\varepsilon>0$ B may be decomposed

$$B=B'_\varepsilon+B''_\varepsilon \quad\text{with}\quad B'_\varepsilon: W_2^l\to W_2^{l+N},\ B''_\varepsilon: W_2^{l-1}\to W_2^{l+N}$$

and for the norms we have

$$\|B'_\varepsilon\varphi\|_{l+N}\leqslant (K_1+\varepsilon)\|\varphi\|_l,\quad \|B''_\varepsilon\varphi\|_{l+N}\leqslant K_3(\varepsilon)\|\varphi\|_{l-1}.$$

If B is independent of l, then B'_ε and B''_ε may also be chosen independent of l. In particular we have

$$B''^{(l-1)}_\varepsilon|_l=B''_\varepsilon(l): W_2^l(\mathbb{R}^r)\to W_2^{l+1+N}(\mathbb{R}^r).$$

Proof. We apply the Fourier transformation and transpose the operator B to the H^l-space (see Theorem 5.2)

$$\begin{array}{ccc} W_2^l & \xrightarrow{B} & W_2^{l+N} \\ \wr\| & & \wr\| \\ H^l & \xrightarrow{\mathscr{F}^{-1}\circ B\circ\mathscr{F}} & H^{l+N}. \end{array}$$

Let $\psi_\rho(t)$ be a function from $C^\infty(\mathbb{R}^1_+)$ with

$$\psi_\rho(t)=\begin{cases}1 & \text{for } 0\leqslant t\leqslant\rho\\ 0 & \text{for } t\geqslant\rho+1\\ \text{between 0 and 1} & \text{for } \rho\leqslant t\leqslant\rho+1\end{cases}$$

where ρ must still be chosen dependent on ε. We put

$$B''_\varepsilon=B\circ\mathscr{F}^{-1}\psi_\rho(|\xi|)\mathscr{F},\quad B'_\varepsilon=B-B''_\varepsilon.$$

The norm property of B''_ε follows immediately from the fact that the support of $\psi_\rho(|\xi|)$ is compact:

$$\psi_\rho(|\xi|)(1+|\xi|^2)^{1/2} \leqslant C \quad \text{for } \xi \in \mathbb{R}^r.$$

If we write $v = \mathscr{F}^{-1}(1-\psi_\delta(|\xi|))\mathscr{F}\varphi$ we have

$$\begin{aligned}\|B'_\varepsilon\varphi\|_{l+N} = \|Bv\|_{l+N} \leqslant K_1\|v\|_l + K_2\|v\|_{l-1} \leqslant K_1\|\varphi\|_l \\ + K_2(1+\rho^2)^{-1/2}\|\varphi\|_l,\end{aligned} \tag{$*$}$$

where we have used the inequality

$$[1-\psi_\delta(|\xi|)](1+|\xi|^2)^{(l-1)/2} \leqslant (1+\rho^2)^{-1/2}(1+|\xi|^2)^{1/2}.$$

If we make the ρ in ($*$) so large that $K_2(1+\rho^2)^{-1/2} \leqslant \varepsilon$, then we have proved everything. ■

We still need a lemma for step IV, this time for the half space $\mathbb{R}^r_+$.

Lemma 13.3 *Let B be a continuous operator from*

$$W^l_2(\mathbb{R}^r_+) \to W^{l+N}(\mathbb{R}^r_+), \quad \text{where} \quad 1 \leqslant l \leqslant L < \infty,\ l+N \geqslant 0,$$

(N can be negative), which satisfies the estimate

$$\|B\varphi\|_{l+N} \leqslant K_1\|\varphi\|_l + K_2\|\varphi\|_{l-1}, \quad \varphi \in W^l_2(\mathbb{R}^r_+).$$

Then for each $\varepsilon > 0$ B may be decomposed as $B = B'_\varepsilon + B''_\varepsilon$ with

$$B'_\varepsilon : W^l_2 \to W^{l+N}_2, \quad B''_\varepsilon : W^{l-1}_2 \to W^{l+N}_2$$

and for the norms we have

$$\|B'_\varepsilon\varphi\|_{l+N} \leqslant (K_1+\varepsilon)\|\varphi\|_l, \quad \|B''_\varepsilon\varphi\|_{l+N} \leqslant K_3(\varepsilon)\|\varphi\|_{l-1}.$$

If B is independent of l, then B'_ε and B''_ε can also be chosen to be independent of l.

The proof comes from the diagram

$$\begin{array}{ccc} W^l_2(\mathbb{R}^r_+) & \xrightarrow{\ B\ } & W^{l+N}_2(\mathbb{R}^r_+) \\ M \uparrow \downarrow F & & M \uparrow \downarrow F \\ W^l_2(\mathbb{R}^r) & \xrightarrow{\ B\ } & W^{l+N}_2(\mathbb{R}^r). \end{array}$$

where F is the extension operator from Theorem 5.5, for $L < \infty$ it is chosen fixed and independent of l; M is the restriction operator. We have $\tilde{B} = F \circ B \circ M = \tilde{B}'_\varepsilon + \tilde{B}''_\varepsilon$, and obtain the desired decomposition by means of

$$B'_\varepsilon = M \circ \tilde{B}'_\varepsilon \circ F, \quad B''_\varepsilon = M \circ \tilde{B}''_\varepsilon \circ F,$$

where if $1 \leqslant l \leqslant L$ the operators B'_ε and B''_ε are independent of l. ∎

For the proof of the implication (a)⇒(b) we must show that the operator given by (2) is smoothable, in the sense of Definition 12.4. For this we must construct the regularisers R_r, R_l, also the smoothing operators $g_X = T_1$, $g_Y = T_2$, and show their continuity. We do this stepwise, beginning with the simplest case: $A(\mathrm{D})$ homogeneous with constant coefficients and $\Omega = \mathbb{R}^r$, hence no boundary conditions. In the second step we consider the half space $\Omega = \mathbb{R}^r_+$ and take $A(\mathrm{D}), b_j(\mathrm{D}), j = 1, \ldots, m$ homogeneous and with constant coefficients; here lies the crux of the argument. Step III involves small perturbations on $\Omega = \mathbb{R}^r$, without boundary conditions; in other words, the coefficients $A(x, \mathrm{D})$, are not constant, though they vary only slightly, see (37). Step IV involves small perturbations on the half space $\Omega = \mathbb{R}^r_+$; this time boundary conditions are involved. In the course of concluding the proof in Step V we cover $\bar{\Omega}$ with small neighbourhoods U, map U to $\mathbb{R}^r$ if $U \subset \Omega$ and $U \cap \Omega$ to $\mathbb{R}^r_+$ if $U \cap \partial\Omega \neq \emptyset$, use the regularisers from Step III (respectively IV) and with the help of partitions of unity construct the smoothing regularisers for $\bar{\Omega}$, $A(x, \mathrm{D}), b_j(x, \mathrm{D}), j = 1, \ldots, m$.

Step I. Let $A(\mathrm{D})$ be a homogeneous, elliptic differential operator of order $2m$ with constant coefficients. We consider the problem

$$A(\mathrm{D})u = f \quad \text{in} \quad \mathbb{R}^r,$$

(without boundary values). We construct the regulariser $R = R_l = R_r$ by

$$Rf = \mathscr{F}^{-1}(|\xi|^{2m}(1 + |\xi|^{2m})^{-1} A^{-1}(\xi)\mathscr{F}(f)) \tag{20}$$

and consider the scheme (2) with

$$\Omega = \mathbb{R}^r, \quad H^l = W^l_2(\mathbb{R}^r), \quad L = A(\mathrm{D}).$$

The continuity of

$$R: W^l_2(\mathbb{R}^r) \to W^{2m+1}_2(\mathbb{R}^r), \quad l \geqslant 0$$

is clear from Definition (20), see Theorem 5.2, and from $|A^{-1}(\xi)| \leqslant c|\xi|^{-2m}$ (ellipticity). We note for later (see Step III) that $R: H^l \to H^{l+2m}$ (see §5) is also continuously defined for negative l. Furthermore we have the decompositions

$$A \circ Rf = If + Tf, \quad R \circ Au = Iu + Tu,$$

where because of (20)

$$Tf := -\mathscr{F}^{-1}((1 + |\xi|^{2m})^{-1}\mathscr{F}f) \tag{21}$$

holds, and the same for Tu. The operator T is a smoothing operator, see Definition 12.4, it acts continuously

$$T: W_2^{2m+l}(\mathbb{R}^r) \to W_2^{2m+l+1}(\mathbb{R}^r),$$

respectively

$$T: W_2^l(\mathbb{R}^r) \to W_2^{l+1}(\mathbb{R}^r),$$

since because of (21), see also Theorem 5.2, and $2m > 1$, we have

$$T: W_2^l(\mathbb{R}^r) \to W_2^{l+2m}(\mathbb{R}^r) \subsetneq W_2^{l+1}(\mathbb{R}^r).$$

Therefore we have proved that the elliptic operator $A(\mathrm{D})$ is smoothable on $\mathbb{R}^r$. All operators considered, now and later, do not depend, as one easily checks, on the subscript l: these operators act on the argument and on the function value, the W_2^l are function spaces, hence the independence of l is clear. ■

Step II. We consider the half space $\mathbb{R}^r_+$, and suppose that the assumptions of Lemma 13.1 hold for the operators $A(\mathrm{D})$, $b_j(\mathrm{D})$, $j = 1, \ldots, m$. We consider the elliptic boundary value problem

$$\begin{aligned} A(\mathrm{D})u &= f \quad \text{on} \quad \mathbb{R}^r_+(x_r > 0) \\ b_j(\mathrm{D})u &= g_j \quad \text{on} \quad \mathbb{R}^{r-1}(x_r = 0), j = 1, \ldots, m \end{aligned} \tag{22}$$

and wish to show that the operator

$$L:(A(\mathrm{D}), b_1(\mathrm{D}), \ldots, b_m(\mathrm{D}))$$

is smoothable, in scheme (2) we put $\Omega = \mathbb{R}^r_+$, $\partial\Omega = \mathbb{R}^{r-1}$. For this we construct the left and right regularisers and the smoothing operators.

Let $F: W_2^l(\mathbb{R}^r_+) \to W_2^l(\mathbb{R}^r)$ be the continuous extension operator from Theorem 5.5, $M: W_2^l(\mathbb{R}^r) \to W_2^l(\mathbb{R}^r_+)$ the continuous restriction operator and $\mathscr{T}_0: W_2^l(\mathbb{R}^r_+) \to W_2^{l-1/2}(\mathbb{R}^{r-1})$ the continuous trace operator, see Theorem 8.1, we put $\mathscr{T}_0 := T_0 \circ F$. Let $\omega_j(\xi', t)$ be the fundamental solution of

$$A\left(\xi', \frac{1}{\mathrm{i}}\frac{\mathrm{d}}{\mathrm{d}t}\right)\omega_j(\xi', t) = 0, \quad b_\mu\left(\xi', \frac{1}{\mathrm{i}}\frac{\mathrm{d}}{\mathrm{d}t}\right)\omega_j(\xi', 0) = \delta_{j,\mu}, \ j, \mu = 1, \ldots, m, \quad \text{in } \mathscr{M}^+.$$

By Condition 11.2 $\omega_j(\xi', t)$ is uniquely determined and by (19) we obtain the following estimate (the sum on the left-hand side of (19) estimates the individual summands):

$$|\xi'|^{2(2m-i)}\left\|\frac{d^i\omega_j(\xi',\cdot)}{dt^i}\right\|_0^2 \leqslant c|\xi'|^{2(2m-m_j)-1}\cdot 1,$$

respectively

$$|\xi'|^{2(2m-i-\lambda)}I_\lambda\left(\frac{d^i\omega_j(\xi',\cdot)}{dt^i}\right) \leqslant c|\xi'|^{2(2m-m_j)-1}\cdot 1, \quad i=0,\ldots,[l].$$

The factor $|\xi'|^{-2i}$, respectively $|\xi'|^{-2(i+\lambda)}$, carried over to the right-hand side and estimated by $(1+|\xi'|^2)^l$ gives after summation $i=0,\ldots,[l]$

$$\|\omega_j(\xi',t)\|_{l,t>0}^2 \leqslant c(1+|\xi'|^2)^l|\xi'|^{-2m_j-1}, \quad l\geqslant 0.$$

We extend the function $\omega_j(\xi',t)$ from $t\in\mathbb{R}^1_+$ to $\mathbb{R}^1$ by Hestenes' Theorem 5.5, to obtain $F\omega_j(\xi',t)$, where of course we retain dependence on parameters ξ', and estimates. By Fourier transformation in t we have, see Theorem 5.2,

$$\int_{-\infty}^{\infty}|\bar\omega_j(\xi',\tau)|^2(1+\tau^2)^l d\tau \leqslant c(1+|\xi'|^2)^l|\xi'|^{-2m_j-1}, \quad l\geqslant 0, \tag{23}$$

where we have written $\bar\omega_j(\xi',\tau)=\mathscr{F}_t\circ F\omega_j(\xi',t)$. We define the regularisers – here R is the regulariser from Step I (20) –

$$\begin{aligned} R_0 f &:= M\circ R\circ Ff = M\circ\mathscr{F}^{-1}(|\xi|^{2m}(1+|\xi|^{2m})^{-1}A^{-1}(\xi)\mathscr{F}\circ Ff) \\ R_j g_j &= M\mathscr{F}^{-1}(|\xi'|^{m_j+1}(1+|\xi'|^{m_j+1})^{-1}\bar\omega_j(\xi',\tau)\mathscr{F}'g_j), \quad j=1,\ldots,m. \end{aligned} \tag{24}$$

$\mathscr{F}'$ is the Fourier transformation with respect to the variables $x'=(x_1,\ldots,x_{r-1})$, $\mathscr{F}$ with respect to $x=(x_1,\ldots,x_r)$. As right (and left) regulariser for the problem (22) we take the operator

$$R(f,g_1,\ldots,g_m) := R_0 f + \sum_{j=1}^m R_j(g_j-\mathscr{F}_0\circ b_j\circ R_0 f) \tag{25}$$

(a) We show continuity. We have

$$R_0\colon W_2^l(\mathbb{R}^r_+)\to W_2^{2m+l}(\mathbb{R}^r_+) \text{ continuous}, \; l\geqslant 0, \tag{26}$$

see Step I.

Now the continuity of

$$R_j\colon W_2^{2m+l-m_j-1/2}(\mathbb{R}^{r-1})\to W_1^{2m+l}(\mathbb{R}^r_+), \quad l\geqslant 0. \tag{27}$$

The properties of the Fourier transformation (see the theorems from §5) and the continuity of M have the effect that

$$\|R_j g_j\|_{2m+l,\mathbb{R}^r_+}^2 \leqslant c_1\int_{\mathbb{R}^r}\left|\frac{|\xi'|^{m_j+1}}{(1+|\xi'|^{m_j+1})}\bar\omega_j(\xi',\tau)\mathscr{F}'g_j\right|^2(1+|\xi|^2)^{2m+l}d\xi.$$

Because

$$(1+|\xi|^2)^{2m+l}=(1+|\xi'|+\tau^2)^{2m+l}\leqslant c_2[(1+|\xi'|^2)^{2m+l}+(1+\tau^2)^{2m+l}],$$

Fubini's theorem and (23), with $l=0$ once and $2m+l$ a second time, we can further estimate

$$\begin{aligned}
&\|R_jg_j\|^2_{2m-l,\mathbb{R}^r_+}\\
&\leqslant c_3\int_{\mathbb{R}^{r-1}}\mathrm{d}\xi'\int_{-\infty}^{\infty}\mathrm{d}\tau\frac{|\xi'|^{2(m_j+1)}((1+|\xi'|^2)^{2m+l}+(1+\tau^2)^{2m+l})}{(1+|\xi'|^{m_j+1})^2}\\
&\quad\cdot|\bar{\omega}_j(\xi',\tau)|^2\cdot|\mathscr{F}'g_j|^2\\
&=c_3\int_{\mathbb{R}^{r-1}}\mathrm{d}\xi'\frac{|\xi'|^{2(m_j+1)}|\mathscr{F}'g_j|^2}{(1+|\xi'|^{m_j+1})^2}\\
&\quad\times\left(\int_{-\infty}^{\infty}|\bar{\omega}_j(\xi',\tau)|^2\mathrm{d}\tau\cdot(1+|\xi'|^2)^{2m+l}+\int_{-\infty}^{\infty}|\bar{\omega}_j(\xi',\tau)|^2(1+\tau^2)^{2m+l}\mathrm{d}\tau\right)\\
&\leqslant c_4\int_{\mathbb{R}^{r-1}}\frac{|\xi'|^{2(m_j+1)}}{(1+|\xi'|^{m_j+1})^2}|\mathscr{F}'g_j|^2(1+|\xi'|^2)^{2m+l-m_j-1/2}\mathrm{d}\xi'\\
&\leqslant c_5\int_{\mathbb{R}^{r-1}}|\mathscr{F}'g_j|^2(1+|\xi'|^2)^{2m+l-m_j-1/2}\mathrm{d}\xi'\leqslant c_6\|g_j\|^2_{2m+l-m_j-1/2,\mathbb{R}^{r-1}}
\end{aligned}$$

which is (27). The continuity of

$$R:W^l_2(\mathbb{R}^r_+)\mathop{\times}_{j=1}^{m}W_2^{2m+l-m_j-1/2}(\mathbb{R}^{r-1})\to W_2^{2m+l}(\mathbb{R}^r_+),\quad l\geqslant 0,$$

may be read off from definition (25) because of (26) and (27). We note further that $w_j(x)=R_j(g_j)$ is the solution of the boundary value problem

$$\begin{aligned}
&A(\mathrm{D})w_j=0,\quad x_r>0,\\
&b_k(\mathrm{D})w_j=0,\quad k\neq j=1,\ldots,m;\\
&b_j(\mathrm{D})w_j(x',0)=\mathscr{F}'^{-1}(|\xi'|^{m_j+1}(1+|\xi'|^{m_j+1})^{-1})\mathscr{F}'(g_j).
\end{aligned}\tag{28}$$

(b) R is a right regulariser for $L=(A(\mathrm{D}),b_1(\mathrm{D}),\ldots,b_m(\mathrm{D}))$. Because of (28) we have $A\circ R_j=0$ for $j=1,\ldots,m$ from which it follows that

$$A\circ R\{f,g_1,\ldots,g_m\}=A\circ R_0f=f+V_0f,$$

with

$$V_0f=-M\circ\mathscr{F}^{-1}(1+|\xi|^{2m})^{-1}\mathscr{F}Ff,$$

where

$$V_0:W^l_2(\mathbb{R}^r_+)\to W_2^{l+2m}(\mathbb{R}^r_+)\subsetneq W_2^{l+1}(\mathbb{R}^r_+)\tag{29}$$

is continuous and so a smoothing operation. Furthermore by (28)

$$b_jR_jg_j=g_j+V_jg_j\quad\text{with}\quad V_jg_j:=-\mathscr{F}'^{-1}((1+|\xi'|^{m_j+1})^{-1}\mathscr{F}'g_j);$$

and

$$V_j: W_2^{2m+l-m_j-1/2}(\mathbb{R}^{r-1}) \to W_2^{2m+l+1/2}(\mathbb{R}^{r-1}) \hookrightarrow W_2^{2m+l-m_j+1/2}(\mathbb{R}^{r-1}) \quad (30)$$

is continuous, hence smoothing. Altogether we have

$$L \circ R\{f, g_1, \ldots, g_m\} = \{f, g_1, \ldots, g_m\} + T_2\{f, g_1, \ldots, g_m\}$$

with

$$T_2\{f, g_1, \ldots, g_m\} = \{V_0 f; V_j(g_j - \mathscr{F}_0 \circ b_j \circ R_0 f), j = 1, \ldots, m\}, \quad (31)$$

where, from (29) and (30), the operator

$$T_2: W_2^l(\mathbb{R}^r_+) \times_{j=1}^{m} W_2^{2m+l-m_j-1/2}(\mathbb{R}^{r-1}) \to W_2^{l+1}(\mathbb{R}^r_+) \times_{j=1}^{m} W_2^{2m+l-m_j+1/2}(\mathbb{R}^{r-1}), \quad (32)$$

in short,

$$T_2: H^l(\mathbb{R}^r_+, \mathbb{R}^{r-1}) \to H^{l+1}(\mathbb{R}^r_+, \mathbb{R}^{r-1}) \quad \text{is smoothing.}$$

(c) R is a left regulariser for L. We decompose

$$R \circ Lu = u + T_1 u$$

and must show that T_1 is a smoothing operator, respectively that for $l \geqslant 0$

$$T_1: W_2^{2m+l}(\mathbb{R}^r_+) \to W^{2m+l+1}(\mathbb{R}^r_+) \quad (33)$$

is continuous. For this we consider the expression $L \circ R \circ Lu$. Because of (31) we have $L \circ R \circ Lu = (I + T_2)Lu = L(I + T_1)u$, from which it follows that

$$T_2 \circ Lu = L \circ T_1 u,$$

and Lemma 13.1 gives the estimate

$$\| T_1 u \|_{2m+l+1, \mathbb{R}^r_+} \leqslant c_1 \| L T_1 u \|_{H^{l+1}(\mathbb{R}^r_+, \mathbb{R}^{r-1})}. \quad (34)$$

We substitute the expression $T_2 Lu$ for $LT_1 u$, apply (32) and estimate further

$$\| LT_1 u \|_{H^{l+1}} = \| T_2 Lu \|_{H^{l+1}(\mathbb{R}^r_+, \mathbb{R}^{r-1})} \leqslant c_2 \| Lu \|_{H^l(\mathbb{R}^r, \mathbb{R}^{r-1})} \leqslant c_3 \| u \|_{W_2^{2m+l}(\mathbb{R}^r_+)}; \quad (35)$$

(34) and (35) give (33). ■

Step III. Let $A(x, \mathrm{D})$ be a differential operator on $\mathbb{R}^r$ with variable coefficients, which satisfy the differentiability and boundedness assumptions of the main Theorem 13.1: $a_j(x) \in C^{k+1}(\mathbb{R}^r)$.

We decompose

$$A(x, \mathrm{D}) = A^{\mathrm{H}}(0, \mathrm{D}) + A_1(x, \mathrm{D}) + A_2(x, \mathrm{D}), \quad (36)$$

where $A_1(x, \mathrm{D}) = A^{\mathrm{H}}(x, \mathrm{D}) - A^{\mathrm{H}}(0, \mathrm{D})$ and $A_2(x, \mathrm{D}) = A(x, \mathrm{D}) - A^{\mathrm{H}}(x, \mathrm{D})$.

The order of $A_2(x, \mathrm{D})$ is less than $2m$. We assume that $A(x, \mathrm{D})$ is elliptic at the point $0 \in \mathbb{R}^r$, and that for the coefficients of $A_1(x, \mathrm{D}) = \sum_{|s|=2m}(a_s(x) - a_s(0))\mathrm{D}^s$ we have

$$\sup_x \sum_{|s|=2m} |a_s(x) - a_s(0)| < \frac{1}{3\|R\|_l}, \tag{37}$$

where $R: W_2^l(\mathbb{R}^r) \to W_2^{2m+l}(\mathbb{R}^r)$ is the regulariser from Step I, $\|R\|_l$ is its norm and $0 \leqslant l \leqslant k + \kappa - 1$.

We wish to show that under these assumptions the operator $A(x, \mathrm{D})$ is smoothable. Let R be the (left and right) regulariser of $A^{\mathrm{H}}(0, \mathrm{D})$ from Step I. Because of (36) we have

$$\begin{aligned} A(x, \mathrm{D}) \circ Rf &= f + A_1(x, \mathrm{D})Rf + A_2(x, \mathrm{D})Rf + Tf, \\ R \circ A(x, \mathrm{D})u &= u + R \circ A_1(x, \mathrm{D})u + R \circ A_2(x, \mathrm{D})u + Tu. \end{aligned} \tag{38}$$

Since $\operatorname{ord} A_2 \leqslant 2m - 1$ we have (here we use the hypothesis that $a_s \in C^{k+1}(\mathbb{R}^r)$, whereas elsewhere $a_s \in C^k(\mathbb{R}^r)$ suffices)

$$\begin{aligned} A_2 \circ R &: W_2^l \to W_2^{l+2m} \to W_2^{l+1}, \\ R \circ A_2 &: W_2^{2m+l} \to W_2^{l+1} \to W_2^{2m+l+1}, \end{aligned}$$

that is, $A_2 \circ R$ and $R \circ A_2$ are smoothing operators and (38) reduces to

$$\begin{aligned} A(x, \mathrm{D}) \circ Rf &= f + A_1(x, \mathrm{D})Rf + T_1 f, \\ R \circ A(x, \mathrm{D})u &= u + R \circ A_1(x, \mathrm{D})u + T_2 u. \end{aligned} \tag{39}$$

We estimate using the Leibniz product rule and apply (37)

$$\begin{aligned} \|A_1(x, \mathrm{D})Rf\|_l &= \left\| \sum_{|s|=2m} (a_s(x) - a_s(0))\mathrm{D}^s Rf \right\|_l \\ &\leqslant \sup_x \sum_{|s|=2m} |a_s(x) - a_s(0)| \cdot \|Rf\|_{l+2m} + C_1 \|Rf\|_{l+2m-1} \\ &\leqslant \frac{1}{3\|R\|_l} \|Rf\|_{l+2m} + C_1 \|Rf\|_{l+2m-1} \\ &\leqslant \frac{1}{3\|R\|_l} \|R\|_l \|f\|_l + C_1 \|R\|_{l-1} \|f\|_{l-1} \\ &\leqslant \frac{1}{3} \|f\|_l + c\|f\|_{l-1}. \end{aligned} \tag{40}$$

Here $l - 1$ is also allowed to be negative, since by means of the Fourier transformation the operator $R: H^l \to H^{l+2m}$ is also defined for negative l.

We now apply Lemma 13.2 to the operator $A_1 \circ R = B$ and put $N = 0, \varepsilon = \frac{1}{3}$. Then because of (40) we obtain $A_1 \circ R = B_1 + B_2$ with $B_2 \in L(W_2^{l-2}, W_2^l)$ and also $\in L(W_2^l, W_2^{l+1})$. This follows from Lemma 13.2 since $A_1 \circ R$ is independent of l. We also obtain $\| B_1 \|_l \leqslant \frac{2}{3}$. (39) takes the form

$$A(x, \mathrm{D}) \circ Rf = f + B_1 f + B_2 f + T_1 f, \tag{41}$$

and B_2 is a smoothing operator. Because $\| B_1 \|_l \leqslant \frac{2}{3}$ the operator $I + B_1$ is invertible (see Lemma 12.1) and one sees that

$$R_r = R \circ (I + B_1)^{-1}$$

is a right regulariser for A, for from (41) it follows that

$$\begin{aligned} A(x, \mathrm{D}) \circ R \circ (I + B_1)^{-1} f &= (I + B_1)(I + B_1)^{-1} f + B_2 (I + B_1)^{-1} f \\ &\quad + T_1 (I + B_1)^{-1} f \\ &= f + T_3 f T_3 : W_2^l \to W_2^{l+1}. \end{aligned}$$

The same method shows, starting from the second equation (39), that there is also a left regulariser for $A(x, \mathrm{D})$. ■

Step IV. Let $A(x, \mathrm{D}), b_j(x, \mathrm{D}), j = 1, \ldots, m$ be differential operators on the half space $\mathbb{R}_+^r$ with variable coefficients, which satisfy the differentiability and boundedness assumptions of the main Theorem 13.1,

$$a_s(x) \in C^{k+1}(\bar{\mathbb{R}}_+^r), \quad b_{j,s}(x) \in C^{2m - m_j + k + 1}(\bar{\mathbb{R}}_+^r), \quad j = 1, \ldots, m.$$

In the half space $\mathbb{R}_+^r$ we consider the elliptic boundary value problem

$$\begin{aligned} A(x, \mathrm{D}) u &= f, \quad x \in \mathbb{R}_+^r \\ b_j(x, \mathrm{D}) u &= g_j, \quad x \in \mathbb{R}^{r-1}, \quad j = 1, \ldots, m. \end{aligned}$$

We decompose

$$\begin{aligned} A(x, \mathrm{D}) &= A^{\mathrm{H}}(0, \mathrm{D}) + A_1(x, \mathrm{D}) + A_2(x, \mathrm{D}), \\ b_j(x, \mathrm{D}) &= b_j^{\mathrm{H}}(0, \mathrm{D}) + b_{1,j}(x, \mathrm{D}) + b_{2,j}(x, \mathrm{D}), \quad j = 1, \ldots, m, \end{aligned} \tag{42}$$

where we again write

$$\begin{aligned} A_1(x, \mathrm{D}) &= A^{\mathrm{H}}(x, \mathrm{D}) - A^{\mathrm{H}}(0, \mathrm{D}), \\ A_2(x, \mathrm{D}) &= A(x, \mathrm{D}) - A^{\mathrm{H}}(x, \mathrm{D}), \quad \operatorname{ord} A_2 \leqslant 2m - 1, \\ b_{1,j}(x, \mathrm{D}) &= b_j^{\mathrm{H}}(x, \mathrm{D}) - b_j^{\mathrm{H}}(0, \mathrm{D}), \\ b_{2,j}(x, \mathrm{D}) &= b_j(x, \mathrm{D}) - b_j^{\mathrm{H}}(x, \mathrm{D}), \quad \operatorname{ord} b_{2,j} \leqslant m_j - 1, j = 1, \ldots, m. \end{aligned}$$

We assume that the boundary value problem is elliptic at the point $0 \in \bar{\mathbb{R}}_+^r$, that is, $A^{\mathrm{H}}(0, \mathrm{D})$ is elliptic and that the L.Š. condition is satisfied for

$A^{\mathrm{H}}(0,\mathrm{D})$, $b_j^{\mathrm{H}}(0,\mathrm{D})$, $j=1,\ldots,m$. Let $\mathcal{R}: H^l(\mathbb{R}^r_+,\mathbb{R}^{r-1}) \to W_2^{2m+l}(\mathbb{R}^r_+)$ be the regulariser for $A^{\mathrm{H}}(0,\mathrm{D})$, $b_j^{\mathrm{H}}(0,\mathrm{D})$, $j=1,\ldots,m$ from Step II, suppose now that $0 \leqslant l \leqslant k+\kappa-1$. We assume in addition that

$$\begin{aligned} &\sup_{x\in\mathbb{R}^r_+} \sum_{|s|=2m} |a_s(x)-a_s(0)| < \frac{1}{3\|\mathcal{R}\|_l}, \\ &\sup_{x\in\mathbb{R}^r_+} \sum_{|s|=m_j} |b_{j,s}(x)-b_{j,s}(0)| < \frac{1}{3\|\mathbb{R}\|_l \|\mathscr{F}_0\|_l}, \quad j=1,\ldots,m. \end{aligned} \tag{43}$$

We abbreviate the decomposition (42) as

$$L(x,\mathrm{D}) = L^{\mathrm{H}}(0,\mathrm{D}) + L_1(x,\mathrm{D}) + L_2(x,\mathrm{D}). \tag{42'}$$

Let $\mathcal{R}$ be the regulariser from Step II with the smoothing operators T_1 and T_2. From (42′) with $g=(g_1,\ldots,g_m)$ we obtain

$$\begin{aligned} L(x,\mathrm{D})\circ\mathcal{R}\{f,g\} &= \{f,g\} + L_1(x,\mathrm{D})\mathcal{R}\{f,g) + L_2(x,\mathrm{D})\mathcal{R}\{f,g) \\ &\quad + T_1\{f,g\}\mathcal{R}\circ L(x,\mathrm{D})u = u + \mathcal{R}\circ L_1(x,\mathrm{D})u \\ &\quad + \mathcal{R}\circ L_2(x,\mathrm{D})u + T_2 u. \end{aligned} \tag{44}$$

The operators $L_2\circ\mathcal{R}$ and $\mathcal{R}\circ L_2$ are again smoothing operators, since $\operatorname{ord} L_2 < (2m, m_1,\ldots,m_m)$. We construct a left regulariser $\mathcal{R}_1$ for $L(x,\mathrm{D})$. For this we estimate ($\mathscr{T}_0$ is the trace operator)

$$\begin{aligned} &\|\mathcal{R}\circ L_1 u\|_{2m+l} \\ &\leqslant \|\mathcal{R}\|_l \|L_1 u\|_l \\ &= \|\mathcal{R}\|_l \left\{ \|A_1(x,\mathrm{D})u\|_l + \sum_{j=1}^m \|\mathscr{T}_0\circ b_{1,j}(x,\mathrm{D})u\|_{l+2m-m_j-1/2,\mathbb{R}^{r-1}} \right\} \\ &\leqslant \|\mathcal{R}\|_l \left\{ \|A_1(x,\mathrm{D})u\|_l + \sum_{j=1}^m \|\mathscr{T}_0\|_l \|b_{1,j}(x,\mathrm{D})u\|_{l+2m-m_j,\mathbb{R}^r_+} \right\}. \end{aligned}$$

We use the Leibniz product rule and the assumptions (43), similarly as in (40), and finally obtain

$$\|\mathcal{R}\circ L_1 u\|_{2m+l} \leqslant \tfrac{1}{3}\|u\|_{2m+l} + c\|u\|_{2m+l-1} \quad \text{on} \quad \mathbb{R}^r_+.$$

We apply Lemma 13.3 to this estimate with $B=\mathcal{R}\circ L_1$, $N=0$, $\varepsilon=\frac{1}{3}$, $L=k+\kappa+2m$, $l\to 2m+l$ and obtain the decomposition

$$\mathcal{R}\circ L_1 = B_1 + B_2, \quad \|B_1\|_{l+2m} < \tfrac{2}{3},$$

B_2 independent of l and smoothing:

$$W_2^{2m+l-1}(\mathbb{R}^r_+) \to W_2^{2m+l}(\mathbb{R}^r_+) \text{ respectively } W_2^{2m+l}(\mathbb{R}^r_+) \to W_2^{2m+l+1}(\mathbb{R}^r_+).$$

We take as left regulariser for $L(x, D)$

$$\mathscr{R}_l = (I + B_1)^{-1}\mathscr{R},$$

use the second equation of (44) and check that

$$\begin{aligned} R_1 \circ L(x, D)u &= (I + B_1)^{-1}u + (I + B_1)^{-1}\mathscr{R}\circ L_1 u + (I + B_1)^{-1}(\mathscr{R}\circ L_2 u + T_2 u) \\ &= (I + B_1)^{-1}u + (I + B_1)^{-1}(B_1 u + B_2 u) + (I + B_1)^{-1}(\mathscr{R}\circ L_2 u + T_2 u) \\ &= u + (I + B_1)^{-1}B_2 u + (I + B_1)^{-1}(\mathscr{R}\circ L_2 u + T_2 u) = u + T_3 u, \end{aligned}$$

here $T_3 : W_2^{l+2m}(\mathbb{R}^r_+) \to W_2^{l+2m+1}(\mathbb{R}^r_+)$ is a smoothing operator.

The construction of a right regulariser R_r for $L(x, D)$ is somewhat harder. We consider the operators $A_1(x, D)$, $b_{1,j}(x, D)$, $j = 1, \ldots, m$ on $\mathscr{R}^r_+$, apply the Leibniz product rule, and estimate with the help of the assumptions (43)

$$\|A_1(x, D)v\|_l \leqslant \frac{1}{3\|R\|_l}\|v\|_{l+2m} + c\|v\|_{l+2m-1},$$

$$\|b_{1,j}(x, D)v\|_{l+2m-m_j} \leqslant \frac{1}{3\|R\|_l\|\mathscr{T}\|_l}\|v\|_{l+2m} + c\|v\|_{l+2m-1}, \quad j = 1, \ldots, m.$$

We again apply Lemma 13.3, once with $B = A_1$, $N = -2m$, $\varepsilon = \frac{1}{3}\|\mathscr{R}\|_l$ $L = k + \kappa + 2m$, $l \to l + 2m$, and a second time with $B = b_{i,j}$, $N = -m_j$, $\varepsilon = \frac{1}{3}\|\mathscr{R}\|_l\|\mathscr{T}_0\|_l$, $L = k + \kappa + 2m$, $l \to l + 2m$, and obtain the decompositions (independent of l)

$$A_1 = B' + B'', \qquad b_{1,j} = B'_j + B''_j, \quad j = 1, \ldots, m,$$

where

$$\|B'\| < \frac{2}{3\|\mathscr{R}_l\|}, \quad \|B'_j\| < \frac{2}{3\|\mathscr{R}_l\|\,\|\mathscr{T}_0\|_l},$$

and B'', B''_j an smoothing

$$B'' : W_2^{l+2m}(\mathbb{R}^r_+) \to W_2^{l+1}(\mathbb{R}^r_+), \qquad B''_j : W_2^{l+2m}(\mathbb{R}^r_+) \to W_2^{l+2m-m_j+1}(\mathbb{R}^r_+)$$

set

$$\tilde{B}_1 := (B', \mathscr{T}_0\circ B'_1, \ldots, \mathscr{T}_0\circ B'_m)\circ\mathscr{R}, \quad \tilde{B} := (B'', \mathscr{T}_0\circ B''_1, \ldots, \mathscr{T}_0\circ B''_2)\circ\mathscr{R}$$

where $\mathscr{T}_0$ is the trace operator, and have

$$L_1\circ\mathscr{R} = \tilde{B}_1 + \tilde{B}_2 \quad \text{with } \tilde{B}_1 : H^l \to H^l \quad \text{and } \tilde{B}_2 : H^l \to H^{l+1} \text{ smoothing},$$

where for the norm of $\tilde{B}_0$ we have

$$\|\tilde{B}_1\| \leqslant \max(\|B'\|, \|\mathscr{T}_0\|_l\|B'_1\|, \ldots, \|\mathscr{T}_0\|_l\|B'_m\|)\cdot\|\mathscr{R}\|_l \leqslant \tfrac{2}{3}.$$

The operator $(I + \tilde{B}_1)^{-1}$ again exists and we take a right regulariser for $L(x, D)$

$$\mathscr{R}_r = \mathscr{R}\circ(I + \tilde{B}_1)^{-1}.$$

We use the first equation of (44) and check that

$$
\begin{aligned}
&L(x,\mathrm{D})\circ\mathscr{R}_r\{f,g\}\\
&\quad= L(x,\mathrm{D})\mathscr{R}\circ(I+\tilde{B}_1)^{-1}\{f,g\} = (I+\tilde{B}_1)^{-1}\{f,g\} + L_1(x,\mathrm{D})\mathscr{R}(I+\tilde{B}_1)^{-1}\{f,g\}\\
&\qquad+(L_2(x,\mathrm{D})\mathscr{R}+T_1)(I+\tilde{B}_1)^{-1}\{f,g\} = (I+\tilde{B}_1)^{-1}\{f,g\}\\
&\qquad+(\tilde{B}_1+\tilde{B}_2)(I+\tilde{B}_1)^{-1}\{f,g\} + (L_2(x,\mathrm{D})\mathscr{R}+T_1)(I+\tilde{B}_1)^{-1}\{f,g\}\\
&\quad= \{f,g\} + \tilde{B}_2(I+\tilde{B}_1)^{-1}\{f,g\} + (L_2(x,\mathrm{D})\mathscr{R}+T_1)(I+\tilde{B}_1)^{-1}\{f,g\}\\
&\quad= \{f,g\} + T_4\{f,g\},
\end{aligned}
$$

here $T_4: H^l \to H^{l+1}$ is a smoothing operator. We have therefore regularised the boundary value problem in the half space $\mathbb{R}^r_+$ under the conditions (43). ■

Step V (end of the proof). Suppose that the assumptions made before the main Theorem 13.1 are satisfied for Ω and $L(x,\mathrm{D}) = (A(x,\mathrm{D}), b_1(x,\mathrm{D}), \ldots, b_m(x,\mathrm{D}))$. Let the boundary value problem (1) be elliptic, we show that (2) $L(x,\mathrm{D})$ is smoothable.

First an auxiliary concept. Let U_i be open and $U_i \subset \Omega$. We say that $L(x,\mathrm{D})$ is locally smoothable on $(U_i, \partial U_i \cap \partial\Omega)$, that R^i_1, R^i_r are local regularisers, and that T^i_1, T^i_2 are local smoothing operators if

$$
\begin{aligned}
&\varphi L\circ R^i_r\{f^i,g^i\} = \varphi\{f^i,g^i\} + T^i_1\{f^i,g^i\},\\
&\varphi R^i_1\circ Lu^i = \varphi u^i + T^i_2 u^i,
\end{aligned}
\tag{45}
$$

for $\{f^i,g^i\}\in H^l(U_i,\partial U_i\cap\partial\Omega)$, $u^i\in W_2^{l+1m}(U_i)$, with

$$
\begin{aligned}
&R^i_r, R^i_1: H^l(U_i,\partial U_i\cap\partial\Omega)\to W_2^{l+2m}(U_i), \text{ continuous},\\
&T^i_1: H^l(U_i,\partial U_i\cap\partial\Omega)\to H^{l+1}(U_i,\partial U_i\cap\partial\Omega), \text{ continuous}\\
\text{and}\quad &T^i_2: W_2^{l+2m}(U_i)\to W_2^{l+1+2m}(U_i), \text{ continuous}.
\end{aligned}
$$

where φ is a function from $C^\infty(U_i)$, which is the restriction to $U_i = V_i\cap\Omega$ of a function from $\mathscr{D}(V_i)$, for V_i see Lemma 13.4 below. If $\partial U_i\cap\partial\Omega = \varnothing$, we write $g^i = 0$ in the first equation of (45), and for the H^l-spaces we then take $W_2^l(U_i)$. ■

Lemma 13.4 *Let $\{V_i\}$ be an open cover of $\bar{\Omega}$, we write $U_i = V_i\cap\Omega$. If $L(x,\mathrm{D})$ is locally smoothable on each $(U_i,\partial U_i\cap\partial\Omega)$, then $L(x,\mathrm{D})$ is 'globally' smoothable on Ω.*

Proof. Let $\{\varphi_i\}$ be a partition of unity subordinate to the cover $\{V_i\}$:

$$\varphi_i\in\mathscr{D}(V_i),\quad \operatorname{supp}\varphi_i\subset V_i,\quad \sum_i \varphi_i = 1.$$

We take further functions $\psi_i\in\mathscr{D}(V_i)$ with the properties $\operatorname{supp}\varphi_i\subset\operatorname{supp}\psi_i\subset V_i$,

$\psi_i = 1$ on $\operatorname{supp}\varphi_i$, from which it follows that

$$\varphi_i = \psi_i \cdot \varphi_i. \tag{46}$$

We define the 'global regularisers' by

$$R_r = \sum_i \varphi_i R_r^i \psi_i, \quad R_l = \sum_i \varphi_i R_l^i \psi_i \tag{47}$$

and check the smoothing properties. We write $\{f,g\} = \{f,g_1,\dots,g_m\} = F$, and have (applying the Leibniz rule)

$$LR_r F = \sum_i L\varphi_i R_r^i \psi_i F = \sum_i \varphi_i L R_r^i \psi_i F + \sum_i \sum_{|\alpha|>0} \frac{1}{\alpha!}(\mathrm{D}^\alpha \varphi_i) L^{(\alpha)} R_r^i \psi_i F$$

$$= \sum_i \varphi_i L R_r^i \psi_i F + T_1 F; \tag{48}$$

T_1 is a smoothing operator, for the order of $L^{(\alpha)}$ is less than $(2m, m_1, \dots, m_m)$. We now use (46) and the local smoothability (45), obtain

$$LR_r F = \sum_i \varphi_i \cdot \psi_i F + \sum_i T_1^i \psi_i F + T_1 F = \sum_i \varphi_i F + TF = F + TF,$$

where T is a smoothing operator. The proof for R_l: $R_l L u = u + \tilde{T}u$ proceeds in the same way.

Back to Step V. Let $x \in \bar{\Omega}$, we consider first the case when x lies in the interior of Ω, therefore $x_1 \in \Omega$. We take a ball $B(x_l, 2\varepsilon_l) \subset \Omega$ so small, that on it the estimate (37) for $A(x,\mathrm{D})$ in Step III holds

$$\sup_x \sum_{|s|=2m} |a_s(x) - a_s(x_l)| < \frac{1}{3\|R(x_l)\|_l}, \tag{37'}$$

where $R(x_l)$ is the regulariser of $A^{\mathrm{H}}(x_l, \mathrm{D})$ which exists by Step I for fixed x_l. We extend the coefficients of $A(x,\mathrm{D})$ from $U_l = B(x_l, \varepsilon_l)$ to $\mathbb{R}^r$, maintaining condition (37), see Theorem 5.7, obtaining in this way an extended differential operator, which by Step III is smoothable:

$$R_l^l A_l u = u + T_1^l u, \quad A_l R_r^l f = f + T_2^l f.$$

From this, after multiplication by $\varphi \in \mathscr{D}(U_l)$ and restriction to U_l, we obtain the local smoothability of $A(x,\mathrm{D})$ on U_l, which is (45). Suppose now that x is a frontier point, thus $x_\alpha \in \partial\Omega$. Since the frontier $\partial\Omega$ belongs by assumption to the class $C^{2m+k,\kappa}$, $k + \kappa \geqslant 1$, by Theorem 2.11 there exists a neighbourhood V'_α of x_α and an admissible transformation $\Phi_\alpha \in C^{2m+k,\kappa}$, which maps

$$V'_\alpha \leftrightarrow W^r, \quad V'_\alpha \cap \Omega \leftrightarrow W^r_+, \quad V'_\alpha \cap \partial\Omega \leftrightarrow W^{r-1}, \quad x_\alpha \leftrightarrow 0.$$

Here, see also Theorem 2.11, the admissible transformation can be so chosen that $(x_1,\ldots,x_{r-1})=\Phi_\alpha^{r-1}$ are the coordinates of $V'_\alpha\cap\partial\Omega$. By Corollary 10.1 and Theorem 11.3 the transformed differential operators $\tilde{L}=(\tilde{A},\tilde{B}_1,\ldots,\tilde{B}_m)$ again form an elliptic boundary value problem for $0\in\bar{W}^r_+$, W^{r-1}, that is, $\tilde{A}(0,\mathrm{D})$ is elliptic and $(\tilde{A}(0,\mathrm{D}),\ \tilde{B}_1(0,\mathrm{D}),\ldots,\tilde{B}_m(0,\mathrm{D}))$ satisfy the L.Š. Condition 11.1. We now choose the half ball $\bar{B}_+(0,2\varepsilon_\alpha)\subset W^r_+$ to be so small that the estimates (43) of Step IV hold on $\bar{B}_+(0,2\varepsilon_\alpha)$

$$
\begin{aligned}
&\sup_y \sum_{|s|=2m} |\tilde{a}_s(y)-\tilde{a}_s(0)| < \frac{1}{3\,\|\tilde{R}(0)\|_l},\\
&\sup_y \sum_{|s|=m_j} |\tilde{b}_{j,s}(y)-\tilde{b}_{j,s}(0)| \leqslant \frac{1}{3\,\|\tilde{R}(0)\|_l\,\|\mathscr{T}_0\|_l}, \quad j=1,\ldots,m,
\end{aligned}
\tag{49}
$$

where $\tilde{R}(0)$ is the regulariser of $\tilde{L}(0,\mathrm{D})$ existing by Step II. Again we extend $\tilde{L}(y,\mathrm{D})$ from $\bar{B}_+(0,\varepsilon_\alpha)$ to $\bar{\mathbb{R}}^r_+$, maintaining conditions (49); this is possible by Theorem 5.7. The operator $\tilde{L}_\alpha(y,\mathrm{D})$ extended in this way is smoothable by Step IV

$$R_1^\alpha\circ\tilde{L}_\alpha u=u+T_1^\alpha u,\quad \tilde{L}_\alpha\circ R_r^\alpha\{f,g\}=\{f,g\}+T_2^\alpha\{f,g\},$$

from which by multiplication by $\varphi\in\mathscr{D}(B(0,\varepsilon_\alpha))$ and restriction to $B_+(0,\varepsilon_\alpha)$ we obtain the local smoothability (45) of $\tilde{L}(y,\mathrm{D})$ on $(B_+(0,\varepsilon_\alpha),\ W^{r-1}\partial B_+(0,\varepsilon_\alpha))$. The properties 'smoothable' and 'locally smoothable' are invariant under admissible coordinate transformations

$$\Phi_\alpha:\Omega\cap V'_\alpha\leftrightarrow W^r_+,\ \Phi_\alpha^{r-1}:V'_\alpha\cap\partial\Omega\leftrightarrow W^{r-1}$$

$(x_1,\ldots,x_{r-1})=\Phi_\alpha^{r-1}$ are the coordinates of $\partial\Omega$. Now the operators in question arise by left and right multiplication by the appropriate pullback operators ${}^*\Phi_\alpha$, ${}^*\Phi_\alpha^{-1}$ and ${}^*\Phi_\alpha^{r-1}$, ${}^*(\Phi_\alpha^{r-1})^{-1}$ using the diagrams (see also Theorem 4.1).

$$
\begin{array}{ccc}
W(V_\alpha\cap\Omega) & \xrightarrow{\ B\ } & W(V_\alpha\cap\partial\Omega)\\
{}^*\Phi_\alpha\ \Big\uparrow\Big\downarrow\ {}^*\Phi_\alpha^{-1} & & {}^*\Phi_\alpha^{r-1}\ \Big\uparrow\Big\downarrow\ {}^*(\Phi_\alpha^{r-1})^{-1}\quad \text{etc.}\\
W(W^r_+) & \xrightarrow{\ B\ } & W(W^{r-1})
\end{array}
$$

that is, $B={}^*(\Phi_\alpha^{r-1})^{-1}\circ B\circ{}^*\Phi_\alpha$ etc.

If we write $V_\alpha:=\Phi_\alpha^{-1}(B(0,\varepsilon_\alpha))$, the image of $B(0,\varepsilon_\alpha)$ under the inverse map Φ_α^{-1}, we see that $L(x,\mathrm{D})$ is locally smoothable on $U_\alpha:=V_\alpha\cap\Omega$, $\partial\Omega\cap U_\alpha$, where x_α was chosen to be arbitrary. Since $\bar{\Omega}$ is compact, we can cover $\bar{\Omega}$ by finitely many U_l and V_α, and apply Lemma 13.4, by which (2) $L(x,\mathrm{D})$ is smoothable on $(\Omega,\partial\Omega)$. ∎

13.2 The index and spectrum of elliptic boundary value problems

We next deduce some theorems on the index of elliptic operators, and in the proofs will, in part, anticipate §15 and §17.

Theorem 13.2 (Lawruk [1]) *The index of a boundary value problem* (1) *depends only on the principal parts of the operators* $A, b_1, \ldots, b_m$, *in other words, if we augment the operator* A *by an operator* A_1 *of order* $\leqslant 2m-1$ *and* b_j by $b_{1,j}$ *of order* $\leqslant m_j - 1$ (*if the order of* $b_j = 0$ *we set* $b_{1,j} = 0$), *then the index is unchanged.*

Proof. We write $L_1 = (A_1, b_{1,1}, \ldots, b_{1,m})$ and in the notation of (2) have

$$L_1 : W_2^{2m+l}(\Omega) \to H^{l+1}(\Omega, \partial\Omega) \subsetneq H^l(\Omega, \partial\Omega)$$

where because of the boundedness of Ω the embedding $\subsetneq$ is compact. We can therefore regard L_1 as a compact perturbation, apply Theorem 12.8 and obtain the result. ■

We consider the spectral value problem (eigenvalue problem):

$$\begin{aligned} &A(x, \mathrm{D})u - \lambda u = f \quad \text{on} \quad \Omega, \lambda \in \mathbb{C}, \\ &b_1(x, \mathrm{D})u = g_1, \ldots, b_m(x, \mathrm{D})u = g_m \quad \text{on} \quad \partial\Omega. \end{aligned} \tag{50}$$

From Theorem 13.2 we immediately obtain:

Theorem 13.3 *For all* $\lambda \in \mathbb{C}$ *the problem* (50) *has the same index*:

$$\operatorname{ind} L = \operatorname{ind}(L - \lambda) \quad \text{for all } \lambda \in \mathbb{C}.$$

Proof. We write $L_1 = (-\lambda \mathrm{D}^0, 0, \ldots, 0)$ in Theorem 13.2. ■

From Theorem 13.3 we obtain statements about the spectrum of (50): if ind $L \neq 0$, then the spectrum of (50) consists of the entire complex plane, Sp $L = \mathbb{C}$. If $\lambda_0 \in \mathbb{C}$ were regular, then $L - \lambda_0$ would be an isomorphism (ind $(L - \lambda_0) = 0$), hence $0 = \operatorname{ind}(L - \lambda_0) = \operatorname{ind} L \neq 0$, a contradiction. For $\operatorname{ind} L > 0$ all complex values $\lambda \in \mathbb{C}$ are eigenvalues of (50) with finite dimensional eigenspaces $E(\lambda)$, because $\operatorname{ind}(L - \lambda) = \operatorname{ind} L > 0$ implies that $\infty > \alpha(L - \lambda) = \dim E(\lambda) > \beta(L - \lambda) \geqslant 0$, while for $\operatorname{ind} L < 0$ besides the eigenvalues, residual spectral values λ can also enter the spectrum $\operatorname{Sp} L = \mathbb{C}$ of (50).

A spectral value λ is called *residual*, if the map $L - \lambda : W_2^{2m}(\Omega) \to \operatorname{im}(L - \lambda)$ is injective, and $\overline{\operatorname{im}(L - \lambda)} \neq H^l$; a spectral value λ is called continuous, if $\overline{\operatorname{im}(L - \lambda)} = H^l$ and $(L - \lambda)^{-1}$ is unbounded. Since by Theorem 13.3 for each $\lambda \in \mathbb{C}$ $(L - \lambda)$ is a Fredholm operator, then $\operatorname{im}(L - \lambda)$ is closed, no continuous spectral values can appear in the spectrum of (50) (open mapping theorem!) and we have

$$\mathrm{Sp}L = \mathrm{Sp}_p L \cup \mathrm{Sp}_r L,$$

where by $\mathrm{Sp}_p L$ we mean the eigenvalues of (50) and by $\mathrm{Sp}_r L$ the residual values.

We can expect more interesting statements about the spectrum of (50) only for $\mathrm{ind}\,L = 0$. In this case there are two possibilities, either $\mathrm{Sp}\,L = \mathbb{C}$, that is, every $\lambda \in \mathbb{C}$ is an eigenvalue of (50) (because $\mathrm{ind}\,L = 0$ we have $\mathrm{Sp}_r L = \varnothing$, hence $\mathrm{Sp}\,L = \mathrm{Sp}_p L = \mathbb{C}$), or there exists at least one $\lambda_0 \in \mathbb{C}$, which is not an eigenvalue of (50). For the first possibility see Example 13.1, while for the second we have the following theorem:

Theorem 13.4 *Let the boundary value operator L (2) be elliptic, let* $\mathrm{ind}\,L = 0$. *We consider problem* (50) *and assume that at least one value* $\lambda_0 \in \mathbb{C}$ *exists, which is not an eigenvalue of* (50), *that is, from*

$$Au - \lambda_0 u = 0, \quad b_1 u = 0, \ldots, b_m u = 0,$$

it should follow that $u = 0$.

Then (50) *satisfies the conclusions of the Riesz–Schauder spectral theorem* §12.1, *with the spectrum there transformed by* $1/\lambda$. *This means that the spectrum of* (50) *consists of at most countably many eigenvalues, possibly with accumulation points,* $|\lambda_n| \to \infty$ *only, each eigenspace* $E(\lambda_n)$ *is finite dimensional, all other values of* λ *are regular, that is, the problem* (50) *is uniquely solvable for these* λ.

Proof. Without loss of generality we may take $\lambda_0 = 0$, otherwise we consider the operator $A - \lambda_0 J$ instead of A. The assumption for $\lambda_0 = 0$ implies that

$$\ker L = (0),$$

which combined with $\mathrm{ind}\,L = 0$ shows that

$$\mathrm{im}\,L = H^l.$$

The operator $L: W_2^{2m+l} \to H^l$ (see (2)) is therefore continuous injective and surjective, hence by the open mapping theorem an isomorphism. We write (50) in the form $L - \lambda J$, where $J = (J, 0. \ldots, 0)$ (identification and boundary values 0) acts as

$$J: W_2^{2m+l} \to H^{2m+l} \hookrightarrow H^l .$$

Because $2m > 0$ and Ω is bounded, the embedding $\hookrightarrow$ is compact, therefore J is also compact, and for (50) we can write

$$L - \lambda J = L(I - \lambda L^{-1} \circ J),$$

with $L^{-1} \circ J: W_2^{2m+l} \to H^l \to W_2^{2m+l}$ compact and $I: W_2^{2m+l} \to W_2^{2m+l}$ the identity. $I - \lambda L^{-1} \circ J$ is a Riesz–Schauder operator to which we can apply Riesz–Schauder theory; the assertions of our theorem now follow from the spectral theorem §12.1.

Here in contrast to the spectral Theorem 12.1 one observes that in $(I - \lambda L^{-1} \circ J)$ λ stands in front of the compact operator. That is, we transform the spectrum by $1/\lambda$, whereby the critical value 0 is mapped to ∞, which for our problem (50) implies that the spectrum of (50) consists only of eigenvalues λ_n. ■

We now show by means of an example that the assumption that there is at least one regular $\lambda_0 \in \mathbb{C}$ in (50) is essential. The example is due to Seeley (quoted in Morrey [1], 252).

Example 13.1 Let Ω be the annulus $\pi < \rho < 2\pi$ in polar coordinates (ρ, φ) for $\mathbb{R}^2$. Let

$$A = e^{-2i\varphi}\left[\frac{\partial^2}{\partial \varphi^2} + \frac{\partial^2}{\partial \rho^2} + I\right]$$

and as boundary problem we take the Dirichlet problem (by Theorem 13.5 the index therefore vanishes, ind = 0)

$$u|_{\partial\Omega} = 0;$$

here the frontier $\partial\Omega$ consists of the two circles $\rho = \pi$ and $\rho = 2\pi$. Let

$$J_0(x) = \frac{1}{2\pi}\int_0^{2\pi} e^{ix\sin\theta}\,d\theta$$

be the 0th Bessel function, we write $u(\rho, \varphi; \lambda) = J_0(\lambda^{1/2} e^{i\varphi}) \sin \rho$, and have

$$Au - \lambda u = 0, \quad u|_{\partial\Omega} = 0, \quad u \neq 0,$$

that is, each $\lambda \in \mathbb{C}$ is an eigenvalue.

Let A be an operator of order $2m$: as boundary value problem we consider the Dirichlet problem

$$u|_{\partial\Omega} = g_1, \left.\frac{\partial u}{\partial n}\right|_{\partial\Omega} = g_2, \ldots, \left.\frac{\partial^{m-1} n}{\partial n^{m-1}}\right|_{\partial\Omega} = g_m, \tag{51}$$

where the $\partial^j/\partial n^j$ are derivatives in the inner normal direction $\mathbf{n}$ to $\partial\Omega$. If A is properly elliptic, then $(A, I, \partial/\partial n, \ldots, \partial^{m-1}/\partial n^{m-1})$ abbreviated as (A, D), always satisfies the L.Š. condition, because the Dirichlet problem (51) corresponds (see §11(2)) to the classical initial value problem for ordinary (linear) differential equations

$$v(0) = 0, \frac{dv(0)}{dt} = 0, \dots, \frac{d^{m-1}}{dt^{m-1}} = 0,$$

which is known (Vandermonde's determinant) to be uniquely solvable in $\mathscr{M}^+$. We have:

Theorem 13.5 *Let A be an elliptic operator (for $r = 2$ we must assume properly elliptic). Then the Dirichlet problem* (51) *always has index zero, that is,*

$$\operatorname{ind}(A, \mathrm{D}) = 0.$$

For the proof we write down some formulae. Let $\bar{A}$ be the operator, which one obtains from A, if one replaces all the coefficients $a_s(x)$ by their complex conjugates $\overline{a_s(x)}$, similarly for $\bar{b}_1$, etc. We have

$$\operatorname{ind}(\bar{A}, \bar{b}_1, \dots, \bar{b}_m) = \operatorname{ind}(A, b_1, \dots, b_m) \tag{52}$$

for from

$$Au = f, \quad b_1 u = g_1, \dots, b_m u = g_m,$$

it follows that

$$\bar{A}\bar{u} = \bar{f}, \quad \bar{b}_1 \bar{u} = \bar{g}_1, \dots, \bar{b}_m \bar{u} = \bar{g}_m, \tag{53}$$

and conversely. Suppose further that A^* is the (formal) adjoint operator defined by §10(3). The adjoint boundary problem to (A, D), see Definition 14.6, is by Theorem 14.9 (52) equal to (A^*, D), and by Theorem 15.7 we have

$$\operatorname{ind}(A^*, \mathrm{D}) = -\operatorname{ind}(A, \mathrm{D}). \tag{54}$$

From formula §10(3) we see that $A^*u - \bar{A}u$ is a differential operator of order $\leqslant 2m - 1$; applying the perturbation Theorem 13.2 we obtain

$$\operatorname{ind}(A^*, \mathrm{D}) = \operatorname{ind}(\bar{A}, \mathrm{D}). \tag{55}$$

For the proof of Theorem 13.5 we use all the index formula (52), (54), (55) derived above:

$$-\operatorname{ind}(A, \mathrm{D}) \underset{(54)}{=} \operatorname{ind}(A^*, \mathrm{D}) \underset{(55)}{=} \operatorname{ind}(\bar{A}, \mathrm{D}) = \operatorname{ind}(\bar{A}, \bar{\mathrm{D}}) \underset{(52)}{=} \operatorname{ind}(A, \mathrm{D}),$$

that is, $2 \cdot \operatorname{ind}(A, \mathrm{D}) = 0$, or $\operatorname{ind}(A, \mathrm{D}) = 0$. ■

Exercises

13.1 Study the eigenvalues and the eigenfunctions for the Laplace operator $-\Delta$ on the disc $B=\{(x,y)\in\mathbb{R}^2 : x^2+y^2<1\}$. Attention–Bessel functions!

13.2 Let

$$y''+\frac{1}{x}y'+\left(1-\frac{m^2}{x^2}\right)y=0$$

be the Bessel differential equation. Show that the regular, for $x=0$, solutions are given by

$$y=c\cdot T_m(x)$$

where J_m is the mth Bessel function

$$J_m(x):=\left(\frac{x}{2}\right)^m\sum_{p=0}^{\infty}\frac{(-1)^p}{\Gamma(p+1)\cdot\Gamma(m+p+1)}\left(\frac{x}{2}\right)^{2p}.$$

For negative values of m we write

$$J_m(x):=(-1)^m J_{-m}(x).$$

13.3 Prove the formula

$$\mathrm{e}^{iz\sin\theta}=\sum_{m=-\infty}^{+\infty}J_m(z)\mathrm{e}^{\mathrm{I}m\theta},$$

from which for real x it follows that

$$|J_m(x)|\leqslant\frac{1}{\sqrt{2}},\quad m\neq 0,\quad\text{and}\quad |J_0(x)|\leqslant 1.$$

13.4 Prove

$$J_m(z)=\frac{1}{\pi}\int_0^{\pi}\cos(m\theta-z\sin\theta)\mathrm{d}\theta.$$

§14 Green's formulae

We introduce the important concepts of normal and Dirichlet boundary value systems, and show that normal boundary conditions can always be satisfied, see Theorem 14.1. We prove the two Green formulae, Theorem 14.2 and Theorem 14.8, and introduce $(A,b_1,\ldots,b_m)^*$, the boundary value problem adjoint to $(A,b_1,\ldots,b_m)$. By means of the first Green formula one sees that $(A,b_1,\ldots,b_m)^*$ is realisable by boundary values, therefore

$$(A,b_1,\ldots,b_m)^*=(A^*,B'_1,\ldots,B'_m),$$

and that if $(A,b_1,\ldots,b_m)$ satisfies the L.Š. condition, then the adjoint problem also satisfies this condition. If $b_1, \ldots, b_m$ is a Dirichlet system of order m, then for the adjoint problem we have, Theorem 14.9, that

$$(A, b_1, \ldots, b_m)^* = (A^*, b_1, \ldots, b_m).$$

With the help of Theorem 14.9 we can also explicitly determine many other adjoint problems – see the examples at the end of §16.

At the close of this section we examine the connection between the antidual operator L' and the adjoint boundary value problem.

As already stated, in this section we wish to prove the two Green formulae for an elliptic differential operator $A(x, \mathrm{D})$ of order $2m$. For both formulae, as also for the other theorems we need different differentiability assumptions, which we shall describe precisely in each case. The C^∞-theory certainly spares us writing down the assumptions, but since the proofs remain the same, we gain no essential simplification.

Let Ω be a bounded region in $\mathbb{R}^r$, which belongs to class $C^{2m,1}$ (the 1 because we work with admissible coordinates), let $A(x, \mathrm{D})$ be an elliptic operator defined on $\bar{\Omega}$ with coefficients $a_s \in C(\bar{\Omega})$, so that

$$A(x, \mathrm{D}): W_2^{2m}(\Omega) \to L_2(\Omega)$$

is continuous, let $b_j, j = 1, \ldots, n$ be boundary value operators on $\partial\Omega$ (see §11); we interpret b_j as

$$b_j(x, \mathrm{D}) := T_0\left(\sum_{|s| \leqslant m_j} b_{j,s}(x)\mathrm{D}^s\right) = \sum_{|s| \leqslant m_j} b_{js}(x) T_0(\mathrm{D}^s), \quad j = 1, \ldots, n,$$

where T_0 is the trace operator $\Omega \to \partial\Omega$ from §8. For the coefficients $b_{js}(x)$ of $b_j(x, \mathrm{D})$ we assume that they are extendable from $\partial\Omega$ to $\bar{\Omega}$ and belong to the class $C^{2m-m_j}(\bar{\Omega})$, $|s| \leqslant m_j$, $j = 1, \ldots, n$. Under these assumptions

$$b_j(x, \mathrm{D}): W_2^{2m}(\Omega) \to W_2^{2m-m_j-1/2}(\partial\Omega), \quad j = 1, \ldots, n,$$

acts continuously. Here as in the earlier sections (§11 and §13)

$$0 \leqslant \operatorname{ord} b_j = m_j \leqslant 2m - 1, \quad j = 1, \ldots, n.$$

14.1 Normal boundary value operators and Dirichlet systems

Definition 14.1 *Let Γ be a subset of $\partial\Omega$. We say that the boundary value operators $b_j(x, \mathrm{D}), j = 1, \ldots, n$ are* normal *on Γ, if the following conditions are satisfied:*

1. $m_j \neq m_i$ *for* $j \neq i = 1, \ldots, n$.
2. $b_j^{\mathrm{H}}(x, \xi) \neq 0$ *for all* $x \in \Gamma$ *and all* $0 \neq \xi \in \mathbb{R}^r$, *where* ξ *is normal to* $\partial\Omega$ *at* x.

Condition 2 implies that the highest purely normal derivative $\partial^{m_j}/\partial n^{m_j}$ appears in $b_j(x, \mathrm{D})$ with a coefficient different from zero.

From the L.Š. condition 11.1 it follows that

$$b_j^{\mathrm{H}}(x, (\xi', \xi_n)) \neq 0 \quad \text{for } \xi' \neq 0 \text{ tangential to } \partial\Omega \text{ at } x \text{ and } \xi_n \neq 0 \text{ normal to } \partial\Omega \text{ at } x, \tag{1}$$

and condition (2) forms the supplement to (1) for $\xi' = 0$, because $b_j^{\mathrm{H}}(x, (0, \xi_n)) = b_j^{\mathrm{H}}(x, \xi_n)$, ξ_n normal.

Definition 14.2 *We say that the boundary value operators $b_j(x, \mathrm{D})$, $j = 1, \ldots, n$ form a Dirichlet system on Γ, if in addition to conditions* 1 *and* 2 *condition* 3 *is satisfied:*

3. *The orders m_j, $j = 1, \ldots, n$ run through all numbers $0, 1, \ldots, n-1$. Hence without loss of generality $m_j = j - 1$, $j = 1, \ldots, n$. The number n is called the order of the Dirichlet system.*

It is clear that the properties 'normal' and 'Dirichlet system' are invariant under admissible transformations (Definition 2.9), therefore in particular under the transformations of Theorem 2.12.

We consider the half cube W^r_+, $\Gamma = W^{r-1}$, with the coordinates $(x_1, \ldots, x_r)$; here $(x_1, \ldots, x_{r-1}) = x'$ are the coordinates of W^{r-1}, and x_r is normal to W^{r-1}. We prove a lemma for (W^r_+, W^{r-1}).

Lemma 14.1 *Let $\Omega = W^r_+$, $\Gamma = W^{r-1}$,* let *$\{F_j(x, \mathrm{D})\}_{j=1}^n$ and $\{F'_j(x, \mathrm{D})\}_{j=1}^n$ be two Dirichlet systems of order n on Γ, where the coefficients of $F_j(x, \mathrm{D})$ and $F'_j(x, \mathrm{D})$, $j = 1, \ldots, n$ belong to the class $C^{M-j}(\bar{W}^r_+)$, $M \geqslant n$. Then we have the decompositions*

$$F_j = \sum_{s=1}^{i} \Lambda'_{js} F'_s, \quad j = 1, \ldots, n, \tag{2}$$

$$F'_j = \sum_{s=1}^{j} \Lambda_{js} F_s, \quad j = 1, \ldots, n, \tag{3}$$

where Λ'_{jj}, Λ_{jj} are functions from the class $C^{M-j}(\bar{W}^r_+)$ not vanishing on Γ, and Λ_{js}, Λ'_{js} $j \neq s$ are tangential differential operators (that is, they consist of derivatives with respect to the variables $x_1, \ldots, x_{r-1}$) of order $j - s$ with coefficients again belonging to the class $C^{M-j}(W^r_+)$, $M \geqslant n$.

Proof. On grounds of symmetry it is enough to prove one of the formulae (2) or (3); we show (3). If we can show (3) for $F'_j = \mathrm{D}_{x_r}^{j-1}$, then we have already proved everything. Since F'_j is a Dirichlet system of order n, we can write, using Definition 14.1.2,

$$F'_j = \sum_{i=1}^{j} \Theta'_{ji} D_{x_r}^{i-1}, \quad j = 1, \ldots, n, \tag{4}$$

where the Θ'_{jj} are functions from $C^{M-j}(\bar{W}^r_+)$ not vanishing on Γ, and the Θ'_{ji}, $i \neq j$ are tangential operators of order $j-1$ with coefficients from $C^{M-j}(\bar{W}^r_+)$. If now we have

$$D_{x_r}^{j-1} = \sum_{s=1}^{j} \Phi_{js} F_s, \quad j = 1, \ldots, n \tag{5}$$

with $0 \neq \Phi_{jj} \in C^{M-j}(\bar{W}^r_+)$ and $\Phi_{js} \in C^{M-j}(\bar{W}^r_+)$, then after substitution of (5) in (4) we obtain

$$F'_j = \sum_{i=1}^{j} \Theta'_{ji} D_{x_r}^{i-1} = \sum_{i=1}^{j} \Theta'_{ji} \sum_{s=1}^{i} \Phi_{is} F_s = \sum_{s=1}^{j} \Lambda_{js} F_s,$$

which is (3) with

$$\Lambda_{jj} = \Theta'_{jj} \Phi_{jj} \quad \text{and} \quad \Lambda_{js} = \sum_{i=s}^{j} \Theta'_{ji} \circ \Phi_{is}, \quad \operatorname{ord} \Lambda_{js} \leqslant j - s,$$

where the coefficients of Λ_{js} belong to the class $C^{M-j}(\bar{W}^r_+)$, since

$$\Theta'_{ji} \circ \Phi_{is} \in C^{M-i-(j-i)} = C^{M-j}$$

Therefore we only need to prove (5). (5) is correct for $j = 1$. We carry out the proof by induction; let (5) hold for $j < k$, we must show that (5) then also holds for $j = k$. Since $\{F_j\}_{j=1}^n$ is a Dirichlet system, we can decompose F_k as

$$F_k = \sum_{i=1}^{k} \Theta_{ki} D_{x_r}^{i-1},$$

where the Θ_{ki} have the same properties as the Θ'_{ki}. Therefore

$$\Theta_{kk} D_{x_r}^{k-1} = F_k - \sum_{i=1}^{k-1} \Theta_{ki} D_{x_r}^{i-1} = F_k - \sum_{i=1}^{k-1} \Theta_{ki} \sum_{s=1}^{i} \Phi_{is} F_s$$

$$= F_k - \sum_{s=1}^{k-1} \left(\sum_{i=s}^{k-1} \Theta_{ki} \circ \Phi_{is} \right) F_s,$$

from which by division by $\Theta_{kk} \neq 0$ we obtain (5) for $j = k$. We check the fact that the coefficients of Φ_{js} belong to the class $C^{M-j}(\bar{W}^r_+)$. Θ_{kk} belongs to C^{M-k} and is different from zero in some neighbourhood u of W^{r-1}, therefore $1/\Theta_{kk} \in C^{M-k}(\bar{W}^r_+ \cap U)$ or $\Phi_{kk} := 1/\Theta_{kk} \in C^{M-k}(\bar{W}^r_+)$, since outside W^{r-1} we can extend arbitrarily. We have

$$\Phi_{ks} = -\frac{1}{\Theta_{kk}} \sum_{i=s}^{k-1} \Theta_{ki} \circ \Phi_{is}, \quad 1 \leqslant s \leqslant k-1;$$

because $\Theta_{ki} \circ \Phi_{is} \in C^{M-i-(k-i)} = C^{M-k}$ and $1/\Theta_{kk} \in C^{M-k}$ it follows immediately that $\Phi_{ks} \in C^{M-k}$. ■

We remark that the conclusion of the lemma remains invariant under admissible coordinate transformations.

With the help of the inverse trace theorem 8.8 we now want to prove an important theorem, which states that normal boundary conditions are always satisfiable.

Theorem 14.1 *Let Ω be a bounded region in $\mathbb{R}^r$, which belongs to the class $C^{M,1}$ (the 1 because we need normal coordinates). Let $b_j(x, \mathrm{D})$, $j = 1, \ldots, n$, be boundary value operators on $\partial\Omega$ with coefficients $b_{js}(x) \in C^{M+1-j}(\bar{\Omega})$, $j = 1, \ldots, n$. Suppose that the boundary value operators form a Dirichlet system of order n; here for the sake of simplicity we have put* $\operatorname{ord} b_j = j - 1, j = 1, \ldots, n$. *Let $1 \leqslant n \leqslant M$. Let $(\varphi_1, \ldots, \varphi_n)$ be a given n-tuple of functions on $\partial\Omega$ with $\varphi_j \in W_2^{M-j+1/2}(\partial\Omega)$, $j = 1, \ldots, n$. Then there exists a function $u \in W_2^M(\Omega)$ with*

$$b_j(x, \mathrm{D})u = \varphi_j \quad \text{on } \partial\Omega, j = 1, \ldots, n, \tag{6}$$

and such that the map

$$(\varphi_1, \ldots, \varphi_n) \to u, \quad \textit{from} \ \mathop{\times}_{j=1}^{n} W_2^{M-j+1/2}(\partial\Omega) \to W_2^M(\Omega)$$

is linear and continuous.

Since one can always extend normal boundary value systems to Dirichlet systems (one simply adds the missing normal derivatives $\partial^j/\partial n^j$, see also the proof of Theorem 14.2) it suffices to assume that the system $(b_1, \ldots, b_n)$ is normal.

Addendum *If the supports of φ_j, $j = 1, \ldots, n$ lie in $\Gamma \subset \partial\Omega$, we can so choose u that* $\operatorname{supp} u \subset U_\varepsilon \cap \Omega$, *where U_ε, $\varepsilon > 0$ is an ε-neighbourhood of Γ.*

Proof. We cover $\partial\Omega$ with finitely many $\{U_i\}$ and by Theorem 2.12 introduce normal coordinates on U_i. Let $\Phi_i : U_i \to W^r$ be the transformation from $C^{M-1,1}$, which leaves tangents and normals invariant. Let $\{\alpha_i\}$ be a partition of unity subordinate to the cover U_i. By Lemma 14.1 on U_i we have:

$$b_j(x, \mathrm{D}) = \sum_{s=1}^{j} \Lambda^i_{js} \circ \frac{\partial^{s-1}}{\partial n^{s-1}}, \quad \frac{\partial^{j-1}}{\partial n^{j-1}} = \sum_{s=1}^{j} \Phi^i_{js} \circ b_{js}(x, \mathrm{D}), \quad j = 1, \dots, n, \tag{7}$$

where $\operatorname{ord} \Lambda^i_{js} = \operatorname{ord} \Phi^i_{js} = j - s$, and the coefficients of Λ^i_{js}, Φ^i_{js} belong to C^{M+1-j}, and $\partial/\partial n$ is the derivative in the direction of the inner normal to $\partial\Omega$. On $U_i \cap \partial\Omega$ we consider the equations

$$\frac{\partial^{j-1}}{\partial n^{j-1}} v_i = \sum_{s=1}^{j} \Phi^i_{js}(\alpha_i \varphi_s), \quad j = 1, \dots, n. \tag{8}$$

Since $\Phi^i_{js}(\alpha_i \varphi_s) \in W_2^{M-j+1/2}(U_i \cap \partial\Omega)$, by Theorem 8.8, respectively Theorem 8.6 (with subsequent localisation, see Remark 8.1) we can find a solution $v_i \in W_2^M(\Omega)$ of (8). Indeed we even have

$$v_i = Z^i_n\left(\sum_{s=1}^{j} \Phi^i_{js}(\alpha_i \varphi_s)\right), \quad j = 1, \dots, n \tag{9}$$

where Z^i_n is the continuous extension operator of Theorem 8.8,

$$Z^i_n \colon \mathop{\times}_{j=1}^{n} W_2^{M-j+1/2}(U_i \cap \partial\Omega) \to W_2^M\left(\bigcup_i U_i \cap \Omega\right) = W_2^M(\Omega).$$

If we apply, $b_j(x, \mathrm{D})$ to v_i from (9), then because

$$\sum_{s=1}^{j} \Lambda^i_{js} \circ \Phi^i_{sl} = \delta_{jl}, \quad 1 \leqslant l \leqslant j \leqslant n,$$

we obtain – see (7) – that

$$b_j(x, \mathrm{D}) v_i = \alpha_i \varphi_j, \quad j = 1, \dots, n.$$

This means that $u = \sum_i v_i$ is a solution of (6):

$$b_j(x, \mathrm{D}) u = b_j(x, \mathrm{D}) \sum_i v_i = \sum_i b_j(x, \mathrm{D}) v_i = \sum_i \alpha_i \varphi_j = \varphi_j,$$

and because of (9), the map

$$(\varphi_1, \dots, \varphi_n) \mapsto u = \sum_i Z'_n\left(\sum_{s=1}^{j} \Phi^i_{js}(\alpha_i \varphi_s)\right), \quad j = 1, \dots, n,$$

acts linearly and continuously

$$\mathop{\times}_{j=1}^{n} W_2^{M-j+1/2}(\partial\Omega) \to \mathop{\times}_{j=1}^{n} W_2^{M-j+1/2}(U_i \cap \partial\Omega) \to W_2^M(\Omega). \qquad \blacksquare$$

We obtain the proof of the addendum by covering only Γ with U_is, setting

$U_\varepsilon = \bigcup_i U_i$, and with this change repeating the proof; outside U_ε we set v_i equal to 0.

14.2 The first Green formula

We come to the first Green formula. We need higher differentiability assumptions on the region Ω and on the coefficients of $A(x, \mathrm{D})$, $b_j(x, \mathrm{D})$, $j = 1, \dots, m$. We require that

$$a_\alpha(x) \in C^{|\alpha| + k}(\bar{\Omega}), |\alpha| \leqslant 2m, \quad \text{and} \quad b_{js}(x) \in C^{4m + k - 1 - m_j}(\bar{\Omega}), \quad j = 1, \dots, m. \tag{10}$$

On the region Ω we impose the requirement

$$\Omega \in C^{4m + k - 1, 1}. \tag{11}$$

Theorem 14.2 *Let $A(x, \mathrm{D}) = \sum_{|s| \leqslant 2m} a_s(x) \mathrm{D}^s$ be an elliptic differential operator on $\bar{\Omega}$, let $b_j(x, \mathrm{D}), j = 1, \dots, m$ be a normal boundary value system on $\partial\Omega$, whose coefficients satisfy* (10). *Then on $\partial\Omega$ one can find another normal boundary value system $\{S_j\}_{j=1}^m$ with the properties*

$$0 \leqslant \operatorname{ord} S_j =: \mu_j \leqslant 2m - 1, \quad S_j \in C^{4m + k - 1 - \mu_j}(\bar{\Omega}), \quad j = 1, \dots, m,$$

and $\{b_1, \dots, b_m, S_1, \dots, S_m\}$ is a Dirichlet system of order $2m$ on $\partial\Omega$. The choice of the S_j is not unique, there clearly exist many S_j which have these properties.

We can further construct $2m$ boundary value operators B'_j, T_j, $j = 1, \dots, m$, with the properties:

(a) $\operatorname{ord} B'_j = 2m - 1 - u_j$, $\operatorname{ord} T_j = 2m - 1 - m_j$, $j = 1, \dots, m$.
(b) $B'_j \in C^{2m + k - \operatorname{ord} B'_j}(\bar{\Omega})$, $T_j \in C^{2m + k - \operatorname{ord} T_j}(\bar{\Omega})$, $j = 1, \dots, m$.
(c) $\{B'_1, \dots, B'_m, T_1, \dots, T_m\}$ *is a Dirichlet system of order $2m$ on $\partial\Omega$.*
(d) *For all $u, v \in W_2^{2m+k}(\Omega)$ respectively from $W_2^{2m}(\Omega)$ we have the Green formula*

$$\int_\Omega A u \bar{v} \, \mathrm{d}x - \int_\Omega u \overline{A^* v} \, \mathrm{d}x = \sum_{j=1}^m \left[\int_{\partial\Omega} S_j u \cdot \overline{B'_j v} \, \mathrm{d}\sigma - \int_{\partial\Omega} b_j u \cdot \overline{T_j v} \, \mathrm{d}\sigma \right], \tag{12}$$

where A^* is the formal adjoint operator to A defined in Definition 10.1.

Remark 14.1 Here the differentiability assumptions are so strong because Theoem 14.2 holds for all boundary value systems. As we shall see later in §16, for special boundary value systems Green formulae exist under far less restrictive differentiability assumptions.

Proof. The choice of the $S_j, j = 1, \dots, m$ is simple to make, we take for $S_j = (\partial/\partial n)^{\mu_j}$ derivatives in the direction of the inner normal, where the μ_j are the numbers missing from the sequence $0 \leqslant m_1 < \cdots < m_m \leqslant 2m - 1$. It is enough to prove formula (12) for $u, v \in C^{2m+k}(\bar{\Omega})$, functions from $C^{2m+k}(\bar{\Omega})$ are dense in $W_2^{2m+k}(\Omega)$ (and in $W_2^{2m}(\Omega)$) by Theorem 3.6. The operators $A, A^*, b_j, \ldots, T_j$ act continuously into $L^2(\Omega)$, respectively into $L_2(\partial\Omega)$, because of (10) and (b). We cover $\bar{\Omega}$ with finitely many U_i, introduce normal coordinates in U_i by Theorem 2.12, and carry out the transformation $\Phi_i: U_i \to W^r(U_i \cap \Omega \to W^r_+,\ U_i \cap \partial\Omega \to W^{r-1})$. By Theorem 2.12 Φ_j is also admissible at each point of U_i and belongs to $C^{4m+k-2,1}$. Let $\{\varphi_j\}$ be a partition of unity subordinate to the cover $\{U_i\}$. Since those U_i which lie completely in Ω $(U_i \subset \Omega)$ make no contribution to formula (12), indeed we have

$$\int_{U'} Au\cdot\bar{v}\,dx - \int_{U'} u\cdot\overline{A^*v}\,dx = 0 \quad \text{for } u, v \in \mathscr{D}(U_i), \quad U_i \subset \Omega,$$

it follows that we may assume $U_i \cap \partial\Omega \neq \varnothing$. It is sufficient to prove (12) for the half cube W^r_+ and for functions u, v whose support does not meet $\partial W^r_+ \setminus W^{r-1}$, that is,

$$\int_{W^r_+} \mathscr{A}u\cdot\bar{v}\,dx - \int_{W^r_+} u\cdot\overline{\mathscr{A}^*v}\,dx = \sum_{j=1}^{m}\left[\int_{W^{r-1}} \mathscr{S}_j u\cdot\overline{\mathscr{B}'_j v}\,d\sigma - \int_{W^{r-1}} \mathscr{B}_j u\cdot\overline{\mathscr{F}_j v}\cdot d\sigma\right].$$

Here $\mathscr{A}, \ldots, \mathscr{S}_j$ are the operators transformed by Φ_i. The inverse transformation Φ_i^{-1} and the partition of unity, together with (13), give that

$$\int_{\Omega\cap U_i\cap U_{i'}} A\varphi_i u\overline{\varphi_{i'} v}\,dx - \int_{\Omega\cap U_i\cap U_{i'}} \varphi_i u\overline{A^*\varphi_{i'} v}\,dx$$
$$= \sum_{j=1}^{m}\left[\int_{\partial\Omega\cap U_i\cap U_{i'}} S_j\varphi_i u\overline{B_j'^{i'}\varphi_{i'} v}\,d\sigma - \int_{\partial\Omega\cap U_i\cap U_{i'}} b_j\varphi_i u\overline{T_j^{i'}\varphi_{i'} v}\,d\sigma\right]$$

and by summation $\sum_i, \sum_{i'}$

$$\int_\Omega Au\bar{v}\,dx - \int_\Omega u\overline{A^*v}\,dx = \sum_i\sum_{i'}\int_\Omega (A\varphi_i u\overline{\varphi_{i'} v} - \varphi_i u\overline{A^*\varphi_{i'} v})\,dx$$
$$= \sum_i\sum_{i'}\sum_{j=1}^{m}\int_{\partial\Omega}(S_j\varphi_i u\overline{B_j'^{i'}\varphi_{i'} v} - b_j\varphi_i u\overline{T_j^{i'}\varphi_{i'} v})\,d\sigma$$
$$= \sum_{j=1}^{n}\left[\int_{\partial\Omega} S_j u\cdot\overline{B'_j v}\,d\sigma - \int_{\partial\Omega} b_j u\overline{T_j v}\,d\sigma\right],$$

which is (12), where we have put

$$B'_j = \sum_i B_j'^i\varphi_i, \quad T_j = \sum_i T^i_j\varphi_i.$$

We now prove (13) on (W^r_+, W^{r-1}), that is, for $(\mathscr{B}_1, \ldots, \mathscr{B}_m, \mathscr{S}_1, \ldots, \mathscr{S}_m)$ we must find a Dirichlet system $(\mathscr{B}'_1, \ldots, \mathscr{B}'_m, \mathscr{T}_1, \ldots, \mathscr{T}_m)$ which satisfies (a), (b), (c) and (13). In order not to make the writing unnecessarily complicated we put

$$\mathscr{A} = \sum_{|\alpha| \leqslant 2m} a_\alpha(x) \mathrm{D}^\alpha, \quad x = (x', x_r), \quad \alpha = (\alpha', \alpha_r).$$

We partially integrate first according to x' and then according to x_r, and obtain (the functions u and v are zero in the neighbourhood of $\partial W^r_+ \setminus W^{r-1}$):

$$\int_{W^r_+} \mathscr{A} u \bar{v} \,\mathrm{d}x_r = \int_{W^r_+} u \cdot \overline{\mathscr{A}^* v} \,\mathrm{d}x' \mathrm{d}x_r$$
$$+ \sum_{j=1}^{2m} \int_{W^{r-1}} [\mathrm{D}_{x_r}^{j-1} u(x', x_r) \overline{N_{2m-j} v(x', x^r)}]_{x_r=0} \,\mathrm{d}x',$$

where we have written

$$N_{2m-j} v := \sum_{\substack{|\alpha| \leqslant 2m \\ \alpha_r \geqslant j}} (-1)^{|\alpha|-j} \mathrm{D}_{x_r}^{\alpha_r - j} \mathrm{D}_{x'}^{\alpha'} (\overline{a_\alpha(x', x_r)} v(x', x_r)). \tag{15}$$

Since $\mathscr{A}$ is elliptic, the $\{N_{2m-j}\}_{j=1}^{2m}$ form a Dirichlet system of order $2m$, and the coefficients of N_{2m-j} belong to $C^{j+k}(\bar{W}^r_+)$. Since the coordinate transformation $\Phi_i: U_i \cap \Omega \to W^r_+$ is admissible at each point of $\partial\Omega \cap U_i \to W^{r-1}$ (see Theorem 2.12), the transformed boundary value operators $\{\mathscr{B}_1, \ldots, \mathscr{B}_m, \mathscr{S}_1, \ldots, \mathscr{S}_m\}$ again form a Dirichlet system of order $2m$ on W^{r-1} (for short we denote it by $\{F_j\}_{j=1}^{2m}$) and by the assumptions (10) the coefficients of F_j belong to $C^{4m+k-j}(\bar{W}^r_+)$.

By Lemma 14.1 with $M = 4m + K$, we can write

$$\mathrm{D}_{x_r}^{j-1} = \sum_{s=1}^{j} \Phi_{js} F_s, \quad j = 1, \ldots, 2m, \tag{16}$$

where the Φ_{js} are tangential differential operators of order $j - s$ with coefficients from $C^{4m+k-j}(\bar{W}^r_+)$. We denote by Φ^*_{js} the formally adjoint tangential differential operators on W^{r-1}, that is,

$$\int_{W^{r-1}} \Phi_{js} \varphi \cdot \bar{\psi} \,\mathrm{d}x' = \int_{W^{r-1}} \varphi \cdot \overline{\Phi^*_{js} \psi} \,\mathrm{d}x', \quad \varphi, \psi \in \mathscr{D}(W^{r-1}). \tag{17}$$

For the coefficients of Φ^*_{js} we have that they belong to $C^{4m+k-j-(j-s)} \subset C^{s+k}$, since $4m + k - 2j + s \geqslant s + k$. (16) substituted in (15) gives for (14)

$$\int_{W^r_+} \mathscr{A} u v \,\mathrm{d}x - \int_{W^r_+} u \cdot \overline{\mathscr{A}^* v} \,\mathrm{d}x$$
$$= \int_{W^{r-1}} \sum_{s=1}^{2m} [F_s u(x', x_r)]_{x_r=0} \overline{\sum_{j=s}^{2m} \Phi^*_{js} \circ [N_{2m-j} v(x', x_r)]_{x_r=0}} \,\mathrm{d}x'. \tag{18}$$

The operators

$$F'_s = \sum_{j=s}^{2m} \Phi^*_{js} \circ N_{2m-j} = \overline{\Phi_{js}} \circ N_{2m-s} + \sum_{j=s+1}^{2m} \Phi^*_{js} \circ N_{2m-j}, \quad s = 1, \ldots, 2m, \tag{19}$$

form a Dirichlet system of order $2m$ on W^{r-1} and for the coefficients of F'_s we have

$$F_s \in C^{s+k}(\bar{W}^r_+) = C^{2m+k-\operatorname{ord} F'_s}(\bar{W}^r_+). \tag{20}$$

We read off from (19) that $\bar{\Phi}_{ss} \neq 0$, $\operatorname{ord} F'_s = \operatorname{ord} N_{2m-s} = 2m - s$, N_{2m-s} attains the highest order in the normal derivative $D^{2m-s}_{x_r}$, where because of the normality of N_{2m-s} the coefficient is different from zero. Therefore we have established the Dirichlet property of $\{F'_s\}^{2m}_{s=1}$. We also deduce (20) from (19):

$$N_{2m-j} \in C^{j+k}, \operatorname{ord} \Phi^*_{js} = j - s, \Phi^*_{js} \in C^{s+k},$$

hence

$$\Phi^*_{js} \circ N_{2m-j} \in C^{j-(j-s)+k} \cap C^{s+k} = C^{s+k},$$

from which (20) follows immediately.

We designate by $\mathscr{B}'_j$ those F'_s for which $F_s = \mathscr{o}_j$ and $-\mathscr{S}'_j$ that F'_s for which $F_s = \mathscr{B}_j$. With this designation (18) becomes (13) and the orders satisfy (a). (b) follow from the redesignation of (20), and we have also proved (c), since $\{\mathscr{B}'_1, \ldots, \mathscr{B}'_m, \mathscr{T}_1, \ldots, \mathscr{T}_m\} = \{F'_s\}^{2m}_{s=1}$. ■

14.3 Adjoint boundary value operators and boundary value spaces

Definition 14.3 *An intermediate closed subspace V*

$$\mathring{W}^{2m}_2(\Omega) \subset V \subset W^{2m}_2(\Omega)$$

is called a boundary space. We say that the boundary values $B = (b_1(x, \mathrm{D}), \ldots, b_m(x, \mathrm{D}))$ *determine the space V if*

$$\varphi \in V \text{ is equivalent to } b_1(x, \mathrm{D})\varphi = 0, \ldots, b_m(x, \mathrm{D})\varphi = 0.$$

Definition 14.4 *Let* $\Omega \in C^{2m,1}$, *let* $b_j(x, \mathrm{D})$, $j = 1, \ldots, m$ *be boundary value operators with coefficients*

$$b_{js} \in C^{2m-m_j}(\bar{\Omega}), j = 1, \ldots, m, \text{ where } 0 \leqslant \operatorname{ord} b_j = m_j \leqslant 2m - 1.$$

We denote the subspace of $W^{2m}_2(\Omega)$ *determined by* $b_j(x, \mathrm{D})\varphi = 0, j = 1, \ldots, m$ *by*

$$W^{2m}(B) := W^{2m}(\{b_j\}^m_1).$$

Since $b_j: W_2^{2m}(\Omega) \to W_2^{2m-m_j-1/2}(\partial\Omega)$ is continuous, we see immediately that $W^{2m}(B)$ is a closed subspace of $W_2^{2m}(\Omega)$.

We now define the adjoint subspace to $W^{2m}(B)$ abstractly.

Definition 14.5 *Let $A(x, \mathrm{D})$ be a linear differential operator on $\bar{\Omega}$ with coefficients*

$$a_\alpha(x) \in C^{|\alpha|}(\bar{\Omega}), \text{ then}$$

$$A, A^*: W_2^{2m}(\Omega) \to L_2(\Omega) \tag{21}$$

act continuously, where A^* is the (formal) adjoint operator, see Definition 10.1. *Let V be a boundary space; we say that $v \in V^*$, if for each $u \in V$ we have*

$$\int_\Omega Au \cdot \bar{v}\, \mathrm{d}x = \int_\Omega u \cdot \overline{A^* v}\, \mathrm{d}x. \tag{22}$$

We call V^* the adjoint space to V. V^* is a closed subspace of $W_2^{2m}(\Omega)$, for let $v_n \in V^*$ and let $v_n \to v$ in $W_2^{2m}(\Omega)$. Then given the continuity of (21) and of the scalar product in $L_2(\Omega)$ we have

$$\begin{array}{ccc} \displaystyle\int_\Omega Au \cdot \bar{v}_n\, \mathrm{d}x & = & \displaystyle\int_\Omega u \cdot \overline{A^* v_n}\, \mathrm{d}x \\ \downarrow & & \downarrow \\ \displaystyle\int_\Omega Au \cdot \bar{v}\, \mathrm{d}x & = & \displaystyle\int_\Omega u \cdot \overline{A^* v}\, \mathrm{d}x, \end{array}$$

thus proving that V^* is closed. An easy check also gives $\mathring{W}_2^{2m}(\Omega) \subset V^* \subset W_2^{2m}(\Omega)$, so V^* is once more a boundary space.

$$A^{**} = A \text{ gives } V^{**} \supset V. \tag{23}$$

Definition 14.6 *Let the boundary value operators b_j, $j = 1, \ldots, m$ and the boundary space $W^{2m}(B)$ be given. If the boundary value operators $(B'_1, \ldots, B'_m)$ determine the adjoint space $W^{2m}(B)^*$, in the sense of Definition* 14.3, *hence $W^{2m}(B)^* = W^{2m}(B')$, we say that the boundary value problem $(A^*, B'_1, \ldots, B'_m)$ is adjoint to $(A, b_1, \ldots, b_m)$, in short:*

$$(A, b_1, \ldots, b_m)^* = (A^*, B'_1, \ldots, B'_m).$$

Since the $B'_1, \ldots, B'_m$ mostly arise from Green's formula (12), they are not uniquely determined by $(A, b_1, \ldots, b_m)$.

Let A be a differential operator and V a boundary space, we say that (A, V) is self-adjoint, if: $A = A^*$ and $V = V^*$. We call the boundary value problem $A, b_1, \ldots, b_m$ self-adjoint, if $A = A^*$ and $W^{2m}(B) = W^{2m}(B)^*$. It is important to note that the definition of self-adjointness is independent of Green's formula, and there is no notion of adjoint boundary values.

We wish to examine the properties of the adjoint boundary value problem more closely, and we begin with the connection with the Green formula.

Theorem 14.3 *Let* $\Omega \in C^{2m,1}$, *suppose given* $A(x, \mathrm{D})$, $b_j(x, \mathrm{D})$, $S_j(x, \mathrm{D})$, $j = 1, \ldots, m$ *with coefficients*

$$a_\alpha(x) \in C^{|\alpha|}(\bar{\Omega}), \quad b_j \in C^{2m - \operatorname{ord} b_j}(\Omega), \quad S_j \in C^{2m - \operatorname{ord} S_j}(\bar{\Omega}), \quad j = 1, \ldots, m,$$

where $\operatorname{ord} b_j$, $\operatorname{ord} S_j \leqslant 2m - 1$. *We assume further that* $b_1, \ldots, b_m$, $S_1, \ldots, S_m$ *is a Dirichlet system of order* $2m$. *Let* $B'_1, \ldots, B'_m$, $T_1, \ldots, T_m$ *be some other boundary value system of order* $2m$ *(not necessarily normal) with* $\operatorname{ord} B'_j$, $\operatorname{ord} T_j \leqslant 2m - 1$, *and*

$$B'_j \in C^{2m - \operatorname{ord} B'_j}(\bar{\Omega}), \quad T_j \in C^{2m - \operatorname{ord} T_j}(\bar{\Omega}), \quad j = 1, \ldots, m,$$

for the corresponding coefficients. We assume further that Green's formula (12) *holds for* A, A^*, $\{b_1, \ldots, b_m, S_1, \ldots, S_m\}$, $\{B'_1, \ldots, B'_m, \ldots, T_m\}$. *If we write* $B = (b_1, \ldots, b_m)$ *and* $B' = (B'_1, \ldots, B'_m)$, *then we have*

$$W^{2m}(B)^* = W^{2m}(B'), \tag{24a}$$

that is, the boundary values $B' = (B'_1, \ldots, B'_m)$ *determine, in the sense of Definition* 14.5, *the adjoint space* $V^* = W^{2m}(B)^*$, *in short*

$$(A, b_1, \ldots, b_m)^* = (A^*, B'_1, \ldots, B'_m).$$

If we interchange $A, b_1, \ldots, b_m$ and $A^*, B'_1, \ldots, B'_m$, and assume in addition that $(B'_1, \ldots, B'_m, T_1, \ldots, T_m\}$ is also a Dirichlet system, we find that

$$W^{2m}(B')^* = W^{2m}(B), \tag{24b}$$

which altogether gives

$$W^{2m}(B)^{**} = W^{2m}(B) \quad \text{or} \quad (A, b_1, \ldots, b_m)^{**} = (A, b_1, \ldots, b_m). \tag{24c}$$

Proof. Let $v \in W^{2m}(B')$, that is, $v \in W_2^{2m}(\Omega)$ and $B'_1 v = 0, \ldots, B'_m v = 0$, if we take $u \in W^{2m}(B)$, that is, $u W_2^{2m}(\Omega)$ and $b_1 u = 0, \ldots, b_m u = 0$, and substitute everything in (12), then we obtain (22):

$$\int_\Omega A u \bar{v} \, \mathrm{d}x = \int_\Omega u \cdot \overline{A^* v} \, \mathrm{d}x,$$

with which, by Definition 14.5, we have proved that $v \in W^{2m}(B)^*$. Suppose conversely that $v \in W^{2m}(B)^*$, then by Definition 14.5, (22) holds for all $u \in W_2^{2m}(\Omega)$ with $b_1 u = 0, \ldots, b_m u = 0$, which substituted in (12) gives

$$\sum_{j=1}^{m} \int_{\partial\Omega} S_j u \overline{B'_j v} \, d\sigma = 0 \quad \text{for all} \quad u \in W^{2m}(B). \tag{25}$$

We fix j_0 and solve the system of equations

$$b_1 u_0 = 0, \ldots, b_m u_0 = 0, \quad S_1 u_0 = 0, \ldots, S_{j_0} u_0 = \varphi_{j_0}, \ldots, S_n u_0 = 0, \tag{26}$$

where

$$\varphi_{j_0} \in C^{2m+1}(\partial\Omega) \subset W_2^{2m - \operatorname{ord} S_{j_0} - 1/2}(\partial\Omega).$$

By Theorem 14.1 with $M = 2m$, there exists a solution $u_0 \in W_2^{2m}(\Omega)$ of the equations above, which because $b_1 u_0 = 0, \ldots, b_m u_0 = 0$ belongs to $W^{2m}(B)$. Therefore we can substitute (26) in (25) and obtain

$$\int_{\partial\Omega} \varphi_j \overline{B'_j v} \, d\sigma = 0 \quad \text{for all} \quad \varphi_j \in C^{2m+1}(\partial\Omega). \tag{27}$$

Since $B'_j v \in W_2^{2m - \operatorname{ord} B'_j - 1/2}(\partial\Omega) \subset L_2(\partial\Omega)$ and because $C^{2m+1}(\partial\Omega)$ is dense in $L_2(\partial\Omega)$ (see Theorem 4.3) it follows from (27) that

$$B'_j v = 0 \quad \text{for } j = 1, \ldots, m$$

which is $v \in W^{2m}(B')$. ■

The second part of Theorem 14.3 is obvious.

Corollary 14.1 *If the boundary value problem $A, b_1, \ldots, b_m$ is symmetric, i.e., Green's formula* (12) *holds with $A^* = A$, $B'_1 = b_1, \ldots, B'_m = b_m$, and the system B is normal, then the boundary problem $A, b_1, \ldots,$ is self-adjoint in the sense of Definition* 14.6.

Remark 14.1 If we make the wrong choice of $S_1, \ldots, S_m$ in Green's formula (12), it may happen that $B \neq B'$, though the boundary value problem is self-adjoint in the sense of Definition 14.6. We come now to an existence theorem.

Theorem 14.4 *Let $\Omega \in C^{4m-1,1}$, let $A(x, D)$ be an elliptic differential operator on $\bar{\Omega}$ with $a_\alpha(x) \in C^{|\alpha|}(\Omega)$, let $B = (b_j(x, D), j = 1, \ldots, m)$ be a normal boundary value system on $\partial\Omega$, where the coefficients*

$$b_{js}(x) \in C^{4m-1-m_j}(\bar{\Omega}), \quad j = 1, \ldots, m, m_j \leqslant 2m - 1.$$

Then on $\partial\Omega$ there exists an adjoint normal boundary value system $B' = (B_j'(x, \mathrm{D}), j = 1, \ldots, m)$ with $\operatorname{ord} B_j' \leqslant 2m - 1, j = 1, \ldots, m$, *and*

$$B_{js}'(x) \in C^{2m - \operatorname{ord} B_j'}(\bar{\Omega}), \quad j = 1, \ldots, m,$$

for which

$$W^{2m}(B)^* = W^{2m}(B')$$

that is, the boundary values $B' = (B_1', \ldots, B_m')$ determine, in the sense of Definition 14.3, *the* adjoint space $V^* = W^{2m}(B)^*$, *in short*

$$(A, b_1, \ldots, b_m)^* = (A^*, B_1', \ldots, B_m').$$

Proof. We apply Theorem 14.2 with $k = 0$, and obtain the existence of

$$S_j \in C^{4m - 1 - \operatorname{ord} S_j}(\bar{\Omega}), \quad B_j' \in C^{2m - \operatorname{ord} B_j'}(\Omega), \quad T_j \in C^{2m - \operatorname{ord} T_j}(\bar{\Omega}), \quad j = 1, \ldots, m,$$

where both systems $\{b_1, \ldots, b_m, S_1, \ldots, S_m\}, \{B_1', \ldots, B_m', T_1, \ldots, T_m\}$ are Dirichlet systems of order $2m$, so B' is normal, and the Green formula is valid (12). Therefore we are in the range of validity of the assumptions for Theorem 14.3 and we have (24). ∎

We call B' the boundary value system adjoint to $B = (b_j)$.

Our next theorems are concerned with the transferability of the L.Š. condition on B to the adjoint boundary value system B'.

Theorem 14.5 *Let Ω be bounded and from $C^{2m,1}$, suppose given $A(x, \mathrm{D})$ elliptic on $\bar{\Omega}$ and two boundary value systems $B = (b_1(x, \mathrm{D}), \ldots, b_m(x, \mathrm{D}))$, $B' = (B_1'(x, \mathrm{D}), \ldots, B_m'(x, \mathrm{D}))$ which are coupled together by means of a Green formula* (12). *For orders and coefficients we assume*: $\operatorname{ord} A = 2m$, $\operatorname{ord} b_j$, $\operatorname{ord} B_j'$, $\operatorname{ord} S_j$, $\operatorname{ord} T_j \leqslant 2m - 1$,

$$a_\alpha(x) \in C^{|\alpha|}(\bar{\Omega}), \qquad b_j \in C^{2m - \operatorname{ord} b_j}(\bar{\Omega}),$$
$$B_j' \in C^{2m - \operatorname{ord} B_j'}(\bar{\Omega}), \quad S_j \in C^{2m - \operatorname{ord}} S_j(\bar{\Omega}), \quad T_j \in C^{2m - \operatorname{ord} T_j}(\bar{\Omega}).$$

If $A(x, \mathrm{D})$, $b_1(x, \mathrm{D}), \ldots, b_m(x, \mathrm{D})$ satisfy the L.Š. *condition* 11.1 *at $x_0 \in \partial\Omega$, then the system $A^*(x, \mathrm{D})$, $B_1'(x, \mathrm{D}), \ldots, B_m'(x, \mathrm{D})$ also satisfies the* L.Š. *condition* 11.1 *at $x_0 \in \partial\Omega$.*

We emphasise that this theorem is true without assuming normality, also B' need not determine the space $W^{2m}(B)^*$.

Proof. Our first problem is to modify the Green formula (12), so that it fits in with the L.Š. condition. We take a small neighbourhood U of x_0

and transform U admissibly by Theorem 2.12 onto W^r. Under this transformation $\Phi(\in C^{2m-1,1})$ (12) becomes (see the proof of Theorem 14.2)

$$\int_{\mathbb{R}^r_+} \mathscr{A}u\bar{v}\,\mathrm{d}x'\,\mathrm{d}x_r - \int_{\mathbb{R}^r_+} u\cdot\overline{\mathscr{A}^*v}\,\mathrm{d}x'\,\mathrm{d}x_r$$
$$= \sum_{j=1}^{m}\left[\int_{\mathbb{R}^{r-1}} \mathscr{S}_j u\overline{\mathscr{B}'_j v}\,\mathrm{d}x' - \int_{\mathbb{R}^{r-1}} \mathscr{B}_j\overline{\mathscr{T}_j v}\,\mathrm{d}x'\right], \tag{28}$$

where (28) holds for all $u, v \in W_2^{2m}(\mathbb{R}^r_+)$, which have supports in $W^r_+ \cup W^{r-1}$. Let $\lambda > 1$, the functions $u(\lambda x', \lambda x_r)$, $v(\lambda x', \lambda x_r)$ again belong to $W_2^{2m}(\mathbb{R}^r_+)$ and have supports in $W^r_+ \cup W^{r-1}$, we can therefore substitute them in (28), and by change of coordinates and simple manipulations we obtain

$$\int_{\mathbb{R}^r_+}\left(\mathscr{A}^{\mathrm{H}}\left(\frac{x}{\lambda},\mathrm{D}\right)u(x)\right)\overline{v(x)}\,\mathrm{d}x - \int_{\mathbb{R}^r_+} u(x)\overline{\left(\mathscr{A}^{*\mathrm{H}}\left(\frac{x}{\lambda},\mathrm{D}\right)v(x)\right)}\,\mathrm{d}x$$
$$= \sum_{j=1}^{m}\left[\int_{\mathbb{R}^{r-1}} \mathscr{S}_j^{\mathrm{H}}\left(\frac{x'}{\lambda},\mathrm{D}\right)u(x',x_r)\overline{\mathscr{B}_j'^{\mathrm{H}}\left(\frac{x'}{\lambda},\mathrm{D}\right)v(x',x_r)}\,\mathrm{d}x'|_{x_r=0}\right.$$
$$\left. - \int_{\mathbb{R}^{r-1}} \mathscr{B}_j^{\mathrm{H}}\left(\frac{x'}{\lambda},\mathrm{D}\right)u(x',x_r)\overline{\mathscr{F}_j^{\mathrm{H}}\left(\frac{x'}{\lambda},\mathrm{D}\right)v(x',x_r)}\,\mathrm{d}x'|_{x_r=0}\right] + O\left(\frac{1}{\lambda}\right)$$

For $\lambda \to +\infty$ we obtain

$$\int_{\mathbb{R}^r_+}(\mathscr{A}^{\mathrm{H}}(0,\mathrm{D})u(x))\overline{v(x)}\,\mathrm{d}x - \int_{\mathbb{R}^r_+} u(x)\overline{(\mathscr{A}^{*\mathrm{H}}(0,\mathrm{D})v(x))}\,\mathrm{d}x$$
$$= \sum_{j=1}^{m}\left[\int_{\mathbb{R}^{r-1}} \mathscr{S}_j^{\mathrm{H}}(0,\mathrm{D})u(x',x_r)\overline{\mathscr{B}_j'^{\mathrm{H}}(0,\mathrm{D})v(x',x_r)}\,\mathrm{d}x'|_{x_r=0}\right.$$
$$\left. - \int_{\mathbb{R}^{r-1}} \mathscr{B}_j^{\mathrm{H}}(0,\mathrm{D})u(x',x_r)\overline{\mathscr{T}_j^{\mathrm{H}}(0,\mathrm{D})v(x',x_r)}\,\mathrm{d}x'|_{x_r=0}\right]. \tag{29}$$

Here the differential expressions

$$\mathscr{A}^{\mathrm{H}}, \mathscr{A}^{*\mathrm{H}}, \mathscr{B}_j^{\mathrm{H}}, \mathscr{B}_j'^{\mathrm{H}}, \mathscr{S}_j^{\mathrm{H}}, \mathscr{T}_j^{\mathrm{H}}, \quad j = 1, \ldots, m,$$

are homogeneous and have constant coefficients. Since (29) is invariant under the stretching of axes, and functions with bounded support are dense in $W_2^{2m}(\mathbb{R}^r_+)$, see Theorem 3.6, (29) also holds for all $u, v \in W_2^{2m}(\mathbb{R}^r_+)$. We submit (29) to a further manipulation. Let $\varphi(x_r)$, $\psi(x_r) \in W_2^{2m}(\mathbb{R}^1_+)$ and $\chi(x') \in W_2^{2m}(\mathbb{R}^{r-1})$, we take $\lambda > 0$, $0 \neq \xi' \in \mathbb{R}^{r-1}$, and write

$$\begin{aligned} u(x', x_r) &= \mathrm{e}^{\mathrm{i}(\xi', x')}\cdot\chi(\lambda x')\varphi(x_r), \\ v(x', x_r) &= \mathrm{e}^{\mathrm{i}(\xi', x')}\cdot\chi(\lambda x')\psi(x_r). \end{aligned} \tag{30}$$

We have $u,v\in W_2^{2m}(\mathbb{R}^r_+)$ by Fubini. We substitute (30) in (29), change the coordinates $t=x_r$, $y'=\lambda x'$, exploit homogeneity, carry out easy manipulations, write $\mathscr{A}^{\mathrm{H}}(0,\mathrm{D})=\mathscr{A}^{\mathrm{H}}(0,\mathrm{D}_{y'},\mathrm{D}_t)$ etc. Here we recall the convention in §11 that derivatives carry the factor 1/i. We obtain

$$\left[\int_0^\infty\left(\mathscr{A}^{\mathrm{H}}\left(0,\xi',\frac{1}{\mathrm{i}}\frac{\mathrm{d}}{\mathrm{d}t}\right)\varphi(t)\right)\overline{\psi(t)}\,\mathrm{d}t-\int_0^\infty\varphi(t)\cdot\overline{\mathscr{A}^{*\mathrm{H}}\left(0,\xi',\frac{1}{\mathrm{i}}\frac{\mathrm{d}}{\mathrm{d}t}\right)\psi(t)}\,\mathrm{d}t\right]$$

$$\cdot\int_{\mathbb{R}^{r-1}}|\chi(y')|^2\,\mathrm{d}y'-\left[\sum_{j=1}^m\mathscr{S}_j^{\mathrm{H}}\left(0,\xi',\frac{1}{\mathrm{i}}\frac{\mathrm{d}}{\mathrm{d}t}\right)\varphi(t)|_{t=0}\overline{\mathscr{B}_j'^{\mathrm{H}}\left(0,\xi',\frac{1}{\mathrm{i}}\frac{\mathrm{d}}{\mathrm{d}t}\right)\psi(t)|_{t=0}}\right.$$

$$\left.-\mathscr{B}_j^{\mathrm{H}}\left(0,\xi',\frac{1}{\mathrm{i}}\frac{\mathrm{d}}{\mathrm{d}t}\right)\varphi(t)|_{t=0}\cdot\overline{\mathscr{T}_j^{\mathrm{H}}\left(0,\xi',\frac{1}{\mathrm{i}}\frac{\mathrm{d}}{\mathrm{d}t}\right)\psi(t)|_{t=0}}\right]$$

$$\cdot\int_{\mathbb{R}^{r-1}}|\chi(y')|^2\,\mathrm{d}y'=O(\lambda).$$

We allow λ to tend to zero, and obtain

$$\int_0^\infty\left(\mathscr{A}^{\mathrm{H}}\left(0,\xi',\frac{1}{\mathrm{i}}\frac{\mathrm{d}}{\mathrm{d}t}\right)\varphi(t)\right)\overline{\psi(t)}\,\mathrm{d}t-\int_0^\infty\varphi(t)\overline{\mathscr{A}^{*\mathrm{H}}\left(0,\xi',\frac{1}{\mathrm{i}}\frac{\mathrm{d}}{\mathrm{d}t}\right)\varphi(t)}\,\mathrm{d}t$$

$$=\sum_{j=1}^m\left[\mathscr{S}_j^{\mathrm{H}}\left(0,\xi',\frac{1}{\mathrm{i}}\frac{\mathrm{d}}{\mathrm{d}t}\right)\varphi(t)|_{t=0}\overline{\mathscr{B}_j'^{\mathrm{H}}\left(0,\xi',\frac{1}{\mathrm{i}}\frac{\mathrm{d}}{\mathrm{d}t}\right)\psi(t)|_{t=0}}\right.$$

$$\left.-\mathscr{B}_j^{\mathrm{H}}\left(0,\xi',\frac{1}{\mathrm{i}}\frac{\mathrm{d}}{\mathrm{d}t}\right)\varphi(t)|_{t=0}\cdot\overline{\mathscr{T}_j^{\mathrm{H}}\left(0,\xi',\frac{1}{\mathrm{i}}\frac{\mathrm{d}}{\mathrm{d}t}\right)\psi(t)|_{t=0}}\right]\tag{31}$$

for $\varphi,\psi\in W_2^{2m}(\mathbb{R}^1_+)$. (31) is the form of the Green formula, as we need it for the L.Š. condition. Let

$$\begin{aligned}&\mathscr{A}^{*\mathrm{H}}\left(0,\xi',\frac{1}{\mathrm{i}}\frac{\mathrm{d}}{\mathrm{d}t}\right)\psi(t)=0,\qquad t>0,\ 0\neq\xi'\in\mathbb{R}^{r-1},\\&\mathscr{B}_j'^{\mathrm{H}}\left(0,\xi',\frac{1}{\mathrm{i}}\frac{\mathrm{d}}{\mathrm{d}t}\right)\psi(t)|_{t=0}=0,\quad j=1,\ldots,m,\end{aligned}\tag{32}$$

for $\psi(t)\in W_2^{2m}(\mathbb{R}^1_+)$, we must show (Condition 11.1) that then $\psi\equiv0$. Since by assumption $\mathscr{A}^{\mathrm{H}},\mathscr{B}_1^{\mathrm{H}},\ldots,\mathscr{B}_m^{\mathrm{H}}$ satisfy Condition 11.1, which is equivalent to 11.3, we obtain that for each $f\in L_2(\mathbb{R}^1_+)$ the following initial value problem has a solution φ in $W_2^{2m}(\mathbb{R}^1_+)$.

$$\begin{aligned}&\mathscr{A}^{\mathrm{H}}\left(0,\xi',\frac{1}{\mathrm{i}}\frac{\mathrm{d}}{\mathrm{d}t}\right)\varphi(t)=f(t)\\&\mathscr{B}_j'^{\mathrm{H}}\left(0,\xi',\frac{1}{\mathrm{i}}\frac{\mathrm{d}}{\mathrm{d}t}\right)\psi(t)|_{t=0}=0,\quad j=1,\ldots,m.\end{aligned}\tag{33}$$

Substitution of (32) and (33) in (31) gives

$$\int_0^\infty f\bar{\psi}\,dt = 0 \quad \text{for each } f \in L_2(\mathbb{R}^1_+),$$

from which it follows that $\Psi = 0$, which is Condition 11.1 for $\mathscr{A}^*, \mathscr{B}'_1, \ldots, \mathscr{B}'_m$. Transforming back (admissible coordinates, Theorem 11.3) gives Condition 11.1 for A^*, $B'_1, \ldots, B'_m$ at $x_0 \in \partial\Omega$. ■

We now show that the L.Š. condition is less a property of the special boundary value conditions $b_1, \ldots, b_m$, but more a property of the boundary space $V = W^{2m}(B)$.

Theorem 14.6 *Let Ω be bounded and from $C^{2m,1}$, let $A(x, D)$ be an elliptic differential operator on $\bar{\Omega}$ with coefficients $a_\alpha(x) \in C^1(\bar{\Omega})$, and let $b_1(x, D), \ldots, b_m(x, D)$ (in short, B) be boundary value operators of order $m_j \leqslant 2m - 1$, $j = 1, \ldots, m$, whose coefficients belong to $C^{2m+1-m_j}(\bar{\Omega})$. If $A, b_1, \ldots, b_m$ satisfy Condition 11.1 on $\partial\Omega$, then each normal boundary value system $\tilde{B} = (\tilde{b}_1(x, D), \ldots, \tilde{b}_m(x, D))$ determining $W^{2m}(B)$ also satisfies Condition 11.1. It suffices to assume that the coefficients of $\tilde{B}$ are such that*

$$\tilde{b}_j \in C^{2m-\tilde{m}_j}(\bar{\Omega}), \quad \tilde{m}_j \leqslant 2m - 1, \quad j = 1, \ldots, m. \tag{34}$$

For the proof we use the main theorem 13.1. By the assumptions made in Theorem 14.6 the implication (a)⇒(d) of the main theorem is valid and we have the *a priori* estimate

$$\|\varphi\|_{2m} \leqslant c\left\{\|A\varphi\|_0 + \sum_{j=1}^m \|b_j\varphi\|_{2m-m_j-1/2,\partial\Omega} + \|\varphi\|_{2m-1}\right\}, \tag{35}$$

for all $\varphi \in W_2^{2m}(\Omega)$.

Let $w \in W^{2m}(B)$, that is $b_1 w = 0, \ldots, b_m w = 0$, then (35) becomes

$$\|w\|_{2m} \leqslant c_1\{\|Aw\|_0 + \|w\|_{2m-1}\} \quad \text{for all } w \in W^{2m}(B). \tag{36}$$

Again let φ from $W_2^{2m}(\Omega)$ be arbitrary. Since the system $\tilde{B} = (\tilde{b}_1, \ldots, \tilde{b}_m)$ is normal, we can enlarge it to a Dirichlet system of order $2m$ and apply Theorem 14.1, that is, find some $\varphi_0 \in W_2^{2m}(\Omega)$ which depends continuously on $(\psi_1, \ldots, \psi_m)$ and which satisfies the equations

$$\tilde{b}_j(x, D)\varphi_0 = \tilde{b}_j(x, D)\varphi =: \psi_j \in W_2^{2m-\tilde{m}_j-1/2}(\partial\Omega), \tag{37}$$

where because of the continuity in Theorem 14.1

$$\|\varphi_0\|_{2m} \leqslant c_2 \sum_{j=1}^m \|\tilde{b}_j\varphi\|_{2m-\tilde{m}_j-1/2,\partial\Omega}. \tag{38}$$

Since $\tilde{B}$ determines the space $W^{2m}(B)$, it follows from (37) that

$$w = \varphi - \varphi_0 \in W^{2m}(B).$$

For this w, applying (36), (38) and the estimates

$$\| A\varphi_0 \|_0 \leqslant c \| \varphi_0 \|_{2m}, \; \| \varphi_0 \|_{2m-1} \leqslant \| \varphi_0 \|_{2m},$$

we have

$$\begin{aligned} \| \varphi \|_{2m} &\leqslant \| w \|_{2m} + \| \varphi_0 \|_{2m} \leqslant c_3(\| Aw \|_0 + \| w \|_{2m-1} + \| \varphi_0 \|_{2m}) \\ &= c_3(\| A\varphi - A\varphi_0 \|_0 + \| \varphi - \varphi_0 \|_{2m-1} + \| \varphi_0 \|_{2m}) \\ &\leqslant c_4(\| A\varphi \|_0 + \| \varphi_0 \|_{2m} + \| \varphi \|_{2m-1} + \| \varphi_0 \|_{2m} + \| \varphi_0 \|_{2m}) \\ &\leqslant c_5\left(\| A\varphi \|_0 + \sum_{j=1}^{m} \| \tilde{b}_j \varphi \|_{2m - \tilde{m}_j - 1/2, \partial\Omega} + \| \varphi \|_{2m-1} \right), \; \varphi \in W_2^{2m}(\Omega), \end{aligned}$$

that is, the system $A, \tilde{b}_1, \ldots, \tilde{b}_m$ satisfies the *a priori* estimate (d) of the main theorem 13.1. We can therefore apply the implication (d)$\Rightarrow$(a) of the main theorem – here the assumptions (34) suffice – and obtain that the system $A, \tilde{b}_1, \ldots, \tilde{b}_m$ satisfies Condition 11.1 on $\partial\Omega$. ■

We now combine Theorems 14.4, 14.5 and 14.6 and have:

Theorem 14.7 *Let Ω be bounded and from $C^{4m,1}$, let $A(x, \mathrm{D})$ be an elliptic differential operator on $\bar{\Omega}$ with coefficients $a_\alpha(x) \in C^{1+|\alpha|}(\bar{\Omega})$, let $B = (b_j(x, \mathrm{D}), j = 1, \ldots, m)$ be a normal boundary value system on $\partial\Omega$, where the coefficients*

$$b_{js}(x) \in C^{4m - m_j}(\bar{\Omega}), \quad |s| \leqslant m_j, \; j = 1, \ldots, m, \; m_j \leqslant 2m - 1;$$

for $(A, b_1, \ldots, b_m)$ suppose that the Condition 11.1 is satisfied on $\partial\Omega$. Then there exists an adjoint, normal, boundary value system $B' = (B'_j(x, \mathrm{D}), j = 1, \ldots, m)$ with the following properties.

(a) *The coefficients satisfy*

$$B'_{js}(x) \in C^{2m+1-\operatorname{ord} B'_j}(\bar{\Omega}), \quad |s| \leqslant \operatorname{ord} B'_j \leqslant 2m - 1, j = 1, \ldots, m.$$

(b) *B' determines the space $W^{2m}(B)^*$, that is,*

$$W^{2m}(B)^* = W^{2m}(B').$$

(c) *on $\partial\Omega$ $A^*, B'_1, \ldots, B'_m$ satisfies the L.Š. condition* 11.1.

Furthermore every other normal boundary value system $\tilde{B}' = (\tilde{B}'_1, \ldots, \tilde{B}'_m)$, determining the boundary space $V^* = W^{2m}(B)^*$ also satisfies Condition 11.1, *where for the coefficients of $\tilde{B}$ we assume*

$$\tilde{B}_j \in C^{2m - \operatorname{ord} \tilde{B}'_j}(\bar{\Omega}).$$

Proof. By Theorem 14.4 with the differentiability assumptions raised by (a) we deduce the existence of some B' with (a) and (b), where B' and B are coupled by Green's formula (12). Therefore we can apply Theorem 14.5 and obtain (c). We obtain the final part of our assertion by applying Theorem 14.6 to

$$A^*, B'_1, \ldots, B'_m \quad \text{and} \quad A^*, \tilde{B}'_1, \ldots, \tilde{B}'_m. \qquad \blacksquare$$

Because $A^{**} = A$ and $W^{2m}(B)^{**} = W^{2m}(B)$, Theorem 14.7 is basically an equivalence theorem

$$\left.\begin{array}{l}\text{(a) } B \text{ determines } V \subset W_2^{2m}(\Omega)\\ \text{(b) } B \text{ is normal on } \partial\Omega\\ \text{(c) } B \text{ satisfies the L.Š.}\\ \quad\ \text{condition 11.1}\end{array}\right\} \Leftrightarrow \left\{\begin{array}{l}\text{(a) } B' \text{ determines } V^*\\ \text{(b) } B' \text{ is normal}\\ \text{(c) } B' \text{ satisfies the L.Š.}\\ \quad\ \text{condition 11.1}\end{array}\right.$$

We leave the precise statement of this equivalence theorem – it is a matter of precise differentiability assumptions – to the reader.

14.4 The second Green formula

For later use – V-elliptic differential equations – we need the second Green formula. It is useful here to write the operator $A(x, \mathrm{D})$ in one of several possible different ways

$$A(x, \mathrm{D})u = \sum_{|p|,|q| \leqslant m} (-1)^q \mathrm{D}^q(a_{pq}\mathrm{D}^p u). \tag{39}$$

we associate a sesquilinear form with the operator (39)

$$a(u, v) = \int_\Omega \sum_{|p|,|q| \leqslant m} a_{pq}\mathrm{D}^p u \cdot \overline{\mathrm{D}^q v}\,\mathrm{d}x. \tag{40}$$

Theorem 14.8 *Let Ω be bounded and from $C^{2m+k,1}$, $k \geqslant 0$, let $A(x, \mathrm{D})$ (39) be an elliptic differential operator on $\bar{\Omega}$ with coefficients*

$$a_{pq} \in C^{|q|+k}(\bar{\Omega}). \tag{41}$$

Let $b_j(x, \mathrm{D})$, $j = 1, \ldots, m$ be a Dirichlet boundary value system of order m on $\partial\Omega$, hence $0 \leqslant m_j \leqslant m - 1$; for the coefficients we suppose

$$b_{js}(x) \in C^{2m+k-m_j}(\bar{\Omega}), \quad s \leqslant m_j, j = 1, \ldots, m, \tag{41'}$$

or, after eventual reordering, $b_j \in C^{2m+1+k-j}(\bar{\Omega})$. Then on $\partial\Omega$ there exists a normal boundary value system $C_1(x, \mathrm{D}), \ldots, C_m(x, \mathrm{D})$ with coefficients

$$C_{js}(x) \in C^{2m+k-\operatorname{ord} C_j}(\bar{\Omega}), \quad |s| \leqslant \operatorname{ord} C_j, j = 1, \ldots, m,$$

such that the second Green formula holds

$$a(u, v) = \int_\Omega (Au)\bar{v}\,\mathrm{d}x - \sum_{j=1}^m \int_{\partial\Omega} C_j u \cdot \overline{b_j v}\,\mathrm{d}\sigma; \quad u, v \in W_2^{2m}(\Omega), \tag{42}$$

where for the orders of the C_j *we have*

$$\operatorname{ord} C_j = 2m - 1 - \operatorname{ord} b_j, \quad j = 1, \ldots, m.$$

We begin with the proof as in Theorem 14.2: we cover $\bar{\Omega}$, take a partition of unity, transform admissibly onto the cube W^r, Theorem 2.12, and have to prove (42) in the form

$$a(u,v) = \int_{W^r_+} (\mathscr{A}u)\bar{v}\,\mathrm{d}x - \sum_{j=1}^{m} \int_{W^{r-1}} \mathscr{C}_j u \cdot \overline{\mathscr{B}_j v}\,\mathrm{d}x', \tag{42'}$$

for functions u, v from $C^{2m}(\bar{W}^r)$ or $C^\infty(\bar{W}^r)$, whose support does not meet $\partial W^r_+ \setminus W^{r-1}$. Again we label the coefficients of $\mathscr{A}$ as a_{pq}. We write $x = (x', x_r)$, $q = (q', q_r)$, integrate partially according to x', and because of the support properties of u and v have

$$\begin{aligned} a(u,v) &= \int_{W^r_+} \sum_{|p|,|q| \leqslant m} a_{pq} \mathrm{D}^p u \mathrm{D}^q \bar{v}\,\mathrm{d}x \\ &= \int_{W^r_+} \sum_{|p|,|q| \leqslant m} (-1)^{|q'|} \mathrm{D}^{q'}(a_{pq}\mathrm{D}^p u) \cdot \mathrm{D}^{q_r}_{x_r} \bar{v}\,\mathrm{d}x. \end{aligned} \tag{43}$$

We integrate partially according to x_r, and have further

$$\begin{aligned} a(u,v) &= \int_{W^r_+} \sum_{|p|,|q| \leqslant m} (-1)^{|q'| + q_r} \mathrm{D}^{q_r}_{x_r} \mathrm{D}^{q'}(a_{pq}\mathrm{D}^p u)\bar{v}\,\mathrm{d}x \\ &\quad - \int_{W^{r-1}} \sum_{|p|,|(q',q_r)| \leqslant m} \sum_{i=1}^{q_r} (-1)^{|q'| + i - 1} \mathrm{D}^{q_r - i}_{x_r} \mathrm{D}^{q'}(a_{pq}\mathrm{D}^p u) \mathrm{D}^{i-1}_{x^r} \bar{v}\,\mathrm{d}x' \\ &= \int_{W^r_+} (\mathscr{A}u) \cdot \bar{v}\,\mathrm{d}x - \sum_{j=1}^{m} \int_{W^{r-1}} N_j(u) \cdot \mathrm{D}^{j-1}_{x_r} \bar{v}\,\mathrm{d}x', \end{aligned} \tag{43'}$$

where we have written

$$N_j(u) := \sum_{\substack{|p|,|q| \leqslant m \\ q = (q', q_r)}} \sum_{q_r \geqslant j} (-1)^{|q'| + j - 1} \mathrm{D}^{q_r - j}_{x_r} \mathrm{D}^{q'}_{x'}(a_{pq}\mathrm{D}^p u). \tag{44}$$

From (44) we read off: $\operatorname{ord} N_j = 2m - j$, the highest order is attained at $a_{(0,m)(0,m)} \mathrm{D}^{2m-j}_{x_r} u$, where because of the ellipticity of $\mathscr{A}(x, \mathrm{D})$ $a_{(0,m)(0,m)} \neq 0$, in $\mathscr{A}^{\mathrm{H}}(x,\xi) \neq 0$ we substitute $\xi = (0, \xi_r) \neq 0$ and have $\mathscr{A}^{\mathrm{H}}(x, (0, \xi_r)) = a_{(0,m)(0,m)} \cdot \xi_r^{2m} \neq 0$. For the coefficients of $N_j(u)$ we have, see (41),

$$N_j \in C^{|q| + k - (|q| - j)}(\bar{W}^r_+) = C^{j+k}(\bar{W}^r_+). \tag{45}$$

The system N_j, $j = 1, \ldots, m$ is therefore normal on W^{r-1}. Since $\mathscr{B}_1, \ldots, \mathscr{B}_m$ is a Dirichlet system, by Lemma 14.1 we have

$$D_{x_r}^{j-1} = \sum_{s=1}^{j} \Phi_{js} \circ \mathscr{B}_s, \quad j = 1, \ldots, m, \tag{46}$$

where because of (41′) the coefficients of Φ_{js} satisfy

$$\Phi_{js} \in C^{2m+1+k-j}(\overline{W^r_+}), \tag{47}$$

ord $\Phi_{js} = j - s$ (Φ_{js} tangential). We denote by Φ^*_{js} the formally adjoint tangential differential operator on W^{r-1}, that is,

$$\int_{W^{r-1}} \Phi_{js}\varphi \cdot \bar{\psi}\, dx' = \int_{W^{r-1}} \varphi \cdot \overline{\Phi^*_{js}\psi}\, dx', \quad \varphi, \psi \in \mathscr{D}(W^{r-1}). \tag{48}$$

Because of (47) for the coefficients of Φ^*_{js} we have

$$\Phi^*_{js} \in C^{2m+1+k-j-(j-s)}(\overline{W^r_+}) \subset C^{k+s}(\overline{W^r_+}); \tag{49}$$

since $2m + 1 + k - j - (j - s) = 2m + 1 + k - 2j + s \geqslant k + 1 + s$ for $j = 1, \ldots, m$. We substituate (46) in (43) and because of (48) we have

$$\begin{aligned} a(u, v) &= \int_{W^r_+} (\mathscr{A}u)\bar{v}\, dx - \sum_{j=1}^{m} \int_{W^{r-1}} N_j(u) \cdot \overline{\sum_{s=1}^{j} \Phi_{js}\mathscr{B}_s v}\, dx' \\ &= \int_{W^r_+} (\mathscr{A}u)\bar{v}\, dx - \sum_{j=1}^{m} \int_{W^{r-1}} \left(\sum_{s=j}^{m} \Phi^*_{sj} \circ N_s(u) \right) \cdot \overline{\mathscr{B}_j v}\, dx. \end{aligned}$$

we have found the $\mathscr{C}$s in (42):

$$\mathscr{C}_j := \sum_{s=j}^{m} \Phi^*_{sj} \circ N_s. \tag{50}$$

Again we read off from (50): because $\Phi^*_{jj} = \bar{\Phi}_{jj} \neq 0$ and $a_{(0,m)(0,m)} \neq 0$, ord $\mathscr{C}_j$ = ord N_j = ord $a_{(0,m)(0,m)} D_{x_r}^{2m-j} = 2m - j$, and the $\mathscr{C}_j$, $j = 1, \ldots, m$ are normal on W^{r-1}. For the coefficients of $\mathscr{C}_j$ (45) and (49) give

$$\mathscr{C}_j \in C^{k+j}(\overline{W^r_+}) \cap C^{k+s-(s-j)}(\overline{W^r_+}) = C^{k+j}(\overline{W^r_+}) = C^{2m+k-\operatorname{ord}\mathscr{C}_j}(\overline{W^r_+}).$$

Transforming back we obtain the assertions of our theorem. ■

For Dirichlet boundary value systems $b_1, \ldots, b_m$ we can recover the Green formula (12) by Theorem 14.8, and determine adjoint boundary value problems.

Theorem 14.9 *Suppose that the assumptions of Theorem* 14.8 *are satisfied, suppose in particular that* $b_1, \ldots, b_m$ *form a Dirichlet system of order m on* $\partial\Omega$

then the first Green formula holds

$$\int_\Omega Au\bar{v}\,\mathrm{d}x - \int_\Omega u\cdot\overline{A^*v}\,\mathrm{d}x = \sum_{j=1}^m \int_{\partial\Omega} [C_ju\cdot\overline{b_jv} - b_ju\cdot\overline{C_j'v}]\,\mathrm{d}\sigma, \tag{51}$$

where C_j (respectively C_j') are determined by $(A, b_1, \ldots, b_m)$ (respectively $(A^*, b_1, \ldots, b_m)$) in the sense of Theorem 14.8.

From (51) one obtains (adjoint boundary value problems, see Definition 14.6 and Theorem 14.3)

$$(A, b_1, \ldots, b_m)^* = (A^*, b_1, \ldots, b_m) \quad (A, C_1, \ldots, C_m)^* = (A^*, C_1', \ldots, C_m') \tag{52}$$

therefore in particular for the Dirichlet problem $\mathrm{D} = (\partial^0/\partial n^0, \ldots, \partial^{m-1}/\partial n^{m-1})$

$$(A, \mathrm{D})^* = (A^*, \mathrm{D}).$$

Proof. We write the operator $A(x, \mathrm{D})$ in the form (39), then the adjoint operator A^* has the form

$$A^*(x, \mathrm{D})u = \sum_{|p|,|q| \leqslant m} (-1)^q \mathrm{D}^q(\overline{a_{qp}}\mathrm{D}^p u)$$

and the sesquilinear form associated with it by (40) is equal to

$$a^*(u, v) = \int_\Omega \sum_{|p|,|q| \leqslant m} \overline{a_{qp}}\mathrm{D}^p u\cdot\overline{\mathrm{D}^q v}\,\mathrm{d}x = \int_\Omega \sum_{|p|,|q| \leqslant m} \overline{a_{pq}\mathrm{D}^p v\cdot\overline{\mathrm{D}^q u}}\,\mathrm{d}x = \overline{a(v, u)}.$$

We apply Theorem 14.8 once to $A, b_1, \ldots, b_m$ and again to A^*, $b_1, \ldots, b_m$, obtaining

$$a(u, v) = \int_\Omega Au\bar{v}\,\mathrm{d}x - \sum_{j=1}^m \int_{\partial\Omega} C_ju\cdot\overline{b_jv}\,\mathrm{d}\sigma,$$

$$a^*(v, u) = \int_\Omega A^*v\cdot\bar{u}\,\mathrm{d}x - \sum_{j=1}^m \int_{\partial\Omega} C_j'v\cdot\overline{b_ju}\,\mathrm{d}\sigma, \tag{54}$$

or

$$\overline{a^*(v, u)} = \int_\Omega u\cdot\overline{A^*v}\,\mathrm{d}x - \sum_{j=1}^m \int_{\partial\Omega} b_ju\cdot\overline{C_j'v}\,\mathrm{d}\sigma.$$

Since by (53) $a^*(v, u) = \overline{a(u, v)}$, we obtain formula (51) from (54) by setting the quantities equal. ■

Proposition 14.1 *Let $b_1, \ldots, b_m$ be a Dirichlet system of order m and A a properly elliptic operator of order 2m. Then $A, b_1, \ldots, b_m$ fulfils the L.Š condition* 11.1.

Proof. We have to show that the initial value problem

$$A^{\mathrm{H}}\left(x;\xi',\frac{1}{\mathrm{i}}\frac{\mathrm{d}}{\mathrm{d}t}\right)v(t)=0,\quad t\geqslant 0,$$

$$b_1^{\mathrm{H}}(x;\xi',(1/\mathrm{i})(\mathrm{d}/\mathrm{d}t))v(0)=0,\ldots,b_m^{\mathrm{H}}(x;\xi',(1/\mathrm{i})(\mathrm{d}/\mathrm{d}t))v(0)=0,$$

admits only the solution $v(t)\equiv 0$ in $\mathscr{M}^+$. Now, observing that the principal parts b_j^{H} of b_j also form a Dirichlet system, we can apply Lemma 14.1 to b_j^{H}, $\mathrm{d}^{j-1}/\mathrm{d}t^{j-1}$, $j=1,\ldots,m$ (t-normal!), obtaining that the initial value problem above is equivalent to the 'Dirichlet Problem'

$$A^{\mathrm{H}}\left(x;\xi',\frac{1}{\mathrm{i}}\frac{\mathrm{d}}{\mathrm{d}t}\right)v(t)=0,\ t\geqslant 0,\ \frac{1}{\mathrm{i}}v(0)=0,\ldots,\left(\frac{1}{\mathrm{i}}\frac{\mathrm{d}}{\mathrm{d}t}\right)^{m-1}v(0)=0,$$

which by Example 13.2 satisfies the L.Š. condition 11.1. So in $\mathscr{M}^+$ both initial value problems only admit the solution $v(t)\equiv 0$. We can also apply the proof procedure of Theorem 13.5 to $(A,b_1,\ldots,b_m)$ and obtain from (52)

$$\operatorname{ind}(A,b_1,\ldots,b_m)=0.$$

We may correspondingly mix the Bs and Cs, and in this way determine other adjoint boundary value problems, see examples in §16.

If A is (formally) self-adjoint, $A=A^*$, we can put $C_j'=C_j$ and simplify formula (51) to

$$\int_\Omega Au\bar{v}\,\mathrm{d}x-\int_\Omega u\overline{Av}\,\mathrm{d}x=\sum_{j=1}^{m}\int_{\partial\Omega}[C_ju\overline{b_jv}-b_ju\overline{C_jv}]\,\mathrm{d}\sigma,$$

which means that both boundary value problems

$$(A,b_1,\ldots,b_m)\quad\text{and}\quad(A,C_1,\ldots,C_m)$$

are symmetric (see Corollary 14.1). Since by Theorem 14.8 the system $C_1,\ldots,C_m$ is normal, we can apply Corollary 14.1 and obtain:

Proposition 14.2 *Let $b_1,\ldots,b_m$ be a Dirichlet system and A formally self-adjoint, $A=A^*$. Then the boundary value problems*

$$(A,b_1,\ldots,b_m)\quad\textit{and}\quad(A,C_1,\ldots,C_m)$$

are self-adjoint in the sense of Definition 14.6, *or*

$$(A,b_1,\ldots,b_m)^*=(A,b_1,\ldots,b_m),\quad(A,C_1,\ldots,C_m)^*=(A,C_1,\ldots,C_m).$$

14.5 The antidual operator L' and the adjoint boundary value problem

In conclusion we indicate the connection between the antidual operator L' and the adjoint boundary value problem.

Let V be a boundary space and V^* its adjoint (see Definition 14.5). We denote by L_V the restriction of $A(x,\mathrm{D})$ to V, by $L^*_{V^*}$ the restriction of the adjoint $A^*(x,\mathrm{D})$ to V^*. If V is determined by $B=(b_1,\ldots,b_m')$, and V^* by $B^*=B'_1,\ldots,B'_{m'}$, that is,

$$V=W^{2m}(B),\quad V^*=W^{2m}(B^*),$$

we abbreviate $L_V=L_0$, $L^*_{V^*}=L^*_0$. We have the continuous maps

$$L_V:V\to L_2(\Omega),$$
$$L^*_{V^*}:V^*\to L_2(\Omega).$$

We consider the Gelfand triple, see §17,

$$V\subset L_2(\Omega)\subset V'$$

with the mappings

$$V\xrightarrow[L_V]{}L_2(\Omega)\xrightarrow[L'_V]{}V',$$

where L'_V is the antidual to L_V. We presuppose the existence of the inverses

$$L_V^{-1}:L_2(\Omega)\to V=\operatorname{im}L_V^{-1},$$
$$L^{*-1}_{V^*}:L_2(\Omega)\to V^*=\operatorname{im}L^{*-1}_{V^*},$$

and call the compositions

$$\begin{aligned} G_V&:L_2(\Omega)\xrightarrow[L_V^{-1}]{}V\subsetneq L_2(\Omega),\\ G_{V^*}&:L_2(\Omega)\xrightarrow[L^{*-1}_{V^*}]{}V^*\subsetneq L_2(\Omega), \end{aligned}\tag{55}$$

Green solution operators. Because

$$V\underset{\text{continuous}}{\subsetneq}W_2^{2m}(\Omega)\underset{\text{compact}}{\subsetneq}L_2(\Omega),$$
$$V^*\subsetneq W_2^{2m}(\Omega)\subsetneq L_2(\Omega),$$

they are compact $L^2\to L^2$ operators.

Theorem 14.10 *We have*

$$L'_V=L^*_{V^*}\quad on\ V^*\cap L_2(\Omega).$$

Let $G':L_2\to L_2$ be the antidual ($=$ adjoint) of $G:L_2\to L_2$. We have

$$G'_V=G_{V^*}.\tag{56}$$

If therefore (A,V) is self-adjoint: $A=A^*$, $V=V^*$, we have $G'_V=G_V$, thus the Green solution operator G_V is compact and self-adjoint in the sense of L_2-theory.

Proof. We take as duality in the Gelfand triple $V \subset L_2 \subset V'$ the scalar product $(\,,\,)_0$ from L_2, see §17, and because of (22) for $\varphi \in V^*$ and $\psi \in V$ have

$$(L'_V\varphi, \psi)_0 = (\varphi, L_V\psi)_0 = \int_\Omega \varphi \cdot A\psi \,\mathrm{d}x = \int_\Omega A^*\varphi\bar{\psi}\,\mathrm{d}x = (L^*_{V^*}\varphi, \psi)_0,$$

that is, $L'_V = L^*_{V^*}$ on V^*. Since $\mathscr{D}(\Omega) \subset V^*$ is dense in $L_2(\Omega)$, the dual operator L'_V is uniquely determined by $L^*_{V^*}$.

We now substitute $\varphi = G_{V^*}g$, $\psi = G_V g$ in

$$(L^*_{V^*}\varphi, \psi)_0 = (\varphi, L_V\psi)_0$$

and obtain

$$(f, G_V g)_0 = (L^*_{V^*} \circ G_V f, G_V g)_0 = (G_{V^*}f, L_V G_V g)_0 = (G_{V^*}f, g)_0$$

for all $f, g \in L_2(\Omega)$, which is $(G_V)' = G_{V^*}$. ■

We can also answer the question of the existence of a Green solution operator G_V.

Theorem 14.11 *Let $V = W^{2m}(B)$, with $B = (b_1, \ldots, b_m)$ normal. The Green solution operator $G_{L_0} = G_{W^{2m}(B)}$ (55) exists if and only if $\lambda = 0$ is not an eigenvalue of the boundary value problem* §13, (50) *and* $\operatorname{ind} L = 0$. *By Theorem* 13.4 *we again find ourselves in the realm of the Riesz–Schauder spectral theorem, see* §12.1. *Similarly for the existence of $G_{\lambda_0} = G_{(L_0 - \lambda_0)}$ if $\lambda = \lambda_0$ is not an eigenvalue.*

Proof. That the condition is sufficient for the existence of G_{L_0} one sees immediately from the proof of Theorem 13.4.

Necessity: if G_{L_0} exists, thus

$$L_0^{-1} : L_2(\Omega) \to W^{2m}(B) \tag{57}$$

represents an isomorphism, the boundary value problem

$$Au = 0, \quad b_1 u = 0, \ldots, \quad b_m u = 0,$$

can only have the trivial solution $u = 0$. Hence $\lambda = 0$ cannot be an eigenvalue, from which it follows that $\alpha(L) = 0$. If we can show that the boundary value problem §13, (1) is solvable for each $(f, g_1, \ldots, g_m)$ $H^0(\Omega, \partial\Omega)$, then $\beta(L) = 0$, hence $\operatorname{Ind} L = \alpha(L) - \beta(L) = 0$, and we are finished. Since the boundary value operators $b_j(x, \mathrm{D})$ are normal on $\partial\Omega$, then for $(g_1, \ldots, g_m) \in \times\, {}_2^{2m - m_j - 1/2}(\Omega)$ there is some $u_0 \in W_2^{2m}(\Omega)$ by Theorem 14.1 with $b_1 u_0 = g_1, \ldots, b_m u_0 = g_m$.

We put $\tilde{f} := f - Au_0 \in L_2(\Omega)$, we have $v := L_0^{-1}(\tilde{f}) \in W^{2m}(B)$, and see that

$u := v + u_0$ is a solution of §13, (1) for

$$Au = Av + Au_0 = f - Au_0 + Au_0 = f,$$

and

$$b_j u = b_j v + b_j u_0 = 0 + b_j u_0 = g_j, \quad j = 1, \ldots, m. \qquad \blacksquare$$

Since later (§17) we will once more return to the Green operator, we give the definition of a Green solution operator G (see Theorem 14.10) somewhat differently:

Definition 14.7 *We say that $G: L_2(\Omega) \to L_2(\Omega)$ is a Green solution operator for $L = (A, b_1, \ldots, b_m)$, if $u = Gf \in W_2^{2m}(\Omega)$ is the unique solution of the boundary value problem*

$$Au = f, \quad b_1 u = 0, \ldots, b_m u = 0.$$

We immediately have that $\operatorname{Im} G = W^{2m}(B)$ and that $G: L_2(\Omega) \to \operatorname{Im} G = W^{2m}(B)$ is inverse to L_0, see Theorem 14.10. By the open mapping theorem the continuity of $L_0^{-1}: L_2(\Omega) \to W^{2m}(B)$ follows from that of $L_0: W^{2m}(B) \to L_2(\Omega)$, therefore $G: L_2(\Omega) \xrightarrow[L_0^{-1}]{} W^{2m}(B) \subsetneq L_2(\Omega)$ is also continuous. In this way we see the equivalence with the definition of G (55) in Theorem 14.10.

In this section we have come to know three different definitions of a self-adjoint boundary value problem; assuming the existence of a Green operator G, we can show that they are all equivalent.

Theorem 14.12 *Let (A, V) be self-adjoint in the sense of Definition 14.6, that is, $A^* = A$, $V^* = V$, then for the operator L'_V antidual to L_V* we have

$$L'_V = L_V \text{ on } V = V^*. \tag{58}$$

If we assume the existence of a Green operator G, from (58) we can deduce

$$G' = G \text{ on } L_2(\Omega), \tag{59}$$

that is, G is continuous (also compact) and self-adjoint as a Hilbert space operator $L_2 \to L_2$; from (59) it again follows that

$$V^* = V. \tag{60}$$

Proof. By Theorem 14.10 we have $L'_v = L^*_{V^*} = L_V$, hence (58). Also (56) follows from Theorem 14.10, so that

$$G' = G_{V^*} = G_V = G,$$

with which we have proved (59). Now for the implication (59)⇒(60). We have by (56) and (55)

$$\operatorname{im} G' = \operatorname{im} G_{V^*} = V^*, \quad \operatorname{im} G = \operatorname{im} G_V = V,$$

hence if we use (59)

$$V = \operatorname{im} G = \operatorname{im} G' = V^*,$$

which is (60). ■

§15 The adjoint boundary value problem and the connection with the image space of the original operator

By the main theorem 13.1 an elliptic operator L is Fredholm, that is, the kernel of L is finite dimensional and the image space $\operatorname{im} L$ is closed and has finite codimension. This implies that in order for the boundary value problem $Lu = F$, or written out more fully

$$A(x, \mathrm{D})u = f \text{ on } \Omega, \quad b_j(x, \mathrm{D})u = g_j, \quad j = 1, \ldots, m, \text{ on } \partial\Omega$$

to be solvable, the given functions $(f, g_1, \ldots, g_m) =: F$ must satisfy finitely many conditions. The aim of this section is to make these conditions explicit. As a result we discover something which is already familiar to us for ordinary differential operators, that F must be 'orthogonal' to the kernel of the adjoint boundary value problem.

For this we recall that the general elliptic boundary value theory should be regarded as a C^∞-theory, that is, for the individual theorems we need (if they are stated in full generality, see below) high differentiability assumptions on the boundary $\partial\Omega$ and on the coefficients $A(x, \mathrm{D})$, $b_j(x, \mathrm{D})$, $j = 1, \ldots, m$. The reasons for these assumptions lie partly in Theorem 15.1 and partly in the Green formulae, Theorem 14.2, respectively Theorem 14.4, which give us the adjoint boundary conditions B'. In individual cases, for example in the Dirichlet problem, or in the Neumann problem for the Laplace operator, we know the adjoint boundary value problem, we do not need to follow the general path, and far weaker differentiability assumptions suffice, see also §16.

We first give the general assumptions under which all the theorems in this section hold. For the individual theorems we also give the minimal assumptions under which they hold. Here we understand 'minimal' in the context of C-spaces; it is of course possible to weaken the assumptions further, but everything becomes more complicated since we must work with other classes of spaces and apply different methods of proof. We assume:

$$\Omega \text{ is bounded and belongs to } C^{6m-2,1}. \tag{1}$$

Let $A(x, D)$ be an elliptic differential operator of order $2m$ on $\bar{\Omega}$ with coefficients

$$a_\alpha(x) \in C^{2m+|\alpha|}(\bar{\Omega}). \tag{2}$$

Let $B = (b_j(x, D), j = 1, \ldots, m)$ be a normal boundary value system on $\partial\Omega$, which with A satisfies the L.Š. condition 11.1. Let the coefficients of B be such that

$$b_{j,s}(x) \in C^{6m-2-m_j}(\tilde{\Omega}), \quad |s| \leqslant m_j \leqslant 2m-1, j = 1, \ldots, m. \tag{3}$$

By Theorem 14.7 there then exists an adjoint boundary value system $B' = (B'_j(x, D), j = 1, \ldots, m)$ with the following properties:

1. The coefficients of B' satisfy

$$B'_{j,s}(x) \in C^{4m-1-\operatorname{ord} B'_j}(\tilde{\Omega}), \quad |s| \leqslant \operatorname{ord} B'_j \leqslant 2m-1, \quad j = 1, \ldots, m. \tag{4}$$

2. B' is normal and determines the space $W^{2m}(B)^*$, see Definitions 14.3 and 14.5, therefore $W^{2m}(B)^* = W^{2m}(B')$.
3. $A^*, B'_1, \ldots, B'_m$ satisfies the L.Š. condition 11.1 on $\partial\Omega$.

For this section we fix the boundary value operators

$$B'_1(x, D), \ldots, B'_m(x, D),$$

and similarly the boundary value operators

$$S_1, \ldots, S_m, \quad T_1, \ldots, T_m$$

which we have obtained by means of Theorems 14.2 and 14.7. We call

$$A^* v = h \text{ on } \Omega, \quad B'_1 v = \psi_1, \ldots, B'_m v = \psi_m, \text{ on } \partial\Omega, \tag{5}$$

the adjoint boundary value problem. Since A^* is elliptic on $\bar{\Omega}$ ($a^*_s \in C^{2m}(\bar{\Omega})$) and because of the L.Š. condition 3, the adjoint problem is again elliptic on $(\Omega, \partial\Omega)$ and we can apply the main Theorem 13.1 (with $k = 2m-2$, $\kappa = 1$) to (5). In this way we obtain statements (b), (c) and (d) of the main Theorem 13.1, in particular therefore the nullspace N^* of (5) is finite dimensional. We consider the operator

$$L^* = (A^*, B'_1, \ldots, B'_m), \quad m'_j = \operatorname{ord} B'_j;$$

$$L^* : W_2^{2m}(\Omega) \to L_2(\Omega) \bigtimes_{j=1}^{m} W_2^{2m-m'_j-1/2}(\partial\Omega) \text{ is continuous,} \tag{6}$$

see (2) and (4). We need the following subspaces of $W_2^{2m}(\Omega)$:

$$N := \ker L = \{u \in W_2^{2m}(\Omega): Au = 0, b_1 u = 0, \ldots, b_m u = 0\},$$
$$N^* := \ker L^* = \{v \in W_2^{2m}(\Omega): A^* v = 0, B'_1 v = 0, \ldots, B'_m v = 0\},$$

$$\left.\begin{aligned} M &:= \left\{ v \in W_2^{2m}(\Omega): \int_\Omega u\bar{v}\,\mathrm{d}x = 0, \text{ for all } u \in N \right\}, \\ M^* &:= \left\{ w \in W_2^{2m}(\Omega): \int_\Omega w\bar{v}\,\mathrm{d}x = 0, \text{ for all } v \in N^* \right\}, \end{aligned}\right. \tag{7}$$

$$W^{2m}(B) := \{u \in W_2^{2m}(\Omega): b_1 u = 0, \ldots, b_m u = 0\},$$
$$W^{2m}(B') = W^{2m}(B)^* = \{v \in W_2^{2m}(\Omega): B'_1 v = 0, \ldots, B'_m v = 0\}.$$

All these subspaces are closed; this is easy to see; N and N^* are finite dimensional, this follows from the main theorem 13.1(c) applied once to $(A, b_1, \ldots, b_m)$ and again to $(A^*, B'_1, \ldots, B'_m)$.

Because $W_2^{2m}(\Omega) \subset L_2(\Omega)$ we can embed N and N^* in $L_2(\Omega)$; as finite dimensional subspaces N and N^* are also closed in $L_2(\Omega)$, and we have the orthogonal decompositions

$$L_2(\Omega) = N \oplus N^\perp, \quad L_2(\Omega) = N^* \oplus N^{*\perp}, \tag{7a}$$

by which the spaces $N^\perp$ and $N^{*\perp}$ are defined. We have

$$M = N^\perp \cap W_2^{2m}(\Omega), \quad M^* = N^{*\perp} \cap W_2^{2m}(\Omega). \tag{7b}$$

We consider the sesquilinear form

$$[u, v] =: \int_\Omega A^* u \overline{A^* v}\,\mathrm{d}x + \sum_{j=1}^m (B'_j u, B'_j v)_j, \tag{8}$$

where $(\ ,\ )_j$ is the scalar product on $W_2^{2m-m_j-1/2}(\partial\Omega)$. We see that $[u, v]$ is continuous on $W_2^{2m}(\Omega) \times W_2^{2m}(\Omega)$. Since, as already stated, by our assumptions $A^*, B'_1, \ldots, B'_m$ is elliptic, we can apply the main Theorem 13.1 and find the *a priori* estimate (d) (with $l = 1$), which with the help of (8) may be written as

$$c_1 \|u\|_{2m}^2 \leqslant [u, u] + \|u\|_0^2, \quad u \in W_2^{2m}(\Omega), \ c_1 > 0,$$

or, together with the continuity $(0 \leqslant [u, u] = \mathrm{Re}(u, u])$,

$$c_1 \|u\|_{2m}^2 \leqslant \mathrm{Re}(u, u] + \|u\|_0^2 \leqslant c_2 \|u\|_{2m}^2, \quad u \in W_2^{2m}(\Omega). \tag{9}$$

(9) means that $[u, v]$ is V-coercive for the space pair $W_2^{2m}(\Omega) \subsetneq L_2(\Omega)$, see Definition 17.4.

Solution of V-coercive equations have regularity properties; we have:

Theorem 15.1 *Let $u \in W_2^{2m}(\Omega)$ and $f \in L_2(\Omega)$, let the equation*

$$[u, v] = \int_\Omega f \bar{v} \, dx = (f, v)_0 \tag{10}$$

be satisfied for all $v \in W_2^{2m}(\Omega)$. Then $u \in W_2^{4m}(\Omega)$.

As assumptions for this theorem we need: $\Omega \in C^{4m,1}$, the 1 because of normal coordinates, see Theorem 2.12, $a_s^*(x) \in C^{2m}(\bar{\Omega})$, $B'_{j,s}(x) \in C^{4m-1-\operatorname{ord} B'_j}(\Omega)$; it suffices to assume $B'_j \in C^{2m}(\bar{\Omega})$. However, in order to obtain $B'_j \in C^{4m-m_j-1}(\bar{\Omega})$, we must apply Theorem 14.2 and there write $k = 2m-1$, that is, $\Omega \in C^{6m-2,1}, b_j \in C^{6m-2-m_j}$. We will present the proof of Theorem 15.1 later in §20, where we consider the regularity of V-coercive equations, see Theorem 20.5.

If we restrict u, v to the subspace M^* of $W_2^{2m}(\Omega)$, see definition (7), then $[u, v]$ is even V-elliptic.

Theorem 15.2 *For all $u \in M^*$* (7) *we have the inequality*

$$\|u\|_{2m}^2 \leqslant c[u, u], \tag{11}$$

that is, $[u, v]$ is V-elliptic for the space pair $M^* \subsetneq L_2(\Omega)$, see Definition 17.3.

Proof. We argue by contradiction. Let (11) be false, then there exists a constant c_1 and a sequence $u_n \in M^*$ with

$$\|u_n\|_{2m} \to \infty \quad \text{and} \quad [u_n, u_n] \leqslant c_1 \quad \text{for all } n.$$

We put

$$w_n = \frac{u_n}{\|u_n\|_{2m}} \in M^*$$

and have

$$[w_n, w_n] \to 0 \quad \text{and} \quad \|w_n\|_{2m} = 1 \quad \text{for all } n. \tag{12}$$

Since by Theorem 7.2 the space $W_2^{2m}(\Omega)$ is compactly embedded in $L_2(\Omega)$, there exists some subsequence of w_n, we again denote it simply by w_n, with the property

$$\|w_n - w_k\|_0 \to 0 \quad \text{as } n, k \to \infty. \tag{13}$$

Substituting (12) and (13) in (9) gives

$$\|w_k - w_n\|_{2m} \to 0 \quad \text{as } n, k \to \infty.$$

But a closed subspace of a complete space M^* is again complete, and there exists some $w \in M^*$ with

$$\| w_n \to w \|_{2m} \to 0 \quad \text{as } n \to \infty. \tag{14}$$

Since $[u, v]$ is continuous, it follows from (13) with (12) that

$$[w, w] = 0,$$

which with definition (8) gives $w \in N^*$.

On the other hand $w \in M^*$, see (7),

$$\int_\Omega w \cdot \bar{w} \, dx = 0,$$

that is, $w = 0$, contradicting (12): $\| w \|_{2m} = 1$. ∎

Theorem 15.3 *Let $f \in L_2(\Omega)$. The boundary value problem*

$$Au = f, \text{ on } \Omega, \ b_1 u = 0, \ldots, b_m u = 0, \text{ on } \partial\Omega, \tag{15}$$

has at least one solution $u \in W_2^{2m}(\Omega)$ if and only if $f \in N^{*\perp}$, see (7) and (7a). Here we understand the equation $Au = f$ in the distributional sense $\mathscr{D}'(\Omega)$.

For this theorem we first need the assumptions of Theorem 15.1

$$\Omega \in C^{4m,1}, \quad a_s^*(x) \in C^{2m}(\bar{\Omega}), \quad B'_{j,s}(x) \in C^{4m-1-\operatorname{ord} B'_j}(\bar{\Omega}),$$

and secondly, since we wish to apply Theorem 14.3, we must assume for (15) that

$$a_s(x) \in C^{|s|}(\bar{\Omega}), \quad b_{j,s}(x) \in C^{2m-m_j}(\bar{\Omega}).$$

All these conditions are of course satisfied, if the premises stated at the beginning of this section are satisfied.

Proof. Green's formula §14, (12) shows that the condition $f \in N^{*\perp}$ is necessary. We show the reverse implication. Let $f \in N^{*\perp}$, we consider the weak equation

$$[g, \varphi'] = (f, \varphi')_0 \quad \text{for all } \varphi' \in M^*. \tag{16}$$

Since by Theorem 15.2 $[u, v]$ is M^*-elliptic, by Theorem 17.11 (16) has a solution g for each f. We can decompose each $\varphi \in W_2^{2m}(\Omega)$ as $\varphi = \psi' + \varphi''$ where $\varphi' \in M^*$ and $\varphi'' \in N^*$; for we have (7a) $L_2(\Omega) = N^* \oplus N^{*\perp}$; since, however, in addition $\varphi' = \varphi - \varphi'' \in W_2^{2m}(\Omega)$, we have

$$\varphi' \in N^{*\perp} \cap W_2^{2m}(\Omega) = M^*, \text{ see (7b).}$$

By definition of the form $[u, v]$, see (8), for $\varphi'' \in N^*$ $[g, \varphi''] = 0$ and $f \in N^{*\perp}$

gives $(f, \varphi'')_0 = 0$. Therefore we have

$$[g, \varphi] = [g, \varphi'] = (f, \varphi')_0 = (f, \varphi)_0 \quad \text{for all } \varphi \in W_2^{2m}(\Omega), \tag{17}$$

and we can apply Theorem 15.1, by which $g \in W_2^{4m}(\Omega)$. We put $u := A^* g \in W_2^{2m}(\Omega)$ and show that u is a solution of (15). For $\varphi \in \mathcal{D}(\Omega)$ partial integration and (17) give

$$\int_\Omega Au\bar{\varphi}\,dx = \int_\Omega u\overline{A^*\varphi}\,dx = \int_\Omega A^*g\overline{A^*\varphi}\,dx = [g, \varphi] = \int_\Omega f\bar{\varphi}\,dx$$

which is $Au = f$ in the distributional sense $\mathcal{D}'(\Omega)$.

We must still check the boundary conditions: again because of (17) we have, putting $f = Au$

$$\int_\Omega u\overline{A^*\varphi}\,dx = \int_\Omega A^*g\overline{A^*\varphi}\,dx + 0 = [g, \varphi] = \int_\Omega f\bar{\varphi}\,dx = \int_\Omega Au\bar{\varphi}\,dx$$

for all $\varphi \in W^{2m}(B')$.

By Definition 14.5 this implies that

$$u \in W^{2m}(B')^* = W^{2m}(B), \tag{18}$$

the last equation because of (24b) from Theorem 14.3. However, (18) means that the boundary conditions

$$b_1 u = 0, \dots, b_m u = 0$$

are satisfied on $\partial\Omega$, so we have shown (15). ∎

We can dualise Theorem 15.3; we leave the straightforward proof to the reader; it is entirely analogous to that of Theorem 15.3.

Theorem 15.4 *Let $f \in L_2(\Omega)$. The boundary value problem*

$$A^*v = f \ (on\ \Omega), \quad B'_1 v = 0, \dots, B'_m v = 0 \ (on\ \partial\Omega), \tag{19}$$

has at least one solution v in $W_2^{2m}(\Omega)$ if and only if $f \in N^\perp$.

The operator

$$L: W_2^{2m}(\Omega) \to L_2(\Omega) \times_{j=1}^{m} W_2^{2m-m_j-1/2}(\partial\Omega), \quad \text{continuous,}$$

always possesses a formally dual operator L'

$$L': \left[L_2(\Omega) \times_{j=1}^{m} W_2^{2m-m_j-1/2}(\partial\Omega) \right]' \to [W_2^{2m}(\Omega)]'. \tag{20}$$

We study the connection between L' and the operator L^* from (6).

Theorem 15.5 *Let N' be the kernel of* (20), $N' = \ker L'$. *We have*

$$N' = \{(v, T_1 v, \ldots, T_m v) : v \in N^*\}, \tag{21}$$

where $T_1, \ldots, T_m$ are the boundary value operators from the Green formula. As differentiability assumptions for this theorem we can take the assumptions of Theorem 15.3, *or simply the following global assumptions*:

$$\Omega \in C^{6m-2,1}, \quad a_\alpha \in C^{2m+|\alpha|}(\bar{\Omega}), \quad b_j \in C^{6m-2-m_j}(\bar{\Omega}).$$

Proof. Let $\langle , \rangle$ be the duality brackets between the spaces $W_2^{2m}(\Omega)$ and $W_2^{2m}(\Omega)'$, see §17; and correspondingly between

$$L_2(\Omega) \mathop{\times}_{j=1}^{m} W_2^{2m-m_j-1/2}(\partial\Omega) \quad \text{and} \quad \left[L_2(\Omega) \mathop{\times}_{j=1}^{m} W_2^{2m-m_j-1/2}(\partial\Omega) \right]',$$

then L' is defined by

$$\langle L'\Phi, \Psi \rangle = \langle \Phi, L\Psi \rangle. \tag{22}$$

We take $\Phi = (v, T_1 v, \ldots, T_m v)$, $v \in N^*$ and show that $\langle \Phi, L\Psi \rangle = 0$ for all $\Psi \in W_2^{2m}(\Omega)$. We will then have shown that $\Phi \in N'$ and hence one inclusion of (21). We have

$$\langle \Phi, L\Psi \rangle := \int_\Omega v \overline{A\psi} \, dx + \sum_{j=1}^{m} \langle T_j v, b_j \psi \rangle_j. \tag{23}$$

Since the three spaces $W_2^{2m-m_j-1/2}(\partial\Omega)$, $L_2(\partial\Omega)$ and $[W_2^{2m-m_j-1/2}(\partial\Omega)]'$ form a Gelfand triple on $\partial\Omega$, see Theorem 17.4, we can express the duality brackets $\langle , \rangle_j$ by means of the scalar product in $L_2(\partial\Omega)$, and have

$$\langle \Phi, L\Psi \rangle = \int_\Omega v \overline{A\psi} \, dx + \sum_{j=1}^{m} \int_{\partial\Omega} T_j v \cdot \overline{b_j \varphi} \, d\sigma. \tag{23a}$$

An application of the Green formula §14, (12) gives

$$\langle \Phi, L\Psi \rangle = \int_\Omega A^* v \cdot \bar{\psi} \, dx + \sum_{j=1}^{m} \int_{\partial\Omega} B_j' v \overline{S_j \psi} \, d\sigma = 0, \quad \text{hence } v \in N^*. \tag{23b}$$

and so we have proved that $\langle \Phi, L\Psi \rangle = 0$.

Suppose now that $\Phi = (v, \varphi_1, \ldots, \varphi_m) \in N'$. Here

$$v \in [L_2(\Omega)]' = L_2(\Omega) \quad \text{and} \quad \varphi_j \in [W_2^{2m-m_j-1/2}(\partial\Omega)]'.$$

We show first that in $\mathscr{D}'(\Omega)$ we have

$$A^* v = 0. \tag{24}$$

By definition (22) we have

$$0 = \langle \Phi, L\Psi \rangle = \int_\Omega v\overline{A\psi} + \sum_{j=1}^m \langle \varphi_j, b_j\psi \rangle_j \quad \text{for all } \psi \in W_2^{2m}(\Omega), \tag{25}$$

or

$$\int_\Omega v\overline{A\psi}\,\mathrm{d}x = 0 \quad \text{for all } \psi \in \mathscr{D}(\Omega)\ (b_j\psi = 0!),$$

which is (24). We decompose v in $L_2(\Omega) = N^* \oplus N^{*\perp}$

$$v = v_1 + v_2, \quad v_1 \in N^*, \quad v_2 \perp N^*. \tag{26}$$

An application of the Green formula §14, (12) to the functions $u \in W^{2m}(B)$ and $v_1 \in N^*$ gives

$$\int_\Omega Au\bar{v}_1\,\mathrm{d}x = 0 \quad \text{for all } u \in W^{2m}(B). \tag{27}$$

From (25) it follows that

$$\int_\Omega Au\bar{v}\,\mathrm{d}x = 0 \quad \text{for all } u \in W^{2m}(B),$$

and with (27)

$$\int_\Omega Au\cdot\bar{v}_2\,\mathrm{d}x = 0 \quad \text{for all } u \in W^{2m}(B). \tag{28}$$

We show now that $v_2 = 0$, with which we will have proved

$$v = v_1 \in N^* \subset W_2^{2m}(\Omega). \tag{29}$$

Let h be arbitrary in $L_2(\Omega)$. Then there exists some $h_1 \in N^*$ and some $u_1 \in W^{2m}(B)$ with

$$h = Au_1 + h_1, \tag{30}$$

for $h - h_1 \in N^{*\perp}$, and by Theorem 15.3 the equation $Au_1 = h - h_1$ is solvable in $W^{2m}(B)$. With (28) and (26) the decomposition (30) gives

$$\int_\Omega h\bar{v}_2 = \int_\Omega Au_1\bar{v}_2\,\mathrm{d}x + \int_\Omega h_1\bar{v}_2\,\mathrm{d}x = 0 \quad \text{for all } h \in L_2(\Omega),$$

and so we have proved (29). We now apply the Green formula §14, (12) to $\int_\Omega v\overline{A\varphi}$ in (25), where we use (29); we obtain

$$\sum_{j=1}^m \langle b_j\psi, \varphi_j - T_jv \rangle_j + \sum_{j=1}^m \int_{\partial\Omega} S_j\psi\overline{B_j'v}\,\mathrm{d}\sigma = 0 \quad \text{for all } \psi \in \psi W^{2m}(\Omega). \tag{31}$$

We interpret the functional φ_j in the sense of the Riesz theorem, therefore $\varphi_j \in W_2^{2m-m_j-1/2}(\partial\Omega)$, and we solve the system of equations (use Theorem 14.1)

$$b_1\psi_0 = \varphi_1 - T_1 v, \ldots, b_m\psi_0 = \varphi_m - T_m v, \quad S_1\psi_0 = B'_1 v, \ldots, S_m\psi_0 = B'_m v,$$

here $(b_1, \ldots, b_m, S_1, \ldots, S_m)$ is a Dirichlet system. ψ_0 substituted in (31) gives $\varphi_1 = T_1 v, \ldots, \varphi_m = T_m v, v \in N^*$, and so we have completely proved our theorem. ■

We can characterise explicitly the image of the operator L, that is, say precisely when the boundary value problem $Lu = F = (f, g_1, \ldots, g_m)$ is solvable.

Theorem 15.6 *Suppose that the following assumptions are satisfied*

$$\Omega \in C^{6m-2,1}, \quad a_\alpha(x) \in C^{2m+|\alpha|}(\bar{\Omega}), \quad b_{j,s}(x) \in C^{6m-2-m_j}(\bar{\Omega}).$$

The boundary value problem

$$Au = f \quad (on\ \Omega), \quad b_1 u = g_1, \ldots, b_m u = g_m \quad (on\ \partial\Omega),$$

where $f \in L^2(\Omega)$, $g_j \in W_2^{2m-m_j-1/2}(\partial\Omega)$, $j = 1, \ldots, m$, *possesses at least one solution* $u \in W_2^{2m}(\Omega)$ *if and only if*

$$\int_\Omega f\bar{v}\,dx + \sum_{j=1}^m \int_{\partial\Omega} g_j \overline{T_j v}\,d\sigma = 0 \quad \text{for all } v \in N^*. \tag{32}$$

Here N^* *is the nullspace* (7) *of the adjoint boundary value problem* (5), *that is,*

$$A^* v = 0,\ on\ \Omega; \quad B'_1 v = 0, \ldots, B'_m v = 0,\ on\ \partial\Omega.$$

If $(v_1, \ldots, v_n)$ *is a basis of* N^* *then we can replace* (32) *by finitely many conditions*

$$\int_\Omega f\bar{v}_k\,dx + \sum_{j=1}^m \int_{\partial\Omega} g_j \overline{T_j v_k}\,d\sigma = 0, \quad k = 1, \ldots, n.$$

Proof. By theorems from functional analysis (see §12, Theorem 12.3) we have $\operatorname{im} L = (\ker L')^\perp = N'^\perp$, or $(f, g_1, \ldots, g_m) \in L$ if and only if

$$\langle (f, g_1, \ldots, g_m), (v, \varphi_1, \ldots, \varphi_m) \rangle = 0 \quad \text{for all } (v, \varphi_1, \ldots, \varphi_m) \in N', \tag{33}$$

here $\langle , \rangle$ is the duality bracket of (22). Writing out (33) gives

$$(f, v)_0 + \sum_{j=1}^m \langle g_j, \varphi_j \rangle_j = 0,$$

and applying Theorem 15.5 gives

$$(f, v)_0 + \sum_{j=1}^m \langle g_j, T_j v \rangle_j = 0 \quad \text{for all } v \in N^*.$$

Since, as already used, we are dealing with Gelfand triples, we can replace the duality brackets $\langle , \rangle_j$ by scalar products from $L_2(\partial\Omega)$, and have

$$\int_\Omega f\bar{v}\,\mathrm{d}x + \sum_{j=1}^m \int_{\partial\Omega} g_j\overline{T_j v}\,\mathrm{d}\sigma = 0 \quad \text{for all } v\in N^*,$$

which is condition (32). ■

With Theorem 15.5 we have actually proved the following:

Theorem 15.7 *We have*

$$\dim\ker L' = \dim\ker L^*, \tag{34}$$

$$\dim\operatorname{coker} L' = \dim\operatorname{coker} L^*, \tag{35}$$

from which it follows that

$$\operatorname{ind} L^* = \operatorname{ind} L' = -\operatorname{ind} L, \tag{36}$$

that is, L^* completely realises the dual operator L', see also Theorem 14.10.

Proof. (34) was proved by means of Theorem 15.5, and we need only show (35) – everything else is clear. Because of $L^{**} = L$ (Green's formula §14, (24c)) and (34) we have

$$\dim\operatorname{coker} L' = \dim\ker L = \dim\ker(L^*)^* = \dim\ker(L^*)' = \dim\operatorname{coker} L^*. \quad ■$$

Corollary 15.8 *If the boundary value problem is self-adjoint, $L^* = (A, b_1, \ldots, b_m)^* = (A, b_1, \ldots, b_m) = L$, the index is equal to zero*: $\operatorname{ind} L = 0$.

This follows immediately from (36).

For normal, self-adjoint boundary value problems $(A, b_1, \ldots, b_m)$ it is possible to prove the existence of a Green solution operator and also the full spectral theorem. We begin with a general fact.

Proposition 15.1 *Suppose that the boundary problem $A, b_1, \ldots, b_m$ satisfies the L.Š. condition* 11.1. *Denoting by $L_V = L_0$ the restriction of A to the boundary space $V = W^{2m}(B)$, we assert that the Hilbert space operator $L_0: L_2(\Omega) \to L_2(\Omega)$ (it is not continuous!) with the domain of definition $D(L_0) = V = W^{2m}(B)$, is closed, and $D(L_0)$ is dense in $L_2(\Omega)$.*

Proof. Let $W^{2m}(B) \ni f_n \to f$ in L_2 and $Af_n \to g$ in L_2, then by the *a priori* estimate (d) from Theorem 13.1 we have

$$\|f_n - f_k\|_{2m} \leqslant C\{\|Af_n - Af_k\|_0 + \|f_n - f_k\|_0\},$$

so f_n is a Cauchy sequence in $W_2^{2m}(\Omega)$ and we obtain $f \in V = W^{2m}(B)$ because $W^{2m}(B)$ is closed in $W_2^{2m}(\Omega)$, and $Af = g$ by continuity of A: $W_2^{2m} \to L_2$. Since $\mathscr{D}(\Omega) \subset W^{2m}(B)$, $D(L_0)$ obviously is dense in $L_2(\Omega)$.

Theorem 15.9 *Let the boundary value problem $A, b_1, \ldots, b_m$ fulfil the L.Š. condition* 11.1, *let the boundary system $b_1, \ldots, b_m$ be normal and let $(A, W^{2m}(B))$ be self-adjoint (see Definition* 14.6). *Then:*

(a) *There exists a $\lambda_0 \in \mathbb{R}$ not belonging to the spectrum of L_0.*
(b) *$L_0 : L_2(\Omega) \to L_2(\Omega)$, $D(L_0) = W^{2m}(B)$, is closed and L_2-self-adjoint in the sense of von Neumann.*
(c) *The Green solution operator $G_\lambda : L_2(\Omega) \to L_2(\Omega)$ exists for each $\lambda \notin \mathrm{Sp} L_0$ and is compact.*

Therefore the spectrum $\mathrm{Sp} L_0$ *of L_0 is real, discrete, and consists of countably many eigenvalues $\{\lambda_n\}$ with $|\lambda_n| \to \infty$, the eigenspaces $E(\lambda_n)$ are finite-dimensional, orthogonal to each other and complete in $L_2(\Omega)$. Also all the other assertions of the Riesz–Schauder spectral theorem* 12.1 *are true.*

Proof. By identification we see that

$$L_V = L_0 = (A, 0, \ldots, 0) = (A, B_1, \ldots, B_m)|_{W^{2m}(B)} = L|_{W^{2m}(B)},$$

and that the operator

$$L_0 - \lambda : W^{2m}(B) \to L_2(\Omega)$$

is continuous, where λ is real and $W^{2m}(B)$ has the topology of $W_2^{2m}(\Omega)$. We are going to show, that $L_0 - \lambda$ is Fredholm with index 0. In general, the restriction of a Fredholm operator is not necessarily Fredholm.

We have $\ker(L_0 - \lambda) = \ker(L - \lambda)$, $(D(L_0) = W^{2m}(B)!)$, therefore

$$d = \dim \ker(L_0 - \lambda) < \infty.$$

To get the dimension of the cokernel of $L_0 - \lambda$ we shall use Theorem 15.6. The system $b_1, \ldots, b_m$ being normal, we can apply Theorem 14.4 obtaining a normal system $B'_1, \ldots, B'_m$ and Green's formula §14, (12), it may happen that $B \neq B'$, see Remark 14.1. By definition, the dimension of the cokernel of $L_0 - \lambda$ is the number of constraints on f such that

$$L_0 u - \lambda u = f, \text{ has a solution } u \in W^{2m}(B),$$

or that

$$Au - \lambda u = f, \quad b_1 u = 0, \ldots, b_m u = 0, \quad u \in W_2^{2m}(\Omega)$$

is solvable. By Theorem 15.6 those constraints are

$$0=\int_{\Omega}\overline{v_l}\cdot f\,\mathrm{d}x=\int_{\Omega}\overline{v_l}\cdot f\,\mathrm{d}x+\sum_{j=1}^{m}\int_{\partial\Omega}\overline{T_jv_l}\cdot 0\,\mathrm{d}\sigma,\quad l=1,\ldots,d^*,\tag{37}$$

where v_l, $l=1,\ldots,d^*$, is a basis of $\ker(L_0^*-\bar{\lambda})$, that is,

$$(A^*-\bar{\lambda})v_l=(A-\lambda)v_l=0,\quad B'_1v_l=0,\ldots,B'_mv_l=0,\tag{38}$$

here we use our assumption that $\lambda\in\mathbb{R}$. Writing (38) as

$$(A-\lambda)v_l=0,\quad v_l\in W^{2m}(B'),\quad l=1,\ldots,d^*,$$

and using Theorem 14.4 and Definition 14.6, that is, $W^{2m}(B')=W^{2m}(B)^*=W^{2m}(B)$, we obtain, that problem (38) is equivalent to

$$(A-\lambda)v_l=0,\quad b_1v_l=0,\ldots,b_mv_l=0.$$

Thus we get $d^*=d$ in (37), which means

$$d^*=\dim\operatorname{coker}(L_0-\lambda)=d<\infty,$$

and

$$\operatorname{ind}(L_0-\lambda)=0\quad\text{for }\lambda\in\mathbb{R},$$

thus showing that $L_0-\lambda$ is Fredholm.

We decompose the spectrum of L_0 into

$$\mathrm{Sp}L_0=\mathrm{Sp}_pL_0\cup\mathrm{Sp}_rL_0\cup\mathrm{Sp}_cL_0,$$

see §13.2, where Sp_p is the point spectrum (eigenvalues), Sp_c the continuous spectrum and Sp_r the residual. Now, $L_0-\lambda$ for $\lambda\in\mathbb{R}$ being Fredholm, we have

$$\mathrm{Sp}_cL_0\cap\mathbb{R}=\varnothing,$$

and from $\operatorname{ind}(L_0-\lambda)=0$ we obtain

$$\mathrm{Sp}_rL_0\cap\mathbb{R}=\varnothing,$$

concluding that

$$\mathrm{Sp}L_0\cap\mathbb{R}=\mathrm{Sp}_pL_0\cap\mathbb{R}.$$

Since $A=A^*$, for eigenvalues λ and eigenfunctions φ, we have

$$\lambda(\varphi,\varphi)_0=(A\varphi,\varphi)_0=(\varphi,A\varphi)_0=\bar{\lambda}(\varphi,\varphi)_0,$$

where $\varphi\in W^{2m}(B)=W^{2m}(B)^*$, thus $\lambda=\bar{\lambda}$; and we have

$$\mathrm{Sp}L_0\cap\mathbb{R}=\mathrm{Sp}_pL_0.$$

Now, there must exist some $\lambda_0\in\mathbb{R}$ not belonging to the spectrum of L_0:

$\lambda_0 \notin \mathrm{Sp} L_0$. Suppose to the contrary, that

$$\mathrm{Sp} L_0 \cap \mathbb{R} = \mathrm{Sp}_p L_0 = \mathbb{R};$$

the normed eigenvectors $\varphi_\lambda, \varphi_\mu$ belonging to different real eigenvalues $\lambda \neq \mu$ and being orthogonal, we have

$$\lambda(\varphi_\lambda, \varphi_\mu)_0 = (A\varphi_\lambda, \varphi_\mu)_0 = (\varphi_\lambda, A\varphi_\mu)_0 = \mu(\varphi_\lambda, \varphi_\mu)_0 \Rightarrow (\varphi_\lambda, \varphi_\mu)_0 = 0,$$

and we obtain an uncountable, orthonormal system $(\varphi_\lambda)_{\lambda \in \mathbb{R}}$, in $L_2(\Omega)$, which is a contradiction because $L_2(\Omega)$ has a countable, orthonormal basis. In establishing $\lambda_0 \notin \mathrm{Sp} L_0$, thus verifying (a), we proved the existence of a continuous inverse, or Green solution operator

$$G_{\lambda_0} = (L_0 - \lambda_0)^{-1} : L_2(\Omega) \to W^{2m}(B).$$

Compactly embedding $W_2^{2m}(\Omega)$ into $L_2(\Omega)$, we obtain

$$G_{\lambda_0} : L_2(\Omega) \to W^{2m}(B) \subsetneq L_2(\Omega) \quad \text{compact.}$$

Now, from the presupposed self-adjointness of $(A, b_1, \ldots, b_m)$ it follows that

$$L_0^* = (A^*, 0, \ldots, 0) = (A, 0, \ldots 0) = L_0,$$

since

$$D(L_0^*) = W^{2m}(B)^* = W^{2m}(B) = D(L_0),$$

and by Theorem 14.10 we obtain

$$G'_{\lambda_0} = G^*_{\lambda_0} = G_{\lambda_0}.$$

Having at our disposal a compact L_2-self-adjoint resolvent, or Green solution operator G_{λ_0} of L_0 we obtain the conclusions (c) and (b) from well-known theorems (see Weidmann [1]). Proposition 15.1 implies that the operator L_0 is closed. ■

We can also prove a partial converse to Theorem 15.9.

Theorem 15.10 *Let $b_1, \ldots, b_m$ be normal, $V = W^{2m}(B)$ and the operator $L_0: V \to$ L_2-self-adjoint. Then $(A, b_1, \ldots, b_m)$ satisfies the L.Š. condition, so by Theorem 15.9 it has a compact resolvent and its spectrum is discrete.*

Proof. By von Neumann's spectral theorem, see Weidmann [1], we have $\mathrm{Sp} L_0 \subset \mathbb{R}$, and for $\lambda \in \mathbb{C}$, $\mathrm{Im}\, \lambda \neq 0$, there exists a continuous resolvent or Green solution operator

$$G_\lambda : L_2(\Omega) \to V \subset L_2(\Omega). \tag{39}$$

We consider the operator $L-\lambda=(A-\lambda, b_1,\ldots,b_m)$:

$$W_2^{2m}(\Omega)\to L_2(\Omega)\times\prod_{j=1}^{m} W_2^{2m-m_j-1/2}(\partial\Omega).$$

Existence of the resolvent G_λ means that

$$\ker(L-\lambda)=(0),$$

and we now show that $L-\lambda$ is surjective, or

$$\operatorname{coker}(L-\lambda)=(0).$$

Considering the boundary value problem

$$(A-\lambda)u=f,\quad b_1u=g_1,\ldots,b_mu=g_m, \tag{40}$$

we first solve for some u_0 with $b_1u_0=g_1,\ldots,b_mu_0=g_m$, which by the normality of $b_1,\ldots,b_n$ and Theorem 14.1 is possible. Secondly we solve the problem

$$(A-\lambda)v=f-(A-\lambda)u_0,\quad b_1v=0,\ldots,b_mv=0,$$

which is possible by (39), and lastly by putting $u=v+u_0$ we find a solution of (40). Having thus proved that $L-\lambda$ is Fredholm, we deduce from Theorem 13.1 that this is equivalent to the L.Š. condition 11.1 for $(A, b_1,\ldots,b_m)$. ∎

§16 Examples

We first consider the Dirichlet problem for the Laplace operator $-\Delta$.

Example 16.1 Suppose that for

$$-\Delta u=f \text{ on } \Omega,\quad u=g_1 \text{ on } \partial\Omega, \tag{1}$$

the differentiability assumptions for the main theorem 13.1 are satisfied, thus in particular $k=0, \kappa=1$,

$$\Omega\in C^{2,1}. \tag{2}$$

By Theorem 13.5 (1) is elliptic, and has index 0. We show that $\lambda=0$ is not an eigenvalue of (1). Suppose therefore

$$-\Delta u=0 \text{ on } \Omega;\quad u=0 \text{ on } \partial\Omega,\quad u\in W_2^2(\Omega). \tag{3}$$

By Theorem 8.9, $u=0$ on $\partial\Omega$ implies that

$$u\in W_2^2(\Omega)\cap \mathring{W}_2^1(\Omega), \tag{4}$$

and we can apply the first Poincaré inequality (see Theorem 7.6)

$$\|u\|_1^2 \leqslant c \sum_{j=1}^{r} \int_\Omega \left|\frac{\partial u}{\partial x_j}\right|^2 \mathrm{d}x, \quad u \in \mathring{W}_2^1(\Omega) \cap W_2^2(\Omega). \tag{5}$$

Because of (4) we may partially integrate inside (5) (first for $u \in C^2$, then pass to the limit) and obtain

$$\|u\|_1^2 \leqslant c \sum_{j=1}^{r} \int_\Omega \left|\frac{\partial u}{\partial x_j}\right|^2 \mathrm{d}x = c \int_\Omega (-\Delta u)\bar{u}\,\mathrm{d}x \quad \text{for } u \in \mathring{W}_2^1(\Omega) \cap W_2^2(\Omega),$$

so that substituting $-\Delta u = 0$ from (3) gives $u = 0$ as the unique solution of (3), that is, $\lambda = 0$ is not an eigenvalue of (1). Therefore the map $Lu = (-\Delta u, u|_{\partial\Omega})$

$$L: W_2^2(\Omega) \to L_2(\Omega) \times W_2^{2-1/2}(\partial\Omega)$$

is an isomorphism, which implies that for each $f \in L_2(\Omega)$, $g_1 \in W_2^{3/2}(\partial\Omega)$ the problem (1) is uniquely solvable in $W_2^2(\Omega)$.

Since $\lambda = 0$ is not an eigenvalue and $\operatorname{ind} L = 0$, we can also apply Theorem 13.4, which says that the conclusions of the Riesz–Schauder theorem are valid for the spectral problem

$$-\Delta u - \lambda u = f \text{ on } \Omega, \quad u|_{\partial\Omega} = g_1 \text{ on } \partial\Omega.$$

The adjoint problem to (1) (see 14) is easily read off from Green's formula ($\Delta^* = \Delta$)

$$\int_\Omega v \cdot \overline{\Delta u}\,\mathrm{d}x - \int_\Omega \Delta v \cdot \bar{u}\,\mathrm{d}x = \int_{\partial\Omega} v \frac{\partial \bar{u}}{\partial n}\mathrm{d}\sigma - \int_{\partial\Omega} \frac{\partial v}{\partial n}\bar{u}\,\mathrm{d}\sigma, \tag{6}$$

the adjoint boundary value problem for $A = -\Delta, b_1 = I|_{\partial\Omega}$ is again the Dirichlet problem

$$A^* = -\Delta, \quad b_1' = I|_{\partial\Omega},$$

that is, the Dirichlet problem is self-adjoint for the Laplace operator, and since the Dirichlet problem is normal, the conclusions of Theorem 15.9 also hold.

Now some words on the region of validity of (6). (6) is the so-called second Green formula, of course here we have no need of the higher differentiability assumptions of §14; (2) is more adequate, we can even manage with $\Omega \in N^{0,1}$, see Nečas [1].

We next consider the Neumann problem for the Laplace operator $-\Delta$.

Example 16.2 Let

$$-\Delta u = f \text{ in } \Omega, \quad \frac{\partial u}{\partial n} = g_1 \text{ on } \partial\Omega,$$

$$f \in L_2(\Omega), \quad g_1 \in W_2^{1/2}(\partial\Omega), \quad u \in W_2^2(\Omega).$$

By Example 11.2 and (2) the assumptions of the main theorem 13.1 are satisfied by (7). We see that $\lambda = 0$ is an eigenvalue of (7), since $c \in \mathbb{C}$ trivially satisfies $-\Delta c = 0$, $\partial c/\partial n = 0$.

We want to show that the eigenspace $E(\lambda = 0) = \ker L$, $L = (-\Delta, \partial/\partial n|_{\partial\Omega})$ consists only of these constant functions, that is,

$$E(\lambda = 0) = \mathbb{C}, \quad \dim \ker L = 1.$$

For this we take the orthogonal decomposition

$$W_2^1(\Omega) = \mathbb{C} \oplus \mathbb{C}^\perp,$$

with the projection map

$$\operatorname{proj}_{\mathbb{C}} \varphi = \frac{1}{\operatorname{mes} \Omega} \int_\Omega \varphi(x)\,\mathrm{d}x, \quad \varphi \in W_2^1(\Omega).$$

Therefore the functions φ from $\mathbb{C}^\perp$ are characterised by

$$\int_\Omega \varphi(x)\,\mathrm{d}x = 0. \tag{8}$$

We take the second Poincaré inequality. Theorem 7.7.

$$\|\varphi\|_1^2 \leqslant c\left\{\left|\int_\Omega \varphi(x)\,\mathrm{d}x\right|^2 + \sum_{j=1}^r \int_\Omega \left|\frac{\partial \varphi}{\partial x_j}\right|^2 \mathrm{d}x\right\}, \quad \varphi \in W_2^1(\Omega) \cap W_2^2(\Omega), \tag{9}$$

which because of (8) reduces on $\mathbb{C}^\perp$ to

$$\|u\|_1^2 \leqslant c \sum_{j=1}^r \int_\Omega \left|\frac{\partial u}{\partial x_j}\right|^2 \mathrm{d}x, \quad u \in \mathbb{C}^\perp \cap W_2^2(\Omega).$$

Because of (2) Green's formula holds

$$\sum_{j=1}^r \int_\Omega \left|\frac{\partial u}{\partial x_j}\right|^2 \mathrm{d}x = \int_\Omega (-\Delta u)\bar{u} + \int_{\partial\Omega} \frac{\partial u}{\partial n} \bar{u} \cdot \mathrm{d}\sigma, \quad u \in W_2^2(\Omega),$$

and for (9) we obtain

$$\|u\|_1^2 \leqslant c \sum_{j=1}^r \int_\Omega \left|\frac{\partial u}{\partial x_j}\right|^2 = c\left\{\int_\Omega (-\Delta u)\bar{u} + \int_{\partial\Omega} \frac{\partial u}{\partial n} \bar{u}\,\mathrm{d}\sigma\right\}, \quad u \in \mathbb{C}^\perp \cap W_2^2(\Omega). \tag{10}$$

(10) shows that the only solution in $\mathbb{C}^{\perp} \cap W_2^2(\Omega)$ of

$$-\Delta u = 0, \quad \left.\frac{\partial u}{\partial n}\right|_{\partial\Omega} = 0, \tag{11}$$

is the null solution $u = 0$, or that $u = \text{constant}$ are all the solutions of (11) in $W_2^2(\Omega)$.

From (6) we see that the Neumann problem is self-adjoint, that is, $(-\Delta, \partial/\partial n|_{\partial\Omega})^* = (-\Delta, \partial/\partial n|_{\partial\Omega}) = L$, and so we have $\operatorname{Ind} L = 0$, (dim coker $L = \dim\ker L^* = \dim\ker L = 1$).

From (6) we see, in the notation of §15, that we have

$$b_1 = \frac{\partial}{\partial n}, \quad S_1 = I, \quad B_1' = \frac{\partial}{\partial n}, \quad T_1 = I, \quad \ker L^* = \ker L = \mathbb{C},$$

and by Theorem 15.6 the condition

$$\int_\Omega f\bar{c}\,\mathrm{d}x + \int_{\partial\Omega} g_1\bar{c}\,\mathrm{d}\sigma = 0 \quad \text{for all } c\in\mathbb{C},$$

$$\int_\Omega f\,\mathrm{d}x + \int_{\partial\Omega} g_1\,\mathrm{d}\sigma = 0, \tag{12}$$

is necessary and sufficient for the solvability of (7), that is, if $f\in L_2(\Omega)$, $g_1\in W_2^{1/2}(\partial\Omega)$ satisfy condition (12), then there exists at least one solution of (7), $u_0\in W_2^2(\Omega)$, and all solutions of (7) have the form

$$u = u_0 + c, \quad c\in\mathbb{C}, \text{ arbitrary}.$$

Since the Neumann condition $\partial/\partial n$ is normal, and $(-\Delta, \partial/\partial n)$ is self-adjoint Theorem 15.9 is applicable. We will later show, Example 21.2, that in the half plane $\operatorname{Re}\lambda < 0$ there are no eigenvalues.

Theorem 16.1 *If the coefficients of the general boundary value operator $b(x, D)$ (see §11, (32)) are real, $r \geqslant 3$, then for the Laplace operator $-\Delta$ all elliptic boundary value problems have index* 0.

Proof. As in §11 we distinguish between two cases:

Case 1 ord $b(x, \mathrm{D}) = 0$. The boundary value problem is elliptic (= condition 11.1 is satisfied) if and only if (see §11, (34))

$$b_0(x) \neq 0 \text{ on } \partial\Omega.$$

Multiplication of the boundary values by $1/b_0(x)$ shows that we are dealing with the Dirichlet problem, hence the index = 0.

Case 2 ord $b(x, \mathrm{D}) = 1$. Because $r \geqslant 3$ by §11, (35) $(-\Delta, b(x, \mathrm{D}))$ is elliptic if and only if

$$b_r(x) \neq 0 \text{ on } \partial\Omega. \tag{13}$$

If we introduce the vector field $\mu = (b_1(x), \ldots, b_r(x))$ on $\partial\Omega$, then we have

$$b^{\mathrm{H}}(x, \mathrm{D}) = \frac{\partial}{\partial\mu} \quad \text{(directional derivative)},$$

and (13) implies that μ never lies tangentially. Therefore we can always continuously deform $\partial/\partial\mu$ to $\pm\partial/\partial n$ inside the class of Fredholm operators, that is, without becoming tangential, and by Theorem 12.10 and Example 16.2

$$\operatorname{ind}\left(-\Delta, \frac{\partial}{\partial\mu}\right) = \operatorname{ind}\left(-\Delta, \frac{\partial}{\partial n}\right) = 0.$$

Since by Theorem 13.2 the index depends only on the principal parts of the operators in question, we have proved our theorem. ■

The situation in the plane is quite different, that is, for $r = 2$. Instead of (13) we have the condition

$$\mu = (b_1(x), b_2(x)) \neq 0 \text{ on the bounding curve } \partial\Omega,$$

which is necessary and sufficient for the ellipticity of $(-\Delta, \partial/\partial\mu)$, and for a simple closed curve $\partial\Omega$ we obtain, see Hörmander [1] or Vekua [2]

$$\operatorname{ind} L = \operatorname{ind}\left(-\Delta, \frac{\partial}{\partial\mu}\right) = 2 - 2p, \tag{14}$$

where p is the rotation of the vector field μ on the curve $\partial\Omega$, see Krasnoselskiĭ [1]. Formula (14) gives a first glimpse into the marvellous index theorem of Atiyah–Singer, see Palais [1], which shows that the index is a topological invariant.

Example 16.3 We now want to work out adjoint boundary value problems for a general second-order elliptic operator:

$$A(x, \mathrm{D}) = -\sum_{j,k} a_{jk} \frac{\partial}{\partial x_j} \frac{\partial}{\partial x_k} + \sum_{j=1}^{r} a_i \frac{\partial}{\partial x_j} + a_0. \tag{15}$$

The general first-order boundary condition may always be written in the form

$$b_1(x, \mathrm{D})u = b_r \cdot N_A u + Tu + b_0 \cdot u, \tag{16}$$

where

$$N_A u = \sum_{j,k} a_{jk} \frac{\partial u}{\partial x_j} \cos(\mathbf{n}, x_k)$$

is the conormal derivative, and Tu is a first-order differential operator tangential to $\partial\Omega$. If we take coordinates $(x_1, \ldots, x_{r-1})$ tangential to $\partial\Omega$ and x_r normal, we have

$$N_A u = \sum_{j=1}^{r-1} a_{jr} \frac{\partial}{\partial x_j} + a_{rr} \frac{\partial}{\partial x_r},$$

where because of the ellipticity of A, $a_{rr} \neq 0$, that is, N_A is normal in the sense of Definition 14.1.

We recognise that the boundary operator $b_1(x, \mathrm{D})$ is then normal, see Definition 14.1, if and only if $b_r(x) \neq 0$ for all $x \in \partial\Omega$.

In order to obtain the boundary value operator adjoint to (16) we begin with the Gauss–Stokes formula

$$\int_\Omega u \frac{\partial v}{\partial x_j} \mathrm{d}x = -\int_\Omega v \frac{\partial u}{\partial x_j} \mathrm{d}x + \int_{\partial\Omega} uv \cdot \cos(\mathbf{n}, x_j) \, \mathrm{d}\sigma, \tag{17}$$

and apply (17) to the individual terms of the expression $\int_\Omega Au \cdot \bar{v} \, \mathrm{d}x$, obtaining

$$\int_\Omega Au \cdot \bar{v} \cdot \mathrm{d}x = \int_\Omega \sum_{j,k} a_{jk} \frac{\partial u}{\partial x_j} \frac{\partial \bar{v}}{\partial x_k} \mathrm{d}x + \int_\Omega \sum_{j=1}^{r} \left(a_j - \sum_{k=1}^{r} \frac{\partial a_{jk}}{\partial x_k} \right) \frac{\partial u}{\partial x_j} \bar{v} \, \mathrm{d}x$$
$$+ \int_\Omega a_0 u \bar{v} - \int_{\partial\Omega} \sum_{j,k} a_{jk} \frac{\partial u}{\partial x_j} \cos(\mathbf{n}, x_k) \bar{v} \cdot \mathrm{d}\sigma.$$

Doing the same to $\int_\Omega u \cdot \overline{A^* v}$ and subtracting the two formulae from each other gives

$$\int_\Omega Au\bar{v} \, \mathrm{d}x - \int_\Omega u\overline{A^* v} \, \mathrm{d}x = \int_{\partial\Omega} u\overline{N_{\bar{A}} v} \, \mathrm{d}\sigma - \int_{\partial\Omega} N_A u \bar{v} \, \mathrm{d}\sigma$$
$$+ \int_{\partial\Omega} \sum_{j=1}^{r} \left(a_j - \sum_{k=1}^{r} \frac{\partial a_{jk}}{\partial x_k} \right) \cos(\mathbf{n}, x_j) u\bar{v} \, \mathrm{d}\sigma, \tag{18}$$

where $\bar{A}$ is the operator with complex conjugate coefficients. In order to abridge equation (18) we write

$$\beta(x) := \sum_{j=1}^{r} \left(a_j - \sum_{k=1}^{r} \frac{\partial a_{jk}}{\partial x_k} \right) \cos(\mathbf{n}, x_j), \quad x \in \partial\Omega.$$

since $\partial\Omega$ has no frontier, (17) gives

$$\int_{\partial\Omega} Tu\bar{v} \, \mathrm{d}\sigma = \int_{\partial\Omega} u \cdot \overline{T^+ v} \, \mathrm{d}\sigma, \tag{19}$$

where T^+ is the formally adjoint tangential differential operator to T on $\partial\Omega$.

We can write (18) and (19) in terms of the boundary operator (16) (first put $b_r = 1$ in (16))

$$\int_\Omega Au\bar{v}\,dx - \int_\Omega u\overline{A^*v}\,dx = \int_{\partial\Omega} u\cdot\overline{[N_{\bar{A}}v + T^+v + \bar{\beta}(x)v + \bar{b}_0\bar{v}]}\,d\sigma$$

$$- \int_{\partial\Omega} [N_A u + Tu + b_0 u]\cdot\bar{v}\,d\sigma. \tag{20}$$

Comparison with Green's formula §14, (12) gives

$$b = N_A + T + b_0, \quad S_1 = I, \quad B_1' = N_{\bar{A}} + T^+ + \bar{b}_0 + \bar{\beta}, \quad T_1 = I,$$

that is, (16) ($b_r = 1$) has the adjoint boundary operator

$$B' = N_{\bar{A}} + T^+ + \bar{b}_0 + \bar{\beta}.$$

If we write (16) as

$$b(x, D) = b_r\cdot\left[N_A u + \frac{Tu}{b_r} + \frac{b_0}{b_r}u\right]$$

(here we have used the normality $b_r \neq 0$) an easy substitution in (20) gives

$$b = b_r N_A + T + b_0, \quad S_1 = \frac{1}{b_r}.$$

$$B' = \bar{b}_r\cdot N_{\bar{A}} + b_r\left(\frac{T}{b_r}\right)^+ + \bar{b}_0 + \bar{b}_r\cdot\bar{\beta}, \quad T_1 = \frac{1}{\bar{b}_r}. \tag{21}$$

Therefore we have determined all adjoint boundary value problems to $(A, b(x, D))$. We see that for the derivation of (20), respectively (21), the following differentiability assumptions suffice:

$$\Omega\in C^{2,1}, \quad a_{jk}\in C^2(\bar{\Omega}), \quad a_j\in C^1(\bar{\Omega}), \quad a_0\in C(\bar{\Omega}), \quad b(x, D)\in C^2(\bar{\Omega}).$$

Indeed then

$$S_1\in C^2(\bar{\Omega}), \quad T_1\in C^2(\bar{\Omega}), \quad B'\in C^1(\bar{\Omega}),$$

and Theorem 14.3 is applicable (= definition of the adjoint problem).

The same trick as for the calculation of the index of the Dirichlet problem (see Theorem 13.5) also allows us to give the index of the general Neumann problem.

Let A (15) be of second order and properly elliptic; we consider the Neumann problem.

Example 16.4

$$Au = f \text{ in } \Omega, \quad N_A u = g_1 \text{ on } \partial\Omega,$$

we have $\operatorname{ind}(A, N_A) = 0$, for

$$-\operatorname{ind}(A, N_A) = \operatorname{ind}(A^*, N'_A) = \operatorname{ind}(\bar{A}, N_{\bar{A}}) = \operatorname{ind}(\tilde{A}, \bar{N}_A) = \operatorname{ind}(A, N_A).$$

Here the first equation follows from Theorem 15.7, the second from formula (21) and Theorem 13.2, A^* and $\bar{A}$ differ from each other by a differential term of order 1, and the remaining equations are clear.

Theorem 13.2 applied once more also gives the index of the third boundary value problem.

Example 16.5

$$Au = f \text{ in } \Omega, \quad N_A u + b_0(x) u = g_1 \text{ on } \partial\Omega,$$
$$\operatorname{ind}(A, N_A + b_0) = \operatorname{ind}(A, N_A) = 0.$$

Example 16.6 We now consider boundary value problems for the biharmonic operator $= \Delta^2$ and apply Theorem 14.9 to various possibilities. We start from Green's formula (6), substitute there for $v \to \Delta v$ and obtain

$$\int_\Omega \Delta u \cdot \Delta \bar{v} \, dx = \int_\Omega u \cdot \Delta^2 \bar{v} \, dx + \int_{\partial\Omega} \frac{\partial u}{\partial n} \cdot \Delta \bar{v} \, d\sigma - \int_{\partial\Omega} u \cdot \frac{\partial \Delta \bar{v}}{\partial n} \, d\sigma, \tag{22}$$

Interchanging u and v and taking complex conjugates leads to

$$\int_\Omega \Delta u \cdot \Delta \bar{v} \, dx = \int_\Omega u \cdot \Delta^2 \bar{v} \, dx + \int_{\partial\Omega} \frac{\partial u}{\partial n} \cdot \Delta \bar{v} \, d\sigma - \int_{\partial\Omega} u \cdot \frac{\partial \Delta \bar{v}}{\partial n} \, d\sigma,$$

both formulae together give

$$\int_\Omega \Delta^2 u \cdot \bar{v} \, dx - \int_\Omega u \cdot \Delta^2 \bar{v} \, dx = \int_{\partial\Omega} u \frac{-\partial \Delta \bar{v}}{\partial n} \, d\sigma + \int_{\partial\Omega} \frac{\partial u}{\partial n} \Delta \bar{v} \, d\sigma$$
$$- \int_{\partial\Omega} \frac{-\partial \Delta u}{\partial n} \bar{v} \, d\sigma - \int_{\partial\Omega} \Delta u \frac{\partial \bar{v}}{\partial n} \, d\sigma. \tag{23}$$

This is formula (51) from Theorem 14.9, Δ^2 is (formally) self-adjoint: $(\Delta^2)^* = \Delta^2$. One sees immediately, that the boundary value operators

$$I, \frac{\partial}{\partial n}, \Delta, -\frac{\partial \Delta}{\partial n}, \tag{24}$$

form a fourth-order Dirichlet system, for – perhaps after a change of coordinates – the highest order derivative will be in a purely normal direction with coefficients different from zero. Therefore we can apply Theorem 14.3 and compare (23) with Green's formula §14, (12), where we can group the boundary conditions (24) differently. For the adjoint boundary value problems of Examples 16.1–16.6 we obtain

$$\left(\Delta^2; I, \frac{\partial}{\partial n}\right)^* = \left(\Delta^2; I, \frac{\partial}{\partial n}\right), \quad (\Delta^2; I, \Delta)^* = (\Delta^2; I, \Delta),$$

$$\left(\Delta^2; \frac{\partial}{\partial n}, \Delta\right)^* = \left(\Delta^2, I, -\frac{\partial \Delta}{\partial n}\right), \quad \left(\Delta^2; I, -\frac{\partial \Delta}{\partial n}\right)^* = \left(\Delta^2; \frac{\partial}{\partial n}, \Delta\right),$$

$$\left(\Delta^2; \frac{\partial}{\partial n}, -\frac{\partial \Delta}{\partial n}\right)^* = \left(\Delta^2; \frac{\partial}{\partial n}, -\frac{\partial \Delta}{\partial n}\right), \quad \left(\Delta^2; \Delta, -\frac{\partial \Delta}{\partial n}\right)^* = \left(\Delta^2; \Delta, -\frac{\partial \Delta}{\partial n}\right).$$

The boundary value problems 1, 2, 5, 6 are self-adjoint, and for the index we obtain:

$$1 \quad \operatorname{ind}\left(\Delta^2; I, \frac{\partial}{\partial n}\right) = 0; \; 2 \quad \operatorname{ind}(\Delta^2; I, \Delta) = 0; \; 5 \quad \operatorname{ind}\left(\Delta^2; \frac{\partial}{\partial n}, -\frac{\partial \Delta}{\partial n}\right) = 0,$$

1, 2, 5 satisfy the L.Š. condition 11.1 – see Example 11.9. Since they are normal, we can apply Theorem 15.9 and find that the Green solution operator G_λ exists for all $\lambda \in \mathbb{C}$, except for a discrete countable number of real eigenvalues λ_n. Later – see Example 21.6 – we shall see, with the help of other methods, that the eigenvalues λ_n lie on the half-line $\lambda \geqslant -k_0$.

Problem 6 $(\Delta^2; \Delta, -\partial\Delta/\partial n)$ is certainly self-adjoint, however it possesses no index, since it is not elliptic – by Example 11.9 $(\Delta^2; \Delta, -\partial\Delta/\partial n)$ does not satisfy the L.Š. condition 11.1.

The reader may supply the required minimal differentiability assumptions needed for the validity of these assertions.

III

Strongly elliptic differential operators and the method of variations

§17 Gelfand triples, the Lax–Milgram theorem, V-elliptic and V-coercive operators

In this section we introduce functional analytic tools for treating strongly elliptic differential operators and abstractly develop the so-called method of variations, also called the direct method. We begin with Gelfand triples, give a collection of examples, Theorem 17.4, and study the dual spaces to Sobolev spaces. Then we deal with the important Lax–Milgram theorem, define V-elliptic and V-coercive operators and use the Lax–Milgram theorem to solve V-elliptic equations (method of variations). The general definition of a Green operator, Definition 17.5, allows us to prove a far-reaching spectral theorem for V-coercive operators, Theorem 17.12, and to extract the *a priori* estimate from the Gårding inequality. Finally we formulate V-ellipticity and V-coercion as it is usually done for differential operators.

17.1 Gelfand triples

Let X be a Banach space. We denote its antidual space by X', see §12, again we shall work – as everywhere in this book – with the antidual space X' and with antidual operators A'.

Theorem 17.1 *Let X and Y be Banach spaces and $A: X \to Y$ a continuous linear operator from X into Y. Then the following statement holds:*

$$\operatorname{im} A \text{ dense in } Y \Leftrightarrow A' \text{ is injective.}$$

Proof. Necessity. Let $A'y' = 0$, which is $\langle y', Ax \rangle = \langle A'y', x \rangle = 0$ for all $x \in X$, and thus $\langle y', y \rangle = 0$ for all $y \in Y$, since $\operatorname{im} A$ is dense in Y; this implies that $y' = 0$, with which we have proved injectivity.

Sufficiency: Let $\operatorname{im} A$ not be dense in Y, then by the Hahn–Banach theorem there exists $0 \neq y' \in Y'$ with $\langle y', Ax \rangle = 0$ for all $x \in X$. Therefore $\langle A'y', x \rangle = \langle y', Ax \rangle = 0$ for all $x \in X$, that is, $A'y' = 0$ and $y' \neq 0$, so that A' cannot be injective. ■

In general $A'' \supset A$; however, for (anti)reflexive Banach spaces ($X = X''$) we have $A'' = A$, and we can express Theorem 17.1 in the form

$$A \text{ injective} \Leftrightarrow \operatorname{im} A' \text{ dense in } X'.$$

We recall that we write Riesz' theorem in the form $H = H'$, where H is a Hilbert space. Every Hilbert space H is also (anti)reflexive, because $H = H' = H''$.

Definition 17.1 *Let V be an (anti)reflexive Banach space and H a Hilbert space. Suppose that $V \underset{i}{\subsetneq} H$, and that the embedding i is continuous, injective and that* $\operatorname{im} i$ *is dense in H. By Theorem* 17.1, *applied once to i and once to i', $i': H \to V'$ is continuous, injective and* $\operatorname{im} i'$ *is dense in V'. Altogether we have*

$$V \underset{i}{\subsetneq} H \underset{i'}{\subsetneq} V', \tag{1}$$

where both embeddings i, i' are continuous, injective and have dense images in H and V'.

A scheme of this kind is called a Gelfand triple. So long as there is no danger of confusion we omit the symbols i and i'.

Because of the continuity of the map i we have

$$\| ix \|_H \leqslant c \| x \|_V \quad \text{for all } x \in V.$$

By renorming V, by means of an equivalent norm, we achieve

$$\| ix \|_H \leqslant \| x \|_V \quad \text{for all } x \in V, \text{ that is, } \| i \| \leqslant 1.$$

Then it is also true that $\| i' \| = \| i \| \leqslant 1$ and $\| i'h \|_{V'} \leqslant \| h \|_H$ for all $h \in H$. Altogether

$$\| i'ix \|_{V'} \leqslant \| ix \|_H \leqslant \| x \|_V \quad \text{for all } x \in V, \tag{2}$$

or, by omitting i and i',

$$\| x \|_{V'} \leqslant \| x \|_H \leqslant \| x \|_V \quad \text{for all } x \in V. \tag{2a}$$

Denoting by $(\cdot,\cdot)_H$ the scalar product in H, by definition of i' we have $\langle i'h, x \rangle_V = (h, ix)_H$ and

$$|(h, ix)_H| = |\langle i'h, x \rangle_V| \leqslant \| i'h \|_{V'} \| x \|_V \leqslant \| h \|_H \| x \|_V. \tag{3}$$

Therefore we may consider each element $h \in H$ as an antilinear, continuous functional on V; $\operatorname{im} i' = i'H$ is dense in V' (with respect to the functional norm $\| \cdot \|_{V'}$), that is, we can uniformly approximate each functional $\langle x', \cdot \rangle_V$

on the unit ball of V by the scalar product $(h, i\cdot)_H$:

$$\langle x', x\rangle_V = \lim_{i'h \to x'} (h, ix)_H. \tag{4}$$

This means that we can regard the continuous extension of $(\cdot,\cdot)_H$ on $V' \times \dot{V}$ as a new representation formula for the functionals from V'. By an example we show that this new representation (4) is in no way trivial.

Example 17.1 We consider the Gelfand triple

$$\mathring{W}_2^1(\Omega) \subsetneq L_2(\Omega) \subsetneq (\mathring{W}_2^1(\Omega))'.$$

Since by Definition 3.2 and Theorem 3.4 $\mathscr{D}(\Omega)$ is dense in $\mathring{W}_2^1$ and L_2, the conditions of Definition 17.1 for a Gelfand triple are satisfied. By the Riesz theorem there exists an isomorphism

$$R:(\mathring{W}_2^1(\Omega))' \to \mathring{W}_2^1(\Omega)$$

Let $v = Rf$ for some $f \in (\mathring{W}_2^1(\Omega))'$, then (we denote functionals by angle brackets and scalar products by round)

$$\langle f, \varphi\rangle = (v, \varphi)_1 = \int_\Omega \left(\sum_{j=1}^r \frac{\partial v}{\partial x_j}\frac{\partial \bar{\varphi}}{\partial x_j} + v\cdot\bar{\varphi}\right)\mathrm{d}x$$

$$= -\int_\Omega (-\Delta v + v)\bar{\varphi}\,\mathrm{d}x = (-\Delta v + v, \varphi)_0, \quad \varphi \in \mathring{W}_2^1(\Omega).$$

Here the integrals are interpreted in the distributional sense $\mathscr{D}'(\Omega)$. By extension of $(\cdot, \varphi)_0$ to $(\mathring{W}_2^1(\Omega))'$ we obtain

$$\langle f, \varphi\rangle = (f, \varphi)_0,$$

formally, since f does not need to belong to L_2. Hence

$$(-\Delta v + v, \varphi)_0 = (f, \varphi)_0, \quad \text{that is, } f = -\Delta v + v = R^{-1}v.$$

Theorem 17.2 (second definition of the Gelfand triple) *Let X_1, X_{-1} be (anti)-reflexive Banach spaces, H a Hilbert space and suppose that the maps*

$$X_1 \underset{i_1}{\subsetneq} H \underset{i_{-1}}{\subsetneq} X_{-1} \tag{5}$$

are continuous, injective and dense. Suppose further that

$$|(h, i_1 x)_H| \leqslant c_1 \|x\|_1 \cdot \|i_{-1}h\|_{-1} \quad \textit{for all } x \in X_1 \textit{ and } h \in H, \tag{6}$$

where $(\cdot,\cdot)_H$ is the scalar product in H, and that

$$\|i_{-1}h\|_{-1} \leqslant c_2 \sup_{0 \neq x \in X_1} \frac{|(h, i_1 x)|}{\|x\|_1} \quad \textit{for all } h \in H. \tag{7}$$

In this situation we have $X_{-1}=(X_1)'$ *up to equivalence of norms and* $i_{-1}=i'_1$.

Remark It is easy to see that a Gelfand triple in the sense of Definition 17.1 satisfies the conditions of Theorem 17.2: we put $X_1=V, X_{-1}=V'$, $i_1=i$, $i_{-1}=i'$, so that (5) is satisfied, and the first estimate of (3) shows the validity of (6). We obtain (7) from the definition of the V'-norm by application of (4)

$$\|i'h\|_{V'}:=\sup_{0\neq x\in V}\frac{|\langle i'h,x\rangle|}{\|x\|_V}=\sup_{0\neq x\in V}\frac{|(h,ix)|}{\|x\|_V}.$$

Theorem 17.2 now asserts that the converse is also true.

Proof. We embed H in $X_{-1}, i_{-1}H\subsetneq X_{-1}$ is dense. The elements from $i_{-1}H=\operatorname{im} i_{-1}$ then form functionals from X'_1 by means of the definition $\langle i_{-1}h,x\rangle:=(h,i_1x)_H$. Because of condition (6) these functionals are continuous, and we have therefore defined the desired correspondence

$$X_{-1}\to X'_1 \tag{8}$$

on the dense subset $i_{-1}H$ of X_{-1}. We now introduce a second norm on $i_{-1}H$

$$\|i_{-1}h\|_{,}:=\sup_{0\neq x\in X_1}\frac{|\langle i_{-1}h,x\rangle|}{\|x\|_1}=\sup_{0\neq x\in X_1}\frac{|(h,i_1x)|}{\|x\|_1},$$

which is the X'_1-norm. Then by (6) we have

$$\|i_{-1}h\|_{,}\leqslant c_1\|i_{-1}h\|_{-1}, \tag{9}$$

and by (7)

$$\|i_{-1}h\|_{-1}\leqslant c_2\sup_{0\neq x\in X_1}\frac{|(h,i_1x)|}{\|x\|_1}=c_2\|i_{-1}h\|_{,}, \tag{10}$$

so that both norms $\|\ \ \|_{-1}$ and $\|\ \ \|_{,}$ are equivalent on $i_{-1}H$. We are now in a position to define the correspondence (8) for all $y\in X_{-1}$. Let $y\in X_{-1}$, then there exists a sequence $h_n\in H$, so that

$$i_{-1}h_n\to y \text{ in } (X_{-1},\|\cdot\|_{-1}) \quad \text{as } n\to\infty.$$

Thus $i_{-1}h_n$ is a Cauchy sequence in $(X'_1,\|\cdot\|_{-1})$, and hence because of (9) also a Cauchy sequence in $(X'_1,\|\cdot\|_{,})$. There then exists some $y'\in X'_1$ with

$$i_{-1}h_n\to y' \quad \text{as } n\to\infty.$$

We associate the $y'\in X'_1$ found in this way with $y\in X_{-1}$,

$$X_{-1}\ni y\mapsto y'\in X'_1, \tag{11}$$

and by (9), (10) and passing to the limit we have

$$\|y'\|_{,}\leqslant c_1\|y\|_{-1},\ \|y\|_{-1}\leqslant c_2\|y'\|_{,}\quad \text{for all } y\in X_{-1}. \tag{12}$$

The correspondence (11) is independent of the choice of the approximating sequence. In order to show this we take $i_{-1}\bar{h}_n \in i_{-1}H$ to be another sequence such that

$$i_{-1}\bar{h}_n \to y \quad \text{in } (X_{-1}, \|\cdot\|_{-1}) \quad \text{as } n \to \infty$$

and

$$i_{-1}\bar{h}_n \to \bar{y}' \quad \text{in } (X'_1, \|\cdot\|_,) \quad \text{as } n \to \infty.$$

Then

$$i_{-1}(h_n - \bar{h}_n) \to 0 \quad \text{in } (X_{-1}, \|\cdot\|_{-1}) \quad \text{as } n \to \infty.$$

$$i_{-1}(h_n - \bar{h}_n) \to y' - \bar{y}' \quad \text{in } (X'_1, \|\cdot\|_,) \quad \text{as } n \to \infty$$

and by (12) it follows that

$$\|y' - \bar{y}'\|_, = 0, \text{ that is, } y' = \bar{y}'.$$

The embedding $i'_1 : H \subsetneq X'_1$ is defined by

$$\langle i'_1 h, x\rangle = (h, i_1 x) \quad \text{for all } x \in X_1, \quad \text{for all } h \in H.$$

Since by definition it is also true that

$$\langle i_{-1} h, x\rangle = (h, i_1 x) \quad \text{for all } x \in X_1, \quad \text{for all } h \in H,$$

(see above) i_{-1} and i'_1 agree on H,

$$i_{-1} = i'_1, \tag{13}$$

and we have proved the second statement of our theorem. In order to prove the first statement we have still to show that the correspondence (11) is surjective. Injectivity and continuity in both directions follow immediately from (12).

X_1 is (anti)reflexive and $i_1 : X_1 \to H$ is injective, so that by Theorem 17.1 $i'_1 H = \operatorname{im} i'_1$ is dense in X'_1; therefore because of (13) $i_{-1}H$ is also dense in X'_1.

The construction of the correspondence Z (11) gives

$$i_{-1}H \subset ZX_{-1} \subset X'_1,$$

and so we have shown that ZX_{-1} is dense in X'_1.

We must distinguish two cases:

(a) ZX_{-1} closed in X'_1,
(b) ZX_{-1} not closed in X'_1.

Case (b) leads as follows to a contradiction: let ZX_{-1} not be closed in X'_1, then ZX_{-1} is not complete with respect to $\|\cdot\|_,$, because of (12)X_{-1} is also not complete with respect to $\|\cdot\|_{-1}$ contradicting the assumption. In case (a), because ZX_{-1} is dense in X'_1, $ZX_{-1} = X'_1$, and we have shown the surjectivity of Z (11).

Theorem 17.3 (third definition of the Gelfand triple) *Let X_1 be an (anti)-reflexive Banach space, H a Hilbert space, and let the embedding $i_1 \colon X_1 \hookrightarrow H$ be dense continuous and injective. We introduce a second norm on H*

$$\|h\|_{-1} := \sup_{0 \neq x \in X_1} \frac{|(h, i_1 x)|}{\|x\|_1}, \tag{14}$$

and label the completion of H in the norm $\|\cdot\|_{-1}$ as X_{-1}. Then

$$X_1 \underset{i_1}{\hookrightarrow} H \underset{i_{-1}}{\hookrightarrow} X_{-1}$$

form a Gelfand triple, where i_{-1} is the canonical embedding of H in the completion, and in particular we have

$$X_{-1} = X'_1 \quad \text{and} \quad i_{-1} = i'_1.$$

If in addition X_1 is a Hilbert space, then X_{-1} is also a Hilbert space.

Proof. We want to apply Theorem 17.2 and must therefore check conditions (5), (6) and (7):

(5): as canonical embedding in the completion i_{-1} is dense, continuous and injective.
(7): this follows immediately from definition (14).
(6): this also follows immediately from (14); we have only to omit the sup sign.

Statement (15) is then the assertion of Theorem 17.2.

We note that here in (15) – in contrast to Theorem 17.2 – we have actual equality and not just equivalence, since the relevant norms are the same.

Now for the Hilbert space property of X_{-1}. By definition of i'_1 we have

$$(h, i_1 x)_H = \langle i'_1 h, x \rangle_{X_1},$$

and (14) becomes

$$\|h\|_{-1} = \sup_{0 \neq x \in X_1} \frac{|\langle i'_1 h, x \rangle_{X_1}|}{\|x\|_1} = \|h\|_{X'_1}.$$

However, by the Riesz theorem the functional norm $\|\cdot\|_{X'_1}$ is a Hilbert space norm, that is, it satisfies the parallelogram equation, and $X'_1 = X_{-1}$ is thus a Hilbert space with the Hilbert space norm $\|\cdot\|_{-1}$. ■

Warning If V is a Hilbert space, we have the Riesz isomorphism

$$V \underset{i}{\hookrightarrow} H \underset{i'}{\hookrightarrow} V', \qquad R \colon V \to V'$$

But in Gelfand triples we do not represent functionals from V' by the scalar product $(\cdot,\cdot)_V$ in V, (Riesz theorem) but by the scalar product $(\cdot,\cdot)_H$ in H, see (4).

We can easily generalise the third definition of a Gelfand triple (Theorem 17.3) to scales: let $X_1,\ldots,X_m$ be (anti)reflexive Banach spaces, H a Hilbert space and let the embeddings

$$X_m \underset{i_m}{\hookrightarrow} \cdots \underset{i_2}{\hookrightarrow} X_1 \underset{i_1}{\hookrightarrow} H$$

be continuous, injective and dense. If we define the so-called negative norms on H by

$$\|h\|_{-n} = \sup_{0\neq x\in X_n} \frac{|(h, i_1\circ\cdots\circ i_n x)_H|}{\|x\|_n} \quad \text{for } 1\leqslant n\leqslant m,$$

and if we complete H with respect to these negative norms, we obtain

$$X_m \underset{i_m}{\hookrightarrow} \cdots \underset{i_2}{\hookrightarrow} X_1 \underset{i_1}{\hookrightarrow} H \underset{i_{-1}}{\hookrightarrow} X_{-1} \underset{i_{-2}}{\hookrightarrow} \cdots \underset{i_{-m}}{\hookrightarrow} X_{-m}$$

where $i_{-n} = i_n'$ and $X_{-n} = X_n'$ for $1\leqslant n\leqslant m$.

In the next theorem we give a sequence of concrete examples for Gelfand triples.

Theorem 17.4 *The following imbedding maps are injective, continuous and dense; in the sense of Definition* 17.1 *they may therefore be extended to Gelfand triples:*

(a) $H^l \hookrightarrow H^0$ *(see §5), $l\in\mathbb{R}_+$ arbitrary.*
(b) $H^l(\Omega) \hookrightarrow H^0(\Omega)$, *$l\in\mathbb{R}_+$ and Ω arbitrary.*
(c) $\mathring{H}^l(\Omega) \hookrightarrow H^0(\Omega)$, *$l\in\mathbb{R}_+$ and Ω arbitrary.*
(d) $\mathring{W}^l_2(\Omega) \hookrightarrow L_2(\Omega)$, *$l\in\mathbb{R}_+$ and Ω arbitrary.*
(e) $W^l_2(\Omega) \hookrightarrow L_2(\Omega)$, *$l\in\mathbb{R}_+$ and Ω arbitrary.*
(f) *Let M be a compact $C^{k,\kappa}$-manifold and $l\leqslant k+\kappa$ ($\kappa = 0, 1$), see* §4. *Then* $W^l_2(M) \hookrightarrow L_2(M)$, $l\in\mathbb{R}_+$.

Proof. By Theorem 6.11 all six maps are injective and continuous; we have only to check the density property. By Theorem 5.1 $\mathscr{D}$ is dense in each H^l, therefore the map (a) is dense. Since $L_2(\Omega)\simeq H^0(\Omega)$ and by definition $\mathscr{D}(\Omega)$ is dense in each $\mathring{H}^l(\Omega)$, we have shown density for maps (b) and (c).

The density of map (d) follows from map 3 since $\mathring{W}^l_2(\Omega) = \mathring{H}^l(\Omega)$. Map (e) follows from map (d), since $\mathscr{D}(\Omega)$ is dense in $L_2(\Omega)$ and $\mathscr{D}(\Omega)\subset W^l_2(\Omega)$. Finally to map (f): by Theorem 4.3 $C^{k,\kappa}(M)$ is dense in $W_2(M)$ and in $L_2(M)$, with which we have proved the density of the embedding (f). ■

Example 17.2 We have the scales, where we put $W_2^{-l}(\Omega) := \mathring{W}_2^l(\Omega)'$

$$\cdots \subsetneq \mathring{W}_2^2(\Omega) \subsetneq \mathring{W}_2^1(\Omega) \subsetneq L_2(\Omega) \subsetneq W_2^{-1}(\Omega) \subsetneq W_2^{-2}(\Omega) \subsetneq \cdots$$

and (let Ω have the segment property, Definition 2.1)

$$\cdots \subsetneq W_2^2(\Omega) \subsetneq W_2^1(\Omega) \subsetneq L_2(\Omega) \subsetneq W_2^1(\Omega)' \subsetneq W_2^2(\Omega)' \subsetneq \cdots$$

The embeddings are dense, first because $\mathscr{D}(\Omega)$ is dense in all the spaces of the first scale and secondly by Theorem 3.6.

17.2 Representations for functionals on Sobolev spaces

For Gelfand triples $V \subsetneq H \subsetneq V'$ the dual spaces V' play an essential role; it is therefore important either to characterise these precisely, or at least to give a representation of the functionals $f \in V'$. We can do this in some cases. We begin with $V = W_2^l(\Omega)$, $l \in \mathbb{N}$.

In §3.1 we embedded the space $W_2^l(\Omega)$, $l \in \mathbb{N}$, as a closed subspace in the Cartesian product $\times_{|s| \leqslant l} L_2(\Omega)$ by means of the map

$$W_2^l(\Omega) \ni \varphi \mapsto \{D^s\varphi;\, |s| \leqslant l\} \in \underset{|s| \leqslant l}{\times} L_2(\Omega).$$

In general let $l = [l] + \lambda \in \mathbb{R}_+$, then $W_2^l(\Omega)$ is the subspace of the Cartesian product

$$\underset{|s| \leqslant [l]}{\times} L_2(\Omega) \times \underset{|s| \leqslant [l]}{\times} L_2(\Omega \times \Omega),$$

via the embedding

$$W_2^l(\Omega) \ni \varphi \mapsto \left\{ D^s\varphi(x),\, |s| \leqslant [l];\, \frac{D^s\varphi(x) - D^s\varphi(y)}{|x-y|^{r/2+\lambda}},\, |s| \leqslant [l] \right\}$$

$$\in \underset{|s| \leqslant [l]}{\times} L_2(\Omega) \times \underset{|s| \leqslant [l]}{\times} L_2(\Omega \times \Omega). \tag{16}$$

The embedding (16) immediately provides a representation of the functionals f in $W_2^l(\Omega)'$.

Theorem 17.5 *Let $f \in W_2^l(\Omega)'$, then there exist functions $f_s^1(x) \in L_2(\Omega)$, $f_s^2(x, y) \in L_2(\Omega \times \Omega)$, $|s| \leqslant [l]$ with*

$$f(\varphi) = \sum_{|s| \leqslant [l]} \left[\int_\Omega f_s^1(x) D^s\bar{\varphi}(x)\, dx + \iint_{\Omega \times \Omega} f_s^2(x, y) \frac{D^s\bar{\varphi}(x) - D^s\bar{\varphi}(y)}{|x-y|^{r/2+\lambda}}\, dx\, dy \right] \tag{17}$$

and

$$\|f\|_l' = \left[\sum_{|s| \leqslant [l]} \left(\int_\Omega |f_s^1(x)|^2 dx + \iint_{\Omega \times \Omega} |f_s^2(x, y)|^2 dx\, dy \right) \right]^{1/2}. \tag{18}$$

We also abbreviate (17) and (18) as

$$f(\varphi)=\sum_{|s|\leq[l]}\left[(f_s^1,\mathrm{D}^s\varphi(x))_{0,x}+\left(f_s^2(x,y),\frac{\mathrm{D}^s\varphi(x)-\mathrm{D}^s\varphi(y)}{|x-y|^{r/2+\lambda}}\right)_{0,x,y}\right] \tag{17a}$$

$$\|f\|_l'=\left[\sum_{|s|\leq[l]}(\|f_s^1\|_{0,x}^2+\|f_s^2\|_{0,x,y}^2)\right]^{1/2}. \tag{18a}$$

The proof follows immediately from (16) using the Riesz theorem and the theorem on orthogonal decomposition applied to the Cartesian product

$$\mathop{\times}_{|s|\leq[l]} L_2(\Omega)\times \mathop{\times}_{|s|\leq[l]} L_2(\Omega\times\Omega). \qquad \blacksquare$$

Since $\mathscr{D}(\Omega)\subsetneq W_2^l(\Omega)$ we can consider each functional $f\in W_2^l(\Omega)'$ as a distribution (we restrict $f(\varphi)$ to $\mathscr{D}(\Omega)$) only, in general the correspondence $W_2^l(\Omega)'\to\mathscr{D}'(\Omega)$ is not injective, since usually $\mathscr{D}(\Omega)$ is not dense in $W_2^l(\Omega)$. For the same reason a distribution of the type (17) may be extended in several ways to $W_2^l(\Omega)$. It is not so in the case of $\mathring{W}_2^l(\Omega)$, which we now study for $l\in\mathbb{N}$.

Theorem 17.6 *Suppose given $\mathring{W}_2^l(\Omega)$, $l\in\mathbb{R}_+$. We have $\mathring{W}_2^l(\Omega)'\subsetneq\mathscr{D}'(\Omega)$ continuously and injectively. Let $l\in\mathbb{N}$, we consider the distribution*

$$T=\sum_{|s|\leq[l]}(-1)^{|s|}\mathrm{D}^s f_s\in\mathscr{D}'(\Omega),\quad \text{where } f_s\in L_2(\Omega). \tag{19}$$

The distribution (19) *$T(\varphi)$ admits exactly one continuous extension to $\mathring{W}_2^l(\Omega)$ and we have*

$$\|\tilde{T}\|_l'\leq\left[\sum_{|s|\leq[l]}\|f_s\|_0^2\right]^{1/2}. \tag{20}$$

Proof. Let $f\in W_2^l(\Omega))'$, from $\mathscr{D}(\Omega)\subsetneq\mathring{W}_2^l(\Omega)$ we find that $f(\varphi)|_{\mathscr{D}(\Omega)}$ is a distribution in $\mathscr{D}'(\Omega)$ and we have defined the embedding

$$(\mathring{W}_2^l(\Omega))'\subsetneq\mathscr{D}'(\Omega). \tag{21}$$

Since by Definition 3.2 $\mathscr{D}(\Omega)$ is dense in $\mathring{W}_2^l(\Omega)$, if $f(\varphi)=0$ for $\varphi\in\mathscr{D}(\Omega)$ and $f\in\mathring{W}_2^l(\Omega)'$, then $f=0$. Therefore we have shown that the map (21) is injective. Now if $f_n\to f$ in $\mathring{W}_2^l(\Omega)'$, then (weakly) $f_n(\varphi)\to f(\varphi)$ for $\varphi\in\mathring{W}_2^l(\Omega)$; if we take $\varphi\in\mathscr{D}(\Omega)$, then from $f_n(y)\to f(y)$ it follows that $f_n\to f$ in $\mathscr{D}'(\Omega)$, and so we have also proved the continuity of the map (21).

Let T be as in (19). By Definition 1.4 of the distributional derivatives we have

$$T(\varphi)=\sum_{|s|\leq l}(f_s,\mathrm{D}^s\varphi)_0,\quad \varphi\in\mathscr{D}(\Omega)$$

and we can estimate

$$|T(\varphi)| \leqslant \sum_{|s|\leqslant l} \|f_s\|_0 \cdot \|D^s\varphi\|_0 \leqslant \left(\sum_{|s|\leqslant l} \|f_s\|_0^2 \right)^{1/2} \cdot \|\varphi\|_{\mathring{W}_2^l(\Omega)}. \tag{22}$$

Therefore we have shown that $T(\varphi)$ is continuous on $\mathscr{D}(\Omega)$ in the topology of $\mathring{W}_2^l(\Omega)$. Since $\mathscr{D}(\Omega)$ is dense in $\mathring{W}_2^l(\Omega)$, by the (continuous) extension theorem T has a unique extension $\tilde{f}$ on $\mathring{W}_2^l(\Omega)$. The estimate (20) follows immediately from (22). ■

Theorem 17.6 suggests the definition of 'negative' Sobolev spaces $W_2^{-l}(\Omega)$, $l \in \mathbb{N}$.

Definition 17.2 *We define the space $W_2^{-l}(\Omega)$, $l \in \mathbb{N}$, to be the space of all distributions $T \in \mathscr{D}'(\Omega)$, which admit a representation in the form of (19), where $f_s \in L_2(\Omega)$. As norm $\|\cdot\|_{-l}$ we take*

$$\|T\|_{-l} = \inf_{(19)} \left[\sum_{|s|\leqslant l} \|f_s\|_0^2 \right]^{1/2}, \tag{23}$$

We write here the infimum as taken over all possible representations

$$W_2^{-l}(\Omega) := \left\{ T \in \mathscr{D}'(\Omega) \colon T = \sum_{|s|\leqslant l} (-1)^{|s|} D^s f_s, f_s \in L_2(\Omega) \right\}.$$

Theorems 17.5 and 17.6 give:

Theorem 17.7 $W_2^{-l}(\Omega) = (\mathring{W}_2^l(\Omega))'$, $l \in \mathbb{N}$.

Proof. Everything is clear, since $\mathring{W}_2^l(\Omega)$ is a closed subspace of $W_2^l(\Omega)$. The equality of norms $\|f\|_l' = \|T\|_{-l}$ $(f(\varphi) = T(\varphi))$ is a consequence of the Riesz theorem, which says in particular that the infimum (23) over all representations (19) is actually attained. ■

The last theorems may be carried over unchanged to compact manifolds M. There, since $W_2^l(M) = \mathring{W}_2^l(M)$, we have for $l \in \mathbb{N}$ ($W_2^{-l}(M)$ correspondingly defined)

$$(W_2^l(M))' \subsetneq \mathscr{D}'(M), \quad (W_2^l(M))' = W_2^{-l}(M),$$

and we can obtain the representation (17) in terms of local coordinates.

If we employ the dual theory of H^l-spaces, see Volevič & Panejach [1], we can characterise almost all the dual spaces in Theorem 17.4

(a) $(H^l)' = H^{-l}$.
(b) $[H^l(\Omega)]' = \mathring{H}^{-l}(\Omega)$,

(c) $[\mathring{H}^l(\Omega)]' = H^{-l}(\Omega)$,
(d) $[\mathring{W}_2^l(\Omega)]' = H^{-l}(\Omega)$, $l \in \mathbb{R}_+$, and for $l \in \mathbb{N}$: $[\mathring{W}_2^l(\Omega)]' = W_2^{-l}(\Omega)$,
(e) $[W_2^l(\Omega)]' = \mathring{H}^{-l}(\Omega)$,

where Ω has the extension property. Finally

(f) $[W_2^l(M)]' = W_2^{-l}(M)$, $l \in \mathbb{N}$ and M is a compact $C^{k,\kappa}$-manifold, $l \leqslant k + \kappa$, $\kappa = 0, 1$.

17.3 The Lax–Milgram theorem

In order to handle strongly elliptic equations we shall work with bilinear forms, or more precisely with sesquilinear forms. Let H_1 and H_2 be two Hilbert spaces; a map b: $H_1 \times H_2 \to \mathbb{C}$ is called a sesquilinear form if it is linear in the first variable and antilinear in the second. We define the continuity of b by means of the estimate

$$|b(x, y)| \leqslant c \|x\|_1 \|y\|_2, \quad x \in H_1, \ y \in H_2. \tag{24}$$

We easily see the equivalence with other notions of continuity. Estimate (24) makes it possible to define the (functional) norm of b,

$$\|b\| = \inf\{c : c \text{ from } (24)\} = \sup_{\substack{0 \neq x \in H_1 \\ 0 \neq y \in H_2}} \frac{|b(x, y)|}{\|x\|_1 \|y\|_2}. \tag{25}$$

Continuous sesquilinear forms on Hilbert spaces may be represented by linear maps.

Theorem 17.8 *Let H_1 and H_2 be pre-Hilbert spaces, and let $b: H_1 \times H_2 = \mathbb{C}$ be a continuous sesquilinear form. Then we have:*

(a) *If H_1 is complete, then b may be represented by*

$$b(x, y) = (x, Ay)_1, \quad x \in H_1, \ y \in H_2,$$

where the continuous linear map $A: H_2 \to H_1$ is uniquely determined by b, and

$$\|b\| = \|A\|.$$

(b) *If H_2 is complete, then b may be represented by*

$$b(x, y) = (Bx, y)_2, \quad x \in H_1, \ y \in H_2,$$

where the continuous linear map $B: H_1 \to H_2$ is again uniquely determined by b, and $\|b\| = \|B\|$.

Proof. (a) For fixed $y \in H_2$ the functional b_y defined by $b_y(x) := b(x, y)$ belongs to H_1'. By the Riesz theorem there exists a uniquely determined element $z_y =: Ay \in H_1$ such that

$$b(x, y) = b_y(x) = (x, z_y)_1 = (x, Ay)_1.$$

The operator A is uniquely determined by b. Indeed were $b(x, y) = (x, A_1 y)_1 = (x, A_2 y)_1$ for all $x \in H_1$, $y \in H_2$, then we would have

$$(x, A_1 y - A_2 y)_1 = 0 \quad \text{for all } x \in H_1. \tag{26}$$

In particular, if in (26) we put $x := A_1 y - A_2 y$, it follows that $A_1 y = A_2 y$ for all $y \in H_2$, that is, $A_1 = A_2$. In the same way the linearity of A can be shown. If we use the Schwarz inequality, then by (25) we have

$$\| b \| = \sup \frac{|b(x, y)|}{\| x \|_1 \| y \|_2} = \sup \frac{|(x, Ay)_1|}{\| x \|_1 \| y \|_2} \leqslant \sup \frac{\| Ay \|}{\| y \|_2} = \| A \|.$$

On the other hand

$$\| b \| = \sup \frac{|(x, Ay)_1|}{\| x \|_1 \| y \|_2} \geqslant \sup \frac{|(Ay, Ay)_1|}{\| Ay \|_1 \| y \|_2} = \sup \frac{\| Ay \|_1}{\| y \|_2} = \| A \|.$$

From this the continuity of A follows, and $\| b \| = \| A \|$.

(b) Put $\beta(y, x) := \overline{b(x, y)}$ and interchange the roles of H_1 and H_2. ■

The Lax–Milgram theorem is important for the existence of solutions:

Theorem 17.9 *Let H be a Hilbert space and $b: H \times H \to \mathbb{C}$ a sesquilinear map, for which*

$$|b(x, y)| \leqslant \gamma \| x \| \, \| y \|, \quad x, y \in H \ (\text{continuity})$$

and

$$|b(x, x)| \geqslant \delta \| x \|^2, \quad x \in H \ (\text{strictly positive})$$

where $\gamma, \delta > 0$. Then there exists a unique bijective linear map $B: H \to H$, continuous in both directions and uniquely determined by b, with

$$b(x, y) = (Bx, y), \quad b(B^{-1}x, y) = (x, y), \quad x, y \in H,$$

and for the norms we have $\| B \| \leqslant \gamma$, $\| B^{-1} \| \leqslant 1/\delta$.

Proof. By Theorem 17.8 there exists a (unique) continuous, linear map $B: H \to H$ with $b(x, y) = (Bx, y)$ and with $\| B \| \leqslant \gamma$. We show that B is injective. If $Bx = 0$, then $b(x, y) = (Bx, y) = 0$ for all $y \in H$, so from $0 = |b(x, x)| \geqslant \delta \| x \|^2$ it follows that $x = 0$. Thus B^{-1} exists as a map $B^{-1}: \operatorname{im} B \to H$; we show the continuity of B^{-1}. We have $\| x \|^2 \cdot \delta \leqslant |b(x, x)| = |(Bx, x)| \leqslant \| Bx \| \cdot \| x \|$, and therefore $\|x\|\delta \leqslant \|Bx\|$, or $\|B^{-1}y\| \leqslant (1/\delta)\|y\|$, $y \in \operatorname{im} H$, which is $\|B^{-1}\| \leqslant 1/\delta$. It remains to show that B is surjective. From the continuity of B^{-1} it follows that $\operatorname{im} B$ is closed in H; if now $\operatorname{im} B \neq H$, then by the orthogonal decomposition theorem there would exist some $y_0 \neq 0$ with $(Bx, y_0) = 0$ for all $x \in H$, and we would have

$$0 = |(By_0, y_0)| = |b(y_0, y_0)| \geqslant \delta \|y_0\|^2,$$

which would imply that $y_0 = 0$, a contradiction. ■

17.4 V-elliptic and V-coercive forms, solution theorems

We are coming to the definition of V-ellipticity. Let $V \subsetneq H \subsetneq V'$ be a Gelfand triple, where V and H are Hilbert spaces (by Theorem 17.3 V' is then also a Hilbert space) and let $a(x, y)$ be a sesquilinear form on V.

Definition 17.3 *We say that $a(x, y)$ is V-elliptic, if the following conditions are satisfied:*

1. $|a(x, y)| \leqslant c_1 \|x\|_V \|y\|_V$ *for all* $x, y \in V$,
2. $|a(x, x)| \geqslant c_2 \|x\|_V^2$ *for all* $x \in V$ *(Gårding's inequality),*

where $c_1, c_2 > 0$ *are independent of* x, y.

Since a is sesquilinear, hence can be extended, it suffices that 1 and 2 be satisfied for all $x, y \in \mathscr{D}$, where $\mathscr{D}$ is a dense subspace of V.

We ought to say (V, H)-elliptic; since, however, for elliptic equations $H = L_2(\Omega)$ always, and only V is varied, it is sufficient to say V-elliptic.

Let $(x', x)_H$ be the representation (4) of functionals from V' by means of the scalar product $(\cdot,\cdot)_H$ in the Gelfand triple. By the Riesz theorem we have

$$(x', x)_H = (Rx', x)_V, \tag{27}$$

where R is the Riesz isomorphism $R: V' \to V$. (27) in combination with Theorem 17.8 gives

$$a(x, y) = (Ax, y)_V = (R^{-1}Ax, y)_H, \quad x, y \in V,$$

or spelt out in detail: condition 1 in Definition 17.3 is equivalent to the existence and continuity of a representation operator $L = R^{-1} \circ A$,

$$L: V \to V', \quad a(x, y) = (Lx, y)_H, \quad x, y \in V, \tag{28}$$

and we say that $L: V \to V'$ is V-elliptic, if $a(x, y)$ is.

As we shall see later, boundary conditions are included in the operator $L: V \to V'$, as was also the case for the operator L of §13. It is useful to consider so-called weak equations, respectively weak solvability. We interpret the weak equation

$$Lx = f,\ f \in V' \text{ as } a(x, y) = (f, y)_H \quad \text{for all } y \in V. \tag{29}$$

Theorem 19.9 (Lax–Milgram) immediately gives:

Theorem 17.10 (solution theorem) *Let $a(x, y)$ be V-elliptic, then $L: V \to V'$ is a linear topological isomorphism between the spaces V and V', and for the norms we have*

$$\|L\| \leqslant c_1, \quad \|L^{-1}\| \leqslant 1/c_2,$$

where the constants c_1, c_2 are from Definition 17.3. Put otherwise, the weak equation (29) $Lx = f$ possesses for each $f \in V'$ a unique solution $x \in V$, and this solution depends continuously on f.

Proof. We have $L = R^{-1} \circ A$, by Theorem 17.9 $A: V \to V$ is an isomorphism and by the Riesz theorem $R: V' \to V$ is an isomorphism. Hence $L: V \to V'$ is also an isomorphism. ■

Definition 17.4 *We say that $a(x, y)$ is V-coercive if the following conditions are satisfied:*

1. $|a(x, y)| \leqslant c_1 \|x\|_V \|y\|_V$ *for* $x, y \in V$,
2. *There exist constants $k \in \mathbb{C}$ and $c_2 > 0$ with $|a(x, x) + k\|x\|_H^2| \geqslant c_2 \|x\|_V^2$ for all $x \in V$ (Gårding's inequality).*

Because $|(x, y)_H| \leqslant \|x\|_V \|y\|_V$, V-coercive implies that the sesquilinear form $A(x, y) = a(x, y) + k(x, y)_H$ is V-elliptic, or that the operator $(L + kJ)$ is V-elliptic, where J denotes the embedding $V \to V'$ of the Gelfand triple $V \subsetneq H \subsetneq V'$.

If in addition the embedding $V \subsetneq H$ is compact (it was assumed to be continuous, injective and to have dense image) then with the help of the Riesz–Schauder theory for compact operators we obtain both a solution and a spectral theorem.

Theorem 17.11 *Let $V \subsetneq H \subsetneq V'$ be a Gelfand triple, suppose in addition that the embedding $V \subsetneq H$ is compact. Suppose that the sesquilinear form $a(x, y)$ is V-coercive, then the operator*

$$L - \lambda J: V \to V'$$

is a linear topological isomorphism between the spaces V and V' for all $\lambda \notin \mathrm{Sp}L$. For the spectrum $\mathrm{Sp}L$ we have:

(a) *The spectrum* SpL *consists of at most countably many eigenvalues* λ_n *which do not finitely accumulate; the adjoint operator* $L':V''\to V'$, $V''=V$, *has the spectrum* Sp$L'=\{\bar{\lambda}_n\}$.
(b) *The equation* $Lx-\lambda_n x=0$ *and the adjoint equation* $L'y-\bar{\lambda}_n y=0$ *have the same finite number of linearly independent solutions.*
(c) *The equation* $Lx-\lambda_n x=f$ *possesses at least one solution* x *if and only if* f *is orthogonal to the kernel of* $L'-\bar{\lambda}_n$, *that is,* $(f,v)_H=0$ *for* $v\in\ker(L'-\bar{\lambda}_n)$.
(d) *The coercivity constant* k (*Definition* 17.4.2) *does not belong to* SpL.

Proof. We have $L-\lambda J=(L+kJ)-\mu J$, by the preliminary remark $L+kJ$ is an isomorphism – thus (d) is proved – and by assumption J is compact. If we apply the map $(L+kJ)^{-1}$ to $(L+kJ)-\mu J$, we obtain the operator

$$I-\mu K:=I-(L+kJ)^{-1}\circ\mu J:V\to V,\quad \mu=k+\lambda,$$

where $K:V\to V$ is a compact operator and $I:V\to V$ is the identity operator. $I-\mu K$ is therefore a Riesz–Schauder operator and the remaining statements of our theorem follow from the Riesz–Schauder theory of §12.1 for the operator $\mu((1/\mu)I-K)$, where we have to modify the spectrum of §12.1 by means of the transformation $1/\mu$. ■

17.5 The Green operator

For special problems – here boundary value problems – it is useful to give the solution theorems in different versions. The notion of a Green operator or resolvent is particularly valuable – see also Definition 14.7. We define:

Definition 17.5 *We say that the continuous operator* $G:H\to H$ *is a Green solution operator for the weak equation*

$$Lx=f,\ f\in H \text{ preassigned},\quad x\in V,$$

that is, for

$$a(x,y)=(Lx,y)_H=(f,y)_H,\quad y\in V,$$

if $Gf=x(f\in H')$ *is the unique solution in* V *of the weak equation* $Lx=f$.

We have:

Theorem 17.12

(a) *Let* $a(x,y)$ *be* V*-elliptic, then a Green solution operator* G_L *exists for the weak equation* $Lx=f$. *The dual problem* $L'y=g$ *also possesses a Green* *solution operator* $G_{L'}$ *and we have*

$$G_{L'} = (G_L)' : H \to H.$$

(b) *If in addition the embedding* $V \subsetneq H$ *is compact, then* $G : H \to H$ *is a compact operator, to which we may directly apply the Riesz–Schauder theory of* §12.1. *For the spectrum of L we obtain*

$$\operatorname{Sp} L = \left\{ \lambda \in \mathbb{C} : \lambda = \frac{1}{\Lambda},\ 0 \neq \Lambda \in \operatorname{Sp} G \right\}, \tag{30}$$

it consists only of eigenvalues. Of course statements (a), (b) *and* (c) *of Theorem* 17.11 *again hold, and we see that* G and L *possess the same eigenfunctions.*

(c) *We also consider the self-adjoint case: in order that the Green operator* $G : H \to H$ *be self-adjoint,* $G = G'$, *it is necessary and sufficient that* $a(x, y)$ *be antisymmetric, that is,*

$$a(x, y) = \overline{a(y, x)} \quad \textit{for all } x, y \in V;$$

$L = L'$ *is also necessary and sufficient. If therefore G is self-adjoint and compact, the spectrum of L, consisting of countably infinitely many eigenvalues* λ_n *with* $|\lambda_n| \to \infty$, *lies on the real axis, and the eigenfunctions of L (suitably normed) form an orthonormal basis for H.*

Proof. (a) Let $L^{-1} : V' \to V$ be the solution operator from Theorem 17.10, then $G_L := i \circ L^{-1} \circ i' : H \subsetneq V' \to V \subsetneq H$ is a Green solution operator with the properties desired in Definition 17.5. By definition of the antidual operator $L' : V'' = V \to V'$ we have

$$(Lx, y)_H = (x, L'y)_H = \overline{(L'y, x)_H},$$

therefore $\tilde{a}(x, y) = \overline{a(y, x)}$ is the sesquilinear form belonging to L'. It is again elliptic, and so we have shown the existence of $G_{L'} : H \to H$. Substituting $x = G_L f$ and $y = G_{L'} g$ in $(Lx, y)_H = (x, L'y)_H$ gives

$$(G_L f, g)_H = (f, G_{L'} g)_H \quad \text{for all } f, g \in H,$$

that is, $G_{L'} = (G_L)'$.

(b) With $i : V \subsetneq H$ compact the composition of operators $i \circ L^{-1} \circ i' = G : H \to H$ is compact. For $\lambda \neq 0$ the equations

$$Lx - \lambda x = 0 \quad \text{and} \quad Gx - (1/\lambda)x = 0 \tag{31}$$

are equivalent, as one sees by application of G (respectively L). G and L

therefore have the same eigenfunctions for $\lambda \neq 0$, that is, for the eigenspaces we have

$$E(\lambda_n, L) = E(1/\lambda_n, G), \quad \lambda_n \neq 0. \tag{32}$$

Now for the spectrum of L. Since $L: V \to V'$ is an isomorphism, Theorem 17.10, and the embedding $J: V \subsetneq V'$ is compact, by Theorem 12.8, for all $\lambda \in \mathbb{C}$, the operator $L - \lambda = L - \lambda J$ has the index

$$\operatorname{ind}(L - \lambda) = 0,$$

and the spectrum of L can only consist of eigenvalues

$$\operatorname{Sp} L = \operatorname{Sp}_p L,$$

where $0 \notin \operatorname{Sp} L$ (L is an isomorphism!). The equivalences (31) therefore give formula (30) for the spectrum. We can now apply the spectral theorem from §12.1 to $G: H \to H$, use (31) and (32) and obtain in this way the statements of (b). Thus we can for example check (c) of Theorem 17.11: let $\lambda_n \neq 0$, then the equation $Lx - \lambda_n x = f$ is equivalent to $x - \lambda_n Gx = Gf$, the last equation is solvable if and only if (see §12.1) $Gf \perp \ker(I - \bar{\lambda}_n G')$, or written out explicitly, if $(Gf, v)_H = 0$ for all v with $v - \bar{\lambda}_n G'v = 0$, that is,

$$0 = (Gf, v)_H = (f, G'v)_H = \left(f, \frac{v}{\bar{\lambda}_n}\right)_H = \frac{1}{\lambda_n}(f, v)_H.$$

Since $\lambda_n \neq 0$ the equation $v - \bar{\lambda}_n G'v = 0$ is equivalent to $L'v - \bar{\lambda}_n v = 0$, and the condition $f \perp \ker(L' - \bar{\lambda}_n)$ (im H) is necessary and sufficient for the solvability of the original equation $Lx - \lambda x - \lambda_n x = f$.

(c) We have $(Lx, y)_H = a(x, y) = \overline{a(y, x)} = (x, Ly)_H$, that is, the asymmetry of $a(x, y)$ is necessary and sufficient for $L = L'$, from which if follows that

$$G = G_L = G_{L'} = G'.$$

We must still prove that $L = L'$ follows from $G = G'$. Since H is dense in V', the solution operator $L^{-1}: V' \to V$ is completely determined by G (see Theorem 17.10); $L^{-1} = L^{-1'}$ follows therefore from $G = G'$, and also $L = L'$.

The other statements of (c) follow from the spectral theorem §12.1, if we show that im G is dense in H. Namely we then have (see §12.1, assertion (h) and the formulae (30), (32))

$$H = \overline{\operatorname{im} G} = \bigoplus_n E(\Lambda_n, G) = \bigoplus_n E\left(\frac{1}{\lambda_n}, G\right) = \bigoplus_n E(\lambda_n, L),$$

where $0 \neq \Lambda_n = 1/\lambda_n$ are the eigenvalues of G and λ_n the eigenvalues of L. Because $G: H \to \operatorname{im} G \subset V$, and V is dense in H, it is enough to show that

$\operatorname{im} G$ is dense in V. Let $v \in V$, then there exists some $v' \in V'$ with $L^{-1}v' = v$. Since H is dense in V', we can approximate v' by $h_m \in H$, that is, $h_m \to v'$ in V'; and the continuity of L^{-1} guarantees that $L^{-1}h_m \to v$ in V. On H, however, L^{-1} agrees with G, which has the consequence that $L^{-1}h_m \in \operatorname{im} G$ and so we have proved everything. ■

We can also formulate and prove Theorem 17.12 for V-coercive forms $a(x, y)$. We have only to take the operator $i \circ (L + k)^{-1} \circ i'$ as the Green operator, that is, shift the spectrum by k. We can use the Green operator to obtain a Friedrichs inequality, or *a priori* estimate – see §13.

We write down the assumptions which we need for this. Suppose given:

1. a Gelfand triple $V \subsetneq H \subsetneq V'$,
2. a V-elliptic form $a(x, y) = (Lx, y)_H$, $L: V \to V'$ continuous,
3. an additional Hilbert space $W \subset H$, the embedding $\subset$ needs to be neither dense nor continuous,
4. suppose that $L: W \to H$ is continuous, $L: V \to H$ is usually not definable.

We equip $V \cap W$ with the norm $\|\cdot\|_V + \|\cdot\|_W$; we see immediately that

$$L: V \cap W \to H \text{ is continuous,} \tag{33}$$

since $\|Lx\|_H \leqslant c\|x\|_W \leqslant c\{\|x\|_W + \|x\|_V\}$. If we use the Green operator $G: H \to V$ from Theorem 17.12, we see that

$$V \cap W \subset \operatorname{im} G. \tag{34}$$

Let $x \in V \cap W$, by (33) (here we do not need the continuity of (33)) $Lx =: f$ is in H, therefore $x = Gf$ (unique solution in V) in $\operatorname{im} G$, and so we have demonstrated (34). We next assume a 'Weyl's lemma'

5. $\operatorname{im} G \subset V \cap W$,

and finally

6. let $V \cap W$ be complete in the $\|\cdot\|_V + \|\cdot\|_W$-norm.

We have:

Theorem 17.13 *Under the assumptions* 1–6 *the Friedrichs inequality*

$$\|x\|_{V \cap W} \leqslant c\|Lx\|_H \textit{ holds for all } x \in V \cap W. \tag{35}$$

Proof. (34) and assumption 5 give

$$\operatorname{im} G = V \cap W,$$

and the properties of G, Theorem 17.12, show that $L: V \cap W \to H$ acts bijectively, the open mapping theorem applied to (33) ($V \cap W = \operatorname{im} G$ is complete, see 6!) gives the continuity of $L^{-1}: H \to V \cap W$, which is (35).

If instead of assumption 2 we assume that $a(x, y)$ is V-coercive, or that $(L+k)$ is V-elliptic, and argue with the operator $L+k$, we obtain the Friedrichs inequality in the form

$$\|x\|_{V \cap W} \leqslant c\{\|Lx\|_H + \|x\|_H\} \quad \text{for all } x \in V \cap W. \tag{36}$$

It is easily seen that $\lambda = 0$ fails to be an eigenvalue of L if and only if the Friedrichs inequality (36) may be written in the simplified form (35). ■

17.6 The concepts V-elliptic and V-coercive for differential operators

We wish to restrict the definitions of V-elliptic (respectively V-coercive) for differential operators. Let $H = L_2(\Omega)$ and let V be a closed subspace (with $\|\cdot\|_m$-norm) lying between $\mathring{W}_2^m(\Omega)$ and $W_2^m(\Omega)$.

For differential operators we define:

Definition 17.6 *We say that $a(x, y)$ is V-elliptic if the following conditions are satisfied:*

1. $|a(x, y)| \leqslant c_1 \|x\|_m \cdot \|y\|_m$ *for all* $x, y \in W_2^m(\Omega)$,
2. $\operatorname{Re} a(x, x) \geqslant c_2 \|x\|_m^2$ *for all* $x \in V$.

Here $c_1, c_2 > 0$ and are independent of x and y.

Definition 17.7 *We say that $a(x, y)$ is V-coercive, if the following conditions are satisfied*

1. $|a(x, y)| \leqslant c_1 \|x\|_m \cdot \|y\|_m$ *for all* $x, y \in W_2^m(\Omega)$,
2. *There exist real constants k and $c_2 > 0$ with*

$$\operatorname{Re} a(x, x) + k(x, x)_{L^2} \geqslant c_2 \|x\|_m^2 \quad \textit{for all } x \in V \tag{37}$$

(Gårding's inequality)

Of course the narrower definitions imply the wider ones (17.3 and 17.4) and all the theorems which we have formulated in this section hold.

For the spectrum Definition 17.7 has an important consequence.

Theorem 17.14 *In the half-plane*

$$\operatorname{Re} \lambda \leqslant -k \tag{38}$$

there are no values of the spectrum $\operatorname{Sp} L$. We deduce (38) directly from (37).

§18 Agmon's condition

The aim of this and the following section is to prove Agmon's theorem (Agmon [2]), that is, to give conditions for V-coercion, where V is a subspace determined by boundary values (see Definition 14.3) lying between $W_2^m(\Omega)$ and $\mathring{W}_2^m(\Omega)$,

$$\mathring{W}_2^m(\Omega) \subset V \subset W_2^m(\Omega)$$

We continue with the notation of §11; thus for example equipping the basic derivatives $\partial/\partial x_1, \ldots, \partial/\partial x_r$ always with the factor $1/\mathrm{i} = -\sqrt{(-1)}$, so that under the Fourier transformation $\mathscr{F}$

$$\left(\frac{1}{\mathrm{i}}\frac{\partial}{\partial x_1}, \ldots, \frac{1}{\mathrm{i}}\frac{\partial}{\partial x_r}\right) \text{ becomes } (\xi_1, \ldots, \xi_r) = \xi.$$

We now formulate Agmon's condition, which plays the same role for his theorem as the L.Š. condition for the main theorem in §13.

Let Ω be an open set in $\mathbb{R}^r$ with sufficiently smooth frontier – since in this section the investigation is really one-dimensional, we postpone the exact smoothness conditions until the next section.

Let $A(x, \mathrm{D})$ be a linear differential operator of order $2m$; as in §14 (Theorem 14.8 *et seq.*) it is useful to write $A(x, \mathrm{D})$ in the form

$$A(x, \mathrm{D})\varphi = \sum_{|\alpha|, |\beta| \leqslant m} (-1)^{|\alpha|} \mathrm{D}^\alpha (a_{\alpha\beta}(x) \mathrm{D}^\beta(\varphi)). \tag{1}$$

We suppose that A is strongly elliptic on $\bar{\Omega}$, that is,

$$\operatorname{Re} A^{\mathrm{H}}(x, \xi) = \operatorname{Re} \sum_{|\alpha| = |\beta| = m} a_{\alpha\beta} \xi^{\alpha+\beta} > 0 \quad \text{for all } 0 \neq \xi \in \mathbb{R}^r, x \in \bar{\Omega}. \tag{2}$$

Suppose given normal (Definition 14.1) boundary value operators

$$b_j(x, \mathrm{D}), \quad j = 1, \ldots, p,$$

of order $0 \leqslant m_s = \operatorname{ord} b_s \leqslant m - 1$, let $0 \leqslant p \leqslant m$, where the case $p = 0$ implies that no boundary conditions are preassigned, therefore $V = W_2^m(\Omega)$ (we also consider this case). We fix some point $x_0 \in \partial\Omega$, take x_0 as origin of coordinates and choose the coordinate axis $x_r = t$ in the direction of the inward pointing normal and $(x_1, \ldots, x_{r-1})$ as local coordinates on $\partial\Omega$. We apply the Fourier transformation $\mathscr{F}_{r-1}$ to $((1/\mathrm{i})(\partial/\partial x_1), \ldots, (1/\mathrm{i})(\partial/\partial x_{r-1}))$ to obtain $(\xi_1, \ldots, \xi_{r-1}) = \xi \in \mathbb{R}^{r-1}$ and we consider the linear differential equation with constant coefficients

$$\tilde{A}^{\mathrm{H}}\left(x_0; \xi', \frac{1}{\mathrm{i}}\frac{\mathrm{d}}{\mathrm{d}t}\right) v(t) = 0, \tag{3}$$

with the initial conditions

$$b_j^{\mathrm{H}}\left(x_0;\xi',\frac{1}{\mathrm{i}}\frac{\mathrm{d}}{\mathrm{d}t}\right)v(t)|_{t=0}=0,\quad j=1,\ldots,p. \tag{4}$$

Here the operator $\tilde{A}$ is distinguished from A by means of the coefficients

$$\tilde{a}_{\alpha\beta}=\frac{a_{\alpha\beta}+\bar{a}_{\beta\alpha}}{2} \tag{5}$$

As in §11 we use $\mathscr{M}^+$ to denote the solutions of (3) which lie in $W_2^m(\mathbb{R}_+^1)$; for equivalent characterisations see §11. For short we write

$$\mathrm{D}_t=\frac{1}{\mathrm{i}}\frac{\mathrm{d}}{\mathrm{d}t} \tag{6}$$

and formulate Agmon's condition for $x_0\in\partial\Omega$.

Condition 18.1 *For all* $0\neq\xi'\in\mathbb{R}^{r-1}$, *we suppose that the non-zero solutions* $v(t)\in W_2^m(\mathbb{R}_+^1)$ (*or* $\in\mathscr{M}^+$) *of the initial value problem* (3) *and* (4) *satisfy the inequality*

$$\operatorname{Re}\int_0^\infty \sum_{\substack{|\alpha'|+k=m\\|\beta'|+l=m}} a_{(\alpha',k),(\beta',l)}(x_0)\cdot\xi'^{\alpha'}\cdot\xi'^{\beta'}\cdot\mathrm{D}_t^k v(t)\cdot\overline{\mathrm{D}_t^l v(t)}\cdot\mathrm{d}t>0. \tag{7}$$

The integral in (7) *converges because* $v\in\mathscr{M}^+$.

Remarks If $p=0$ in Condition 18.1 we require that all solutions v of (3) with $0\not\equiv v\in\mathscr{M}^+$ satisfy the inequality (7). If $p=m$, and it follows from (3) and (4) that $0\equiv v\in\mathscr{M}^+$ (corresponding to the L.Š. condition), then in this case Condition 18.1 is never satisfied.

We next give Agmon's condition in an equivalent formulation which corresponds to version 3 (Condition 11.3) of the L.Š. condition, and which may be applied directly to the proof of Agmon's theorem. We need three lemmas on ordinary differential equations with constant coefficients.

We consider the ordinary differential operators on $\mathbb{R}_+^1$

$$L(\mathrm{D}_t)=\sum_{k,l=0}^{m}(-1)^l a_{kl}\mathrm{D}_t^{k+l} \tag{8}$$

of order $\leqslant 2m$, and the associated sesquilinear form

$$A[\varphi,\psi]=\int_0^\infty\sum_{k,l=0}^{m}a_{kl}\mathrm{D}_t^k\varphi(t)\cdot\overline{\mathrm{D}_t^l\psi(t)}\cdot\mathrm{d}t \tag{9}$$

for $\varphi, \psi \in W_2^m(\mathbb{R}_+^1)$. For us the Hermitian part is important

$$\tilde{A}[\varphi,\psi] = \tfrac{1}{2}(A[\varphi,\psi] + \overline{A[\psi,\varphi]}), \tag{10}$$

as also the operator

$$\tilde{L}(D_t) = \sum_{k,l=0}^{m} (-1)^l \tilde{a}_{kl} D_t^{k+l}, \quad \tilde{a}_{kl} = \frac{a_{kl} + \overline{a_{lk}}}{2}. \tag{11}$$

Here $\tilde{L}$ is the operator associated with the form $\tilde{A}$ (the relation between A and L, respectively between $\tilde{A}$ and $\tilde{L}$, follows by the second Green formula, see Theorem 14.8).

Suppose given p linear normal (Definition 14.1) boundary value operators with constant coefficients

$$b_j(D_t)|_{t=0}, \quad j = 1,\ldots,p, \tag{12}$$

of order $0 \leqslant m_j = \operatorname{ord} b_j \leqslant m-1$, let $0 \leqslant p \leqslant m$. Let V be the subspace of $W_2^m(\mathbb{R}_+^1)$ determined by the boundary conditions b_j, $j = 1,\ldots,p$, that is,

$$V = \{\varphi \in W_2^m(\mathbb{R}_+^1) : b_j\varphi = 0, j = 1,\ldots,p\}, \quad V := W_2^m(\mathbb{R}_+^1) \text{ if } p = 0. \tag{13}$$

Since the functions $\varphi \in \mathscr{D}(\mathbb{R}_+^1)$ always satisfy the boundary conditions $b_j\varphi = 0$, $j = 1,\ldots,p$, and because of the continuity of the boundary value operators $b_j : W_2^m(\mathbb{R}_+^1) \to \mathbb{C}$, it follows that

$$\mathring{W}_2^m(\mathbb{R}_+^1) \subset V \subset W_2^m(\mathbb{R}_+^1). \tag{14}$$

We equip V with the topology induced from W_2^m, V is then a closed subspace of $W_2^m(\mathbb{R}_+^1)$. We say – following Definition 17.6 – that $A[\varphi,\psi]$ is V-elliptic, if

$$\operatorname{Re} A[\varphi,\varphi] \geqslant c\|\varphi\|_m^2 = c\int_0^\infty \sum_{k=0}^m |D_+^k\varphi|^2 \, dt \quad \text{for all } \varphi \in V = W^m(\{b_j\}_1^p). \tag{15}$$

We first prove a one-dimensional result of Gårding.

Lemma 18.1 *Two conditions are necessary and sufficient for the* $\mathring{W}_2^m(\mathbb{R}_+^1)$-*ellipticity of* $A[\varphi,\varphi]$:

(a) $\operatorname{ord} \tilde{L}(\xi) = \operatorname{ord} \operatorname{Re} L(\xi) = 2m$,
(b) $\tilde{L}(\xi) = \operatorname{Re} L(\xi) > 0$ *for all* $0 \neq \xi \in \mathbb{R}^1$.

Proof. Sufficiency: It follows from (a) and (b) that there exists some positive constant $c > 0$ with

$$\tilde{L}(\xi) \geqslant c \sum_{k=0}^m \xi^{2k} \quad \text{for all } \xi \in \mathbb{R}^1. \tag{16}$$

Let $\tilde{\varphi}\in \mathring{W}_2^m(\mathbb{R}_+^1)$, by Lemma 3.4 we can extend φ by zero to some $\tilde{\varphi}\in W_2^m(\mathbb{R}^1)$, let $\tilde{\phi}$ be the Fourier transform of $\tilde{\varphi}$. Then using the Parseval formula, see Theorem 1.24, we have

$$\operatorname{Re} A[\varphi,\varphi]=\frac{1}{2\pi}\int_{-\infty}^{\infty}\tilde{L}(\xi)\cdot|\phi(\xi)|^2\,\mathrm{d}\xi,$$

and with (16)

$$\operatorname{Re} A[\varphi,\varphi]\geqslant\frac{c}{2\pi}\sum_{k=0}^{m}\int_{-\infty}^{\infty}\xi^{2k}|\phi(\xi)|^2\,\mathrm{d}\xi\geqslant c_1\|\varphi\|_m^2,$$

which is (15) for $\varphi\in \mathring{W}_2^{2m}(\mathbb{R}_+)$.

Necessity: We fix a function different from zero $\varphi(t)$ in $\mathring{W}_2^m(\mathbb{R}_+)$. If either (a) or (b) is not satisfied, then there exists some $0\neq\xi_0\in\mathbb{R}^1$ with $\tilde{L}(\xi_0)=0$ or the order of $\tilde{L}(\xi)$ is smaller than $2m$. In the first case for $\varepsilon>0$ we write

$$f_\varepsilon(t):=\varepsilon^{1/2}\cdot\mathrm{e}^{\mathrm{i}\xi_0 t}\cdot\varphi(\varepsilon t),$$

and in the second case

$$f_\varepsilon(t):=\varepsilon^{m-1/2}\cdot\varphi\left(\frac{x}{\varepsilon}\right).$$

We have that $f_\varepsilon\in \mathring{W}_2^m(\mathbb{R}_+^1)$ and in both cases we can verify that

$$\lim_{\varepsilon\to 0}\|f_\varepsilon\|_m\quad\text{exists and }>0,$$

and

$$\lim_{\varepsilon\to 0}\operatorname{Re} A[f_\varepsilon,f_\varepsilon]=0.$$

This says that (15) cannot hold for f_ε, and so we have shown the necessity of both conditions. ■

Lemma 18.2 *Let* $V=\{\varphi\in W_2^m(\mathbb{R}_+^1):b_j\varphi=0,\quad j=1,\dots,p\}$, *respectively* $V=W_2^m(\mathbb{R}_+^1)$ *if* $p=0$. *The following three conditions (together) are necessary and sufficient for the V-ellipticity of* $A[\varphi,\psi]$:

(a) $\operatorname{ord}\tilde{L}(\xi)=2m$,
(b) $\tilde{L}(\xi)=\operatorname{Re}L(\xi)>0$ for all $0\neq\xi\in\mathbb{R}^1$,
(c) *for all solutions* $v(t)\not\equiv 0$ *of the differential equation*

$$\tilde{L}(\mathrm{D}_t)v(t)=0,\ t\in\mathbb{R}_+^1 \tag{17}$$

in V(that is, $v\in V$) we have

$$\operatorname{Re} A[v,v]>0. \tag{18}$$

(18) is Agmon's condition (7).

Proof. Necessity. Since $V \supset \mathring{W}_2^m(\mathbb{R}_+^1)$ the necessity of conditions (a) and (b) follows from Lemma 18.1, while condition (c), that is, (18), follows immediately from (15). Before proving the sufficiency we make a remark. Let $\mathcal{M}^+(b)$ be the subspace of solutions of (17) in V, $\mathcal{M}^+(b)$ is finite dimensional, let $v_1, \ldots, v_N$ be a basis, then each $0 \not\equiv v \in \mathcal{M}^+(b)$ may be represented as

$$v = \sum_{i=1}^{N} d_i v_i, \quad \text{with} \quad (d_1, \ldots, d_N) \neq 0, \tag{19}$$

and condition (c) may be written as

$$\operatorname{Re} A[v,v] = \tilde{A}[v,v] = \sum_{i,j=1}^{N} \tilde{A}[v_i, v_j] d_i \bar{d}_j > 0. \tag{20}$$

For $\tilde{A}$ see (10). This implies that the Hermitian quadratic form $\sum_{i,j} \tilde{A}[v_i, v_j] d_i \bar{d}_j$ is positive definite in $(d_1, \ldots, d_N)$. From (20) it follows that

$$\operatorname{Re} A[v,v] \geqslant c_1 \sum_{i=1}^{N} |d_i|^2 \geqslant c_2 \| v \|_m^2 \quad \text{for all } v \in \mathcal{M}^+(b), \tag{21}$$

where $c_1, c_2 > 0$.

Now we prove that conditions (a)–(c) are sufficient. The characteristic polynomial $\tilde{L}(\lambda)$ of $L(\mathrm{D}_t)$ has real coefficients, degree $\tilde{L} = 2m$ (condition (a)), and there are no roots on the real axis (condition (b)). Therefore m characteristic roots lie in the upper half plane $\operatorname{Im} \lambda > 0$ and m in the lower half plane, therefore $\dim \mathcal{M}^+ = m$. Since we have assumed that the boundary operators b_j, $j = 1, \ldots, p$ are normal, they are linearly independent, and we calculate the dimension of $\mathcal{M}^+(b)$ as

$$N = \dim \mathcal{M}^+(b) = m - p.$$

We now complete the normal boundary operator system $b_j, j = 1, \ldots, p$ to a Dirichlet system $B_j, j = 1, \ldots, m$ of order m. By Lemma 14.1 the system B_j $j = 1, \ldots, m$ is equivalent to D_t^{j-1}, $j = 1, \ldots, m$ which by Theorem 8.9 implies that we have

$$\begin{aligned} &\{\varphi \in W_2^m(\mathbb{R}_+^1) : B_j \varphi = 0, j = 1, \ldots, m\} \\ &\quad = \{\varphi \in W_2^m(\mathbb{R}_+^1) : \mathrm{D}_t^{j-1} \varphi = 0, j = 1, \ldots, m\} = \mathring{W}_2^m(\mathbb{R}_+^1). \end{aligned} \tag{22}$$

We show now that each function $\varphi \in V$, (13), possesses the unique decomposition

$$\varphi = \varphi_0 + v, \tag{23}$$

where $v \in \mathcal{M}^+(b)$ and $\varphi_0 \in \mathring{W}_2^m(\mathbb{R}_+^1)$. We determine v from (23)

$$v = \sum_{i=1}^{m-p} d_i v_i, \quad N = m - p, \quad \text{see (19)},$$

by calculating the numbers d_i from the system of equations

$$\sum_{i=1}^{m-p} d_i[B_{p+j}(\mathrm{D}_t)v_i]_{t=0} = [B_{p+j}(\mathrm{D}_t)\varphi]_{t=0}, \quad j=1,\dots,m-p. \tag{24}$$

We assert that the matrix

$$[B_{p+j}(\mathrm{D}_t)v_i]_{t=0}, \quad i,j=1,\dots,m-p$$

is non-singular. Otherwise a non-trivial solution would satisfy the homogeneous system of equations (24), from which would follow the existence of a function $v \not\equiv 0$ with $v \in \mathscr{M}^+$ and $B_j v|_{t=0}=0$ for $j=1,\dots,m$, (Lemma 14.1). Therefore $\mathrm{D}_t^{j-1}v|_{t=0}=0, j=1,\dots,m$ which is impossible.

We have therefore proved that the system of equations (24) is uniquely solvable for (d_i), and we obtain the desired decomposition (23) as

$$v := \sum_{i=1}^{m-p} d_i v_i, \quad \varphi_0 := \varphi - v.$$

we still wish to assert that $\varphi_0 \in \mathring{W}_2^m(\mathbb{R}_+^1)$. We have $\varphi, v \in V$, which is

$$b_j(\mathrm{D}_t)(\varphi - v)|_{t=0} = 0 \quad \text{for } j=1,\dots,p.$$

The conditions

$$B_{p+j}(\mathrm{D}_t)(\varphi - v)|_{t=0} = 0 \quad \text{for } j=1,\dots,m-p,$$

come from (24) and (22) gives $\varphi_0 = \varphi - v \in \mathring{W}_2^m(\mathbb{R}_+^1)$. The decomposition (23) possesses the orthogonality property

$$\tilde{A}[v,\varphi_0] = 0. \tag{25}$$

We write the second Green formula (see Theorem 14.8) as

$$\tilde{A}[\varphi,\psi] = \int_0^\infty \tilde{L}\varphi\cdot\bar{\psi}\,\mathrm{d}t - \sum_{j=1}^m \mathscr{C}_j\varphi\cdot\overline{\mathrm{D}_t^{j-1}\psi}|_{t=0},$$

put $\varphi = v$, and $\psi = \varphi_0$, obtaining

$$\tilde{A}[v,\varphi_0] = \int_0^\infty 0\cdot\bar{\psi}\,\mathrm{d}t - \sum_{j=1}^m \mathscr{C}_j v\cdot 0 = 0.$$

Now for the actual proof. Let $\varphi \in V$, we decompose according to (23) and because of (25) have

$$\begin{aligned}\operatorname{Re} A[\varphi,\varphi] &= \tilde{A}[\varphi,\varphi] = \tilde{A}[\varphi_0+v,\varphi_0+v]\\ &= \tilde{A}[\varphi_0,\varphi_0] + 2\operatorname{Re}\tilde{A}[\varphi_0,v] + \tilde{A}[v,v]\\ &= \tilde{A}[\varphi_0,\varphi_0] + \tilde{A}[v,v].\end{aligned} \tag{26}$$

Since $\varphi_0 \in \mathring{W}_2^m(\mathbb{R}_+^1)$ we can apply Lemma 18.1 and have

$$\tilde{A}[\varphi_0, \varphi_0] \geqslant c_1 \|\varphi_0\|_m^2,$$

and for $v \in \mathcal{M}^+(b)$ by (21) we have

$$\tilde{A}[v, v] \geqslant c_2 \|v\|_m^2.$$

Together with (26) the last inequalities give

$$\operatorname{Re} A[\varphi, \varphi] \geqslant c\{\|\varphi_0\|_m^2 + \|v\|_m^2\} \geqslant c\|\varphi_0 + v\|_m^2 = c\|\varphi\|_m^2 \quad \text{for all } \varphi \in V,$$

which is (15). ∎

Lemma 18.3 *Suppose once more that* $V = W_2^m(\{b_j\}_1^p) = \{\varphi \in W_2^m(\mathbb{R}_+)$: $b_j\varphi = 0,\ j = 1, \ldots, p\}$. *The estimate*

$$\operatorname{Re} A[\varphi, \varphi] \geqslant c_1 \|\varphi\|_m^2 - c_2 \sum_{j=1}^{p} |[b_j(\mathrm{D}_t)\varphi]_{t=0}|^2 \quad \text{for all } \varphi \in W_2^m(\mathbb{R}_+^1),$$

$c_1, c_2 > 0$, *is necessary and sufficient for the V-ellipticity of* $A[\varphi, \psi]$.

Proof. That (27) suffices follows immediately, since for $\varphi \in V$ (27) becomes (15). We prove the necessity. Since the boundary value operators $b_j(\mathrm{D}_t)$, $j = 1, \ldots, p$, are normal, we can complete them to a Dirichlet system (Definition 14.2), $B_j(\mathrm{D}_t)$, $j = 1, \ldots, m$, and see that the matrix $\{B_j(\mathrm{i}k)\}, k$, $j = 1, \ldots, m$, $\mathrm{i} = \sqrt{(-1)}$, is not singular. Let $\varphi \in W_2^m(\mathbb{R}_+^1)$, the linear system of equations

$$\sum_{k=1}^{m} d_k B_j(\mathrm{i}k) = [b_j(\mathrm{D}_t)\varphi]_{t=0}, \quad j = 1, \ldots, p,$$

$$\sum_{k=1}^{m} d_k B_j(\mathrm{i}k) = 0, \qquad j = p+1, \ldots, m, \tag{28}$$

possesses a uniquely determined solution

$$d_k = d_k(\varphi), \quad k = 1, \ldots, m,$$

and we define the map P of $W_2^m(\mathbb{R}_+^1)$ onto the subspace $E = [\mathrm{e}^{-kt}, k = 1, \ldots, m]$ of $W_2^m(\mathbb{R}_+^1)$ spanned by the functions e^{-kt}, $k = 1, \ldots, m$ by means of

$$P\varphi = \sum_{k=1}^{m} d_k(\varphi)\mathrm{e}^{-kt}. \tag{29}$$

We easily see that the norms $\|\psi\|_m$ and $(\sum_{j=1}^m |\mathrm{D}_t^{j-1}\psi|_{t=0}^2)^{1/2}$ are equivalent on E. By Lemma 14.1 the Dirichlet system $\{B_j(\mathrm{D}_t), j = 1, \ldots, m\}$ is equivalent to $\{\mathrm{D}_t^{j-1}, j = 1, \ldots, m\}$, therefore from (29) and (28) we have the estimate

$$\|P\varphi\|_m^2 \leqslant c_1 \sum_{j=1}^{m} |D_t^{j-1}P|_{t=0}^2 \leqslant c_2 \sum_{j=1}^{m} |B_j(D_t)P\varphi|_{t=0}^2$$

$$= \sum_{j=1}^{m} \left| \sum_{k=1}^{m} d_k B_j(\mathrm{i}k)\mathrm{e}^{-kt} \right|_{t=0}^2 = \sum_{j=1}^{m} \left| \sum_{k=1}^{m} d_k B_j(\mathrm{i}k) \right|^2 = \sum_{j=1}^{p} |b_j(D)\varphi|_{t=0}^2. \quad (30)$$

The system of equations (28) was chosen that we have – see (29) –

$$\varphi = P\varphi \in V = W_2^m(\{b_j\}_1^p),$$

and the assumption about V-ellipticity gives

$$\operatorname{Re} A[\varphi - P\varphi, \varphi - P\varphi] \geqslant c_4 \|\varphi - P\varphi\|_m^2. \quad (31)$$

Since $A[\varphi, \psi]$ satisfies the continuity estimate,

$$|A[\varphi, \psi]| \leqslant c_5 \|\varphi\|_m \|\psi\|_m,$$

see definition (a), we have

$$\operatorname{Re} A[\varphi, \varphi] \geqslant \operatorname{Re} A[\varphi - P\varphi, \varphi - P\varphi] - 2c_6 \|\varphi\|_m \cdot \|P\varphi\|_m - c_6^2 \|P\varphi\|_m^2. \quad (32)$$

If we use the inequality $2rs \leqslant \varepsilon r^2 + s^2/\varepsilon$, we obtain from (31) and (32) that

$$\operatorname{Re} A[\varphi, \varphi] \geqslant c_4 \|\varphi - P\varphi\|_m^2 - \varepsilon \|\varphi\|_m^2 - c_6^2 \left(1 + \frac{1}{\varepsilon}\right) \|P\varphi\|_m^2. \quad (33)$$

The inequality $\|\varphi - P\varphi\|_m^2 \geqslant (1-\varepsilon)\|\varphi\|_m^2 - (1/\varepsilon)\|P\varphi\|_m^2$ together with (33) and (30) gives, if we write $\varepsilon := \frac{1}{2}c_4(c_4+1)$, that

$$\operatorname{Re} A[\varphi, \varphi] \geqslant \tfrac{1}{2}c_4 \|\varphi\|_m^2 - c_7 \|P\varphi\|_m^2 \geqslant c_1 \|\varphi\|_m^2 - c_2 \sum_{j=1}^{p} |[b_j(D_t)\varphi]_{t=0}|^2,$$

which is (27). ∎

We are now in a position to reformulate Agmon's Condition 18.1. Let $A(x, D)$ be a differential operator given by (1), and

$$a(\varphi, \psi) = \int_\Omega \sum_{\substack{|\alpha| \leqslant m \\ |\beta| \leqslant m}} a_{\alpha\beta}(x) D^\beta \varphi \overline{D^\alpha \psi} \, dx,$$

the associated sesquilinear form. We are interested in the principal parts

$$A^H(x, D)\varphi = \sum_{\substack{|\alpha| = m \\ |\beta| = m}} (-1)^{|\alpha|} D^\alpha (a_{\alpha\beta}(x) D^\beta \varphi),$$

$$a^H(\varphi, \psi) = \int_\Omega \sum_{\substack{|\alpha| = m \\ |\beta| = m}} a_{\alpha\beta}(x) D^\beta \varphi \cdot \overline{D^\alpha \psi} \, dx.$$

Suppose we are given p normal boundary operators

$$b_j(x,\mathrm{D}), \quad j=1,\ldots,p,$$

where $0\leqslant p\leqslant m$ ($p=0$ means that no boundary conditions are preassigned).

Condition 18.2 *Let $x_0\in\partial\Omega$, let the operator $A(x_0,\mathrm{D})$ be strongly elliptic of order $2m$, let $0\neq\xi'\in\mathbb{R}^{r-1}$. We consider the sesquilinear form* ($t\in\mathbb{R}^1_+$)

$$a^{\mathrm{H}}(x_0,\xi';\varphi,\psi)=\int_0^\infty \sum_{\substack{|\alpha'|+k=m\\|\beta'|+l=m}} a_{(\alpha',k)(\beta',l)}(x_0)\xi'^{\alpha'}\cdot\xi'^{\beta'}\cdot \mathrm{D}_t^k\varphi(t)\cdot\mathrm{D}_t^l\overline{\psi(t)}\,\mathrm{d}t.$$

we have the estimate

$$\operatorname{Re}a^{\mathrm{H}}(x_0,\xi';\varphi,\varphi)\geqslant c_1\|\varphi\|_m^2-c_2\sum_{j=1}^{p}|[b_j^{\mathrm{H}}(x_0,\xi',\mathrm{D}_t)\varphi]_{t=0}|^2, \tag{34}$$

for all $\varphi\in W_2^m(\mathbb{R}^1_+)$, and the constants $c_1,c_2>0$ can be chosen independent of ξ', as ξ' varies on the unit sphere of $\mathbb{R}^{r-1}$, that is, $|\xi'|_{r-1}=1$.

We show the equivalence of Conditions 18.1 and 18.2.

It is easy to see that Condition 18.1 follows from 18.2; let $v\not\equiv 0$ be a solution of (3) and (4). Then because of (4) the estimate (34) becomes

$$\operatorname{Re}a^{\mathrm{H}}(x_0,\xi',v,v)\geqslant c_1\|v\|_m^2>0,$$

and so we have checked (7).

In order to prove the converse we write, in the sense of (8) and (9)

$$\begin{aligned}
&L_{\xi'}(\mathrm{D}_t)=A^{\mathrm{H}}(x_0,\xi';\mathrm{D}_t), && 0\neq\xi'\in\mathbb{R}^{r-1},\\
&A_{\xi'}[\varphi,\psi]=a^{\mathrm{H}}(x_0,\xi';\varphi,\psi), && 0\neq\xi'\in\mathbb{R}^{r-1},\\
&b_{j\xi'}(\mathrm{D}_t)=b_j^{\mathrm{H}}(x_0,\xi';\mathrm{D}_t), && j=1,\ldots,p, 0\neq\xi'\in\mathbb{R}^{r-1}.
\end{aligned}$$

Since $A(x_0,\mathrm{D})$ is strongly elliptic, for all $\eta\in\mathbb{R}^1$ we have

$$\tilde{L}_{\xi'}(\eta)=\operatorname{Re}L_{\xi'}(\eta)=\operatorname{Re}A^{\mathrm{H}}(x_0,\xi',\eta)\neq 0,$$

since, because $\xi'\neq 0$, the vector $(\xi',\eta)\neq 0$ and belongs to $\mathbb{R}^r$. Also

$$\operatorname{ord}\tilde{L}_{\xi'}(\eta)=\operatorname{ord}\operatorname{Re}L_{\xi'}(\eta)=2m \quad \text{for all } \eta\in\mathbb{R}^1.$$

Therefore conditions (a) and (b) of Lemma 18.2 are satisfied. We check condition (c). We have

$$\tilde{L}_{\xi'}(\mathrm{D}_t)v(t)=\tilde{A}(x_0,\xi',\mathrm{D}_t)v(t)=0$$

which is (3), and $v\in V$ implies (4), where we must write

$$b_{j\xi'}(\mathrm{D}_t):=b_j^{\mathrm{H}}(x_0,\xi',\mathrm{D}_t).$$

Condition 18.1 reads $\operatorname{Re} A_{\xi'}[v,v] = \operatorname{Re} a^{\mathrm{H}}(x_0,\xi',v,v) > 0$, which is (18), and therefore all three conditions of Lemma 18.2 are satisfied. We have that the form $A_{\xi'}[\varphi,\psi] = a^{\mathrm{H}}(x_0,\xi',\varphi,\psi)$ is V-elliptic, and Lemma 18.3 gives us the estimate

$$\operatorname{Re} A_{\xi'}[\varphi,\varphi] \geqslant c_1(\xi')\|\varphi\|_m^2 - c_2(\xi')\sum_{j=1}^{p}|b_j^{\mathrm{H}}(x_0,\xi',\mathrm{D}_t)\varphi(0)|^2,$$

for all $\varphi \in W_2^m(\mathbb{R}_+)$, hence, given that $A_{\xi'}[\varphi,\varphi] = a^{\mathrm{H}}(x_0,\xi',\varphi,\varphi)$ we have checked (34).

We have still to show that $c_1(\xi')$, $c_2(\xi')$ depend continuously on $\xi' \in \mathbb{R}^{r-1}\setminus\{0\}$ we write (34) in the equivalent form, $c_1 \neq 0$!

$$\|\varphi\|_m^2 \leqslant c_1(\xi')\cdot\operatorname{Re} A_{\xi'}[\varphi,\varphi] + c_2(\xi')\sum_{j=1}^{p}|b_j^{\mathrm{H}}(x_0,\xi',\mathrm{D}_t)\varphi(0)|^2, \tag{35}$$

and understand by continuous dependence, that if $\bar{\xi}'$ is near to ξ', we can also choose $c_i(\bar{\xi}')$ near to $c_i(\xi')$, $i = 1, 2$, so that

$$\|\varphi\|_m^2 \leqslant c_1(\xi')\cdot\operatorname{Re} A_{\xi'}[\varphi,\varphi] + c_2(\xi')\sum_{j=1}^{p}|b_j^{\mathrm{H}}(x_0,\xi',\mathrm{D}_t)\varphi(0)|^2, \tag{36}$$

holds, for all $\varphi \in W_2^m(\mathbb{R}_+)$. We choose $\bar{\xi}'$ so near to ξ' that

$$|\operatorname{Re} A_{\xi'}[\varphi,\varphi] - \operatorname{Re} A_{\xi'}[\varphi,\varphi]| \leqslant \varepsilon\|\varphi\|_m^2,$$

$$\left|\sum_{j=1}^{p}|b_j^{\mathrm{H}}(x_0,\xi',\mathrm{D}_t)\varphi(0)|^2 - \sum_{j=1}^{p}|b_j^{\mathrm{H}}(x_0,\bar{\xi}',\mathrm{D}_t)\varphi(0)|^2\right| \leqslant \varepsilon\|\varphi\|_m^2, \tag{37}$$

which is always possible because of the continuous dependence of the expressions $A^{\mathrm{H}}(\ldots\xi'\ldots)$, $b_j^{\mathrm{H}}(\ldots\xi'\ldots)$ on $\xi' \in \mathbb{R}^{r-1}\setminus\{0\}$. Here $\varepsilon > 0$ (independent of φ) is arbitrarily preassigned. (35) in conjunction with (37) gives

$$\|\varphi\|_m^2 \leqslant c_1(\xi')\cdot\operatorname{Re} A_{\bar{\xi}'}[\varphi,\varphi] + c_2(\xi')\sum_{j=1}^{p}|b_j^{\mathrm{H}}(x_0,\bar{\xi}',\mathrm{D}_t)\varphi(0)|^2$$
$$+ [\varepsilon c_1(\xi') + \varepsilon c_2(\xi')]\|\varphi\|_m^2.$$

If we impose the restriction $1 - \varepsilon c_1(\xi') - \varepsilon c_2(\xi) > 0$ on $\varepsilon > 0$ and if we put $c_i(\bar{\xi}'){:}c_i(\xi')/(1 - \varepsilon c_1(\xi') - \varepsilon c_2(\xi'))$, $i = 1, 2$, then we see that (36) holds, where $c_i(\bar{\xi}')$ can be arbitrarily close to $c_i(\xi')$, $i = 1, 2$. A compactness argument ends the proof, and we obtain the estimate (34) with constants c_1, c_2 independent of ξ', if ξ' varies on the unit sphere of $\mathbb{R}^{r-1}$.

Theorem 18.1 *Agmon's condition is invariant under admissible coordinate transformations.*

For the proof we use Condition 18.2. Let $\Phi: x=(x',t)\mapsto(y',\tau)=y$ be an admissible coordinate transformation, by Definition 2.9 its Jacobi matrix decomposes as

$$\frac{\partial\Phi}{\partial x}(x_0)=\begin{bmatrix} A & 0 \\ 0 & \sigma \end{bmatrix},\quad \sigma>0, x_0\in\partial\Omega. \tag{38}$$

By Theorem 10.3 we have

$$\tilde{a}^{\mathrm{H}}(y_0,\eta',\tilde{\varphi},\tilde{\bar{\psi}})=ca^{\mathrm{H}}(x_0(y_0),{}^{\mathrm{T}}A\eta',\varphi,\psi), c>0,$$

and

$$\tilde{b}_j^{\mathrm{H}}(y_0,\eta',\mathrm{D}_\tau)\tilde{\varphi}(\tau)|_{\tau=0}=\sigma b_j^{\mathrm{H}}(x_0(y_0),{}^{\mathrm{T}}A\eta',\sigma\mathrm{D}_\tau)\varphi(\tau)|_{\tau=0}. \tag{39}$$

By Theorem 4.1 (transformation theorem)

$$\|\tilde{\varphi}\|_m\simeq\|\varphi\|_m. \tag{40}$$

With (39) and (40), (34) becomes

$$\operatorname{Re}\tilde{a}^{\mathrm{H}}(y_0,\eta',\tilde{\varphi},\tilde{\varphi})\geqslant\tilde{c}_1\|\tilde{\varphi}\|_m-\tilde{c}_2\sum_{j=1}^{p}|\tilde{b}_j^{\mathrm{H}}(y_0,\eta',\mathrm{D}_\tau)\tilde{\varphi}|^2_{\tau=0},$$

where because of (38) $\mathbb{R}^{r-1}\ni{}^{\mathrm{T}}A\eta'\neq0$, if $\eta'\neq0$. Therefore we have proved the invariance of Agmon's condition 18.2 under admissible transformations. ■

§19 Agmon's theorem: conditions for V-coercion of strongly elliptic differential operators

19.1 The theorems of Gårding and Agmon

Let $A(x,\mathrm{D})$ be a linear differential operator of the form

$$A(x,\mathrm{D})\varphi:=\sum_{|\alpha|,|\beta|\leqslant m}(-1)^{|\alpha|}\mathrm{D}^\alpha(a_{\alpha\beta}(x)\mathrm{D}^\beta\varphi),\quad \operatorname{ord}A=2m, \tag{1}$$

and let $a(\varphi,\psi)$ be the associated (see Green's Theorem 14.8) sesquilinear form

$$a(\varphi,\psi):=\int_\Omega\sum_{\substack{|\alpha|\leqslant m\\|\beta|\leqslant m}}a_{\alpha\beta}(x)\cdot\mathrm{D}^\beta\varphi\cdot\mathrm{D}^\alpha\bar{\psi}\,\mathrm{d}x. \tag{2}$$

If the coefficients $a_{\alpha\beta}(x)$ are continuous and bounded on Ω, that is, $a_{\alpha\beta}\in C(\Omega)$, then $a(\varphi,\psi)$ satisfies the estimate

$$|a(\varphi,\psi)|\leqslant c\|\varphi\|_m\cdot\|\psi\|_m\quad\text{for all }\varphi,\psi\in W_2^m(\Omega). \tag{3}$$

Let V be a closed subspace of $W_2^m(\Omega)$, $V\subset W_2^m(\Omega)$, equipped with the Hilbert space structure induced from W_2^m. We take $H=L_2(\Omega)$, and say – following Definition 17.7 – that $a(\varphi,\psi)$ is V-coercive, if in addition to (3) Gårding's inequality

$$\operatorname{Re} a(\varphi, \varphi) + k(\varphi, \varphi)_0 \geq c_1 \|\varphi\|_m^2 \quad \text{holds for all } \varphi \in V, \tag{4}$$

where the constants $c_1 > 0, k$ are independent of φ.

We begin with some general remarks. It is clear that if $a(\varphi, \psi)$ is V-coercive and $W \subset V$, then $a(\varphi, \psi)$ is also W-coercive.

Theorem 19.1 *Let $b(\varphi, \psi)$ be a second continuous, sesquilinear form which, when $\varphi = \psi$, satisfies the stronger estimate, instead of* (3)

$$|b(\varphi, \varphi)| \leq c_2 \|\varphi\|_m \|\varphi\|_{\tilde{m}}, \quad \varphi \in W_2^m(\Omega) \tag{5}$$

where $\tilde{m} < m$; let V and Ω be such, that an Ehrling inequality holds

$$\|\varphi\|_{\tilde{m}} \leq \varepsilon \|\varphi\|_m + c(\varepsilon) \|\varphi\|_0 \quad \textit{for all } \varphi \in V, \ \varepsilon \textit{ arbitrary}, \tag{6}$$

then if the form $a(\varphi, \psi)$ is V-coercive, the same holds for the sesquilinear form

$$A(\varphi, \psi) = a(\varphi, \psi) + b(\varphi, \psi).$$

Proof. It is clear that $A(\varphi, \psi)$ is again continuous. We estimate $|b(\varphi, \varphi)|$ with the help of (5), (6) and $2rs \leq \bar{\varepsilon} r^2 + (1/\bar{\varepsilon}) s^2$, $\bar{\varepsilon} > 0$:

$$\begin{aligned} |b(\varphi, \varphi)| &\leq c_2 \|\varphi\|_m \|\varphi\|_{\tilde{m}} \leq c_2 \varepsilon \|\varphi\|_m^2 + c_2 c(\varepsilon) \|\varphi\|_m \|\varphi\|_0 \\ &\leq c_2 \varepsilon \|\varphi\|_m^2 + \frac{c_2 c(\varepsilon) \bar{\varepsilon}}{2} \|\varphi\|_m^2 + \frac{c_2 c(\varepsilon)}{2\bar{\varepsilon}} \|\varphi\|_0^2, \quad \varphi \in V. \end{aligned} \tag{7}$$

We choose $\varepsilon > 0 : c_2 \varepsilon = \frac{1}{4} c_1$, $0 < c_1$ is the constant from (4), and $\bar{\varepsilon} > 0$: $\frac{1}{2} c_2 \cdot c(\varepsilon) \bar{\varepsilon} = \frac{1}{4} c_1$. The inequality (7) then takes the form

$$|b(\varphi, \varphi)| \leq \tfrac{1}{2} c_1 \|\varphi\|_m^2 + c_3 \|\varphi\|_0^2 \tag{8}$$

and with the help of (8) we can show that $A(\varphi, \psi)$ is V-coercive:

$$\begin{aligned} \operatorname{Re} A(\varphi, \varphi) &= \operatorname{Re} a(\varphi, \varphi) + \operatorname{Re} b(\varphi, \varphi) \geq \operatorname{Re} a(\varphi, \varphi) - |b(\varphi, \varphi)| \\ &\geq c_1 \|\varphi\|_m^2 - k \|\varphi\|_0^2 - \tfrac{1}{2} c_1 \|\varphi\|_m^2 - c_3 \|\varphi\|_0^2 \\ &= \tfrac{1}{2} c_1 \|\varphi\|_m^2 - (k + c_3) \|\varphi\|_0^2 \quad \text{for all } \varphi \in V. \end{aligned}$$

■

We start with the smallest subspace, $V = \mathring{W}_2^m(\Omega)$, and prove Gårding's theorem [1].

Theorem 19.2 *Let the region Ω be bounded in $\mathbb{R}^r$, let the coefficients $a_{\alpha\beta}(x)$ of* (2) *be continuous on $\bar{\Omega}$*, that is,

$$a_{\alpha\beta} \in C(\bar{\Omega}), \quad |\alpha| \leq m, |\beta| \leq m.$$

Strong ellipticity of $A(x, \mathrm{D})$, (1), *on $\bar{\Omega}$ is necessary and sufficient for the $\mathring{W}_2^m(\Omega)$-coercion of $a(\varphi, \psi)$,* (2).

We remark that Gårding's theorem holds without any regularity requirements on the frontier $\partial\Omega$ of Ω.

Proof. Sufficiency: Because of Theorems 19.1 and 7.4 we only need to consider the principal parts A^{H} and $a^{\mathrm{H}}(\varphi, \psi)$. Let $A(x, \mathrm{D})$ be strongly elliptic on $\bar{\Omega}$. By Addendum 10.2 we have

$$\operatorname{Re} A^{\mathrm{H}}(x, \xi) \geqslant c_0 |\xi|^{2m} \quad \text{for all } x \in \bar{\Omega} \text{ and } \xi \in \mathbb{R}^r. \tag{9}$$

Here for the sake of simplicity we have taken the ellipticity constant $\gamma = 1$.

(a) We fix some point $x_0 \in \bar{\Omega}$, therefore have constant coefficients, and apply the Fourier transformation. Let $\varphi \in \mathscr{D}(\Omega)$, by (9) and the Parseval equation §1, (24) we have

$$\begin{aligned}\operatorname{Re} a_x^{\mathrm{H}}(\varphi, \varphi) &= \operatorname{Re} \int_\Omega \sum_{\substack{|\alpha|=m\\|\beta|=m}} a_{\alpha\beta}(x_0) \mathrm{D}^\alpha \varphi \mathrm{D}^\beta \bar{\varphi}\, \mathrm{d}x \\ &= \operatorname{Re} \frac{1}{(2\pi)^r} \int_{\mathbb{R}^r} \sum_{\substack{|\alpha|=m\\|\beta|=m}} a_{\alpha\beta}(x_0) \xi^\alpha \xi^\beta |\hat{\varphi}|^2\, \mathrm{d}\xi \\ &= \operatorname{Re} \frac{1}{(2\pi)^r} \int_{\mathbb{R}^r} A^{\mathrm{H}}(x_0, \xi) |\hat{\varphi}|^2\, \mathrm{d}\xi \geqslant \frac{c_0}{(2\pi)^r} \int_{\mathbb{R}^r} |\xi|^{2m} |\hat{\varphi}|^2\, \mathrm{d}\xi \\ &\geqslant c_1 \|\varphi\|_m^2. \end{aligned} \tag{10}$$

In order to obtain the last inequality we have applied Theorem 7.6, the first Poincaré inequality. Because of (9) c_1 is not dependent on x_0, but only on Ω, see Theorem 7.6.

(b) Suppose as before that $x_0 \in \bar{\Omega}$ is fixed, we take some small neighbourhood of x_0 so that

$$\sum_{\substack{|\alpha|=m\\|\beta|=m}} |a_{\alpha\beta}(x) - a_{\alpha\beta}(x_0)| \leqslant \frac{c_1}{2}, \quad x \in \omega \cap \bar{\Omega}, c_1 \text{ from (10)}. \tag{11}$$

for $\varphi \in \mathscr{D}(\omega \cap \Omega)$, because of (10) and (11) we have

$$\begin{aligned}\operatorname{Re} a^{\mathrm{H}}(\varphi, \varphi) &= \operatorname{Re} \int_\Omega \sum_{\substack{|\alpha|=m\\|\beta|=m}} a_{\alpha\beta}(x) \mathrm{D}^\alpha \varphi(x) \mathrm{D}^\beta \overline{\varphi(x)}\, \mathrm{d}x \\ &= \operatorname{Re} \int_\Omega \sum_{|\alpha|=|\beta|=m} a_{\alpha\beta}(x_0) \mathrm{D}^\alpha \varphi \mathrm{D}^\beta \bar{\varphi}\, \mathrm{d}x \\ &\quad + \operatorname{Re} \int_\Omega \sum_{\substack{|\alpha|=m\\|\beta|=m}} (a_{\alpha\beta}(x) - a_{\alpha\beta}(x_0)) \mathrm{D}^\alpha \varphi \mathrm{D}^\beta \bar{\varphi}\, \mathrm{d}x \end{aligned}$$

$$\geqslant c_1 \|\varphi\|_m^2 - \int_\Omega \sum_{\substack{|\alpha|=m\\|\beta|=m}} |a_{\alpha\beta}(x) - a_{\alpha\beta}(x_0)| D^\alpha\varphi D^\beta\bar{\varphi}\, dx$$

$$\geqslant c_1 \|\varphi\|_m^2 - \frac{c_1}{2}\|\varphi\|_m^2 = \frac{c_1}{2}\|\varphi\|_m^2. \tag{12}$$

(c) We cover $\bar{\Omega}$ (compact!) with finitely many small balls $\omega = B(x_0, \varepsilon)$, so that on $\omega \cap \bar{\Omega}$ estimate (11) is valid, let $\{\omega_j\}$, $j = 1, \ldots, n$, be this cover. We choose functions $h_j \in \mathscr{D}(\omega_j)$ with

$$\sum_{j=1}^n h_j^2(x) = 1 \text{ on } \bar{\Omega}, \tag{13}$$

and after application of the Leibniz product rule and (13) for $\varphi \in \mathscr{D}(\Omega)$ we have

$$\begin{aligned} \operatorname{Re} a^{\mathrm{H}}(\varphi, \varphi) &= \operatorname{Re} \int_\Omega \sum_{\substack{|\alpha|=m\\|\beta|=m}} a_{\alpha\beta}(x) D^\alpha\varphi D^\beta\bar{\varphi}\, dx \\ &= \operatorname{Re} \int_\Omega \sum_{\substack{|\alpha|=m\\|\beta|=m}} \sum_{j=1}^n h_j^2 a_{\alpha\beta}(x) D^\alpha\varphi D^\beta\bar{\varphi}\, dx \\ &= \operatorname{Re} \sum_{j=1}^n \int_\Omega \sum_{\substack{|\alpha|=m\\|\beta|=m}} a_{\alpha\beta}(x) D^\alpha(h_j\varphi)\overline{D^\beta(h_j\varphi)}\, dx \\ &\quad + \operatorname{Re} \int_\Omega \sum_{\substack{|\alpha|,|\beta|\leqslant m\\|\alpha|+|\beta|<2m}} b_{\alpha\beta}(x) D^\alpha\varphi D^\beta\bar{\varphi}\, dx. \end{aligned} \tag{14}$$

Because

$$\|\varphi\|_m^2 = \left\|\sum_{j=1}^n h_j^2\varphi\right\|_m^2 \leqslant \left[\sum_{j=1}^n \|h_j^2\varphi\|_m\right]^2 \leqslant n\sum_{j=1}^n \|h_j^2\varphi\|_m^2 \leqslant nc\sum_{j=1}^n \|h_j\varphi\|_m^2$$

it follows from (12), $h_j\varphi \in \mathscr{D}(\omega_j \cap \Omega)$, that the first form in the decomposition (14) is coercive:

$$\operatorname{Re} a_1(\varphi, \varphi) \geqslant \sum_{j=1}^n \frac{c_1}{2}\|h_j\varphi\|_m^2 \geqslant \frac{c_1}{2nc}\|\varphi\|_m^2,$$

while the second $b(\varphi, \varphi)$ satisfies the assumptions of Theorem 19.1. Therefore $a^{\mathrm{H}}(\varphi, \varphi)$ is also $\mathring{W}_2^m(\Omega)$-coercive, since $\mathscr{D}(\Omega)$ is dense in $\mathring{W}_2^m(\Omega)$.

Necessity: We show the necessity of strong ellipticity. Let x_0 be an arbitrary point from $\bar{\Omega}$, we first reduce the inequality (4) to the principal part a^{H} and to constant coefficients $a_{\alpha\beta}(x_0)$, hence to $a^{\mathrm{H}}_{x_0}$, and at the same time localise, that is, take $\varphi\in\mathscr{D}(\omega)$. We have

$$\operatorname{Re} a^{\mathrm{H}}_{x_0}(\varphi,\varphi)=\operatorname{Re} a(\varphi,\varphi)+k(\varphi,\varphi)_0+\operatorname{Re}[a^{\mathrm{H}}_{x_0}(\varphi,\varphi)-a^{\mathrm{H}}(\varphi,\varphi)]+\operatorname{Re} b(\varphi,\varphi), \tag{15}$$

where the remainder form b has order smaller than m, therefore

$$|b(\varphi,\varphi)|\leqslant c_0\|\varphi\|_m\cdot\|\varphi\|_{m-1}. \tag{16}$$

We now choose the ball $B(x_0,\varepsilon)=:\omega$ so that on it, first the estimate (11) holds with c_1 from (4), from which it follows that

$$|a^{\mathrm{H}}_{x_0}(\varphi,\varphi)-a^{\mathrm{H}}(\varphi,\varphi)|\leqslant\frac{c_1}{2}\|\varphi\|_m^2\quad\text{for all }\varphi\in\mathscr{D}(\omega), \tag{17}$$

and secondly the Poincaré inequality holds, Theorem 7.6,

$$\|\varphi\|_{m-1}\leqslant c(\varepsilon)\|\varphi\|_m,\quad \varphi\in\mathscr{D}(\omega), \tag{18}$$

with

$$c(\varepsilon)\leqslant\frac{c_1}{4c_0}. \tag{19}$$

(19) and (18) substituted in (16) give

$$|b(\varphi,\varphi)|\leqslant(c_1/4)\|\varphi\|_m^2,\quad \varphi\in\mathscr{D}(\omega). \tag{20}$$

For (15) the estimates (17) and (20) in conjunction with (4) give

$$\operatorname{Re} a^{\mathrm{H}}_{x_0}(\varphi,\varphi)\geqslant c_1\|\varphi\|_m^2-\frac{c_1}{2}\|\varphi\|_m^2-\frac{c_1}{4}\|\varphi\|_m^2=\frac{c_1}{4}\|\varphi\|_m^2\quad\text{for all }\varphi\in\mathscr{D}(\omega), \tag{21}$$

which is the desired localisation of (4).

If $A(x,\mathrm{D})$ were not strongly elliptic at some point $x_0\in\bar{\Omega}$, then there would exist some $0\neq\xi_0\in\mathbb{R}^r$ with

$$\operatorname{Re} A^{\mathrm{H}}(x_0,\xi_0)=\operatorname{Re}\sum_{\substack{|\alpha|=m\\|\beta|=m}}a_{\alpha\beta}(x_0)\xi_0^{\alpha+\beta}=0. \tag{22}$$

We choose some $0\neq u_0\in\mathscr{D}(\omega)$ and write

$$\varphi_0(x)=u_0(x)\mathrm{e}^{\mathrm{i}\lambda\xi_0x},\quad \lambda\in\mathbb{R}^1.$$

Then $0\neq\varphi_0\in\mathscr{D}(\omega)$ and because of (22) we have

$$\begin{aligned}\operatorname{Re} a^{\mathrm{H}}_x(\varphi_0,\varphi_0)&=\operatorname{Re}\int_\omega\sum_{\substack{|\alpha|=m\\|\beta|=m}}a_{\alpha\beta}(x_0)\mathrm{D}^\alpha\varphi_0(x)\cdot\mathrm{D}^\beta\overline{\varphi_0(x)}\,\mathrm{d}x\\&=\lambda^{2m}\|u_0\|_0^2\left(\operatorname{Re}\sum_{\substack{|\alpha|=m\\|\beta|=m}}a_{\alpha\beta}(x_0)\xi_0^{\alpha+\beta}\right)+P_{2m-1}(\lambda)=P_{2m-1}(\lambda),\end{aligned} \tag{23}$$

where $P_{2m-1}(\lambda)$ is a real polynomial in λ, with degree at most $2m-1$. Furthermore

$$\|\varphi_0\|_m^2 = \sum_{|s|\leqslant m} \int_\omega |D^s(u_0(x)e^{i\lambda\xi_0 x})|^2 \, dx$$

$$= \lambda^{2m}\|u_0\|_0^2\left(\sum_{|s|=m} \xi_0^{2s}\right) + \cdots = Q_{2m}(\lambda), \tag{24}$$

where $Q_{2m}(\lambda)$ is a real polynomial in λ, for which the leading coefficient $= \|u_0\|_0^2(\sum_{|s|=m}\xi_0^{2s}) \neq 0$, hence for which the degree is precisely $2m$. Substituting (23) and (24) in (21) gives

$$P_{2m-1}(\lambda) \geqslant (c_1/4)Q_{2m}(\lambda) \quad \text{for all } \lambda \in \mathbb{R}^1, \tag{25}$$

a contradiction. Hence $A(x, D)$ is strongly elliptic for each $x_0 \in \bar{\Omega}$. ■

We give the differentiability assumptions for Agmon's theorem:

1. The region $\Omega \subset \mathbb{R}^r$ is bounded and belongs to class $C^{m-1,1}$.
2. $A(x, D)$ is a linear differential operator of order $2m$, written in the form (1) with continuous coefficients $a_{\alpha\beta}(x)$, $x \in \bar{\Omega}$,

$$a_{\alpha\beta} \in C(\bar{\Omega}), \quad |\alpha| \leqslant m, |\beta| \leqslant m. \tag{26}$$

3. Suppose given p linear normal boundary value operators

$$b_j(x, D), \quad j = 1, \ldots, p,$$

where $0 \leqslant m_j = \operatorname{ord} b_j \leqslant m-1$ and $0 \leqslant p \leqslant m$, $p = 0$ implies that no boundary values are preassigned. For the coefficients $b_{s,j}(x)$ of $b_j(x, D)$ we require that

$$b_{s,j}(x) \in C^{m-m_j}(\bar{\Omega}), \quad |s| \leqslant m_j, j = 1, \ldots, p. \tag{27}$$

We write

$$V := W^m(\{b_j\}_1^p) = \{\varphi \in W_2^m(\Omega) : b_j(x, D)\varphi = 0 \text{ on } \partial\Omega, j = 1, \ldots, p\}.$$

V is a closed subspace of $W_2^m(\Omega)$, we give it the W_2^m-topology. If $p = 0$ we put $V := W_2^m(\Omega)$. Since $\mathscr{D}(\Omega)$ is in V, we have

$$\mathring{W}_2^m(\Omega) \subset V \subset W_2^m(\Omega).$$

We can now give conditions for the V-coercion of $a(\varphi, \psi)$ (2) for subspaces V, which are determined by boundary values.

Theorem 19.3 (Agmon [2]) *Suppose that the differentiability assumptions* 1, 2 *and* 3 *are satisfied. Let the operator* $A(x, D)$ (1) *be strongly elliptic for all* $x \in \bar{\Omega}$

and for $A(x,D)$, $b_1(x,D),\dots,b_p(x,D)$ *let Agmon's Condition* 18.1 *be satisfied for all* $x\in\partial\Omega$. *Then the sesquilinear form* $a(\varphi,\psi)$ (2) *is* V*-coercive,* $V=W_2^m(\{b_j\}_1^p)$.

Since $\mathring{W}_2^m(\Omega)\subset V$, strong ellipticity is necessary for V-coercion, see Theorem 19.2; by Agmon [2] Condition 18.1 is also necessary for V-coercion.

Remark Agmon's theorem implies (see Theorems 17.11 and 17.14) that there exists some k_0 (coercion constant) such that for each $\lambda\in\mathbb{C}$ with $\operatorname{Re}\lambda\leqslant -k_0$, the weak equation

$$a(u,\varphi)-\lambda(u,\varphi)_0=(f,\varphi)_0,\ \varphi\in V$$

is uniquely solvable in $u\in V$ for each $f\in V'$. We can also fully apply the spectral Theorem 17.12 to the weak equation. We concern ourselves in §21 with the realisation of the weak equation as a boundary value problem for differential equations.

For the proof of Theorem 19.3 we give two preliminary propositions.

Proposition 19.1 *Let* $A(D)$, $b_1(D),\dots,b_p(D)$ *be homogeneous differential operators with constant coefficients. Let* $A(D)$ *be strongly elliptic and let Agmon's Condition* 18.1 *be satisfied (for* $x_0=0$). *We consider the half space* $\Omega=\mathbb{R}^r_+$, $\partial\Omega=\mathbb{R}^{r-1}$. *Then for all* $\varphi\in W_2^m(\mathbb{R}^r_+)$ *we have the estimate*

$$\operatorname{Re}a(\varphi,\varphi)\geqslant c_1\|\varphi\|_m^2-c_2\sum_{j=1}^p\|b_j(D_{x'},D_t)|_{t=0}\varphi\|^2_{m-m_j-1/2,\mathbb{R}^{r-1}},\tag{28}$$

where $0\leqslant m_j=\operatorname{ord}b_j\leqslant m-1$, $x=(x',t)\in\mathbb{R}^r_+$, $0\leqslant p\leqslant m$, $c_1,c_2>0$. *If* $p=0$ *the second term on the right-hand side of* (28) does not occur.

Proof. We apply the Fourier transformation $\mathscr{F}_{r-1}$ at $x'\in\mathbb{R}^{r-1}$ and for $\varphi\in C_0(\overline{\mathbb{R}}^r_+)$ we have

$$\begin{aligned}\operatorname{Re}a(\varphi,\varphi)&=\operatorname{Re}\int_{\mathbb{R}^r_+}\sum_{\substack{|\alpha|=m\\|\beta|=m}}a_{\alpha\beta}D^\alpha\varphi D^\beta\bar\varphi\,dx\\&=\operatorname{Re}\int_{\mathbb{R}^{r-1}}d\xi'\cdot\int_0^\infty dt\sum_{\substack{|\alpha'|+k=m\\|\beta'|+l=m}}a_{(\alpha',k)(\beta',l)}\xi'^{\alpha'}\cdot\xi'^{\beta'}D_t^k\phi(\xi',t)\cdot D_t^l\phi(\xi',t)\\&=\operatorname{Re}\int_{\mathbb{R}^{r-1}}d\xi' A_{\xi'}[\phi(\xi',t),\phi(\xi',t)].\end{aligned}\tag{29}$$

We change the variable by $t\mapsto t/|\xi'|$ in the integral $\int_0^\infty dt$, obtaining

$$A_{\xi'}[f(t),f(t)]=|\xi'|^{2m-1}A_{\xi'/|\xi'|}\left[f\left(\frac{t}{|\xi'|}\right),f\left(\frac{t}{|\xi'|}\right)\right],$$

and can rewrite (29) as

$$\operatorname{Re} a(\varphi,\varphi)=\operatorname{Re}\int_{\mathbb{R}^{r-1}} d\xi'|\xi'|^{2m-1}A_{\xi'/|\xi'|}\left[\phi\left(\xi',\frac{t}{|\xi'|}\right),\phi\left(\xi',\frac{t}{|\xi'|}\right)\right]. \quad (29')$$

Since Condition 18.2 follows from Condition 18.1, we can apply the estimate §18, (34), to (29′) (we put $\varphi(t)=\phi(\xi',t/|\xi'|)$) and obtain

$$\begin{aligned}\operatorname{Re} a(\varphi,\varphi)&\geqslant c_1'\sum_{l=0}^{m}\int_{\mathbb{R}^{r-1}}|\xi'|^{2m-1}\int_0^\infty\left|D_t^l\phi\left(\xi',\frac{t}{|\xi'|}\right)\right|^2 dt\,d\xi'\\&\quad-c_c'\sum_{j=1}^{p}\int_{\mathbb{R}^{r-1}}|\xi'|^{2m-1}\left|b_j\left(\frac{\xi'}{|\xi'|},D_t\right)\phi\left(\xi',\frac{t}{|\xi'|}\right)\right|_{t=0}^2 d\xi'\\&=c_1'\sum_{l=0}^{m}\int_{\mathbb{R}^{r-1}}\int_0^\infty|\xi'|^{2(m-l)}|D_t^l\phi(\xi',t)|^2\,dt\,d\xi'\\&\quad-c_2'\sum_{j=1}^{p}\int_{\mathbb{R}^{r-1}}|\xi'|^{2(m-m_j)-1}|b_j(\xi',D_t)\phi(\xi',t)|_{t=0}^2\,d\xi'\\&\geqslant c_1\|\varphi\|_m^2-c_2\sum_{j=1}^{p}\|[b_j(D_{x'},D_t)]_{t=0}\varphi\|_{m-m_j-1/2,\mathbb{R}^{r-1}}^2.\end{aligned}$$

Here in order to establish the last inequality we have applied the inverse Fourier transformation $\mathscr{F}_{r-1}^{-1}$. Therefore we have proved (28) for $\varphi\in C_0^\infty(\bar{\mathbb{R}}_+^r)$, and our Proposition 19.1 follows from Theorem 3.6, according to which $C_0^\infty(\bar{\mathbb{R}}_0^r)$ is dense in $W_2^m(\mathbb{R}_+^r)$. ∎

Proposition 19.2 *Let $A(x,D)$, $b_1(x,D),\ldots,b_p(x,D)$ be differential operators with variable coefficients defined on the half cube $\bar{W}_+^r$, which satisfy the differentiability assumptions (26) and (27). Let $A(x,D)$ be strongly elliptic for $x_0=0$ and suppose that Condition 18.1 is satisfied for $x_0=0$. Then there exists a small cube $W^r(\delta)$ (edge length 2δ) about the origin $x_0=0$, such that for all $\varphi\in W_2^m(W_+^r(\delta))$ we have the estimate*

$$Re\, a(\varphi,\varphi)\geqslant c_1\|\varphi\|_m^2-c_2\|\varphi\|_0^2-c_3\sum_{j=1}^{p}\|b_j(x;D_{x'},D_t)|_{t=0}\varphi\|_{m-m_j-\frac{1}{2},W^{r-1}(\delta)}^2. \quad (30)$$

Proof. We decompose

$$\begin{aligned}a(\varphi,\varphi)&=\int_{W_+^r(\delta)}\sum_{|\alpha|=|\beta|=m}a_{\alpha\beta}(0)D^\alpha\varphi\cdot D^\beta\bar{\varphi}\,dx\\&\quad+\int_{W_+^r(\delta)}\sum_{|\alpha|=|\beta|=m}[a_{\alpha\beta}(x)-a_{\alpha\beta}(0)]D^\alpha\varphi D^\beta\bar{\varphi}\,dx+A'(\varphi,\varphi)\\&=a_0^H(\varphi,\varphi)+a_0^H(\varphi,\varphi)+A'(\varphi,\varphi),\end{aligned} \quad (31)$$

where the form A' has at most order $(m, m-1)$. By Proposition 19.1 we have

$$\operatorname{Re} a_0^{\mathrm{H}}(\varphi, \varphi) \geqslant c_1 \| \varphi \|_m^2 - c_2 \sum_{j=1}^{p} \| b_j^{\mathrm{H}}(0, \mathrm{D}_{x'}, \mathrm{D}_t)\varphi \|_{m-m_j-1/2}^2. \tag{32}$$

For the forms a_1^{H} and A' we have the estimates

$$|a_1^{\mathrm{H}}(\varphi, \varphi)| \leqslant c_1(\delta) \| \varphi \|_m^2, \quad \varphi \in W_2^m(W_+^r(\delta)). \tag{33}$$

with $c(\delta) \to 0$ for $\delta \to 0$. ($c(\delta)$ is the continuity modulus of $a_{\alpha\delta}(x) - a_{\alpha\beta}(0)$) and

$$|A'(\varphi, \varphi)| \leqslant c \| \varphi \|_m \| \varphi \|_{m-1} \leqslant \frac{c\varepsilon}{2} \| \varphi \|_m^2 + \frac{c}{2\varepsilon} \| \varphi \|_{m-1}^2. \tag{34}$$

In the same way we decompose the boundary value operators

$$b_j(x, \mathrm{D}) = b_j^{\mathrm{H}}(0, \mathrm{D}) + [b_j^{\mathrm{H}}(x, \mathrm{D}) - b_j^{\mathrm{H}}(0, \mathrm{D})] + b_j'(x, \mathrm{D}), \quad j = 1, \ldots, p, \tag{35}$$

where the order of b_j' is at most $m_j - 1$. We estimate further by applying the Leibniz product rule; T_0 is the trace operator of §8

$$\begin{aligned} &\| [b_j^{\mathrm{H}}(x, \mathrm{D}) - b^{\mathrm{H}}(0, \mathrm{D})]\varphi \|_{m-m_j-1/2, W^{r-1}(\delta)}^2 \\ &\leqslant \| T_0 \|^2 \| [b_j^{\mathrm{H}}(x, \mathrm{D}) - b^{\mathrm{H}}(0, \mathrm{D})]\varphi \|_{m-m_j, W_+^r(\delta)} \\ &\leqslant \| T_0 \|^2 c_2(\delta) \| \varphi \|_m^2 + c_3 \| \varphi \|_{m-1}^2. \end{aligned} \tag{36}$$

Here $c_2(\delta)$ is the continuity modulus of the coefficients

$$b_{s,j}(x) - b_{s,j}(0), \; |s| = m_j, \quad \text{hence } c_2(\delta) \to 0 \text{ as } \delta \to 0.$$

For b_j' we have

$$\| b_j'(x, \mathrm{D})\varphi \|_{m-m_j-1/2, W^{r-1}}^2 \leqslant \| T_0 \|^2 c \| \varphi \|_{m-1}^2. \tag{37}$$

We estimate (31) with the help of (32), (33) and (34) and obtain

$$\begin{aligned} \operatorname{Re} a(\varphi, \varphi) \geqslant{}& c_1 \| \varphi \|_m^2 - c_2 \sum_{j=1}^{p} \| b_j^{\mathrm{H}}(0, \mathrm{D})\varphi \|_{m-m_j-1/2, W^{r-1}(\delta)}^2 \\ &- c_1(\delta) \| \varphi \|_m^2 - \frac{c\varepsilon}{2} \| \varphi \|_m^2 - \frac{c}{2\varepsilon} \| \varphi \|_{m-1}^2. \end{aligned} \tag{38}$$

We replace $b_j^{\mathrm{H}}(0, \mathrm{D})\,\varphi$ from (38) by (35), and apply the estimates (36) and (37), by which (38) takes the form

$$\begin{aligned} \operatorname{Re} a(\varphi, \varphi) \geqslant{}& \| \varphi \|_m^2 \left[c_1 - c_1(\delta) - \frac{c\varepsilon}{2} - p \| T_0 \|^2 c_2(\delta) c_2 \right] \\ &- c_2 \sum_{j=1}^{p} \| b_j(x, \mathrm{D})\varphi \|_{m-m_j-1/2, W^{r-1}(\delta)}^2 \\ &- \| \varphi \|_{m-1}^2 \left[\frac{c}{2\varepsilon} + pc_3c_2 + c_2pc \| T_0 \|^2 \right]. \end{aligned} \tag{39}$$

We choose δ and ε so small that

$$\frac{c_1}{4} = c_1(\delta) + \frac{c\varepsilon}{2} + p\|T_0\|^2 c_2(\delta)c_2$$

and apply Theorem 7.4, Ehrling's inequality, to the last term of (39), $\eta > 0$ arbitrary:

$$\|\varphi\|_{m-1}^2 \leqslant \eta\|\varphi\|_m^2 + c(\eta)\|\varphi\|_0^2, \quad \varphi \in W_2^m(W_+^r(\delta)).$$

Next we choose η so small that

$$\eta\left(\frac{c}{2\varepsilon} + pc_3c_2 + c_2pc\|T_0\|^2\right) = \frac{c_1}{4}.$$

For $\varphi \in W_2^m(W_+^r(\delta))$ we obtain from (39) that

$$\operatorname{Re} a(\varphi, \varphi) \geqslant \frac{c_1}{2}\|\varphi\|_m^2 - c_2 \sum_{j=1}^p \|b_j(x, \mathrm{D})\varphi\|_{m-m_j-1/2, W^{r-1}(\delta)}^2$$

$$- c(\eta)\left(\frac{c}{2\varepsilon} + pc_3c_2 + c_2pc\|T_0\|^2\right)\|\varphi_0\|^2, \tag{40}$$

which is (30). ■

We are now in the position to give the proof of Theorem 19.3 by means of a partition of unity. We show that for all $\varphi \in W_2^m(\Omega)$ we have the estimate

$$\operatorname{Re} a(\varphi, \varphi) \geqslant c_1\|\varphi\|_m^2 - c_2 \sum_{j=1}^p \|b_j(x, \mathrm{D})\varphi\|_{m-m_j-1/2, \partial\Omega}^2 - c_3\|\varphi\|_0^2. \tag{41}$$

We fix a point $x_0 \in \partial\Omega$, take some neighbourhood $U \ni x_0$ and an admissible coordinate transformation $\Phi: U \to W^r$, $x_0 \to 0$; this exists, since $\partial\Omega$ belongs to $C^{m-1,1}$.

Strong ellipticity and Agmon's Condition 18.1 are invariant under Φ (see Theorem 18.1), by Proposition 19.2 we can find a small $\delta > 0$, so that for $\tilde{\varphi} \in W_2^m(W_+^r(\delta))$ we have

$$\operatorname{Re} \tilde{a}(\tilde{\varphi}, \tilde{\varphi}) \geqslant \tilde{c}_1\|\tilde{\varphi}\|_m^2 - \tilde{c}_2 \sum_{j=1}^p \|\tilde{b}_j(\tilde{x}, \mathrm{D})\tilde{\varphi}\|_{m-m_j-1/2}^2 - \tilde{c}_3\|\tilde{\varphi}\|_0^2,$$

or after transforming back $\Phi^{-1}: U(\delta) \leftarrow W^r(\delta)$

$$\operatorname{Re} a(\varphi, \varphi) \geqslant c_1\|\varphi\|_m^2 - c_2 \sum_{j=1}^p \|b_j(x, \mathrm{D})\varphi\|_{m-m_j-1/2}^2 - c_3\|\varphi\|_0^2 \tag{42}$$

for all $\varphi \in W_2^m(U(\delta) \cap \Omega)$. Since the frontier $\partial\Omega$ is compact, we can cover it with finitely many $U_j(\delta)$ with property (42), $i = 1, \ldots, M$. The remaining part $\Omega\setminus\bigcup_{i=1}^{M} U_i(\delta)$ may be covered by $U_0(\subset\Omega)$. To the cover $\{U_i\}_{i=0,\dots,M}$ we associate a 'partition' of unity h_i^2 – as in the proof of Theorem 19.2: $h_i\in\mathscr{D}(U_i)$, $\sum_{i=0}^{M} h_i^2(x)=1$ on $\bar{\Omega}$. We decompose, as in (14), for $\varphi\in W_2^m(\Omega)$

$$\operatorname{Re} a(\varphi,\varphi)=\operatorname{Re}\sum_{i=0}^{M} a(h_i\varphi,h_i\varphi)+\operatorname{Re} b(\varphi,\varphi), \tag{43}$$

where for $b(\varphi,\varphi)$ we have the estimate

$$|b(\varphi,\varphi)|\leqslant c_4\|\varphi\|_m\|\varphi\|_{m-1}\leqslant\frac{c_4\varepsilon}{2}\|\varphi\|_m^2+\frac{c_4}{2\varepsilon}\|\varphi\|_{m-1}^2. \tag{44}$$

By construction of the h_i we have $h_0\varphi\in\mathring{W}_2^m(U_0)$ and $h_i\varphi\in W_2^m(U_i(\delta)\cap\Omega)$, $j=1,\dots,M$ and we can apply Theorem 19.2 or the estimate (42) which for (43) gives

$$\begin{aligned}\operatorname{Re} a(\varphi,\varphi)\geqslant{}& c_1\sum_{i=0}^{M}\|h_i\varphi\|_m^2-\frac{c_4\varepsilon}{2}\|\varphi\|_m^2\\ &-c_2\sum_{j=1}^{p}\sum_{i=1}^{M}\|b_j(x,\mathrm{D})h_i\varphi\|_{m-m_j-1/2}^2\\ &-c_3\sum_{i=0}^{M}\|h_i\varphi\|_0^2-\frac{c_4}{2\varepsilon}\|\varphi\|_{m-1}^2.\end{aligned} \tag{45}$$

We estimate (see the lines following (14))

$$\sum_{i=0}^{M}\|h_i\varphi\|_m^2=\frac{1}{(M+1)c_5}\|\varphi\|_m^2,\ \|b_j(x,\mathrm{D})h_i\varphi\|_{m-m_j-1/2}$$
$$\leqslant c_6\|b_j(x,\mathrm{D})\varphi\|_{m-m_j-1/2}+c_7\|\varphi\|_{m-1},\ \|h_i\varphi\|_0^2\leqslant c_8\|\varphi\|_0^2,$$

and for (45) obtain

$$\begin{aligned}\operatorname{Re} a(\varphi,\varphi)\geqslant{}&\left[\frac{c_1}{(M+1)c_5}-\frac{c_4}{2}\varepsilon\right]\|\varphi\|_m^2\\ &-\bar{c}_2\sum_{j=1}^{p}\|b_j(x,\mathrm{D})\varphi\|_{m-m_j-1/2,\partial\Omega}^2\\ &-c_3'\|\varphi\|_0^2-\left[\frac{c_4}{2\varepsilon}+Mc_7\right]\|\varphi\|_{m-1}^2.\end{aligned} \tag{46}$$

We choose ε so that

$$-\frac{c_4}{2}\varepsilon+\frac{c_1}{2(M+1)c_5}=:\frac{\bar{c}_1}{2}>0,$$

and use Ehrling's inequality (see Theorem 7.4)

$$\|\varphi\|_{m-1}^2 \leqslant \eta\|\varphi\|_m^2 + c(\eta)\|\varphi\|_0^2$$

with η so small that

$$\eta\left(\frac{c_4}{2} + Mc_7\right) = \frac{\bar{c}_1}{4}.$$

Then (46) takes the form

$$\operatorname{Re} a(\varphi, \varphi) \geqslant \frac{\bar{c}_1}{4}\|\varphi\|_m^2 - \bar{c}_2 \sum_{j=1}^{p} \|b_j(x, \mathrm{D})\varphi\|_{m-m_j-1/2,\partial\Omega}^2 - c_3\|\varphi\|_0^2,$$

and thus we have proved (41). From (41) it follows immediately that $a(\varphi, \varphi)$ is V-coercive, for $\varphi = V$ is characterised by $b_j(x, \mathrm{D})\varphi = 0$, $j = 1, \ldots, p$. Therefore we have proved Agmon's theorem. ■

As a corollary of Agmon's theorem we deduce a result of Aronszajn [1].

Theorem 19.4 *Let the quadratic form* (2) *be formally positive, that is,*

$$a(\varphi, \varphi) = \int_{\Omega} \sum_{k=1}^{n} |A_k(x, \mathrm{D})\varphi|^2 \,\mathrm{d}x, \quad A(x, \mathrm{D})\varphi = \sum_{k=1}^{n} A_k^*(x, \mathrm{D}) \circ A_k(x, \mathrm{D})\varphi, \tag{47}$$

where $\operatorname{ord} A_k \leqslant m$ *and* (26) *holds for the coefficients of* A_k. *Let* Ω *be bounded and* $\partial\Omega \in C^{m-1,1}$. *Let* $\tilde{A}_k(x, \mathrm{D})$ *be the sum of the terms of* $A_k(x, \mathrm{D})$, *which have order m. We assume further that*

$$\sum_{k=1}^{n} |\tilde{A}_k(x, \xi)|^2 > 0 \quad \text{for all } x \in \bar{\Omega} \text{ and } 0 \neq \xi \in \mathbb{R}^r. \tag{48}$$

As in Agmon's condition we write

$$\tilde{A}_k(x; \mathrm{D}) = \tilde{A}_k(x; \mathrm{D}_{x'}, \mathrm{D}_t),$$

where x' are the coordinates of $\partial\Omega$ and t points in the direction of the inner normal. We consider the polynomials in z

$$\tilde{A}_k(x_0; \xi', z) = 0, \quad k = 1, \ldots, n, x_0 \in \partial\Omega, 0 \neq \xi' \in \mathbb{R}^{r-1}, \tag{49}$$

and require that they do not possess a common zero with positive imaginary part. Then the form $a(\varphi, \varphi)$ given by (47) *is $W_2^m(\Omega)$-coercive; hence also coercive for all $V \subset W_2^m(\Omega)$.*

Proof. We have

$$\operatorname{Re} A^{\mathrm{H}}(x;\xi)=\sum_{k=1}^{n}|\tilde{A}_k(x,\xi)|^2>0,$$

so that $A(x,\mathrm{D})$ is strongly elliptic, and it remains to check Agmon's condition 18.1. We consider the ordinary differential equation (3) from §18 associated with the operator $A(x,\mathrm{D})$ (47). Its characteristic polynomial is

$$A^{\mathrm{H}}(x_0,\xi',z)=\sum_{k=1}^{n}|\tilde{A}_k(x_0,\xi',z)|^2. \tag{50}$$

The requirement of a solution v different from zero for (3) (§18) in $W_2^m(\mathbb{R}_+^1)$ is equivalent to the existence of a root z of (50) with non-negative imaginary part. This is excluded by (48) and (49); Condition 18.1 is therefore (vacuously) satisfied. Agmon's Theorem 19.3 ends the proof. ■

19.2 Examples, including the Dirichlet problem for strongly elliptic differential operators

We wish to test Agmon's Condition 18.1 against examples, that is, determine V-coercion for different spaces V. For differential operators of the second order we do not need Agmon's theorem, indeed we have:

Example 19.1 Let

$$A(x,\mathrm{D})=-\sum_{j,k=1}^{r}a_{jk}\frac{\partial}{\partial x_j}\frac{\partial}{\partial x_k}+\sum_{j=1}^{r}a_j\frac{\partial}{\partial x_j}+a_0 \tag{51}$$

be a strongly elliptic second-order differential operator, hence

$$\operatorname{Re}\sum_{j,k=1}^{r}a_{jk}\xi_j\xi_k\geqslant c_0|\xi|^2\quad\text{for }\xi\in\mathbb{R}^r, \tag{52}$$

Then A is W_2^1-coercive, that is, V-coercive for each V with $\mathring{W}_2^1\subsetneq V\subsetneq W_2^1(\Omega)$.

Here is the simple proof. From (52) we obtain the inequality

$$\operatorname{Re}\sum_{j,k=1}^{r}a_{jk}z_j\bar{z}_k\geqslant c_0|z|^2\quad\text{for all }z\in\mathbb{C}^r, \tag{53}$$

indeed because of the symmetry $a_{jk}=a_{kj}$ we have

$$\begin{aligned}\operatorname{Re}a_{jk}z_j\bar{z}_k+\operatorname{Re}a_{kj}z_k\bar{z}_j&=\operatorname{Re}a_{jk}[z_j\bar{z}_k+\bar{z}_jz_k]\\&=\operatorname{Re}a_{jk}2[\operatorname{Re}z_j\cdot\operatorname{Re}z_k+\operatorname{Im}z_j\cdot\operatorname{Im}z_k],\end{aligned}$$

which is

$$\begin{aligned}\operatorname{Re}\sum_{j,k=1}^{r}a_{jk}z_j\bar{z}_k&=2\operatorname{Re}\sum_{j,k=1}^{r}a_{jk}[\operatorname{Re}z_j\cdot\operatorname{Re}z_k+\operatorname{Im}z_j\cdot\operatorname{Im}z_k]\\&\geqslant c_0(|\operatorname{Re}z|^2+|\operatorname{Im}z|^2)=c_0|z|^2.\end{aligned}$$

We write $z_j = \partial\varphi/\partial x_j$ for $\varphi \in W_2^1(\Omega)$ in (53), integrate over Ω according to x and adjoin $c_0 \| \varphi \|_0^2$

$$\operatorname{Re} \int_\Omega \sum_{j,k=1}^r a_{jk} \frac{\partial\varphi}{\partial x_j} \cdot \frac{\partial\bar{\varphi}}{\partial x_k} \,dx + c_0 \|\varphi\|_0^2 \geqslant c_0 \int_\Omega \sum_{j=1}^r \left| \frac{\partial\varphi}{\partial x_j} \right|^2 dx + c_0 \|\varphi\|_0^2$$
$$= c_0 \|\varphi\|_1^2.$$

Therefore the principle part of (51) is W_2^1-coercive and Theorem 19.1 concludes the proof. ■

We next give all* boundary value spaces V, which satisfy Agmon's Condition 18.1 for the biharmonic operator Δ^2, that is, for which Δ^2 is V-coercive.

Example 19.2 Let $A = \Delta^2$, since ord $\Delta^2 = 4$, hence $m = 2$, p may equal 0, 1 or 2. For the sesquilinear-form we take

$$a(\varphi, \psi) = \int_\Omega \Delta\varphi \cdot \overline{\Delta\psi} \,dx, \quad \varphi, \psi \in W_2^2(\Omega),$$

so for the quadratic form (7) in Condition 18.1 we have

$$\operatorname{Re} \tilde{a}^{\mathrm{H}}(v) = \int_0^\infty \left| \left(|\xi'|^2 - \frac{d^2}{dt^2} \right) v(t) \right|^2 dt, \quad v \in \mathcal{M}^+, \xi' \in \mathbb{R}^{r-1}, \tag{54}$$

when by Example 11.8

$$\mathcal{M}^+ = \{ c_1 e^{-|\xi'|t} + c_2 t e^{-|\xi'|t} : c_1, c_2 \in \mathbb{C} \}. \tag{55}$$

1. The case $p = 0$, here $V = W_2^2(\Omega)$ and v runs through the whole space $\mathcal{M}^+$. If we take $0 \not\equiv v = c_1 e^{-|\xi'|t}$, $c_1 \neq 0$, then because $(|\xi'|^2 - d^2/dt^2)v(t) \equiv 0$ the form (54) $\operatorname{Re} \tilde{a}^{\mathrm{H}}(v) = 0$ for all $0 \neq \xi' \in \mathbb{R}^{r-1}$, that is, Condition 18.1 is not satisfied and Δ^2 is not $W_2^2(\Omega)$-coercive.

2. The case $p = 1$. Suppose first that ord $b_1(x, \mathrm{D}) = 0$, the boundary condition b_1 then has the form $b_1(x, \mathrm{D}) = b_0(x)$, where because of normality $b_0(x) \neq 0$. For §18, (4) we obtain

$$b_1\left(x, \xi', \frac{1}{i} \frac{d}{dt}\right) v(0) = b_0(x) c_1 = 0, \quad v \in \mathcal{M}^+, \text{ therefore } c_1 = 0,$$

and in order to check Condition 18.1 we must show that in (54) $\operatorname{Re} \tilde{a}^{\mathrm{H}}(v) > 0$ holds for all $v \in \mathcal{M}^+$ of the form

$$0 \not\equiv v(t) = c_2 t e^{-|\xi'|t}, \quad c_2 \neq 0.$$

* That is, for a fixed chosen sesquilinear form $a(\varphi, \psi) = \int_\Omega \Delta\varphi \cdot \overline{\Delta\psi} \,dx$.

Since, as is easily calculated,

$$\left(|\xi'|^2 - \frac{d^2}{dt^2}\right) te^{-|\xi'|t} \neq 0 \quad \text{for all } 0 \neq \xi' \in \mathbb{R}^{r-1},$$

we have

$$\operatorname{Re} \tilde{a}^H(v) = |c_2|^2 \int_0^\infty \left| \left(|\xi'|^2 - \frac{d^2}{dt^2}\right) te^{-|\xi'|t} \right|^2 dt > 0, \quad 0 \neq \xi' \in \mathbb{R}^{r-1},$$

that is, Δ^2 is V_1-coercive, where

$$V_1 = W_2^2(\{b_0\}) = \{\varphi \in W_2^2(\Omega) | \varphi|_{\partial\Omega} = 0\} = W_2^2(\Omega) \cap \mathring{W}_2^1(\Omega). \tag{56}$$

Now let $\operatorname{ord} b_1(x, D) = 1$, that is,

$$b_1^H(x, D) = \sum_{j=1}^{r-1} b_j(x) \frac{\partial}{\partial x_j} + b_r \frac{\partial}{\partial x_r}, \quad b_r \neq 0,$$

$$b_1^H\left(x; \xi', \frac{1}{i}\frac{d}{dt}\right) = i \sum_{j=1}^{r-1} b_j(x)\xi'_j + b_r \frac{d}{dt},$$

where again the coordinates $x_1, \ldots, x_{r-1}$ are tangentially directed and the coordinate x_r is normally directed. For §18, (4) we obtain

$$b_1^H v(0) = c_1 i \sum_{j=1}^{r-1} b_j(x)\xi'_j - b_r |\xi'| c_1 + b_r c_2 = 0, \quad v \in \mathcal{M}^+,$$

and must check §18, (7) for all $v \in \mathcal{M}^+$, which may be written in the form

$$v(t) = c\left[b_r + b_r t|\xi'| - it \sum_{j=1}^{r-1} b_j(x)\xi'_j \right] \cdot e^{-t|\xi'|}. \tag{57}$$

Let the coefficients of $b_1(x, D)$ be real, since $b_r \neq 0$ (normality), we have

$$b_r(1 + t|\xi'|) - it \sum_{j=1}^{r-1} b_j \xi'_j \neq 0,$$

and $v(t) \not\equiv 0$ means $c \neq 0$. Substitution of (57) in (54) gives, for all $0 \neq \xi' \in \mathbb{R}^{r-1}$

$$\tilde{a}^H(v) = |c|^2 \left| b_r|\xi'| - i \sum_{j=1}^{r-1} b_j \xi'_j \right|^2 \int_0^\infty \left| \left(|\xi'|^2 - \frac{d^2}{dt^2}\right) te^{-t|\xi'|} \right|^2 dt > 0$$

since, because $b_r \neq 0$, the expression

$$b_r|\xi'| - i \sum_{j=1}^{r-1} b_j \xi'_j \neq 0, \quad \text{if } \xi' \neq 0.$$

Therefore Δ^2 is V_2-coercive for all spaces

$$V_2 = W_2^2(\{b_1(x, D)\}) = \left\{ \varphi \in W_2^2(\Omega) : b_0 \varphi + \sum_{j=1}^{r} b_j \frac{\partial \varphi}{\partial x_j} = 0 \quad \text{on } \partial\Omega \right\}. \tag{58}$$

3. The case $p=2$. Here we must handle two boundary conditions $b_1(x,\mathrm{D})$, $b_2(x,\mathrm{D})$; because of the assumed normality one must have

$$\operatorname{ord} b_1 = 0, \operatorname{ord} b_2 = 1 \quad \text{and} \quad b_0^{(1)} \cdot b_2^{(2)} \neq 0. \tag{59}$$

Comparison with Example 11.8, (41) shows: (59) implies that the L.Š. Condition 11.1 is satisfied. Therefore from

$$\left.\begin{aligned} b_1^{\mathrm{H}}\left(x;\xi',\frac{1}{\mathrm{i}}\frac{\mathrm{d}}{\mathrm{d}t}\right)v(0)=0 \\ b_2^{\mathrm{H}}\left(x,\xi',\frac{1}{\mathrm{i}}\frac{\mathrm{d}}{\mathrm{d}t}\right)v(0)=0 \end{aligned}\right. \quad v\in\mathcal{M}^+,$$

if follows that $v\equiv 0$, that is, Condition 18.1 is (vacuously) satisfied. Δ^2 is therefore V_3-coercive for all spaces

$$V_3 = W_2^2\begin{pmatrix} b_1 \\ b_2 \end{pmatrix} = \left\{\varphi\in W_2^2(\Omega): b_0^{(1)}\varphi = 0 \text{ on } \partial\Omega, \text{ and } b_0^{(2)}\varphi + \sum_{j=1}^{r} b_j^{(2)}\frac{\partial\varphi}{\partial x_j} = 0 \text{ on } \partial\Omega\right\}. \tag{60}$$

By Theorem 8.9 and Lemma 14.1 we have $V_3 = \mathring{W}_2^2(\Omega)$.

As a conclusion to this section we want to show what we have achieved for the Dirichlet problem by means of Gårding's theorem. The importance of Agmon's theorem for boundary value problems will be discussed later in §21; in order to apply the Green formulae we need regularity properties of the solutions, which we will consider in §20.

Theorem 19.5 *Let the region Ω be open and bounded in $\mathbb{R}^r$, let $A(x,\mathrm{D})$ (1) be strongly elliptic on $\bar{\Omega}$. We impose the stronger condition (see Theorem 19.2)*

$$a_{\alpha\beta}\in C^m(\bar{\Omega}) \tag{61}$$

On the coefficients of $A(x,\mathrm{D})$. By Theorem 19.2 A is $\mathring{W}_2^m$-coercive, let k_0 be the coercion constant. Then the conclusions of Theorems 17.11 and 17.12 hold, and in the notation of Theorem 17.11 we have

$$L = A \quad \text{and} \quad L' = A^* \ (\text{on } V = \mathring{W}_2^m(\Omega)).$$

For the eigenvalue problem

$$Au - \lambda u = f, \quad u\in\mathring{W}_2^m(\Omega) \tag{62}$$

the conclusions of the Riesz–Schauder spectral theorem (Theorem 17.12) hold, and in the half plane $\operatorname{Re}\lambda \leqslant -k_0$ there are no eigenvalues of (62) (Theorem 17.14). Here we can interpret (62) in the distributional sense $\mathcal{D}'(\Omega)$. If $A = A^*$ is formally self-adjoint, then the Dirichlet problem is L_2-self-adjoint, and by Theorem 17.12 there exists a self-adjoint, compact Green solution operator

$G: L_2(\Omega) \to L_2(\Omega)$; the spectrum of (62) therefore consists of infinitely many, discrete, real eigenvalues $\lambda_n > -k_0$ with $\lambda_n \to +\infty$, and the eigenfunctions of (62) form an orthonormal basis for $L_2(\Omega)$. Thus all the conclusions of Theorem 17.12 hold.

Proof. We must first show that $L = A$ and that $Au - \lambda u = f$ (in the sense of $\mathscr{D}'(\Omega)$) is equivalent to $Lu - \lambda u = f$. Let $\varphi, \psi \in \mathscr{D}(\Omega)$, then by partial integration (a from (21)) we obtain

$$a(\varphi, \psi) = \int_\Omega A\varphi \cdot \bar{\psi} \, dx. \tag{63}$$

The assumption (61) has a consequence that

$$A: \mathring{W}_2^m(\Omega) \to W_2^{-m}(\Omega) = [\mathring{W}_2^m(\Omega)]', \tag{64}$$

is continuous, see Theorem 17.7. If we interpret $\int_\Omega \varphi\bar{\psi}\, dx$ as a scalar product on the Gelfand triple $\mathring{W}_2^m \subsetneq L_2(\Omega) \subsetneq W_2^{-m}(\Omega)$ (Theorem 17.4), then by a density argument we conclude that (63) also holds for $\varphi, \psi \in \mathring{W}_2^m(\Omega)$, from which it follows that $L = A$. Writing (63) as $a(\varphi, \psi) = \int_\Omega \varphi \cdot \overline{A^*\psi}\, dx$ also gives $L' = A^*$ on $\mathring{W}_2^m(\Omega)$, and in the case $A = A^*: a(\varphi, \psi) = \overline{a(\psi, \varphi)}$, or $L = L'$, which is the criterion 17.12(c) for self-adjointness.

The interpretation of (62) as a distributional equation follows from the fact that if we fix $\varphi \in \mathring{W}_2^m(\Omega)$, the distribution $A\varphi$ is defined by $\int_\Omega A\varphi \cdot \bar{\psi}\, dx: \mathscr{D}(\Omega) \to \mathbb{C}$, here we again use (64). All the remaining conclusions follow from theorems already stated.

In order to obtain boundary values from $u \in \mathring{W}_2^m(\Omega)$ we must impose regularity assumptions on the boundary $\partial\Omega$ on Ω, for example

$$\Omega \in C^{m,1} \quad \text{for } m \geq 2, \text{ respectively } \Omega \in C^{0,1} \quad \text{for } m = 1. \tag{65}$$

By Theorem 8.9 $u \in \mathring{W}_2^m(\Omega)$ is then equivalent to the Dirichlet boundary values

$$u|_{\partial\Omega} = 0, \ldots, \left.\frac{\partial^{m-1} u}{\partial n^{m-1}}\right|_{\partial\Omega} = 0$$

and (62) is equivalent to the boundary eigenvalue problem

$$Au - \lambda u = f \text{ in } \Omega, \quad u \in W_2^m(\Omega)$$

$$\left.\frac{\partial^j u}{\partial n^j}\right|_{\partial\Omega} = 0, \quad j = 0, \ldots, m-1.$$

The assumptions (65) are also sufficient for the validity of the trace Theorem 8.7 and the inverse Theorem 8.8. Therefore we can for example show (k_0 = the coercion constant): let $f \in W_2^{-m}(\Omega)$, $g_1 \in W_2^{m-1/2}(\partial\Omega), \ldots, g_m \in W_2^{1/2}(\partial\Omega)$ be preassigned, then there exists a unique solution $u \in W_2^m(\Omega)$ of the Dirichlet problem

$$Au + k_0 u = f \text{ in } \Omega, \quad \left.\frac{\partial^j u}{\partial n^j}\right|_{\partial\Omega} = g_{j+1}, \quad j = 0, \ldots, m-1. \tag{66}$$

Indeed by Theorem 8.8 there exists $u_0 \in W_2^m(\Omega)$ with

$$\left.\frac{\partial^j u_0}{\partial n^j}\right|_{\partial\Omega} = g_{j+1}, \quad j = 0, \ldots, m-1.$$

We now solve the equations

$$Aw + k_0 w = f - Au_0 - k_0 u_0, \quad \left.\frac{\partial^j w}{\partial n^j}\right|_{\partial\Omega} = 0, \quad j = 0, \ldots, m-1, \tag{67}$$

where $f - Au_0 - k_0 u_0 \in W_2^{-m}(\Omega) = [\mathring{W}_2^m(\Omega)]'$ (assumptions). By what has been said above the equations are equivalent to the weak equation

$$a(w, \varphi) + k_0(w, \varphi)_0 = (f - Au_0 - k_0 u_0, \varphi)_0, \quad \varphi, w \in \mathring{W}_2^m(\Omega),$$

which by Gårding's Theorem 19.2 is $\mathring{W}_2^m$-elliptic, therefore possesses a unique solution $w \in \mathring{W}_2^m(\Omega)$, so that (67) is also uniquely solvable. With the help of the trace Theorem 8.7 one now easily checks that

$$u := w + u_0$$

is the unique solution of (66) in $W_2^m(\Omega)$. ■

Gårding's Theorem 19.2 allows us to solve the Dirichlet problem in $\mathring{W}_2^m(\Omega)$, the main Theorem 13.1 deals with solvability in $\mathring{W}_2^m(\Omega) \cap W_2^{2m}(\Omega)$, that both methods of solution agree follows from the regularity Theorem 20.4.

Exercise

19.1 We consider the operator Δ^n. For $n = 3, 4, \ldots$ give all boundary conditions which satisfy Agmon's condition.

§20 Regularity of the solutions of strongly elliptic equations

As in §19 let $A(x, D)$ be a strongly elliptic differential operator of the type (1), §19, with the sesquilinear form $a(\varphi, \psi)$ §19, (2), and assume that $a(\varphi, \psi)$ is V-coercive, see Definition 17.7. We consider the weak equation (see §17, (29))

$$a(u, \varphi) = (f, \varphi)_0 \quad \text{for all } \varphi \in V, \tag{1}$$

and our aim is to prove regularity theorems for the solution u – these theorems are named after H. Weyl (Weyl's Lemma).

We begin with $V=\mathring{W}_2^m(\Omega)$, and prove regularity in the interior of Ω (as we shall see later, there are additional difficulties on the boundary $\partial\Omega$). Regularity theorems are local: for the interior of this implies that they hold for small $\omega\subset\bar{\omega}\subset\subset\Omega$, therefore for regularity inside Ω we can manage without any assumptions on $\partial\Omega$, and the boundedness of Ω is also redundant.

Theorem 20.1 *Let Ω be open in $\mathbb{R}^r$ and let $a(\varphi,\psi)$ be $\mathring{W}_2^m(\Omega)$-coercive. Let $k\geqslant 1$ be preassigned, for the coefficients of a we assume*

$$a_{\alpha\beta}(x)\in C^k(\bar{\Omega}). \tag{2}$$

Let $f\in W_2^{-m}(\Omega)$ (see Definition 17.2) be such that $D^\alpha f\in W_2^{-m+1}(\Omega)$ for all $|\alpha|\leqslant k-1$, for example let $f\in W_2^{k-m}(\Omega)$ (see also Example 17.2). Let ω,ω', ω'',ω''' be arbitrary open bounded sets with

$$\omega\subset\subset\omega'\subset\omega'_\delta\subset\omega''\subset\omega''_\delta\subset\omega'''\subset\subset\Omega,\quad \textit{where } \delta>0. \tag{3}$$

Let $u\in W_2^m(\omega''')$ be a weak solution of (1). Then $u\in W_2^{m+k}(\omega)$.

Since $\omega,\omega',\omega'',\omega'''$ were arbitrary, by means of a partition of unity we obtain

$$u\in W_2^{m+k}(\Omega')\quad \text{for each } \Omega'\subset\subset\Omega,$$

if u were assumed to be in $W_2^m(\Omega)$.

For the proof we first give a Lemma:

Lemma 20.1 *Let the assumptions of Theorem 20.1 be satisfied. Let $\chi\in\mathscr{D}(\omega')$ with $\chi\equiv 1$ on ω, and let $|h|<\delta$. We write*

$$\Delta_h^i\varphi:=\frac{\varphi(\ldots,x_i+h,\ldots)-\varphi(\ldots,x_i,\ldots)}{h},\quad i=1,\ldots r,$$

then for all $\varphi\in\mathscr{D}(\omega'')$,

$$a(\Delta_h^i(\chi u),\varphi)=-a(u,\chi\Delta_h^i\varphi)+I(u,\varphi),\quad i=1,\ldots,r, \tag{4}$$

where for $I(u,\varphi)$ we have the estimate

$$|I(u,\varphi)|\leqslant c\,\|u\|_m\cdot\|\varphi\|_m, \tag{5}$$

with $c\geqslant 0$, independent of h,

Proof. We have ($h_i=(0,\ldots,h,\ldots 0)$)

$$\Delta_h^i(\varphi\cdot\psi)=(\Delta_h^i\varphi)\cdot\psi+\varphi(x+h_i)\cdot\Delta_h^i\psi \tag{6}$$

and

$$\int_\Omega \Delta_h^i \varphi \, dx = 0, \tag{7}$$

if the supports of φ and $\varphi(x+h_i)$ lie in Ω. From these simple rules we deduce

$$\begin{aligned} a(\Delta_h^i(\chi u), \varphi) &= \int_\Omega \sum_{|\alpha|,|\beta| \leqslant m} a_{\alpha\beta}(x) D^\alpha(\Delta_h^i(\chi u)) \cdot D^\beta \bar{\varphi} \, dx \\ &= -\int_\Omega \sum_{\alpha,\beta} \Delta_h^i a_{\alpha\beta} \cdot D^\alpha(\chi u) \cdot D^\beta \overline{\varphi(x+h_i)} \, dx \\ &\quad - \int_\Omega \sum_{\alpha,\beta} a_{\alpha\beta}(x) D^\alpha(\chi u) D^\beta(\Delta_h^i \bar{\varphi}) \, dx, \end{aligned}$$

where because of assumption (2) for $k=1$, the first integral satisfies the estimate (5). Here and elsewhere we use the fact that the translation operator $\tau_{h_i}\varphi := \varphi(x+h_i)$ is uniformly continuous in the W_2^m-spaces; this follows from the Kolmogorov compactness criterion §1.1. In the second integral we apply the Leibniz product rule $(D^\alpha(\chi u) = \chi D^\alpha u + \cdots + \chi D^\beta \Delta_h^i \bar{\varphi} = D^\beta(\chi \Delta_h^i \bar{\varphi}) - \cdots)$ and we obtain

$$\begin{aligned} \int_\Omega \sum_{\alpha,\beta} a_{\alpha\beta}(x) D^\alpha(\chi u) D^\beta(\Delta_h^i \bar{\varphi}) \, dx &= \int_\Omega \sum_{\alpha,\beta} a_{\alpha\beta}(x) D^\alpha u D^\beta(\chi \Delta_h^i \bar{\varphi}) \, dx \\ &\quad + \int_\Omega \sum_{\substack{|\alpha| \leqslant m \\ |\beta| \leqslant m-1}} b_{\alpha\beta}(x) D^\alpha u D^\beta(\Delta_i^h \bar{\varphi}) \, dx \\ &= a(u, \chi \Delta_h^i \varphi) + I_1(u, \Delta_h^i \varphi). \end{aligned}$$

For I_1 by Theorem 9.3 we have the estimate

$$I_1(u, \Delta_h^i \varphi) \leqslant c' \|u\|_m \cdot \|\Delta_h^i \varphi\|_{m-1} \leqslant c \|\varphi\|_m \|u\|_m,$$

and so we have proved (4).

Now for the proof of Theorem 20.1. Let $k=1$. For $\varphi \in \mathscr{D}(\omega'')$ we have, see (4),

$$a(\Delta_h^i(\chi u), \varphi) = -a(u, \chi \Delta_h^i \varphi) + I(u, \varphi) = -(f, \chi \cdot \Delta_h^i \varphi)_0 + I(u, \varphi). \tag{8}$$

Since by assumption $f \in W_2^{-(m-1)}(\Omega)$ we can estimate

$$|(f, \chi \Delta_h^i \varphi)_0| \leqslant c \|f\|_{-(m-1)} \|\chi \Delta_h^i \varphi\|_{m-1} \leqslant c \|\varphi\|_m \|f\|_{-m+1}, \tag{9}$$

where we interpret $(f, \varphi)_0$ as a scalar product on the Gelfand triple $\mathring{W}_2^{m-1} \subsetneq L_2(\Omega) \subsetneq W_2^{-m+1}$, and we use the second property (3), §17 of a Gelfand triple, as well as Theorem 9.3. By the usual density argument (5), (8) and (9) hold for all $\varphi \in \mathring{W}_2^m(\omega'')$. We put $\varphi = \Delta_h^i(\chi u)$ in (8), use the $\mathring{W}_2^m$-coercion of a and

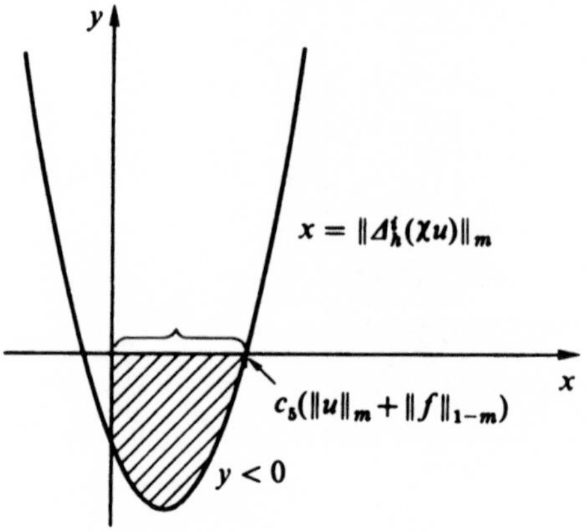

Fig. 20.1

the estimates (5) and (9); we obtain

$$\begin{aligned} c_1 \|\Delta_h^i(\chi u)\|_m^2 - c_2' \|\Delta_h^i(\chi u)\|_0^2 &\leqslant \operatorname{Re} a(\Delta_h^i(\chi u), \Delta_h^i(\chi u)) \\ &\leqslant c_3 \|\Delta_h^i(\chi u)\|_m \cdot \|f\|_{-m+1} + c_4 \|u\|_m \|\Delta_h^i(\chi u)\|_m, \end{aligned}$$

and again applying Theorem 9.3 to $\|\Delta_h^i(\chi u)\|_0^2 \leqslant c \|\Delta_h^i(\chi u)\|_{m-1}^2 \leqslant c \|u\|_m^2$ gives

$$c_1 \|\Delta_h^i(\chi u)\|_m^2 \leqslant c_2 \|u\|_m^2 + c_3 \|\Delta_h^i(\chi u)\|_m \|f\|_{-m+1} + c_4 \|u\|_m \|\Delta_h^i(\chi u)\|_m. \tag{10}$$

Comparison with the parabola (see Fig. 20.1)

$$y = c_1 x^2 - (c_3 \|f\|_{-m+1} + c_4 \|u\|_m) x - c_2 \|u\|_m^2$$

gives an estimate for $\|\Delta_h^i(\chi u)\|_m^2$ from (10)

$$\|\Delta_h^i(\chi u)\|_m \leqslant c_5(\|u\|_m + \|f\|_{-m+1}) \tag{11}$$

(we look for the x-interval, where the parabola is negative, see (10)). Since f and u were chosen fixed, by Theorem 9.6 the estimate (11) implies that $\chi u \in \mathring{W}_2^{m+1}(\omega')$, respectively that $u \in W_2^{m+1}(\omega)$, with which we have completed the proof for $k = 1$. Given (11), Theorem 9.6 also gives the estimate

$$\|u\|_{m+1,\omega} \leqslant c_5(\|u\|_m + \|f\|_{-m+1}). \tag{12}$$

Suppose now that $k \geqslant 2$. We consider in succession the expressions

$$a\left(\frac{\partial u}{\partial x_i}, \varphi\right), \quad i = 1, \ldots, r, \quad a(D^\gamma u, \varphi), \quad 1 < |\gamma| \leqslant k - 1; \quad \varphi \in \mathscr{D}(\omega'').$$

By partial integration we have

$$a\left(\frac{\partial u}{\partial x_i}, \varphi\right) = -a\left(u, \frac{\partial \varphi}{\partial x_i}\right) - \int_\Omega \sum_{\substack{|\alpha| \leqslant m \\ |\beta| \leqslant m}} \frac{\partial a_{\alpha\beta}}{\partial x_i} \mathrm{D}^\alpha u \cdot \mathrm{D}^\beta \bar{\varphi}\, \mathrm{d}x$$

$$= -a\left(u, \frac{\partial \varphi}{\partial x_i}\right) + I_i(\varphi), \quad i = 1, \dots, r, \tag{13}$$

$$a(\mathrm{D}^\gamma u, \varphi) = (-1)^{|\gamma|} a(u, \mathrm{D}^\gamma \varphi) + I_\gamma(\varphi), \quad |\gamma| \leqslant k-1,$$

and with (1)

$$a\left(\frac{\partial u}{\partial x_i}, \varphi\right) = -\left(f, \frac{\partial \varphi}{\partial x_i}\right)_0 + I_i(\varphi), \quad i = 1, \dots, r,$$

$$a(\mathrm{D}^\gamma u, \varphi) = (-1)^{|\gamma|}(f, \mathrm{D}^\gamma \varphi)_0 + I_\gamma(\varphi), \quad |\gamma| \leqslant k-1.$$

We write the functionals

$$(-1)^{|\gamma|}(f, \mathrm{D}^\gamma \varphi)_0 + I_\gamma(\varphi) =: F_\gamma(\varphi), \quad 1 \leqslant |\gamma| \leqslant k-1$$

and instead of (1) we consider in succession the equations

$$a(\mathrm{D}^\gamma u, \varphi) = F_\gamma(\varphi), \quad 1 \leqslant |\gamma| \leqslant k-1. \tag{14}$$

If instead of the first estimate in (9) we can prove the estimates

$$|F_\gamma(\varphi)| \leqslant c \|\varphi\|_{m-1}, \quad 1 \leqslant |\gamma| \leqslant k-1, \tag{15}$$

then we have finished, for the proof above (for $k = 1$) applied successively to the equations (14) gives

$$\frac{\partial u}{\partial x_i} \in W_2^{m+1}(\omega), \quad i = 1, \dots, r, \quad \mathrm{D}^\gamma u \in W_2^{m+1}(\omega), \quad 1 < |\gamma| \leqslant k-1,$$

which is $u \in W_2^{m+k}(\omega)$. Also without difficulty we have an estimate corresponding to (12)

$$\|u\|_{m+k,\omega} \leqslant c_5 \left[\|u\|_m + \sum_{|\alpha| \leqslant k-1} \|\mathrm{D}^\alpha f\|_{-m-1} \right]. \tag{16}$$

We check (15), for $\varphi \in \mathscr{D}(\omega'')$ we have (we partially integrate and use the hypothesis $\mathrm{D}^\gamma f \in W_2^{1-m}$ for $|\gamma| \leqslant k-1$)

$$|(-1)^{|\gamma|}(f, \mathrm{D}^\gamma \varphi)_0| = |(\mathrm{D}^\gamma f, \varphi)_0| \leqslant c' \|\mathrm{D}^\gamma f\|_{-m+1} \cdot \|\varphi\|_{m-1} \leqslant c \|\varphi\|_{m-1}. \tag{17}$$

Now for the functionals $I_\gamma(\varphi)$, we must show

$$|I_\gamma(\varphi)| \leqslant c \|\varphi\|_{m-1}. \tag{18}$$

We have

$$I_\gamma(\varphi)=(-1)^{|\gamma|-1}\int_\Omega \sum_{\substack{|\alpha|\leqslant m\\|\beta|\leqslant m}}\sum_{|s|\leqslant|\gamma|-1}\binom{s}{\gamma}\mathrm{D}^{\gamma-s}a_{\alpha\beta}\cdot\mathrm{D}^\alpha u\cdot\mathrm{D}^{\beta+s}\bar{\varphi}\,\mathrm{d}x \tag{19}$$

with

$$u\in W^{m+|\gamma|}. \tag{20}$$

Expression (20) is the reason why we must proceed in stages. We estimate the individual terms of (19), distinguishing two cases; first $|\beta+s|\leqslant m-1$, here (18) is apparent, second $|\beta+s|\geqslant m$. In order to obtain (18), we partially integrate $(|\beta+s|-m+1)$ times away from φ. In doing this $\mathrm{D}^\alpha u$ may be increased to $\mathrm{D}^{\alpha_1}u$, where

$$|\alpha_1|=|\alpha|+|\beta|+|s|-m+1\leqslant 2m+|\gamma|-1-m+1=m+|\gamma|,$$

which is permitted because of (20). $\mathrm{D}^{\gamma-s}a_{\alpha\beta}$ may be increased to $\mathrm{D}^{\gamma_1}a_{\alpha\beta}$, where

$$\begin{aligned}|\gamma_1|&=|\gamma|-|s|+|\beta|+|s|-m+1=|\gamma|+|\beta|-m+1\\&\leqslant k-1+m+m+1=k,\end{aligned}$$

which by assumption (2) is also permitted. We see therefore that after partial integration we again have the estimate (18); (18) and (17) give (15). ■

Remark Careful manipulation of (19) shows that for $k>m$ we can weaken assumption (2) to

$$a_{\alpha\beta}\in C^{\max(m,|\beta|+k-m)}(\bar{\Omega}),$$

a result which goes back to Nirenberg [1].

We now proceed to prove the regularity of a solution $u\in V$ of (1) up to the frontier $\partial\Omega$. In order to include Theorem 15.1 in the regularity Theorem 20.4, we must adjoin a boundary form $c(\varphi,\psi)$ to the sesquilinear form $a(\varphi,\psi)$. We consider

$$c(\varphi,\psi)=\sum_{|\alpha|+|\beta|\leqslant 2m-1}\int_{\partial\Omega} c_{\alpha\beta}\mathrm{D}^\alpha\varphi\cdot\mathrm{D}^\beta\bar{\psi}\cdot\mathrm{d}x. \tag{21}$$

Definition 20.1 *We say that the boundary form* (21) *is* $(m-1)$*-transversal if the derivatives in the normal direction, whether arising in* D^α *or in* D^β, *are at most of order* $m-1$.

Theorem 20.2 *Let* Ω *be bounded and* $\in C^{2m,1}$, *let* $c_{\alpha\beta}\in C^1(\bar{\Omega})$ *if* $|\alpha|\leqslant m-1$ *and* $|\beta|\leqslant m-1$, *let* $c_{\alpha\beta}\in C^{|\alpha|-m+2}(\bar{\Omega})$ *for* $|\alpha|>m-1$ *and* $c_{\alpha\beta}\in C^{|\beta|-m+2}(\bar{\Omega})$ *for* $|\beta|>m-1$. *If the boundary form* (21) *is* $(m-1)$*-transversal, then it is continuous on*

$W_2^m(\Omega) \times W_2^m(\Omega)$. *If in addition for the boundary form* (21) *we have*

$$|\alpha| + |\beta| \leqslant 2m - 2 \tag{22}$$

then $c(\varphi, \psi)$ *satisfies the assumptions of Theorem* 19.1:

$$|c(\varphi, \varphi)| \leqslant c \|\varphi\|_m \cdot \|\varphi\|_{m-1} \tag{23}$$

from which it follows that along with $a(\varphi, \psi)$, $a(\varphi, \psi) + c(\varphi, \psi)$ *is again V-coercive.*

Proof. We carry out an elementary proof without using trace operators. By means of a partition of unity and normal coordinate transformations we need only to consider the case $\partial\Omega = W^{r-1}, \Omega = W^r_+$, and $\varphi, \psi \in C^\infty(\overline{W^r_+})$, where the supports of φ, ψ cut the frontier of W^r_+ in W^{r-1}. In order to estimate we consider the individual terms

$$\int_{W^{r-1}} c_{\alpha\beta} \mathrm{D}^\alpha \varphi \cdot \mathrm{D}^\beta \bar{\psi} \, \mathrm{d}x_1 \cdots \mathrm{d}x_{r-1}, \tag{24}$$

of (21). If $|\alpha| \leqslant m - 1$ and $|\beta| \leqslant m - 1$, we change nothing in (24), and write (24) in the form

$$\int_{W^{r-1}} c w \bar{v} \, \mathrm{d}x_1 \cdots \mathrm{d}x_{r-1}, \tag{25}$$

where

$$w := \mathrm{D}^\alpha \varphi, \quad |\alpha| \leqslant m - 1, \quad v := \mathrm{D}^\beta \psi, \quad |\beta| \leqslant m - 1, \quad c \in C^1(\overline{W^r}). \tag{26}$$

Let $|\alpha| > m - 1$, because of $(m-1)$-transversality the $|\alpha| - m + 1$ superfluous differentiations in $\mathrm{D}^\alpha \varphi$ must be tangential (that is, $\partial/\partial x_1, \ldots, \partial/\partial x_{r-1}$), and we can, because of the support properties of φ and ψ, by means of partial integration (without extra terms) bring them into $\mathrm{D}^\beta \psi$. The integral (24) then takes the form

$$\int_{W^{r-1}} c w \frac{\overline{\partial v}}{\partial x_1} \cdot \mathrm{d}x_1 \cdots \mathrm{d}x_{r-1} \tag{27}$$

where (26) is again valid. We can also bring the integral (24) into the form (27) if $|\beta| > m - 1$. These are all the possible cases, since $|\alpha| \geqslant m$ and $|\beta| \geqslant m$ are impossible given that $|\alpha| + |\beta| \leqslant 2m - 1$. For (25) we have (support properties of φ and ψ!):

$$\begin{aligned} \int_{W^{r-1}} c w \bar{v} \, \mathrm{d}x &= -\int_0^1 \mathrm{d}t \int_{W^{r-1}} \frac{\partial}{\partial t}(c w \bar{v}) \, \mathrm{d}x \\ &= -\int_{W^r_+} \left[\frac{\partial c}{\partial t} w \bar{v} + c \frac{\partial w}{\partial t} \bar{v} + c w \frac{\partial \bar{v}}{\partial t} \right] \mathrm{d}y, \end{aligned} \tag{28}$$

from which because of (26) we have the estimate

$$\left|\int_{W^{r-1}} cw\bar{v}\,dx\right| \leqslant c_1[\|\varphi\|_{m-1}\|\psi\|_{m-1} + \|\varphi\|_m\|\psi\|_{m-1} + \|\varphi\|_{m-1}\|\psi\|_m]$$
$$\leqslant c_2\|\varphi\|_m\|\psi\|_m. \tag{29}$$

For (27) we have

$$\int_{W^{r-1}} cw\frac{\partial\bar{v}}{\partial x_1}dx = -\int_0^1 dt\int_{W^{r-1}}\frac{\partial}{\partial t}\left[cw\frac{\partial\bar{v}}{\partial x_1}\right]dx$$
$$= -\int_{W^r_+}\left[\frac{\partial c}{\partial t}w\frac{\partial\bar{v}}{\partial x_1} + c\frac{\partial w}{\partial t}\frac{\partial\bar{v}}{\partial x_1} + cw\frac{\partial^2\bar{v}}{\partial t\partial x_1}\right]dy.$$

In the last term we integrate partially according to x_1 and obtain

$$-\int_{W^r_+} cw\frac{\partial^2\bar{v}}{\partial t\partial x_1}dy = \int_{W^r_+}\frac{\partial(cw)}{\partial x_1}\frac{\partial\bar{v}}{\partial t}dy = \int_{W^r_+}\left[\frac{\partial c}{\partial x_1}w\frac{\partial\bar{v}}{\partial t} + c\frac{\partial w}{\partial x_1}\frac{\partial\bar{v}}{\partial t}\right]dy,$$

by which for $\int_{W^{r-1}} cw(\partial\bar{v}/\partial x_1)\,dx$ we obtain

$$\int_{W^{r-1}} cw\frac{\partial\bar{v}}{\partial x_1}dx = \int_{W^r_+}\left[\frac{\partial c}{\partial x_1}w\frac{\partial\bar{v}}{\partial t} - c\frac{\partial w}{\partial t}\frac{\partial\bar{v}}{\partial x_1} + c\frac{\partial w}{\partial x_1}\frac{\partial\bar{v}}{\partial t} - \frac{\partial c}{\partial t}w\frac{\partial\bar{v}}{\partial x_1}\right]dy,$$

from which follows the estimate

$$\left|\int_{W^{r-1}} cw\frac{\partial\bar{v}}{\partial x_1}\right| \leqslant c_2\|\varphi\|_m\cdot\|\psi\|_m. \tag{30}$$

(29) and (30) prove the continuity of (21) on $W_2^m(\Omega)\times W_2^m(\Omega)$. If in addition $|\alpha|+|\beta|\leqslant 2m-2$, we can always put (24) in the form (25), and for $\varphi=\psi$ the first estimate of (29) gives

$$\left|\int_{W^{r-1}} cw\bar{v}\,dx\right| \leqslant c_3\|\varphi\|_m\|\varphi\|_{m-1}$$

which is (23). ■

We next give the assumptions for the regularity Theorem 20.4. Let $k\geqslant 1$ (k can also be equal to zero, however we then obtain nothing new).

1. Let Ω be bounded and belong to $C^{m+k,1}$.
2. For the coefficients of the sesquilinear form $a(\varphi,\psi)$ we assume

$$a_{\alpha\beta}(x)\in C^k(\bar{\Omega}). \tag{31}$$

3. Suppose that on the frontier $\partial\Omega$ we are given p normal boundary

value operators $b_j(x, \mathrm{D})$, $j = 1, \ldots, p$, with $0 \leqslant m_j = \operatorname{ord} b_j \leqslant m - 1$. Let

$$V := W^m(\{b_j\}_1^p);$$

if no boundary value operators are preassigned, hence $p = 0$, we put

$$V := W_2^m(\Omega).$$

For the coefficients $b_j(x, \mathrm{D})$ we assume

$$b_{j,s}(x) \in C^{m+k-m_j}(\bar{\Omega}), \quad |s| \leqslant m_j, j = 1, \ldots, p. \tag{32}$$

4. Let an $(m-1)$-transversal boundary form $c(\varphi, \psi)$ be preassigned. Since on $\partial\Omega$ we can always partially integrate tangentially, without introducing additional terms, $\partial\Omega$ has no frontier since $\partial\partial\Omega = \emptyset$, we can – as we have shown in the proof of Theorem 20.2 – modify $c(\varphi, \psi)$ to

$$c(\varphi, \varphi) = \sum_{\substack{|\alpha| \leqslant m-1 \\ |\beta| \leqslant m}} \int_{\partial\Omega} c_{\alpha\beta} \cdot \mathrm{D}^\alpha \varphi \cdot \mathrm{D}^\beta \bar{\psi} \lambda\sigma, \tag{33}$$

and we assume for the coefficients of $c(\varphi, \psi)$ in (33) that

$$c_{\alpha\beta}(x) \in C^k(\bar{\Omega}), \quad |\alpha| \leqslant m - 1, |\beta| \leqslant m. \tag{34}$$

5. Let the sesquilinear form

$$\mathscr{A}(\varphi, \psi) = a(\varphi, \psi) + c(\varphi, \psi) \tag{35}$$

be V-coercive. By Theorem 20.2 it suffices, if for (33) $|\alpha| + |\beta| \leqslant 2m - 2$ holds for all α, β, to require that $a(\varphi, \psi)$ is V-coercive.

6. We consider the weak equation

$$a(u, \varphi) + c(u, \varphi) = (f, \varphi)_0, \quad \varphi \in V, \tag{36}$$

where the solution u should be from V, and for the preassigned functional f we should have

$$f \in [W_2^{m-k}(\Omega)]' \quad \text{for } k < m, \quad f \in W_2^{k-m}(\Omega) \quad \text{for } k \geqslant m. \tag{37}$$

We have, see Example 17.2, that $[W_2^{m-k}]'$, respectively $W_2^{k-m} \subset [W_2^{m-1}]' \subset [W_2^m]'$. Here for $k < m$ we interpret $(f, \varphi)_0$ as functional representation on the Gelfand triple

$$W_2^{m-k}(\Omega) \subsetneq L_2(\Omega) \subsetneq [W_2^{m-k}(\Omega)]',$$

see §17, with the estimate

$$|(f, \varphi)_0| \leqslant \|f\|_{k-m} \|\varphi\|_{m-k} \leqslant \|f\|_{k-m} \|\varphi\|_m, \tag{38}$$

that is, $(f, \varphi)_0$ can also be interpreted as a continuous functional on the

Gelfand triple

$$V \subsetneq L_2(\Omega) \subsetneq V',$$

so that the weak equation (36) makes sense. For $k \geqslant m$ we write

$$(f, \varphi)_0 := \int_\Omega f \cdot \bar{\varphi} \, \mathrm{d}x$$

and have the estimate

$$|(f, \varphi)_0| \leqslant \| f \| \, \| \varphi \|_0 \leqslant \| f \| \, \| \varphi \|_m \tag{38'}$$

by which $(f, \varphi)_0$ again becomes a continuous functional on the triple

$$V \subsetneq L_2(\Omega) \subsetneq V'.$$

In order to obtain the regularity of a solution $u \in V$ of (36) up to the boundary $\partial\Omega$, we must overcome the following difficulties:

(a) We must straighten out the boundary $\partial\Omega$ locally, this occurs by means of an admissible coordinate transformation, that is, we apply Theorem 2.12.

(b) In the proof of Theorem 20.1 we have worked with very good differentiable functions from $\mathscr{D}$, here we show that $W_2^{m+k}(\Omega) \cap V$ is dense in V (Theorem 20.3). Under the assumption (32) this is the best possible. If the coefficients of the boundary operators b_j, $j = 1, \ldots, p$, are better than (32) differentiable, we can draw better conclusions:

$$W_2^{m+l}(\Omega) \cap V \text{ is dense in } V \text{ for } l > k.$$

(c) Again in the proof we use the difference operators Δ_h, this time, however, only in the tangential direction ($\Omega = W^r_+$, $\partial\Omega = W^{r-1}$, h tangential to W^{r-1}). Here it can happen that $\Delta_h u$ no longer belongs to V, although $u \in V$ (that is, $b_j u = 0$, $j = 1, \ldots, p$, but $b_{j_0} \Delta_h u \neq 0$). However, we can introduce a correction operator Z_h, such that $\Delta_h u - Z_h u \in V$ for $u \in V$ (Proposition 20.1). The same can be done for the tangential derivatives $\mathrm{D}_x u$, $x \in W^{r-1}$, see Proposition 20.2.

(d) If the displacement h takes place in the normal direction, it diverges from the frontier and the boundary conditions, so we must handle the normal derivatives differently, this occurs in Proposition 20.3 and 20.4.

(e) Regularity theorems are local; we localise by means of multiplication by a function $\chi \in \mathscr{D}(\omega)$, where $\omega \cap \partial\Omega \neq \varnothing$, see also Lemma 20.1. Here it can also happen that $u \in V$, but χu is no longer in V; however, again there exists a correction R with $\chi u - Ru \in V$. For the sake of brevity we consider multiplication by χ together with the difference operators Δ_h, respectively with D_x^α (see Proposition 20.1 and 20.2).

We begin with:

Theorem 20.3 *Let Ω satisfy assumption* 1 *and the boundary value operators $b_j, j = 1,\dots,p$ assumption* 3, *that is,* (32). *We consider the space V determined by the boundary values b_j, $V = W^m(\{b_j\}_{j=1}^p)$ (see also* §14). *We have*

$$V \cap W_2^{m+k}(\Omega) \text{ is dense in } V.$$

Proof. Let $v \in V \subset W_2^m(\Omega)$, by Theorem 3.6 there exists a sequence $\varphi_n \in C^\infty(\Omega)$ with $\varphi_n \to v$ in $W_2^m(\Omega)$ and $b_j\varphi_n \to b_jv = 0$ in $W_2^{m-m_j-1/2}(\partial\Omega)$. Since $b_j\varphi_n$ also belongs to $W_2^{m+k-m_j-1/2}(\partial\Omega)$, $j = 1,\dots,p$, we can consider the equations

$$b_1\psi_n = b_1\varphi_n, \dots, b_p\psi_n = b_p\varphi_n. \tag{39}$$

By Theorem 14.1, with $M = k + m$, the equations (19) possess a solution ψ_n in $W_2^{m+k}(\Omega)$, which depends continuously in the W_2^m-topology on $b_j\varphi_n$, that is, $\psi_n \to 0$ in $W_2^m(\Omega)$. We put $v_n := \varphi_n - \psi_n$ and by (39) have $b_jv_n = b_j\varphi_n - b_j\psi_n = 0$, which is $v_n \in V \cap W_2^{m+k}(\Omega)$, as also $v_n = (\varphi_n - \psi_n) \to (v - 0)$ in $W_2^m(\Omega)$, and thus we have proved our theorem. ■

We come now to the correction operators.

Proposition 20.1 *We consider the half cube W_+^r. Suppose on W^{r-1} that the normal boundary value operators $b_j(x, \mathrm{D}), j = 1,\dots,p$ are given, where $0 \leqslant m_j \leqslant m - 1$ and that* (32) *holds, that is,*

$$b_{j,s}(x) \in C^{m+k-m_j}(\overline{W_+^r}), |s| \leqslant m, \quad j = 1,\dots,p.$$

We denote by $\mathscr{V}$ the space of all functions φ from $W_2^m(W_+^r)$, which vanish in the neighbourhood of $\partial W_+^r \setminus W^{r-1}$, and which satisfy the boundary conditions $b_j(x, \mathrm{D})\varphi = 0$ on W^{r-1}. For brevity we write

$$W_\varepsilon^m := W_2^m(W_+^r(1 - \varepsilon)), V_\varepsilon := \mathscr{V} \cap W_\varepsilon^m \tag{40}$$

where $1 - \varepsilon$ is half the edge length of the cube $W^r(1 - \varepsilon)$. Let $\chi \in \mathscr{D}(\mathbb{R}^r)$ be given with $\operatorname{supp}\chi \cap W_+^r \neq \varnothing$, let $|h| \leqslant \varepsilon/2$, we consider the tangential difference operators Δ_h^τ, $\tau = 1,\dots,r-1$. Then there exist correction operators Z_h^τ, $\tilde{Z}_h^\tau$, $\tau = 1,\dots,r-1$, with the following properties: for $v \in V_{3\varepsilon}$

$$\Delta_h^\tau(\chi v) - Z_h^\tau v \in V_\varepsilon, \quad \tau = 1,\dots,r-1,$$

and

$$\chi(\Delta_h^\tau v) - \tilde{Z}_h^\tau v \in V_\varepsilon, \quad \tau = 1,\dots,r-1, \tag{41}$$

Z_h and $\tilde{Z}_h: V_{2\varepsilon} \to W_\varepsilon^m$ are linear and continuous, and for their norms we have

$$\|Z_h^\tau\|_m, \|\tilde{Z}_h^\tau\|_m \leqslant c < \infty, \quad \tau = 1,\dots,r-1 \tag{42}$$

where c is independent of h.

Proof. We again use Theorem 14.1, respectively the extension operator Z from §8. Let $\varphi_j^h := b_j(\Delta_h^\tau(\chi v))$. By application of formula (6) and the Leibniz product rule we have

$$\begin{aligned}\varphi_j^h &= b_j(\Delta_h^\tau(\chi v)) = b_j[(\Delta_h^\tau \chi)v + \chi(x+h)\Delta_h^\tau v] \\ &= \Delta_h^\tau \chi \cdot b_j v + b_j' v + \chi(x+h)\cdot b_j(\Delta_h^\tau v) + b_j''(\Delta_h^\tau v),\end{aligned} \tag{43}$$

where the orders satisfy

$$\operatorname{ord} b_j', \operatorname{ord} b_j'' \leqslant m_j - 1. \tag{44}$$

Because of the boundary conditions $b_j v = 0$ and $b_j(x+h, \mathrm{D})v(x+h) = 0$ we have

$$b_j(\Delta_h^\tau v) = \sum_{|s| \leqslant m^j} \Delta_h^\tau b_{j,s}(x) \mathrm{D}^s v(x+h) \tag{45}$$

and the assumptions (32), that is, $\partial b_{j,s}/\partial x_\tau \in C^{m+k-1-m_j}(\overline{W_+^r})$, $\tau = 1, \ldots, r-1$, give the estimate for (45)

$$\| b_j(\Delta_h^\tau v) \|_{m-m_j-1/2} \leqslant c \| v \|_m, \tag{46}$$

where c is independent of h. Altogether – looking back to (44) and (46) – we obtain the estimate for (43)

$$\| \varphi_j^h \|_{m-m_j-1/2} \leqslant c \| v \|_m, \tag{47}$$

where c is independent of h. By Theorem 14.1 the equations

$$b_j u^h = \varphi_j^h, \quad j = 1, \ldots, p, \tag{48}$$

possess a solution $u^h \in W_2^m$, which depends continuously on φ_j^h. We define $Z_h^\tau v := u^h$, and again (48) immediately deduce (41), while (47) together with continuous dependence gives us (42). The proof for $\tilde{Z}_h^\tau$ proceeds in the same way, instead of (43) as initial formula we only have to take the following

$$\tilde{\varphi}_j^h := b_j(\chi(\Delta_h^\tau v)) = \chi \cdot b_j(\Delta_h^\tau v) + \tilde{b}_j \Delta_h^\tau v,$$

with $\operatorname{ord} \tilde{b}_j \leqslant m_j - 1$ and $b_j(\Delta_h^\tau v)$ as in (45). For $\tilde{b}_j \Delta_h^\tau v$ Theorem 9.4 again gives the estimate

$$\| \tilde{b}_j \Delta_h^\tau v \|_{m-m_j-1/2} = \| \tilde{b}_j \Delta_h^\tau v \|_{(m-1)-(m_j-1)-1/2} \leqslant \| \tilde{b}_j \| \, \| \Delta_h^\tau v \|_{m-1} \leqslant c \| v \|_m,$$

and the rest of the proof proceeds as above. ∎

The dependence on ε (support!) is also given by Theorem 14.1.

Warning Again we note that Propositions 1 and 2 only hold in the tangential direction.

Proposition 20.2 *We use the assumptions and notation of Proposition* 20.1, *let* D_x^s *be tangential derivatives, as always we write*

$$y=(x,t)=(x_1,\ldots,x_{r-1},t)\in W_+^r, \quad x\in W^{r-1}, t\geqslant 0.$$

We consider $\Delta_h^\tau D_x^s(\chi v)$ *and* $\chi[D_x^s\Delta_h^\tau v]$ *for* $v\in V_{3\varepsilon}\cap W_2^{m+l}$, *where* $1\leqslant l\leqslant k-1$ *and* $|s|=l$. *There exist correction operators* $Z_h^{s,\tau}$, $\tilde{Z}_h^{s,\tau}$ *with the following properties*:

(a) *For* $v\in V_{3\varepsilon}\cap W_2^{m+l}$ *we have*

$$\begin{aligned}&\Delta_h^\tau D_x^s(\chi v)-Z_h^{s,\tau}v\in V_\varepsilon\\ &\chi[D_x^s\Delta_h^\tau v]-\tilde{Z}_h^{s,\tau}v\in V_\varepsilon\end{aligned} \quad \tau=1,\ldots,r-1. \tag{49}$$

(b) *We have the representations*

$$\begin{aligned}Z_h^{s,\tau}v&=\sum_{|\alpha|\leqslant|s|}D_x^\alpha Z_h^\alpha v\\ \tilde{Z}_h^{s,\tau}v&=\sum_{|\alpha|\leqslant|s|}D_x^\alpha \tilde{Z}_h^\alpha v,\end{aligned} \tag{50}$$

where the operators Z_h^α, $\tilde{Z}_h^\alpha$ *act linearly and continuously*

$$Z_h^\alpha, \tilde{Z}_h^\alpha:\ \begin{matrix}W_2^m\cap V_{2\varepsilon} & \to W_2^m\\ \vdots & \vdots\\ W_2^{m+1}\cap V_{2\varepsilon} & \to W_2^{m+1}\end{matrix} \tag{51}$$

and for the corresponding operator norms we have

$$\|Z_h^\alpha\|_m, \|\tilde{Z}_h^\alpha\|_m,\ldots,\|Z_h^\alpha\|_{m+l}, \|\tilde{Z}_h^\alpha\|_{m+l}\leqslant c<\infty, \tag{52}$$

where c is independent of h.

Remark In this proof we shall work with the Leibniz formulae in several places. Multipliers in front of the differential D can be brought behind D.

$$\chi Dv=D(\chi v)-D^0(D\chi)v, D^0=I=\text{identity},$$

and similarly with higher derivatives. The last statement and the Leibniz product rule give

$$D^\alpha(b(x,D)v)=b(x,D)D^\alpha v+\sum_{|\beta|\leqslant|\alpha|-1}D^\beta(b_\beta(x,D)v),$$

or written differently

$$b(x,D)(D^\alpha v)=D^\alpha(b(x,D)v)+\sum_{|\beta|\leqslant|\alpha|-1}D^\beta(b_\beta(x,D)v), \tag{53}$$

where the orders satisfy

$$\operatorname{ord} b_\beta\leqslant\operatorname{ord} b.$$

The coefficients of b_β are obtained from the coefficients of b by differentiation and arithmetic operations, without division.

Proof of Proposition 20.2. We start from the representations

$$\begin{aligned} b_j\{\Delta_h^\tau \mathrm{D}_x^s(\chi v)\} &= \sum_{|\alpha|\leqslant|s|} \mathrm{D}_x^\alpha W_{h,j}^\alpha, \\ b_j\{\chi[\mathrm{D}_x^j\Delta_h^\tau v]\} &= \sum_{|\alpha|\leqslant|s|} \mathrm{D}_x^\alpha \tilde{W}_{j,h}^\alpha, \end{aligned} \quad j=1,\dots,p \tag{54}$$

with the estimates

$$\| W_{j,h}^\alpha \|_{m+n-m_j-1/2} \leqslant c\| v\|_{m+n}, \quad n=0,1,\dots,l \tag{55}$$

(similarly for $\tilde{W}_{j,h}^\alpha$), where c is independent of h. We prove (54) and (55) at the end. We look for corrections in the form

$$Z_h^{s,\tau}v =: g_h^{s,\tau} = \sum_{|\alpha|\leqslant|s|} \mathrm{D}_x^\alpha g_h^\alpha = \sum_{|\alpha|\leqslant s} \mathrm{D}_x^\alpha Z_h^\alpha v \tag{56}$$

and must determine g_h^α by (54). In order for (49) to be satisfied, we must have

$$b_j\Bigg(\sum_{|\alpha|\leqslant|s|} \mathrm{D}_x^\alpha g_h^\alpha\Bigg) = \sum_{|\alpha|\leqslant|s|} \mathrm{D}_x^\alpha W_{h,j}^\alpha, \quad j=1,\dots,p. \tag{57}$$

In order to satisfy (57) and to prove our proposition, we determine the g_h^α successively. In doing this we face the difficulty that $b_j(x, \mathrm{D})$ does not commute with D_x^α.

We begin with the highest order α, $|\alpha|=|s|$. We determine g_h^α, $|\alpha|=|s|$ by the equations

$$b_j g_h^\alpha = W_{h,j}^\alpha, \quad j=1,\dots,p. \tag{58}$$

By Theorem 14.1 and because of the estimate (55) there exists a solution $g_h^\alpha =: Z_h^\alpha v$ of (58), which satisfies (51) and (52). We substitute the solution g_h^α found for $|\alpha|=|s|$ in (57) and apply (53)

$$b_j(\mathrm{D}_x^\alpha g_h^\alpha) = \mathrm{D}_x^\alpha(b_j g_h^\alpha) + \sum_{|\beta|\leqslant|\alpha|-1} \mathrm{D}_x^\beta(b_{j,\beta} g_h^\alpha),$$

where $\operatorname{ord} b_{j,\beta}\leqslant m_j$. (57) becomes

$$b_j\Bigg(\sum_{|\alpha|\leqslant|s|-1} \mathrm{D}_x^\alpha g_h^\alpha\Bigg) = \sum_{|\alpha|\leqslant|s|-1} \mathrm{D}_x^\alpha W_{h,j}^\alpha - \sum_{\substack{|\alpha|\leqslant|s|-1\\ |\alpha'|=|s|}} \mathrm{D}_x^\alpha b_{j,\alpha} g_h^{\alpha'}. \tag{59}$$

We now determine g_h^α with $|\alpha|=|s|-1$ by

$$b_j(x,\mathrm{D}) g_h^\alpha = W_{h,j}^\alpha - b_{j,\alpha}(x,\mathrm{D}) g_h^{\alpha'}, \quad j=1,\dots,p. \tag{60}$$

Since $\operatorname{ord} b_{j,\alpha}\leqslant m_j$ we have the estimate

$$\| b_{j,\alpha}(x,\mathrm{D}) g_h^{\alpha'} \|_{m+n-m_j-1/2} \leqslant c\| g_h^{\alpha'}\|_{m+n} \leqslant c\| v\|_{m+n}, \tag{61}$$

where because of (52) ($|\alpha'| = |s|$) c is independent of h. By Theorem 14.1 the estimates (55) and (61) give the properties (51) and (52) for $g_h^\alpha =: Z_h^\alpha v$, $|\alpha| = |s| - 1$. If we continue with this process, we can determine all the g_h^α in (56), where (57), that is, (49) is satisfied, together with the properties (51) and (52). The representation (50) is (56) written in another way. In the same way we derive the properties of the $\tilde{Z}$-operators from the second formula of (54) with the corresponding estimate in (55).

Now for the proof of (54) with the estimate (55). Since the b_j are normal, we can without any loss of generality, assume that the coefficient of $\partial^{m_j}/\partial t^{m_j}$ is equal to 1. Because of (6) the Leibniz product rule and (53) we have

$$\begin{aligned} b_j(\Delta_h^\tau \mathrm{D}_x^s(\chi v)) &= b_j(\mathrm{D}_x^s \Delta_h^\tau(\chi v)) = b_j(\mathrm{D}_x^s[(\Delta_h^\tau \chi)v + \chi(x+h)\Delta_h^\tau v]) \\ &= \Delta_h^\tau \chi \cdot b_j(\mathrm{D}_x^s v) + b'_{j,h}(\mathrm{D}_x^s v) + \chi(x+h) b_j(\mathrm{D}_x^s \Delta_h^\tau v) \\ &\quad + b''_j(\mathrm{D}_x^s \Delta_h^\tau v), \end{aligned} \tag{62}$$

where ord $b'_{j,h}$, ord $b''_j \leqslant m_j - 1$ and the coefficients of $b'_{j,h}$ are bounded, with a bound independent of h, and by (32) satisfy the necessary differentiability assumptions. With the help of (53) we can already represent in (62) and estimate

$$\Delta_h^\tau \chi \cdot b_j(x, \mathrm{D})\mathrm{D}_x^s v = \sum_{|\alpha| \leqslant |s|} \mathrm{D}_x^\alpha b_{j,h}^\alpha v, \quad \text{ord } b_{j,h}^\alpha \leqslant m_j,$$

$$\| b_{j,h}^\alpha v \|_{m+n-m_j-1/2} \leqslant c \| v \|_{m+n},$$

$$b'_{j,h}(\mathrm{D}_x^s v) = \sum_{|\alpha| \leqslant |s|} \mathrm{D}_x^\alpha b'^{\alpha}_{j,h} v, \quad \text{ord } b'^{\alpha}_{j,h} \leqslant m_j - 1,$$

$$\| b'^{\alpha}_{j,h} v \|_{m+n-m_j-1/2} \leqslant c \| v \|_{m+n},$$

$$b''_j(x, \mathrm{D})(\mathrm{D}_x^s \Delta_h^\tau v) = \sum_{|\alpha| \leqslant |s|} \mathrm{D}_x^\alpha b''^{\alpha}_j(\Delta_h^\tau v), \quad \text{ord } b''_j \leqslant m_j - 1,$$

$$\| b''^{\alpha}_j(\Delta_h^\tau v) \|_{m+n-1-(m_j-1)-1/2} \leqslant c \| \Delta_h^\tau v \|_{m+n-1} \leqslant c \| v \|_{m+n},$$

since by assumption (32) the coefficients of $b_{j,n}^\alpha$, $b'^{\alpha}_{j,n}$ and b''^{α}_j are multipliers in the corresponding W-spaces. It remains to represent $\chi(x+h) \cdot b_j(\mathrm{D}_x^s \Delta_h^\tau v)$ in (62) and to estimate it. Again because of (53) we have

$$\chi(x+h) b_j(x, \mathrm{D})(\mathrm{D}_x^s \Delta_h^\tau v) = \chi(x+h) \mathrm{D}_x^s b_j \Delta_h^\tau v + \sum_{|\alpha| \leqslant |s|-1} \mathrm{D}_x^\alpha \tilde{b}_j^\alpha(\Delta_h^\tau v). \tag{63}$$

Since $v \in V_{3\varepsilon}$ satisfies the boundary conditions, $b_j(x, \mathrm{D})\Delta_h^\tau v$ has the form (45) and can be estimated by (46):

$$\| b_j(x, \mathrm{D})\Delta_h^\tau v \|_{m+n-m_j-1/2} \leqslant c \| v \|_{m+n}, \quad n = 0, \ldots, l.$$

With the assumption 'the coefficient of $\partial^{m_j}/\partial t^{m_j}$ equals 1', $\tilde{b}_j(x, \mathrm{D})$ either has order $m_j - 1$ or at least one tangential differentiation appears, which we can move to the front. In each case we can write

$$\sum_{|\alpha|\leqslant|s|-1} D_x^{\alpha}\bar{b}_j^{\alpha}\Delta_h^{\tau}v=\sum_{|\beta|\leqslant|s|} D_x^{\beta}\bar{b}_j^{\beta}(\Delta_h^{\tau}v),\quad \operatorname{ord}\bar{b}_j^{\beta}\leqslant m_j-1,$$

and estimate

$$\|\bar{b}_j^{\beta}(\Delta_h^{\tau}v)\|_{m+n-1-(m_j-1)-1/2}\leqslant c\|\Delta_h^{\tau}v\|_{m+n-1}\leqslant c\|v\|_{m+n}.$$

As already remarked in the derivation of (53), the prefactor $\chi(x+h)$ in (63) plays no role, and we have represented (62) by (54), where the estimates (55) hold. The proof for the second representation in (54) proceeds in the same way and may be left to the reader. ■

We come now to the derivatives $\partial/\partial t$ in the normal direction.

Proposition 20.3 *Again* let W^r *be the unit cube* $W_+^r=\{(x,t)\in W^r:t>0\}$. *Let* $f\in L_2(W_+^r)$ *with* $\partial f/\partial x_\tau\in L_2(W_+^r)$ *for* $\tau=1,\ldots,r-1$. *We consider the* (*anti*)-*linear functional*

$$f(\varphi)=\int_{W_+^r} f\frac{\partial^m\bar{\varphi}}{\partial t^m}\,dy\quad \text{on } \mathring{W}_2^{m-1}(W_+^r). \tag{64}$$

If the estimate

$$|f(\varphi)|\leqslant c\|\varphi\|_{m-1} \text{ for } \varphi\in\mathscr{D}(W_+^r) \text{ respectively } \varphi\in\mathring{W}_2^{m-1}(W_+^r) \text{ holds,} \tag{65}$$

then also $\partial f/\partial t\in L_2(W_+^r)$, *and we have*

$$\left\|\frac{\partial f}{\partial t}\right\|_0\leqslant c\left[\sum_{\alpha+|\beta|\leqslant m-1}\|w_{\alpha\beta}\|_0+\sum_{\tau=1}^{r-1}\left\|\frac{\partial f}{\partial x_\tau}\right\|_0\right]. \tag{66}$$

for the $w_{\alpha\beta}$*s see the proof*.

Proof. Because of the estimate (65) f belongs to $(\mathring{W}_2^{m-1})'$, and by Theorem 17.6 we have the representation

$$f(\varphi)=\int_{W_+^r} f\frac{\partial^m\bar{\varphi}}{\partial t^m}\,dy=\int_{W_+^r}\sum_{\alpha+|\beta|\leqslant m-1} w_{\alpha\beta}\frac{\partial^\alpha}{\partial t^\alpha}D_x^\beta\bar{\varphi}\,dy, \tag{67}$$

where we have written $y=(x,t)=(x_1,\ldots,x_{r-1},t)$. We write (67) as

$$E(\varphi)=\int_{W_+^r} f\frac{\partial^m\bar{\varphi}}{\partial t^m}\,dy-\int_{W_+^r}\sum_{\alpha+|\beta|\leqslant m-1} w_{\alpha\beta}\frac{\partial^\alpha}{\partial t^\alpha}D_x^\beta\bar{\varphi}\,dy=0 \tag{68}$$

and by Hestenes extend f and $w_{\alpha\beta}$ to W^r:

$$\tilde{f}(x,t)=\begin{cases} f(x,t) & \text{for } t>0,\\ \sum_{j=1}^{m+1}\lambda_j f(x,-jt) & \text{for } t<0,\end{cases}$$

$$\tilde{w}_{\alpha\beta}(x,t)=\begin{cases} w_{\alpha\beta}(x,t) & \text{for } t>0,\\ \sum_{j=1}^{m+1}\lambda_j(-j)^{m-\alpha}w_{\alpha\beta}(x,-jt) & \text{for } t<0.\end{cases}$$

Here we have so chosen the λ_j that

$$\sum_{j=1}^{m+1}(-j)^{m-1-\alpha}\lambda_j=1 \quad \text{for } \alpha=0,\ldots,m. \tag{69}$$

We consider the extended functional

$$\tilde{E}(\varphi)=\int_{W^r}\tilde{f}\frac{\partial^m\bar{\varphi}}{\partial t^m}\,\mathrm{d}y-\int_{W^r}\sum_{\alpha+|\beta|\leqslant m-1}\tilde{w}_{\alpha\beta}\frac{\partial^\alpha}{\partial t^\alpha}\mathrm{D}_x^\beta\bar{\varphi}\,\mathrm{d}y,$$

and claim that if

$$\varphi\in\mathscr{D}(W^r),\ \text{then } \tilde{E}(\varphi)=0. \tag{70}$$

Indeed we have

$$\begin{aligned}\tilde{E}(\varphi)&=E(\varphi)+\sum_{j=1}^{m+1}\lambda_j\int_{W^r_+}f(x,jt)\frac{\partial^m}{\partial t^m}\overline{\varphi(x,-t)}\,\mathrm{d}y\\ &\quad+\sum_{\alpha+|\beta|\leqslant m-1}\sum_{j=1}^{m+1}\lambda_j\cdot j^{-1}(-j)^{m-\alpha}\int_{W^r_+}w_{\alpha\beta}(x,t)\mathrm{D}_x^\beta\frac{\partial^\alpha}{\partial t^\alpha}\overline{\varphi\left(x,-\frac{t}{j}\right)}\,\mathrm{d}y\\ &=E(\varphi)+\sum_{j=1}^{m+1}\lambda_j j^{-1}\int_{W^r_+}f(x,t)\frac{\partial^m}{t^m}\overline{\varphi\left(x,-\frac{t}{j}\right)}\,\mathrm{d}y\\ &\quad+\sum_{\alpha+|\beta|\leqslant m-1}\sum_{j=1}^{m+1}\lambda_j j^{-1}(-j)^{m-\alpha}\int_{W^r_+}w_{\alpha\beta}(x,t)\mathrm{D}_x^\beta\frac{\partial^\alpha}{\partial t^\alpha}\overline{\varphi\left(x,-\frac{t}{j}\right)}\,\mathrm{d}y.\end{aligned} \tag{71}$$

We write

$$\varphi_0(x,t)\mapsto-\sum_{j=1}^{m+1}\lambda_j(-j)^{m-1}\varphi\left(x,-\frac{t}{j}\right),\quad v=\varphi+\varphi_0 \tag{72}$$

and have $v=\varphi+\varphi_0\in C_0^m(W^r_+)\subset \mathring{W}_2^{m-1}(W^r_+)$, that is, all derivatives up to order m vanish on W^{r-1}, this follows immediately from (69). If we use definition (72) we can write (71) as

$$\tilde{E}_{W^r}(\varphi)=E_{W^r_+}(\varphi)+E_{W^r_+}(\varphi_0)=E_{W^r_+}(\varphi+\varphi_0)=E_{W^r_+}(v)=0,$$

the last because of (68). In this way we have proved (70). In order to obtain (66) we can use the Fourier transformation, or Fourier series. To make a change we shall for once work with Fourier series. We take $W^r(1) = W^r$ as 'Fourier interval' and in order to have no difficulties with the Fourier coefficients (that is, with (73)), we assume that $\operatorname{supp} f \subset W^r_+(1-\varepsilon)$, $0<\varepsilon$ small – this involves no loss of generality. We then also have

$$\operatorname{supp} w_{\alpha\beta} \subset W^r_+(1-\varepsilon) \quad \text{and} \quad \operatorname{supp} \tilde{f} \subset W^r(1-\varepsilon), \quad \operatorname{supp} \tilde{w}_{\alpha\beta} \subset W^r(1-\varepsilon).$$

Let $\psi \in \mathscr{D}(W^r)$ with $\psi \equiv 1$ on $W^r(1-\varepsilon)$, we take

$$\varphi = i^m \pi^m \cdot \psi \cdot e^{-i\pi(k,x)}, \quad 0 \neq k = (l_1, \ldots, l_{r-1}, \tau) \in \mathbb{Z}^r,$$

substitute this in (70) and obtain

$$\tau^m \int_{W^r} \tilde{f} e^{-i\pi(k,x)}\,dx = \sum_{|s| \leqslant m-1} \int_{W^r} \tilde{w}_s(x) k^s \cdot e^{-i\pi(k,x)}\,dx, \tag{73}$$

where the functions $\tilde{f}$ and $\tilde{w}_s$ belong to $L^2(W^r)$. If we denote the Fourier coefficients of $\tilde{f}$ and $\tilde{w}_s$ by c_k and γ_{sk}, then we can write (73) as

$$\tau^m \cdot c_k = \sum_{|s| \leqslant m-1} k^s \cdot \gamma_{sk}, \quad k \neq 0. \tag{74}$$

We have the obvious estimate (where $k = (l_1, \ldots l_{r-1}, \tau) = (l, \tau)$)

$$|\tau c_k| \leqslant 2^m \frac{|\tau|^{2m+1}}{|k|^{2m}} |c_k| + 2^m \frac{|\tau|\,|l|^{2m}}{|k|^{2m}} |c_k|,$$

$$\leqslant \frac{2^m}{|k|^{m-1}} \sum_{|s|<m-1} |k|^{m-1} |\gamma_{sk}| + 2^m |l|\,|c_k|,$$

or finally

$$|\tau c_k| \leqslant 2^m \Bigg(\sum_{|s|<m-1} |\gamma_{sk}| + |l|\,|c_k| \Bigg). \tag{75}$$

We use the assumption that $\partial f/\partial x_\tau \in L^2$, $\tau = 1, \ldots, r-1$, the Hestenes extension gives $\partial \tilde{f}/\partial x_\tau \in L^2(W^r)$, that is, the Fourier series $\sum_k |l|^2 |c_k|^2$ converges, and hence (75), there also $\sum_k |\gamma_{sk}|^2 < \infty$, gives

$$\sum_k |\tau c_k|^2 < \infty,$$

which is equivalent to $\partial \tilde{f}/\partial t \in L^2(W^r)$. It follows that $\partial f/\partial t \in L^2(W^r_+)$, and (66) is an immediate consequence of (75).

Proposition 20.4 *Let the assumptions* 2–6 *be satisfied for* $\Omega = W^r_+$, $\partial\Omega = W^{r-1}$ *and* V_ε, *let* u *be a solution of* (36) *from* $V_\varepsilon \subset W^m_\varepsilon$, *let* $k \geqslant 1$. *If* $u \in W^m_\varepsilon$

and $D_x^s u \in W_\varepsilon^m$ *for all tangential derivatives* $|s| \leqslant k$, *then*

$$u \in W_\varepsilon^{m+k} \tag{76}$$

Proof. The assumptions imply that

$$\frac{\partial^\alpha}{\partial t^\alpha} D_x^s u \in L_{2,\varepsilon} := L_2(W_+^r(1-\varepsilon)) \tag{77}$$

for all $\alpha + |s| \leqslant k + m$, with $\alpha \leqslant m$, and we must prove that

$$\frac{\partial^\gamma}{\partial t^\gamma} D_x^s u \in L_{2,\varepsilon} \quad \text{for all } \gamma + |s| \leqslant k + m, \tag{78}$$

this time without restriction on γ, hence $0 \leqslant \gamma \leqslant k + m$. We prove (78) by induction on γ. For $\gamma \leqslant m$ everything is correct because of (77). Therefore for $\gamma \geqslant m$ suppose that (78) is correct; we then prove the correctness of (78) for $\gamma + 1$. We consider

$$u' := \frac{\partial^\gamma}{\partial t^\gamma} D_x^s u = \frac{\partial^m}{\partial t^m}\left[\frac{\partial^{\gamma-m}}{\partial t^{\gamma-m}} D_x^s u\right] \quad \text{for } \gamma + |s| \leqslant m + k - 1, \tag{79}$$

estimate the integral

$$\int_{W_+^r} a_{(0,\ldots,m)(0,\ldots,m)}(y) \frac{\partial^m}{\partial t^m}\left[\frac{\partial^{\gamma-m}}{\partial t^{\gamma-m}} D_x^s u\right] \frac{\partial^m \bar{\varphi}}{\partial t^m}\, dy \tag{80}$$

for $\varphi \in \mathscr{D}_\varepsilon = \mathscr{D}(W_+^r(1+\varepsilon))$ and use Proposition 20.3.

Remark $\varphi \in \mathscr{D}_\varepsilon$ implies that in the proof all partial integrations may be carried out without boundary terms entering.

Now to estimate (80); we have

$$\begin{aligned} &\int_{W_+^r} a_{(0,\ldots,m)(0,\ldots,m)}(y) \frac{\partial^m}{\partial t^m}\left[\frac{\partial^{\gamma-m}}{\partial t^{\gamma-m}} D_x^s u\right] \frac{\partial^m \bar{\varphi}}{\partial t^m}\, dy \\ &\quad = a\left(\frac{\partial^{\gamma-m}}{\partial t^{\gamma-m}} D_x^s u, \varphi\right) - \sum_{\substack{|\alpha| \leqslant m \\ |\beta| \leqslant m}}' \int_{W_+^r} a_{\alpha\beta}(y) D^\alpha\left[\frac{\partial^{\beta-m}}{\partial t^{\beta-m}} D_x^s u\right] D^\beta \bar{\varphi}\, dy, \end{aligned} \tag{81}$$

where $\sum'$ denotes that the term with $a_{(0,\ldots,m)(0,\ldots,m)}$ is missing from the sum $\sum_{\alpha\beta}$. Since $\varphi \in \mathscr{D}_\varepsilon$, we can apply the manipulation (13) and obtain

$$a\left(\frac{\partial^{\gamma-m}}{\partial t^{\gamma-m}} D_x^s u, \varphi\right) = (-1)^{\gamma-m+|s|} a\left(u, \frac{\partial^{\gamma-m}}{\partial t^{\gamma-m}} D^s \varphi\right) + I_{\gamma,s}(\varphi), \tag{82}$$

where because of the inductive assumption (78) for $I_{\gamma,s}(\varphi)$ (see the derivation of

(18)) we have

$$|I_{\gamma,s}(\varphi)| \leqslant c \| \varphi \|_{m-1}. \tag{83}$$

By assumption u satisfies equation (38), since $\varphi \in \mathscr{D}_\varepsilon$ (36) takes the simpler form

$$a(u, \varphi) = (f, \varphi)_0.$$

Substituting this in (82) gives for (81)

$$\int_{W'_+} a_{(0,m)(0,m)}(y) \frac{\partial^m}{\partial t^m}\left[\frac{\partial^{\gamma-m}}{\partial t^{\gamma-m}} D_x^s u\right] \frac{\partial^m \bar{\varphi}}{\partial t^m} \, dy$$

$$= (-1)^{\gamma - m + |s|} \cdot \left(f, \frac{\partial^{\gamma-m}}{\partial t^{\gamma-m}} D^s \varphi\right)_0$$

$$+ I_{\gamma,s}(\varphi) - \sum_{\substack{|\alpha| \leqslant m \\ |\beta| \leqslant m}}' \int_{W'_+} a_{\alpha\beta}(y) D^\alpha \left[\frac{\partial^{\beta-m}}{\partial t^{\beta-m}} D_x^s u\right] D^\beta \bar{\varphi} \, dy, \tag{84}$$

and after partial integration the assumption (37) gives the estimate

$$\left|\left(f, \frac{\partial^{\gamma-m}}{\partial t^{\gamma-m}} D_x^s \varphi\right)_0\right| \leqslant c \| f \|_{k-m} \cdot \| \varphi \|_{m-1}, \tag{85}$$

for $\gamma - m + |s| \leqslant k - 1$ and $\varphi \in \mathscr{D}_\varepsilon$.

It remains to estimate the sum $\sum'$ in (84). The terms with $|\beta| \leqslant m - 1$ immediately give the estimate $\leqslant c \| \varphi \|_{m-1}$, if we use the inductive assumption (78). Therefore suppose that $|\beta| = m$, if there is a tangential $(\partial/\partial x)$-derivative occurring in D^β, by means of partial integration we bring it into the term $a_{\alpha\beta} D^\alpha[(\partial^{\gamma-m}/\partial t^{\gamma-m}) D_x^s u]$ where $\gamma - m + |s| \leqslant k - 1$. After this manipulation we thus have

$$|\alpha| + 1 + \gamma - m + |s| \leqslant m + 1 + k - 1 = m + k,$$

and are once more in the domain of the assumption (78), where we can estimate with $\leqslant c \| \varphi \|_{m-1}$. If $D^\beta = \partial^m / \partial t^m$, then at least one tangential $(\partial/\partial x)$-derivative must occur in D^α (the term with $a_{(0,m)(0,m)}$ is missing in $\sum'$), and integrating partially once with respect to dt gives with (78) that

$$\frac{\partial}{\partial t} a_{\alpha\beta} D^\alpha \frac{\partial^{\gamma-m}}{\partial t^{\gamma-m}} D_x^s u \in L_{2,\varepsilon}. \tag{86}$$

This is so because

$$\frac{\partial}{\partial t} D^\alpha \frac{\partial^{\gamma-m}}{\partial t^{\gamma-m}} D_x^s u = D^{\alpha-1} \frac{\partial^{\gamma+1-m}}{\partial t^{\gamma+1-m}} D_x^{s+1} u,$$

with

$$|\alpha|-1+\gamma+1-m\leqslant m-1+\gamma+1-m=\gamma$$

and

$$|\alpha|-1+\gamma+1-m+|s|+1\leqslant m+\gamma-m+|s|+1\leqslant k-1+m+1.$$

(86) shows that we may once again estimate by $\leqslant c\|\varphi\|_{m-1}$, and therefore for the $\sum'$ sum we finally obtain

$$\left|\sum_{\substack{|\alpha|\leqslant m\\|\beta|\leqslant m}}' \int_{W_+^r} a_{\alpha\beta}(y)\mathrm{D}^\alpha\left[\frac{\partial^{\gamma-m}}{\partial t^{\gamma-m}}\mathrm{D}_x^s u\right]\cdot \mathrm{D}^\beta\bar{\varphi}\,\mathrm{d}y\right|\leqslant c\|\varphi\|_{m-1},\quad \varphi\in\mathscr{D}_\varepsilon. \tag{87}$$

(83), (85) and (87) give the desired estimate for (84) (see (79)):

$$\left|\int_{W_+^r} a_{(0,m)(0,m)}(y)\frac{\partial^\gamma}{\partial t^\gamma}\mathrm{D}_x^s u\frac{\partial^m\bar{\varphi}}{\partial t^m}\mathrm{d}y\right|\leqslant c\|\varphi\|_{m-1},$$

for $\gamma+|s|\leqslant m+k-1$, $\varphi\in\mathscr{D}_\varepsilon$.

Since by the inductive assumption (78)

$$\frac{\partial}{\partial x_\tau}\frac{\partial^\gamma}{\partial t^\gamma}\mathrm{D}_x^s u=\frac{\partial^\gamma}{\partial t^\gamma}\mathrm{D}_x^{s+1}u\in L_{2,\varepsilon},\quad \tau=1,\ldots,r-1,$$

($\gamma+|s|+1\leqslant m+(k-1)+1=m+k$!) and also (see (31))

$$\frac{\partial}{\partial x_\tau}\left[a_{(0,m)(0,m)}\frac{\partial^\gamma}{\partial t^\gamma}\mathrm{D}_x^s u\right]\in L_{2,\varepsilon},\quad \tau=1,\ldots,r-1,$$

Proposition 20.3 gives

$$\frac{\partial}{\partial t}\left[a_{(0,m)(0,m)}\frac{\partial^\gamma}{\partial t^\gamma}\mathrm{D}_x^s u\right]=\frac{\partial a}{\partial t}\frac{\partial^\gamma}{\partial t^\gamma}\mathrm{D}_x^s u+a\frac{\partial^{\gamma+1}}{\partial t^{\gamma+1}}\mathrm{D}_x^s u\in L_{2,\varepsilon}. \tag{88}$$

However, the $\mathring{W}_2^m$-coercion of $a(\varphi,\psi)+c(\varphi,\psi)=a(\varphi,\psi)+0$ ($c(\varphi,\psi)$ is a boundary form, thus $=0$ on $\mathring{W}_2^m(\Omega)$) now follows from V-coercion. By Theorem 19.2 $a(\varphi,\psi)$ is strongly elliptic on $\overline{W_+^r}$, which implies that $a_{(0,m)(0,m)}\neq 0$ on $\overline{W_+^r}$, and therefore $1/a_{(0,m)(0,m)}$ belongs to $C(\overline{W_+^r})$. Using the assumptions

$$\frac{\partial a_{(0,m)(0,m)}}{\partial t}\frac{\partial^\gamma}{\partial t^\gamma}\mathrm{D}_x^s u\in L_2,$$

by (88) we obtain

$$\frac{\partial^{\gamma+1}}{\partial t^{\gamma+1}}\mathrm{D}_x^s u=\frac{1}{a_{(0,m)(0,m)}}a_{(0,m)(0,m)}\frac{\partial^{\gamma+1}}{\partial t^{\gamma+1}}\mathrm{D}_x^s u\in L_{2,\varepsilon},$$

which is (78) for $\gamma+1$.

Proposition 20.3 also provides the estimate

$$\|u\|_{m+k} \leqslant c\left[\|u\|_m + \sum_{|s|\leqslant k} \|D_x^s u\|_m + \sum_{n=0}^{k} \|f\|_{-m+n}\right] \tag{89}$$

Theorem 20.4 (regularity theorem) *Suppose that assumptions* 1–6 *are satisfied and that u is a solution of* (36) *in V. Then*

$$u \in W_2^{m+k} \cap V. \tag{90}$$

Proof. Since Ω is bounded, we can localise the theorem by means of some finite partition of unity, that is, we must only prove that $\chi u \in W_2^{m+k}(U)$, where U is a small neighbourhood of an arbitrary point $x \in \bar{\Omega}$, supp $\chi \subset U$. If x belongs to the interior of $\bar{\Omega}$, $x \in \Omega$, and we have already proved everything in Theorem 20.1, so it only remains to consider the frontier points $x \in \partial\Omega$. We fix some frontier point $x_0 \in \partial\Omega$ and take so small a neighbourhood U of x_0, that admissible coordinates may be introduced in $U \cap \Omega$. By Theorem 2.12 and assumption 1, there exists an admissible transformation

$$\Phi: U \leftrightarrow W^r, \quad U \cap U \cap \Omega \leftrightarrow W_+^r, \quad U \cap \partial\Omega \leftrightarrow W^{r-1}$$

which belongs to the class C^{m+k}. We now take $\chi \in \mathscr{D}(U)$ with $\chi = 1$ on U', where $x_0 \in U' \subset U$. We consider the sesquilinear form (35)

$$\mathscr{A}(\varphi, \psi) = a(\varphi, \psi) + c(\varphi, \psi),$$

and only take such φ (and ψ)$\in V$, supp $\varphi \subset U$ which are mapped by the transformation $^*\Phi$ into $\mathscr{V}$, that is, $^*\Phi\varphi = 0$ on $\partial W_+^r \setminus W^{r-1}$, (see Proposition 20.1). We apply the transformation Φ to $\mathscr{A}$ and easily see, by Theorem 4.1, that $\mathscr{A}$ becomes a $\mathscr{V}$-coercive form $\mathscr{A}_\Phi$; indeed a and c are integral forms, which behave correctly under the admissible transformation Φ. The continuity estimate

$$\|\mathscr{A}|(\varphi, \psi)\| \leqslant c\|\varphi\|_m \|\psi\|_m \tag{91}$$

is also coordinate invariant, again because of Theorem 4.1. Since $\Phi \in C^{m+k}$ the assumptions 2–6 remain unchanged; in short by means of the transformation Φ and restriction to $\varphi, \psi \in \mathscr{V}$ we have reduced the regularity problem from $(\Omega, \partial\Omega)$ to (W_+^r, W^{r-1}), where the transformed form $\mathscr{A}_\Phi$ – we again denote it simply by $\mathscr{A}$ – satisfies assumptions 2–6. The boundary conditions $b_{j\Phi}$, $j = 1, \dots, p$, are now only given on W^{r-1}, while on $\partial W_+^r \setminus W^{r-1}$ all functions are zero.

From now on we need only to carry out our arguments on (W_+^r, W^{r-1}). Let $k = 1$, we show that under the assumptions 2–6 $\chi u \in W_2^{m+1}(W_+^r)$. We begin with the tangential derivatives and it is apparent that we may proceed analogously to the proof of Theorem 20.1. For the approximation of the tangential derivatives $\partial/\partial x_\tau$, $\tau = 1, \dots, r-1$, we again use the difference operators Δ_h^τ, $\tau = 1, \dots, r-1$, so that we may apply the results from §9. Since with a tangential translation through $|h| \leqslant \varepsilon/2$ we remain in W_+^r, respectively W^{r-1}, if the support of φ was in $W_+^r(1-\varepsilon)$, for $\varphi \in V_\varepsilon$ the formulae (7) are valid

$$\int_{W_+^r} \Delta_h^\tau \varphi \, \mathrm{d}y = 0, \quad \int_{W^{r-1}} \Delta_h^\tau \varphi \, \mathrm{d}x = 0, \quad \tau = 1, \dots, r-1,$$

and the same proof as for Lemma 20.1 shows that for $\tau = 1, \dots, r-1$, $|h| \leqslant \varepsilon/2$, $\chi u, \varphi \in W_\varepsilon^m$ we have

$$\mathscr{A}(\Delta_h^\tau(\chi u), \varphi) = -\mathscr{A}(u, \chi \cdot \Delta_h^\tau \varphi) + I(u, \varphi), \tag{92}$$

with the estimate

$$|I(u, \varphi)| \leqslant c \| u \|_m \cdot \| \varphi \|_m, \quad c \text{ independent of } h \text{ and } \varphi. \tag{93}$$

Let Z_h^τ and $\tilde{Z}_h^\tau$ be the correction operators from Proposition 20.1, let u be a solution of (36), let $|h| \leqslant \varepsilon/2$, and to guarantee good behaviour we take $\varphi \in V_{3\varepsilon}$ and $\chi u \in V_{5\varepsilon}$. For $\tau = 1, \dots, r-1$ because of (92), (41) and (36) we have that

$$\begin{aligned}
&\mathscr{A}(\Delta_h^\tau(\chi u) - Z_h^\tau u, \varphi) \\
&\quad = -\mathscr{A}(u, \chi \Delta_h^\tau \varphi) + I(u, \varphi) - \mathscr{A}(Z_h^\tau u, \varphi) \\
&\quad = -\mathscr{A}(u, \chi \Delta_h^\tau \varphi - \tilde{Z}_h^\tau \varphi) - \mathscr{A}(u, \tilde{Z}_h^\tau \varphi) + I(u, \varphi) - \mathscr{A}(Z_h^\tau u, \varphi) \\
&\quad = -(f, (\chi \Delta_h^\tau \varphi - \tilde{Z}_h^\tau \varphi))_0 - \mathscr{A}(u, \tilde{Z}_h^\tau \varphi) + I(u, \varphi) - \mathscr{A}(Z_h^\tau u, \varphi).
\end{aligned} \tag{94}$$

We can estimate all terms on the right-hand side of (94): by assumption (37) and Theorem 9.4 we have

$$|(f, (\Delta_h^\tau \varphi))_0| \leqslant c \| f \|_{-m+1} \cdot \| \Delta_h^\tau \varphi \|_{m-1} \leqslant c \| \varphi \|_m, \tag{95}$$

while (37) and (42) give

$$|(f, (\tilde{Z}_h^\tau \varphi))_0| \leqslant c \| f \|_{-m} \cdot \| \tilde{Z}_h^\tau \varphi \|_m \leqslant c \| \varphi \|_m. \tag{96}$$

(42) gives once more

$$|\mathscr{A}(u, \tilde{Z}_h^\tau \varphi)| \leqslant c \| u \|_m \cdot \| \tilde{Z}_h^\tau \varphi \|_m \leqslant c \| \varphi \|_m, \tag{97}$$

and

$$|\mathscr{A}(Z_h^\tau u, \varphi)| \leqslant c \| Z_h^\tau u \|_m \cdot \| \varphi \|_m \leqslant c \| \varphi \|_m, \tag{98}$$

where all constants c are independent of h. Taken together the estimates (93), (95), (96), (97) and (98) give the estimate for $\mathscr{A}(\Delta_h^\tau \chi u - Z_h^\tau u, \varphi)$

$$|\mathscr{A}(\Delta_h^\tau \chi u - Z_h^\tau u, \varphi)| \leqslant c \| \varphi \|_m, \quad c \text{ independent of } h. \tag{99}$$

Here we have considered u and f to be fixed elements.

We now substitute $\varphi = \Delta_h^\tau(\chi u) - Z_h^\tau u$, by Proposition 20.1 we have $\Delta_h^\tau \chi u - Z_h^\tau u \in V_{3\varepsilon}$ and use assumption 5, that is, $\mathscr{V}$-coercion, and obtain

$$c_1 \|\Delta_h^\tau(\chi u) - Z_h^\tau u\|_m^2 - c_2' \|\Delta_h^\tau \chi u - Z_h^\tau u\|_0^2 \leqslant \operatorname{Re} \mathscr{A}(\Delta_h^\tau \chi u - Z_h^\tau u, \Delta_h^\tau \chi u - Z_h^\tau u),$$

which together with (99) gives

$$c_1 \|\Delta_h^\tau(\chi u) - Z_h^\tau u\|_m^2 \leqslant c_3 \|\Delta_h^\tau(\chi u) - Z_h^\tau u\|_m + c_2' \|\Delta_h^\tau \chi u - Z_h^\tau u\|_0^2. \tag{100}$$

An application of Theorem 9.4 and Proposition 20.1 (42) gives

$$\|\Delta_h^\tau(\chi u) - Z_h^\tau u\|_0 \leqslant \|\Delta_h^\tau(\chi u)\|_0 + \|Z_h^\tau u\|_0 \leqslant c' \|u\|_m + c\|u\|_m$$

and estimate (100) takes the form

$$c_1 \|\Delta_h^\tau(\chi u) - Z_h^\tau u\|_m^2 \leqslant c_3 \|\Delta_h^\tau(\chi u) - Z_h^\tau u\|_m + c_2 \|u\|_m^2. \tag{101}$$

From (101) it follows immediately (u is fixed, see also the transition from (10) to (11)) that

$$\|\Delta_h^\tau(\chi u) - Z_h^\tau u\|_m \leqslant K_1 < \infty$$

or by Proposition 20.1 (42) that

$$\|\Delta_h^\tau(\chi u)\|_m \leqslant K < \infty, \quad \text{with } K \text{ independent of } h.$$

By Theorem 9.6 this implies that the derivatives $\partial(\chi u)/\partial x_\tau$, $\tau = 1, \ldots, r-1$ together with χu belong to $W_2^m(W_+^r)$. Therefore we can apply Proposition 20.4 for $k = 1$, and obtain $\partial(\chi u)\partial t \in W_\varepsilon^m$, that is, $\chi u \in W_\varepsilon^{m+1}$, with which we have proved our regularity theorem for $k = 1$.

If we take more care with the estimates, starting with (99), we obtain as in (12)

$$\left\|\frac{\partial u}{\partial x_\tau}\right\|_m \leqslant c\{\|u\|_m + \|f\|_{-m+1}\} \quad \text{for } \tau = 1, \ldots, r-1,$$

which together with (66) gives the estimate

$$\|u\|_{m+1} \leqslant c\{\|u\|_m + \|f\|_{-m+1} + \|f\|_{-m}\}. \tag{102}$$

Suppose now that $k \geqslant 2$. We carry out the further proof by induction on k. Suppose therefore that

$$u \in W_2^{m+k-1} \cap V_\varepsilon =: W_\varepsilon^{m+k-1} \tag{103}$$

which as above we know holds for $k = 2$; this starts the induction. We must show that it is then true that $u \in W_\varepsilon^{m+k}$ or, since we may localise that $\chi u \in W_\varepsilon^{m+k}$. Since we have Proposition 20.4 to hand, we need only look at the tangential derivatives

$$\mathrm{D}_x^s(\chi u) \quad \text{for } |s| \leqslant k, k \geqslant 2.$$

By assumption (103) we have that

$$D_x^s(\chi u) \in W_\varepsilon^m \quad \text{for } |s| \leqslant k-1, \tag{104}$$

and we prove (104) for $|s| = k$.

In order to simplify this proof we will not look at the dependence of the supports on ε; the reader can easily fill in the missing details. We rely on Proposition 20.2 and first establish the estimate

$$|\mathscr{A}(\Delta_h^\tau D_x^s(\chi u) - Z_h^{s,\tau} u, \varphi)| \leqslant c \| \varphi \|_m \quad \text{for } \varphi \in \mathscr{V}. \tag{105}$$

Since by Theorem 20.3 $\mathscr{V} \cap W_2^{m+k}$ is dense in $\mathscr{V}$ it is enough to prove (105) for $\varphi \in \mathscr{V} \cap W_2^{m+k}$. We again have as for (92), by partial integration in the tangential direction, and by taking note of (6) and (7)

$$\mathscr{A}(\Delta_h^\tau D_x^s(\chi u), \varphi) = (-1)^{|s|+1} \mathscr{A}(u, \chi[D_x^s \Delta_h^\tau \varphi]) + I(u, \varphi), \tag{106}$$

with the estimate

$$|I(u, \varphi)| \leqslant c' \| u \|_{m+k-1} \| \varphi \|_m \leqslant c \| \varphi \|_m, \quad c \text{ independent of } h \text{ and } \varphi. \tag{107}$$

For the sake of completeness we derive the transformation (106) for the principal form $a(u, \varphi)$; the boundary form $c(u, \varphi)$ is handled similarly. We have

$$\int_\Omega \sum_{\alpha,\beta} a_{\alpha\beta}(x) D^\alpha[\Delta_h^\tau D_x^s(\chi u)] \cdot D^\beta \bar{\varphi} \, dx$$

$$= -\int_\Omega \sum_{\alpha,\beta} \Delta_h^\tau a_{\alpha\beta} D^\alpha [D_x^s(\chi u)] D^\beta \overline{\varphi(x + h_\tau)} \, dx$$

$$- \int_\Omega \sum_{\alpha,\beta} a_{\alpha\beta} D^\alpha [D_x^s(\chi u)] \cdot D^\beta (\Delta_h^\tau \bar{\varphi}) \, dx$$

$$= I_1(u, \varphi) - \int_\Omega \sum_{\alpha,\beta} a_{\alpha\beta} D^\alpha [D_x^s(\chi u)] D^\beta (\Delta_h^\tau \bar{\varphi}) \, dx,$$

where for I_1 the estimate (107) is immediately apparent. We transform further, that is, integrate partially according to D_x^s to obtain

$$\int_\Omega \sum_{\alpha,\beta} a_{\alpha\beta}(x) D^\alpha[\Delta_h^\tau D_x^s(\chi u)] \cdot D^\beta \bar{\varphi} \, dx$$

$$= I_1(u, \varphi) + (-1)^{|s|+1} \int_\Omega \sum_{\alpha,\beta} D^\alpha(\chi u) \cdot D_x^s [a_{\alpha\beta} D^\beta (\Delta_h^\tau \bar{\varphi})] \, dx$$

$$= I_1(u, \varphi) + (-1)^{|s|+1} \int_\Omega \sum_{\alpha,\beta} D^\alpha(\chi u) \sum_{|\gamma| \leqslant |s|-1} D_x^{s-\gamma} a_{\alpha\beta} \cdot D_x^\gamma D^\beta (\Delta_h^\tau \varphi) \, dx$$

$$+ (-1)^{|s|+1} \int_\Omega \sum_{\alpha,\beta} a_{\alpha\beta} D^\alpha(\chi u) D^\beta (D_x^s \Delta_h^\tau \bar{\varphi}) \, dx.$$

The integral

$$I_2(u,\varphi)=(-1)^{|s|+1}\int_\Omega \sum_{\alpha,\beta} D^\alpha(\chi u)\sum_{|\gamma|\leqslant|s|-1} D^{s-\gamma}a_{\alpha\beta}\cdot D^\beta(D_x^\gamma \Delta_h^\tau\bar{\varphi})\,dx$$

can be brought into a form which admits an estimate (107). To this end we integrate partially according to D_x^γ (away from $\bar{\varphi}$), use (6) and (7) (Δ_h^τ away from $\bar{\varphi}$), Theorem 9.4 and $|\gamma|\leqslant|s|-1\leqslant k-2$. In this way we obtain

$$\int_\Omega \sum_{\alpha,\beta} a_{\alpha\beta}(x)D^\alpha[\Delta_h^\tau D_x^s(\chi u)]\cdot D^\beta\bar{\varphi}\,dx$$

$$=I_1(u,\varphi)+I_2(u,\varphi)+(-1)^{|s|+1}\int_\Omega \sum_{\alpha,\beta} a_{\alpha\beta}D^\alpha(\chi u)D^\beta(D_x^s\Delta_h^\tau\bar{\varphi})\,dx$$

and the Leibniz product rule gives

$$\int_\Omega \sum_{\alpha,\beta} a_{\alpha\beta}(x)D^\alpha[\Delta_h^\tau D_x^s(\chi u)]\cdot D^\beta\bar{\varphi}\,dx$$

$$=I_1(u,\varphi)+I_2(u,\varphi)$$

$$+(-1)^{|s|+1}\int_\Omega \sum_{\alpha,\beta} a_{\alpha\beta}\sum_{|\gamma|\leqslant|\alpha|-1} D^{\alpha-\gamma}\chi\cdot D^\gamma u\cdot D^\beta(D_x^s\Delta_h^\tau\bar{\varphi})\,dx$$

$$+(-1)^{|s|+1}\int_\Omega \sum_{\alpha,\beta} D^\alpha u\cdot\chi D^\beta(D_x^s\Delta_h^\tau\bar{\varphi})\,dx$$

$$=I_1(u,\varphi)+I_2(u,\varphi)+I_3(u,\varphi)+(-1)^{|s|+1}\int_\Omega \sum_{\alpha,\beta} a_{\alpha\beta}D^\alpha u\cdot\chi D^\beta(D_x^s\Delta_h^\tau\bar{\varphi})\,dx.$$

I_3 can be modified and estimated in the same way as I_2. Once more applying the Leibniz rule to $D^\beta(\chi\cdot D_x^s\Delta_h^\tau\bar{\varphi})=\cdots$ we obtain

$$\int_\Omega \sum_{\alpha,\beta} a_{\alpha\beta}(x)D^\alpha[\Delta_h^\tau D_x^s(\chi u)]\cdot D^\beta\bar{\varphi}\,dx=I_1(u,\varphi)+I_2(u,\varphi)+I_3(u,\varphi)+I_4(u,\varphi)$$

$$+(-1)^{|s|+1}\int_\Omega \sum_{\alpha,\beta} a_{\alpha\beta}D^\alpha u\cdot D^\beta[\chi D_x^s\Delta_h^\tau\bar{\varphi}]\,dx,$$

where I_4 can again be modified and estimated in the same manner as I_2. In this way we have proved (106) with the estimate (107). For $\tau=1,\ldots,r-1$, the transformation (106) gives

$$\mathscr{A}(\Delta_h^\tau D_x^s(\chi u)-Z_h^{s,\tau}u,\varphi)=\mathscr{A}(\Delta_h^\tau D_x^s(\chi u),\varphi)-\mathscr{A}(Z_h^{s,\tau}u,\varphi)$$

$$=(-1)^{|s|+1}\mathscr{A}(u,\chi[D_x^s\Delta_h^\tau\varphi])+I(u,\varphi)-\mathscr{A}(Z_h^{s,\tau}u,\varphi)$$

$$=(-1)^{|s|+1}\{\mathscr{A}(u,\chi[D_x^s\Delta_h^\tau\varphi]-\tilde{Z}_h^{s,\tau}\varphi)+\mathscr{A}(u,\tilde{Z}_h^s\varphi)+I(u,\varphi)-\mathscr{A}(Z_h^{s,\tau}u,\varphi)\}$$

$$= (-1)^{|s|+1}(f, (\chi[\mathrm{D}_x^s \Delta_h^\tau \varphi] - \tilde{Z}_h^{s,\tau}\varphi))_0 + (-1)^{|s|+1}\mathscr{A}(u, \tilde{Z}_h^s \varphi)$$
$$+ l(u, \varphi) - \mathscr{A}(Z_h^{s,\tau}u, \varphi), \tag{108}$$

where $Z_h^{s,\tau}$ and $\tilde{Z}_h^{s,\tau}$ are the correction operators from Proposition 20.2, and we have taken note of (36) and (49). The assumptions (37) and Proposition 20.2 allow us to estimate all terms on the right hand side of (108). We begin with $(f, (\chi[\mathrm{D}_x^s \Delta_h^\tau \varphi]))_0$, we have because of (37) the estimate

$$|(f, (\chi[\mathrm{D}_x^s \Delta_h^\tau \varphi]))_0| \leqslant \|\chi f\|_{k-m} \cdot \|\Delta_h^\tau \varphi\|_{m-1}.$$

In proving this estimate we distinguish two cases, first with $k \leqslant m$, then for all $|s| \leqslant m-1$ (see assumption 6) we obtain

$$|(f, (\mathrm{D}^s \varphi))_0| \leqslant \|f\|_{k-m} \|\mathrm{D}^s \varphi\|_{m-k} \leqslant \|f\|_{k-m} \|\varphi\|_{m-k+|s|}$$
$$\leqslant \|f\|_{k-m} \|\varphi\|_{m-1}, \quad \text{since } m-k+|s| \leqslant m-1;$$

and secondly with $k > m$. Then $f \in W_2^{k-m}$ and $(f, (\chi \mathrm{D}_x^s \varphi))_0 = \int \chi f \cdot \overline{\mathrm{D}_x^s \varphi}\, \mathrm{d}x$, see assumption 6. Because of the support properties of χ we can partially integrate $(k-m)$ times in the tangential direction x, without introducing boundary terms, thus

$$(f, (\chi \mathrm{D}_x^s \varphi))_0 = (-1)^{|\gamma|} \int \mathrm{D}_x^\gamma(\chi f) \mathrm{D}_x^{s-\gamma} \bar{\varphi}\, \mathrm{d}x, \quad \text{with } |\gamma| \leqslant k-m,$$

and we obtain the estimate

$$|(f, (\chi \mathrm{D}_x^s \varphi))_0| \leqslant \|\mathrm{D}_x^\gamma(\chi f)\|_0 \|\mathrm{D}_x^{s-\gamma}\varphi\|_0 \leqslant \|\chi f\|_{k-m} \|\varphi\|_{|s-\gamma|}$$
$$\leqslant \|\chi f\|_{k-m} \cdot \|\varphi\|_{m-1},$$

since $|s-\gamma| \leqslant |s| - k + m \leqslant m-1$. Substituting Δ_h^τ instead of φ again gives the estimate above. Using Theorem 9.4 we can therefore estimate

$$|(f, (\chi[\mathrm{D}_x^s \Delta_h^\tau \varphi]))_0| \leqslant \|\chi f\|_{k-m} \cdot \|\Delta_h^\tau \varphi\|_{m-1} \leqslant c\|f\|_{-m+k} \|\varphi\|_m, \tag{109}$$

with c independent of h and φ. For $\tilde{Z}_h^{s,\tau}\varphi$ we use the representation (50) with (52)

$$|(f, (\tilde{Z}_h^{s,\tau}\varphi))_0| = \left|\left(f, \left(\sum_{|\alpha| \leqslant |s|} \mathrm{D}_x^\alpha \tilde{Z}_h^\alpha \varphi\right)\right)_0\right| \leqslant \sum_{|\alpha| \leqslant |s|} |(f, (\mathrm{D}_x^\alpha \tilde{Z}_h^\alpha \varphi))_0|$$
$$= \sum_{|\alpha| \leqslant |s|} |(\mathrm{D}_x^\alpha f, \tilde{Z}_h^\alpha \varphi)_0| \leqslant c\|f\|_{-m+k} \cdot \|\varphi\|_m. \tag{110}$$

We also use the representation (50) for $\mathscr{A}(u, \tilde{Z}_h^{s,\tau}\varphi)$ and integrate partially according to D_x^α

$$\begin{aligned}\mathscr{A}(u, \tilde{Z}_h^{s,\tau}\varphi) &= \mathscr{A}\Big(u, \sum_{|\alpha|\leqslant|s|} D_x^\alpha \tilde{Z}_h^\alpha \varphi\Big)\\ &= \int_{W_+^r} \sum_\alpha \sum_{\beta,\gamma} a_{\beta\gamma} D^\beta u \cdot D^\gamma \overline{(D_x^\alpha \tilde{Z}_h^\alpha \varphi)}\, dy\\ &\quad + \int_{W^{r-1}} \sum_\alpha \sum_{\beta,\gamma} c_{\beta\gamma} D^\beta u D^\gamma \overline{(D_x^\alpha \tilde{Z}_h^\alpha \varphi)}\, dx\\ &= \int_{W_+^r} \sum_\alpha \sum_{\beta,\gamma} (-1)^{|\alpha|} D_x^\alpha (a_{\beta\gamma} D^\beta u) D^\gamma \overline{\tilde{Z}_h^\alpha \varphi}\, dy\\ &\quad + \int_{W^{r-1}} \sum_\alpha \sum_{\beta,\gamma} (-1)^{|\alpha|} D_x^\alpha (c_{\beta\gamma} D^\beta u) D^\gamma \overline{\tilde{Z}_h^\alpha \varphi}\, dx\\ &= \sum_{|\alpha|\leqslant|s|} \mathscr{A}_\alpha(D_x^\alpha u, \tilde{Z}_h^\alpha \varphi). \end{aligned} \tag{111}$$

We can estimate and obtain

$$|\mathscr{A}(u, \tilde{Z}_h^{s,\tau}\varphi)| \leqslant \sum_{|\alpha|\leqslant|s|} |\mathscr{A}_\alpha(D_x^\alpha u, \tilde{Z}_h^\alpha \varphi)| \leqslant c\,\| D_x^\alpha u\|_m \|\tilde{Z}_h^\alpha \varphi\|_m \leqslant c\,\|u\|_{m+k-1}\|\varphi\|_m,$$

with c independent of h and φ because of (52). For $\mathscr{A}(Z_h^{s,\tau}u, \varphi)$ we use the first representation from (50) and because of (52) obtain directly that

$$\begin{aligned}|\mathscr{A}(Z^{s,\tau}u, \varphi)| &= \Big|\mathscr{A}\Big(\sum_{|\alpha|\leqslant s} D_x^\alpha Z_h^\alpha u, \varphi\Big)\Big|\\ &= \leqslant \sum_{|\alpha|\leqslant|s|} |\mathscr{A}(D_x^\alpha Z_h^\alpha u, \varphi)| \leqslant \sum_{|\alpha|\leqslant|s|} c\,\|D_x^\alpha Z_h^\alpha u\|_m \cdot \|\varphi\|_m,\\ &\leqslant \sum_{|\alpha|\leqslant|s|} c\,\|Z_h^\alpha u\|_{m+k-1}\|\varphi\|_m \leqslant c\,\|u\|_{m+k-1}\|\varphi\|_m. \end{aligned} \tag{112}$$

For (108) we may use the estimates (107), (109), (110), (111) and (112) thus obtaining the sought for estimate (105). The remainder of the proof now proceeds as for the case $k = 1$. We substitute $\varphi = \Delta_h^\tau D_x^s(\chi u) - Z_h^{s,\tau}u \in \mathscr{V}$, use the $\mathscr{V}$-coercion (assumption 5) and with (105) obtain

$$\begin{aligned} &c_1\|\Delta_h^\tau D_x^s(\chi u) - Z_h^{s,\tau}u\|_m^2 - c_2'\|\Delta_h^\tau D_x^s(\chi u) - Z_h^{s,\tau}u\|_0^2\\ &\leqslant \operatorname{Re}\mathscr{A}(\Delta_h^\tau D_x^s(\chi u) - Z_h^{s,\tau}u, \Delta_h^\tau D_x^s(\chi u) - Z_h^{s,\tau}u)\\ &\leqslant c_3\|\Delta_h^\tau D_x^s(\chi u) - Z_h^{s,\tau}u\|_m. \end{aligned} \tag{113}$$

An application of Theorem 9.4 and the representation (50) has the effect that

$$\begin{aligned}\|\Delta_h^\tau D_x^s(\chi u) - Z_h^{s,\tau}u\|_0 &\leqslant \|\Delta_h^\tau D_x^s(\chi u)\|_0 + \|Z_h^{s,\tau}u\|_0\\ &\leqslant c\|\Delta_h^\tau D_x^s(\chi u)\|_{m-1} + c\left\|\sum_{|\alpha|\leqslant|s|} D_x^\alpha Z_h^\alpha u\right\|_m\\ &\leqslant c\|\chi D_x^s u\|_m + c\sum_{|\alpha|\leqslant|s|}\|Z_h^\alpha u\|_{m+k-1}\\ &\leqslant c\|u\|_{m+k-1} + c\|u\|_{m+k-1},\end{aligned}$$

which applied to (113) gives

$$c_1\|\Delta_h^\tau D_x^s(\chi u) - Z_h^{s,\tau}u\|_m^2 \leqslant c_3\|\Delta_h^\tau D_x^s(\chi u) - Z_h^{s,\tau}u\|_m + c_3\|u\|_{m+k-1}$$

from which it follows (see also the passage from (10) to (11)) that

$$\|\Delta_h^\tau D_x^s(\chi u) - Z_h^{s,\tau}u\|_m \leqslant K_1 < \infty,$$

and again with (50) and (52) that

$$\|\Delta_h^\tau D_x^s(\chi u)\|_m \leqslant K < \infty, \tag{113a}$$

where K is independent of h. By Theorem 9.6 (113) implies that the tangential derivatives $D_x^s(\chi u)$ of order $|s| = k$ also belong to W_2^m, which by application of Proposition 20.4 gives that $\chi u \in W_2^{m+k}$. ■

With (102) and (89) we obtain by induction the estimate

$$\|u\|_{m+k} \leqslant c\left[\|u\|_m + \sum_{n=0}^{k}\|f\|_{-m+n}\right] \leqslant c[\|u\|_m + \|f\|_{k-m}]. \tag{114}$$

We now prove Theorem 15.1. We recall the assumptions. Let $(A^*, B_1', \ldots, B_m')$ be elliptic on $(\Omega, \partial\Omega)$, where for the coefficients we have

$$a_s^*(x) \in C^{2m}(\bar{\Omega}), \quad |s| \leqslant 2m, \quad B_{j,\alpha}'(x) \in C^{4m-1-\operatorname{ord} B_j'}(\bar{\Omega}), |\alpha| \leqslant 2m-1; \tag{115}$$

it suffices to assume $B_{j,\alpha}' \in C^{2m}(\Omega)$ – see (34). For the region Ω we presuppose that

$$\Omega \in C^{4m,1}. \tag{116}$$

We consider the sesquilinear form

$$\begin{aligned}[u,v] &:= \int_\Omega A^*u\cdot\overline{A^*v}\,dx + \sum_{j=1}^{m}\int_{\partial\Omega} B_j'u\cdot\overline{B_j'v}\,d\sigma\\ &= (A^*u, A^*v)_0 + \sum_{j=1}^{m}(B_j'u, B_j'v)_j, \quad u, v \in W_2^m(\Omega),\end{aligned}$$

where $(\cdot,\cdot)_j$ is the scalar product on $W_2^{2m-m_j-1/2}(\partial\Omega)$, see also Theorem 17.4. We have:

Theorem 20.5 *Let* $u\in W_2^{2m}(\Omega)$, $f\in L_2(\Omega)$ *and* $[u,v]=(f,v)_0$ *for all* $v\in W_2^{2m}(\Omega)$. *Then* $u\in W_2^{4m}(\Omega)$.

Proof. We check the assumptions 1–6. The sesquilinear form $[u,v]$ has rank $(2m,2m)$, therefore in assumptions 1–6 we must put

$$m\mapsto 2m,\quad k\mapsto 2m. \tag{117}$$

We see that because of (116) assumption 1 is satisfied. (31) is (115), which equals assumption 2. For 3: $V=W_2^{2m}(\Omega)$, therefore $p=0$, so that 3 is vacuously satisfied. For 4 we have

$$c(\varphi,\psi)=\sum_{j=1}^{m}\int_{\partial\Omega}B'_j\varphi\cdot\overline{B'_j\psi}\,\mathrm{d}\sigma$$

and the order of c amounts to $(2m-1,2m-1)$, so that Theorem 20.2 is applicable. The coefficients of c satisfy (34), because of (115). For 5: we have shown that $[u,v]$ is V-coercive for the pair of spaces $W_2^{2m}(\Omega)$, $L_2(\Omega)$ in §15 – see the text after formula (8). For 6: because of (117) (37) has the form $f\in L_2(\Omega)$, which is the assumption for Theorem 20.5. Since the assumptions 1–6 are satisfied, we can apply Theorem 20.4, which immediately gives us Theorem 20.5. ∎

§21 The solution theorem for strongly elliptic equations and examples

In order to deal with examples we must consider which boundary value problems we can solve by means of the weak equations §20, (36). Let $A(x,\mathrm{D})$ be a strongly elliptic differential operator §19, (1) with the sesquilinear form $a(\varphi,\psi)$ §19, (2).

We assume the following with notation as in §20.

1. Green's formula (Theorem 14.8). Let $\Omega\in C^{m,1}$, suppose given boundary value operators

$$b_1,\dots,b_p,B'_{p+1},\dots,B'_m\in C^{2m-m_j}(\bar\Omega),\quad m_1=\operatorname{ord}b_1,\dots,m_m=\operatorname{ord}B'_m, \tag{1}$$

which form a Dirichlet system of order m. Suppose given m further boundary value operators (ord $\leqslant 2m-1$)

$$\mathrm{D}_1,\dots,\mathrm{D}_p,B_{p+1},\dots,B_m\in C^{2m-\mu_j}(\bar\Omega),\quad \mu_1=\operatorname{ord}\mathrm{D}_1,\dots,\mu_m=\operatorname{ord}B_m, \tag{2}$$

so that Green's formula holds

$$a(u,\varphi)=\int_\Omega Au\bar{\varphi}\,\mathrm{d}x+\sum_{j=1}^{p}\int_{\partial\Omega}\mathrm{D}_j u\cdot\overline{b_j\varphi}\,\mathrm{d}\sigma+\sum_{j=p+1}\int_{\partial\Omega}B_j u\overline{B'_j\varphi}\,\mathrm{d}\sigma, \qquad (3)$$

for all $u\in W_2^{2m}(\Omega)$, $\varphi\in W_2^m(\Omega)$.

Remark Since the assumptions of Theorem 14.8 are satisfied by $(A, b_1,\ldots,b_p, B'_{p+1},\ldots,B'_m)$, the existence of at least one system $(\mathrm{D}_1,\ldots,\mathrm{D}_p, B_{p+1},\ldots,B_m)$ satisfying (3) is guaranteed. Since, however, only $(b_1,\ldots,b_p)$ is fixed beforehand, and $(B'_{p+1},\ldots,B'_m)$ can be widely varied, the system $(\mathrm{D}_1,\ldots,\mathrm{D}_p, B_{p+1},\ldots,B_m)$ is in no way uniquely determined. Therefore in the first assumption we have fixed both systems $(b_1,\ldots,b_p, B'_{p+1},\ldots,B'_m)$ and $(\mathrm{D}_1,\ldots,\mathrm{D}_p, B_{p+1},\ldots,B_m)$.

2. Regularity theorem. Suppose that the assumptions 1–6 in §20 for the regularity Theorem 20.4 are satisfied, with $k=m$. Here the boundary form $c(\varphi,\psi)$ 20, (33) has the form

$$c(\varphi,\psi)=\sum_{j=p+1}^{m}\int_{\partial\Omega}C_j\varphi\cdot\overline{B'_j\psi}\,\mathrm{d}\sigma, \qquad (4)$$

with $C_j(x,\mathrm{D})\in C^{2m-\mathrm{ord}\,C_j}(\bar{\Omega})$ and $C_j(x,\mathrm{D})$ $(m-1)$-transversal to $\partial\Omega$. By assumption B'_j is $(m-1)$-transversal, since (1) is a Dirichlet system of order m. For the orders we suppose

$$\mathrm{ord}\,C_j+\mathrm{ord}\,B'_j\leqslant 2m-1.$$

For the functionals f §20, (37) we take $f\in L_2(\Omega)$. We consider also the sesquilinear form $\mathscr{A}(\varphi,\psi):=a(\varphi,\psi)+c(\varphi,\psi)$ and the weak equation $\mathscr{A}(u,\varphi)=(f,\varphi)_0$ for all $\varphi\in V$.

3. Assumptions for the boundary values. We take $b_1,\ldots,b_p$ from (1) and for $j>p$ put $b_j=B_j+C_j$ $(j=p+1,\ldots,m)$, where B_j is from (2) and C_j from (4). We require that for each tuple $(g_1,\ldots,g_m)$ with $g_j\in W_2^{2m-\mathrm{ord}\,b_j-1/2}(\partial\Omega)$, $j=1,\ldots,m$, at least one $u_0\in W_2^{2m}(\Omega)$ exists with $b_1u_0=g_1,\ldots,b_mu_0=g_m$ on $\partial\Omega$. By Theorem 14.1 this is for example the case if the system $(b_1,\ldots,b_m)$ is normal. If $C_j=0$ and (2) was constructed by means of Theorem 14.8, then the system $(b_1,\ldots,b_p, B_{p+1},\ldots,B_m)$ is normal, for each system $(b_1,\ldots,b_p)$ $(B_{p+1},\ldots,B_m)$ considered individually is normal. If the combined system $(b_1,\ldots,b_p, B_{p+1},\ldots,B_m)$ were not normal, then for at least one pair (h_0,k_0) we would have

$$\mathrm{ord}\,b_{j_0}=\mathrm{ord}\,B_{k_0}=2m-1-\mathrm{ord}\,B'_{k_0},$$

the last equation by Theorem 14.8, or $\operatorname{ord} b_{j_0} + \operatorname{ord} B'_{k_0} = 2m - 1$. This is impossible since $(b_1, \dots, b_p, B'_{p+1}, \dots, B'_m)$ is a Dirichlet system of order m, that is,

$$\operatorname{ord} b_j + \operatorname{ord} B'_k \leqslant 2m - 2.$$

We again put

$$V = W^m(\{b_j\}_1^p), \quad W^{2m}(B) = W^{2m}(\{b_j\}_1^m).$$

Remark If we only consider zero boundary conditions, $b_j u = 0$, $j = 1, \dots, m$, the assumption 3 is superfluous.

Corollary 21.1 *Let* $\mathscr{L}$ *be the representation operator* (§17, (28)) *of the sesquilinear form*

$$\mathscr{A}(\varphi, \psi) = a(\varphi, \psi) + c(\varphi, \psi), \quad \varphi, \psi \in V.$$

Then on

$$W^{2m}(B) = W^{2m}(\{b_j\}_1^m) = V \cap W^{2m}(\{b_j\}_{p+1}^m)$$

$\mathscr{L}$ *agrees with* $A(x, \mathrm{D})$ *and with* L, *where* (§13) $L := (A, b_1, \dots, b_m)$, *or stated otherwise the weak equation*

$$\mathscr{A}(u, \varphi) = (f, \varphi), \quad \varphi \in V$$

is for $u \in W^{2m}(B)$, $f \in L_2(\Omega)$ *equivalent to*

$$Au = f, \quad b_1 u = 0, \dots, b_m u = 0.$$

Proof. By (3) and (4) we have

$$\begin{aligned}(\mathscr{L}\varphi, \psi)_0 &= a(\varphi, \psi) + c(\varphi, \psi)\\ &= (A\varphi, \psi)_0 + \sum_{j=1}^{p} \int_{\partial\Omega} \mathrm{D}_j \varphi \cdot \overline{b_j \psi}\, \mathrm{d}\sigma + \sum_{j=p+1}^{m} \int_{\partial\Omega} (B_j + C_j)\varphi \cdot \overline{B'_j \psi}\, \mathrm{d}\sigma.\end{aligned}$$

Since $\psi \in V$ the $\sum_{j=1}^{p}$-sum equals zero, and because $\varphi \in W^{2m}(B)$ implies $b_j \varphi = (B_j + C_j)\varphi = 0$ for $j = p + 1, \dots, m$, the $\sum_{j=p+1}^{m}$-sum also equals zero. We therefore have

$$(\mathscr{L}\varphi, \psi)_0 = (A\varphi, \psi)_0 \quad \text{for } \varphi \in W^{2m}(B), \psi \in V,$$

and since $\mathscr{D}(\Omega) \subset V \subset W_2^m(\Omega)$ is dense in $L_2(\Omega)$

$$\mathscr{L}\varphi = A\varphi \quad \text{for all } \varphi \in W^{2m}(B).$$

Suppose conversely that $Au = f$, $b_1 u = 0, \dots, b_m u = 0$, then because $u \in W^{2m}(B) \subset W_2^{2m}(\Omega)$, $\varphi \in V \subset W_2^{2m}(\Omega)$, Green's formula (4) holds and with (4) we obtain

$$\begin{aligned}\mathscr{A}(u,\varphi) &= a(u,\varphi) + c(u,\varphi)\\ &= (Au,\varphi)_0 + \sum_{j=1}^{p} \int_{\partial\Omega} \mathrm{D}_j u \overline{b_j \varphi}\, \mathrm{d}\sigma + \sum_{j=p+1}^{m} \int_{\partial\Omega} (B_j + C_j) u \overline{B'_j \varphi}\, \mathrm{d}\sigma\\ &= (Au,\varphi)_0 + \sum_{j=1}^{p} \int_{\partial\Omega} \mathrm{D}_j u \overline{b_j \varphi}\, \mathrm{d}\sigma + \sum_{j=p+1}^{m} \int_{\partial\Omega} b_j u \overline{B'_j \varphi}\, \mathrm{d}\sigma,\end{aligned}$$

which after substitution gives

$$\mathscr{A}(u,\varphi) = (Au,\varphi)_0 + 0 = (f,\varphi)_0 \quad \text{for all } \varphi \in V. \qquad \blacksquare$$

In order to formulate theorems about the solvability of the weak equation, instead of §20, (30) we consider the equation

$$a(u,\varphi) + k_0(u,\varphi)_0 + c(u,\varphi) = (f,\varphi)_0, \quad \varphi \in V, \tag{5}$$

where $k_0 \in \mathbb{R}$ is the coercion constant of $\mathscr{A}(u,\varphi) = a(u,\varphi) + c(u,\varphi)$, that is,

$$\operatorname{Re} a(u,u) + \operatorname{Re} c(u,u) + k_0(u,u)_0 \geqslant c\,\|u\|_m^2, \quad u \in V, \tag{6}$$

and $V = W_2^m(\{b_j\}_1^p)$.

We gain the advantage of being able to talk about V-ellipticity, instead of V-coercion.

Theorem 21.1 (solution theorem) *Let $A(x,\mathrm{D})$ be a strongly elliptic differential operator of the form* §19, (1), *with the sesquilinear form $a(\varphi,\psi)$,* §19, (2); *let the assumptions* 1, 2(6) *and* 3 *be satisfied, then the boundary value problem*

$$\begin{aligned}&(A + k_0)u = f \text{ in } \Omega,\\ &b_1 u = g_1, \ldots, b_p u = g_p,\\ &b_{p+1} u = (B_{p+1} + C_{p+1})u = g_{p+1}, \ldots, b_m u = (B_m + C_m)u = g_m \text{ on } \partial\Omega\end{aligned} \tag{7}$$

possesses exactly one solution $u \in W_2^{2m}(\Omega)$ for each $(f, g_1, \ldots, g_m)$ in $H^0(\Omega, \partial\Omega)$ (*see* §13), *that is, the operator*

$$L + k_0 = (A + k_0, b_1, \ldots, b_m): W_2^{2m}(\Omega) \to H^0(\Omega, \partial\Omega)$$

is a (topological) isomorphism. By Theorem 13.3 *$L = (A, b_1, \ldots, b_m)$ is therefore a Fredholm operator of index zero,*

$$\operatorname{ind}(A, b_1, \ldots, b_m) = 0.$$

All the conclusions of the Riesz–Schauder spectral theorem hold for the spectral value problem

$$Au - \lambda u = f, \quad b_1 u = g_1, \ldots, b_m u = g_m, \quad \lambda \in \mathbb{C},$$

where, because of (6), there are no spectral values in the half plane $\operatorname{Re}\lambda \leqslant -k_0$ *(see Theorem 17.4). If in addition the system* $b_1,\ldots,b_m$ *is normal, then the (anti)dual operator* L' *can be realised by means of the adjoint elliptic boundary value problem* $(A^*, b'_1,\ldots,b'_m) = L^*$.

Corollary 21.2

(a) *The system* $(A, b_1,\ldots,b_m)$ *from* (7) *satisfies the L.Š. Condition* 11.1, *this follows from Theorem* 13.1.

(b) *The Green solution operator* G_{k_0} *for* (5) *(see Definition* 17.5) *acts continuously and surjectively*

$$G_{k_0}: L_2(\Omega) \to W_2^{2m}(\Omega) \cap V \cap W^{2m}(\{b_j\}_{p+1}^m) = W^{2m}(B).$$

The restriction of $L + k_0$ *to* $W_2^{2m}(\Omega) \cap V \cap W^{2m}(\{b_j\}_{p+1}^m) = W^{2m}(B)$ *is again a topological isomorphism:*

$$L + k_0: W_2^{2m}(\Omega) \cap V \cap W^{2m}(\{b_j\}_{p+1}^m) \to L_2(\Omega) \times 0 \cdots \times 0 \simeq L_2(\Omega),$$

and because of Corollary 21.1 *we have* $L + k_0 = \mathscr{L} + k_0 = A + k_0$, *hence also*

$$G_{k_0} = (L + k_0)^{-1} = (\mathscr{L} + k_0)^{-1} = (A + k_0)^{-1},$$

where $\mathscr{L}$ *is the representation operator of the sesquilinear form* $\mathscr{A}(\varphi, \psi)$.

If we use Theorem 20.4 and there write $k = m + q$, then we can also show the continuity of

$$G_{k_0}: W_2^p(\Omega) \to W_2^{p+2m}(\Omega) \cap W^{2m}(B)$$

and

$$L + k_0: W_2^{q+2m}(\Omega) \cap W^{2m}(B) \to W_2^q(\Omega).$$

Proof of Theorem 21.1. We show first that for $u \in V$, $f \in L_2(\Omega)$ and $\lambda \in \mathbb{C}$ the weak equation

$$\mathscr{A}_\lambda(u, \varphi) := a(u, \varphi) + c(u, \varphi) - \lambda(u, \varphi)_0 = (f, \varphi)_0 \quad \text{for all } \varphi \in V \tag{8}$$

is equivalent to the elliptic boundary value problem

$$A(x, D)u - \lambda u = f, \quad b_1 u = 0, \ldots, b_m u = 0, \quad u \in W_2^{2m}(\Omega). \tag{9}$$

Proof. (8)⇒(9): We use the individual assumptions. Let $u \in V$, that is,

$$u \in W_2^m(\Omega) \quad \text{with } b_1 u = 0, \ldots, b_p u = 0. \tag{10}$$

By assumption 2 we can apply the regularity Theorem 20.4 with $k = m$ and have

$$u \in W_2^{2m}(\Omega) \cap V.$$

We now apply the Green formula (3) (assumption 1) to (8), where we allow φ to vary

$$\varphi \in W_2^{2m}(\Omega) \cap V.$$

For these φ the expression $\sum_{j=1}^{p} \int_{\partial\Omega} D_j u \cdot \overline{b_j \varphi}\, d\sigma = 0$ in (3), and we obtain

$$\begin{aligned} a(u, \varphi) &- \lambda(u, \varphi)_0 + c(u, \varphi) \\ &= \int_\Omega (Au - \lambda u) \cdot \bar{\varphi}\, dx + \sum_{j=p+1}^{m} \int_{\partial\Omega} B_j u \cdot \overline{B_j' \varphi}\, d\sigma + \sum_{j=p+1}^{m} \int_{\partial\Omega} C_j u \cdot \overline{B_j' \varphi}\, d\sigma \\ &= \int_{\text{L}\Omega} f \cdot \bar{\varphi}\, dx, \end{aligned}$$

that is

$$\int_\Omega (Au - \lambda u)\bar{\varphi}\, dx + \sum_{j=p+1}^{m} \int_{\partial\Omega} (B_j + C_j)u \cdot \overline{B_j' \varphi}\, d\sigma = \int_\Omega f \cdot \bar{\varphi}\, dx. \tag{11}$$

By Theorem 20.3 $W_2^{2m}(\Omega) \cap V$ is dense in V, and a density argument shows that (11) holds for all $\varphi \in V$. We take $\varphi \in \mathscr{D}(\Omega) \subset V$, then (11) reduces to

$$\int_\Omega (Au - \lambda u)\bar{\varphi}\, dx = \int_\Omega f \bar{\varphi}\, dx, \quad \varphi \in \mathscr{D}(\Omega)$$

and we have obtained the first equation $(A - \lambda)u = f$ of (9). From (11) there remains

$$\sum_{j=p+1}^{m} \int_{\partial\Omega} (B_j + C_j)u \cdot \overline{B_j' \varphi}\, d\sigma = 0 \quad \text{for all } \varphi \in V, \tag{12}$$

where $u \in W_2^{2m}(\Omega) \cap V$. For the orders we have by assumption that

$$\text{ord}\, B_j, \text{ord}\, C_j \leqslant 2m - 1, \quad \text{ord}\, B_j' \leqslant m - 1.$$

Since by assumption 1 the system $(b_1, \ldots, b_p, \ldots, B'_{p+1}, \ldots, B'_m)$ is normal, by Theorem 14.1 we can solve the equations

$$b_1 v = 0, \ldots, b_p v = 0, \quad B'_{p+1} v = \varphi_{p+1}, \ldots, B'_m v = \varphi_m \quad \text{for } v \in W_2^m(\Omega),$$

where we write

$$\varphi_j = \delta_{jk}(B_k + C_k)u, \quad j, k = p+1, \ldots, m,$$

or

$$B'_{p+1} v = \varphi_{p+1}, \ldots, B'_m v = \varphi_m \quad \text{in } V (v \in V). \tag{13}$$

Substituted in (12) gives

$$b_j u = (B_j + C_j)u = 0 \quad \text{for } j = p+1, \ldots, m. \tag{14}$$

(14) and (10) give the boundary conditions of (9).

The reverse implication (9)$\Rightarrow$(8) follows as in Corollary 2.1.

The unique solvability of (7) for the special case $(f, g_1, \ldots, g_m) = (f, 0, \ldots, 0)$ is an immediate consequence of the equivalence of (5) and (7) (substitute $\lambda = -k_0$ in (8) and (9)) and the V-ellipticity of (5) (Theorem 17.10). We show the general solvability of (7) by the usual homogenisation trick. Let $(g_1, \ldots, g_m)$ be given, by assumption 3 we choose u_0 so that

$$b_1 u_0 = g_1, \ldots, b_m u_0 = g_m,$$

and solve the boundary value problem

$$(A + k_0)w = f - (A + k_0)u_0 = \tilde{f}, b_1 w = 0, \ldots, b_m w = 0.$$

If we put $u := w + u_0$, then we have satisfied (7) and also proved the uniqueness of the solution – see above.

Now for the spectral assertion. Since (8) and (9) are equivalent, and we can apply the spectral Theorem 17.12 to (8), we also have the spectral statements for the partially homogeneous eigenvalue problem (9). It remains only forms to give conditions for solving the inhomogeneous problem

$$Au - \lambda u = f, \quad b_1 u = g_1, \ldots, b_m u = g_m.$$

We again apply the homogenisation process, by which the solvability of the inhomogeneous problem is equivalent to the solvability of

$$(A - \lambda)w = f - (A - \lambda)u_0, \quad b_1 w = 0, \ldots, b_m w = 0, \tag{15}$$

where $w = u - u_0$ and u_0 is a solution, see assumption 3, of

$$b_1 u_0 = g_1, \ldots, b_m u_0 = g_m.$$

By Theorem 17.12 (15) is solvable if and only if

$$(f - (A - \lambda)u_0, v)_0 = 0 \quad \text{for all } v \in \ker(L' - \bar{\lambda}).$$

We can give this condition in the accustomed form, in which the given functions $g_1, \ldots, g_m$ appear, if the system $b_1, \ldots, b_m$ is normal. Thus, let $(b_1, \ldots, b_m)$ be normal; since by Corollary 21.2 $(A, b_1, \ldots, b_m)$ satisfies the L.Š. condition, we can apply Theorem 15.7, which says that the (anti)dual operator L' is realised by an adjoint elliptic boundary value problem $L^* = (A^*, b'_1, \ldots, b'_m)$, and that Green's formula also holds §14, (12). With §14, (12) we have for all $v \in \ker(L^* - \bar{\lambda}) = \ker(L' - \bar{\lambda})$ that

$$\begin{aligned} 0 &= (f - (A - \lambda)u_0, v)_0 = (f, v)_0 - ((A - \lambda)u_0, v)_0 \\ &= (f, v)_0 - (u_0, (A^* - \bar{\lambda})v)_0 - \sum_{j=1}^{m} \int_{\partial\Omega} S_j u_0 \overline{b'_j v}\, d\sigma + \sum_{j=1}^{m} \int_{\partial\Omega} b_j u_0 \overline{T_j v}\, d\sigma \\ &= \int_\Omega f \bar{v}\, dx + \sum_{j=1}^{m} \int_{\partial\Omega} g_1 \cdot \overline{T_j v}\, d\sigma, \end{aligned}$$

which is the already known solvability condition §15, (32) for the inhomogeneous problem. ■

Remark From the equivalence shown at the beginning of the proof it follows that the regularity Theorem 20.4 only applies to the half-homogeneous boundary value problem

$$Au = f, \quad b_1 u = 0, \dots, b_m u = 0.$$

In order to establish the regularity of a solution of the full homogeneous problem

$$Au = f, \quad b_1 u = g_1, \dots, b_m u = g_m,$$

we may go back to the main Theorem 13.1 and Corollary 13.1 ($k \geqslant m$). By Corollary 21.2 the L.Š. condition is indeed satisfied. For the same purpose we can also use the homogenisation process and Theorem 14.1.

Remark The solution method for elliptic boundary value problems described here by means of weak equations §20, (36) is also called the direct or variational method. By Theorem 21.1 it is not applicable to such elliptic problems, where ind $L \neq 0$ or such that all $\lambda \in \mathbb{C}$ are eigenvalues, see Example 13.1, or such that $\lim \operatorname{Re} \lambda_{n_k} = -\infty$ (λ_n eigenvalues).

We now have a closer look at the self-adjoint case. By Theorem 17.12 the antisymmetry of $\mathscr{A}$ on $V \times V$, that is,

$$\mathscr{A}(\varphi, \psi) = \overline{\mathscr{A}(\psi, \varphi)} \quad \text{for all } \varphi, \psi \in V,$$

in necessary and sufficient for the Green solution operator $G_{k_0}: L_2(\Omega) \to L_2(\Omega)$ to be self-adjoint in the L_2-sense, see Corollary 21.2. Here we want to give additional (sufficient) conditions, under which the boundary value problem $(A, b_1, \dots, b_m)$ is self-adjoint in the sense of Definition 14.6.

Theorem 21.2 *Let the sesquilinear form*

$$\mathscr{A}(u, v) := a(u, v) + c(u, v)$$

be antisymmetric on $S_2^{2m}(\Omega) \times W_2^{2m}(\Omega)$, *that is,*

$$\mathscr{A}(u, v) = \overline{\mathscr{A}(v, u)}, u, v \in W_2^{2m}(\Omega).$$

Then the boundary value problem $(A, b_1, \dots, b_m)$ *associated with* $\mathscr{A}$ *by Theorem* 21.1 *is self-adjoint in the sense of Definition* 14.6

$$(A, b_1, \dots, b_m)^* = (A, b_1, \dots, b_m).$$

Proof. For $u, v \in \mathscr{D}(\Omega)$ the antisymmetry gives

$$(Au, v)_0 = \overline{(Av, u)_0} = (u, Av)_0,$$

that is, $A = A^*$. Therefore the differential operator A is formally self-adjoint. Formulae (3) and (4) give

$$\begin{array}{c}\mathscr{A}(u, v) = (Au, v)_0 + \sum_{j=1}^{p} \int_{\partial\Omega} \mathrm{D}_j u \overline{b_j v}\, \mathrm{d}\sigma + \sum_{j=p+1}^{m} \int_{\partial\Omega} b_j u \cdot \overline{B'_j v}\, \mathrm{d}\sigma,\\ \| \\ \overline{\mathscr{A}(v, u)} = (u, Av)_0 + \sum_{j=1}^{p} \int_{\partial\Omega} b_j u \overline{\mathrm{D}_j v}\, \mathrm{d}\sigma + \sum_{j=p+1}^{m} \int_{\partial\Omega} B'_j u \overline{b_j v}\, \mathrm{d}\sigma.\end{array}$$

If we write $\mathrm{D}_1 = -C_1, \ldots, \mathrm{D}_p = -C_p$, $B'_{p+1} = C_{p+1}, \ldots, B'_m = C_m$, then both formulae above give

$$\int_\Omega Au \cdot \bar{v}\, \mathrm{d}x - \int_\Omega u \cdot \overline{Au}\, \mathrm{d}x = \sum_{j=1}^{m} \int_{\partial\Omega} [C_j u \overline{b_j v} - b_j u \overline{C_j v}]\, \mathrm{d}\sigma,$$

which is Green's formula (51) from Theorem 14.9. Therefore Theorem 14.3 is applicable, and we have

$$(A, b_1, \ldots, b_m)^* = (A, b_1, \ldots, b_m)$$

as also

$$(A, C_1, \ldots, C_m)^* = (A, C_1, \ldots, C_m)$$ ■

Theorem 21.3 *Let the boundary value operators $b_1, \ldots, b_m$ be normal, let the sesquilinear form $\mathscr{A}(\varphi, \psi) = a(\varphi, \psi) + c(\varphi, \psi)$ be antisymmetric on $V \times V$. Then by Theorem* 21.1 *the boundary value problem $(A, b_1, \ldots, b_m)$ associated with $\mathscr{A}$ is self-adjoint in the sense of Definition* 14.6.

Remark Since by Theorem 20.3 $V \cap W_2^{2m}(\Omega)$ is dense in V, the requirement of antisymmetry on $V \times V$ is certainly weaker than antisymmetry on $W_2^{2m}(\Omega) \times W_2^{2m}(\Omega)$. However, here we must (in contrast to Theorem 21.2) assume in addition the normality of $b_1, \ldots, b_m$.

Proof. Because $\mathscr{D}(\Omega) \subset V$ for the differential operator A we have, as in Theorem 21.2, that $A = A^*$. Since the system $B = (b_1, \ldots, b_m)$ is normal, by Theorem 14.4 there exists a system $B' = (B'_1, \ldots, B'_m)$ with $W^{2m}(B)^* = W^{2m}(B')$. By Theorem 14.10 we have maps

$$G_{k_0}: L_2(\Omega) \to W^{2m}(B) = \operatorname{Im} G_{k_0},$$
$$G'_{k_0}: L_2(\Omega) \to W^{2m}(B') = \operatorname{Im} G'_{k_0},$$

and by Theorem 17.12 we have $G'_{k_0} = G_{k_0}$ from which it follows that $\operatorname{Im} G_{k_0} =$

$\operatorname{Im} G'_{k_0}$, that is,

$$W^{2m}(B) = W^{2m}(B') = W^{2m}(B)^*$$

which is self-adjointness in the sense of Definition 14.6. ■

We have added an additional boundary form $c(\varphi, \psi)$ (4) to the form $a(\varphi, \psi)$; this enables us to consider more boundary value operators (see assumption 3), than just those given by Green's formula (3); namely $b_j = B_j + C_j$, $j = 1 + p, \ldots, m$, instead of $b_j = B_j$, $j = 1 + p, \ldots, m$, for this see Example 21.3. A further possibility of altering the boundary value operators $b_j, j = 1 + p, \ldots, m$, is obtained by adding a skew-symmetric form $\tilde{a}(\varphi, \psi)$ to $a(\varphi, \psi)$, see Example 21.4. $\tilde{a}(\varphi, \psi)$ is called skew-symmetric, if the components of the principal part are antisymmetric

$$\tilde{a}_{\alpha\beta} = -\tilde{a}_{\beta\alpha} \quad \text{for } |\alpha| = |\beta| = m.$$

Here and everywhere we have implicitly assumed that the original form $a(\varphi, \psi)$ has a symmetric coefficients $a_{\alpha\beta} = a_{\beta\alpha}$, for if we take the derivatives D^α, D^β in the distributional sense, we have interchangeability $D^\alpha D^\beta = D^\beta D^\alpha$. The addition of $\tilde{a}(\varphi, \psi)$ has no effect on the ellipticity of a, for because of the skew-symmetry we have

$$\sum_{|\alpha| = |\beta| = m} \tilde{a}_{\alpha\beta} \xi^\alpha \xi^\beta = 0 \quad \text{for all } \xi \in \mathbb{R}^r,$$

and

$$\tilde{A}^{\mathrm{H}}(x, \mathrm{D}) = \sum_{|\alpha| = |\beta| = m} (-1)^m \tilde{a}_{\alpha\beta} \mathrm{D}^\alpha \mathrm{D}^\beta \equiv 0,$$

so the principal part of $A(x, \mathrm{D})$ remains unchanged. For operators A of order 2 this already implies that V-coercion remains unchanged; by Example 19.1 strongly elliptic differential operators of the second order are always $W_2^1(\Omega)$-coercive. In this way we can submit the problem with the so-called skew derivative to the variational method; see Example 21.4 and Fichera [1], where he works through the general case of a second-order operator A.

We have seen in §16 that for the Laplace operator $-\Delta$, $r \geqslant 3$, all elliptic boundary value problems have index equal to zero, the reason for this is that they may all be handled by the variational method, therefore Theorem 21.1 is applicable. We show this by examples.

Example 21.1 The Dirichlet problem for the Laplace operator $-\Delta$. For the form a associated with the Laplace operator $-\Delta$ we take

$$a(\varphi, \psi) = \sum_{j=1}^{r} \int_\Omega \frac{\partial \varphi}{\partial x_j} \cdot \overline{\frac{\partial \psi}{\partial x_j}} \, \mathrm{d}x, \tag{16}$$

and for the boundary value space associated with the Dirichlet problem we take $V = \mathring{W}_2^1(\Omega)$.

We have: let Ω be an arbitrary open and bounded set (no assumptions on the frontier $\partial\Omega$), then (16) is $\mathring{W}_2^1$-elliptic for we have the estimates

$$|a(\varphi, \psi)| = \left|\sum_{j=1}^{r}\int_{\Omega}\frac{\partial\varphi}{\partial x_j}\cdot\overline{\frac{\partial\psi}{\partial x_j}}\,\mathrm{d}x\right| \leqslant \|\varphi\|_1\cdot\|\psi\|_1 \quad \text{for all } \varphi, \psi \in W_2^1,$$

and

$$\operatorname{Re} a(\varphi, \psi) = \sum_{j=1}^{r}\int_{\Omega}\left|\frac{\partial\varphi}{\partial x_j}\right|^2 \mathrm{d}x \geqslant c\|\varphi\|_1^2, \quad \varphi \in \mathring{W}_2^1(\Omega),$$

the latter because of the Poincaré inequality (Theorem 7.6). By Theorem 17.10 the equation

$$a(u, \varphi) = (f, \varphi)_0, \quad \varphi \in \mathring{W}_2^1(\Omega) \tag{17}$$

(called the generalised Dirichlet problem) has for each $f \in W_2^{-1}(\Omega)$ a unique solution $u \in \mathring{W}_2^1(\Omega)$, and by Theorem 17.4 the half plane $\operatorname{Re}\lambda \leqslant 0$ contains no eigenvalues of the Dirichlet problem. Suppose now in addition that

$$\Omega \in C^{0,1}, \tag{18}$$

then by Theorem 8.9 the space $\mathring{W}_2^1(\Omega)$ is characterised by the boundary condition

$$T_0 u =: u|_{\partial\Omega} = 0$$

in the sense of the trace operator T_0. If $u, \varphi \in \mathring{W}_2^1(\Omega)$, Green's formula holds

$$a(u, \varphi) = \sum_{j=1}^{r}\int_{\Omega}\frac{\partial u}{\partial x_j}\cdot\frac{\partial\bar{\varphi}}{\partial x_j}\,\mathrm{d}x = \int_{\Omega}(-\Delta u)\bar{\varphi}\,\mathrm{d}x. \tag{19}$$

We prove it by first taking $\int_\Omega(-\Delta u)\bar{\varphi}\,\mathrm{d}x = (-\Delta u, \varphi)_0$, $u, \varphi \in \mathscr{D}(\Omega)$ as scalar product in the Gelfand triple

$$\mathring{W}_2^1(\Omega) \subsetneq L_2(\Omega) \subsetneq W_2^{-1}(\Omega),$$

see Definitions 17.1 and 17.2, then carrying out the usual density argument. Suppose now that $g \in W_2^{1/2}(\Omega)$, because of (18) we can apply Theorem 8.8 and find some $u_0 \in W_2^1(\Omega)$ with

$$u_0|_{\partial\Omega} := T_0 u_0 = g_0 \quad \text{on } \partial\Omega. \tag{20}$$

We consider the weak equation

$$-\Delta w = f + \Delta u_0, \quad w \in \mathring{W}_2^1(\Omega),$$

that is, because of (19) and (17),

$$\int_\Omega (-\Delta w)\bar{\varphi}\,dx = a(w,\varphi) = (f + \Delta u_0, \varphi)_0$$

$$= \int_\Omega (f + \Delta u_0)\bar{\varphi}\,dx \quad \text{for all } \varphi \in \mathring{W}_2^1(\Omega).$$

Since ord $\Delta = 2$ we have $\Delta u_0 \in W_2^{-1}(\Omega)$ and the solution Theorem 17.10 is again applicable, that is, $w \in \mathring{W}_2^1(\Omega)$ is uniquely determined by (21). We write

$$u := w + u_0$$

and because of (21) have

$$u \in W_2^1(\Omega), \quad -\Delta u = -\Delta w - \Delta u_0 = f + \Delta u_0 - \Delta u_0 = f.$$

Because of (20) the trace Theorem 8.7 (use (18) again here) gives

$$u|_{\partial\Omega} = Tu_0 = T_0 w + T_0 u_0 = 0 + g.$$

Hence we have proved: for each pair $f \in W_2^{-1}(\Omega)$, $g \in W_2^{1/2}(\Omega)$,

$$-\Delta u = f \text{ in } \Omega,\ u|_{\partial\Omega} = g \text{ on } \partial\Omega,$$

possesses a uniquely determined solution $u \in W_2^1(\Omega)$ – this is much more than one can obtain from the main Theorem 13.1. It is interesting that if we strengthen the assumption (18) only a little, for example to $\Omega \in N^{0,1}$ or to (18) plus the uniform cone condition, we can make far-reaching regularity assertions about the solution u. For example let $f \in C^\infty(\bar{\Omega})$, and let g be extendable to $\bar{\Omega}$ in such a way that $g \in C^\infty(\bar{\Omega})$, then for the solution u of $-\Delta u = f, u|_{\partial\Omega} = g$, we have that $u \in C^\infty(\bar{\Omega})$, and this in spite of the fact that the region $\Omega \in N^{0,1}$ may have corners. The reader may carry out the proof for himself, it is a simplified version of Theorem 20.4 – we need no correction operators, but rather a multiple application of the Calderon–Zygmund extension Theorem 5.4 and the Sobolev lemma (Theorem 6.2). Both hold under minimal assumptions on Ω, for example $\Omega \in N^{0,1}$.

Since the Dirichlet problem is self-adjoint, for example by Theorem 21.3, and the half plane $\operatorname{Re}\lambda \leqslant 0$ does not belong to the spectrum, the spectral conclusions of Theorems 15.9, 17.12, respectively 21.1 also hold for the Dirichlet problem.

Example 21.2 The Neumann problem for the Laplace operator $-\Delta$. Again for the form $a(\varphi, \psi)$ we take (16) and for $k > 0$, $\varphi \in W_2^1(\Omega)$ we find that

$$\operatorname{Re} a(\varphi,\varphi) + k\|\varphi\|_0^2 = \sum_{j=1}^{r} \int_\Omega \left|\frac{\partial\varphi}{\partial x_j}\right|^2 dx + k\int_\Omega |\varphi|^2\,dx \geqslant c_k\|\varphi\|_1^2,$$

with which we have shown that $-\Delta$ is $W_2^1(\Omega)$-coercive, respectively that $-\Delta+k$ is $W_2^1(\Omega)$-elliptic for each $k>0$. Since we can choose $k>0$ arbitrarily small, it follows that the half plane $\operatorname{Re}\lambda<0$ contains no eigenvalues of the Neumann problem ($\lambda=0$ is an eigenvalue), this we know from Example 16.2.

By the Neumann problem we wish once more to illustrate the method of proof of Theorem 21.1. By Theorem 17.10 the equation

$$a(u,\varphi)+k(u,\varphi)_0=(f,\varphi)_0,\ \varphi W_2^1(\Omega), \tag{22}$$

(*the generalised Neumann problem*), possesses for each $f\in[W_2^1(\Omega)]'$ a unique solution $u\in W_2^1(\Omega)$. Here we only require of Ω that it be open and bounded. In order to be able to make any interpretation of the Neumann boundary condition $\partial u/\partial n|_{\partial\Omega}$, for example in the sense of the trace operator of §8, we need sharper assumptions on Ω, for example

$$\Omega\in C^{2,1}, \tag{23}$$

and on u, for example $u\in W_2^2(\Omega)$. This is for the Neumann problem, as distinct from the Dirichlet problem, where (18) is sufficient. We have Green's formula

$$\sum_{j=1}^{r}\int_\Omega \frac{\partial u}{\partial x_j}\frac{\partial\bar\varphi}{\partial x_j}\,\mathrm{d}x+k\int_\Omega u\cdot\bar\varphi\,\mathrm{d}x=\int_\Omega(-\Delta+k)u\cdot\bar\varphi\,\mathrm{d}x+\int_{\partial\Omega}\frac{\partial u}{\partial n}\cdot\bar\varphi\,\mathrm{d}\sigma, \tag{24}$$

for $u\in W_2^2(\Omega)$ and $\varphi\in W_2^1(\Omega)$. We need assumption (23) for the existence of the integral $\int_{\partial\Omega}(\partial u/\partial n)\bar\varphi\,\mathrm{d}\sigma$: because of (23) we can define the map $(\partial/\partial n)|_{\partial\Omega}=T_0(\partial/\partial n)$, it is continuous, see Theorem 8.7,

$$\left.\frac{\partial}{\partial n}\right|_{\partial\Omega}: W_2^2(\Omega)\to W_2^{1/2}(\partial\Omega)\subset L_2(\partial\Omega)$$

and we have

$$\left|\int_{\partial\Omega}\frac{\partial u}{\partial n}\bar\varphi\,\mathrm{d}\sigma\right|=\left|\int_{\partial\Omega}T_0\left(\frac{\partial u}{\partial n}\right)\cdot T_0\bar\varphi\,\mathrm{d}\sigma\right|\leqslant\left\|T_0\frac{\partial u}{\partial n}\right\|_{0,\partial\Omega}\cdot\|T_0\bar\varphi\|_{0,\partial\Omega}$$

$$\leqslant c\left\|T_0\frac{\partial u}{\partial n}\right\|_{1/2,\partial\Omega}\cdot\|T_0\bar\varphi\|_{1/2,\partial\Omega}\leqslant c\|u\|_2\cdot\|\varphi\|_1,$$

with which we have guaranteed the existence of the surface integral. For the proof of (24) we employ the usual method, we first prove (24) for $u,\varphi\in C^\infty(\Omega)$ and then carry out a density argument. Suppose now that $f\in L_2(\Omega)$, by Theorem 17.10 and the regularity Theorem 20.4, (22) possesses a uniquely determined solution $u\in W_2^2(\Omega)$; if we apply (24) to (22) we obtain

$$\int_\Omega(-\Delta+k)u\cdot\bar\varphi\,\mathrm{d}x+\int_{\partial\Omega}\frac{\partial u}{\partial n}\bar\varphi\,\mathrm{d}\sigma=\int_\Omega f\cdot\bar\varphi\,\mathrm{d}x,\quad \varphi\in W_2^1(\Omega). \tag{25}$$

Restricting φ to $\mathscr{D}(\Omega)$ we obtain $-\Delta u + ku = f$ on Ω. Substituting this in (25) – written with the trace operator T_0 – we find that

$$\int_{\partial\Omega} \frac{\partial u}{\partial n}\cdot T_0\bar{\varphi}\,\mathrm{d}\sigma = 0 \quad \text{for all } \varphi\in W_2^1(\Omega). \tag{26}$$

By Theorem 8.8 $T_0: W_2^1(\Omega)\to W_2^{1/2}(\partial\Omega)$ is surjective and by Theorem 4.3 $W_2^{1/2}(\partial\Omega)$ is dense in $L_2(\partial\Omega)$, hence it follows from (26) that

$$\left.\frac{\partial u}{\partial n}\right|_{\partial\Omega} = 0,$$

that is, the weak equation (22) is equivalent to

$$-\Delta u + ku = f \text{ in } \Omega,$$

$$\frac{\partial u}{\partial n} = 0 \text{ on } \partial\Omega, \quad u\in W_2^2(\Omega).$$

We can now solve the inhomogeneous equation: let $g\in W_2^{1/2}(\partial\Omega)$ and $f\in L_2(\Omega)$ be preassigned, then by Theorem 8.8 there exists some $u_0\in W_2^2(\Omega)$ with $(\partial u_0/\partial n)|_{\partial\Omega} = g$. We again solve

$$-\Delta w + kw = f + \Delta u_0 - ku_0, \quad \frac{\partial w}{\partial n} = 0 \quad \text{on } \partial\Omega,$$

respectively

$$a(w,\varphi) + k(w,\varphi)_0 = (f + \Delta u_0 - ku_0, \varphi)_0$$

and obtain $u := w + u_0$ as solution of the inhomogeneous problem

$$-\Delta u + ku = f \quad \text{in } \Omega \quad \frac{\partial u}{\partial n} = g \quad \text{on } \partial\Omega.$$

By Example 16.2 the Neumann problem is self-adjoint (see also Theorem 21.2), previously we have shown that $\operatorname{Re}\lambda < 0$ does not form part of the spectrum, therefore the Theorems 15.9, 17.12 and 21.1 are applicable to the Neumann problem and we obtain a far-reaching spectral theorem.

By Green's formula (24) the Neumann problem (see §17, (28)) has representation operator $\mathscr{L}$ given by

$$(\mathscr{L}\varphi,\psi)_0 = a(\varphi,\psi) = \int_\Omega (-\Delta\varphi)\bar{\psi}\,\mathrm{d}x + \int_{\partial\Omega}\frac{\partial\varphi}{\partial n}\bar{\psi}\,\mathrm{d}\sigma,$$

therefore

$$\mathscr{L}\varphi = -\Delta\varphi + \frac{\partial\varphi}{\partial n}\cdot\delta_{\partial\Omega}.$$

where $\delta_{\partial\Omega}$ is the δ-distribution concentrated on the frontier $\partial\Omega$, see Gelfand & Silov [1]. We observe that $\mathscr{L}$ agrees with $-\Delta$ only on $W^2(B) = \{\varphi \in W_2^2(\Omega) : \partial\varphi/\partial n = 0 \text{ on } \partial\Omega\}$. If $\varphi \in W_2^2(\Omega)$ but $\varphi \notin W^2(B)$, then $\mathscr{L}\varphi$ can be a distribution not belonging to $L_2(\Omega)$ – compare also Corollary 21.1.

Example 21.3 The third boundary value problem for the Laplace operator $-\Delta$. We continue with the notation and assumptions of Example 21.2. To the form (16) we add the boundary form

$$c(\varphi, \psi) = \int_{\partial\Omega} b_0(x) \cdot \varphi(x) \cdot \overline{\psi(x)}\, d\sigma.$$

In the sense of Definition 20.1 c is $m - 1 = 0$ transversal and degree $c = 2m - 2 = 0$, we can therefore apply Theorem 20.2, by which $a(\varphi, \psi) + c(\varphi, \psi)$ is again $W_2^1(\Omega)$-coercive, let k_0 be the coercion constant. By Theorem 17.10 the weak equation

$$a(u, \varphi) + c(u, \varphi) + k_0(u, \varphi)_0 = (f, \varphi)_0, \quad \varphi \in W_2^1(\Omega) \tag{27}$$

then possesses a unique solution $u \in W_2^1(\Omega)$ for each $f \in [W_2^1(\Omega)]'$, and there are no eigenvalues of the third boundary value problem lying in the half plane $\operatorname{Re}\lambda \leqslant -k_0$. Suppose now that $f \in L_2(\Omega)$, then by Theorem 20.4 $u \in W_2^2(\Omega)$ and we can apply Green's formula (24) to (27), obtaining

$$\int_\Omega (-\Delta + k_0)u \cdot \bar{\varphi}\, dx + \int_{\partial\Omega} \left(b_0 u + \frac{\partial u}{\partial n} \right) \bar{\varphi}\, d\sigma = \int_\Omega f \bar{\varphi}\, dx, \quad \varphi \in W_2^1(\Omega).$$

From this it follows as in Example 21.2 that

$$-\Delta u + k_0 u = f \quad \text{in } \Omega, \quad b_0 u + \frac{\partial u}{\partial n} = 0 \quad \text{on } \partial\Omega.$$

For $f \in L_2(\Omega)$, $g \in W_2^{1/2}(\partial\Omega)$ we can in $W_2^2(\Omega)$ uniquely solve the third boundary value problem

$$-\Delta u + ku = f \quad \text{in } \Omega, \quad b_0 u + \frac{\partial u}{\partial n} = g \quad \text{on } \partial\Omega.$$

Here also we can directly apply the solution Theorem 21.1.

If $b_0(x)$ is real valued, by Example 16.3 or by Theorem 21.2 the third boundary value problem is self-adjoint for the Laplace operator; since the half plane $\operatorname{Re}\lambda \leqslant -k_0$ does not belong to the spectrum, the spectral conclusions of Theorems 15.9, 17.12 and 21.1 are again valid.

Example 21.4 The boundary value problem with skew derivative. We work through the case $r = 2$ here. We first add the skew-symmetric form below to the form (16)

$$\mathcal{s}(\varphi, \psi) = \int_\Omega \left[a \frac{\partial \varphi}{\partial x_1} \overline{\frac{\partial \psi}{\partial x_2}} - a \frac{\partial \varphi}{\partial x_2} \overline{\frac{\partial \psi}{\partial x_1}} \right] \mathrm{d}x.$$

This does not alter the strong ellipticity, by Example 19.1 strongly elliptic forms of order 2 are always $W_2^1(\Omega)$-coercive. Then we add the form

$$\mathcal{b}(\varphi, \psi) = - \int_\Omega \left[\frac{\partial a}{\partial x_1} \frac{\partial \varphi}{\partial x_2} \bar{\psi} - \frac{\partial a}{\partial x_2} \frac{\partial \varphi}{\partial x_1} \bar{\psi} \right] \mathrm{d}x,$$

since order $b = (1, 0)$, by Theorem 19.1 $W_2^1(\Omega)$-coercion is maintained, and the form

$$\mathscr{A}(u, \varphi) = \mathcal{a}(\varphi, \psi) + \mathcal{s}(\varphi, \psi) + \mathcal{b}(\varphi, \psi)$$

is W_2^1-coercive. Let $f \in L_2(\Omega)$, we consider the weak equation

$$\mathscr{A}(u, \varphi) = (f, \varphi)_0, \quad \varphi \in W_2^1(\Omega), u \in W_2^2(\Omega),$$

and apply Green's formula (24) and the Gauss formula §16, (77) to it

$$\begin{aligned}
\int_\Omega a \frac{\partial u}{\partial x_1} \frac{\partial \bar{\varphi}}{\partial x_2} \mathrm{d}x &= - \int_\Omega a \frac{\partial^2 u}{\partial x_1 \partial x_2} \bar{\varphi} \,\mathrm{d}x - \int_\Omega \frac{\partial a}{\partial x_2} \frac{\partial u}{\partial x_1} \bar{\varphi} \,\mathrm{d}x \\
&\quad + \int_{\partial\Omega} a \frac{\partial u}{\partial x_1} n_2 \bar{\varphi} \,\mathrm{d}\sigma - \int_\Omega a \frac{\partial u}{\partial x_2} \frac{\partial \bar{\varphi}}{\partial x_1} \mathrm{d}x \\
&= \int_\Omega a \frac{\partial^2 u}{\partial x_1 \partial x_2} \bar{\varphi} \,\mathrm{d}x + \int_\Omega \frac{\partial a}{\partial x_1} \frac{\partial u}{\partial x_2} \bar{\varphi} \,\mathrm{d}x - \int_{\partial\Omega} a \frac{\partial u}{\partial x_2} n_1 \bar{\varphi} \,\mathrm{d}\sigma,
\end{aligned}$$

($\mathbf{n} = (n_1, n_2)$ is the unit vector in the direction of the inner normals.) This gives

$$\begin{aligned}
\mathscr{A}(u, \varphi) &= \mathcal{a}(u, \varphi) + \mathcal{s}(u, \varphi) + \mathcal{b}(u, \varphi) \\
&= \int_\Omega - \Delta u \bar{\varphi} \,\mathrm{d}x + \int_{\partial\Omega} \left[\frac{\partial u}{\partial x_1} n_1 + \frac{\partial u}{\partial x_2} n_2 \right] \bar{\varphi} \,\mathrm{d}\sigma \\
&\quad + \int_{\partial\Omega} \left[a \frac{\partial u}{\partial x_1} n_2 - a \frac{\partial u}{\partial x_2} n_1 \right] \bar{\varphi} \,\mathrm{d}\sigma \\
&\quad + \mathcal{b}(u, \varphi) - \mathcal{b}(u, \varphi) = \int_\Omega f \cdot \bar{\varphi} \,\mathrm{d}x.
\end{aligned}$$

With this we have submitted Example 21.4 to the variational method (Theorem 21.1), where

$$- \Delta u = f \quad \text{in } \Omega,$$

and we have the boundary value

$$\frac{\partial u}{\partial x_1}(n_1 + an_2) + \frac{\partial u}{\partial x_2}(n_2 - an_1) = 0 \quad \text{on } \partial\Omega \tag{28}$$

Let the direction $\mu = (\mu_1, \mu_2)$ of the skew derivation be given, where 'skew' means that

$$\mu_1 \cdot n_1 + \mu_2 \cdot n_2 > 0. \tag{29}$$

Then we can always so determine a that

$$n_1 + an_2 = \rho \cdot \mu_1, \quad n_2 - an_1 = \rho \cdot \mu_2, \quad \text{where } \rho > 0. \tag{30}$$

We can therefore write (28) in the form

$$\frac{\partial u}{\partial \mu} = \frac{\partial u}{\partial x_1}\mu_1 + \frac{\partial u}{\partial x_2}\mu_2 = 0 \quad \text{on } \partial\Omega,$$

which justifies the name. Now for the solvability of (30); we multiply the first equation by n_1, the second by n_2, and after addition obtain, because of (29) that

$$\rho = \frac{1}{\mu_1 n_1 + \mu_2 n_2} > 0.$$

If $n_2 \neq 0$, we obtain a from the first equation (30)

$$a = \frac{\rho\mu_1 - n_1}{n_2} = \frac{\mu_1 n_2 - \mu_2 n_1}{\mu_1 n_1 + \mu_2 n_2},$$

and this a also satisfies the second equation. If $n_1 \neq 0$ we determine a by the second equation (30), obtaining the same result. The calculations for the general case $r \geqslant 2$ run similarly, see Fichera [1], where the calculations for the boundary value problem with skew derivative are carried out for a strongly elliptic second-order differential operator.

The boundary value problem with skew derivative does not need to be self-adjoint.

Example 21.5 Let

$$A(x, \mathrm{D}) = -\sum_{j,k}^{r} a_{jk}\frac{\partial}{\partial x_j}\frac{\partial}{\partial x_k} + \sum_{j=1}^{r} a_j \frac{\partial}{\partial x_j} + a_0 \tag{31}$$

be a strongly elliptic second-order differential operator, let

$$a(\varphi, \psi) = \int_\Omega \left[\sum_{j,k} a_{jk}\frac{\partial\varphi}{\partial x_j}\frac{\partial\bar{\psi}}{\partial x_k} + \sum_{j=1}^{r}\left(a_j + \sum_{k=1}^{r}\frac{\partial a_{jk}}{\partial x_k}\right)\frac{\partial\varphi}{\partial x_j}\bar{\psi} + a_0 \cdot \varphi \cdot \bar{\psi}\right]\mathrm{d}x$$

be an associated sesquilinear form, note that (31) is not written in the form §19, (1). We want to show that we can fit the Neumann problem

$$Au = f \text{ on } \Omega, \quad N_A u = g \text{ on } \partial\Omega$$

into the framework of the variational method, where (see §16)

$$N_A u = \sum_{j,k} a_{jk} \frac{\partial u}{\partial x_j} \cos(\mathbf{n}, x_k).$$

For this we need only check the assumptions 1, 2 and 3 for Theorem 21.1. For assumption 1: we apply the Gauss formula §16, (17) to the individual terms of the expression $\int_\Omega Au\bar{\varphi}\,\mathrm{d}x$ and obtain

$$a(u, \varphi) = \int_\Omega Au\bar{\varphi}\,\mathrm{d}x + \int_{\partial\Omega} N_A u \cdot \bar{\varphi}\,\mathrm{d}\sigma,$$

(see the formula after §16, (17)), that is, assumption 1 holds with $p = 0$, $B_1 = N_A$, $B_1' = I$.
For assumption 2: by Example 19.1 $a(\varphi, \psi)$ is W_2^1-coercive, hence $p = 0$.
For assumption 3: we have noted the normality of N_A in Example 16.3.

Example 21.6 For the biharmonic operator Δ^2 we consider the boundary value problems

$$(\Delta^2; I, \Delta) \quad \text{and} \quad \left(\Delta^2; \frac{\partial}{\partial n}, -\frac{\partial \Delta}{\partial n}\right) \tag{32}$$

and show that Theorem 21.1 is applicable to them, that is, that they can be solved by the variational method. We have $A = \Delta^2$, $a(\varphi, \psi) = \int_\Omega \Delta\varphi \cdot \Delta\bar{\psi}\,\mathrm{d}x$, and for assumption 1 we take Green's formula §16, (22)

$$a(u, \varphi) = \int_\Omega \Delta u \cdot \Delta\bar{\varphi} = \int_\Omega \Delta^2 u \cdot \bar{\varphi} + \int_{\partial\Omega} \Delta u \cdot \frac{\partial\bar{\varphi}}{\partial n}\,\mathrm{d}\sigma + \int_{\partial\Omega} -\frac{\partial \Delta u}{\partial n} \cdot \bar{\varphi}\,\mathrm{d}\sigma. \tag{33}$$

Assumption 2, V-coercion, is satisfied (Example 19.2), once with §19, (56)

$$V_1 = \{\varphi \in W_2^2(\Omega) : I\varphi|_{\partial\Omega} = 0\},$$

the second time with §19, (58)

$$V_2 = \left\{\varphi \in W_2^2(\Omega) : \left.\frac{\partial\varphi}{\partial n}\right|_{\partial\Omega} = 0\right\}.$$

We do not need a boundary form, therefore $c = 0$. Because of the normality of both boundary value problems, assumption 3 is also satisfied.

Consider $(\Delta^2; I, \Delta)$; Δ^2 is V_1-coercive and from Green's formula (33) (with the notation of (3)) it is immediate that

$$b_1 = I, \quad B_2 = \Delta, \quad \left(\mathrm{D}_1 = -\frac{\partial\Delta}{\partial n}, B_2' = \frac{\partial}{\partial n}\right).$$

Consider $(\Delta^2; \partial/\partial n, -\partial\Delta/\partial n)$, Δ^2 is V_2-coercive. We again read off from (33) that

$$b_1 = \frac{\partial}{\partial n}, \quad B_2 = -\frac{\partial\Delta}{\partial n} \quad (\mathrm{D}_1 = \Delta, B_2' = I).$$

Therefore both boundary value problems

$$\Delta^2 u = f \text{ in } \Omega, \quad u|_{\partial\Omega} = g_1, \quad \Delta u|_{\partial\Omega} = g_2,$$

and

$$\Delta^2 u = f \text{ in } \Omega, \left.\frac{\partial u}{\partial n}\right|_{\partial\Omega} = g_1, -\left.\frac{\partial \Delta u}{\partial n}\right|_{\partial\Omega} = g_2,$$

are amenable to the variational method, and the conclusions of Theorem 21.1 hold. In particular the half plane $\mathrm{Re}\,\lambda \leqslant -k_0$ does not belong to the spectrum, which together with the self-adjointness of both problems (32), see Example 16.6 or Theorem 21.3, gives: for both problems (32) the spectral conclusions of Theorems 15.9, 17.12 and 21.1 hold.

By the variational method we can also handle boundary value problems, which do not fall into the scheme of §13.

Example 21.7 The mixed problem for the Laplace operator $-\Delta$. Let Ω be open, connected and bounded, we decompose the frontier $\partial\Omega$ into two subsets $\Gamma_1 \cup \Gamma_2$, and look for a solution $u(x)$ of

$$-\Delta u = f \text{ in } \Omega, \quad u = 0 \quad \text{on } \Gamma_1 \quad \text{and} \quad \frac{\partial u}{\partial n} = 0 \quad \text{on } \Gamma_2,$$

respectively $u = g_1$ on Γ_1 and $\partial u/\partial n = g_2$ on Γ_2.

We show that for $\Omega \in N^{0,1}$, $\mu_{r-1}\Gamma_1 \neq 0$, where μ_{r-1} is the surface measure on $\partial\Omega$, the mixed problem is uniquely solvable in its variational method version. We set

$$a(\varphi, \psi) := \sum_{j=1}^{r} \int_\Omega \frac{\partial\varphi}{\partial x_j} \cdot \frac{\partial\bar{\psi}}{\partial x_j} \mathrm{d}x, \quad \varphi, \psi \in \tilde{V}$$

$$\tilde{V} := \{\varphi \in W_2^1(\Omega) : T_0\varphi|_{\Gamma_1} = 0\}, \quad H = L_2(\Omega),$$

because of the continuity of T_0 (for this we need $\Omega \in C^{0,1}$), $\tilde{V}$ is a closed subspace of $W_2^1(\Omega)$ and we have

$$\mathring{W}_2^1(\Omega) \subset \tilde{V} \subset W_2^1(\Omega).$$

We have the estimate

$$|a(\varphi, \psi)| \leqslant \|\varphi\|_1 \cdot \|\psi\|_1 \quad \text{for } \varphi, \psi \in \tilde{V} \tag{34}$$

and need Gårding's inequality

$$\operatorname{Re} a(\varphi, \varphi) \geqslant c \|\varphi\|_1^2 \quad \text{for } \varphi \in \tilde{V}. \tag{35}$$

On $W_2^1(\Omega)$ we establish the inequality

$$\|\varphi\|_1^2 \leqslant c \left[\sum_{j=1}^{r} \left\| \frac{\partial \varphi}{\partial x_j} \right\|_0^2 + \| T_0 \varphi \|_{1/2, \Gamma}^2 \right]. \tag{36}$$

Suppose that (36) is not correct, then there exists a sequence $\varphi_n \in W_2^1(\Omega)$ with $\|\varphi_n\|_1 = 1$ and

$$1 = \|\varphi_n\|_1^2 > n \left[\sum_{j=1}^{r} \left\| \frac{\partial \varphi_n}{\partial x_j} \right\|_0^2 + \| T_0 \varphi_n \|_{1/2, \Gamma_1}^2 \right], \tag{37}$$

from which it follows that

$$\frac{\partial \varphi_n}{\partial x_j} \to 0 \quad \text{in } L_2(\Omega) \quad \text{for } j = 1, \ldots, r. \tag{38}$$

By Theorem 7.2 (here we use $\Omega \in N^{0,1}$) $\{\varphi_n\}$ is relatively compact in $L_2(\Omega)$, therefore there exists a subsequence – for the sake of simplicity we again call it φ_n – with $\varphi_n \to \varphi$ in $L_2(\Omega)$, which together with (38) gives

$$\varphi_n \to \varphi \quad \text{in } W_2^1(\Omega) \quad \text{and} \quad \frac{\partial \varphi}{\partial x_j} = 0, \quad j = 1, 2, \ldots, r, \tag{39}$$

φ is therefore constant on Ω. Also from (37) we have $T_0 \varphi_n \to 0$ and from (39)

$$T_0 \varphi_n \to T_0 \varphi_0 = T_0(\text{const}) = \text{const}.$$

To recapitulate: the constant vanishes on Γ_1 ($\mu_{r-1} \Gamma_1 \neq 0$), that is, $\varphi \equiv 0$ contradicting $\|\varphi\|_1 = 1$ (see above). (36) considered on $\tilde{V}$ is (35).

By Definition 17.5, (34) and (35) imply that $a(\varphi, \psi)$ is $\tilde{V}$-elliptic, this in turn implies that the weak equation

$$a(u, \varphi) = (f, \varphi)_0, \quad \varphi \in \tilde{V}, \tag{40}$$

is uniquely solvable in $\tilde{V}$ for $f \in \tilde{V}'$, for example $f \in L_2(\Omega) \subset \tilde{V}'$. By Theorem 17.14 no values of the spectrum of (40) lie in the half plane $\operatorname{Re} \lambda \leqslant 0$. Because

$\mathscr{D}(\Omega) \subset \tilde{V}$ we obtain the equation

$$-\Delta u = f \text{ in } \Omega$$

as an immediate consequence of (40), and if we formally apply Green's formula (24) to (40) we also obtain the boundary conditions

$$u = 0 \quad \text{on } \Gamma_1 \quad \text{and} \quad \frac{\partial u}{\partial n} = 0 \quad \text{on } \Gamma_2.$$

In order to apply Green's formula honestly, we need global regularity theorems (in the sense of Theorem 20.4) which are, however, very difficult to obtain, there are problems with the points of $\Gamma_1 \cap \Gamma_2$, see also Fichera [1].

It is also possible to solve uniquely the inhomogeneous problem

$$-\Delta u = f, \quad u|_{\Gamma_1} = g_1, \quad \left.\frac{\partial u}{\partial n}\right|_{\Gamma_2} = g_2 \tag{41}$$

for $\Omega \in C^{2,1}$, $f \in L_2(\Omega)$, $g_1 \in W_2^{3/2}(\Gamma_1)$, $g_2 \in W_2^{1/2}(\Gamma_2)$. We extend g_1 and g_2 to $\partial\Omega$, to obtain

$$g_1 \in W_2^{3/2}(\partial\Omega), \quad g_2 \in W_2^{1/2}(\partial\Omega),$$

by Theorem 14.1 (here we use $\Omega \in C^{2,1}$). We then determine a function $g \in W_1^2(\Omega)$ with

$$g|_{\Gamma_1} = g_1 \quad \text{and} \quad \left.\frac{\partial g}{\partial n}\right|_{\Gamma_2} = g_2.$$

and solve the equation (for $w \in \tilde{V}$)

$$a(w, \varphi) = (f - \Delta g, \varphi)_0, \quad \varphi \in \tilde{V},$$

and put

$$u := w + g.$$

We easily see that u satisfies (41). Since $a(\varphi, \psi)$ acts antisymmetrically on $\tilde{V}$, by Theorem 17.12 the Green solution operator G for the mixed problem is L_2-self-adjoint, and all the conclusions of Theorem 17.12 hold.

Example 21.8 The transmission problem for the Laplace operator $-\Delta$. Let Ω be open and bounded, we decompose Ω into two subsets $\Omega_1 \cup \Omega_2$. Let $\Gamma_1 := \bar{\Omega}_1 \cap \partial\Omega$, $\Gamma_2 := \bar{\Omega}_2 \cap \partial\Omega$, $\Gamma := \bar{\Omega}_1 \cap \bar{\Omega}_2$ (see Fig. 21.1). We look for solution $u_i \in W_2^1(\Omega_i)$, $i = 1, 2$ of

$$-a\Delta u_1 = f \quad \text{in } \Omega_1, \quad -b\Delta u_2 = f \quad \text{in } \Omega_2$$
$$u_1 = g \quad \text{on } \Gamma_1, \quad u_2 = g \quad \text{on } \Gamma_2, \tag{42}$$

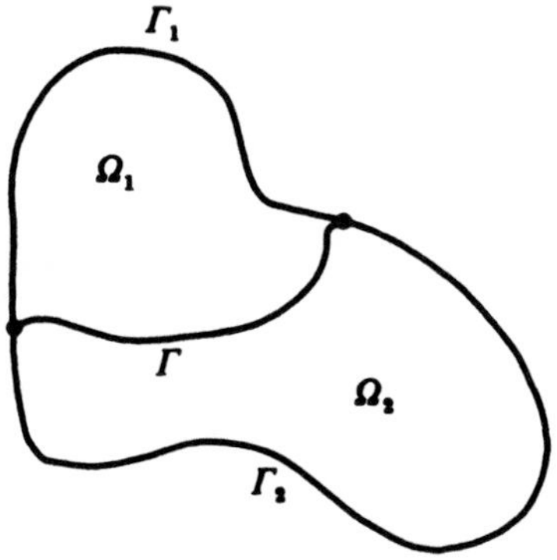

Fig. 21.1

which satisfy the transmission condition:

$$u_1 = u_2, \quad a\frac{\partial u_1}{\partial n} = b\frac{\partial u_2}{\partial n} \quad \text{on } \Gamma, \tag{43}$$

where $a, b > 0$ are positive constants. As $V = W$ we take the closed subspace of $W_2^1(\Omega_1) \times W_2^1(\Omega_2)$ defined by

$$W: \{(\varphi_1, \varphi_2): \varphi_i \in W_2^1(\Omega_i), \quad i = 1, 2,$$
$$T_0 u_1 = 0 \quad \text{on } \Gamma_1, \quad T_0 u_2 = 0 \quad \text{on } \Gamma_2, \quad T_0 u_1 = T_0 u_2 \quad \text{on } \Gamma\}.$$

Here we suppose that $\Omega_i \in N^{0,1} \subset C^{0,1}$, hence that the trace operator is defined and continuous, $i = 1, 2$. We have

$$\mathring{W}_2^1(\Omega_1) \times \mathring{W}_2^1(\Omega_2) \subset W \subset W_2^1(\Omega_1) \times W_2^1(\Omega_2).$$

As sesquilinear form we take

$$a((\varphi_1, \varphi_2), (\psi_1, \psi_2)) = \sum_{j=1}^{r} \left[a \int_{\Omega_1} \frac{\partial \varphi_1}{\partial x_j} \cdot \frac{\partial \bar{\psi}_1}{\partial x_j} \mathrm{d}x + b \int_{\Omega_2} \frac{\partial \varphi_2}{\partial x_j} \cdot \frac{\partial \bar{\psi}_2}{\partial x_j} \mathrm{d}x \right] \tag{44}$$

and immediately have the estimate

$$|a((\varphi_1, \varphi_2), (\psi_1, \psi_2))| \leqslant c_1 \|(\varphi_1, \varphi_2)\|_1 \cdot \|(\psi_1, \psi_2)\|_1 \quad \text{for } (\varphi_1, \varphi_2), (\psi_1, \psi_2) \in W. \tag{45}$$

Here we have furnished $W_2^1(\Omega_1) \times W_2^1(\Omega_2)$ with the Hilbert space norm

$$\|(\varphi_1, \varphi_2)\|_1^2 = \|\varphi_1\|_1^2 + \|\varphi_2\|_1^2,$$

and W carries the topology induced from $W_2^1(\Omega_1) \times W_2^1(\Omega_2)$. In order to prove Gårding's inequality for (44), we proceed as in Example 21.7. On

$W_2^1(\Omega_1) \times W_2^1(\Omega_2)$ we have the estimate

$$\|(\varphi_1, \varphi_2)\|_1^2 \leqslant c_2 \left\{ \sum_{j=1}^{r} \left(a \left\| \frac{\partial \varphi_1}{\partial x_j} \right\|_0^2 + b \left\| \frac{\partial \varphi_2}{\partial x_j} \right\|_0^2 \right) \right.$$

$$\left. + \| T_0(\varphi_1 - \varphi_2) \|_{1/2, \Gamma}^2 + \| T_0 \varphi_1 \|_{1/2, \Gamma_1}^2 + \| T_0 \varphi_2 \|_{1/2, \Gamma_2} \right\}; \tag{46}$$

we obtain the proof by means of a compactness argument, as in (36). Here we need the assumptions $\Omega_i \in N^{0,1}$, $i = 1, 2$. On W (46) takes the form

$$\|(\varphi_1, \varphi_2)\|_1^2 \leqslant c_2 \sum_{j=1}^{r} \left(a \left\| \frac{\partial \varphi_1}{\partial x_j} \right\|_0^2 + b \left\| \frac{\partial \varphi_2}{\partial x_j} \right\|_0^2 \right),$$

with which we have proved Gårding's inequality for (44)

$$\operatorname{Re} a \| (\varphi_1, \varphi_2), (\varphi_1, \varphi_2) \|_1^2 \geqslant c_3 \| (\varphi_1, \varphi_2) \|_1^2, \quad (\varphi_1, \varphi_2) \in W. \tag{47}$$

Let $H = L_2(\Omega_1) \times L_2(\Omega_2) \simeq L_2(\Omega_1 \cup \Omega_2) = L_2(\Omega)$, $V = W$, then (45) and (47) imply that (44) is (H, V)-elliptic in the sense of Definition 17.5. Thus the weak equation

$$a((u_1, u_2), (\varphi_1, \varphi_2)) = (f, (\varphi_1, \varphi_2))_0, \quad (\varphi_1, \varphi_2) \in W, \tag{48}$$

is uniquely solvable in $W((u_1, u_2) \in W)$ for $f \in L_2(\Omega_1 \cup \Omega_2) = L_2(\Omega)$. Also by Theorem 17.14 there are no values of the spectrum of (48) in the half plane $\operatorname{Re} \lambda \leqslant 0$, and all conclusions of the spectral Theorem 17.12 hold. The Green solution operator G for the transmission problem is L_2-self-adjoint, since (44) is antisymmetric on W. Because $\mathscr{D}(\Omega_1) \times \mathscr{D}(\Omega_2) \subset W$ from (48) we obtain the equations

$$-a\Delta u_1 = f \quad \text{in } \Omega, \quad -b\Delta u_2 = f \quad \text{in } \Omega_2.$$

If we formally apply Green's formula (24) twice to (48), we obtain the transmission condition (43), while the boundary conditions $u_1 = 0$ on Γ_1, $u_2 = 0$ on Γ_2, $u_1 = u_2$ on Γ belongs to the definition of W.

Therefore we have uniquely solved the homogeneous transmission problem $(g = 0)$. We remark that for the actual application of Green's formula (24) on (48) we need global regularity theorems for the solution $(u_1, u_2) \in W$ – here also we have problems with the points of $\Gamma_1 \cap \Gamma_2 \cap \Gamma$.

In order to solve the inhomogeneous transmission problem, we take $g \in W_2^{1/2}(\Gamma_1)$, $g \in W_2^{1/2}(\Gamma_2)$, and use Theorem 8.8 to determine

$$u_1^0 \in W_2^1(\Omega_1), \quad u_2^0 \in W_2^1(\Omega_2)$$

with

$$T_0 u_1^0 = g \quad \text{on } \Gamma_1, \quad \text{and} \quad T_0 u_2^0 = g \quad \text{on } \Gamma_2,$$

(here the assumptions $\Omega_i \in C^{0,1}$, $i=1,2$ are sufficient). We solve the weak equation ($(w_1, w_2) \in W$)

$$a((w_1, w_2), (\varphi_1, \varphi_2)) = (f - \Delta u^0, (\varphi_1, \varphi_2))_0$$
$$:= (f - \Delta u_1^0, \varphi_1)_{0,\Omega_1} + (f - \Delta u_2^0, \varphi_2)_{0,\Omega_2}$$

and put $(u_1, u_2) := (w_1, w_2) + (u_1^0, u_2^0)$. We easily see that (u_1, u_2) formally satisfies (42) and (43).

Generalisations to arbitrary strongly elliptic differential operators of the second order lie to hand, the reader may carry out the necessary steps for himself.

Exercises

21.1 We consider the Dirichlet problem for the Laplace operator

$$-\Delta u = f \quad \text{in } \Omega \quad \text{and} \quad u = g \quad \text{on } \partial\Omega. \tag{1}$$

By a Green function $G(x, y)$, $x \in \bar{\Omega}$, $y \in \Omega$ for the Dirichlet problem we understand the following function:

1. $r \geq 3$:

$$G(x; y) = \frac{\Gamma(\frac{r}{2})}{(r-2)2\pi^{r/2}|x-y|^{r-2}} + v(x; y),$$

where for each $y \in \Omega$ v satisfies the boundary value problem

$$\Delta_x v(x, y) = 0, x \in \Omega; \quad v(x; y) = -\frac{\Gamma(\frac{r}{2})}{(r-2)2\pi^{r/2}|x-y|^{r-2}} \quad \text{for } x \in \partial\Omega;$$

2. For $r = 2$ we take

$$G(x; y) = \frac{1}{2\pi} \ln \frac{1}{|x-y|} + v(x; y)$$

where for each $y \in \Omega$ v satisfies

$$\Delta_x v(x; y) = 0; \quad x \in \Omega; \quad v(x; y) = -\frac{1}{2\pi} \ln \frac{1}{|x-y|}, \quad x \in \partial\Omega.$$

Show that for the solution u of (1) we have the representation

$$u(y) = \int_{\partial\Omega} g(x) \frac{\partial G(x; y)}{\partial n_x} \, d\sigma_x + \int_\Omega f(x) G(x; y) \, dx.$$

This means among other things that the Green operator $G: L_2(\Omega) \to L_2(\Omega)$ (from Definition 14.7 or from Definition 17.5) is an integral operator of the form

$$u(y) = Gf = \int_\Omega f(x) G(x; y) \, dx.$$

Hint: see Exercise 1.14 and apply Green's formula (see older textbooks on partial differential equations).

21.2 Find the Green function for the ball $\Omega = B(0, a)$, $r = 3$.

Answer:

$$G(x; y) = \frac{1}{4\pi}\left[\frac{1}{|x-y|} - \frac{a}{|x|}\cdot\frac{1}{\left|y - x\dfrac{a^2}{|x|^2}\right|}\right]$$

21.3 Find the Green function for the disc $\Omega = B(0, a)$, $r = 2$.

Answer:

$$G(x; y) = \frac{1}{2\pi}\left[\ln\frac{1}{|x-y|} - \ln\left(\frac{a}{|x|}\cdot\frac{1}{\left|y - x\dfrac{a^2}{|x|^2}\right|}\right)\right]$$

21.5 Find the Green function for the half ball $\Omega = B_+(0, a)$, $r = 3$.

Answer:

$$G(x; y) = \frac{1}{4\pi}\left[\frac{1}{|x-y|} - \frac{a}{|y|}\frac{1}{|x-y^1|} - \frac{1}{|x-y^2|} + \frac{a}{|y|}\frac{1}{|x-y^3|}\right].$$

Here we have obtained y^1 from y by inversion on the sphere $S(0, a)$ $(y^1 = a^2 y/|y|^2)$, y^2 by reflection in the plane $x_3 = 0$ $(y^2 = (y_1, y_2, -y_3))$, and y^3 from y^2 by inversion $(y^3 = (a^2 y^2/|y|^2)$.

21.6 Find a Green function for the three-dimensional quadrant $\Omega = \{x : x_1, x_2 \geq 0\}$

Answer:

$$G(x; y) = \frac{1}{4\pi}\left[\frac{1}{|x-y|} - \frac{1}{|x-y^2|} + \frac{1}{|x-y^3|} - \frac{1}{|x-y^4|}\right]$$

where

$$y = (y_1, y_2, y_3), y^2 = (-y_1, y_2, y_3), y^3 = (-y_1, -y_2, y_3) \quad \text{and} \quad y^4 = (y_1, -y_2, y_3).$$

21.7 By the Green function for the Neumann problem

$$-\Delta u = f \quad \text{in } \Omega, \quad \frac{\partial u}{\partial n} = g \quad \text{on } \partial\Omega, r = 3, \tag{2}$$

we understand the function $G(x, y)$, $x \in \bar{\Omega}$, $y \in \Omega$ with the following properties:

1. $G(x; y) = 1/4\pi|x-y| + v(x; y)$, where v is harmonic, that is, for all $y \Delta_x v(x; y) = 0$;
2. $\partial G(x; y)/\partial n_x = 1/S_0$ for $x \in \partial\Omega$, $y \in \Omega$, $S_0 = \text{area}(\partial\Omega)$ is the surface measure of $\partial\Omega$;
3. $\int_{\partial\Omega} G(x; y)\, d\sigma_x = 0$, $y \in \Omega$.

Show that G is uniquely determined by the requirements 1, 2 and 3, and that for the solutions u of (2) we have the representation

$$u(y) = \int_\Omega f(x)G(x; y)\,dx - \int_{\partial\Omega} g(x)G(x; y)\,d\sigma_x + c,$$

assuming that the integrability condition of (2) is satisfied

$$\int_{\Omega} f(x)\,\mathrm{d}x + \int_{\partial\Omega} g(x)\,\mathrm{d}\sigma_x = 0.$$

21.8 Find the Green function of the Neumann problem for the ball $\Omega = B(0, a)$, $r = 3$.
Answer:

$$G(x; y) = \frac{1}{4\pi}\left[\frac{1}{|x-y|} + \frac{a}{|y||x-y^1|} + \frac{1}{a}\ln\frac{2}{1 - \dfrac{|x^1|}{|y^1|} + \dfrac{|x-y^1|}{|y^1|}}\right] - \frac{1}{2\pi a}$$

Here $y^1 = a^2 y/|y|^2$, and x^1 is the base point of the perpendicular through x onto the line $[0, y]$.

21.9 We consider the Dirichlet problem for the biharmonic operator ($r = 2$)

$$\Delta^2 u = f \quad \text{in } \Omega; \quad u|_{\partial\Omega} = 0, \quad \left.\frac{\partial u}{\partial n}\right|_{\partial\Omega} = 0 \quad \text{on } \partial\Omega \tag{3}$$

Show that the Green operator $G: L_2(\Omega) \to L_2(\Omega)$ is an integral operator with the representation

$$u(y) = \int_{\Omega} f(x) G(x; y)\,\mathrm{d}x,$$

where the 'Green function' $G(x, y)$ is (uniquely) determined by the conditions
1. $G(x; y) = (1/8\pi)|x-y|^2 \ln|x-y| + v(x; y)$, where for each $y \in \Omega$ $v(x; y)$ satisfies the equation $\Delta_x^2 v(x; y) = 0$ (see also Exercise 1.15);
2. $G(x; y)|_{\partial\Omega_x} = 0$, $\partial G(x; y)/\partial n_x|_{\partial\Omega_x} = 0$ for all $y \in \Omega$. Find the Green formula for the solution u of

$$\Delta^2 u = f, \quad u|_{\partial\Omega} = g_1, \quad \left.\frac{\partial u}{\partial n}\right|_{\partial\Omega} = g_1.$$

21.10 Find the Green function of (3) for the disc $\Omega = B(0, a)$, $r = 2$.
Answer:

$$G(x; y) = \frac{a^2}{16\pi}\left[\frac{|x-y|^2}{a^2}\ln\frac{a^2|x-y|^2}{|y|^2|x-y^1|^2} + \left(1 - \frac{|y|^2}{a^2}\right)\left(1 - \frac{|x|^2}{a^2}\right)\right]$$

Here again $y^1 = a^2 y/|y|^2$.

§22 The Schauder fixed point theorem and a non-linear problem

We begin with a special case, the Brouwer theorem.

Theorem 22.1 *Let $\bar{B}(0, 1)$ be the closed unit ball in $\mathbb{R}^r$ and $T: \bar{B} \to \bar{B}$ a C^∞-map,* *that is,* $T(x)=(t_1(x),\dots,t_r(x))$ *with* $t_i\in C^\infty(\bar{B})$, $i=1,\dots,r$. *Then* T *possesses at least one fixed point* $x_0\in\bar{B}$, *that is,* $Tx_0=x_0$.

We need a lemma.

Lemma 22.1 *Let* f *be a* C^∞*-function of* $r+1$ *variables with values in* $\mathbb{R}^r$, $f=(f_1(x_0,x_1,\dots,x_r),\dots,f_r(x_0,x_1,\dots,x_r))$. *Let* $D_0,D_1,\dots,D_r$ *be the cofactors of the first row of the matrix*

$$\begin{bmatrix} e_0 & e_1 & \cdots & e_r \\ \dfrac{\partial f_1}{\partial x_0} & \dfrac{\partial f_1}{\partial x_1} & \cdots & \dfrac{\partial f_1}{\partial x_r} \\ \cdots & \cdots & & \cdots \\ \dfrac{\partial f_r}{\partial x_0} & \dfrac{\partial f_r}{\partial x_1} & \cdots & \dfrac{\partial f_r}{\partial x_r} \end{bmatrix}$$

that is,

$$D_0=\det\begin{bmatrix} \dfrac{\partial f_1}{\partial x_1} & \cdots & \dfrac{\partial f_1}{\partial x_r} \\ \cdots & \cdots & \cdots \\ \dfrac{\partial f_r}{\partial x_1} & \cdots & \dfrac{\partial f_r}{\partial x_r} \end{bmatrix},\quad D_1=-\det\begin{bmatrix} \dfrac{\partial f_1}{\partial x_0} & \dfrac{\partial f_1}{\partial x_2} & \cdots \\ \cdots & \cdots & \cdots \\ \dfrac{\partial f_r}{\partial x_0} & \dfrac{\partial f_r}{\partial x_2} & \cdots \end{bmatrix},\ \text{etc.}$$

We have the identity

$$\frac{\partial D_0}{\partial x_0}+\frac{\partial D_1}{\partial x_1}+\cdots+\frac{\partial D_r}{\partial x_r}=0. \tag{1}$$

Proof. We use the differentiation theorem for determinants, we differentiate by columns, and see that for each term in the expansion of (1)

$$\pm\frac{\partial f_1}{\partial x_i}\cdot\ldots\cdot\frac{\partial^2 f_j}{\partial x_k\partial x_l}\cdot\ldots\cdot\frac{\partial f_r}{\partial x_q},\quad i,j,\dots,q=0,1,\dots,r$$

exactly one other term with k and l interchanged appears, that is, of the form

$$\mp\frac{\partial f_1}{\partial x_i}\cdot\ldots\cdot\frac{\partial^2 f_j}{\partial x_l\partial x_k}\cdot\ldots\cdot\frac{\partial f_r}{\partial x_q};\quad i,j,\dots,q=0,1,\dots,r.$$

They cancel each other and we have (1). ■

We prove Brouwer's theorem by contradiction. Let $x\neq Tx$ for each point of $\bar{B}$, let y be the intersection point of the line $[x,Tx]$ with the sphere $\Gamma=$

$\{x:|x|=1\}$, we have two intersection points; we take the one closer to x. We have

$$y = x + a(x)(x - T(x)), \quad a(x) \geqslant 0.$$

We show that $a \in C^\infty(\bar{B})$. For this we calculate $a(x)$. We have $(y, y) = |y|^2 = y_1^2 + \cdots + y_r^2 = 1$, or

$$|x|^2 + 2a(x, x - Tx) + a(x)^2|x - Tx|^2 = 1$$

or

$$a(x) = \frac{-(x, x - Tx) + \sqrt{[(x, x - Tx)^2 + |x - Tx|^2(1 - |x|)^2]}}{|x - Tx|^2},$$

which because of our assumption $|x - Tx| \neq 0$, gives

$$a \in C^\infty(\bar{B}).$$

We also see immediately that

$$x \in \Gamma \Leftrightarrow a(x) = 0. \tag{2}$$

We next construct a subsidiary function with values in $\mathbb{R}^r$.

$$f(t, x_1, \ldots, x_r) := x + t \cdot a(x)(x - T(x)). \tag{3}$$

We have

$$f(0, x) = x, \quad f(1, x) = x + a(x)(x - T(x)) = y \quad \text{and } f \in C^\infty. \tag{4}$$

The Jacobi determinant $\partial f/\partial x$ is D_0 from Lemma 22.1 where we put x_0 equal to t, and we have

$$D_0(0, x) = 1 \quad \text{and} \quad D_0(1, x) = 0. \tag{5}$$

Indeed because of (4) we have

$$\left[\frac{\partial f_i}{\partial x_j}(0, x)\right] = E = \text{identity matrix},$$

white from $f(1, x) = y \in \Gamma$, we have

$$f_1^2(1, x) + \cdots + f_r^2(1, x) = 1. \tag{6}$$

Differentiating (6) according to x_j we obtain

$$f_1(1, x)\frac{\partial f_1(1, x)}{\partial x_j} + \cdots + f_r(1, x)\frac{\partial f_r(1, x)}{\partial x_j} = 0, \quad j = 1, \ldots, r.$$

with $f(1, x) \neq 0$, see (6) from which $D_0(1, x) = 0$ follows. We form the integral

$I(t)=\int_B D_0(t,x)\,dx$. By (1) and Stokes' theorem we have

$$\frac{dI(t)}{dt}=\int_B \frac{\partial D_0(t,x)}{\partial t}dx=-\int_B \sum_{i=1}^{r}\frac{\partial D_i}{\partial x_i}dx=-\int_\Gamma \sum_{i=1}^{r} D_i n_i\,d\sigma,$$

where $n=(n_1,\ldots,n_r)$ is the outward pointing normal, $D_j(t,x)$ contains the column $\partial f_i/\partial t=a(x)(x-T(x))_i$, which because of (2) vanishes on Γ. Therefore

$$\frac{\partial I(t)}{\partial t}\equiv 0. \tag{7}$$

On the other hand (5) gives

$$I(0)=\int_B 1\,dx>0 \quad\text{and}\quad I(1)=\int_B D_0(1,x)\,dx=0,$$

in contradiction to (7). ■

The assumption $T\in C^\infty$ in Brouwer's theorem may be weakened to $T\in C(\bar{B})$, if one approximates T by C^∞-functions.

Theorem 22.2 *Let $T:\bar{B}\to\bar{B}$, $T\in C(\bar{B})$ be uniformly approximated by $T_n:\bar{B}\to\bar{B}$, $T_n\in C^\infty(\bar{B})$. Then T also has the fixed point property, that is, there exists $x_0\in\bar{B}$ with $Tx_0=x_0$.*

Proof. Let $x_n\in\bar{B}$ be a fixed point of T_n. Since $\bar{B}$ is compact, there exists a subsequence of $\{x_n\}$ – we again call it simply $\{x_n\}$ – with $x_n\to x_0\in\bar{B}$. We show that x_0 is a fixed point of T. For $\varepsilon>0$ we get

$$|Tx_0-x_0|\leqslant|Tx_0-Tx_n|+|Tx_n-T_nx_n|+|x_n-x_0|\leqslant\varepsilon+\varepsilon+\varepsilon,$$

here we have the first ε-inequality because of the continuity of T, the second because of the uniform convergence $T_n\to T$ on $\bar{B}$, the third because $x_n\to x_0$. $|Tx_0-x_0|\leqslant 3\varepsilon$ implies that $Tx_0=x_0$. ■

For the approximation of a continuous function by C^∞-functions, or polynomials, we have Weierstrass' theorem. However, we must be a little careful: from $|T(x)-P_\varepsilon(x)|\leqslant\varepsilon$, $P_\varepsilon(x)$ a Weierstrass polynomial, one can only conclude that $|P_\varepsilon(x)|\leqslant|Tx|+\varepsilon\leqslant 1+\varepsilon$, that is $P_\varepsilon:\bar{B}(0,1)\to\bar{B}(0,1+\varepsilon)$. However, we need maps $P_\varepsilon:\bar{B}(0,1)\to\bar{B}(0,1)$.

Theorem 22.3 *Let $T:\bar{B}\to\bar{B}$, $T\in C(\bar{B})$. For each $\varepsilon>0$ a polynomial map $P_\varepsilon(x)=(P_{1,\varepsilon}(x),\ldots,P_{r,\varepsilon}(x))$, where the components $P_{i,\varepsilon}$, $i=1,\ldots,r$ are polynomials in $x_1,\ldots,x_r$, may be found with the following properties:*

$$|T(x)-P_\varepsilon(x)|\leqslant\varepsilon \quad\text{for all } x\in\bar{B} \quad\text{and}\quad P_\varepsilon:\bar{B}\to\bar{B}. \tag{8}$$

Proof. First step: we deform the map T so that its image lies in the smaller ball $\bar{B}(0, 1-\varepsilon)$, that is, let $T_\varepsilon(x) := (1-\varepsilon)T(x)$. We see easily that $T_\varepsilon(x)$ is again continuous on $\bar{B}$, and that

$$|T_\varepsilon(x)| \leqslant 1-\varepsilon, \quad \text{that is,}\ T_\varepsilon: \bar{B}(0,1) \to \bar{B}(0, 1-\varepsilon), \tag{9}$$

$$|T_\varepsilon(x) - T(x)| \leqslant |T(x)|\,|1-\varepsilon-1| \leqslant \varepsilon, \quad \text{since}\ |T(x)| \leqslant 1. \tag{10}$$

We now apply Weierstrass' theorem to $T_\varepsilon(x)$, and find a polynomial map $P_\varepsilon(x)$ with

$$|T_\varepsilon(x) - P_\varepsilon(x)| \leqslant \varepsilon \quad \text{for all}\ x \in \bar{B}. \tag{11}$$

The first part of (18) now follows from (10) and (11). (11) and (9) give

$$|P_\varepsilon(x)| \leqslant |T_\varepsilon(x)| + \varepsilon \leqslant 1 - \varepsilon + \varepsilon = 1$$

or

$$P_\varepsilon: \bar{B}(0,1) \to \bar{B}(0,1)$$ ■

Collecting together what we know, we prove the fixed point theorem in $\mathbb{R}^r$.

Theorem 22.4 *Let $K \subset \mathbb{R}^r$ be closed, bounded and convex, and $T: K \to K$ continuous. Then T has at least one fixed point x_0 in K, that is,*

$$x_0 \in K \quad \textit{with} \quad Tx_0 = x_0.$$

Proof. Since K is bounded, we can find some closed ball $\bar{B}(0, R)$ with

$$K \subset \bar{B}(0, R).$$

We define a surjective map $N: \bar{B} \to K$, by taking $N(x)$ to be the point in K nearest to x. Because of the convexity of K, $N(x)$ is uniquely defined. Otherwise we would have $y_1 \neq y_2$, $\operatorname{dist}(x, K) = |x - y_1| = |x - y_2|$, and hence $\frac{1}{2}(y_1 + y_2) \in K$ (convexity). Then

$$\left| x - \frac{y_1 + y_2}{2} \right| < |x - y_1| = \operatorname{dist}(x, K),$$

a contradiction. The map $N: \bar{B} \to K$ is continuous, assuming the opposite let $x_n \to x$ and $N(x_n) \not\to N(x)$. Because of the compactness of K we can choose a subsequence of x_n (we again denote the subsequence by x_n) with

$$x_n \to x \quad \text{and} \quad N(x_n) \to y \neq N(x), \text{ where } y \in K. \tag{12}$$

Because $N(x_n)$ is the 'nearest' point to x_n we have

$$|x_n - N(x_n)| \leqslant |x_n - N(x)|. \tag{13}$$

Taking the limit, in the sense of (12), of both sides of (13) we obtain

$$|x-y| \leqslant |x-N(x)|. \tag{14}$$

Again, since $N(x)$ is the 'nearest' point to x in K, it already follows from (14) that $y=N(x)$, contradicting (12). We now consider the composite continuous map

$$T \circ N : \bar{B} \to K \to K \subsetneq \bar{B}.$$

Because of Theorems 22.1, 22.2, 22.3 the fixed point theorem holds for balls $\bar{B}(0,R)$, the radius is unimportant, and we find some x_0 with $T(N(x_0)) = x_0$. Because $T \circ N : \bar{B} \to K \to K$ it must be that $x_0 \in K$. However, for $x_0 \in K$ we have $x_0 = N(x_0)$ and hence that $T(x_0) = x_0 \in K$. ∎

We now prove a fixed point theorem for Banach spaces X, namely the *Schauder theorem*. First a definition:

Definition 22.1 *A continuous map $f: X \to X$ is called compact, if for each closed, bounded subset $\Omega \subset X$, the set $\overline{f(\Omega)}$ is compact.*

Theorem 22.5 *Let Ω be a closed bounded subset of X. The map $f: \Omega \to X$ is compact if and only if f may be uniformly approximated on Ω by continuous, finite dimensional maps. A map $g: \Omega \to X$ is called finite dimensional if its image is contained in a finite dimensional subspace of X.*

Proof. Let f be compact, then $f(\Omega)$ is relatively compact, hence for each $n \in \mathbb{N}$ there exist finitely many $x_i \in f(\Omega)$, $i = 1, \ldots, m$, such that

$$\min_i \| f(x) - x_i \| < \frac{1}{n} \quad \text{for all } x \in \Omega. \tag{15}$$

The Schauder operator

$$f_n(x) = \frac{\sum_i a_i(x) x_i}{\sum_i a_i(x)} \quad \text{with } a_i(x) = \max\left(0, \frac{1}{n} - \| f(x) - x_i \|\right)$$

possesses the desired properties, since because of (15) for $x \in \Omega$ the continuous $a_i(x)$ are not all simultaneously zero, f_n is therefore continuous, and we have

$$\| f_n(x) - f(x) \| = \left\| \frac{\sum_i a_i(x)(x_i - f(x))}{\sum_i a_i(x)} \right\| \leqslant \frac{\sum_i a_i(x) \frac{1}{n}}{\sum_i a_i(x)} = \frac{1}{n} \quad \text{for all } x \in \Omega.$$

The f_n are also all finite dimensional, since $\operatorname{im} f_n \subset$ linear hull $[x_1, \ldots, x_m]$.

Sufficiency: As a uniform limit of continuous maps f_n, f is itself continuous, and because

$$\|f_n(x) - f(x)\| < 1/n \quad \text{for all } x \in \Omega, \tag{16}$$

$f(\Omega)$ is relatively compact, for (16) means that $f(\Omega)$ possesses for each $n \in \mathbb{N}$ a finite $(2/n)$-mesh $(x_1, \ldots, x_m)$. ■

Theorem 22.6 (Schauder) *Let Ω be a closed, bounded and convex subset of the Banach space X and let $f: \Omega \to \Omega$ be continuous and compact. Then f possesses at least one fixed point x_0 in Ω, that is, there exists some $x_0 \in \Omega$ with $f(x_0) = x_0$.*

Proof. Let $f_\varepsilon(x)$ be an ε-approximation of f, see above, and let N_ε be the finite dimensional subspace of X spanned by the vectors $x_1, \ldots, x_{j(\varepsilon)}$ in $f(\Omega)$. Since Ω is convex, and (as we have seen from the proof of Theorem 22.5) $f_\varepsilon(\Omega)$ is contained in the convex hull of $f(\Omega) \subset \Omega$, we have

$$f_\varepsilon: \Omega \to \Omega \cap N_\varepsilon,$$

or restricted

$$f_\varepsilon: \Omega \cap N_\varepsilon \to \Omega \cap N_\varepsilon.$$

However, $\Omega \cap N_\varepsilon$ is finite dimensional, closed, bounded and convex, thus we can apply Theorem 22.4 to prove that there exists some $x_\varepsilon \in \Omega$ with $f_\varepsilon(x_\varepsilon) = x_\varepsilon$. Now let $\varepsilon \to 0$. From $\overline{f(\Omega)}$ compact it follows that $\overline{CHf(\Omega)}$ is compact, where CH denotes the convex hull (see Bourbaki [1] p. 81, X is complete), therefore $f_\varepsilon(x_\varepsilon)$ possesses a convergent subsequence with limit x_0, which we again denote by $f_\varepsilon(x_\varepsilon)$:

$$x_\varepsilon = f_\varepsilon(x_\varepsilon) \to x_0 \in \Omega.$$

If we pass to the limit as $\varepsilon \to 0$ in $\|x_\varepsilon - f(x_\varepsilon)\| = \|f_\varepsilon(x_\varepsilon) - f(x_\varepsilon)\| < \varepsilon$ we obtain $\|x_0 - f(x_0)\| = 0$ or $f(x_0) = x_0$. ■

As an application of Schauder's theorem we wish to consider a non-linear elliptic problem. We assume the following (see Theorem 13.1): let $\Omega \subset \mathbb{R}^r$ be bounded and $C^{3,0}$-regular. Let $L = \sum_{|s| \leq 2} a_s(x) D^s$ be elliptic in $\bar{\Omega}$, $a_s \in C^1(\bar{\Omega})$. Suppose that zero is not an eigenvalue of L. Let the function g be continuous in all variables

$$g\left(x, u, \frac{\partial u}{\partial x_1}, \ldots, \frac{\partial u}{\partial x_r}\right) = G(u),$$

and satisfy the estimates ($u_1, u_2, u \in \mathring{W}_2^1(\bar{\Omega})$)

$$|G(u_1) - G(u_2)| \leqslant C_2 \sqrt{\left(\sum_{|s| \leqslant 1} |D^s(u_1 - u_2)|^1 \right)}, \tag{17}$$

where C_2 independent of $x, x \in \bar{\Omega}, u_1$ and u_2;

$$|G(u)|^2 \leqslant M \sqrt{\left(\sum_{|s| \leqslant 1} |D^s u|^2 \right)} \tag{18}$$

and M similarly independent of u and $x \in \partial\Omega$.

We consider the non-linear Dirichlet problem

$$Lu = g\left(x, u, \frac{\partial u}{\partial x_1}, \ldots, \frac{\partial u}{\partial x_r} \right) = G(u), \quad u \in \mathring{W}_2^1(\Omega), \text{ (that is, } u = 0 \text{ on } \partial\Omega). \tag{19}$$

We have:

Theorem 22.7 *Under the assumptions above the problem* (19) *has at least one solution* $u_0 \in \mathring{W}_2^1(\Omega) \cap W_2^2(\Omega)$ *and we have the estimate*

$$\| u_0 \|_1 \leqslant (C_1 C_3)^2,$$

where for the constants see the following proof.

Proof. By Theorem 8.9 the space $\{u = 0 \text{ on } \partial\Omega, u \in W_2^2(\Omega)\}$ equals $\mathring{W}_2^1 \cap W_2^2(\Omega)$ and we give it the W_2^2-norm; we easily see that $\mathring{W}_2^1 \cap W_2^2(\Omega)$ is a closed subspace of $W_2^2(\Omega)$. Since the Dirichlet problem always has index equal to zero (Theorem 13.5) and $\lambda = 0$ – by assumption – is not an eigenvalue of L, we can apply Theorem 13.4, by which $L: \mathring{W}_2^1 \cap W_2^2(\Omega) \to L^2(\Omega)$ represents an isomorphism, that is,

$$\| u \|_2 \leqslant C_1 \| Lu \|_0 \quad \text{for all } u \in \mathring{W}_2^1 \cap W_2^2,$$

and

$$\| L^{-1} f \|_2 \leqslant C_1 \| f \|_0 \quad \text{for all } f \in L^2(\Omega), \tag{20}$$

that is, $L^{-1}: L^2(\Omega) \to \mathring{W}_2^1(\Omega) \cap W_2^2(\Omega)$ is continuous.

(19) is equivalent to the fixed point problem

$$u = L^{-1}(G(u)) =: Fu, \quad u \in \mathring{W}_2^1(\Omega). \tag{21}$$

From (17) it follows that

$$G: \mathring{W}_2^1(\Omega) \to L^2(\Omega) \quad \text{is continuous,} \tag{22}$$

for

$$\int_\Omega |G(u_1) - G(u_2)|^2 \leqslant C_2^2 \int_\Omega \sum_{|s| < 1} |D^s u_1 - D^s u_2|^2 \, dx,$$

or

$$\| G(u_1) - G(u_2) \|_0 \leqslant C_2 \| u_1 - u_2 \|_1.$$

Therefore

$$F = L^{-1} \circ G : \mathring{W}_2^1 \to L^2 \to \mathring{W}_2^1(\Omega) \cap W_2^2(\Omega) \to \mathring{W}_2^1(\Omega)$$

is continuous and compact, here the last embedding is compact by Theorem 7.2 and $\mathring{W}_2^1 \cap W_2^2(\Omega)$ is closed in $W_2^2(\Omega)$.

We now want to find a closed ball $\bar{B}(0, R)$ in $\mathring{W}_2^1(\Omega)$, so that

$$F : \bar{B} \to \bar{B}. \tag{23}$$

By means of the Schwartz inequality we deduce from (18) that

$$\int_\Omega |G(u)|^2 \, dx \leqslant M \sqrt{(\text{mes}\, \Omega)} \left(\int_\Omega \sum_{|s| \leqslant 1} |D^s u|^2 \, dx \right)^{1/2}$$

or

$$\| G(u) \|_0 \leqslant C_3 \| u \|_1^{1/2} \quad \text{with } C_3^2 = M \sqrt{(\text{mesh}\, \Omega)} \tag{24}$$

We take $R = (C_1 C_3)^2$, $\bar{B} = \bar{B}(0, R)$ and have:from $\| u \|_1 \leqslant R$ it follows that

$$\begin{aligned} \| F(u) \|_1 \leqslant \| L^{-1} G(u) \|_2 &\underset{(20)}{\leqslant} C_1 \| Gu \|_0 \underset{(24)}{\leqslant} C_1 C_3 \| u \|_1^{1/2} \\ &\leqslant C_1 C_3 R^{1/2} = (C_1 C_3)(C_1 C_3) = (C_1 C_3)^2 = R. \end{aligned}$$

Therefore we do indeed have (23), can apply Theorem 22.6, and find some $u_0 \in \mathring{W}_2^1(\Omega)$ with $Fu_0 = u_0$ or (19). Because of (17) $u \in W_2^1$ implies that $f = g(x, u, \ldots) \in L^2(\Omega)$, so putting $f = g(x, u, \ldots)$ in the main Theorem 13.1, we get that $u = Fu_0$ belongs to $\mathring{W}_2^1(\Omega) \cap W_2^2(\Omega)$. If we make stronger regularity assumptions for the frontier $\partial\Omega$ and the function g, we obtain regular solutions for (19):

Theorem 22.8 *Suppose that the assumptions to Theorem 22.7 are satisfied and that in addition: Ω is $(k+3)$-regular, $a_s \in C^{k+1}(\bar{\Omega})$ and $g(x, u, \partial u/\partial x_1, \ldots, \partial u/\partial x_r) \in W_2^k(\Omega)$ for $u \in W_2^{k+1}(\Omega)$, $k \geqslant 0$. Then each solution u of (19) [$u \in \mathring{W}_2^1(\Omega)$!] necessarily belongs to $W_2^{k+2}(\Omega) \cap \mathring{W}_2^1(\Omega)$.*

The proof is by induction on k, using Corollary 13.1 with $f = G(u)$. In order to solve (19) classically, we apply Theorem 6.2:

Theorem 22.9 *Suppose that the assumptions above are satisfied for $k = [r/2] + 1$, hence $g \in W_2^{[r/2]+1}(\Omega) \subset C(\bar{\Omega})$ for $u \in W_2^{[r/2]+2}(\Omega)$. Then for the solution u of (19) we have:*

$$u \in C^2(\bar{\Omega}),\ Lu = g\left(x, u, \frac{\partial u}{\partial x_1}, \ldots, \frac{\partial u}{\partial x_r} \right) \text{ classically and } \lim_{x \to \partial\Omega} u(x) = u(x) \Big|_{\partial\Omega} = 0.$$

Generalisation of (19) to elliptic differential operators L of order $= 2m$ are obvious, the reader may formulate these and prove them himself, similarly generalisations to other boundary value problems, see also Ladyženskaya & Uralceva [1], Lions [3], where there is a large bibliography.

§23 Elliptic boundary value problems for unbounded regions

Here we consider the so-called exterior space problem for the Laplace operator Δ, that is, the Dirichlet problem for an unbounded region Ω. The properties of the spectrum of Δ show that in the framework of L^2-theory we can no longer expect to have a compact Green solution operator, because for $\Omega = \mathbb{R}^r$, $\lambda = 0$ belongs to the essential spectrum, see Schechter [2] or Kato [1]. Furthermore for the Sobolev spaces – if Ω is unbounded $(\text{mes}(\Omega) = \infty)$ – there no longer exist compact embedding theorems. It has, however, been known for a long time that by imposing growth restrictions at ∞ one can save the unique solvability of $\Delta u = f$, $u|_{\partial\Omega} = 0$. For us the growth restrictions take the form that we consider the Sobolev spaces as being equipped with polynomial weights. In this section we limit ourselves to a description by Mäulen [1]. For other elliptic boundary value problems on unbounded regions see Nečas [1] and Triebel [1]; where there is also a broad guide to the literature.

We introduce some notation and definitions. Let $p_\delta(x) := \frac{1}{2}(1 + |x|^2)^{(1+\delta)/2}$, where $\delta > 0$.

Definition 23.1 *We write* $\frac{1}{2}(1 + \delta) = l$ *and in agreement with* §5 *define the weighted Hilbert space* $L_2^l(\Omega)$ *as*

$$L_2^l(\Omega) := \{\varphi \text{ measurable on } \Omega : p_\delta \varphi \in L_2(\Omega)\},$$

with the scalar product

$$(\varphi, \psi) := \int_\Omega \varphi \cdot \bar{\psi} \cdot p_\delta^2 \, dx. \tag{1}$$

It is easy to check that $L_2^l(\Omega)$ is a Hilbert space; we have already used the spaces L_2^l in §5.

Definition 23.2 *We define the weighted Sobolev space* $W_2^{1,\delta}$ *as*

$$W_2^{1,\delta}(\Omega) := \left\{ \varphi \in \mathscr{D}'(\Omega) : p_\delta^{-1} \varphi \in L_2(\Omega), \frac{\partial \varphi}{\partial x_i} \in L_2(\Omega), i = 1, \ldots, r \right\}$$

equipped with the scalar product

$$(\varphi,\psi)_1=(\varphi,\psi)_{W_2^{1,\delta}}=\int_\Omega \varphi\bar{\psi}p_\delta^{-2}\,\mathrm{d}x+\sum_{i=1}^{r}\int_\Omega \frac{\partial\varphi}{\partial x_i}\frac{\partial\bar{\psi}}{\partial x_i}\,\mathrm{d}x. \tag{2}$$

Theorem 23.1 *$W_2^{1,\delta}(\Omega)$ is a Hilbert space.*

Proof. We have only to check completeness. Let φ_n be a Cauchy sequence in $W_2^{1,\delta}(\Omega)$. Then because $L_2(\Omega)$ is complete and $p_\delta(x)\neq 0$

$$p_\delta^{-1}\varphi_n \text{ converges to } p_\delta^{-1}\varphi \text{ in } L_2(\Omega) \quad\text{and}\quad \frac{\partial\varphi_n}{\partial x_i}\to\psi_i \text{ in } L_2(\Omega),\quad i=1,\dots,r. \tag{3}$$

By the text following Theorem 1.4 we deduce from (3) that

$$p_\delta^{-1}\varphi_n\to p_\delta^{-1}\varphi \quad\text{in}\quad \mathscr{D}'(\Omega) \tag{4}$$

and

$$\frac{\partial\varphi_n}{\partial x_i}\to\psi_i \quad\text{in}\quad \mathscr{D}'(\Omega),\ i=1,\dots,r.$$

By Definition 1.5 we may multiply in $\mathscr{D}'(\Omega)$ by $p_\delta\in C^\infty$, obtain $\varphi_n\to\varphi$ in $\mathscr{D}'(\Omega)$, and by differentiation, Theorem 1.6, obtain $\partial\varphi_n/\partial x_i\to\partial\varphi/\partial x_i$ in $\mathscr{D}'(\Omega)$, which together with (4) gives

$$\frac{\partial\varphi}{\partial x_i}=\psi_i,\quad i=1,\dots,r,$$

and thus we have proved completeness.

Definition 23.3 *We define $\mathring{W}_2^{1,\delta}(\Omega)$ as the closure of $\mathscr{D}(\Omega)$ in the $W_2^{1,\delta}$-topology $\mathring{W}_2^{1,\delta}:=\overline{\mathscr{D}(\Omega)}^{W_2^{1,\delta}}$.*

$\mathring{W}_2^{1,\delta}$ is again a Hilbert space, and we show that the second Poincaré inequality holds in $\mathring{W}_2^{1,\delta}$, see Theorem 7.6.

Theorem 23.2 *Let $r\geqslant 3$, on $\mathring{W}_2^{1,\delta}$ the expression*

$$\left[\int_\Omega\sum_{i=1}^{r}\left|\frac{\partial\varphi}{\partial x_i}\right|^2\mathrm{d}x\right]^{1/2} \textit{ is equivalent to the norm } \|\varphi\|_{W_2^{1,\delta}}.$$

Proof. By (2) we trivially have

$$\left[\int_\Omega\sum_{i=1}^{r}\left|\frac{\partial\varphi}{\partial x_i}\right|^2\mathrm{d}x\right]^{1/2}\leqslant\|\varphi\|_{W_2^{1,\delta}}\quad\text{for } \varphi\in\mathring{W}_2^{1,\delta}(\Omega). \tag{5}$$

For $\varphi\in\mathscr{D}(\mathbb{R}^r)$ we show the inequality

$$\int_{\mathbb{R}^r}|p_\delta^{-1}\varphi|^2\,dx\leqslant\frac{1}{2\delta(r-2)}\int_{\mathbb{R}^r}\sum_{i=1}^{r}\left|\frac{\partial\varphi}{\partial x_i}\right|^2 dx, \tag{6}$$

from which by an addition and restriction to Ω is follows that

$$\|\varphi\|^2_{W_2^{1,\delta}}\leqslant\left[1+\frac{1}{2\delta(r-2)}\right]\int_{\mathbb{R}^r}\sum_{i=1}^{r}\left|\frac{\partial\varphi}{\partial x_i}\right|^2 dx \quad \text{for } \varphi\in\mathring{W}_2^{1,\delta}(\Omega),$$

which, with (5), proves our theorem. Now for the proof of (6). We introduce polar coordinates on $\mathbb{R}^r$.

$$\begin{aligned} x_1&=\rho\cos\vartheta_1,\\ x_2&=\rho\sin\vartheta_1\cdot\cos\vartheta_2,\\ &\vdots\\ x_{r-1}&=\rho\sin\vartheta_1\cdot\sin\vartheta_2\cdots\sin\vartheta_{r-2}\cos\vartheta_{r-1},\\ x_r&=\rho\sin\vartheta_1\cdot\sin\vartheta_2\cdots\sin\vartheta_{r-2}\sin\vartheta_{r-1}, \end{aligned} \tag{7}$$

where $0\leqslant\rho<\infty$, $0\leqslant\vartheta_i\leqslant\pi$, $i=1,\ldots,r-2$, $0\leqslant\vartheta_{r-1}\leqslant 2\pi$. In polar coordinates we have for $\varphi\in\mathscr{D}(\mathbb{R}^r)$ that

$$\varphi(\rho,\vartheta_1,\ldots,\vartheta_{r-1})=-\int_\rho^\infty\frac{\partial}{\partial\tau}\varphi(\tau,\vartheta_1,\ldots,\vartheta_{r-1})\,d\tau.$$

The Schwarz inequality gives

$$\begin{aligned}|\varphi(\rho,\vartheta_1,\ldots,\vartheta_{r-1})|^2&\leqslant\int_\rho^\infty\frac{d\tau}{\tau^{r-1}}\cdot\int_\rho^\infty\left|\frac{\partial}{\partial\tau}\varphi(\tau,\ldots)\right|^2\tau^{r-1}\,d\tau\\ &\leqslant\frac{1}{(r-2)\rho^{r-2}}\int_0^\infty\left|\frac{\partial}{\partial\tau}\varphi(\tau,\vartheta_1,\ldots,\vartheta_{r-1})\right|^2\tau^{r-1}\,d\tau.\end{aligned}$$

we multiply by $\prod_{j=1}^{r-1}(\sin\vartheta_j)^{r-j-1}(\geqslant 0)$, integrate with respect to the angle variables and obtain

$$\begin{aligned}&\int_0^\pi d\vartheta_1\cdots\int_0^\pi d\vartheta_{r-2}\int_0^{2\pi}d\vartheta_{r-1}|\varphi(\rho,\vartheta_1,\ldots,\vartheta_{r-1})|^2\prod_{j=1}^{r-1}(\sin\vartheta_j)^{r-j-1}\\ &\leqslant\frac{1}{(r-2)\rho^{r-2}}J(\varphi),\end{aligned} \tag{8}$$

where we have written

$$\begin{aligned}J(\varphi)=&\int_0^\pi d\vartheta_1\cdots\int_0^\pi d\vartheta_{r-2}\int_0^{2\pi}d\vartheta_{r-1}\int_0^\infty d\tau\cdot\tau^{r-1}\\ &\cdot\left|\frac{\partial}{\partial\tau}\varphi(\tau,\vartheta_1,\ldots,\vartheta_{r-1})\right|^2\prod_{j=1}^{r-1}(\sin\vartheta_j)^{r-j-1}.\end{aligned} \tag{9}$$

We multiply (8) by $\rho^{r-1}/(1+\rho^2)^{1+\delta}$, integrate according to ρ and return to the original coordinates $x_1, \ldots, x_r$,

$$\int_0^\infty \mathrm{d}\rho \int_0^\pi \mathrm{d}\vartheta_1 \cdots \int_0^\pi \mathrm{d}\vartheta_{r-2} \int_0^{2\pi} \mathrm{d}\vartheta_{r-1} \left|\frac{\varphi(\rho, \vartheta_1, \ldots, \vartheta_{r-1})}{(1+\rho^2)^{(1+\delta)/2}}\right|^2 \rho^{r-1} \prod_{j=1}^{r-1} (\sin \vartheta_j)^{r-j-1}$$
$$= \int_{\mathbb{R}^r} \left|\frac{\varphi(x)}{(1+|x|^2)^{(1+\delta)/2}}\right|^2 \mathrm{d}x = \int_{\mathbb{R}^r} |p_\delta^{-1}\varphi|^2 \,\mathrm{d}x$$
$$\leqslant \frac{1}{(r-2)} \int_0^\infty \frac{\rho}{(1+\rho^2)^{1+\delta}} \mathrm{d}\rho \cdot J(\varphi) = \frac{1}{(r-2)2\delta} J(\varphi). \tag{10}$$

We need the assumptions $\delta > 0$ and $r \geqslant 3$ for the convergence of the integral $\int_0^\infty (\rho/(1+\rho^2)^{1+\delta})\mathrm{d}\rho$. We must still estimate the integral $J(\varphi)$ in (10). We have (see (9) and return to the coordinates $(x_1, \ldots, x_r)$)

$$J(\varphi) = \int_{\mathbb{R}^r} \left|\frac{\partial \varphi}{\partial \rho}\right|^2 \mathrm{d}x.$$

We use the chain rule

$$\frac{\partial \varphi}{\partial \rho} = \frac{\partial \varphi}{\partial x_1}\frac{\partial x_1}{\partial \rho} + \cdots + \frac{\partial \varphi}{\partial x_r}\frac{\partial x_r}{\partial \rho}.$$

and obtain

$$J(\varphi) = \int_{\mathbb{R}^r} \left|\sum_{i=1}^r \frac{\partial \varphi}{\partial x_i}\frac{\partial x_i}{\partial \rho}\right|^2 \mathrm{d}x$$
$$\leqslant \int_{\mathbb{R}^r} \left[\sum_{i=1}^r \left|\frac{\partial \varphi}{\partial x_i}\right|^2\right]\left[\sum_{i=1}^r \left|\frac{\partial x_i}{\partial \rho}\right|^2\right] \mathrm{d}x = \int_{\mathbb{R}^r} \sum_{i=1}^r \left|\frac{\partial \varphi}{\partial x_i}\right|^2 \mathrm{d}x, \tag{11}$$

since by (7) $\sum_{k=1}^r |\partial x_i/\partial \rho|^2 = 1$. Substitution of (11) in (10) gives (6). ■

With the help of Theorem 17.9 (Lax–Milgram) we can now solve the exterior space problem. We consider the equation $-\Delta u = f$, and we interpret the Dirichlet boundary condition $u|_{\partial\Omega} = 0$ as $u \in \mathring{W}_2^{1,\delta}(\Omega)$.

Theorem 23.3 *Let $f \in L_2^l(\Omega)$, $l = (1+\delta)/2$, $\delta > 0$ and $r \geqslant 3$. We consider the weak equation*

$$(-\Delta u, \varphi)_0 = \int_\Omega (-\Delta u)\bar{\varphi}\,\mathrm{d}x = \int_\Omega f \cdot \bar{\varphi}\,\mathrm{d}x = (f, \varphi)_0. \tag{12}$$

For all $\varphi \in \mathscr{D}(\Omega)$, or for all $\varphi \in \mathring{W}_2^{1,\delta}(\Omega)$. (12) possesses a unique solution u in $\mathring{W}_2^{1,\delta}(\Omega)$, and u depends continuously on f, in other words, if we put $u = Gf$, then

$$G: L_2^l(\Omega) \to \mathring{W}_2^{1,\delta}(\Omega) \text{ is linear and continuous.} \tag{13}$$

Proof. We take

$$a(\varphi,\psi):=\int_\Omega(-\Delta\varphi)\bar{\psi}\,\mathrm{d}x=\int_\Omega\sum_{i=1}^r\frac{\partial\varphi}{\partial x_i}\cdot\frac{\partial\bar{\psi}}{\partial x_i}\mathrm{d}x, \tag{14}$$

$a(\varphi,\psi)$ is a continuous sesquilinear form on $\mathring{W}_2^{1,\delta}\times\mathring{W}_2^{1,\delta}$, since (2) and the Schwarz inequality give us

$$|a(\varphi,\psi)|\leqslant\|\varphi\|_{W_2^{1,\delta}}\|\psi\|_{W_2^{1,\delta}}$$

$a(\varphi,\varphi)$ satisfies a Gårding inequality on $\mathring{W}_2^{1,\delta}$ – see Theorem 23.2 –

$$\operatorname{Re}a(\varphi,\varphi)=\int_\Omega\sum_{i=1}^r\left|\frac{\partial\varphi}{\partial x_i}\right|^2\geqslant c\|\varphi\|^2_{W_2^{1,\delta}},\quad\varphi\in\mathring{W}_2^{1,\delta}, \tag{15}$$

where we can calculate c

$$c=\frac{2\delta(r-2)}{2\delta(r-2)+1}.$$

Therefore the assumptions of Theorem 17.9 are satisfied, and we have

$$A:\mathring{W}_2^{1,\delta}(\Omega)\leftrightarrow\mathring{W}_2^{1,\delta}(\Omega)\quad\text{is an isomorphism,}$$

where

$$(\varphi,\psi)_1=a(A\varphi,\psi)\quad\text{for all }\varphi,\psi\in\mathring{W}_2^{1,\delta}(\Omega). \tag{16}$$

We now consider $(f,\varphi)_0$, we have

$$|(f,\varphi_0|=|p_\delta f,p_\delta^{-1}\varphi)_0|\leqslant\|p_\delta f\|_0\cdot\|p_\delta^{-1}\varphi\|_0\leqslant\|f\|_{L_2^l}\cdot\|\varphi\|_{\mathring{W}_2^{1,\delta}}.$$

Thus $(f,\cdot)$ is continuous on $\mathring{W}_2^{1,\delta}$ and the Riesz theorem gives us that $F:L_2^l(\Omega)\to\mathring{W}_2^{1,\delta}$ is continuous with

$$(f,\varphi)_0=(Ff,\varphi)_1\quad\text{for all }\varphi\in W_2^{1,\delta}(\Omega). \tag{17}$$

(17) in connection with (16) gives

$$(f,\varphi)_0=(Ff,\varphi)_1=a(A\circ Ff,\varphi),$$

and thus we have shown the existence of a solution u of (12), see also (14): $u:=Gf:=A\circ Ff$. Since A and F are continuous, we have also shown the continuity of (13). The uniqueness of the solution u in $\mathring{W}_2^{1,\delta}$ follows immediately from (15). ∎

One can also prove regularity theorems for the exterior space problem, see Mäulen [1], and in this way show solvability in the classical sense.

Exercises

23.1 Show that one can regard the Laplace operator $-\Delta$ as an unbounded, self-adjoint operator in $L_2(\mathbb{R}^r)$. Show that the spectrum of $-\Delta$ equals $\bar{\mathbb{R}}_+$, but that no eigenvalues lie in $\mathrm{Sp}(-\Delta) = \bar{\mathbb{R}}_+$ ($-\Delta$ has no eigenfunction in $L_2(\mathbb{R}^r)$).

23.2 Show that for each $\lambda \geqslant 0$ there exists some function $0 \neq f \in L^\infty(\mathbb{R}^r) \cap \mathscr{E}(\mathbb{R}^r)$ with $-\Delta f = \lambda f$.

23.3 Let $\lambda \geqslant 0$ be arbitrary. Show that all tempered distributions u (that is, $u \in \mathscr{S}'(\mathbb{R}^2)$), which satisfy the equation

$$(\Delta + \lambda)u = 0 \quad \text{in } \mathbb{R}^2,$$

are given by the series

$$u(z) = \sum_{m=-\infty}^{+\infty} c_m J_m(\sqrt{\lambda}\rho)\mathrm{e}^{\mathrm{i}m\theta}, \quad z = \rho \mathrm{e}^{\mathrm{i}\theta},$$

with c_ms of slow growth, that is,

$$|c_m| \leqslant c(1 + |m|)^c \quad \text{for all } m \in \mathbb{Z}.$$

Here the J_m are the Bessel functions, see the exercises to §13.

23.4 Let $u(x) \in C^2$ (C^2 is rather redundant because of Theorem 20.1) with $u(x) \not\equiv 0$ and $\Delta u = 0$ in $\mathbb{R}^r$. Show that it is then true that

$$\int_{\mathbb{R}^r} |u(x)|^2 \, \mathrm{d}x = +\infty.$$

IV

Parabolic differential operators

§24 The Bochner integral

As a further functional analytic tool we introduce the Bochner integral in Hilbert spaces. We present here complete definitions and proofs, up to measure theory, for which we quote Halmos [1]; the representations in Dunford & Schwartz [1] and Edwards [1] use topological assumptions, we manage without them.

24.1 Pettis' theorem

Definition 24.1 *Let $(S, \mathscr{B}, m)$ be a σ-finite measure space, let H be a Hilbert space with the scalar product $(\cdot,\cdot)_H$ and the norm $\|\cdot\|_H$. The map $s{:}S \to H$ is called:*

1. *weakly measurable (with respect to $\mathscr{B}$) if $(x(s), h)_H$ is a measurable scalar valued function with respect to $(S, \mathscr{B}, m)$ for all $h \in H$.*
2. *countably valued, if*
   (i) $\operatorname{im} x = \{h_n{:}n \in \mathbb{N}\}$, *the image consists of countably many values*
   (ii) $x^{-1}(h_i) = B_i \in \mathscr{B}$ *for all* $i \in \mathbb{N}$,
3. *strongly measurable, if there exists a sequence x_n of countably valued functions $x_n{:}S \to H$ with $x_n(s) \to x(s)$ strongly for H, m almost everywhere. Here we interpret, as usual, strong convergence in H to mean convergence in the norm $\| \quad \|_H$.*
4. *separably valued, if* $\operatorname{im}(x)$ *is separable,*
5. *almost separably valued, if there exists $B_0 \in \mathscr{B}$ with $m(B_0) = 0$ and* $\operatorname{im}(x(s))$ *separable for s running through $S \backslash B_0$.*
6. *finitely valued, if $x(s) = c_i \neq 0$ ($c_i =$ constant) for $s \in B_i$ and $i = 1, \ldots, n$, where $m(B_i) < \infty$, $B_i \cap B_j = \varnothing$ for $i \neq j$ and $x(s) = 0$ on $S \backslash \bigcup_{i=1}^{n} B_i$.*

Theorem 24.1 (Pettis) *The function $x(s)$ is strongly measurable if and only if $x(s)$ is weakly measurable and almost separably valued.*

Proof. Necessity: let $x(s)$ be strongly measurable. Then by definition there exists a sequence $x_n(s)$ of countably valued functions, which strongly converges in H to $x(s)$,

$$x_n(s) \to x(s) \text{ strongly} \quad m\text{-almost everywhere} \quad \text{as } n \to \infty,$$

that is, there exists $B_0 \in \mathscr{B}$ with $m(B_0) = 0$ and $x_n(s) \to x(s)$ strongly as $n \to \infty$ on $S \backslash B_0$.

We show next, quite generally, if $y(s)$ is countably valued with values in H, then $(y(s), h)_H$ is measurable for all $h \in H$. For this let $\operatorname{im}(y) = \{h_n : n \in \mathbb{N}\}$. We set

$$y_n(s) = \sum_{i=1}^{n} h_i \cdot \chi_{B_i}(s) \quad \text{with } B_i = y^{-1}(h_i),$$

so that

$$(y_n(s), h) = \sum_{i=1}^{n} (h_i, h)\chi_{B_i}(s) \text{ is measurable,}$$

and

$$(y_n(s), h) \to (y(s), h) \quad \text{as } n \to \infty.$$

Then by measure theory $(y(s), h)$ is measurable. Now back to the proof of the theorem. Let x_n be countably valued and

$$x_n(s) \to x(s) \quad \text{on} \quad S \backslash B_0 \quad \text{as } n \to \infty,$$

so that also

$$(x_n(s), h) \to (x(s), h) \quad \text{for all } h \in H \quad \text{for all } s \in S \backslash B_0 \quad \text{as } n \to \infty.$$

Since $(x_n(s), h)$ is measurable by our intermediate assertion, it follows that $(x(s), h)$ is measurable, and therefore $x(s)$ is weakly measurable. Moreover

$$\bigcup_{s \in S \backslash B_0} \{x(s)\} \subseteq \overline{\bigcup_{n=1}^{\infty} \operatorname{im}(x_n)},$$

and therefore $x(s)$ is almost separately valued.

Sufficiency: We can assume without loss of generality that H is separable. Otherwise we consider $\overline{[\operatorname{im}_{s \in S \backslash B_0}(x(s))]}$.

First of all we assert that $\|x(s)\|$ is measurable. For the proof let $a \in \mathbb{R}^1$ be chosen arbitrarily. Then it is enough to show

$$A := \{s \in S : \|x(s)\| \leqslant a\}$$

is a measurable set. For this let $h \in H$ and write

$$A_h := \{s \in S : |(x(s), h)| \leqslant a\}.$$

Then by the Schwarz inequality

$$A \subseteq \bigcap_{\|h\| \leqslant 1} A_h.$$

We show the reverse inclusion. Let $s \in \bigcap_{\|h\| \leqslant 1} A_h$ be fixed. We distinguish between two possibilities:

1. $x(s) = 0$; it follows immediately that $s \in A$.
2. Let $x(s) \neq 0$, then it follows that

$$\|x(s)\| = \left(x(s), \frac{x(s)}{\|x(s)\|}\right) \leqslant a, \text{ since } s \in A_{h_0} \text{ with } h_0 = x(s)/\|x(s)\|,\ \|h_0\| = 1,$$

and therefore $s \in A$. With this we have

$$A = \bigcap_{\|h\| \leqslant 1} A_h.$$

H is separable, therefore there exists a sequence $h_n \in H$ with

$$\|h_n\| \leqslant 1 \quad \text{and} \quad \overline{\{h_n : n \in N\}} = \overline{B(0, 1)}.$$

Hence

$$A = \bigcap_{\|h\| \leqslant 1} A_h = \bigcap_{n=1}^{\infty} A_{h_n}.$$

By assumption A_{h_n} is measurable for all $n \in \mathbb{N}$, therefore so is A. With this the first assertion is proved. Suppose now that $n \in \mathbb{N}$ is chosen and fixed. Since H is separable there exists a countable cover by balls

$$S_{j,n} := \{h \in H : \|x_{j,n} - h\| < 1/n\}, H = \bigcup_{j=1}^{\infty} S_{j,n}.$$

As shown in the intermediate assertion, the function

$$\|x(s) - x_{j,n}\| \text{ is measurable.}$$

Therefore

$$B_{j,n} := \{s \in S : x(s) \in S_{j,n}\} = \{s \in S : \|x(s) - x_{j,n}\| < 1/n\}$$

is measurable and $S = \bigcup_{j=1}^{\infty} B_{j,n}$

We set

$$B'_{i,n} = B_{i,n} \setminus \bigcup_{j=1}^{i-1} B_{j,n},\ y_n(s) = x_{i,n} \quad \text{for } s \in B'_{i,n}.$$

Of course $B'_{i,n}$ is measurable and $S = \bigcup_{i=1}^{\infty} B'_{i,n}$, and we have

$$\| x(s) - y_n(s) \| < \frac{1}{n} \quad \text{for all } s \in S.$$

Therefore $x(s)$ is the strong limit of a sequence of countably valued functions $y_n(s)$. ∎

If $m(S) < \infty$, each strongly measurable function can also be approximated by a sequence of finitely valued functions. Let $\mathcal{M}_w$ be the set of weakly measurable functions $x: S \to H$, $\mathcal{M}_s$ the set of strongly measurable functions and $\mathcal{M}_E$ the set of functions that can be approximated almost everywhere by sequences of finitely valued functions in the strong sense. We have

$$\mathcal{M}_E \subset \mathcal{M}_s \subset \mathcal{M}_w,$$

and we show that everywhere there exists equality under the assumption that H is separable and $m(S) < \infty$. (We know already by Pettis' theorem that $\mathcal{M}_s = \mathcal{M}_w$).

Some preliminary assertions. We say that $x_n(s)$ converges almost uniformly to $x(s)$, if for each $\varepsilon > 0$ we can find a measurable set F with $m(F) < \infty$ and $x_n(s)$ converging uniformly to $x(s)$ on $S \backslash F$. Egorov's theorem holds.

Theorem 24.2 *Let $(S, \mathscr{B}, m)$ be a measure space $m(S) < \infty$ and H a separable Hilbert space. Suppose given weakly measurable functions $x_n, x: S \to H$, where*

$$x_n(s) \to x(s) \text{ strongly almost everywhere} \quad \text{as } n \to \infty.$$

Then x_n converges to x almost uniformly.

Proof. Apart from a set of measure zero we may assume that $x_n(s)$ converges to $x(s)$ everywhere on S. Let

$$E_n^m = \bigcap_{i=n}^{\infty} \left\{ s : \| x_i(s) - x(s) - x(s) \| < \frac{1}{m} \right\}$$

($\| x_i(s) - x(s) \|$ is measurable!). We have $E_1^m \subset E_2^m \subset \cdots$, and since $x_n(s)$ tends to $x(s)$ everywhere on S,

$$\lim_n E_n^m = S \quad \text{for } m = 1, 2, \ldots.$$

Therefore $\lim_n m(S \backslash E_n^m) = 0$, here we use $m(s) < \infty$, and there exists some $N_0 = N_{(m)}$ with $m(S \backslash E_{N_0}^m) < \varepsilon / 2^m$ (ε fixed). We set $F = \bigcup_{m=1}^{\infty} (S \backslash E_{N_0}^m)$, have $F \in \mathscr{B}$ and

$$m(F) = m\left(\bigcup_{m=1}^{\infty} (S \backslash E_{N_0}^m) \right) \leqslant \sum_{m=1}^{\infty} m(S \backslash E_{N_0}^m) < \varepsilon.$$

Since

$$S\backslash F = S \cap \bigcap_{m=1}^{\infty} E_{N_0}^m,$$

we have for $n \geqslant N_0$: $s \in S\backslash F \subset E_n^m$, or

$$\| x_n(s) - x(s) \| < \frac{1}{m},$$

which proves uniform convergence on $S\backslash F$. ■

Another important type of convergence is convergence in measure. Let $x_n(s)$, $x(s)$ be weakly measurable functions $S \to H$. We say that $x_n(s)$ converges in measure to $x(s)$, if for each $\varepsilon > 0$ we have

$$\lim_n m\{s : \| x_n(s) - x(s) \| \geqslant \varepsilon\} = 0. \tag{1}$$

Almost uniform convergence implies convergence in measure (also without the hypothesis $m(S) < \infty$): for almost uniform convergence implies that given two positive numbers $\varepsilon, \delta > 0$ we can find $F \in \mathscr{B}$ and $N_0 \in \mathbb{N}$ with $m(F) < \delta$ and $\| x_n(s) - x(s) \| < \varepsilon$ for $s \in S\backslash F$, and (1) follows from this. (2)

If $m(S) < \infty$ convergence in measure is metrisable. Let M be the space of all weakly measurable functions $S \to H$ with the metric (3)

$$d(x, y) := \int_S \min(1, \| x(s) - y(s) \|) \, dm(s). \tag{4}$$

It is easy to see that d is a metric (because $m(S) < \infty$, $d(x, y)$ is always finite), we show now the equivalence of metric convergence and convergence in measure. Let $x_n \to x$ in measure, then because of (1) $\| x_n - x \| \to 0$ in measure as also does $\min(1, \| x_n - x \|)$. Because $\min(,) \leqslant 1$ and $m(S) < \infty$, we can apply the Lebesgue convergence theorem and obtain

$$\lim_n d(x_n, x) = \lim_n \int_S \min(1, \| x_n(s) - x(s) \|) \, dm(s) = \int_S 0 \, dm(s) = 0.$$

Suppose conversely that $d(x_n, x) \to 0$ in M, then for $0 < \varepsilon < 1$, if we set $A_n = \{s : \| x_n(s) - x(s) \| \geqslant \varepsilon\}$, we have

$$d(x_n, x) \geqslant \int_{A_n} \min(1, \| x_n(s) - x(s) \|) \, dm(s) \geqslant \int_{A_n} \varepsilon \, dm(s) = m(A_n)\varepsilon,$$

from which it follows that $m(A_n) \to 0$ or (1).

Furthermore Theorem D of Halmos [1], p. 93, may be carried over to Hilbert space valued functions

Theorem 24.3 *Let $x_n, x: S \to H$ be weakly measurable functions, and suppose that $x_n \to x$ in measure. Then these exists a subsequence x_{n_k} with $x_{n_k} \to x$ almost uniformly.*

Proof. For each k we can find some $\bar{n}(k)$ with

$$m\left\{s: \| x_n(s) - x(s) \| \geqslant \frac{1}{2^k}\right\} < \frac{1}{2^k} \quad \text{for } n \geqslant \bar{n}(k). \tag{5}$$

We put

$$n_1 = \bar{n}(1), \quad n_2 = \max(n_1 + 1, \bar{n}(2)), \quad n_3 = (\max(n_2 + 1, \bar{n}(3)), \ldots$$

and have $n_1 < n_2 < \cdots$, Thus x_{n_k} is a subsequence of x_n. We write

$$E_k := \left\{s: \| x_{n_k}^{(s)} - x(s) \| \geqslant \frac{1}{2^k}\right\},$$

so that because of (5)

$$m(E_k) < \frac{1}{2^k},$$

and for $s \in S \backslash E_k = CE_k$

$$\| x_{n_k}(s) - x(s) \| < \frac{1}{2^k}.$$

In order to conclude the proof we write $F_l := \bigcup_{k=l}^{\infty} E_k$, get

$$m(F_l) \leqslant \sum_{k=l}^{\infty} m(E_k) < \sum_{k=l}^{\infty} \tfrac{1}{2}k = \frac{1}{2^{l-1}},$$

while for $s \in S \backslash F_l = \bigcap_{k=l}^{\infty} CE_k$ we have

$$\| x_{n_k}(s) - x(s) \| < \frac{1}{2^k} \leqslant \frac{1}{2^l} \quad \text{for all } k \geqslant l,$$

with which almost uniform convergence is proved. ∎

After these preliminaries we can prove:

Theorem 24.4 *Let H be separable and $m(S) < \infty$. Then $\mathscr{M}_s = \mathscr{M}_E$.*

Proof. Since H is separable, the concepts of strong and weak measurability coincide. Suppose now that $x \in \mathcal{M}_s$. Then there exists a sequence of countably valued functions x_n with

$$x_n(s) \to x(s) \text{ strongly almost everywhere as } n \to \infty.$$

Because $m(S) < \infty$ Egorov's theorem (24.2) gives

$$x_n(s) \to x(s) \text{ almost uniformly}$$

Then, given (2), $x_n \to x$ in measure also. This convergence in measure is, as we have seen in (3), a metric convergence, and we have

$$d(x_n, x) \to 0 \quad \text{as } n \to \infty.$$

We show now that we can approximate each countably valued function almost uniformly by finite step functions. Suppose given

$$y(s) = \sum_{k=1}^{\infty} y_k \chi_{B_k}(s) \text{ with } B_k = y^{-1}(y_k) \text{ countably valued.}$$

We put $y_n(s) := \sum_{k=1}^{n} y_k \chi_{B_k}(s)$ (finite step functions), and have

$$y_n = y \text{ on } S \setminus \bigcup_{k=n+1}^{\infty} B_k.$$

Since $S = \bigcup_{k=1}^{\infty} B_k$ and $m(S) < \infty$ we conclude that for each $\varepsilon > 0$ there exists some $n_0(\varepsilon) \in \mathbb{N}$, such that for $n \geqslant n_0$

$$m\left(\bigcup_{k=n+1}^{\infty} B_k\right) = \sum_{k=n+1}^{\infty} m(B_k) < \varepsilon.$$

Therefore $y_n(s) \to y(s)$ almost uniformly. Suppose now that $n \in \mathbb{N}$ is chosen fixed. Then there exists a sequence $(y_m^{(n)})$, $m \in \mathbb{N}$ of finite step functions with

$$y_m^{(n)}(s) \xrightarrow[m \to \infty]{} x_n(s) \text{ almost uniformly,}$$

or, because of (2), in measure. We therefore have

$$d(x_n, x) \to 0 \quad (n \to \infty) \quad \text{and} \quad d(y_m^{(n)}, x_n) \to 0 \quad (m \to \infty),$$

that is, for each $k \in \mathbb{N}$ we can find numbers $n_k, m_k \in \mathbb{N}$ with

$$d(x_{n_k}, x) < \frac{1}{2k}, d(y_{m_k}^{(n_k)}, x_{n_k}) < \frac{1}{2k}.$$

From this it follows (and here decisively we use the triangle inequality of the metric) that

$$d(y_{m_k}^{(n_k)}, x) \leqslant d(y_{m_k}^{(n_k)}, x_{n_k}) + d(x_{n_k}, x) < \frac{1}{k}.$$

Therefore $d(y_{m_k}^{(n_k)}, x) \to 0$ as $k \to \infty$, that is, $y_{m_k}^{(n_k)} \to x$ in measure as $k \to \infty$. By Theorem 24.3 there exists a subsequence $(y'_m) \subseteq (y_{m_k}^{(n_k)})$, so that

$$y'_m \to x \text{ almost uniformly} \quad \text{as } m \to \infty,$$

from which it follows that $y'_m \to x$ almost everywhere as $m \to \infty$. Here the y'_m are finite step functions and so $x \in \mathcal{M}_E$.

We prove the last step. Let $F_n \in \mathcal{B}$ with $m(F_n) < 1/n$ and let $y'_m \to x$ uniformly on $S \backslash F_n$. If we put $F := \bigcap_{n=1}^{\infty} F_n$, then $m(F) \leqslant m(F_n) < 1/n$, hence $m(F) = 0$ and $y'_m(s) \to x(s)$ everywhere in $S \backslash F$. ■

In the following we assume always that H is separable.

Corollary 24.1 *Let $x(s)$, $y(s)$ be weakly measurable. Then the scalar valued function $(x(s), y(s))_H$ is measurable.*

Proof. The proof is valid without the assumption $m(S) < \infty$. Let x_n, y_n be sequences of countably valued functions with

$$\left.\begin{array}{l} x_n \to x \\ y_n \to y \end{array}\right. \quad \text{almost everywhere} \quad \text{as } n \to \infty.$$

Then $(x_n(s), y_n(s))$ is countably valued, therefore also measurable. By the continuity of the scalar product it follows that $(x_n(s), y_n(s)) \to (x(s), y(s))$ almost everywhere, and hence $(x(s), y(s))$ is measurable. ■

Corollary 24.2 *Let $x_n(s)$ be weakly measurable for all $n \in \mathbb{N}$. If $x_n(s) \rightharpoonup x(s)$ as $n \to \infty$ (weak convergence in H) it follows that $x(s)$ is weakly measurable.*

Proof. Let $(x_n(s), h) \to (x(s), h)$ for $h \in H$. Then $(x(s), h)$ is measurable, which is equivalent to the weak measurability of $x(s)$. ■

In a similar way we can define L^p spaces ($p \geqslant 1$) for functions with values in a Hilbert space $H, f: S \to H$.

Definition 24.2 *Let $\mathcal{M}_w(S; H) := \{f: S \to H, f$ weakly measurable$\}$. Then for $p \geqslant 1$ we set*

$$L^p(S, H) := \left\{ x : x \in \mathcal{M}_w(S, H) \quad \text{and} \quad \int_S \|x(s)\|^p \, dm(s) < \infty \right\}.$$

Entirely analogously to the usual proof (see for example Wloka [1], p. 52) we show:

Theorem 24.5 *Let $p \geqslant 1$. Then $L^p(S; H)$ is a Banach space. In particular $L^2(S; H)$ is a Hilbert space with respect to the scalar product*

$$[x, y] := \int_S (x(s), y(s))_H \, dm(s).$$

24.2 The Bochner integral

We come now to the definition of the Lebesgue–Bochner integral.

Definitions and Theorem 24.6 *Let H be a separable Hilbert space.*

(a) *Let E denote the set of finitely valued functions $x: S \to H$. E is a linear set and $E \subset L^l(S, H)$. If $x \in E$ we define*

$$\int_S x(s) \, dm(s) := \sum_{i=1}^{n} x_i m(B_i),$$

where $\operatorname{im}(x) = \{x_1, \ldots, x_n, 0\}$ and $B_i = x^{-1}(x_i)$ for $i = 1, \ldots, n$. The integral is linear and

$$\left\| \int_S x(s) \, dm(s) \right\| \leqslant \int_S \| x(s) \| \, dm(s).$$

(b) *We write $B^1(S, H) := \bar{E}^{L^1(S,H)} \subset L^1(S, H)$ and call $B^1(S, H)$ the set of Bochner integrable functions. If $x \in B^1(S, H)$ there exists a sequence $x_n \in E$ with $x_n \to x$ in $L^1(S, H)$ as $n \to \infty$. We put*

$$\int_S x(s) \, dm(s) = \lim_{n \to \infty} \int_S x_n(s) \, dm(s). \tag{6}$$

Proof. (a) is obvious, see Definition 24.1.6.

(b) We have to show that the limit in (6) exists and is independent of the choice of the sequence $x_n \in E$. $x_n \to x$ in $L^1(S, H)$ as $n \to \infty$ means that

$$\lim_{n \to \infty} \int_S \| x(s) - x_n(s) \| \, dm(s) = 0.$$

Given (a) we have

$$\left\| \int_S x_n(s) \, dm(s) - \int_S x_m(s) \, dm(s) \right\|$$

$$= \left\| \int_S (x_n(s) - x_m(s)) \, dm(s) \right\| \leqslant \int_S \| x_n(s) - x_m(s) \| \, dm(s)$$

$$\leqslant \int_S \|x_n(s) - x(s)\| \, dm(s) + \int_S \|x_m(s) - x(s)\| \, dm(s),$$

so that $\int_S x_n(s)\,dm(s)$ forms a Cauchy sequence in H and the limit exists. Suppose now that

$$\begin{array}{l} x_n(s) \to x(s) \\ y_n(s) \to x(s) \end{array} \quad \text{in } L^1(S; H) \quad \text{for } n \to \infty.$$

We define

$$z_n(s) = \begin{cases} x_m(s) & \text{if } n = 2m - 1, \\ y_m(s) & \text{if } n = 2m. \end{cases}$$

Clearly

$$z_n(s) \to x(s) \quad \text{in } L^1(S; H) \quad \text{for } n \to \infty.$$

The same estimate as above shows that $\int_S z_n(s)\,dm(s)$ is a Cauchy sequence in H, hence convergent. Therefore each subsequence converges to the same limit

$$\lim_{n\to\infty} \int_S x_n(s)\,dm(s) = \lim_{n\to\infty} \int_S z_n(s)\,dm(s) = \lim_{n\to\infty} \int_S y_n(s)\,dm(s). \qquad \blacksquare$$

Theorem 24.7 *Let* $x \in B^1(S, H)$, *then*

$$\left\| \int_S x(s)\,dm(s) \right\| \leqslant \int_S \|x(s)\|\,dm(s).$$

Proof. Let $x_n \in E$ be so chosen that $x_n(s) \to x(s)$ in $L^1(S, H)$ as $n \to \infty$. We then have

$$\begin{aligned} \left| \int_S \|x_n\|\,dm(s) - \int_S \|x\|\,dm(s) \right| &\leqslant \int_S \big| \, \|x_n\| - \|x\| \, \big| \, dm(s) \\ &\leqslant \int_S \|x_n - x\|\,dm(s) \xrightarrow[n\to\infty]{} 0. \end{aligned}$$

Therefore

$$\begin{aligned} \left\| \int_S x(s)\,dm(s) \right\| &= \left\| \lim_{n\to\infty} \int_S x_n(s)\,dm(s) \right\| = \lim_{n\to\infty} \left\| \int_S x_n(s)\,dm(s) \right\| \\ &\leqslant \lim_{n\to\infty} \int_S \|x_n(s)\|\,dm(s) = \int_S \|x(s)\|\,dm(s). \end{aligned} \qquad \blacksquare$$

The absolute continuity of the Bochner integral also follows from the last theorem.

Theorem 24.8 *We have $B^1(S,H)=L^1(S,H)$, that is, the function $x(s)$ is Bochner integrable if and only if the number valued function $\|x(s)\|$ is absolutely integrable.*

Proof. Suppose first that $m(S)<\infty$. By Theorem 24.6(b) we have $B^1\subset L^1$. We prove the inverse inclusion. Let $x\in L^1(S,H)$, since $m(S)<\infty$, by Theorem 24.4 there exists a sequence x_n of finitely valued functions with $x_n\to x$ almost everywhere as $n\to\infty$.

We write

$$y_n(s):=\begin{cases}x_n(s) & \text{if } \|x_n(s)\|\leqslant\frac{3}{2}\|x(s)\|,\\ 0 & \text{if } \|x_n(s)\|>\frac{3}{2}\|x(s)\|.\end{cases}$$

It follows that the y_n are again finitely valued,

$$\|y_n(s)\|\leqslant\tfrac{3}{2}\|x(s)\|\leqslant 2\|x(s)\|,$$

and

$$\lim_{n\to\infty}\|y_n(s)-x(s)\|=0 \quad \text{almost everywhere,}$$

for, given $s\in S$ there exists some n_0 with $y_{n_i}(s)=x_{n_0}(s)$. We choose n_0 in such a way that $\|x_{n_0}(s)-x(s)\|\leqslant\frac{1}{2}\|x(s)\|$. By the theorem of Lebesgue

$$\lim_{n\to\infty}\int_S\|x(s)-y_n(s)\|\,\mathrm{d}m(s)=\int_S\lim_{n\to\infty}\|x(s)-y_n(s)\|\,\mathrm{d}m(s)=0.$$

Since $y_n(s)$ is finitely valued it follows from this that $x\in B^1(S,H)$, with which we have proved the equality $B^1(S,H)=L^1(S,H)$ in the case that $m(S)<\infty$. Suppose now that $(S,\mathscr{B},m)$ is a σ-finite measure space. Let

$$S=\bigcup_{i=1}^{\infty}B_i \quad \text{with } B_i\cap B_j=\varnothing \quad \text{for } i\neq j \text{ and } m(B_i)<\infty.$$

For $x\in B^1(S;H)$ and $B\subset S$ measurable, we define

$$\int_B x(s)\,\mathrm{d}m(s):=\int_S x(s)\chi_B(s)\,\mathrm{d}m(s).$$

Let $A_n=\bigcup_{i=1}^n B_i$, then for $x\in B^1(S;H)$ we have

$$\int_{A_n}x(s)\,\mathrm{d}m(s)=\int_S x(s)\chi_{A_n}(s)\,\mathrm{d}m(s)=\int_S x(s)\sum_{i=1}^n\chi_{B_i}(s)\,\mathrm{d}m(s)$$

$$=\sum_{i=1}^n\int_S x(s)\chi_{B_i}(s)\,\mathrm{d}m(s)=\sum_{i=1}^n\int_{B_i}x(s)\,\mathrm{d}m(s).$$

Let $x \in L^1(S; H)$, then $\chi_{B_i} \cdot x \in L^1(B_i; H)$, because

$$\int_{B_i} \| \chi_{B_i} \cdot x \| \, dm(s) = \int_{B_i} \chi_{B_i} \| x \| \, dm(s) = \int_S \chi_{B_i} \cdot \| x \| \, dm(s) \leqslant \int_S \| x \| \, dm < \infty,$$

and by the first part of our proof

$$\chi_{B_i} \cdot x \in B^1(B_i; H) \quad \text{for all } i \in \mathbb{N},$$

hence also

$$\chi_{A_n} \cdot x \in B^1(A_n; H) = B^1\left(\bigcup_{i=1}^{n} B_i; H \right) \quad \text{for all } n \in \mathbb{N}.$$

For each $\varepsilon > 0$ there therefore exists some finitely valued function x_ε with

$$\int_{A_n} \| \chi_{A_n} x - \chi_{A_n} x_\varepsilon \| \, dm(s) < \varepsilon. \tag{7}$$

Suppose now that $\varepsilon > 0$ is given, we choose $n_\varepsilon =: n$ so large that

$$\int_{CA_n} \| x \| \, dm < \varepsilon,$$

which is always possible given the absolute convergence of the series

$$\sum_{i=1}^{\infty} \int_{B_i} \| x \| \, dm = \int_S \| x \| \, dm < \infty,$$

and where we set

$$\int_{CA_n} \| x \| \, dm = \sum_{i=n+1}^{\infty} \int_{B_i} \| x \| \, dm.$$

Using (7) we now choose a finitely valued function $\chi_{A_n} \cdot x_\varepsilon$ and have

$$\int_S \| x - \chi_{A_{n_\varepsilon}} x_\varepsilon \| \, dm = \int_{A_{n_\varepsilon}} \| x - \chi_{A_n} x_\varepsilon \| \, dm + \int_{CA_n} \| x - 0 \| \, dm \leqslant \varepsilon + \varepsilon. \tag{8}$$

(8) implies that x can be approximated by finitely valued functions in L^1, that is, $x \in B^1(S, H)$. ■

Theorem 24.9 *Let $x \in L^p(B, H)$ for $p > 1$ and $m(B) < \infty$. Then $x \in L^1(B, H)$, and also $\chi_B \cdot x \in L^1(S, H)$, so that $\int_B \| x(s) \| \, dm(s) < \infty$, which is local integrability.*

Proof. It follows by the Hölder inequality from the assumptions that

$$\int_B \| x(s) \| \, dm(s) \leqslant \left(\int_B \| x(s) \|^p \, dm(s) \right)^{1/p} \left(\int_B dm(s) \right)^{1/q} < \infty$$

where $1/p + 1/q = 1$. ■

As a conclusion to this introduction to the Bochner integral we present an alternative definition:

We define $\mathscr{M}_E$ somewhat differently, here again suppose that S is σ-finite: $x \in \tilde{\mathscr{M}}_E$ if and only if for each $B \in \mathscr{B}$ with $m(B) < \infty$ there exists a sequence y_n^B of step functions on B with $y_n^B \to x$ almost everywhere on B for $n \to \infty$. Then again

$$\tilde{\mathscr{M}}_E \subset \mathscr{M}_s,$$

for if $S = \bigcup_{i=1}^{\infty} B_i$ with $m(B_i) < \infty$ and $B_i \cap B_j = \emptyset$ for $i \neq j$, then

$$x_n(s) = \sum_{i=1}^{\infty} y_n^{B_i}(s) \chi_{B_i}(s)$$

is countably valued and

$$x_n(s) \to x(s) \text{ almost everywhere} \quad \text{as } n \to \infty.$$

The reverse inclusion is also true, for let $x \in \mathscr{M}_s$, $B \subset S$ with $m(B) < \infty$. Then there exists a sequence x_n of countably valued function with

$$x_n \to x \text{ almost everywhere on } S \quad \text{as } n \to \infty.$$

In particular

$$x_n \to x \text{ almost everywhere on } B \quad \text{as } n \to \infty.$$

In this situation we apply Theorem 24.4 and know that there exists a sequence y_n of finitely valued functions with $y_n \to x$ almost everywhere on B as $n \to \infty$, therefore $x \in \tilde{\mathscr{M}}_E$. Putting everything together we obtain $\mathscr{M}_s = \tilde{\mathscr{M}}_E$.

For $x \in L^1(S, H)$ we define the Bochner integral over B ($m(B) < \infty$) by

$$\int_B x(s)\, dm(s) = \lim_{n \to \infty} \int_B x_n(s)\, dm(s),$$

where x_n is finitely valued and $x_n \to x$ in $L^1(B, H)$ as $n \to \infty$.

This integral exists and is independent of the choice of the sequence x_n. We now define the integral for $x \in L^1(S, H)$ over S in a similar manner by

$$\int_S x(s)\, dm(s) := \sum_{k=1}^{\infty} \int_{B_k} x(s)\, dm(s),$$

where

$$S = \bigcup_{k=1}^{\infty} B_k, \quad m(B_k) < \infty \quad \text{and} \quad B_i \cap B_j = \emptyset \quad \text{for } i \neq j$$

we show that the integral exists. We have

$$\left\|\int_{B_k} x(s)\,\mathrm{d}m(s)\right\| \leqslant \int_{B_k} \|x(s)\|\,\mathrm{d}m(s)$$

and therefore

$$\sum_{k=1}^{\infty}\left\|\int_{B_k} x(s)\,\mathrm{d}m(s)\right\| \leqslant \sum_{k=1}^{\infty}\int_{B_k} \|x(s)\|\,\mathrm{d}m(s) = \int_S \|x(s)\|\,\mathrm{d}m(s) < \infty,$$

that is, the series is absolutely convergent. It remains to show that the integral is independent of the choice of the cover $\{B_k\}$. Let

$$S = \bigcup_{k=1}^{\infty} B_k = \bigcup_{n=1}^{\infty} A_n.$$

We set

$$C_{kn} = B_k \cap A_n.$$

Then because of the absolute convergence of the series, we have

$$\sum_{k\geqslant 1}\int_{B_k} x(s)\,\mathrm{d}m(s) = \sum_{k\geqslant 1}\sum_{n\geqslant 1}\int_{C_{kn}} x(s)\,\mathrm{d}m(s) = \sum_{n\geqslant 1}\sum_{k\geqslant 1}\int_{C_{kn}} x(s)\,\mathrm{d}m(s)$$

$$= \sum_{n\geqslant 1}\int_{A_n} x(s)\,\mathrm{d}m(s).$$

Exercises

24.1 Let X be a Banach space and $f:[0,T]\to X$ a continuous function. Define the Riemann integral $\int_0^T f(t)\,\mathrm{d}t$, and demonstrate its existence and basic properties, for example

$$\left\|\int_0^T f(t)\,\mathrm{d}t\right\| \leqslant \int_0^T \|f(t)\|\,\mathrm{d}t.$$

Show that the linear functional on X', $x'\mapsto\int_0^T\langle x', f(t)\rangle\mathrm{d}t$ defines a unique element $y\in X$ with $y=\int_0^T f(t)\,\mathrm{d}t$.

24.2 Show that if the function $f:[0,1]\to X$ is continuous, then so also are the number valued functions $\langle x', f(t)\rangle$, for all $x'\in X'$, but not conversely. Counterexample: let H be a Hilbert space ($\dim H = +\infty$) with orthonormal basis $e_1,\ldots,e_n,\ldots$, we define the function $g:[0,1]\to H$ in the following way

$$g(0) = 0$$

$$g(t) = (1-\tau)e_{n+1} + \tau e_n \quad \text{for } t = \tau\frac{1}{n} + (1-\tau)\frac{1}{n+1}, \text{ where } 0\leqslant\tau\leqslant 1.$$

Show that the functions $\langle x', g(t)\rangle$, $x'\in H' = H$ are all continuous, but not the function $g:[0,1]\to H$. There is a discontinuity at $t=0$.

24.3 Let X, Y be Banach spaces and $f:[0,1]\to L(X,Y)$ a C^m-function. Show that the transposed function ${}^{T}f:[0,1]\to L(Y',X')$ is also a C^m-function. ($C^m = m$-fold continuously differentiable.)

24.4 Let $f:[0,1]\to X$ be given, $m\geqslant 1$, and $\langle x', f(t)\rangle$ of class C^m for each $x'\in X'$. Show that $f:[0,1]\to X$ is then a function of class C^{m-1}.

24.5 Let H_1, H_2 be Hilbert spaces, let $(S,\mathscr{B},m)$ be a measure space with $m(S)<\infty$. Let the map $A:S\to L(H_1,H_2)$ be measurable in the following sense: for each $h_1\in H_1$ and $h_2\in H_2$ the number valued function $\langle A(s)h_1,h_2\rangle$ is $(S,\mathscr{B},m)$-measurable. If the function $f:S\to H_1$ is measurable, is the function $A(s)f(s)$ also measurable in H_2?

§25 Distributions with values in a Hilbert space H and the space $W(0,T)$

For later use we introduce distributions with values in a Hilbert space H, and corresponding Sobolev spaces.

Definition 25.1 *Let Ω be open in $\mathbb{R}^r$, let H be a Hilbert space. We call a linear map T*

$$T:\mathscr{D}(\Omega)\to H$$

a distribution from $\mathscr{D}'(\Omega,H)$, if for each compact subset $K\subset\subset\Omega$ there exist constants $p,c\geqslant 0$ with

$$\|\langle T,\varphi\rangle\|\leqslant c\cdot\sup_{x\in K}\sum_{|s|\leqslant p}|\mathrm{D}^s\varphi(x)|,\quad \varphi\in\mathscr{D}(\Omega)\ \text{with}\ \operatorname{supp}\varphi\subset K.$$

We define differentiation and convergence of distributions in the known way.

Definition 25.2

1. *The derivative D^sT, $s=(s_1,\ldots,s_r)$ with $s_i\geqslant 0$ of a distribution T is defined by*

$$\langle \mathrm{D}^sT,\varphi\rangle := (-1)^{|s|}\langle T,\mathrm{D}^s\varphi\rangle.$$

2. *Let T_n be a sequence of distributions from $\mathscr{D}'(\Omega,H)$ and $T\in\mathscr{D}'(\Omega,H)$. We say that $T_n\to T$ as $n\to\infty$ in $\mathscr{D}'(\Omega,H)$, if for each $\varphi\in\mathscr{D}(\Omega)$ we have $\langle T_n,\varphi\rangle\to\langle T,\varphi\rangle$ in H as $n\to\infty$.*

We easily prove:

Theorem 25.1

(a) *If $T\in\mathscr{D}'(\Omega,H)$, then $\mathrm{D}^sT\in\mathscr{D}'(\Omega,H)$.*
(b) *If $T_n\to T$ in $\mathscr{D}'(\Omega,H)$ as $n\to\infty$, then $\mathrm{D}^sT_n\to\mathrm{D}^sT$ in $\mathscr{D}'(\Omega,H)$ as $n\to\infty$.*

From now on suppose that H is a separable Hilbert space, Ω open in $\mathbb{R}^r$, and $(\Omega, \mathscr{B}, m = dx)$ the usual Lebesgue measure space.

Let $f: \Omega \to H$ be locally integrable, that is, for all $K \subset\subset \Omega \subset \mathbb{R}^r$, K compact, $f \in L^1(K, H)$. To this f we associate a distribution from $\mathscr{D}'(\Omega, H)$

$$f \to T_f \text{ by } \langle T_f, \varphi \rangle := \int_\Omega f(x)\varphi(x)\,dx, \quad \varphi \in \mathscr{D}(\Omega),$$

where we understand the integral as the Bochner integral. Because

$$\|\langle T_f, \varphi \rangle\| = \left\| \int_\Omega f(x)\varphi(x)\,dx \right\| = \left\| \int_{\operatorname{supp}\varphi} f(x)\varphi(x)\,dx \right\|$$

$$\leqslant \sup_\Omega |\varphi(x)| \cdot \int_{\operatorname{supp}\varphi} \|f(x)\|\,dx = c \cdot \sup_\Omega |\varphi(x)|$$

we have $T_f \in \mathscr{D}'(\Omega; H)$.

Theorem 25.2 *The map $L^2(\Omega, H) \to \mathscr{D}'(\Omega, H)$ is injective and continuous.*

Proof. It is clear that the map $L^2(\Omega, H) \to \mathscr{D}'(\Omega, H)$ is defined from Theorem 24.9 and the definition above. To prove continuity of the map, let $f_n \to f$ in $L^2(\Omega, H)$ as $n \to \infty$ and let $\varphi \in \mathscr{D}(\Omega)$ with $\operatorname{supp}\varphi = K$. Then we have

$$\|\langle T_{f_n} - T_f, \varphi \rangle\| = \left\| \int_K (f_n(x) - f(x))\varphi(x)\,dx \right\| \leqslant \int_K \|f_n(x) - f(x)\|\,|\varphi(x)|\,dx$$

$$\leqslant \sup_{x \in K} |\varphi(x)| \left(\int_K \|f_n(x) - f(x)\|^2\,dx \right)^{1/2} \cdot (\operatorname{mes} K)^{1/2}$$

$$\leqslant \sup_{x \in K} |\varphi(x)| (\operatorname{mes} K)^{1/2} \left(\int_\Omega \|f_n(x) - f(x)\|^2\,dx \right)^{1/2},$$

therefore $T_{f_n} \to T_f$ in $\mathscr{D}'(\Omega; H)$ as $n \to \infty$.

Injectivity of the map: Let $\{h_i : i \in \mathbb{N}\}$ be an orthonormal basis for H. Then

$$f(x) = \sum_{i=1}^\infty (f(x), h_i)h_i.$$

Suppose now

$$\int_\Omega f(x)\varphi(x)\,dx = \int_\Omega \sum_{i=1}^\infty (f(x), h_i)h_i\varphi(x)\,dx = 0 \quad \text{for all } \varphi \in \mathscr{D}(\Omega),$$

then because $\sum$ and $\int$ commute

$$\int_\Omega (f(x), h_i)\varphi(x)\,dx = 0 \quad \text{for all } i \in \mathbb{N} \text{ for all } \varphi \in \mathscr{D}(\Omega)$$

since $(f(x), h_i) \in L^2(\Omega)$ and $\mathscr{D}(\Omega)$ is dense in $L^2(\Omega)$ it follows that

$$(f(x), h_i) = 0 \quad i \in \mathbb{N},$$

and hence $f(x) = 0$. ■

Theorem 25.3 *Let H_1, H_2 be Hilbert spaces, $H_1 \subsetneq H_2$ and suppose that the inclusion is continuous then $\mathscr{D}'(\Omega; H_1) \subsetneq \mathscr{D}'(\Omega; H_2)$ and the inclusion is continuous.*

Proof. We extend $T: \mathscr{D}(\Omega) \to H_1 \to H_2$. As an important tool for the study of the parabolic differential equation we introduce the Sobolev space $W_2^1(0, T)$. Suppose now that $\Omega = (0, T) \subset \mathbb{R}^r$, $0 < T < \infty$, and that V, H are separate Hilbert spaces such that $V \subsetneq H \subsetneq V'$ forms a Gelfand triple. Then V' is also a Hilbert space, see Theorem 17.3. For $f \in L^2((0, T); V)$, because of Theorems 25.3, 25.1, $f \in \mathscr{D}'((0, T), V)$, from which it follows that

$$\frac{\mathrm{d}f}{\mathrm{d}t} \in \mathscr{D}'((0, T); V) \text{ and also } \frac{\mathrm{d}f}{\mathrm{d}t} \in \mathscr{D}'((0, T); V').$$

Definition 25.3 *We define*

$$W_2^1(0, T) := \left\{f : f \in L^2((0, T); V), \frac{\mathrm{d}f}{\mathrm{d}t} \in L^2((0, T); V')\right\}$$

with scalar product given by

$$(f, g)_W := \int_0^T (f(t), g(t))_V \, \mathrm{d}t + \int_0^T \left(\frac{\mathrm{d}f(t)}{\mathrm{d}t}, \frac{\mathrm{d}g(t)}{\mathrm{d}t}\right)_{V'} \mathrm{d}t.$$

We shall omit the affixes in $W_2^1(0, T)$ and simply write $W(0, T)$.

Theorem 25.4 *$W(0, T)$ is a Hilbert space.*

Proof. We need only show that $W(0, T)$ is complete. Let f_n be a Cauchy sequence in $W(0, T)$. By definition of the norm this is equivalent to f_n is a Cauchy sequence in $L^2((0, T); V)$ and $\mathrm{d}f_n/\mathrm{d}t$ is a Cauchy sequence in $L^2((0, T), V)$. It follows that:

(a) $f_n(t) \to f_0(t)$ in $L^2((0, T); V)$

and

(b) $\dfrac{\mathrm{d}f_n}{\mathrm{d}t} \to f_1(t)$ in $L^2((0, T), V')$.

We know that $\mathrm{d}f_0/\mathrm{d}t$ exists in the distributional sense. It remains to show that $\mathrm{d}f_0/\mathrm{d}t = f_1$. From $L^2((0, T); V) \subsetneq \mathscr{D}'((0, T); V) \subsetneq \mathscr{D}'((0, T); V')$ and

$L^2((0, T), V') \subsetneq \mathscr{D}'((0, T); V')$ and because of (a), we have

$$f_n \xrightarrow[(n\to\infty)]{} f_0 \quad \text{in } \mathscr{D}'((0, T); V).$$

Then because of Theorems 25.3 and 25.1

$$\frac{\mathrm{d}f_n}{\mathrm{d}t} \xrightarrow[(n\to\infty)]{} \frac{\mathrm{d}f_0}{\mathrm{d}t} \quad \text{in } \mathscr{D}'((0, T); V').$$

However, because of (b) it is also true that

$$\frac{\mathrm{d}f_n}{\mathrm{d}t} \xrightarrow[(n\to\infty)]{} f_1 \quad \text{in } \mathscr{D}'((0, T); V').$$

From the uniqueness of limits in $\mathscr{D}'((0, T); V')$ it follows that

$$\frac{\mathrm{d}f_0}{\mathrm{d}t} = f_1 \in L^2((0, T); V'). \qquad \blacksquare$$

We prove a lemma:

Lemma 25.1 *Let $\mathscr{D}((-\infty, \infty); V) = \mathscr{D}(V)$ be the space of all functions $f: R \to V$ (with values in V) which are infinitely differentiable and have a compact support. Then the restriction $\mathscr{D}(V)|_{[0,T]}$ is dense in $W(0, T)$.*

Proof. See Theorem 3.6. We take two scalar functions $\alpha, \beta \in \mathscr{D}$ with $\alpha(t) + \beta(t) = 1$ and $\alpha(t) = 0$ in a neighbourhood of T, respectively $\beta(t) = 0$ in a neighbourhood of 0; we can find α, β by a partition of unity. Let $u \in W(0, T)$, we consider the function αu, shift it to the left: $(\alpha u)_h = \alpha(t + h)u(t + h)$, $h > 0$, and extend it by zero on the right. We have

1. $\operatorname{supp}(\alpha u)_h \subset [-h, T - \varepsilon_0 - h]$,
2. $(\alpha u)_h \in W(0, T)$
3. $(\alpha u)_h \to \alpha u$ in $W(0, T)$ as $h \to 0$.

The third property is the continuity in the mean of the L^2-norms, which we obtain by approximation by step functions, see Theorem 24.4.

Thus it is sufficient to approximate functions of the form $(\alpha u)_h$ by $\mathscr{D}(V)$-functions. Let

$$\psi(t) := \begin{cases} 1 \text{ for } t \in [-h/2, +\infty), \\ 0 \text{ for } t \in (-\infty, -h], \\ \geqslant 0 \text{ for } t \in (-h, -h/2), \end{cases} \text{ be a function from } C^\infty(-\infty, +\infty)$$

the function $\psi(\alpha u)_h$ belongs to $W(-\infty, +\infty)$ and we have $\operatorname{supp}[\psi(\alpha u)_h] \subset$

$(-h, T-\varepsilon_0-h)$, and for its restriction

$$\psi(\alpha u)_h|_{[0,T]} = (\alpha u)_h.$$

Regularising by $\varphi_\varepsilon \in \mathscr{D}$ we find, see Theorem 1.3,

$$f_\varepsilon = [\psi(\alpha u)_h]^* \varphi_\varepsilon \in \mathscr{D}(V), \quad \operatorname{supp} f_\varepsilon \subset (-h-, T-\varepsilon_0-h+\varepsilon),$$

$$f_\varepsilon \to \psi(\alpha u)_h \quad \text{in } L^2((-\infty, +\infty); V)$$

and

$$\frac{\mathrm{d}f_\varepsilon}{\mathrm{d}t} = \frac{\mathrm{d}}{\mathrm{d}t}[\psi(\alpha u_h)]^* \varphi_\varepsilon \to \frac{\mathrm{d}}{\mathrm{d}t}[\psi(\alpha u)_h] \quad \text{in } L^2((-\infty, +\infty); V'),$$

which is $f_\varepsilon \to \psi(\alpha u)_h$ in $W(-\infty, \infty)$. Since for the norms in $W(0, T)$ and $W(-\infty, \infty)$ we have $\int_0^T \|\cdot\| \leqslant \int_{-\infty}^{\infty} \|\cdot\|$, by restricting to $[0, T]$ we obtain

$$\mathscr{D}(V)|_{[0,T]} \ni f_\varepsilon \to (\alpha u)_h \quad \text{in } W(0, T).$$

Treating βu analogously, we have proved our lemma. ■

Theorem 25.5 *We have the continuous imbedding*

$$W(0, T) \subsetneq C([0, T], H),$$

or

$$\sup_{t\in[0,T]} \|u(t)\|_H \leqslant c\|u\|_{W(0,T)}, \quad u \in W(0, T), \tag{1}$$

which implies that the functions from $W(0, T)$ are continuous with values in H – eventually after changing each function on a set of measure 0.

Proof. See Theorem 6.2. By Lemma 25.1 it is sufficient to prove estimate (1) for functions $f \in \mathscr{D}(V)|_{[0,T]}$. We have for such fs the formula

$$\frac{\mathrm{d}}{\mathrm{d}t}\|f(t)\|_H^2 = \left(\frac{\mathrm{d}f(t)}{\mathrm{d}t}, f(t)\right)_H + \left(f(t), \frac{\mathrm{d}f(t)}{\mathrm{d}t}\right)_H = 2\mathrm{Re}\left(f(t), \frac{\mathrm{d}f(t)}{\mathrm{d}t}\right)_H$$

or

$$2\mathrm{Re}\int_{t_0}^{t}\left(f(\tau), \frac{\mathrm{d}f(\tau)}{\mathrm{d}\tau}\right)_H \mathrm{d}\tau = \|f(t)\|_H^2 - \|f(t_0)\|_H^2. \tag{2}$$

Let e_T be the following scalar function:

$$e_T(t) = \begin{cases} 1 \text{ for } t = 0, \\ 0 \text{ for } t \in (-\infty, -T/4] \cup [T/4, +\infty), \\ \geqslant 0 \text{ for } t \in (-T/4, T/4). \end{cases}$$

Putting $t_0 := t + T/4$ for $t \in [0, T/2]$ and $t_0 := t - T/4$ for $t \in (T/2, T]$, and multiplying $f(\tau)$ by $e_T(t-\tau)$ we obtain from (2)

$$\|f(t)\|_H^2 = \pm 2\mathrm{Re}\int_{t_0}^{t} \Big(e_T(t-\tau)f(\tau), \frac{\mathrm{d}}{\mathrm{d}\tau}[e_T(t-\tau)f(\tau)]\Big)_H \mathrm{d}\tau. \qquad (3)$$

Using the Gelfand properties of $V \subset H \subset V'$, we see that (3) yields the estimate

$$\begin{aligned}\|f(t)\|_H^2 &\leqslant 2\int_0^T \Big|\Big(e_R f, \frac{\mathrm{d}}{\mathrm{d}\tau}e_R f\Big)_H\Big| \mathrm{d}\tau \\ &= 2\int_0^T \left| |e_R|^2\left(f, \frac{\mathrm{d}f}{\mathrm{d}\tau}\right)_H + |e_R\cdot e_R'|(f,f)_H\right| \mathrm{d}\tau \\ &\leqslant 2c_1\int_0^T \|f\|_V \cdot \left\|\frac{\mathrm{d}f}{\mathrm{d}\tau}\right\|_{V'} \mathrm{d}\tau + 2c_2\int_0^T \|f\|_V^2 \,\mathrm{d}\tau \\ &\leqslant c_3\left[\int_0^T \|f(\tau)\|_V^2\,\mathrm{d}\tau + \int_0^T \left\|\frac{\mathrm{d}f(\tau)}{\mathrm{d}\tau}\right\|_{V'}^2 \mathrm{d}\tau\right] \\ &= c_3\|f\|_{W(0,T)}^2 \quad \text{for } t\in[0,T] \text{ and } f\in\mathscr{D}(V)|_{[0,T]}. \qquad (4)\end{aligned}$$

■

But (4) is (1).

§26 The existence and uniqueness of the solution of a parabolic differential equation

We introduce an abstract solution theorem for parabolic equations, we resume the discussion of §17. Suppose given two Hilbert spaces V, H with $V \subsetneq H$ injective, continuous and dense, let H and V be separable. By §17, see Definition 17.1, we can extend to a Gelfand triple $V \subsetneq H \subsetneq V'$.

Let $\infty > T > 0$. For $t\in[0, T]$ and $\varphi, \psi\in V$ suppose that the form $a(t; \varphi, \psi)$, sesquilinear in φ, ψ, is given. We require:

(a) $a(t; \varphi, \psi)$ is measurable on $[0, T]$, for fixed $\varphi, \psi\in V$.
(b) There exists some $c > 0$, independent of t, with

$$|a(t; \varphi, \psi)| \leqslant c\|\varphi\|_V \cdot \|\psi\|_V \quad \text{for all } t\in[0, T], \quad \varphi, \psi\in V.$$

(c) There exist real $k_0, \alpha \geqslant 0$ independent of t and φ, with

$$\mathrm{Re}\, a(t; \varphi, \varphi) + k_0\|\varphi\|_H^2 \geqslant \alpha\|\varphi\|_V^2 \quad \text{for all } t\in[0, T], \quad \varphi\in V.$$

By Theorem 17.9, (28), (b) shows the existence of a representation operator $L(t(t)$: $a(t; \varphi, \psi) = (L(t)\varphi, \psi)_H$, where $L(t)$: $V \to V'$ is continuous and linear (for fixed t). However, under our assumptions much more is true.

Lemma 26.1 *Suppose that the assumptions* (a) *and* (b) *are satisfied. Then the* *action of the representation operator $L(t)$ of $a(t;\varphi,\psi)$ is continuous and linear as a map*

$$L: L^2((0,T);V)\to L^2((0,T);V'). \tag{1}$$

Here for $f\in L^2((0,T);V)$ we denote by $L(f)$ the function $t\mapsto L(t)(f(t))\in V'$.

Proof. We show first that the function $L(t)(f(t))$ is measurable. Since by assumption V is separable, $T<\infty$ and $f(t)$ $(\in V)$ is measurable, by Theorem 24.4 there exists a sequence $y_n(t)$ of finitely valued functions with $y_n\to f$ almost everywhere in V as $n\to\infty$. Let $y_n(t)=\sum_{k=1}^{m_n} y_k\chi_{B_k}(t)$ with the y_k constants, then for arbitrary $v\in V$ we have

$$(L(t)y_n(t),v)_H=\sum_{k=1}^{m_n}(L(t)y_k,v)_H\chi_{B_k}(t).$$

By (a) $(L(t)y_k,v)_H=a(t;y_k,v)$ is measurable, and hence so is $(L(t)y_n(t),v)_H$. By (b) we have

$$|(L(t)(y_n(t)-f(t)),v)_H|\leqslant c\,\|y_n(t)-f(t)\|_V\cdot\|v\|_V,$$

that is, $(L(t)y_n(t),v)_H\to(L(t)f(t),v)_H$ almost everywhere, so that $(L(t)f(t),v)_H$ is measurable. Therefore by Theorem 24.4 so is $L(t)f(t)$. A similar argument also gives the measurability of $L(t)\in L(V,V')$. Moreover for $L(t)f(t)\in V'$ we have

$$\|L(t)f(t)\|_{V'}=\sup_{\substack{u\in V\\ \|u\|_V\leqslant 1}}|(L(t)f(t),u)_H|\leqslant c\,\|f(t)\|_V,$$

where (b) is used, therefore

$$\int_0^T\|L(t)f(t)\|_{V'}^2\,\mathrm{d}t\leqslant c^2\int_0^T\|f(t)\|_V^2\,\mathrm{d}t,$$

that is, $L(t): L^2((0,T);V)\to L^2((0,T);V')$ is continuous and linear. ■

Now we consider the following problem (parabolic equation). Suppose given $f\in L^2((0,T),V')$ and $y_0\in H$, then we look for a function

$$y\in W(0,T)$$

with

$$L(t)y+\frac{\mathrm{d}y}{\mathrm{d}t}=f\quad \textit{in } V'\textit{, that is, } (L(t)y,v)_H+\left(\frac{\mathrm{d}y}{\mathrm{d}t},v\right)_H=(f,v)_H,\quad v\in V,$$

and

$$y(0)=y_0. \tag{P}$$

We prove the abstract existence theorem.

Theorem 26.1 *Suppose that the hypotheses* (a), (b) *and* (c) *are satisfied. Then if* $T < \infty$ *the problem* (P) *has a unique solution* y, *and this depends continuously on* f and y_0, *that is, the map*

$$(f, y_0) \mapsto y, \quad y \text{ the solution of (P)},$$

is continuous from $L^2((0, T); V') \times H$ *into* $W(0, T)$.

Proof. We observe first of all that if $T < \infty$ we can set $k_0 = 0$ in condition (c). Let $y \in W(0, T)$ be a solution of (P), we set

$$z(t) := y(t) \cdot \exp(-k_0 t) \in W(0, T)$$

and have

$$\frac{dz(t)}{dt} = \frac{dy(t)}{dt} \cdot \exp(-k_0 t) - k_0 z(t),$$

and also

$$(L(t) + k_0 E)z + \frac{dz}{dt} = \exp(-k_0 t) f.$$

We take $L(t) + k_0 E$ instead of $L(t)$ and have

$$\operatorname{Re}((L(t) + k_0 E)\varphi, \varphi)_H \geqslant \alpha \|\varphi\|_V^2 \quad \text{for all } \varphi \in V \quad \text{for all} \quad t \in (0, T).$$

We show first the uniqueness of the solution: Let y be a solution of (P) in $W(0, T)$ with $f = 0$, $y_0 = 0$. For $y(t) \in V$ we have

$$(L(t)y(t), y(t))_H + \left(\frac{dy(t)}{dt}, y(t)\right)_H = 0,$$

from which it follows that

$$\operatorname{Re} \int_0^T a(t; y(t), y(t))\, dt + \operatorname{Re} \int_0^T \left(\frac{dy(t)}{dt}, y(t)\right)_H dt = 0.$$

since

$$\operatorname{Re} \int_0^T \left(\frac{dy(t)}{dt}, y(t)\right)_H dt = \tfrac{1}{2}(\|y(T)\|_H^2 - \|y(0)\|_H^2)$$

see proof of Theorem 25.5 and $y(0) = y_0 = 0$, it follows that

$$\int_0^T \operatorname{Re} a(t; y(t), y(t))\, dt + \tfrac{1}{2}\|y(T)\|_H^2 = 0,$$

and applying (c) one obtains

$$\int_0^T \alpha \|y(t)\|_V^2\, dt + \tfrac{1}{2}\|y(T)\|_H^2 \leqslant 0,$$

from which it follows that

$$y(t) = 0 \text{ almost everywhere,}$$

that is, $y(t) \equiv 0$, since $y \in C^0[0, T]$ by Theorem 25.5.
Existence of the solution. We carry out the proof in several steps:

(i) Approximation of the solution by a sequence $y_m(t)$.
(ii) We show $\| y_m \|_{L^2((0,T),V)} \leqslant K$.

Then by Theorem 9.1 there exists an element $z \in L^2((0, T); V)$ so that without loss of generality y_m weakly converges to z.

(iii) We show that z is a solution of (P).

Finally we show that

(iv) y_m strongly converges to z in $L^2((0, T); V)$.

The details:

Step (i). Let $\{w_m : m \in \mathbb{N}\}$ be linearly independent and total in V, for example an orthonormal basis. We define $y_m(t) = \sum_{i=1}^m g_{im}(t) w_i$, where the $g_{im}(t)$ are so chosen that the ordinary differential equations

$$\left(\frac{d}{dt} y_m(t), w_j\right)_H + (L(t) y_m(t), w_j)_H = (f(t), w_j)_H, \tag{2}$$

are satisfied for $1 \leqslant j \leqslant m$ together with the initial condition

$$y_m(0) = y_{0m} = \sum_{i=1}^m \xi_{im} w_i,$$

where

$$y_{0m} = \sum_{i=1}^m \xi_{im} w_i \to y_0 \quad \text{in } H \text{ as } m \to \infty.$$

(2) is a system of m ordinary linear differential equations, which may be written in the form

$$\mathscr{V}_m \frac{dG_m}{dt} + \mathscr{A}_m(t) G_m = F_m, \quad G_m(0) = (\xi_{1m}, \ldots, \xi_{mm})$$

where $\mathscr{V}_m = ((w_i, w_j)_H)$ is the Gram matrix $1 \leqslant j, i \leqslant m$

$$\mathscr{A}_m(t) = (a(t; w_i, w_j)) \qquad 1 \leqslant i, j \leqslant m$$

$$F_m(t) = ((f(t), w_j)_H) \qquad j = 1, \ldots, m,$$

and

$$G_m(t) = (g_{im}(t)) \qquad i = 1, \ldots, m.$$

Since $\det \mathscr{V}_m \neq 0$, there exists a unique solution G_m. Therefore y_m is uniquely determined.

Step (ii). Multiply (2) by $g_{jm}(t)$ and sum over j for $1 \leqslant j \leqslant m$, obtaining

$$\left(\frac{\mathrm{d}}{\mathrm{d}t} y_m(t), y_m(t)\right)_H + a(t; y_m(t), y_m(t)) = (f(t), y_m(t))_H$$

and by integration

$$\int_0^T \mathrm{Re}\left(\frac{\mathrm{d}}{\mathrm{d}t} y_m(t), y_m(t)\right)_H \mathrm{d}t + \int_0^T \mathrm{Re}\, a(t; y_m(t), y_m(t))\,\mathrm{d}t = \int_0^T \mathrm{Re}\,(f(t), y_m(t))_H\,\mathrm{d}t,$$

from which it follows

$$\tfrac{1}{2}(\| y_m(T)\|_H^2 - \| y_m(0)\|_H^2) + \int_0^T \mathrm{Re}\, a(t; y_m(t), y_m(t))\mathrm{d}t = \int_0^T \mathrm{Re}\,(f(t), y_m(t))_H\,\mathrm{d}t,$$

and rearranged

$$\| y_m(T)\|_H^2 + 2\int_0^T \mathrm{Re}\, a(t; y_m(t), y_m(t))\,\mathrm{d}t \leqslant \| y_m(0)\|_H^2 + 2\int_0^T |(f(t), y_m(t))_H|\,\mathrm{d}t.$$

We now apply (c) and estimate

$$\begin{aligned} \| y_m(T)\|_H^2 + 2\alpha\int_0^T \| y_m(t)\|_V^2\,\mathrm{d}t &\leqslant \| y_m(0)\|_H^2 + 2\int_0^T |(f(t), y_m(t))_H|\,\mathrm{d}t \\ &\leqslant y_{0m}\|_H^2 + 2\int_0^T \| f(t)\|_{V'}\cdot \| y_m(t)\|_V\,\mathrm{d}t \qquad (3) \\ &\leqslant \| y_{0m}\|_H^2 + \alpha\int_0^T \| y_m(t)\|_V^2\,\mathrm{d}t + \frac{1}{\alpha}\int_0^T \| f(t)\|_{V'}^2\,\mathrm{d}t, \end{aligned}$$

since $2|ab| \leqslant \alpha a^2 + \dfrac{1}{\alpha} b^2$. Therefore

$$\int_0^T \| y_m(t)\|_V^2\,\mathrm{d}t \leqslant \frac{1}{\alpha}\| y_{0m}\|_H^2 + \frac{1}{\alpha^2}\int_0^T \| f(t)\|_{V'}^2\,\mathrm{d}t.$$

Because $y_{0m} \to y_0$ in H as $m \to \infty$, there exists some $M > 0$ with

$$\| y_{0m}\|_H \leqslant \| y_0\|_H + M. \qquad (4)$$

Therefore

$$\int_0^T \| y_m(t)\|_V^2\,\mathrm{d}t \leqslant \frac{1}{\alpha}(\| y_0\|_H + M)^2 + \frac{1}{\alpha^2}\int_0^T \| f(t)\|_{V'}^2\,\mathrm{d}t \leqslant K^2 < \infty. \qquad (5)$$

We deduce that

$$y_m(t) \in L^2((0, T); V),$$

and the sequence y_m is bounded in $L^2((0, T); V)$. By Theorem 9.1 there exists a weakly convergent subsequence. We suppose without loss of generality that

$$y_m \rightharpoonup z \text{ (weakly) in } L^2((0, T); V) \text{ as } m \to \infty.$$

Step (iii). Let $\varphi(t) \in C^1[0, T]$ with $\varphi(T) = 0$. We multiply (2) by $\varphi(t)$, integrate over $(0, T)$, set $\varphi_j(t) = \varphi(t)w_j$ and obtain

$$\int_0^T \left[\left(\frac{\mathrm{d}}{\mathrm{d}t} y_m(t), \varphi_j(t)\right)_H + a(t; y_m(t), \varphi_j(t))\right] \mathrm{d}t = \int_0^T (f(t), \varphi_j(t))_H \,\mathrm{d}t.$$

Therefore, since $\varphi(T) = 0$, we obtain by partial integration that

$$\int_0^T [-(y_m(t), \varphi_j'(t))_H + a(t; y_m(t), \varphi_j(t))] \,\mathrm{d}t$$

$$= \int_0^T (f(t), \varphi_j(t))_H \,\mathrm{d}t + (y_m(0), \varphi_j(0))_H.$$

As $m \to \infty$ we obtain because of the weak convergence of y_m to z (we here apply the Riesz representation $(y_m(t), v)_H = (y_m(t), Rv)_V$), that

$$\int_0^T [-(z(t), \varphi_j'(t))_H + a(t; z(t), \varphi_j(t))] \,\mathrm{d}t = \int_0^T (f(t), \varphi_j(t))_H \,\mathrm{d}t + (y_0, \varphi_j(0))_H. \tag{6}$$

In particular this equation holds for each $\varphi \in \mathscr{D}(0, T)$, that is,

$$\frac{\mathrm{d}}{\mathrm{d}t}(z(t), w_j)_H + (L(t)z(t), w_j)_H = (f(t), w_j)_H.$$

Since j was arbitrary, the last equation implies that

$$\frac{\mathrm{d}z}{\mathrm{d}t} + L(t)z = f. \tag{7}$$

From $L(t)u \in L^2((0, T); V')$ for all $u \in L^2((0, T); V)$ and $f \in L^2((0, T); V')$ it therefore follows that

$$\frac{\mathrm{d}z}{\mathrm{d}t} = f - L(t)z \in L^2((0, T); V'),$$

that is, $z \in W(0, T)$. Now we can partially integrate (6) and by (7) and $\varphi(T) = 0$,

$\varphi(0) \neq 0$, we obtain

$$(z(0), w_j)_H \varphi(0) = (y_0, w_j)_H \varphi(0) \quad \text{for all } j \in \mathbb{N},$$

which has the consequence that

$$(z(0), w_j)_H = (y_0, w_j)_H \quad \text{for all } j \in \mathbb{N},$$

and

$$z(0) = y_0.$$

Therefore z is a solution of (P)

Step (iv). Since $\alpha \| y_m(t) - z(t) \|_V^2 \leqslant \operatorname{Re} a(t; y_m(t) - z(t), y_m(t) - z(t))$, it follows that

$$\begin{aligned}
&\alpha \int_0^T \| y_m(t) - z(t) \|_V^2 \, \mathrm{d}t \\
&\leqslant \operatorname{Re} \int_0^T a(t; y_m(t) - z(t), y_m(t) - z(t)) \mathrm{d}t + \tfrac{1}{2} \| y_m(T) - z(T) \|_H^2 \\
&= \operatorname{Re} \Big[\int_0^T a(t; y_m(t), y_m(t)) \, \mathrm{d}t - \int_0^T a(t; y_m(t), z(t)) \, \mathrm{d}t \\
&\quad - \int_0^T a(t; z(t), y_m(t) - z(t)) \, \mathrm{d}t \Big] \\
&\quad - \tfrac{1}{2}(z(T), y_m(T) - z(T))_H - \tfrac{1}{2}(y_m(T), z(T))_H + \tfrac{1}{2} \| y_m(T) \|_H^2.
\end{aligned}$$

We substitute

$$\begin{aligned}
\int_0^T a(t; y_m(t), y_m(t)) \, \mathrm{d}t &= \int_0^T (f(t), y_m(t))_H \, \mathrm{d}t - \int_0^T \left(\frac{\mathrm{d}y_m(t)}{\mathrm{d}t}, y_m(t) \right)_H \mathrm{d}t \\
&= \int_0^T (f(t), y_m(t))_H \, \mathrm{d}t - \tfrac{1}{2}(\| y_m(T) \|_H^2 - \| y_m(0) \|_H^2),
\end{aligned}$$

obtaining

$$\begin{aligned}
&\alpha \int_0^T \| y_m(t) - z(t) \|_V^2 \, \mathrm{d}t \\
&\leqslant \operatorname{Re} \Big[\int_0^T (f(t), y_m(t))_H \, \mathrm{d}t - \int_0^T a(t; y_m(t), z(t)) \, \mathrm{d}t \\
&\quad - \int_0^T a(t; z(t), y_m(t) - z(t)) \, \mathrm{d}t \Big] \\
&\quad - \tfrac{1}{2}(z(T), y_m(T) - z(T))_H - \tfrac{1}{2}(y_m(T), z(T))_H + \tfrac{1}{2} \| y_m(0) \|_H^2. \qquad (8)
\end{aligned}$$

As $m \to \infty$ the right-hand side of the inequality converges to

$$\operatorname{Re}\left[\int_0^T (f(t), z(t))_H \, dt - \int_0^T a(t; z(t), z(t)) \, dt\right] - \tfrac{1}{2}(\| z(T) \|_H^2 - \| y_0 \|_H^2) = 0, \tag{9}$$

and thus we have proved (iv). In passing from (8) to (9) we have used the fact that $y_m(T)$ converges weakly in H to $z(T)$, and now give the proof of this. We show that:

1. $\| y_m(T) \|_H$ is bounded, which follows immediately from (3) when we take note of (4) and (5).
2. For the total set $\{w_j\}$ in H we have

$$(y_m(T), w_j)_H \to (z(T), w_j)_H \quad \text{as } m \to \infty. \tag{10}$$

In order to obtain (10) we integrate (2) and obtain

$$\begin{aligned}(y_m(T), w_j)_H - (y_m(0), w_j)_H &= \int_0^T (f(t), w_j)_H \, dt - \int_0^T (L(t) y_m(t), w_j)_H \, dt \\ &= \int_0^T (f(t), w_j)_H \, dt - \int_0^T (y_m(t), L^*(t) w_j)_H \, dt.\end{aligned} \tag{11}$$

We know that $y_m 0 \to z(0)$ in H and $y_m(t) \to z(t)$ weakly in $L^2((0, T); V)$. Therefore as $m \to \infty$ (11) becomes

$$\lim_{m \to \infty} (y_m(T), w_j)_H - (z(0), w_j)_H = \int_0^T (f(t), w_j)_H \, dt - \int_0^T (L(t) z(t), w_j)_H \, dt. \tag{12}$$

If we integrate (P) from 0 to T, put $v = w_j$, and compare the result with (12), we obtain (10), 1 and 2 give for $h \in H$, $v \in \operatorname{Lin}\{w_j\}$ (linear hull)

$$\begin{aligned}|(y_m(T) - z(T), h)_H| &= |(y_m(T) - z(T), h - v)_H + (y_m(T) - z(T), v)_H| \\ &\leqslant \frac{\varepsilon}{2} + M \| h - v \|_H \leqslant \varepsilon,\end{aligned} \tag{13}$$

since $\operatorname{Lin}\{w_j\}$ is dense in H (totality assumption). (13) shows the weak convergence of $y_m(T)$ to $z(T)$ in H.

By (i) $\{w_j\}$ spans a dense subset in V, the properties '$V \subsetneq H$' and 'V dense in H' of a Gelfand triple imply that $\{w_j\}$ also spans a dense subset in H, from which we departed in 2.

Continuous dependence of the solution; for $y_0 \neq 0$ we can instead of (4) also find some $c_1 > 0$ with

$$\|y_{0_m}\|_H \leqslant c_1 \|y_0\|_H,$$

and we can write (5) in the form

$$\int_0^T \|y_m(t)\|_V^2 \,dt \leqslant c_2\left(\|y_0\|_H^2 + \int_0^T \|f(t)\|_{V'}^2 \,dt\right),$$

from which it follows that

$$\int_0^T \|z(t)\|_V^2 \,dt \leqslant c_2\left(\|y_0\|_H^2 + \int_0^T \|f(t)\|_{V'}^2 \,dt\right),$$

because $y_m \to z$ in $L^2((0, T); V)$ as $m \to \infty$. Therefore, because $dz/dt = f - L(t)z$ we have the estimate (see also (1))

$$\begin{aligned}\int_0^T \left\|\frac{dz(t)}{dt}\right\|_{V'}^2 dt &\leqslant 2\int_0^T (\|f(t)\|_{V'}^2 + \|L(t)z(t)\|_{V'}^2)dt \\ &\leqslant 2\int_0^T \|f(t)\|_{V'}^2 \,dt + 2c^2 \int_0^T \|z(t)\|_V^2 \,dt \\ &\leqslant c_3\left(\|y_0\|_H^2 + \int_0^T \|f(t)\|_{V'}^2 \,dt\right).\end{aligned}$$

Altogether

$$\|z\|_{W(0,T)}^2 = \int_0^T \|z(t)\|_V^2 \,dt + \int_0^T \left\|\frac{dz(t)}{dt}\right\|_{V'}^2 dt \leqslant c_4\left(\|y_0\|_H^2 + \int_0^T \|f(t)\|_{V'}^2 \,dt\right).$$

The last estimate is also known as Hadamard's estimate. ■

§27 The regularity of solutions of a parabolic differential equation

We turn now to the question of the regularity of the solution. We make the same assumptions as at the beginning of §26. We first prove a regularity theorem (Theorem 27.2) for the abstract parabolic equation and then pass to partial differential operators. Since we regard a problem as solved if we have solved it in the distributional sense, or even better in the framework of the Sobolev spaces W_2^l, Theorem 27.6 plays a particularly important role. It gives us the unique solvability of the mixed boundary value problem for parabolic differential equations in the space W_2^l under minimal assumptions. The transition from the space W_2^l to classic solvability involves only an application of Sobolev's lemma, Theorem 6.2.

27.1 An abstract regularity theorem

Definition 27.1 *By the Sobolev space $W_2^k((0,T);V)$, $k\in\mathbb{N}$, we understand the collection of measurable functions (or distributions)*

$$y:(0,T)\to V,$$

with

$$\frac{\mathrm{d}^n}{\mathrm{d}t^n}\in L^2((0,T);V),\quad \textit{for } 0\leqslant n\leqslant k,$$

where the differentiation is in the distributional sense. The norm in $W_2^k((0,T);V)$ is given by

$$\|y\|_k^2=\sum_{n=0}^{k}\int_0^T\left\|\frac{\mathrm{d}^n y(t)}{\mathrm{d}t^n}\right\|_V^2\mathrm{d}t.$$

It is clear that $W_2^k((0,T);V)$ is a Hilbert space with the scalar product correspondingly defined.

Definition 27.2 *Let H_1 and H_2 be Hilbert spaces. Suppose further than for $t\in[0,T]$ $B(t):H_1\to H_2$ is a linear map. We say that $B(t)$ is differentiable in $[0,T]$ if for each $v\in H$, the limit*

$$\lim_{h\to 0}\frac{1}{h}(B(t+h)v-B(t)v)=:\frac{\mathrm{d}}{\mathrm{d}t}B(t)v$$

exists in H_2. For $t=0$ or $t=T$ we take the left or right limit. $(\mathrm{d}/\mathrm{d}t)B(t):H_1\to H_2$ is again linear.

Suppose that the assumptions (a), (b) and (c) of § 26 are satisfied for $a(t,\varphi,\psi)$. We impose the additional requirement on $a(t,\varphi,\psi)$:

(d) for fixed $\varphi,\psi\in V$ *let* $a(t;\varphi,\psi)$ be k-fold differentiable with respect to t in $[0,T]$. We have that $(\mathrm{d}^j/\mathrm{d}t^j)\ a(t;\varphi,\psi)$ is continuous in $[0,T]$ for $j=0,1,\ldots,k-1$, in addition we require the existence and measurability of

$$\frac{\mathrm{d}^k}{\mathrm{d}t^k}a(t;\varphi,\psi)\quad\text{in } [0,T]$$

and

$$\left|\frac{\mathrm{d}^j}{\mathrm{d}t^j}a(t;\varphi,\psi)\right|\leqslant c\|\varphi\|_V\cdot\|\psi\|_V,\quad j=0,1,\ldots,k,$$

with c independent of t.

We see that in the sense of Definition 27.2 we have

$$\frac{\mathrm{d}^j}{\mathrm{d}t^j}a(t;\varphi,\psi)=\left(\frac{\mathrm{d}^j}{\mathrm{d}t^j}L(t)\varphi,\psi\right)_H,\quad j=0,1,\ldots,k,$$

and because of (d) $(\mathrm{d}^j/\mathrm{d}t^j)L(t): V\to V'$ acts linearly and continuously, $j=0,\ldots,k$,
where $\|(\mathrm{d}^j/\mathrm{d}t^j)L(t)\|\leqslant c$, with c the constant independent of T from (d). We again associate

$$f(t)\mapsto L(t)f(t), \tag{1}$$

and then have the generalisation of Lemma 26.1.

Theorem 27.1 *Under the assumptions* (a), (b), (d) *on* $L(t)$ *the map* (1):

$$L: W_2^k((0,T);V)\to W_2^k((0,T);V')$$

is continuous.

Remark 27.1 Corresponding results holds for $\mathrm{d}L/\mathrm{d}t$, $\mathrm{d}^2L/\mathrm{d}t^2$, etc.

Proof. As in Lemma 26.1 we show the measurability of the functions $(\mathrm{d}^j/\mathrm{d}t^j)L(t)f(t)$, $j=0,1,\ldots,k$, assuming that the function $f(t)$ is measurable. The Leibniz formulae also hold for $u\in W_2^k((0,T);V)$

$$\frac{\mathrm{d}}{\mathrm{d}t}(L(t)u(t))=\frac{\mathrm{d}L(t)}{\mathrm{d}t}u(t)+L(t)\frac{\mathrm{d}u(t)}{\mathrm{d}t},$$

$$\frac{\mathrm{d}^2}{\mathrm{d}t^2}(L(t)u(t))=\frac{\mathrm{d}^2L(t)}{\mathrm{d}^2t}u(t)+\cdots.$$

Because of (d) we have

$$\int_0^T\sum_{j=0}^k\left\|\frac{\mathrm{d}^j}{\mathrm{d}t^j}(L(t)u(t))\right\|_{V'}^2\mathrm{d}t=\int_0^T\sum_{j=0}^k\left\|\sum_{i=0}^j\binom{j}{i}\frac{\mathrm{d}^i}{\mathrm{d}t^i}L(t)\frac{\mathrm{d}^{j-i}}{\mathrm{d}t^{j-i}}u(t)\right\|_{V'}^2\mathrm{d}t$$

$$\leqslant\int_0^T\sum_{j=0}^k\sum_{i=0}^j\binom{j}{i}\left\|\frac{\mathrm{d}^i}{\mathrm{d}t^i}L(t)\frac{\mathrm{d}^{j-i}}{\mathrm{d}t^{j-i}}u(t)\right\|_{V'}^2\mathrm{d}t$$

$$\leqslant k^2c^2\int_0^T\sum_{j=0}^k\sum_{i=0}^j\left\|\frac{\mathrm{d}^{j-i}}{\mathrm{d}t^{j-i}}u(t)\right\|_V^2\mathrm{d}t$$

$$\leqslant c_1\int_0^T\sum_{j=0}^k\left\|\frac{\mathrm{d}^j}{\mathrm{d}t^j}u(t)\right\|_V^2\mathrm{d}t,$$

with which we have proved continuity. ■

We are now in a position to prove a far-reaching abstract regularity theorem for parabolic equations.

Theorem 27.2 *We consider the parabolic equation in* V'

$$\frac{dy(t)}{dt} + L(t)y(t) = f(t), \quad t\in(0, T), \tag{2}$$

with the initial condition $y(0) = y_0$. We assume that L satisfies the assumptions (a), (b), (c) of §26 and (d) of this section. Moreover, let

$$f\in W_2^k((0, T); V'), \tag{3}$$

and $y_0 \in H$ for $k = 0$, whereas for $k \geqslant 1$ we suppose

$$y_0\in V, \;\; f(0) - L(0)y_0\in V,\ldots,\in V, \;\; f^{(k-1)}(0) - L(0)f^{(k-2)}(0)\ldots\in H. \tag{4}$$

Then for the solution y of (2) we have:

$$y\in W_2^k((0, T), V) \quad \text{and} \quad \frac{d^{k+1}y(t)}{dt^{k+1}}\in L_2((0, T); V').$$

Remarks 27.2 It is easy to abbreviate formulae (4) by introducing formally the initial values $(d^jy/dt^j)(0)$. Differentiating formally (2) and substituting $t = 0$, we obtain

$$f^{(j-1)}(0) - L(0)f^{(j-2)}(0)\cdots = \frac{d^jy}{dt^j}(0)\in V, \quad j = 0,\ldots, k-1,$$

$$f^{(k-1)}(0) - L(0)f^{(k-2)}(0)\cdots = \frac{d^ky}{dt^k}(0)\in H. \tag{5}$$

'Formally' means here, that we do not know in advance whether the derivatives $(d^jy/dt^j)(0)$ exists; we know this for certain after using (4) and proving our Theorem 27.2. (4), respectively (5) are compatibility conditions. The data y_0, $(d^jf/dt^j)(0)$, $j = 0,\ldots, k-1$, must 'fit' the solution y and the boundary conditions on the lower 'lid' $t = 0$ of the region $\bar{\Omega} \times [0, T]$. An application of Theorem 25.5 shows that the conditions (4) are also necessary for the regularity formulated in Theorem 27.2.

Proof. We use induction on k. If $k = 0$, by Theorem 26.1 and from the assumptions, $f\in L^2((0, T); V')$, $y_0 = y(0)\in H$, it follows that $y\in L^2((0, T), V)$; $y_t\in L^2((0, T); V')$. We show now the transition to $k = 1$; the transition from $(k-1)$ to k follows in a similar way, for this reason we do not write out the argument again.

We differentiate equation (2) formally, that is, we do not know *a priori*, whether the appropriate derivatives exist and lie in the right space, to obtain

$$y_{tt} + L(t)y_t = f_t - L_t(y)y,$$

with $y_t(0) \in H$ because of (4) and (5). This reminds us of the initial value problem

$$v_t + L(t) = f_t - L_t(t)y, \tag{6}$$

$$v(0) = f(0) - L(0)y_0 = [y_t(0)] \in H. \tag{7}$$

Here we have written the value $y_t(0)$ in brackets, because we do not yet know whether it even exists. Because of (3) $f_t \in L^2((0, T); V')$ and by Remark 27.1 L_t maps $L^2((0, T), V')$ continuously into $L^2((0, T); V')$. Hence, $f_t - L_t(t)y$ belongs to $L^2((0, T); V')$ and $v(0) = f(0) - L(0)y_0 = [y_t(0)]$ to H (because of (4)). Therefore we can again apply Theorem 26.1 and obtain $v \in L^2((0, T); V)$, $v_t \in L^2((0, T); V')$. Next we show that $v = y_t$, from which Theorem 27.2 follow for $k = 1$. We form the function: $w(t) = y(0) + \int_0^t v(\tau)\,d\tau$, which is absolutely continuous since $v \in L^2((0, T); V)$, hence $w \in C^0([0, T]; V)$, and we have $w_t = v$. Because $y(0) \in V$ – see (4) – it is also true that $w \in W_2^1((0, T); V)$. If we show $w = y$, we also have $v = y_t$. We integrate (6) from 0 to t, and have

$$\begin{aligned} w_t = v(t) &= v(0) - \int_0^t Lv\,d\tau + f(t) - f(0) - \int_0^t L_t y\,d\tau \\ &= -L(0)y(0) - \int_0^t Lv\,d\tau + f(t) - \int_0^t L_t y\,d\tau \end{aligned} \tag{8}$$

by (7). By partial integration

$$\begin{aligned} \int_0^t Lv\,d\tau = \int_0^t Lw_t\,d\tau &= Lw - L(0)w(0) - \int_0^t L_t w\,d\tau \\ &= Lw - L(0)y(0) - \int_0^t L_t w\,d\tau, \end{aligned}$$

we obtain from (8)

$$w_t = -Lw + f(t) - \int_0^t L_t y\,d\tau + \int_0^t L_t w\,d\tau. \tag{9}$$

If we subtract equation (2) from (9) we have

$$(w - y)_t + L(w - y) = \int_0^t L_t(w - y)\,d\tau, \tag{10}$$

and the initial condition

$$(w - y)(0) = w(0) - y(0) = y(0) - y(0) = 0. \tag{11}$$

Since $w \in W_2^1((0,T);V)$ we have

$$w-y \in L^2((0,T);V), \quad (w-y)_t \in L^2((0,T);V'),$$

and

$$\int_0^t L_t(w-y)\,d\tau \in C([0,T];V') \subset L^2((0,T;V').$$

In Hadamard's estimate of Theorem 26.1 the constants are independent of T, so long as T is finite. We see this if we carry out the proof, or directly (adapted to our purposes): let

$$h_t + Lh = f_1 \quad \text{and} \quad h(0)=0. \tag{12}$$

Then

$$\int_0^{T_0} \|h(t)\|_V^2\,dt \leqslant K^2 \int_0^{T_0} \|f_1(t)\|_{V'}\,dt. \tag{13}$$

We fix $T_0 < \infty$ and take $0 \leqslant T' \leqslant T_0$; let $\tilde{f} \in L^2((0,T');V')$ and h be the solution of

$$\tilde{h}_t + L\tilde{h} = \tilde{f}, \quad \tilde{h}(0)=0, \quad t \in [0,T']. \tag{14}$$

We extend $\tilde{f}$ by 0 to $(0,T_0)$, call the extension f_1, and obtain

$$f_1 \in L^2((0,T);V') \quad \text{with} \int_0^{T'} \|\tilde{f}\|_{V'}^2\,dt = \int_0^{T_0} \|f_1\|_{V'}\,dt.$$

We solve (12) with this f_1, again have (13), and uniqueness gives $\tilde{h}=h$ on $[0,T']$, since the equations (12) and (14) agree on $[0,T']$. Therefore from (13) we obtain

$$\int_0^{T'} \|\tilde{h}\|_V^2\,dt = \int_0^{T'} \|h\|_V^2\,dt \leqslant \int_0^{T_0} \|h\|_V^2\,dt \leqslant K^2 \int_0^{T_0} \|f_1\|_{V'}^2\,dt$$

$$= K^2 \int_0^{T'} \|\tilde{f}\|_{V'}^2\,dt,$$

that is for the solution $\tilde{h}$ of the equation (14) we obtain the estimate on $[0,T']$

$$\int_0^{T'} \|\tilde{h}(t)\|_V^2\,dt \leqslant K^2 \int_0^{T'} \|\tilde{f}(t)\|_{V'}^2\,dt,$$

with K from (13), with which we have shown independence from T'. We return now to (10) and (11). We put $w-y=h$ and

$$f_1 = \int_0^t L_t h\,d\tau. \tag{15}$$

By Theorem 27.1 (Remark 27.1) we have

$$\|f_1(t)\|_{V'} \leqslant \int_0^t \|L_t h\|_V \, d\tau \leqslant \int_0^t K_1 \|h(\tau)\|_V \, d\tau; \tag{16}$$

where again K_1 is independent of T'. Suppose now that T' is the largest number, $0 \leqslant T' \leqslant T \leqslant T_0$; with $h = 0$ in $[0, T']$. (15) shows that

$$f_1 \equiv 0 \quad \text{in} \quad [0, T'],$$

and (13) with (16) gives

$$\int_0^T \|h(t)\|_V^2 \, dt = \int_{T'}^T \|h(t)\|_V^2 \, dt \leqslant K^2 K_1^2 \int_{T'}^T \left\{ \int_{T'}^t \|h(\tau)\|_V \, d\tau \right\}^2 dt$$

$$\leqslant K^2 K_1^2 \int_{T'}^T (t - T') \, dt \int_{T'}^T \|h(t)\|_V^2 \, dt = K^2 K_1^2 \frac{(T - T')^2}{2} \int_{T'}^T \|h(t)\|_V^2 \, dt.$$

We take $T' < T \leqslant T_0$, then the inequality above implies

$$1 \leqslant K^2 K_1^2 \frac{(T - T')^2}{2}.$$

If we allow T to tend to T', then it would follow that $1 \leqslant 0$, a contradiction. Therefore we have proved $y = w$ or $y_1 = v$; and so the value $v(0) = y_t(0)$ also exists, and we obtain it by continuity as $t \to 0$. ■

We now demonstrate one possibility of fulfilling the conditions (4). Let $k = 1$, the conditions (4) reduce to (see (5))

$$y_0 \in V, \quad f(0) - L(0) y_0 \in H. \tag{4a}$$

Because $f \in W_2^1((0, T); V')$ f is continuous in V' by Theorem 25.5 and thus $f(0)$ is defined. In order to satisfy (4) it is enough to require

$$y_0 \in V, \quad f(0) \in H, \quad L(0) y_0 \in H. \tag{17}$$

Let $k = 2$, since $f \in W_2^2((0, T); V')$, by Theorem 25.5 the values $f(0)$, $f_t(0)$ are defined (in V') and the conditions (4) reduce to (see (5))

$$y_t \in V, \quad f(0) - L(0) y_0 \in V, \quad f_t(0) - L(0) f(0) + L^2(0) y_0 - L_t(0) y_0 \in H,$$

and they are then satisfied, if for example

$$y_0 \in V, \quad L(0) y_0 \in V, \quad L^2(0) y_0 \in H, \quad L_t(0) y_0 \in H,$$

$$f(0) \in V, \quad L(0) f(0) \in H, \quad f_t(0) \in H. \tag{18}$$

Continuing in this way we find that the conditions (4) are for example satisfied, if

$$\begin{aligned}
&L^j(0)y_0 \in V, \quad j=0,\dots,k-1, \quad L^k(0)y_0 \in H,\\
&L_t^j(0)y_0 \in V, \quad j=1,\dots,k-2, \quad L_t^{k-1}(0)y_0 \in H,\\
&L_{tt}^j(0)y_0 \in V, \quad j=1,\dots,k-3, \quad L_{tt}^{k-2}(0)y_0 \in H, \quad \text{etc.}\\
&f(0) \in V, \quad L^j(0)f(0) \in V, \quad j=1,\dots,k-2, \quad L^{k-1}(0)f(0) \in H,\\
&L_t^j(0)f(0) \in V, \quad j=1,\dots,k-3, \quad L_t^{k-2}(0)f(0) \in H \quad \text{etc.,}\\
&f_t(0) \in V, \quad L^j(0)f_t(0) \in V, \quad j=1,\dots,k-3, \quad L^{k-2}(0)f_t(0) \in H\\
&L_t^j(0)f_t(0) \in V, \quad j=1,\dots,k-4, \quad L_t^{k-3}(0)f_t(0) \in H \quad \text{etc.,}\\
&f_{\underbrace{t\dots t}_{k-1}}(0) \in H.
\end{aligned} \tag{19}$$

We give yet one more example of how we can satisfy the conditions (4). Let

$$y_0 \in V,$$

and

$$\begin{aligned}
&f(0) - L(0)y_0 = 0,\\
&f_t(0) - L_t(0)y_0 = 0,\\
&f_{tt}(0) - L_{tt}(0)y_0 = 0, \quad \text{etc.}
\end{aligned} \tag{20}$$

Given (5), (20) implies that

$$y_0 \in V, \quad y'(0) = 0, \quad y''(0) = 0, \dots, y^{(k)}(0) = 0.$$

(in shorthand – see Remark 27.2), that is, with (20) the conditions (4) are also satisfied.

Remark 27.3 For parabolic differential operators, that is, $L(t) = \sum_s a_s(x,t) D_x^s =$ some elliptic differential operator plus boundary conditions, one can prove regularity theorems under different compatibility assumptions by other methods (and spaces), see Lions & Magenes [1] and the rich bibliography given there. For parabolic differential equations of the second order and for special boundary value problems Friedman [1] and Ladyženskaya, Solonnikov & Uralceva [2] and others have proved regularity theorems.

We want to proceed further along the path signposted by the conditions (4); this is rewarded by analytic simplicity. The existence Theorem 26.1 and the regularity Theorem 27.2 belong to the framework of §17. In order to formulate theorems about differential operators, we utilise the notation and hypotheses of §20.

27.2 Differentiability with respect to t

Let $t\in[0,T]$, $x\in\Omega\subset\mathbb{R}^r$, Ω open, and let $A(t)$ be a strongly elliptic differential operator given by (see §19, (1))

$$A(t)\varphi=\sum_{|\alpha|,|\beta|\leqslant m}(-1)^{|\alpha|}\mathrm{D}_x^\alpha(a_{\alpha\beta}(x,t)\mathrm{D}_x^\beta\varphi), \tag{21}$$

with the sesquilinear form

$$a(t;\varphi,\psi)=\int_\Omega\sum_{\substack{|\alpha|\leqslant m\\|\beta|\leqslant m}}a_{\alpha\beta}(x,t)\cdot\mathrm{D}_x^\beta\varphi\cdot\mathrm{D}_x^\alpha\bar{\psi}\,\mathrm{d}x. \tag{22}$$

Suppose further given the boundary form (see §20 (33))

$$c(t;\varphi,\psi)=\int_{\partial\Omega}\sum_{\substack{|\alpha|\leqslant m-1\\|\beta|\leqslant m}}c_{\alpha\beta}(x,t)\cdot\mathrm{D}_x^\alpha\varphi\mathrm{D}_x^\beta\bar{\psi}\,\mathrm{d}\sigma.$$

Let $H=L_2(\Omega)$ and let V be a closed subspace of $W_2^m(\Omega)$ with

$$\mathring{W}_2^m(\Omega)\subset V\subset W_2^m(\Omega).$$

V can for example be determined by boundary operators, $V=W_2^m(\{b_j\}_1^p)$, see §19. We consider the sesquilinear form

$$\mathscr{A}(t;\varphi,\psi)=a(t;\varphi,\psi)+c(t;\varphi,\psi), \tag{23}$$

and require that the assumptions 1–6 (at first for $k=0$) of §20 be satisfied, in such a way that the constants appearing, for example the coercion constant k_0, be independent of t. In order for example to satisfy 5, it is enough that Agmon's theorem, 19.3, holds for $a(t;\varphi,\psi)$, V, and that the boundary form $c(t;\varphi,\psi)$ has total order $|\alpha|+|\beta|\leqslant 2m-2$.

In addition, for the coefficients of a and c we require that there exist partial derivatives with respect to t, and that

$$\frac{\partial^j a_{\alpha\beta}(x,t)}{\partial t^j},\frac{\partial^j c_{\alpha\beta}(x,t)}{\partial t^j}\in C(\bar{\Omega}\times[0,T])\quad\text{for } j=0,\ldots,k. \tag{24}$$

It is easily checked that with these conditions the assumptions (a), (b) and (c) of §26 are satisfied and also condition (d) from the beginning of this section. Here we can (see the beginning of the proof of Theorem 26.1) without loss of generality set $k_0=0$ in assumption (c), §26. We consider the parabolic differential equation

$$\frac{\partial y(x,t)}{\partial t}+\mathscr{L}(t)y(x,t)=f(x,t), \tag{25}$$

where $\mathscr{L}(t) = \mathscr{A}(t)$ is the representation operator of (23) (see §17, (28)), or written as a weak equation

$$\frac{\mathrm{d}(y,\varphi)_0}{\mathrm{d}t} + \mathscr{A}(t; y, \varphi) = (f, \varphi)_0, \quad \varphi \in V.$$

If we take $\varphi \in \mathscr{D}(\Omega) \subset V$, from the weak equation we obtain the parabolic differential equation below

$$\frac{\partial y(x,t)}{\partial t} + \sum_{|\alpha|,|\beta| \leqslant m} (-1)^{|\alpha|} \mathrm{D}_x^\alpha (a_{\alpha\beta}(x,t) \mathrm{D}_x^\beta y(x,t))$$
$$= f(x,t), \quad \text{in } \Omega \times (0, T),$$

interpreted in the distributional sense.

However, boundary conditions are also involved in the weak equation. We can now state some regularity theorems. We begin with differentiability with respect to t.

Theorem 27.3 *We consider the parabolic equation (25) with the initial condition*

$$y(x,0) = y_0(x).$$

Suppose that the assumptions above are satisfied, suppose further that

$$f(x,t) \in W_2^k((0,T); V'), \quad k \geqslant 0,$$

and

$$\frac{\partial^j y(x,0)}{\partial t^j} \in V, \quad j = 0, \ldots, k-1, \quad \frac{\partial^k y(x,0)}{\partial t^k} \in L_2(\Omega). \tag{26}$$

Then for the (unique) solution $y(x,t)$ of (25) we have

$$y \in W_2^k((0,T); V) \quad \text{and} \quad \frac{\partial^{k+1} y(x,t)}{\partial t^{k+1}} \in L_2((0,T); V').$$

In particular if $k \geqslant 1$ $y(x,t) \in V$ for all $t \in [0,T]$, that is, the solution $y(x,t)$ satisfies the (homogeneous) boundary conditions given by V.

The proof follows immediately from 27.2.

In order to be able to formulate a regularity theorem for inhomogeneous boundary conditions we must first define what we mean by the acceptance of the boundary conditions V.

Definition 27.3 *Suppose given a function $g(x,t)$. We say that $y(x,t)$ accepts the boundary values determined by V and g, if for each $t \in [0,T]$*

$$y(\cdot,t) - g(\cdot,t) \in V. \tag{27}$$

Theorem 27.4 *Suppose that the assumptions of Theorem 27.3 are satisfied (apart from* (26)*), suppose in addition that a 'boundary'-function $g(x,t)$ is given with*

$$g \in W_2^k((0,T), W_2^m(\Omega)) \quad \textit{and} \quad \frac{\partial^{k+1}}{\partial t^{k+1}} g(x,t) \in L_2((0,T); V'), \tag{28}$$

together with the compatibility conditions

$$\frac{\partial^j y(x,0)}{\partial t^j} - \frac{\partial^j g(x,0)}{\partial t^j} \in V, \quad j = 0, \ldots, k-1$$

$$\frac{\partial^k y(x,0)}{\partial t^k} - \frac{\partial^k g(x,0)}{\partial t^k} \in L_2(\Omega). \tag{29}$$

Then there exists a unique solution $y(x,t)$ of (25) *with $y(x,0) = y_0(x)$ and*

$$y(\cdot,t) - g(\cdot,t) \in V \quad \textit{for all } t \in [0,T]$$

and such that

$$y \in W_2^k((0,T); W_2^m(\Omega)) \quad \textit{and} \quad \frac{\partial^{k+1} y}{\partial t^{k+1}} \in L_2((0,T); V').$$

For $k = 0$ one can prove (27) *only for almost all t.*

For this theorem we sharpen condition (b), we assume the estimate:

$$|\mathscr{A}(t; \varphi, \psi)| \leqslant c \|\varphi\|_m \|\psi\|_m \quad \text{for all } \varphi, \psi \in W_2^m(\Omega), \tag{b$'$}$$

where c is independent of t.

(b′) is usually satisfied for differential operators, see Definition 17.7 (b′) has the consequence that the representation operators $\mathscr{L}(t)$ of $\mathscr{A}(t; \varphi, \psi)$ is not only continuous as a map $V \to V'$ but also as a map from $W_2^m(\Omega)$ to $W_2^m(\Omega)'$.

Proof. First some preliminary remarks. Because $V \subset W_2^m(\Omega)$ we can consider each continuous functional $F \in W_2^m(\Omega)'$ on V also, that is, we have $W_2^m(\Omega)' \subset V'$ (not necessarily injectively) and because

$$\|F\|_{V'} = \sup_{\substack{\varphi \in V \\ \|\varphi\| \leqslant 1}} |F(\varphi)| \leqslant \sup_{\substack{\varphi \in W_2^m(\Omega) \\ \|\varphi\| \leqslant 1}} |F(\varphi)| = \|F\|_{W_2^m(\Omega)'}$$

we have

$$W_2^k((0,T); W_2^m(\Omega)') \subset W_2^k((0,T); V'). \tag{30}$$

further, the sharpened hypothesis (b′) implies that

$$\mathscr{L}(t): W_2^k((0,T); W_2^m(\Omega)) \to W_2^k((0,T); W_2^m(\Omega)'), \quad \text{is continuous,} \tag{31}$$

so that by (30) $\mathscr{L}(t)g$ belongs to $W_2^k((0,T),V')$, if $g\in W_2^k((0,T);\ W_2^m(\Omega))$.

Now for the proof. We apply the well-known trick of homogenisation, that is, we put $y:=u+g$, where u satisfies the equation

$$\frac{\mathrm{d}u}{\mathrm{d}t}+\mathscr{L}u=f-\mathscr{L}g-g_t=:\tilde{f}. \tag{32}$$

Suppose first that $k=0$. We check the hypotheses for the solvability of (32). Because of (29) we have

$$u(0)=y_0-g(0)\in L_2(\Omega). \tag{33}$$

(28), (30) and (31) have the effect that $\tilde{f}=f-\mathscr{L}g-g_t\in L_2((0,T);\ V')$, therefore by Theorem 26.1 there exist uniquely determined u and y with

$$y=u+g\in W_2^0((0,T);\ W_2^m(\Omega)),\quad y_t=u_t+g_t\in L_2((0,T);\ V'),$$

which is the assertion for $k=0$. Since $u\in L^2((0,T),V)$, we have the boundary condition $y(\cdot,t)-g(\cdot,t)=u(\cdot,t)\in V$ for almost all t.

Suppose now that $k\geqslant 1$. We are now looking for a solution u of (33) and (32) with

$$\frac{\partial^j u(0)}{\partial t^j}\in V, j=0,\dots,k-1,\quad \frac{\partial^k u(0)}{\partial t^k}\in L_2(\Omega). \tag{34}$$

However, (34) are the conditions (29), and Theorem 27.3 (given (28), (31) and (30) $\tilde{f}$ also satisfies the assumptions of this theorem) guarantees a solution u with

$$u\in W_2^k((0,T);V)\quad\text{and}\quad \frac{\partial^{k+1}u}{\partial t^{k+1}}\in L_2((0,T);\ V').$$

This, together with (28), leads to the assertion of our theorem

$$y=u+g\in W_2^k((0,T);W_2^m(\Omega)),\quad y^{(k+1)}\in L_2((0,T);\ V'),$$

and to the fact that $y(t)-g(t)\in V$, this time for all $t\in[0,T]$. ■

27.3 Differentiability with respect to *x*, respectively *t*

We make the same assumptions as in §27.2. In Theorems 27.3 and 27.4 we have only considered regularity with respect to t; in order to obtain regularity with respect to x, we must raise the differentiability assumptions with respect to x. We take the assumptions 1 to 6 of §20 for (23) with $k=m+q$, $q\geqslant 0$ (the constants appearing are again independent of t) and then have Theorems 20.4 and 21.1 at our disposal. Since – as already remarked above – we can assume without loss of generality that $\lambda=0$ is not an eigenvalue of $\mathscr{L}(t)$, the map $\mathscr{L}(t): V \to V'$ is a (topological) isomorphism. We consider the (continuous) inverse map $\mathscr{L}^{-1}(t): V' \to V$, and restrict it to $W_2^q(\Omega)$, $q \geqslant 0$. By Corollary 21.2

$$G(t) = \mathscr{L}^{-1}(t): W_2^q(\Omega) \to W_2^{2m+q}(\Omega) \cap V \cap V_N \tag{35}$$

acts continuously and surjectively, in the topologies of W_2^q and W_2^{2m+q}, and we have $G^{-1}(t) = A(t)$ – where $A(t)$ is the differential operator given by (21) – see Corollary 21.1 from Green's Formula. Here, as everywhere in this section, the space V is determined by boundary conditions on $\partial\Omega$, the so-called stable boundary conditions

$$b_1(x, \mathrm{D})\varphi = 0, \ldots, b_p(x, \mathrm{D})\varphi = 0, \ 0 \leqslant p \leqslant m,$$

while V_N is determined by the so-called natural boundary conditions:

$$b_{p+1}(x, \mathrm{D})\varphi = 0, \ldots, b_m(x, \mathrm{D})\varphi = 0, \quad \text{on} \quad \partial\Omega,$$

where we have (see §21, assumption 3) $b_j = B_j + C_j$, $j = p+1, \ldots, m$. B_j originates in Green's formula §21, (3) and C_j in the boundary form $c(t; \varphi, \varphi)$ (22), which as in §21, (4), we write the form

$$c(t; \varphi, \varphi) = \sum_{j=p+1}^{m} C_j(x,t)\varphi \cdot \overline{B'_j \psi} \, \mathrm{d}\sigma.$$

If we use the notation of §14, we have $V \cap V_N = W^{2m}(B) = W^{2m}(\{b_j\}_1^m)$, and for (35)

$$G(t): W_2^q(\Omega) \to W_2^{2m+q}(\Omega) \cap W^{2m}(\{b_j\}_1^m) \text{ (topologically)} \tag{35*}$$

isomorphic, from which it follows that the Friedrichs inequality holds:

$$\| G(t) f \|_{2m+q} \leqslant c \| f \|_q, \quad f \in W_2^q(\Omega), \tag{36}$$

where $c = 1/\alpha$ is independent of t, see §26(c).

But we have a better result (communicated by R. Racke); from assumption (24) we find that the Friedrichs inequality also holds for the derivatives of $G(t)$:

$$\left\| \frac{\partial^n}{\partial t^n} G(t) f \right\|_{2m+q} \leqslant c_n \| f \|_q, \quad t \in [0, T], \quad f \in W_2^q(\Omega), \quad n = 0, \ldots, k, \tag{e}$$

where the constants c_n are independent of t. As we know, see §27.1, condition (24) implies that $(\partial^n/\partial t^n)\mathscr{L}(t)$ exists and $\| \partial^n/\partial t^n \mathscr{L}(t) \| \leqslant c$, $n = 0, \ldots, k$, with c independent of t. We now prove (e) by induction.

$n = 0$, this is (36).

$n = 1$: Let $u(t) := G(t) f$ or $\mathscr{L}(t) u(t) = f$. Differentiating by t we obtain

$$\frac{\partial \mathscr{L}(t)}{\partial t}u(t)+\mathscr{L}(t)\frac{\partial u(t)}{\partial t}=0, \quad \text{or} \quad \frac{\partial u}{\partial t}=-G(t)\frac{\partial \mathscr{L}(t)}{\partial t}u(t).$$

Using this equality we have

$$\begin{aligned}
\left\|\frac{\partial G(t)}{\partial t}f\right\|_{2m+q}=\left\|\frac{\partial (G(t)f)}{\partial t}\right\|_{2m+q}&=\left\|\frac{\partial u}{\partial t}\right\|_{2m+q}\\
&=\left\|G(t)\frac{\partial \mathscr{L}(t)}{\partial t}u(t)\right\|_{2m+q}\\
&\leqslant c_0\left\|\frac{\partial \mathscr{L}(t)}{\partial t}u(t)\right\|_q \quad \text{(by (36))}\\
&\leqslant c_0 c\|u(t)\|_{2m+q} \quad \text{(by Theorem 27.1)}\\
&=c_0 c\|G(t)f\|_{2m+q}\leqslant c_0^2 c\|f\|_q \quad \text{(by (36))},
\end{aligned}$$

and the proof for $n=1$ is complete.

Analogously we obtain the proof for $n\geqslant 2$. Thus by Theorem 27.1 $G(t)$ acts continuously

$$G(t)\colon W_2^k((0,T);W_2^q(\Omega))\to W_2^k((0,T);W_2^{2m+q}\cap W^{2m}(B)). \tag{37}$$

Theorem 27.5 *Let $\mathscr{L}(t)$ be the representation operator of* (23) *and let the assumptions* 1–6 *of* §20 *with $K=m+q$, $H=L_2(\Omega)$, $V=W^m(\{b_j\}_1^p)V_N=W^{2m}(\{b_j\}_{p+1}^m)$, $V\cap V_N=W^{2m}(B)=W^{2m}(\{b_j\}_1^m)$ be satisfied for the sesquilinear form* (23). *Further suppose that* (24) *is satisfied, let y_0 and f be sufficiently regular* (*see Theorem* 27.3) *that for the solution y of* (25) *with $y(0)=y_0$ we have*

$$y\in W_2^k((0,T);V), \quad k\geqslant 1. \tag{38}$$

Suppose in addition that

$$f\in W_2^{k-1}((0,T);W_2^q(\Omega)), \tag{39}$$

then it is also true that

$$\begin{aligned}
&y\in W_2^{k-1}((0,T);W_2^{2m+\min(q,m)}(\Omega)\cap W^{2m}(B)),\\
&y\in W_2^{k-2}((0,T);W_2^{2m+\min(q,m_1)}(\Omega)\cap W^{2m}(B)), \quad m_1=2m+\min(m,q), \textit{ etc.}
\end{aligned} \tag{40}$$

By choosing k and q to be sufficiently large we may always conclude that

$$y\in W_2^l((0,T);W_2^l(\Omega)\cap W^{2m}(B)). \tag{41}$$

Proof. Since $\mathscr{L}(t)y = f - \mathrm{d}y/\mathrm{d}t$, it follows that

$$y = G(t)\left(f - \frac{\mathrm{d}y}{\mathrm{d}t}\right).$$

From (38) it follows that $\mathrm{d}y/\mathrm{d}t \in W_2^{k-1}((0,T); V)$, which with (39) gives

$$f - \frac{\mathrm{d}y}{\mathrm{d}t} \in W_2^{k-1}((0,T); W_2^{\min(m,q)}(\Omega)).$$

we apply $G(t)$ to this, and by (37) have

$$y \in W_2^{k-1}((0,T); W_2^{2m+\min(m,q)}(\Omega) \cap W^{2m}(B)).$$

We can carry on:

$$\frac{\mathrm{d}y}{\mathrm{d}t} \in W_2^{k-2}((0,T); W_2^{2m+\min(m,q)}(\Omega)),$$

$$f - \frac{\mathrm{d}y}{\mathrm{d}t} \in W_2^{k-2}((0,T); W_2^{\min(q,m_1)}(\Omega)), \quad m_1 = 2m + \min(m,q),$$

$$y = G(t)\left(f - \frac{\mathrm{d}y}{\mathrm{d}t}\right) \in W_2^{k-2}((0,T); W_2^{2m+\min(q,m_1)}(\Omega) \cap W^{2m}(B)) \text{ etc.} \qquad \blacksquare$$

By Theorem 27.5 we already know which mixed parabolic problem we have solved.

Theorem 27.6 *In Theorem 27.5 let $k = 1$, $q = 0$, therefore*

$$f \in L_2((0,T); L_2(\Omega)) = L_2(\Omega \times (0,T)); \quad \frac{\partial f(x,t)}{\partial t} \in L_2((0,T), V'),$$

and let $y_0(x) \in W^{2m}(B) \subset V$. We then have that the solution $y(x,t)$ satisfies the (parabolic) differential equation in $\Omega \times (0,T)$

$$\frac{\partial y(x,t)}{\partial t} + \sum_{\substack{|\alpha| \leq m \\ |\beta| \leq m}} (-1)^{|\alpha|} \mathrm{D}_x^\alpha (a_{\alpha\beta}(x,t) \mathrm{D}_x^\beta y(x,t)) = f(x,t),$$

with the initial condition

$$y(x,0) = y_0(x) \quad \text{on} \quad \Omega,$$

and with the boundary conditions ($x \in \partial\Omega$)

$$\begin{aligned} &b_1(x, \mathrm{D}_x)y(x,t) = 0, \ldots, b_p(x, \mathrm{D}_x)y(x,t) = 0, \\ &b_{p+1}(x,t,\mathrm{D}_x)y(x,t) = [B_{p+1}(x,\mathrm{D}_x) + C_{p+1}(x,t,\mathrm{D}_x)]y(x,t) = 0, \ldots, \\ &b_m(x,t,\mathrm{D}_x)y(x,t) = [B_m(x,\mathrm{D}_x) + C_m(x,t,\mathrm{D}_x)]y(x,t) = 0. \end{aligned} \tag{42}$$

Here the first p boundary conditions are satisfied for all $t\in[0,T]$, the last $(m-p)$ for almost all t. If we require that the last $(m-p)$ boundary conditions are also satisfied for all t, then we must take $k=2$ in Theorems 27.3 and 27.5. Since by Corollary 21.1 $\mathscr{L}(t)$ agrees with $A(x,t,\mathrm{D}_x)$ on $W^{2m}(B)$, the converse also holds, that is, under the assumptions above the mixed parabolic problem

$$\frac{\partial y}{\partial t}+A(x,t,\mathrm{D}_x)y=f,\quad y(x,0)=y_0(x),\quad y\in W_2^{2m}(\Omega)$$
$$b_j(x,t,\mathrm{D}_x)y=0,\quad j=1,\ldots,m,\ \text{on}\ \partial\Omega\times[0,T],$$

is equivalent in V' to the weak equation

$$\frac{\mathrm{d}y}{\mathrm{d}t}+\mathscr{L}(t)y=f,\quad y(x,0)=y_0(x),\ y\in V.$$

Proof. We immediately obtain the differential equation from the weak equation (25), if we take $\varphi\in\mathscr{D}(\Omega)\subset V$ there. The initial condition is satisfied in V or in $L_2(\Omega)$, hence on Ω (for almost all x). From the hypotheses $y_0\in W^{2m}(B)\subset V$ it follows from (35) that $\mathscr{L}(0)y_0=G^{-1}(0)y_0$ belongs to $L_2(\Omega)$, which together with $f(x,0)\in L_2(\Omega)$ gives

$$y_0(x)\in V,\quad f(x,0)-\mathscr{L}(0)y_0(x)\in L_2(\Omega),$$

which for $k=1$ are the compatibility conditions (26) in Theorem 27.3. For the first p boundary conditions we obtain

$$y(x,t)\in V\quad\text{for all } t\in[0,T].$$

The assumptions for Theorem 27.5 are also satisfied, and we have (40), that is, for the $(m-p)$ latter boundary conditions

$$y(x,t)\in L_2((0,T);\ W^{2m}(B)).$$

This means that for almost all $t\in[0,T]$, $y(x,t)\in W^{2m}(B)$, which is equivalent to (42). If $k=2$ by Theorem 25.5 we have that

$$y(x,t)\in W_2^1((0,T);\ W^{2m}(B))\subset C^0([0,T];\ W^{2m}(B)),$$

therefore $y(x,t)\in W^{2m}(B)$, this time for all $t\in[0,T]$. The second part of the theorem, the converse, is clear. ■

We can also enunciate a regularity theorem (with respect to x) for inhomogeneous boundary values. We again take the assumption for Theorem 27.5: the sesquilinear form (23) satisfies assumptions 1–6 of §20, with $K=m+q$, $H=L_2(\Omega)$, $V=W^m(\{b_j\}_1^p)$, $W^{2m}(B)=W^{2m}(\{b_j\}_1^m)$, together with (24). In addition (see Theorem 27.3) let $k \geqslant 1$

$$f \in W_2^{k-1}((0,T); W_2^q(\Omega)), \quad \frac{\partial^k f}{\partial t^k} \in L_2((0,T); V'),$$

and let a boundary function $g(x,t)$ be given with

$$g \in W_2^k((0,T); W_2^{2m+\tilde{q}}(\Omega)), \quad \frac{\partial^{k+1} g(x,t)}{\partial t^{k+1}} \in L_2((0,T); V'),$$

where $\tilde{q}$ is the maximum among the numbers $\min(q,m)$, $\min(q,m_1)$, $\min(q,m_2)$, $\min(q,m_2)$, etc., arising in (40). Here we have put $m_1 = 2m + \min(m,q)$, $m_2 = 2m + \min(m_1,q)$, etc. Then for $\tilde{f} = f - A(t)g - g_t$ we have

$$\tilde{f} \in W^{k-1}((0,T); W_2^q(\Omega)). \tag{43}$$

Warning Here $A(t)$ is the differential operator (21) and not the representation operator $\mathscr{L}(t)$. We have that $A(t)$; $W_2^{2m+q}(\Omega) \to W_2^{\tilde{q}}(\Omega)$ is continuous, something which need not be the case for $\mathscr{L}(t)$, for example for the Neumann problem, Example 21.2, where we have worked out the representation operator.

We put $u_0(x) := y_0(x) - g(x,0)$ and impose the following compatibility condition on $\tilde{f}, u_0$

$$\begin{aligned} &u_0(x) \in V, \quad u_t(0) := \tilde{f}(0) - \mathscr{L}(0)u_0 \in V, \\ &u_{tt}(0) := \tilde{f}_t(0) - \mathscr{L}(0)\tilde{f}(0) + \mathscr{L}^2(0)u_0 - \mathscr{L}(0)u_0 \in V, \end{aligned} \tag{44}$$

etc., to $\mathrm{d}^k u/\mathrm{d}t^k(0) := \cdots \in H$.

We obtain (44) from (5), when we there set $y \mapsto u$, $y_0 \mapsto u_0$, $f \mapsto \tilde{f}$. The formulae in (44) are drastically simplified, if we choose g so that $g(x,0) = y_0(x)$, hence $u_0 = 0$.

We have:

Theorem 27.7 *There exists a unique solution $y(x,t)$ of the parabolic differential equation*

$$\frac{\partial y(x,t)}{\partial t} + \sum_{\substack{|\alpha| \leqslant m \\ |\beta| \leqslant m}} (-1)^{|\alpha|} \mathrm{D}_x^\alpha (a_{\alpha\beta}(x,t) \mathrm{D}_x^\beta y(x,t)) = f(x,t) \quad \text{in } \Omega \times (0,T),$$

with $y(x,0) = y_0(x)$, and such that

$$\begin{aligned} &y \in W_2^k((0,T); W_2^m(\Omega)) \\ &y \in W_2^{k-1}((0,T); W_2^{2m+\min(q,m)}(\Omega)), \\ &y \in W_2^{k-2}((0,T); W_2^{2m-\min(q,m_1)}(\Omega)), \quad m_1 = 2m + \min(m,q), \quad \text{etc.} \end{aligned}$$

For all $t \in [0,T]$ *this solution satisfies the boundary conditions belonging to* V:

$$y(\cdot,t) - g(\cdot,t) \in V, \quad t \in [0,T],$$

and if $k > 1$ *also the boundary condition*

$$y(\cdot,t) - g(\cdot,t) \in W^{2m}(B), \quad t \in [0,T].$$

If $k = 1$, *this last boundary condition only holds for almost all* $t \in [0,T]$.

Proof. We put $y := u + g$, where u satisfies the (weak) equation:

$$\frac{\mathrm{d}u}{\mathrm{d}t} + \mathscr{L}u = \tilde{f} = f - Ag - g_t,$$

with the initial condition $u(0) = y_0 - g(\cdot,0)$. (43) with the other hypotheses on f and g imply (see also (30)) that

$$\tilde{f} \in W_2^{k-1}((0,T);\, W_2^q(\Omega)) \subset W_2^{k-1}((0,T);\, V'),$$

$$\frac{\partial^k \tilde{f}}{\partial t^k} \in L_2((0,T);\, V'),$$

which, together with (44), allows the application of Theorem 27.3, hence $u \in W_2^k((0,T);\, V)$. This with (43) and use of Theorem 27.5 gives (40) for $u = y - g$, from which we read off the boundary behaviour (see Theorem 27.6). (40) for $u = y - g$ with the assumption of g yields

$$y \in W_2^{k-1}((0,T);\, W_2^{2m+\min(q,m)}(\Omega)),$$
$$y \in W_2^{k-2}((0,T);\, W_2^{2m+\min(q,m_1)}(\Omega)),\, m_1 = 2m + \min(m,q), \text{ etc.}$$

As in Theorem 27.6 we obtain from the weak equation

$$\left(\frac{\mathrm{d}u}{\mathrm{d}t}, \varphi\right)_0 + (\mathscr{L}u, \varphi)_0 = (\tilde{f}, \varphi)_0 = (f - Ag - g_t, \varphi)_0, \quad \varphi \in V,$$

the partial differential equation

$$\frac{\partial u(x,t)}{\partial t} + A(x,t,\mathrm{D}_x)u(x,t) = f(x,t) - A(x,t,\mathrm{D}_x)g - \frac{\partial g(x,t)}{\partial t},$$

with $u(x,0) = y_0(x) - g(x,0)$, or

$$\frac{\partial y(x,t)}{\partial t} + A(x,t,\mathrm{D}_x)y(x,t) = f(x,t),$$

with $y(x,0) = y_0(x)$. We obtain the uniqueness of the solution from Theorem 27.6, since by Corollary 21.1 $\mathscr{L}(t)$ agrees with $A(x,t,\mathrm{D}_x)$ on $W^{2m}(B)$. ■

Once more by choosing k and q to be sufficiently large we can always arrange that $y \in W_2^l((0,T), W_2^l(\Omega))$.

Our next theorem shows how the functional analytic spaces $W_2^l((0,T); W_2^l(\Omega))$ fit together with the Sobolev spaces $W_2^l(Q)$, $Q = \Omega \times (0,T)$ considered earlier. If Ω has the segment property (see Definition 2.1), then by Theorem 3.6 $C^l(\bar{\Omega})$ is dense in $W_2^l(\Omega)$. We recall that all Sobolev spaces are separable – see Theorem 3.1.

Theorem 27.8 *Let Ω have the segment property. Then*

$$W_2^l((0,T); W_2^l(\Omega)) \subsetneq W_2^l((0,T) \times \Omega), \tag{45}$$

and the inclusion is continuous.

Proof. For the sake of brevity we write $W = W_2^l(\Omega)$. Let $u \in W_2^l((0,T), W)$ and let $\{e_n : n \in \mathbb{N}\}$ be an orthonormal basis of W. Without loss of generality we can require that $e_n \in C^l(\bar{\Omega})$. W is separable, $C^l(\bar{\Omega})$ dense in W, we find $\{C_n\}_{n\in\mathbb{N}}$ dense in W, $C_n \in C^l(\bar{\Omega})$; we take $\{C'_n\}_{n\in N}$ linearly independent and such that the linear hulls of $\{C_n\}$ and $\{C'_n\}$ are the same, and apply the Schmidt orthogonalisation process to $\{C'_n\}$. We have

$$u(t) = \sum_{n=1}^{\infty} u_n(t) e_n \quad \text{with} \quad u_n(t) = (u(t), e_n)_W, \tag{46}$$

is measurable since $u(t)$ was measurable. Also

$$|u_n(t)| \leqslant \| u(t) \|_W,$$

hence

$$\int_0^T |u_n(t)|^2 \, \mathrm{d}t \leqslant \int_0^T \| u(t) \|_W^2 \, \mathrm{d}t < \infty \quad \text{for } n \in \mathbb{N},$$

that is,

$$u_n(t) \in L^2(0,T).$$

Moreover, if $K = 0, \ldots, l$, $\mathrm{D}_t^K u_n(t) = (\mathrm{D}_t^K u(t), e_n)_w \in L^2(0,T)$, hence $u_n(t) \in W_2^l(0,T)$.

We apply the Bessel equation of the Hilbert space $W_2^l((0,T); W)$ to (44) and obtain after easy calculation

$$\sum_{n=1}^{\infty} \| u_n(t) \|_{W_2^l(0,T)}^2 = \| u(t) \|_{W_2^l((0,T);W)}^2 < \infty. \tag{47}$$

Since $(0,T)$ has the segment property (trivial), we can approximate the $u_n(t)$s by $C^l[0,T]$-functions.

$$\|u_n(t)-\tilde{u}_n^m(t)\|_{W_2^l}<\frac{1}{2^n m},\quad \tilde{u}_n^m\in C^l[0,T]. \tag{48}$$

We put

$$w_m(t)=\sum_{n=1}^{m}\tilde{u}_n^m(t)e_n. \tag{49}$$

We have

$$w_m(t)\to u(t)\quad \text{in } W_2^l((0,T);W): \tag{50}$$

Let $\varepsilon>0$ be arbitrary and m_0 so large that

$$\frac{1}{m}<\frac{\varepsilon}{2}\quad\text{and}\quad\sum_{n=m+1}^{\infty}\|u_n(t)\|^2_{W_2^l(0,T)}<\frac{\varepsilon}{2}\quad\text{for}\quad m\geqslant m_0 \tag{51}$$

(see (47)). We then have for $m\geqslant m_0$ – because of (48) and (51) –

$$\|u(t)-w_m(t)\|^2_{W_2^l((0,T),W)}=\sum_{n=1}^{m}\|u_n(t)-\tilde{u}_n^m(t)\|^2_{W_2^l(0,T)}+\sum_{n=m+1}^{\infty}\|u_n(t)\|^2_{W_2^l(0,T)}$$

$$\leqslant\frac{1}{m}\sum_{n=1}^{m}\frac{1}{2^n}+\frac{\varepsilon}{2}<\frac{1}{m}\cdot 1+\frac{\varepsilon}{2}<\frac{\varepsilon}{2}+\frac{\varepsilon}{2}=\varepsilon.$$

By our choice of the functions $e_n(x)$ and $\tilde{u}_n^m(t)$ we have arranged that

$$w_m(t,x)=\sum_{n=1}^{m}\tilde{u}_n^m(t)e_n(x)\in C^l(\overline{(0,T)\times\Omega}), \tag{52}$$

and we now show that $(w_m(t,x))$ is a Cauchy sequence in $W_2^l((0,T)\times\Omega)$.

Let $Q=(0,T)\times\Omega$, then

$$\|w_m-w_k\|^2_{W_2^l(Q)}=\int_0^T\int_\Omega\sum_{|\nu|\leqslant l}\left|\sum_{n=m+1}^{k}D^\nu_{(l,x)}(\tilde{u}_n^m(t)-\tilde{u}_n^k(t))e_n(x)\right|^2\mathrm{d}x\mathrm{d}t \tag{53}$$

$$\leqslant\int_0^T\int_\Omega\sum_{|s|\leqslant l}\left|\sum_{n=m+1}^{k}D_t^r D_x^s(\tilde{u}_n^m(t)-\tilde{u}_n^k(t))e_n(x)\right|^2\mathrm{d}x\,\mathrm{d}t=\|w_m-w_k\|^2_{W_2^1((0,T);W_2^1(\Omega))}$$

There is the inequality sign because more derivatives can appear on the right-hand side. An estimate similar to (53) also shows that $w_m\in W_2^l(\Omega)$. We now map $u\mapsto\tilde{u}=\lim_{m\to\infty}w_m(t,x)\in W_2^l(Q)$, then by completeness $\tilde{u}\in W_2^l(Q)$ and continuity follows from (53) ■

If $l\in\mathbb{N}$ is preassigned, we can always attain by means of appropriate regularity conditions on y_0 and f that the solution y of the parabolic differential equation is such that, see (42),

$$y\in W_2^l((0,T);W_2^l(\Omega)).$$

Theorem 27.8 gives in conjunction with Sobolev's lemma (Theorem 6.2) that for $l-l'>(r+1)/2$

$$y\in W_2^l((0,T);W_2^l(\Omega))\subsetneq W_2^l((0,T)\times\Omega)\subsetneq C^{l'}([0,T]\times\bar{\Omega}).$$

§28 Examples

We begin with the mixed initial value Dirichlet problem.

Example 28.1 Let $A(x,\mathrm{D}_x)$ be a linear, strongly elliptic, differential operator of the form §19, (1) with the sesquilinear form

$$a(\varphi,\psi)=\int_\Omega\sum_{\substack{|\alpha|\leqslant m\\|\beta|\leqslant m}}a_{\alpha\beta}(x)\mathrm{D}^\beta\varphi\cdot\mathrm{D}^\alpha\bar{\psi}\,\mathrm{d}x.$$

Let Ω be bounded, we set $H=L_2(\Omega)$, $V=\mathring{W}_2^m(\Omega)$. If $a_{\alpha\beta}\in C(\bar{\Omega})$ we have the estimates

$$|a(\varphi,\psi)|\leqslant c\,\|\varphi\|_m\cdot\|\psi\|_m\quad\text{for all }\varphi,\psi\in W_2^m(\Omega),\tag{1}$$

and the Gårding inequality, see Theorem 19.2

$$\operatorname{Re}a(\varphi,\varphi)+k_0(\varphi,\varphi)_0\geqslant c\,\|\varphi\|_m^2\quad\text{for all }\varphi\in\mathring{W}_2^m(\Omega).\tag{2}$$

For $f\in L_2((0,T);W_2^{-m}(\Omega))$, $T<\infty$, see Definition 17.2, and the initial function $y_0(x)\in L_2(\Omega)$, there exists by Theorem 26.1 a unique solution $y(x,t)$ of the parabolic equation

$$\frac{\partial y(x,t)}{\partial t}+A(x,\mathrm{D}_x)y(x,t)=f(x,t)\quad\text{in }\Omega\times(0,T).\tag{3}$$

which satisfies the initial condition

$$y(x,0)=y_0(x)\quad\text{on }\Omega,\tag{4}$$

and for almost all $t\in[0,T]$ the boundary condition

$$y(\cdot,t)\in\mathring{W}_2^m(\Omega).\tag{5}$$

For the solution $y(x,t)$ we have

$$y \in L_2((0,T); \mathring{W}_2^m(\Omega)), \frac{\partial y}{\partial t} \in L_2((0,T); W_2^{-m}(\Omega))$$

We have used the fact that if $V = \mathring{W}_2^m(\Omega)$. The weak equation (P) in §26 is equivalent to (3).

We can also solve the inhomogeneous Dirichlet problem, if in addition we are given a boundary function $g(x,t)$ with

$$g \in L_2((0,T); W_2^m(\Omega)), \quad \frac{\partial g}{\partial t} \in L_2((0,T); W_2^m(\Omega)').$$

Since the condition §27, (29) $k=0$ is automatically satisfied because of the assumptions on g and y_0 and because of Theorem 25.5, we can apply Theorem 27.4 by which there exists a unique solution of (3) and (4), which satisfies the boundary conditions

$$y(\cdot,t) - g(\cdot,t) \in \mathring{W}_2^m(\Omega) \quad \text{for almost all } t \in [0,T]. \tag{6}$$

For this solution we have

$$y \in L_2((0,T); W_2^m(\Omega)), \quad \frac{\partial y}{\partial t} \in L_2((0,T); W_2^{-m}(\Omega)).$$

we now strengthen assumptions and take

$$\Omega \in C^{m,1}, \quad a_{\alpha\beta} \in C^m(\bar{\Omega}), \tag{7}$$

$k=1$ in Theorem 27.3 and

$$f \in L_2((0,T), L_2(\Omega)) = L_2(\Omega \times (0,T)); \quad \frac{\partial f}{\partial t} \in L_2((0,T); W_2^{-m}(\Omega)),$$

$$y_0 \in \mathring{W}_2^m(\Omega) \cap W_2^{2m}(\Omega) = W^{2m}(B), \qquad f(x,0) \in L_2(\Omega),$$

from which the compatibility conditions §27, (26) for $k=1$ follow:

$$y_0 \in V = \mathring{W}_2^m(\Omega), \quad f(x,0) - Ay_0 \in L_2(\Omega),$$

and we can apply Theorem 27.6. Hence the unique solution y of (3) and (4) satisfies the boundary conditions (5) for all $t \in [0,T]$, and because of (7) by Theorem 8.7 the trace operator is defined, giving

$$y(x,t) = 0, \frac{\partial y(x,t)}{\partial n} = 0, \ldots, \frac{\partial^{m-1} y(x,t)}{\partial n^{m-1}} = 0 \text{ on } \partial\Omega \times [0,T].$$

Here n points in the direction of the inner normal to the surface $\partial\Omega \times [0,T]$.

By Theorem 27.5 we also have

$$y \in W_2^1((0,T); \mathring{W}_2^m(\Omega)), \quad \frac{\partial^2 y}{dt^2} \in L_2((0,T); W_2^{-m}(\Omega)), \quad y \in L_2((0,T); W^{2m}(B)).$$

There are analogous results for the inhomogeneous Dirichlet problem deriving from Theorem 27.7.

Example 28.2 The mixed initial value Neumann problem. We continue our treatment of Example 21.5. Let

$$A(x, D_x) = -\sum_{j,k}^{r} a_{jk} \frac{\partial}{\partial x_k} \frac{\partial}{\partial x_k} + \sum_{j=1}^{r} a_j \frac{\partial}{\partial x_j} + a_0$$

be a strongly elliptic operator of the second order with the sesquilinear form

$$a(\varphi, \psi) = \int_\Omega \left[\sum_{j,k}^{r} a_{jk} \frac{\partial \varphi}{\partial x_j} \frac{\partial \bar{\psi}}{\partial x_k} + \sum_{j=1}^{t} \left(a_j + \sum_{k=1}^{r} \frac{\partial a_{jk}}{\partial x_k} \right) \frac{\partial \varphi}{\partial x_j} \cdot \bar{\psi} + a_0 \cdot \varphi \cdot \bar{\psi} \right] \mathrm{d}x. \quad (8)$$

We assume that

$$\Omega \in C^{2,1}, \quad a_{j,k} \in C^3(\bar{\Omega}), \quad a_j, a_0 \in C^2(\bar{\Omega}),$$

where $N_A|_{\partial\Omega}$ is defined in the sense of the trace operator of Theorem 8.7 we take $H = L_2(\Omega)$, $V = W_2'(\Omega)$. Estimates (1) and (2) hold, for the latter see Example 19.1. Therefore the assumptions of Theorem 26.1 are satisfied. In order to be able to apply Theorem 27.6 we take $k = 1$,

$$W^2(B) = W^2(N_A) = \{\varphi \in W_2^2(\Omega): N_A \varphi = 0 \quad \text{on } \partial\Omega\},$$

$$f \in L_2(\Omega \times (0,T)), \quad \frac{\partial f}{\partial t} \in L_2((0,T); W_2^1(\Omega)')$$

$$y_0 \in W^2(N_A), \quad f(x,0) \in L_2(\Omega),$$

then

$$y_0 \in V = W_2^1(\Omega), \quad f(x,0) - Ay_0 \in L_2(\Omega) \quad \text{(compatibility)}.$$

On $W^2(N_A)$ the representation operator $\mathscr{L}$ agrees with $A(x, \mathrm{D}_x)$, and we have that there exists a unique solution y of (3) and (4), which for almost all $t \in [0,T]$ satisfies the Neuman boundary condition

$$N_A y(x,t) = 0, \quad x \in \partial\Omega. \quad (9)$$

For this soluton y we have

$$y \in W_2^1((0,T); W_2^1(\Omega)), \quad \frac{\partial^2 y}{\partial t^2} \in L_2((0,T); W_2^1(\Omega)'), \quad y \in L_2((0,T); W^2(N_A)).$$

If one wants to satisfy the boundary condition (9) for all $t \in [0, T]$, one must take $k = 2$ in Theorem 27.5. The inhomogeneous Neumann problem may be solved by means of Theorem 27.7.

Example 28.3 The third boundary value problem for a general parabolic differential operator of the second order. Let $A(x, D_x)$ be as in Example 28.2, we add the boundary form

$$c(\varphi, \psi) = \int_{\partial\Omega} b_0(x)\varphi(x)\bar{\psi}(x)\,d\sigma, \quad \varphi, \psi \in W_2^1(\Omega)$$

to the sesquilinear form (8). Since grad $c = 0 = 2m - 2$, we can apply Theorem 20.2, by which the form

$$\mathscr{A}(\varphi, \psi) = a(\varphi, \psi) + c(\varphi, \psi)$$

is again $W_2^1(\Omega)$-coercive. We make the same assumptions as for Example 28.2, only $W^2(B)$ is now

$$W^2(B) = W^2(N_A + b_0) = \{\varphi \in W_2^2(\Omega); N_A\varphi + b_0\varphi = 0 \text{ on } \partial\Omega\},$$

and by Theorem 27.6 we obtain:

There exists a unique solution y of (3) and (4), which for almost all $t \in [0, T]$ satisfies the so-called third boundary condition

$$N_A y(x, t) + b_0(x)y(x, t) = 0, \quad x \in \partial\Omega.$$

For this solution we have

$$y \in W_2^1((0, T); W_2^1(\Omega)), \quad \frac{\partial^2 y}{\partial t^2} \in L_2((0, T); W_2^1(\Omega)'),$$

$$y \in L_2((0, T); W^2(N_A + b_0)).$$

In all three examples we obtain the heat equation as a special case,

$$A(x, D_x) = -\Delta_x.$$

Example 28.4 The boundary value problem with the skew derivative for the heat equation. We continue our discussion of Example 21.4 and work out – as previously – the case $r = 2$. As sesquilinear form we take

$$\mathscr{A}(\varphi, \psi) = \int_\Omega \left[\frac{\partial\varphi}{\partial x_1}\frac{\partial\bar{\psi}}{\partial x_1} + \frac{\partial\varphi}{\partial x_2}\frac{\partial\bar{\psi}}{\partial x_2} + a\frac{\partial\varphi}{\partial x_1}\frac{\partial\bar{\psi}}{\partial x_2}\right.$$
$$\left. - a\frac{\partial\varphi}{\partial x_2}\frac{\partial\bar{\psi}}{\partial x_1} - \frac{\partial a}{\partial x_1}\frac{\partial\varphi}{\partial x_2}\bar{\psi} + \frac{\partial a}{\partial x_2}\frac{\partial\varphi}{\partial x_1}\bar{\psi}\right]dx,$$

with $H = L_2(\Omega)$, $V = W_2^1(\Omega)$. By Example 21.4 $\mathscr{A}(\varphi, \psi)$ is $W_2^1(\Omega)$-coercive, and we can apply the theorems from §§26, 27 to the weak equation

$$\frac{\mathrm{d}}{\mathrm{d}t}(y, \varphi)_0 + \mathscr{A}(y, \varphi) = (f, \varphi)_0, \quad \varphi \in W_2^1(\Omega) \tag{10}$$

For $\varphi \in \mathscr{D}(\Omega) \subset W_2^1(\Omega)$ (10) gives

$$\frac{\partial y(x,t)}{\partial t} - \Delta_x y(x,t) = f(x,t) \quad \text{in } \Omega \times (0, T). \tag{11}$$

In order to apply Theorem 27.6 we take

$$W^2(B) = W^2\left(\frac{\partial}{\partial \mu}\right) = \left\{\varphi \in W_2^2(\Omega): \frac{\partial \varphi}{\partial \mu} = 0 \text{ on } \partial\Omega\right\},$$

where $\partial/\partial\mu$ is understood to be the directional derivative – see Example 21.4

$$\frac{\partial}{\partial \mu} := \frac{\partial}{\partial x_1}\cdot(n_1 + a n_2) + \frac{\partial}{\partial x_2}\cdot(n_2 - a n_1),$$

$\mathbf{n} = (n_1, n_2)$ being the normal vector to $\partial\Omega$. Let $\Omega \in C^{2,1}$,

$$f \in L_2(\Omega \times (0, T)), \qquad \frac{\partial f}{\partial t} \in L_2((0, T); W_2^1(\Omega)'),$$

$$y_0 \in W^2\left(\frac{\partial}{\partial f}\right), \qquad f(x, 0) \in L_2(\Omega),$$

then for the unique solution $y(x, t)$ of (11) and (4) we have

$$\frac{\partial y(x,t)}{\partial \mu} = 0 \quad \text{for } x \in \partial\Omega, \text{ almost all } t \in [0, T].$$

Similarly

$$y \in W_2^1((0, T); W_2^1(\Omega)), \quad \frac{\partial^2 y}{\partial t^2} \in L_2((0, T); W_2^1(\Omega)'), \quad y \in L_2\left((0, T); W^2\left(\frac{\partial}{\partial \mu}\right)\right).$$

There are corresponding results for the inhomogeneous boundary value problem.

Example 28.5 We wish to present examples for mixed boundary-initial value problems for the parabolic equation with the biharmonic operator Δ_x^2. As form we take $a(\varphi, \psi) = \int_\Omega \Delta\varphi \cdot \overline{\Delta\psi}\, \mathrm{d}x$, and if $\varphi, \psi \in W_2^2(\Omega)$ we have the estimate (1). As boundary values on $\partial\Omega$ we take: (I, Δ), set $H = L_2(\Omega)$, $V = V_1 = \{\varphi \in W_2^2(\Omega): I\varphi = 0 \text{ on } \partial\Omega\}$, and by Example 21.6 have that a is V_1-coercive. In order to apply Theorem 27.6 we take $\Omega \in C^{4,1}$,

$$W^4(B) = W^4(I, \Delta) = \{\varphi \in W_2^4(\Omega);\ \varphi = 0, \quad \Delta\varphi = 0 \text{ auf } \partial\Omega\},$$

$$f \in L_2(\Omega \times (0, T)), \quad \frac{\partial f}{\partial t} \in L_2((0, T); V_1') \quad y_0 \in W^4(I, \Delta), \quad f(x, 0) \in L_2(\Omega),$$

Then there exists a (unique) solution $y(x, t)$ of

$$\frac{\partial y(x,t)}{\partial t} + \Delta_x^2 y(x,t) = f(x,t), \quad y(x,0) = y_0(x), \tag{12}$$

with the boundary conditions

$$y(x,t) = 0, \quad x \in \partial\Omega \quad \text{for all } t \in [0, T];$$
$$\Delta_x y(x,t) = 0, \quad x \in \partial\Omega \quad \text{for almost all } t \in [0, T],$$

for which moreover we have

$$y \in W_2^1((0, T); V_1), \quad \frac{\partial^2 y}{\partial t^2} \in L_2((0, T); V_1'), \quad y \in L_2((0, T); W^4(I, \Delta)). \tag{13}$$

For the boundary value problem $(\partial/\partial n, \partial\Delta/\partial n)$ we take

$$H = L_2(\Omega), \quad V = V_2 = \left\{\varphi \in W_2^2(\Omega): \frac{\partial\varphi}{\partial n} = 0 \quad \text{on} \quad \partial\Omega\right\},$$

by Example 21.6 $a(\varphi, \psi)$ is V_2-coercive. We can again apply Theorem 27.6, for this let

$$\Omega \in C^{4,1}, \; W^4(B) = W^4\left(\frac{\partial}{\partial n}, \frac{\partial\Delta}{\partial n}\right) = \left\{\varphi \in W_2^4(\Omega): \frac{\partial\varphi}{\partial n} = 0, \frac{\partial\Delta\varphi}{\partial n} = 0, \text{ on } \partial\Omega\right\}$$

$$f \in L_2(\Omega \times (0, T)), \quad \frac{\partial f}{\partial t} \in L_2((0, T); V_2'),$$

$$y_0 \in W^4\left(\frac{\partial}{\partial n}, \frac{\partial\Delta}{\partial n}\right), \quad f(x, 0) \in L_2(\Omega).$$

There then exists a unique solution y for (12) and for the boundary conditions on $\partial\Omega \times [0, T]$

$$\frac{\partial y(x,t)}{\partial n} = 0, \quad x \in \partial\Omega \quad \text{for all } t \in [0, T];$$
$$\frac{\partial \Delta_x y(x,t)}{\partial n} = 0, \quad x \in \partial\Omega \quad \text{for almost all } t \in [0, T].$$

The statements corresponding to (13) also hold.

We next consider two examples, for which we do not have the regularity Theorem 27.5 at our disposition. We start from Example 21.7.

Example 28.6 We decompose the frontier $\partial\Omega$ into $\Gamma_1 \cup \Gamma_2$ where $\mu_{r-1}\Gamma_1 \neq 0$. we take $H = L_2(\Omega)$, $V = \tilde{V} = \{\varphi \in W_2^1(\Omega): \varphi = 0 \text{ on } \Gamma_1\}$ and

$$a(\varphi, \psi) = \int_\Omega \sum_{j=1}^r \frac{\partial\varphi}{\partial x_j} \cdot \frac{\partial\bar{\psi}}{\partial x_j} \mathrm{d}x.$$

Since $a(\varphi, \psi)$ is coercive for each V (Example 19.1) and the assumptions of Theorem 26.1 are satisfied, and we have for $f \in L_2((0, T); \tilde{V}')$, $0 < T < \infty$, and the initial function $y_0 \in L_2(\Omega)$, that there exists a unique solution y of the weak equation

$$\frac{\mathrm{d}}{\mathrm{d}t}(y, \varphi)_0 + a(y, \varphi) = (f, \varphi)_0, \quad \varphi \in \tilde{V}, \quad \text{with } y(x, 0) = y_0. \tag{14}$$

This solution y satisfies

$$y \in L_2((0, T); \tilde{V}), \quad \frac{\partial y}{\partial t} \in L_2((0, T); \tilde{V}'). \tag{15}$$

Because $\mathscr{D}(\Omega) \subset \tilde{V}$ from (14) we obtain the heat equation (11) and for almost all $t \in [0, T]$ from (15) the boundary condition

$$y(x, t) = 0, \quad x \in \Gamma_1. \tag{16}$$

Since, however, we have no regularity theorems for Example 21.7, and hence cannot apply Theorem 27.6, we can only regard the boundary condition

$$\frac{\partial y(x, t)}{\partial n} = 0, \quad x \in \Gamma_2, t \in [0, T] \tag{17}$$

as being formally satisfied, it lives in (14). By an application of Theorem 27.3, for example with $k \geqslant 1$, we can certainly improve the statements (15), and also arrange for (16) to be satisfied for all $t \in [0, T]$. However, we gain little for the boundary condition (17).

Example 28.7 The transmission problem for the heat equation. Let $\Omega = \Omega_1 \cup \Omega_2$, $\Gamma_1 = \bar{\Omega}_1 \cap \partial\Omega$, $\Gamma_2 = \bar{\Omega}_2 \cap \partial\Omega$, $\Gamma = \bar{\Omega}_1 \cap \bar{\Omega}_2$, $a, b > 0$. Using the notation of Example 21.8 we can once more solve (uniquely)

$$\frac{\partial y_1(x, t)}{\partial t} - a\Delta_x y_1(x, t) = f(x, t) \quad \text{in } \Omega_1 \times (0, T), \quad y_1(x, 0) = y_0(x) \quad \text{in } \Omega_1,$$

$$\frac{\partial y_2(x, t)}{\partial t} - b\Delta_x y_2(x, t) = f(x, t) \quad \text{in } \Omega_2 \times (0, T), \quad y_2(x, 0) = y_0(x) \quad \text{in } \Omega_2,$$

with the boundary conditions

$$y_1(x,t)=0 \quad \text{on } \Gamma_1 \times [0,T], \quad y_2(x,t)=0 \quad \text{on } \Gamma_2 \times [0,T],$$
$$y_1(x,t)=y_2(x,t) \quad \text{on } \Gamma \times [0,T]$$

while we can only consider the second transmission condition

$$a\frac{\partial y_1(x,t)}{\partial n}=b\frac{\partial y_2(x,t)}{\partial n} \quad \text{on } \Gamma \times [0,T],$$

as being formally satisfied. It is again contained in the appropriate weak equation, see also Example 28.6.

Example 28.8 The Cauchy problem for the heat equation. Let $\Omega = \mathbb{R}^r$, we consider the sesquilinear form

$$a(\varphi,\psi)=\int_{\mathbb{R}^r}\sum_{j=1}^{r}\frac{\partial\varphi}{\partial x_j}\cdot\frac{\partial\overline{\psi}}{\partial x_j}\,dx, \quad \varphi,\psi\in W_2^1(\mathbb{R}^r). \tag{18}$$

Estimate (1) holds and

$$\operatorname{Re} a(\varphi,\varphi)+(\varphi,\varphi)_0=\|\varphi\|_1^2, \quad \varphi\in \mathring{W}_2^1(\mathbb{R}^r)=W_2^1(\mathbb{R}^r),$$

so that the hypotheses for Theorem 26.1 are satisfied, $H=L_2(\mathbb{R}^r)$, $V=W_2^1(\mathbb{R}^r)$. If $f\in L_2((0,T);\ W_2^{-1}(\mathbb{R}^r))$ and the initial function $y_0(x)\in L_2(\Omega)$, by Theorem 26.1 there exists a unique solution $y(x,t)$ of the heat equation

$$\frac{\partial y(x,t)}{\partial t}-\Delta_x y(x,t)=f(x,t) \text{ in } \mathbb{R}^r \times (0,T),$$

with

$$y(x,0)=y_0(x) \text{ in } \mathbb{R}^r,$$

and we have

$$y\in L_2((0,T);\ W_2^1(\mathbb{R}^r)),\ \frac{\partial y}{\partial t}\in L_2((0,T);\ W_2^{-1}(\mathbb{R}^r)).$$

We now use Theorem 27.3 in order to make regularity statements about the solution y of the Cauchy problem. Let $f\in W_2^k((0,T);\ W_2^{-1}(\mathbb{R}^r))$. Here the compatibility conditions §27, (26) have the form (as one easily calculates, see also §27),

$$y_0\in W_2^1,\ \ f(0)+\Delta y_0\in W_2^1,\ \ f'(0)+\Delta f(0)+\Delta^2 y_0\in W_2^1,\ldots,$$
$$y^{(k)}(0)=f^{(k-1)}(0)+\Delta f^{(k-2)}(0)+\cdots+\Delta^{k-1}f(0)+\Delta^k y_0\in L_2.$$

They are satisfied, if one takes

$$y_0 \in W_2^{2k}, \quad f(0) \in W_2^{2(k-1)}, \quad f'(0) \in W_2^{2(k-2)}, \ldots, f^{(k-1)}(0) \in L_2, \tag{19}$$

see also the formulae §27, (19). The hypotheses for Theorem 27.3 are satisfied by taking (18) and (19), and we have

$$y \in W_2^k((0, T); W_2^1(\mathbb{R}^r)). \tag{20}$$

We now bring the Green operator into play. We consider the equation ($u \in W_2^s(\mathbb{R}^r)$)

$$(-\Delta + 1)u(x) = g(x), \quad g \in W_2^{s-2}(\mathbb{R}^r).$$

By application of the Fourier transformation $\mathscr{F}$ we obtain

$$(|\xi|^2 + 1)\hat{u} = \hat{g},$$

and we see that $(-\Delta + 1)$ is an isomorphism between the spaces $W_2^s(\mathbb{R}^r)$ and $W_2^{s-2}(\mathbb{R}^r)$. Hence the Green operator acts continuously

$$G := (-\Delta + 1)^{-1} : W_2^{s-2}(\mathbb{R}^r) \to W_2^s(\mathbb{R}^r).$$

and by Theorem 27.1

$$G : W_2^k((0, T); W_2^{s-2}(\mathbb{R}^r)) \to W_2^k((0, T); W_2^s(\mathbb{R}^r)) \text{ is continuous.} \tag{21}$$

We now impose further conditions on f

$$\begin{aligned} &f \in W_2^{k-1}((0, T); W_2^1(\mathbb{R}^r)), \\ &f \in W_2^{k-2}((0, T); W_2^3(\mathbb{R}^r)), \\ &\vdots \\ &f \in W_2^{k-l}((0, T); W_2^{2l-1}(\mathbb{R}^r)) \end{aligned} \tag{22}$$

We write the heat equation in the form

$$(-\Delta_x + 1)y(x, t) = f(x, t) + y(x, t) - \frac{\partial y(x, t)}{\partial t} = \tilde{f}, \tag{23}$$

Then for the right-hand side of (23) we have because of (20) and (22) that $\tilde{f} \in W_2^{k-1}((0, T); W_2^1(\mathbb{R}^r))$, and the Green operator (21) applied to (23) gives

$$y \in W_2^{k-1}((0, T); W_2^3(\mathbb{R}^r)). \tag{24}$$

(24) with (22; 2) gives in the same way

$$y \in W_2^{k-2}((0, T); W_2^5(\mathbb{R}^r)),$$

and the last assumption on (22) gives

$$y \in W_2^{k-l}((0, T); W_2^{2l+1}(\mathbb{R}^r)). \tag{25}$$

If we take $k = 3l + 1$, $l = 0, 1, \ldots$, from (25) by Theorem 27.8 we obtain

$$y \in W_2^{2l+1}((0, T); W_2^{2l+1}(\mathbb{R}^r)) \hookrightarrow W_2^{2l+1}(\mathbb{R}^r \times (0, T)).$$

Exercises

28.1 Solve the eigenvalue problem on $[0, 1]$

$$y''(x) - \lambda y(x) = 0, \quad y(0) = 0, \quad y(1) = 0,$$

that is, find the eigenvalues λ_k, $k = 1, 2, \ldots$ and the eigenfunctions y_k, $k = 1, 2, \ldots$. Make the eigenfunctions orthonormal and show that they form an orthonormal basis for $L_2(0, 1)$.

28.2 Solve the mixed problem

$$\begin{aligned} &\frac{\partial y(x,t)}{\partial t} - \frac{\partial^2 y(x,t)}{\partial x^2} = f(x,t), \quad 0 < x < 1, t > 0, \\ &y(x,0) = y_0(x), \quad y(0,t) = 0, \quad y(1,t) = 0, \end{aligned} \tag{1}$$

where

$$f(x,t) \in L_2((0,1) \times \mathbb{R}_+), \quad y_0(x) \in L_2(0,1)$$

with the help of the so-called Fourier method, that is, we develop (λ_k and y_k from 28.1)

$$f(x,t) = \sum_{k=1}^{\infty} f_k(t) y_k(x), \quad y_0(x) = \sum_{k=1}^{\infty} c_k y_k(x),$$

and try

$$y(x,t) = \sum_{k=1}^{\infty} e_k(t) y_k(x), \tag{2}$$

as a solution where $e_k(t)$ satisfies the initial value problem

$$e_k'(t) + \lambda_k e_k(t) = f_k(t), \quad k = 1, 2, \ldots, \quad e_k(0) = c_k.$$

28.3 Show that the solution (2) of (1) 'fits' into the framework of §26, that is, carry out the necessary estimates.

28.4 Give regularity theorems for the solution $y(x, t)$ of (1) in terms of the representation (2).

28.5 Solve with the help of the Fourier method the nonhomogeneous mixed problem

$$\begin{aligned} &\frac{\partial y(x,t)}{\partial t} - \frac{\partial^2 y(x,t)}{\partial x^2} = f(x,t), \quad 0 < x < 1, t > 0, \\ &y(x,0) = y_0(x), \quad y(0,t) = g_0(t), \quad y(1,t) = g_1(t), \end{aligned}$$

where the functions f, y_0, g_0 and g_1 are preassigned. State regularity theorems for the solution $y(x, t)$ in terms of the differentiability of the functions f, y_0, g_0 and g_1.

28.6 A generalisation: suppose that the assumptions of §13 and §15 are satisfied. We consider the elliptic, self-adjoint, boundary value problem $(A, b_1, \ldots, b_m) = L$ on $\Omega \subset \mathbb{R}^r$. Suppose the assumptions for Theorem 15.9 are satisfied, let λ_k be the eigenvalues and y_k the eigenfunctions of §13, (50). By Theorem 15.9 the y_k

(suitably orthonormalised) form an orthonormal basis for $L_2(\Omega)$. We consider the mixed problem

$$\frac{\partial y(x,t)}{\partial t} + A(x, D_x)y(x,t) = f(x,t), \quad x \in \Omega, t > 0, \tag{3}$$

$y(x,0) = y_0(x)$; $b_1(x, D_x)y(x,t) = 0, \ldots, b_m(x, D_x)y(x,t) = 0$ for $x \in \Omega$ and $t \geqslant 0$, where for the given functions we have

$$y_0(x) \in L_2(\Omega), \quad f(x,t) \in L_2(\Omega \times \mathbb{R}_+).$$

We solve the mixed problem in terms of

$$y_0(x) = \sum_{k=1}^{\infty} c_k y_k(x), \quad f(x,t) = \sum_{k=1}^{\infty} f_k(t) y_k(x), \tag{4}$$

and for the solution $y(x,t)$ put

$$y(x,t) = \sum_{k=1}^{\infty} e_k(t) y_k(x), \tag{5}$$

where $e_k(t)$ satisfies the initial value problem

$$e_k'(t) + \lambda_k e_k(t) = f_k(t), \quad e_k(0) = c_k, \quad k = 1, 2, \ldots \tag{6}$$

Can one carry over the existence and uniqueness theorem from §26 to (3) and (5)?

28.7 Use the Corollary 13.1, write the solution of (6) down explicitly, state differentiability assumptions for $y_0(x)$ and $f(x,t)$, and in this way prove regularity theorems for the solution (5).

28.8 Solve the Cauchy problem by application of the Fourier transformation $\mathscr{F}_x$:

$$\frac{\partial y(x,t)}{\partial t} - \frac{\partial^2 y(x,t)}{\partial x^2} = 0, \quad y_0(x) = \begin{cases} c & \text{for } |x| < x_0, \\ 0 & \text{for } |x| \geqslant x_0. \end{cases}$$

28.9 Similarly

$$\frac{\partial y(x,t)}{\partial t} - \frac{\partial^2 y(x,t)}{\partial x^2} = 0, \quad y_0(x) = ce^{-b^2x^2}.$$

V

Hyperbolic differential operators

§29 The existence and uniqueness of the solution

As for parabolic equations we first present an abstract solution theorem. Again, let V, H be Hilbert spaces with $V \subsetneq H$ dense; V separable. Then in the familiar notation

$$V \subsetneq H \subsetneq V'$$

we have a Gelfand triple. If $t \in [0, T]$, $0 < T < \infty$, let the sesquilinear form $a(t, \varphi, \psi)$ be continuous, that is,

$$|a(t; \varphi, \psi)| \leqslant c \| \varphi \|_V \| \psi \|_V; \quad \varphi, \psi \in V, \tag{1}$$

where c is independent of t. By Theorem 17.9 there exists a representation operator

$$L(t): V \to V', \tag{2}$$

which for each t is continuous and linear, with $a(t, \varphi, \psi) = (L(t)\varphi, \psi)_H$, see also §26.

We assume further that for all $\varphi, \psi \in V$ the function $t \mapsto a(t; \varphi, \psi)$ (φ, ψ fixed) is continuously differentiable for $t \in [0, T]$; in short

$$a(t; \varphi, \psi) \in C^1[0, T] \quad \text{for all } \varphi, \psi \in V,$$

where

$$\left| \frac{\mathrm{d}}{\mathrm{d}t} a(t; \varphi, \psi) \right| \leqslant c \| \varphi \|_V \| \psi \|_V \quad \text{for all } t \in [0, T], \tag{3}$$

once more independent of t. Furthermore suppose (antisymmetry) that

$$a(t; \varphi, \psi) = \overline{a(t; \psi, \varphi)} \quad \text{for all } \varphi, \psi \in V, \tag{4}$$

We also suppose V-coercion: there exist constants k_0, $\alpha > 0$ with

$$a(t;\varphi,\varphi) + k_0 \|\varphi\|_H^2 \geqslant \alpha \|\varphi\|_V^2 \quad \text{for all } t \in [0, T] \quad \text{for all } \varphi \in V, \tag{5}$$

observe that because of (4) $a(t;\varphi,\psi)$ is real.

(4) in conjunction with (1) and (5) implies by Theorem 17.12 that the representation operator $L(t)$ and the Green operator G_{k_0} are self-adjoint.

Remark (1) is assumption (b) from §26, (5) is (c), while (3) is a sharpened version of (a), it is known that continuity implies measurability. In comparison with §26, here the hypothesis (4) is new. The self-adjointness of $L(t)$ enters in an essential way into the existence-uniqueness proof for hyperbolic equations.

With this preliminary material we consider problem (H): suppose given $f \in L^2((0, T), H)$, $T < \infty$, and initial conditions

$$y_0 \in V, y_1 \in H.$$

We look for a function $y(t)$ with $y \in L^2((0, T); V)$, $\mathrm{d}y/\mathrm{d}t \in L^2((0, T); H)$, so that in V' we have

$$\frac{\mathrm{d}^2 y}{\mathrm{d}t^2} + L(t)y = f \quad \text{for } t \in [0, T], \quad y(0) = y_0, \quad \frac{\mathrm{d}y(0)}{\mathrm{d}t} = y_1. \tag{H}$$

Here we read the equation in V' or – equivalently – as a weak equation in the Gelfand triple $V \subsetneq H \subsetneq V'$, that is, as

$$\left(\frac{\mathrm{d}^2 y}{\mathrm{d}t^2}, \varphi\right)_H + (L(t)y, \varphi)_H = (f, \varphi)_H \quad \text{for all } \varphi \in V.$$

Remark 29.1 Because of (1) and (3) ($a(t;\varphi,\psi)$ is measurable in t) we can apply Lemma 26.1, by which L maps $L^2((0, T); V)$ to $L^2((0, T); V')$ and given our assumptions on f and y, we obtain from (H)

$$\frac{\mathrm{d}^2 y}{\mathrm{d}t^2} = f - L(t)y \in L^2((0, T); V')$$

for each solution y. By Theorem 25.5 we see that

$$\frac{\mathrm{d}y}{\mathrm{d}t}:[0, T] \to V' \quad \text{and} \quad y:[0, T] \to H \text{ are continuous.}$$

The initial conditions are therefore taken continuously in H or V'. However, it is even true (see Lions & Magenes [1] Vol. 1, p. 275) that

$$y:[0,T]\to V \quad \text{and} \quad \frac{\mathrm{d}y}{\mathrm{d}t}:[0,T]\to H \text{ are continuous.}$$

Hence the initial conditions even make sense in V or in H.

We need some integral inequalities.

Lemma 29.1 *Let $g(t)$, $v(t)$, $w(t)$ be continuous functions in $C[0,T]$, let $0\leqslant h(t)\in L_1(0,T)$, and suppose that in $[0,T]$ we have*

$$v(t)\leqslant g(t)+\int_0^t h(\tau)v(\tau)\,\mathrm{d}\tau, \quad w(t)\geqslant g(t)+\int_0^t h(\tau)w(\tau)\,\mathrm{d}\tau, \tag{6}$$

where for each single value $t\in[0,T]$, equality in (6) holds for at most one of the pair. Then

$$v(t)<w(t) \quad \textit{for all } t\in[0,T].$$

Proof. At $t=0$ the assumption implies that $v(0)\leqslant g(0)\leqslant w(0)$, where equality holds at most once, hence $v(0)<w(0)$. If the assertion were false, then there would exist a first point $t_0\in(0,T]$ such that $v(t_0)=w(t_0)$ and $v<w$ for $0\leqslant t<t_0$. With $h(t)\geqslant 0$ this last implies that

$$h(\tau)v(\tau)\leqslant h(\tau)w(\tau) \quad \text{if } 0\leqslant\tau<t_0,$$

from which it follows by (6)

$$v(t_0)\leqslant g(t_0)+\int_0^{t_0} h(\tau)v(\tau)\,\mathrm{d}t\leqslant g(t_0)+\int_0^{t_0} h(\tau)w(\tau)\,\mathrm{d}t\leqslant w(t_0).$$

Again there is strict inequality in at least one place, which gives the contradiction to prove the theorem.

Lemma 29.2 (Gronwall) *Let $g(t)$, $v(t)$ be continuous functions in $C[0,T]$, let $0\leqslant h(t)\in L_1(0,T)$, and suppose that in $[0,T]$ we have*

$$v(t)\leqslant g(t)+\int_0^t h(\tau)v(\tau)\,\mathrm{d}\tau. \tag{7}$$

Then in $[0,T]$ we have

$$v(t)\leqslant g(t)+\int_0^t g(\tau)h(\tau)e^{H(t)-H(\tau)}\,\mathrm{d}\tau=e^{H(t)}\left[g(0)+\int_0^t g'(\tau)e^{-H(\tau)}\,\mathrm{d}\tau\right], \tag{8}$$

with $H(t)=\int_0^t h(\tau)\,\mathrm{d}\tau$ (the second form of the expression in (8) holds for differentiable functions $g(t)$).

The proof uses the fact that the function

$$w(t) := \bar{g}(t) + \int_0^t \bar{g}(\tau)h(\tau)e^{H(t)-H(\tau)}\,d\tau$$

is a solution of the integral equation

$$w(t) = \bar{g}(t) + \int_0^t h(\tau)w(\tau)\,d\tau.$$

The equality of the two integrals follows by differentiation. If we choose $\bar{g} > g$, then besides (7) we also have the inequality

$$w(t) > g(t) + \int_0^t h(\tau)w(\tau)\,d\tau,$$

and we have satisfied both (6) and the assumptions to Lemma 29.1. Therefore $v < w$. The first part of (8) now follows by passage to the limit, $\bar{g} \to g$, while the second part represents the result of carrying out partial integration.

Lemma 29.3 *Let* $v(t) \in C[0, T]$ *be continuous with* $v(t) \geqslant 0$ *and*

$$v(t) \leqslant C \int_0^t v(\tau)\,dt \quad \textit{for } t \in [0, T].$$

Then $v(t) \equiv 0$.

The result follows immediately from Lemma 29.2 by taking $g(t) := 0$, $h(t) := c$.

After these preliminaries we can prove the existence-uniqueness theorem.

Theorem 29.1 *Subject to the assumptions* (1)–(5) *problem* (H) *has a unique solution. The map*

$$\{f, y_0, y_1\} \to \left\{y, \frac{dy}{dt}\right\}$$

is continuous and linear from

$$L^2((0, T); H) \times V \times H \to L^2((0, T); V) \times L^2((0, T); H).$$

Proof. Uniqueness of the solution. Let y be a solution of (H) with $y_0 = 0$, $y_1 = 0$, $f = 0$. Let $s \in (0, T)$ be arbitrary but fixed. We put

$$\psi(t) := \begin{cases} -\displaystyle\int_t^s y(\sigma)\,d\sigma & \text{for } t < s, \\ 0 & \text{for } t \geqslant s. \end{cases}$$

Then

$$\int_0^T \left(\frac{d^2 y(t)}{dt^2} + L(t)y(t), \psi(t) \right)_H dt = 0.$$

Since

$$\frac{d}{dt}(y'(t), \psi(t))_H = (y''(t), \psi(t))_H + (y'(t), \psi'(t))_H,$$

partial integration gives

$$\int_0^T [a(t; y(t), \psi(t)) - (y'(t), \psi'(t))_H] \, dt = 0,$$

hence

$$\operatorname{Re} \int_0^s [a(t; \psi'(t), \psi(t)) - (y'(t), y(t))_H] \, dt = 0.$$

We put

$$\frac{d}{dt} a(t; u, v) = a'(t; u, v) \quad \text{for all } u, v \in V,$$

and because of (4) have

$$\frac{d}{dt} a(t; \psi(t), \psi(t)) = a'(t; \psi(t), \psi(t)) + a(t; \psi'(t), \psi(t)) + a(t; \psi(t), \psi'(t))$$

$$= a'(t; \psi(t), \psi(t)) + 2 \operatorname{Re} a(t; \psi'(t), \psi(t)).$$

Therefore it follows that

$$\int_0^s \frac{d}{dt} [a(t; \psi(t), \psi(t)) - \| y(t) \|_H^2] \, dt - \int_0^s a'(t; \psi(t), \psi(t)) \, dt$$

$$= 2 \operatorname{Re} \int_0^s [a(t; \psi'(t), \psi(t)) - (y'(t), y(t))_H] \, dt = 0,$$

thus

$$a(0; \psi(0), \psi(0)) + \| y(s) \|_H^2 = - \int_0^s a'(t; \psi(t), \psi(t)) \, dt$$

and with (3) and (5)

$$\alpha \| \psi(0) \|_V^2 - k_0 \| \psi(0) \|_H^2 + \| y(s) \|_H^2 \leqslant a(0; \psi(0), \psi(0)) + \| y(s) \|_H^2$$

$$\leqslant \int_0^s |a'(t; \psi(t), \psi(t))| \, dt$$

$$\leqslant c \int_0^s \| \psi(t) \|_V^2 \, dt.$$

Therefore

$$\alpha \| \psi(0) \|_V^2 + \| y(s) \|_H^2 \leqslant c \int_0^s \| \psi(t) \|_V^2 \, dt + k_0 \| \psi(0) \|_H^2,$$

or

$$\| \psi(0) \|_V^2 + \| y(s) \|_H^2 \leqslant c_1 \left(\int_0^s \| \psi(t) \|_V^2 \, dt + \| \psi(0) \|_H^2 \right).$$

If we put

$$w(t) := \int_0^t y(\sigma) \, d\sigma, \quad \psi(t) := w(s) - w(t),$$

we can rewrite the inequality as

$$\| w(s) \|_V^2 + \| y(s) \|_H^2 \leqslant c_1 \left(\int_0^s \| w(t) - w(s) \|_V^2 \, dt + \| w(s) \|_H^2 \right),$$

from which it follows that

$$(1 - 2c_1 s) \| w(s) \|_V^2 + \| y(s) \|_H^2 \leqslant c_2 \int_0^s (\| w(t) \|_V^2 + \| y(t) \|_H^2) \, dt.$$

If we choose $s_0 = 1/4c_1$, we have

$$1 - 2c_1 s_0 = \tfrac{1}{2},$$

then for $0 \leqslant s \leqslant s_0$ it follows that

$$\| w(s) \|_V^2 + \| y(s) \|_H^2 \leqslant c_3 \int_0^s (\| w(t) \|_V^2 + \| y(t) \|_H^2) \, dt.$$

Hence given Lemma 29.3 we have $y \equiv 0$ in $[0, s_0]$. The choice of s_0 was, however, independent of the origin. Arguing with s_0 instead of 0 we obtain $y = 0$ in $[s_0, 2s_0]$, etc. Hence $y = 0$ in $[0, T]$.

Existence of the solution: the proof proceeds analogously to that of the parabolic differential equation, see Theorem 26.1. Again suppose that $\{w_n : n \in \mathbb{N}\}$ is linearly independent and total in V. Let

$$y_{0m} = \sum_{i=1}^m \xi_{im}^0 w_i, \quad y_{1m} = \sum_{i=1}^m \xi_{im}^1 w_i,$$

with $y_{0m} \to y_0$ in V as $m \to \infty$ and $y_{1m} \to y_1$ in H as $m \to \infty$.

We again look for a sequence $y_m(t)$ which approximates the solution,

$$y_m(t) = \sum_{i=1}^m g_{im}(t) w_i$$

with

$$\frac{d^2}{dt^2}(y_m(t), w_j)_H + a(t; y_m(t), w_j) = (f(t), w_j)_H \quad \text{for } 1 \leqslant j \leqslant m,$$

$$y_m(0) = y_{0m}, \quad y_m'(0) = y_{1m}. \tag{9}$$

This system of m linear ordinary differential equations has a unique solution.

If we multiply (9) by $g_{jm}'(t)$ and sum over j, we obtain

$$(y_m''(t), y_m'(t))_H + a(t; y_m(t), y_m'(t)) = (f(t), y_m'(t))_H,$$

and hence because of (4)

$$\frac{d}{dt}[\| y_m'(t) \|_H^2 + a(t; y_m(t), y_m(t))] - a'(t; y_m(t), y_m(t)) = 2\,\mathrm{Re}\,(f(t), y_m'(t))_H.$$

By integration we obtain

$$\| y_m'(t) \|_H^2 + a(t; y_m(t), y_m(t)) = \| y_{1m} \|_H^2 + a(0; y_{0m}, y_{0m}) + \int_0^t a'(\sigma; y_m(\sigma), y_m(\sigma))\,d\sigma + 2\,\mathrm{Re}\int_0^t (f(\sigma), y_m'(\sigma))_H\,d\sigma,$$

from which by (3) and (5) we obtain the estimate

$$\| y_m'(t) \|_H^2 + \alpha \| y_m(t) \|_V^2 \leqslant k_0 \| y_m(t) \|_H^2 + \| y_{1m} \|_H^2 + c \| y_{0m} \|_V^2 + c\int_0^t \| y_m(\sigma) \|_V^2\,d\sigma + 2\int_0^t \| f(\sigma) \|_H \| y_m'(\sigma) \|_H\,d\sigma,$$

that is

$$\| y_m'(t) \|_H^2 + \| y_m(t) \|_V^2 \leqslant c_1(\| y_{0m} \|_V^2 + \| y_{1m} \|_H^2 + \| y_m(t) \|_H^2) + c_1\int_0^t (\| y_m(\sigma) \|_V^2 + \| f(\sigma) \|_H \| y_m'(\sigma) \|_H)\,d\sigma.$$

From

$$\| y_m(t) \|_H \leqslant \| y_{0m} \|_H + \int_0^t \| y_m'(\sigma) \|_H\,d\sigma$$

it follows that

$$\| y_m(t) \|_H^2 \leqslant 2 \| y_{0m} \|_H^2 + 2\left(\int_0^t \| y_m'(\sigma) \|_H\,d\sigma\right)^2 \leqslant 2 \| y_{0m} \|_H^2 + 2T\int_0^t \| y_m'(\sigma) \|_H^2\,d\sigma.$$

If we put

$$\omega_m(t) := \| y_m'(t) \|_H^2 + \| y_m(t) \|_V^2,$$

it follows that

$$\omega_m(t) \leqslant c_1(\|y_{0m}\|_V^2 + \|y_{1m}\|_H^2) + 2c_1\left(\|y_{0m}\|_H^2 + T\int_0^t \|y_m'(\sigma)\|_H^2\,\mathrm{d}\sigma\right)$$

$$+ c_1\int_0^t \|y_m(\sigma)\|_V^2\,\mathrm{d}\sigma + c_1\int_0^t \|f(\sigma)\|_H\|y_m'(\sigma)\|_H\,\mathrm{d}\sigma$$

and

$$\omega_m(t) \leqslant c_2\left(\|y_{0m}\|_V^2 + \|y_{1m}\|_H^2 + \int_0^t \|f(\sigma)\|_H^2\,\mathrm{d}\sigma\right) + c_2\int_0^t \omega_m(\sigma)\,\mathrm{d}\sigma,$$

the last by the Schwarz inequality

$$2|ab| \leqslant a^2 + b^2.$$

If we now apply Gronwall's lemma 29.2, then because $T < \infty$ we have

$$\omega_m(t) \leqslant C, \quad \text{with } C = c_3\left(\|y_0\|_V^2 + \|y_1\|_H^2 + \varepsilon + \int_0^T \|f(\sigma)\|_H^2\,\mathrm{d}\sigma\right). \tag{10}$$

Hence (y_m) is bounded in $L^2((0, T); V)$, and $(\mathrm{d}y_m/\mathrm{d}t)$ is bounded in $L^2((0, T); H)$. We find a subsequence $(y_\mu) \subset (y_m)$ and elements $z \in L^2((0, T); V)$, $\tilde{z} \in L^2((0, T); H)$ with

$$y_\mu \rightharpoonup z \quad \text{in } L^2((0, T); V) \text{ as } \mu \to \infty,$$

and

$$\frac{\mathrm{d}y_\mu}{\mathrm{d}t} \rightharpoonup \tilde{z} \quad \text{in } L^2((0, T); H) \text{ as } \mu \to \infty$$

(see Theorem 9.1). Of course

$$\tilde{z} = \mathrm{d}z/\mathrm{d}t \quad \text{and} \quad y_\mu(0) \rightharpoonup z(0) \quad \text{in } V \text{ as } \mu \to \infty.$$

However, by assumption $y_\mu(0) = y_{0\mu} \to y_0$ in V as $\mu \to \infty$, therefore $z(0) = y_0$. Suppose now that $\varphi \in C^1[0, T]$ with $\varphi(T) = 0$. If we put $\varphi_j(t) = \varphi(t)w_j$ and multiply (9) by $\varphi(t)$, take $m = \mu > j$, and partially integrate we find that

$$\int_0^T [(-y_\mu'(t), \varphi_j'(t))_H + a(t; y_\mu(t), \varphi_j(t))]\,\mathrm{d}t = \int_0^T (f(t), \varphi_j(t))_H\,\mathrm{d}t + (y_{1\mu}, \varphi_j(0))_H$$

and as $\mu \to \infty$ we have

$$\int_0^T [(-z'(t), \varphi_j'(t))_H + a(t; z(t), \varphi_j(t))]\,\mathrm{d}t = \int_0^T (f(t), \varphi_j(t))_H\,\mathrm{d}t + (y_1, \varphi_j(0))_H.$$

Next if we take $\varphi \in \mathscr{D}(0, T)$, then

$$\frac{d^2}{dt^2}(z(t), w_j)_H + a(t; z(t), w_j) = (f(t), w_j)_H$$

and this holds for all $j \in \mathbb{N}$, which means

$$\frac{d^2 z}{dt^2} + L(t)z = f. \tag{12}$$

From (11) and (12) we obtain by partial integration, that

$$(z'(0), w_j)_H \varphi(0) = (y_1, w_j)_H \varphi(0) \quad \text{for all } w_j, \text{ that is, } z'(0) = y_1.$$

Therefore z is a solution of (H).

Continuous dependence of the solution: applying integration to (10) we obtain

$$\int_0^T \| y_m'(t) \|_H^2 \, dt + \int_0^T \| y_m(t) \|_V^2 \, dt \leqslant c_3 T\left(\| y_0 \|_V^2 + \| y_1 \|_H^2 \varepsilon + \int_0^T \| f(t) \|_H^2 \, dt \right),$$

and by the estimate from Theorem 9.1 (weak convergence!)

$$\int_0^T \| z'(t) \|_H^2 \, dt + \int_0^T \| z(t) \|_V^2 \, dt \leqslant c_3 T\left(\| y_0 \|_V^2 + \| y_1 \|_H^2 + \int_0^T \| f(t) \|_H^2 \, dt \right),$$

since $\varepsilon > 0$ was arbitrary.

§30 Regularity of the solutions of hyperbolic differential equations

As for parabolic equations, §27, we first prove an abstract regularity theorem for hyperbolic equations, and then pass to partial differential operators. Our regularity theorems show that we can uniquely solve the mixed boundary-initial value problem in the framework of Sobolev spaces. Here in the minimal case, $k = 2$, $q = 0$, the compatibility conditions are not particularly restrictive, see Theorems 30.4 and 30.5.

30.1 An abstract regularity theorem

For simplicity we assume that the representation operator L, respectively the form $a(\varphi, \psi)$, is not dependent on t.

Theorem 30.1 *We consider the hyperbolic equation*

$$\frac{d^2 y(t)}{dt^2} + Ly(t) = f(t) \quad \textit{in } (0, T), \tag{1}$$

with the initial conditions

$$y(0) = y_0, \quad y'(0) = y_1. \tag{2}$$

Besides the assumptions made in §29 we suppose that

$$f \in W_2^{k-1}((0, T); H), \quad k \geq 1, \tag{3a}$$

and

$$y_0 \in V, \ldots, y_1 \in V, \ldots, f^{(k-3)}(0) - \cdots \in V, \quad f^{(k-2)}(0) - Lf^{(k-3)}(0) + \cdots \in H. \tag{3b}$$

Then the solution y of (1) *and* (2) *satisfies*

$$y \in W_2^{k-1}((0, T); V), \quad \frac{d^k y(t)}{dt^k} \in L^2((0, T); H), \quad \frac{d^{k+1} y(t)}{dt^{k+1}} \in L^2((0, T); V'). \tag{4}$$

Remark See also Remark 27.2. As for parabolic equations it is possible to abbreviate formulae (3b) by introducing formally the initial values $d^j y(0)/dt^j$. Differentiating (1) formally and substituting $t = 0$, we obtain

$$\frac{d^{2n-1} y(0)}{dt^{2n-1}} = f^{(2n-3)}(0) - Lf^{(2n-5)}(0) + \cdots + (-1)^{n-2} L^{n-2} f'(0)$$
$$+ (-1)^{n-1} L^{n-1} y_1,$$

$$\frac{d^{2n} y(0)}{dt^{2n}} = f^{(2n-2)}(0) - Lf^{2n-4}(0) + \cdots + (-1)^{n-1} L^{n-1} f'(0)$$
$$+ (-1)^n L^n y_0,$$

and (3b) becomes

$$\frac{d^j y(0)}{dt^j} \in V, \quad j = 0, \ldots, k-1, \frac{d^k y(0)}{dt^k} \in H. \tag{3b'}$$

Here 'formally' means that we do not know *a priori* whether the derivatives $d^j y/dt^j$ exist; we know this for certain after using (3b) and proving our Theorem 30.1. The formulae (3b) are compatibility conditions: on the lower lid $t = 0$ of the region $\bar{\Omega} \times [0, T]$ the data $(y_0, y_1, d^j f(0)/dt^j)$ must fit both the solution and the boundary conditions V. An application of Lions' theorem quoted in Remark 29.1 shows that the compatibility conditions are also necessary for regularity as formulated here.

Proof. Induction on k. For $k = 1$ it follows by Theorem 29.1 from the assumptions that

$$y \in L^2((0, T); V), \quad y_t \in L^2((0, T); H), \quad y_{tt} \in L^2((0, T); V'). \tag{5}$$

We now show the transition to $k=2$; the general passage from $k-1$ to k follows in the same way, for this reason we do not write it out again. We formally differentiate equation (1) again we do not know *a priori* whether the corresponding derivatives exist and lie in the correct space, and obtain

$$y_{ttt}+Ly_t=f_t(t), \tag{6}$$

with

$$y_t(0)=y_1\in V,\quad y_{tt}(0)=f(0)-Ly_0\in H,\quad f_t\in L^2((0,T);H), \tag{7}$$

because of the assumptions (3), with $k=2$. Therefore we consider the initial value problem

$$y_{tt}+Lv=f_t,\quad v(0)=y_1\in V,\quad v_t(0)=f(0)-Ly_0\in H, \tag{8}$$

and show that $v=y_t$. By Theorem 29.1 for the solution of (8) we have

$$v\in L^2((0,T);V),\quad v_t\in L^2((0,T);H),\quad v_{tt}\in L^2((0,T);V'). \tag{9}$$

We form the auxiliary function

$$w(t):=y(0)+\int_0^t v(\tau)\,\mathrm{d}\tau.$$

Because of (9) we have

$$\begin{aligned} &w\in C([0,T];V)\subset L^2((0,T);V),\quad w_t=v\in L^2((0,T);V)\subset L^2((0,T);H),\\ &w_{tt}=v_t\in L^2((0,T);H)\subset L^2((0,T);V'), \end{aligned} \tag{10}$$

and (see (8))

$$w(0)=y(0)=y_0,\quad w_t(0)=v(0)=y_1. \tag{11}$$

Now back to the auxilliary function $w(t)$. We integrate (8) and because of the initial condition in (8) have

$$w_{tt}=v_t=v_t(0)-\int_0^t Lv+f(t)-f(0)=-Ly(0)-\int_0^t Lv+f(t).$$

Partial integration gives

$$\int_0^t Lv=\int_0^t Lw_t=Lw-Ly(0),$$

that is,

$$w_{tt}=v_t=-Lw+f(t). \tag{12}$$

We subtract equation (1) from (12), and obtain

$$(w-y)_{tt}+L(w-y)=0 \tag{13}$$

and from (11) it follows that

$$(w-y)(0)=0, \quad (w-y)_t(0)=0.$$

Because of (10) and (5) we can apply the uniqueness Theorem 29.1 to (13) and have

$$w=y \quad \text{in } [0,T] \quad \text{or} \quad v=w_t=y_t,$$

with which by (9) we have proved our theorem for $k=2$. ∎

Remark The proof procedure given in Theorem 30.1 also works for time-dependent operators $L(t)$ with

$$\begin{aligned} &\frac{\mathrm{d}L(t)}{\mathrm{d}t}: V\to H, \text{ and} \\ &\frac{\mathrm{d}^{k-1}L(t)}{\mathrm{d}t^{k-1}}: V\to H \text{ continuous,} \end{aligned} \tag{14}$$

when the boundedness constants are independent of t. We cannot manage here – in contrast to the parabolic case – without the strong hypotheses (14), since the Hadamard estimate of Theorem 29.1 only holds for $f\in L^2((0,T);H)$, and not – as in the parabolic case – for $f\in L^2((0,T);V')$.

Up to this point we have argued in the context of §17 – in order to formulate theorems about differential operators, we again exploit the notation and assumptions of §20, see also §27.

30.2 Differentiability with respect to t

We want to simplify the description and take the sesquilinear forms which arise to be independent of t – otherwise we must start from the assumptions (14).

Let $t\in[0,T]$, $T<\infty$, $x\in\Omega\subset\mathbb{R}^r$, Ω open, and let $A(x,\mathrm{D})$ be a strongly elliptic differential operator given by (see §19, (1))

$$A(x,\mathrm{D})\varphi = \sum_{|\alpha|,|\beta|\leqslant m} (-1)^{|\alpha|}\mathrm{D}_x^\alpha(a_{\alpha\beta}(x)\mathrm{D}_x^\beta\varphi), \tag{15}$$

with the sesquilinear form

$$a(\varphi,\psi)=\int_\Omega \sum_{|\alpha|,|\beta|\leqslant m} a_{\alpha\beta}(x)\mathrm{D}_x^\beta\varphi\mathrm{D}_x^\alpha\bar{\psi}\,\mathrm{d}x. \tag{16}$$

Further suppose that the boundary form (see §20, (33)) is preassigned

$$c(\varphi,\psi)=\int_{\partial\Omega} \sum_{\substack{|\alpha|\leqslant m-1\\|\beta|\leqslant m}} c_{\alpha\beta}(x)\mathrm{D}_x^\alpha\varphi\mathrm{D}_x^\beta\bar{\psi}\,\mathrm{d}\sigma \tag{17}$$

also written as

$$c(\varphi, \psi) = \sum_{j=p+1}^{m} \int_{\partial\Omega} C_j\varphi \cdot \overline{B'_j\psi}\, d\sigma, \tag{17a}$$

see §21, (4).

Let $H = L_2(\Omega)$ and let V be a closed subspace of $W_2^m(\Omega)$ with

$$\mathring{W}_2^m(\Omega) \subset V \subset W_2^m(\Omega).$$

For example V can be determined by boundary operators, $V = W_2^m(\{b_j\}_1^p)$, see §19. We consider the sesquilinear form

$$\mathscr{A}(\varphi, \psi) = a(\varphi, \psi) + c(\varphi, \psi), \tag{18}$$

and require that the assumptions 1–6 of §20 for $K = 0$ are satisfied. We also require throughout this section that $\mathscr{A}(\varphi, \psi) = \overline{\mathscr{A}(\psi, \varphi)}$ (antisymmetry), see also Theorems 21.2, 21.3 and 17.12.

Let $\mathscr{L}_x: V \to V'$ be the representation operator of (18) (see §17, (28)). We have (regularity with respect to t):

Theorem 30.2 *We consider the hyperbolic differential equation*

$$\frac{\partial^2 y(x,t)}{\partial t^2} + \mathscr{L}_x y(x,t) = f(x,t), \tag{19}$$

with the initial conditions

$$y(x,0) = y_0(x), \quad \frac{\partial y(x,0)}{\partial t} = y_1(x). \tag{20}$$

Suppose that the conditions above are satisfied, further suppose that

$$f(x,t) \in W_2^{k-1}((0,T); L_2(\Omega)), \quad k \geqslant 1, \tag{21}$$

and that the following compatibility conditions are satisfied,

$$\frac{\partial^j y(x,0)}{\partial t^j} \in V, \quad j = 0, \ldots, k-1, \quad \frac{\partial^k y(x,0)}{\partial t^k} \in L_2(\Omega). \tag{22}$$

Then for the solution y of (19), (20) we have

$$y(x,t) \in W_2^{k-1}((0,T); V), \quad \frac{\partial^k y(x,t)}{\partial t^k} \in L^2((0,T); L^2(\Omega)), \quad \frac{\partial^{k+1} y(x,t)}{\partial t^{k+1}} \in L^2((0,T); V').$$

In particular if $k \geqslant 2$ we have $y(x,t) \in V$ for all $t \in [0,T]$, that is, the solution $y(x,t)$ satisfies the homogeneous boundary conditions given by V.

The proof follows immediately from Theorem 30.1.

30.3 Differentiability with respect to x

We now consider regularity with respect to x, and again rely on Theorems 20.4 and 21.1 for $\mathscr{A}(\varphi, \psi)$ for help (see (18)), with $K = m + q$, $q \geqslant 0$. Once again we divide the boundary conditions into two groups

$$b_1, \ldots, b_p, \quad 0 \leqslant p \leqslant m,$$

are those boundary conditions, which determine V (if $p = 0$, we have $V = W_2^m(\Omega)$) and

$$b_{p+1} = B_{p+1} + C_{p+1}, \ldots, b_m = B_m + C_m$$

are the (natural) boundary conditions arising out of Green's formula §21, (3), B_j, and out of the boundary form (17), C_j. We again write (see §27)

$$V_N = W^{2m}(\{b_j\}_{p+1}^m), \quad W^{2m}(B) = W^{2m}(\{b_j\}_1^m).$$

Let λ_0 be the coercion constant of $\mathscr{A}$, we can also take λ_0 to be some non-eigenvalue. Then by Theorem 17.11 the map $\mathscr{L}_x + \lambda_0 : V \to V'$ is a (topological) isomorphism and by Corollary 21.2 the restriction map $G_{\lambda_0} = (\mathscr{L}_x + \lambda_0)^{-1}|_{W_2^q}$ is continuous:

$$G_{\lambda_0} : W_2^q(\Omega) \to W_2^{2m+q}(\Omega) \cap W^{2m}(\{b_j\}_1^m),$$

that is, the Friedrichs inequality holds,

$$\| G_{\lambda_0} f \|_{2m+q} \leqslant c \| f \|_q, \quad \text{for all} \quad f \in W_2^q(\Omega),$$

from which by Theorem 27.1 it follows that

$$G_{\lambda_0} : W_2^k((0, T); W_2^q(\Omega)) \to W_2^k((0, T); W_2^{2m+q}(\Omega) \cap W^{2m}(B)) \tag{23}$$

is continuous.

We have:

Theorem 30.3 *Let the assumptions of Theorem* 30.2 *be satisfied, let* $k \geqslant 2$, *so that the solution* y *of* (19), (20) *satisfies*

$$y \in W_2^{k-1}((0, T); V), \quad y \in W_2^k((0, T); L_2(\Omega)). \tag{24}$$

Suppose in addition that

$$f \in W_2^{k-2}((0, T); W_2^q(\Omega)). \tag{25}$$

Then it is also true that

$y \in W_2^{k-2}((0, T); W^{2m}(B)),$
$y \in W_2^{k-3}((0, T); W_2^{2m+q_1}(\Omega) \cap W^{2m}(B)),\ q_1 = \min(m, q),$

$y \in W_2^{k-4}((0,T); W_2^{4m}(\Omega) \cap W^{2m}(B))$,
$y \in W_2^{k-5}((0,T); W_2^{2m+q_2}(\Omega) \cap W^{2m}(B))$, $q_2 = \min(m_1, q)$, $m_1 = 2m + q_1$, etc.

By choosing k and q to be sufficiently large we can always arrange that $y \in W_2^l((0,T); W_2^l(\Omega) \cap W^{2m}(B))$.

Proof. We have (19)

$$\mathscr{L}y + \lambda_0 y = f + \lambda_0 y - \frac{\partial^2 y}{\partial t^2}$$

from which follows

$$y = G_{\lambda_0}\left(f + \lambda_0 y - \frac{\partial^2 y}{\partial t^2}\right).$$

The assumptions (24) and (25) have the effect that

$$\begin{aligned} f + \lambda_0 y - \frac{\partial^2 y}{\partial t^2} &\in W_2^{k-2}((0,T); L_2(\Omega)), \\ f + \lambda_0 y - \frac{\partial^2 y}{\partial t^2} &\in W_2^{k-3}((0,T); W_2^{q_1}(\Omega)), \quad q_1 = \min(m,q), \end{aligned} \tag{26}$$

respectively.
Since V is a closed subspace of $W_2^m(\Omega)$.

Applying (23) to (26) gives in one case $y \in W_2^{k-2}((0,T); W^{2m}(B))$ and in the other

$$y \in W_2^{k-3}((0,T); W^{2m+q_1}(\Omega) \cap W^{2m}(B)). \tag{27}$$

If we take (27) instead of (24) and repeat the process, we obtain

$$y \in W_2^{k-4}((0,T); W_2^{4m}(\Omega) \cap W^{2m}(B))$$

and

$$y \in W_2^{k-5}((0,T); W_2^{2m+q_2}(\Omega) \cap W^{2m}(B)), \quad q_2 = \min(m_1, q), m_1 = 2m + q_1$$

etc. ∎

Since by Theorem 27.8 we have

$$W_2^l((0,T); W_2^l(\Omega)) \subsetneq W_2^l(\Omega \times (0,T)),$$

we find for the solution of the hyperbolic equation (19), (20)

$$y(x,t) \in W_2^l(\Omega \times (0,T)) \subsetneq C^{l-(r+1)/2}(\bar{\Omega} \times [0,T]),$$

which is far reaching regularity.

Theorem 30.3 also tells us which mixed hyperbolic problem we have solved.

Theorem 30.4 *Suppose that in Theorems* 30.3 *and* 30.2 *we take* $k=2$, $q=0$, *that is*

$$f\in W_2^1((0,T);L_2(\Omega)).$$

In order to satisfy the compatibility conditions (22), *we assume for* y_0 *and* y_1 *that* $y_0\in W^{2m}(B)$, $y_1\in V$. *By Corollary* 21.1 $\mathscr{L}$ *agrees with* A *on* $W^{2m}(B)$, *hence* $Ay_0=\mathscr{L}y_0\in L_2(\Omega)$, *and for* $f(x,0)\in L_2(\Omega)$ (*this follows from the assumptions above on* f) *we have*

$$y_0\in W^{2m}(B)\subset V,\quad y_1\in V,\quad f(0)-\mathscr{L}y_0\in L_2(\Omega),$$

which are the conditions (22). *Hence by Theorem* 30.2 *the unique solution of* (19), (20) *is such that*

$$y\in W_2^1((0,T),V),\quad y\in W_2^2((0,T);L_2(\Omega)), \tag{28}$$

and by Theorem 30.3,

$$y(x,t)\in L_2((0,T);W^{2m}(B)), \tag{29}$$

that is, in $\Omega\times(0,T)$ $y(x,t)$ *is the unique solution of the hyperbolic differential equation*

$$\frac{\partial^2 y(x,t)}{\partial t^2}+\sum_{\substack{|\alpha|\leq m\\ |\beta|\leq m}}(-1)^{\alpha}D_x^{\alpha}(a_{\alpha\beta}(x)D_x^{\beta}y(x,t))=f(x,t), \tag{30}$$

with the initial conditions

$$y(x,0)=y_0(x),\quad \frac{\partial y(x,0)}{\partial t}=y_1(0)\quad \text{on } \Omega \tag{31}$$

and the boundary conditions ($x\in\partial\Omega$)

$$b_1(x,D)y(x,t)=0,\ldots,b_p(x,D)y(x,t)=0 \qquad \textit{for all } t\in[0,T] \tag{32}$$
$$b_{p+1}(x,D)y(x,t)=0,\ldots,b_m(x,D)y(x,t)=0 \quad \textit{for almost all } t\in[0,T]. \tag{33}$$

Proof. Equation (30), interpreted in the distributional sense, follows immediately from the weak equation (19), since $\mathscr{D}(\Omega)\subset V$. (32) follows from (28) and (33) from (29). If one wishes to satisfy the boundary conditions (33) for all t, one must take $k=3$ in Theorem 30.3. The boundary conditions (32), (33) imply that $y\in W^{2m}(B)$; since by Corollary 21.1 $A(x,D)$ agrees with $\mathscr{L}$ on $W^{2m}(B)$, the set (30), (31), (32), (33) is equivalent to (19), (20) and we have also proved the uniqueness of y, (30)–(33). ∎

As in §27 we can also solve the mixed inhomogeneous problem for hyperbolic differential equations. For the sake of completeness we write down the assumptions. For the sesquilinear form $\mathscr{A}(\varphi, \psi)$ (respectively $\mathscr{L}$) take the assumptions 1–6 of §20, with $K = m + q$, $q \geqslant 0$ there,

$$H = L_2(\Omega), \quad V = W^m(\{b_j\}_1^p), \quad W^{2m}(B) = W^{2m}(\{b_j\}_1^m).$$

Further let $k \geqslant 2$

$$f \in W_2^{k-2}((0, T); W_2^q(\Omega)), \quad f \in W_2^{k-1}((0, T); L_2(\Omega)) \tag{34}$$

and let a boundary function $g(x, t)$ be given

$$g \in W_2^k((0, T); W_2^{2m+\tilde{q}}(\Omega)), \quad g \in W_2^{k+1}((0, T); L_2(\Omega)), \tag{35}$$

where $\tilde{q}$ is the maximum among the numbers $\min(q, m)$, $\min(q, m_1)$, $\min(q, m_2)$, etc. arising in Theorem 30.3. Here we have written

$$m_1 = 2m + \min(m, q), \quad m_2 = 2m + \min(m_1, q), \text{ etc.}$$

If we put

$$\tilde{f} = f - Ag - g_{tt}, \tag{36}$$

we then have

$$\tilde{f} \in W_2^{k-2}((0, T); W_2^q(\Omega), \quad \tilde{f} \in W_2^{k-1}((0, T); L_2(\Omega)), \tag{37}$$

where A is the differential operator (15) and not the representation operator $\mathscr{L}$. We set $u_0(x) := y_0(x) - g(x, 0)$ and $u_1(x) = y_1(x) - g_t(x, 0)$ and impose the following compatibility assumptions on $\tilde{f}, u_0, u_1$

$$u_0 \in V, \quad u_1 \in V, \quad \frac{\partial^2 u(x,0)}{\partial t^2} := \tilde{f}(x, 0) - \mathscr{L} u_0 \in V,$$
$$\frac{\partial^3 u(x,0)}{\partial t^3} := \frac{\partial \tilde{f}(x,0)}{\partial t} - \mathscr{L} u_1 \in V, \ldots, \frac{\partial^k u(x,0)}{\partial t^k} := \cdots \in L_2(\Omega). \tag{38}$$

We obtain (38) from (3) when we put

$$y \mapsto u, \quad y_0 \mapsto u_0, \quad y_1 \mapsto u_1, \quad f \mapsto \tilde{f}.$$

We have:

Theorem 30.5 *There exists a solution $y(x, t)$ of the hyperbolic differential equation*

$$\frac{\partial^2 y(x,t)}{\partial t^2} + \sum_{\substack{|\alpha| \leqslant m \\ |\beta| \leqslant m}} (-1)^{|\alpha|} D_x^\alpha (a_{\alpha\beta}(x) D_x^\beta y(x, t)) = f(x, t) \quad \text{in } \Omega \times (0, T) \tag{39}$$

with the initial conditions

$$y(x,0) = y_0(x), \quad \frac{\partial y(x,0)}{\partial t} = y_1(x) \quad \text{on } \Omega.$$

Furthermore

$$\begin{aligned} &y \in W_2^k((0,T); L_2(\Omega)), && y \in W_2^{k-1}((0,T); W_2^m(\Omega)), \\ &y \in W_2^{k-2}((0,T); W_2^{2m}(\Omega)), && y \in W^{k-3}((0,T); W_2^{2m+q_1}(\Omega)), \\ &y \in W_2^{k-4}((0,T); W_2^{4m}(\Omega)), && y \in W^{k-5}((0,T); W_2^{2m+q_2}(\Omega)), \end{aligned}$$
$$q_1 = \min(m,q), q_2 = \min(m_1,q), m_1 = 2m + q_1, \ldots.$$

This solution satisfies the boundary conditions $b_1, \ldots, b_p$, associated with V.

$$y(\cdot,t) - g(\cdot,t) \in V, \quad \textit{for all } t \in [0,T], \tag{40}$$

and for almost all $t \in [0,T]$ the remaining boundary conditions $b_{p+1}, \ldots, b_m$

$$y(\cdot,t) - g(\cdot,t) \in W^{2m}(B). \tag{41}$$

If $k \geqslant 3$ the boundary conditions (41) are satisfied for all t. Furthermore $y(x,t)$ is the unique solution of (30), (31), (40), (41).

Proof. We write

$$y := u + g$$

where it satisfies the weak equation:

$$\frac{d^2u}{dt^2} + \mathscr{L}u = \tilde{f} = f - Ag - g_{tt}, \tag{42}$$

with the initial conditions

$$u(0) = u_0 = y_0 - g(\cdot,0), \quad \frac{du(0)}{dt} = u_1 = y_1 - g_t(\cdot,0).$$

(37) and (38) imply, that on (42) Theorem 30.2 is applicable, by which

$$u \in W_2^{k-1}((0,T); V), \quad u \in W_2^k((0,T); L_2(\Omega)), \tag{43}$$

and by Theorem 30.3 we have

$$u \in W_2^{k-2}((0,T); W^{2m}(B)), \quad u \in W_2^{k-3}((0,T); W_2^{2m+q_1}(\Omega)), \text{ etc.} \tag{44}$$

(35) with (43) and (44) gives (because $y = u + g$) that

$$\begin{aligned} &y \in W_2^k((0,T); L^2(\Omega)), && y \in W_2^{k-1}((0,T); W_2^m(\Omega)), \\ &y \in W_2^{k-2}((0,T); W_2^{2m}(\Omega)), && y \in W_2^{k-3}((0,T); W_2^{2m+q_1}(\Omega)), \text{ etc.} \end{aligned}$$

Because $u = y - g$ we obtain the boundary conditions (40) and (41) from Theorem 30.4, which we apply to (42). Theorem 30.4 also gives the uniqueness of the solution y of (30), (31), (40), (41) (Corollary 21.1!) ∎

We further remark that for $k = 2$, $q = 0$, the compatibility conditions (38) are not particularly restrictive; it is enough to require

$$u_0 = y_0(x) - g(x,0) \in W^{2m}(B) \subset V, \quad u_1 = y_1(x) - g_t(x,0) \in V, \tag{45}$$

for in this case $A(y_0(x) - g(x,0)) \in L_2(\Omega)$ and also

$$\tilde{f}(x,0) - \mathscr{L}(y_0 - g(x,0)) = f(x,0) - A(y_0(x) - g(x,0)) \in L_2(\Omega), \tag{46}$$

the last because of (34), (35). For

$$\left.\begin{aligned} &f \in W_2^1((0,T); L_2(\Omega)), \\ &g \in W_2^2((0,T); W_2^{2m}(\Omega)), \quad g \in W_2^3((0,T); L_2(\Omega)) \end{aligned}\right. \tag{47}$$

imply

$$\tilde{f} = f - Ag - g_{tt} \in W_2^1((0,T); L_2(\Omega)).$$

Hence the value $\tilde{f}(x,0)$ is taken continuously ($t \to 0$) in $L_2(\Omega)$. However, (45) and (46) are (38) with $k = 2$; and Theorem 30.5, $k = 2$, under conditions (45), (47) presents an existence-uniqueness theorem for the inhomogeneous, mixed, hyperbolic problem, (30), (31), (40), (41).

§31 Examples

We begin with the mixed Dirichlet problem.

Example 31.1 Let $A(x, \mathrm{D})$ be a linear strongly elliptic differential operator of the form §19, (1) with the sesquilinear form

$$a(\varphi, \psi) = \int_\Omega \sum_{\substack{|\alpha| \leq m \\ |\beta| \leq m}} a_{\alpha\beta}(x) \mathrm{D}^\beta \varphi \cdot \overline{\mathrm{D}^\alpha \psi}\, \mathrm{d}x. \tag{1}$$

Let Ω be bounded, we put $H = L_2(\Omega)$, $V = \mathring{W}_2^m(\Omega)$. If $a_{\alpha\beta} \in C(\bar{\Omega})$, we have the estimates

$$|a(\varphi, \psi)| \leq c \, \|\varphi\|_m \|\psi\|_m \quad \text{for all } \varphi, \psi \in W_2^m(\Omega), \tag{2}$$

and the Gårding inequality, see Theorem 19.2

$$\operatorname{Re} a(\varphi, \varphi) + k_0(\varphi, \varphi)_0 \geq c \, \|\varphi\|_m^2 \quad \text{for all } \varphi \in \mathring{W}_2^m(\Omega). \tag{3}$$

If $a_{\alpha\beta}(x) = \overline{a_{\beta\alpha}(x)}$, (1) is antisymmetric and we can apply Theorem 29.1:

If $f \in L_2((0, T); L_2(\Omega)) \simeq L_2(\Omega \times (0, T))$, $T < \infty$, and the initial functions are such that

$$y_0(x) \in \mathring{W}_2^m(\Omega), \quad y_1(x) \in L_2(\Omega)$$

there exists a unique solution $y(x, t)$ of the hyperbolic equation

$$\frac{\partial^2 y(x,t)}{\partial t^2} + A(x, D_x)y(x,t) = f(x,t) \quad \text{in } \Omega \times (0, T), \tag{4}$$

which satisfies the initial conditions

$$y(x, 0) = y_0(x), \quad \frac{\partial y(x,0)}{\partial t} = y_1(0) \quad \text{on } \Omega \tag{5}$$

and for almost all $t \in [0, T]$ the boundary conditions

$$y(\cdot, t) \in \mathring{W}_2^m(\Omega). \tag{6}$$

Here for the solution $y(x, t)$ we have

$$y \in L_2((0, T); \mathring{W}_2^m(\Omega)), \quad \frac{\mathrm{d}y}{\mathrm{d}t} \in L_2(\Omega \times (0, T)).$$

We have used the fact that for $V = \mathring{W}_2^{2m}(\Omega)$ the weak equation §29, (H) is equivalent to (4).

By means of a slight modification of Theorem 30.5 (for $V = \mathring{W}_2^{2m}(\Omega)$, we can also take $k = 1$ there) we can easily solve the inhomogeneous Dirichlet problem. Suppose in addition we are given a boundary function $g(x, t)$ with

$$g(x, t) \in W_2^1((0, T); W_2^{2m}(\Omega)), \quad g(x, t) \in W_2^2((0, T); L_2(\Omega)).$$

Then there exists a unique solution $y(x, t)$ of (4) plus (5), which for almost all $t \in [0, T]$ satisfies the boundary conditions

$$(y \cdot, t) - g(\cdot, t) \in \mathring{W}_2^m(\Omega). \tag{7}$$

For the solution we have

$$y \in L_2((0, T); W_2^m(\Omega)), \quad \frac{\partial y}{\partial t} \in L_2(\Omega \times (0, T)).$$

If we wish to satisfy the boundary conditions (6) (respectively (7)) for all $t \in [0, T]$, then we must make stronger assumptions, and call on Theorem 30.4 (respectively 30.5). Let

$$\begin{aligned} &\Omega \in C^{m,1}, \quad q_{\alpha\beta} \in C^m(\bar{\Omega}), \\ &f \in W_2^1((0, T); L_2(\Omega)), \quad y_1(x) \in \mathring{W}_2^m(\Omega) \cap W_2^{2m}(\Omega), \quad y_1(x) \in \mathring{W}_2^m(\Omega), \end{aligned} \tag{8}$$

then the boundary conditions (6) are satisfied for all $t\in[0,T]$; since because of (8) by Theorem 8.7 the trace operator is defined, and we have

$$y(x,t)=0,\quad \frac{\partial y}{\partial n}(x,t)=0,\ldots,\frac{\partial y^{m-1}}{\partial n^{m-1}}(x,t)=0 \text{ on } \partial\Omega\times[0,T].$$

Here n is taken in the direction of the inner normal to the surface $\partial\Omega\times[0,T]$. There is an analogous result for (7). By Theorem 30.4 we also have

$$y\in W_2^1((0,T);\mathring{W}_2^m(\Omega)),\quad y\in W_2^2((0,T);L_2(\Omega)),\quad y\in L_2((0,T),W_2^{2m}(\Omega)\cap\mathring{W}_2^m(\Omega)).$$

If we take $A(x,\mathrm{D})=-\Delta$, we have solved the mixed Dirichlet problem for the wave equation.

Example 31.2 The mixed Neumann problem for the wave equation. We take $A(x,\mathrm{D})=-\Delta$ and

$$a(\varphi,\psi)=\int_\Omega\sum_{j=1}^{r}\frac{\partial\varphi}{\partial x_j}\frac{\partial\bar{\psi}}{\partial x_j}\,\mathrm{d}x\quad\text{for }\varphi,\psi\in W_2^1(\Omega).$$

We see that a is antisymmetric, and that for $\varphi,\psi\in V=W_2^1(\Omega)$ the estimates (2) and (3) hold (for the latter see Example 19.1). Therefore the assumptions for Theorem 29.1 are satisfied with $H=L_2(\Omega)$, $V=W_2^1(\Omega)$. In order to apply Theorem 30.4 we take $\Omega\in C^{2,1}$ by which $\partial/\partial n|_{\partial\Omega}$ is also defined using Theorem 8.7, and we suppose that

$$y_0(x)\in W^2\left(\frac{\partial}{\partial n}\right)=\left\{\varphi\in W_2^2(\Omega):\frac{\partial\varphi}{\partial n}=0\text{ on }\partial\Omega\right\},$$
$$y_1(x)\in W_2^1(\Omega),\quad f\in W_2^1((0,T);L_2(\Omega)).$$

Using Theorem 30.4 we obtain that there exists a unique solution $y(x,t)$ of the wave equation

$$\frac{\partial^2 y(x,t)}{\partial t^2}-\Delta_x y(x,t)=f(x,t)\quad\text{in }\Omega\times(0,T),$$

which satisfies both initial conditions

$$y(x,0)=y_0(x),\quad \frac{\partial y(x,0)}{\partial t}=y_1(x)\quad\text{in }\Omega$$

and for almost all $t\in[0,T]$ the Neumann boundary condition

$$\frac{\partial y(x,t)}{\partial n}=0,\quad x\in\partial\Omega \tag{9}$$

(n in the direction of the inner normal to $\partial\Omega$). For this solution $y(x,t)$ we have

$$y \in W_2^1((0,T); W_2^1(\Omega)), \quad y \in W_2^2((0,T); L_2(\Omega)), \quad y \in L_2\left((0,T); W^2\left(\frac{\partial}{\partial n}\right)\right).$$

If we want to satisfy the boundary condition (9) for all $t \in [0,T]$, we must take $k = 3$ in Theorem 30.3.

The inhomogeneous Neumann problem may be solved by means of Theorem 30.5.

Example 31.3 The third boundary value problem for the wave equation. We again take $A(x, D_x) = -\Delta$; however, for the sesquilinear form (see Example 21.3) we take

$$\mathscr{A}(\varphi, \psi) = \int_{\Omega} \sum_{j=1}^{r} \frac{\partial \varphi}{\partial x_j} \frac{\partial \bar{\psi}}{\partial x_j} \mathrm{d}x + \int_{\partial\Omega} b_0(x)\varphi(x)\overline{\psi(x)}\, \mathrm{d}\sigma, \quad \varphi, \psi \in W_2^1(\Omega),$$

where $b_0(x)$, $x \in \partial\Omega$ is real valued.

$\mathscr{A}(\varphi, \psi)$ is antisymmetric; with $H = L_2(\Omega)$, $V = W_2^1(\Omega)$ by Example 21.3 $\mathscr{A}$ is V-coercive, and hence the assumptions for Theorem 29.1 are satisfied. In order to apply Theorem 30.4 we take $\Omega \in C^{2,1}$, therefore again $\partial/\partial n|_{\partial\Omega}$ is defined by Theorem 8.7, and we suppose that

$$y_0 \in W^2(B) = \left\{\varphi \in W_2^2(\Omega) : b_0 \cdot \varphi + \frac{\partial \varphi}{\partial n} = 0 \text{ on } \partial\Omega\right\},$$
$$y_1 \in W_2^1(\Omega), \quad f \in W_2^1((0,T); L_2(\Omega)).$$

Using Theorem 30.4 we obtain that there exists a unique solution $y(x,t)$ of the wave equation:

$$\frac{\partial^2 y(x,t)}{\partial t^2} - \Delta_x y(x,t) = f(x,t) \quad \text{in } \Omega \times (0,T),$$

which satisfies the initial conditions

$$y(x,0) = y_0(x), \quad \frac{\partial y(x,0)}{\partial t} = y_1(x) \quad \text{in } \Omega,$$

and for almost all $t \in [0,T]$ also the third boundary condition

$$b_0(x)y(x,t) + \frac{\partial y(x,t)}{\partial n} = 0, \quad x \in \partial\Omega.$$

The reader may formulate the Neumann problem and the third boundary value problem for the general hyperbolic differential equation of second order, and prove its unique solvability.

Example 31.4 We turn to the boundary value problems for the biharmonic operator Δ^2, and wish to solve the corresponding mixed problems for the hyperbolic equation. We take

$$a(\varphi, \psi) = \int_\Omega \Delta\varphi \cdot \Delta\bar{\psi}\, dx,$$

and for $\varphi, \psi \in W_2^2(\Omega)$ have the estimate (2). The form a is antisymmetric and V_1-coercive, where (see Example 21.6)

$$V_1 = \{\varphi \in W_2^2(\Omega): \varphi|_{\partial\Omega} = 0\}$$

and in the case of the second boundary value problem V_2-coercive, where

$$V_2 = \left\{\varphi \in W_2^2(\Omega): \frac{\partial\varphi}{\partial n}\bigg|_{\partial\Omega} = 0\right\}.$$

In order to apply Theorem 30.4 we take

$$\Omega \in C^{4,1}, \quad f \in W_2^1((0, T); L_2(\Omega)),$$

and for the first boundary value problem (I, Δ),

$$y_0 \in W^4(I, \Delta) = \{\varphi \in W_2^4(\Omega): \varphi = 0, \Delta\varphi = 0, \quad \text{on } \partial\Omega\}, \quad y_1 \in V_1,$$

while for the second $(\partial/\partial n, \partial\Delta/\partial n)$ we must take

$$y_0 \in W^4\left(\frac{\partial}{\partial n}, \frac{\partial\Delta}{\partial n}\right) = \left\{\varphi \in W_2^4(\Omega): \frac{\partial\varphi}{\partial n} = 0, \frac{\partial\Delta\varphi}{\partial n} = 0, \quad \text{on } \partial\Omega\right\}, \quad y_1 \in V_2.$$

We obtain a unique solution $y(x, t)$ of the hyperbolic equation

$$\frac{\partial^2 y(x, t)}{\partial t^2} + \Delta_x^2 y(x, t) = f(x, t),$$

which satisfies the initial conditions

$$y(x, 0) = y_0(x), \quad \frac{\partial y(x, 0)}{\partial t} = y_1(0).$$

In one case the equation satisfies the boundary conditions

$$\begin{aligned} y(x, t) &= 0, \quad && x \in \partial\Omega \quad \text{for all } t \in [0, T]; \\ \Delta_x y(x, t) &= 0, \quad && x \in \partial\Omega \quad \text{almost all } t \in [0, T] \end{aligned}$$

and in the second case

$$\begin{aligned} \frac{\partial y(x, t)}{\partial n} &= 0, \quad && x \in \partial\Omega \quad \text{for all } t \in [0, T], \\ \frac{\partial \Delta_x y(x, t)}{\partial n} &= 0, \quad && x \in \partial\Omega \quad \text{almost all } t \in [0, T]. \end{aligned}$$

Example 31.5 We turn to Example 21.7. Let $\Omega \in N^{0,1}$, $\partial\Omega = \Gamma_1 \cup \Gamma_2$, where $\mu_{r-1}\Gamma_1 \neq 0$. We take $H = L_2(\Omega)$, $V = \tilde{V} = \{\varphi \in W_2^1(\Omega) : \varphi|_{\Gamma_1} = 0\}$ and

$$a(\varphi, \psi) = \sum_{j=1}^{r} \int_{\Omega} \frac{\partial \varphi}{\partial x_j} \overline{\frac{\partial \psi}{\partial x_j}} \, dx.$$

By Example 21.7 a is $\tilde{V}$-elliptic, also estimate (2) is satisfied, and a is anti-symmetric. Therefore the assumptions for Theorem 29.1 are satisfied and we have: For

$$f \in L_2((0, T); L_2(\Omega)) = L_2(\Omega \times (0, T))$$

and for the initial functions

$$y_0(x) \in \tilde{V}, \quad y_1(x) \in L_2(\Omega),$$

there exists a unique solution y of the weak equation

$$\frac{\partial^2 (y, \varphi)_0}{\partial t^2} + a(y, \varphi) = (f, \varphi)_0, \quad \varphi \in \tilde{V}, \tag{10}$$

with (5). For this solution y we have

$$y \in L_2((0, T); \tilde{V}), \quad \frac{\partial y}{\partial t} \in L_2(\Omega \times (0, T)). \tag{11}$$

From (10) because $\mathscr{D}(\Omega) \subset \tilde{V}$ we obtain the wave equation

$$\frac{\partial^2 y(x, t)}{\partial t^2} - \Delta_x y(x, t) = f(x, t) \quad \text{in } \Omega \times (0, T),$$

and from (11) for almost all $t \in [0, T]$ the boundary condition

$$y(x, t) = 0, \quad x \in \Gamma_1.$$

Since, however, for Example 21.7 we have no regularity theorems, and therefore cannot apply Theorem 30.4 (Green's formula), we can only regard the boundary condition $\partial y(x, t)/\partial n = 0$, $x \in \Gamma_2$, $t \in [0, T]$ as being formally satisfied, it lives in (10).

Example 31.6 The transmission problem for the wave equation (see Example 21.8). Let

$$\Omega = \Omega_1 \cup \Omega_2, \Gamma_1 = \bar{\Omega}_1 \cap \partial\Omega, \Gamma_2 = \bar{\Omega}_2 \cap \partial\Omega, \Gamma = \bar{\Omega}_1 \cap \bar{\Omega}_2, a, b > 0.$$

In the notation of Example 21.8 we can again (uniquely) solve

$$\frac{\partial^2 y_1(x, t)}{\partial t^2} - a\Delta_x y_1(x, t) = f(x, t) \quad \text{in } \Omega_1 \times (0, T),$$

$$\frac{\partial^2 y_2(x,t)}{\partial t^2} - b\Delta_x y_2(x,t) = f(x,t) \quad \text{in } \Omega_2 \times (0,T),$$

$$y_1(x,0) = y_0(x) \quad \text{in } \Omega_1, \quad y_2(x,0) = y_0(x) \quad \text{in } \Omega_2,$$

$$\frac{\partial y_1(x,0)}{\partial t} = y_1(x) \quad \text{in } \Omega_1, \quad \frac{\partial y_2(x,0)}{\partial t} = y_1(x) \quad \text{in } \Omega_2,$$

$$y_1(x,t) = 0 \quad \text{on } \Gamma_1 \times [0,T], \quad y_2(x,t) = 0 \quad \text{on } \Gamma_2 \times [0,T],$$

$$y_1(x,t) = y_2(x,t) \quad \text{on } \Gamma \times [0,T],$$

for almost all $t \in [0,T]$, while we can regard the second transmission condition

$$a\frac{\partial y_1(x,t)}{\partial n} = b\frac{\partial y_2(x,t)}{\partial n} \quad \text{on } \Gamma \times [0,T]$$

as being only formally satisfied. It is again contained in the appropriate weak equation.

Example 31.7 The Cauchy problem for the wave equation. Let $\Omega = \mathbb{R}^r$, we consider the sesquilinear form

$$a(\varphi, \psi) = \int_{\mathbb{R}^r} \sum_{j=1}^{r} \frac{\partial \varphi}{\partial x_j} \frac{\partial \bar{\psi}}{\partial x_j} \, \mathrm{d}x, \quad \varphi, \psi \in W_2^1(\mathbb{R}^r).$$

Estimate (2) is valid, also antisymmetry and

$$\operatorname{Re} a(\varphi, \varphi) + (\varphi, \varphi)_0 = \|\varphi\|_1^2, \quad \varphi \in \mathring{W}_2^1(\mathbb{R}^r) = W_2^1(\mathbb{R}^r),$$

therefore the assumptions of Theorem 29.1 are satisfied. For $f \in L_2(\mathbb{R}^r \times (0,T))$ and the initial functions $y_0(x) \in W_2^1(\mathbb{R}^r)$, $y_1 \in L_2(\mathbb{R}^r)$, there exists a unique solution $y(x,t)$ of the wave equation

$$\frac{\partial^2 y(x,t)}{\partial t^2} - \Delta_x y(x,t) = f(x,t) \quad \text{in } \mathbb{R}^r \times (0,T)$$

with $y(x,0) = y_0(x)$, $\partial y(x,0)/\partial t = y_1(x)$ in $\mathbb{R}^r$, and we have

$$y \in L_2((0,T); W_2^1(\mathbb{R}^r)), \quad \frac{\partial y}{\partial t} \in L_2(\mathbb{R}^r \times (0,T)).$$

We can use Theorem 30.2 in order to obtain regularity statements for the Cauchy problem; we proceed as in Example 28.8. We begin with the assumption that

$$f \in W_2^{k-1}((0,T); L_2(\mathbb{R}^r)), \quad k \geqslant 2. \tag{12}$$

In order to satisfy the compatibility conditions §30, (22) we take

$$y_0 \in W_2^k(\mathbb{R}^r), \quad y_1 \in W_2^{k-1}(\mathbb{R}^r), \quad f^{(k-j)}(0) \in W_2^{j-2}(\mathbb{R}^r), \quad j = 2, \ldots, k.$$

By Theorem 30.2 we then have

$$y \in W_2^k((0,T); L_2(\mathbb{R}^r)), \quad y \in W_2^{k-1}((0,T); W_2^1(\mathbb{R}^r)). \tag{13}$$

We now write the wave equation in the form

$$(-\Delta_x + 1)y(x,t) = f(x,t) + y(x,t) - \frac{\partial^2 y(x,t)}{\partial t^2} = \tilde{f}. \tag{14}$$

By §28, (21) the Green solution operator $G := (-\Delta + 1)^{-1}$ acts continuously,

$$G: W_2^k((0,T); W_2^{s-2}(\mathbb{R}^r)) \to W_2^k((0,T); W_2^s(\mathbb{R}^r)). \tag{15}$$

(15) applied to (14) gives

$$y \in W_2^{k-2}((0,T); W_2^2(\mathbb{R}^r)). \tag{16}$$

We now place further restrictions on f

$$\begin{aligned} &f \in W_2^{k-3}((0,T); W_2^1(\mathbb{R}^r)) \\ &f \in W_2^{k-4}((0,T); W_2^2(\mathbb{R}^r)) \\ &\vdots \\ &f \in W_2^{k-1}((0,T); W_2^{l-2}(\mathbb{R}^r)). \end{aligned} \tag{17}$$

(15) applied to (14) together with the assumptions (17) gives successively

$$\begin{aligned} &y \in W_2^{k-3}((0,T); W_2^3(\mathbb{R}^r)) \\ &y \in W_2^{k-4}((0,T); W_2^4(\mathbb{R}^r)) \\ &\vdots \\ &y \in W_2^{k-l}(0,T); W_2^l((0,T); W^l(\mathbb{R}^r)). \end{aligned} \tag{18}$$

If we take $k = 2l$, $l = 1, 2, \ldots$, from (18) and by Theorem 27.8 we obtain

$$y \in W_2^l((0,T); W_2^l(\mathbb{R}^r)) \subset W_2^l(\mathbb{R}^r \times (0,T)).$$

Exercises

31.1 Refer to Exercise 28.1. Solve the mixed problem

$$\begin{aligned} &\frac{\partial^2 y(x,t)}{\partial t^2} - \frac{\partial^2 y(x,t)}{\partial x^2} = f(x,t), \quad 0 < x < 1, t > 0, \\ &y(x,0) = y_0(x), \quad \frac{\partial y(x,0)}{\partial t} = y_1(x), \quad y(0,t) = 0, \quad y(1,t) = 0, \end{aligned} \tag{1}$$

by the Fourier method. We expand

$$f(x,t) = \sum_{k=1}^{\infty} f_k(t) y_k(x), \quad y_0(x) = \sum_{k=1}^{\infty} c_k y_k(x), \quad y_1(x) = \sum_{k=1}^{\infty} d_k y_k(x),$$

solve the initial value problem

$$e_k''(t) + \lambda_k e_k(t) = f_k(t), \quad e_k(0) = c_k, \quad e_k'(0) = d_k, \quad k = 1, 2, \ldots,$$

and try as a solution

$$y(x,t)=\sum_{k=1}^{\infty} e_k(t)y_k(x). \tag{2}$$

The reader should write out explicitly the formulae for λ_k, y_k, e_k.

31.2 Show that the solution (2) of (1) 'fits' into the framework of §29, that is, carry through the necessary estimates.

31.3 With the help of the Fourier method solve the general mixed problem

$$\frac{\partial^2 y(x,t)}{\partial t^2}-\frac{\partial^2 y(x,t)}{\partial x^2}=f(x,t), \quad 0<x<1, t>0,$$

$$y(x,0)=y_0(x), \quad \frac{\partial y(x,0)}{\partial t}=y_1(x), \quad y(0,t)=g_0(t), \quad y(1,t)=g_1(t),$$

where the functions f, y_0, g_0 and g_1 are preassigned. Formulate regularity theorems for the solution $y(x,t)$.

31.4 Refer to Exercise 28.6. We consider the mixed hyperbolic problem

$$\frac{\partial^2 y(x,t)}{\partial t^2}+A(x,\mathrm{D}_x)y(x,t)=f(x,t), \quad x\in\Omega, t>0,$$

$$y(x,0)=y_0(x), \quad \frac{\partial y(x,0)}{\partial t}=y_1(x), \quad x\in\Omega, \tag{3}$$

$$b_1(x,\mathrm{D}_x)y(x,t)=0,\ldots,b_m(x,\mathrm{D}_x)y(x,t)=0 \quad \text{for } x\in\partial\Omega, t\geqslant 0.$$

We solve the mixed problem by developing

$$y_0(x)=\sum_{k=1}^{\infty} c_k y_k(x), \quad y_1(x)=\sum_{k=1}^{\infty} d_k y_k(x), \quad f(x,t)=\sum_{k=1}^{\infty} f_k(t)y_k(x), \tag{4}$$

and for the solution $y(x,t)$ write

$$y(x,t)=\sum_{k=1}^{\infty} e_k(t)y_k(x), \tag{5}$$

where $e_k(t)$ solves the initial value problem

$$e_k''(t)+\lambda_k e_k(t)=f_k(t) \quad e_k(0)=c_k, \quad e_k'(0)=d_k, \quad k=1,2,\ldots. \tag{6}$$

Can one carry over the existence-uniqueness theorem from §29 to (3) and (5)?

31.5 Refer to Exercise 28.7. By means of the representation (5) prove regularity theorems for the solution $y(x,t)$.

31.6 One can also solve the mixed problem for the Schrödinger equation (and other equations also) by means of the Fourier method, see Exercise 28.6 for $(A,b_1,\ldots,b_m)$.

$$\frac{\partial y(x,t)}{\partial t}+\mathrm{i}A(x,\mathrm{D}_x)y(x,t)=f(x,t), \quad x\in\Omega, t>0,$$

$$y(x,0) = y_0(x) \in L_2(\Omega), \quad y_0(x) = \sum_{k=1}^{\infty} c_k y_k(x),$$

$$b_1(x, D_x)y(x,t) = 0, \ldots, b_m(x, D_x)y(x,t) = 0 \quad \text{for } x \in \partial\Omega, t \geqslant 0.$$

For the solution $y(x,t)$ we write (see Exercise 31.5)

$$y(x,t) = \sum_{k=1}^{\infty} e_k(t) y_k(x),$$

where we determine the $e_k(t)$ from

$$e_k'(t) + \mathrm{i}\lambda_k e_k(t) = f_k(t), \quad e_k(0) = c_k, \quad k = 1, 2, \ldots .$$

For existence and uniqueness theorems see Lions & Mangenes [1]. Prove regularity theorems, beginning with the simple equation

$$\frac{\partial y(x,t)}{\partial t} - \mathrm{i}\frac{\partial^2 y(x,t)}{\partial x^2} = f(x,t), \quad 0 < x < 1, t > 0.$$

31.7 Solve the Cauchy problem for the equation of a vibrating string

$$\frac{\partial^2 y(x,t)}{\partial t^2} - \frac{\partial^2 y(x,t)}{\partial x^2} = f(x,t), \quad y(x,0) = \varphi(x), \quad \frac{\partial y(x,0)}{\partial t} = \psi(x).$$

Answer:

$$y(x,t) = \tfrac{1}{2}[\varphi(x-t) + \varphi(x+t)] + \frac{1}{2}\int_{x-t}^{x+t} t(\xi)\,\mathrm{d}\xi$$

$$+ \frac{1}{2}\iint_D f(\xi,\tau)\,\mathrm{d}\xi\,\mathrm{d}\tau, \quad \text{where } D$$

is the hatched region in Fig. 31.1.

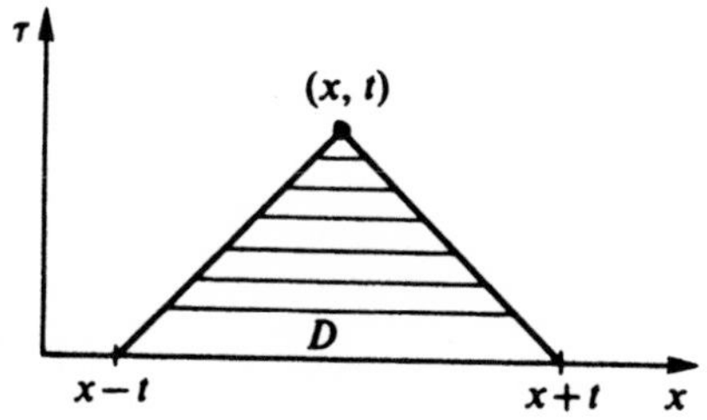

Fig. 31.1

VI

Difference processes in the calculation of the solution of a partial differential equation

§32 Functional analytic concepts for difference processes

In this section we first of all pin down the functional analytic spaces and concepts for difference processes before applying these to concrete boundary problems for partial differential equations.

In particular we formulate as generally as possible notions such as consistency, stability, discretisation, and study their interdependence. The clarification of these logical connections essentially simplifies the convergence proofs for difference methods. Here we follow the lecture notes of Wloka [3] and the thesis of Janssen [1]. For more far reaching descriptions see Kreĭn [1] and Richtmyer & Morton [1]. Let

$B_1, B_2, B_{1,n}, B_{2,n}$ be Banach spaces ($n \in \mathbb{N}$)

$A: B_1 \to B_2$ linear; A need not be continuous nor defined on all on B_1

$A_n: B_{1,n} \to B_{2,n}$ linear, A_n^{-1} exists and is continuous

$$\left.\begin{aligned} D_{1,n}: B_1 \to B_{1,n} \text{ linear} \\ D_{2,n}: B_2 \to B_{2,n} \text{ linear} \end{aligned}\right\} \text{discretisation operators.}$$

Furthermore the discretisation operators need not be defined everywhere. We have thus constructed Fig. 32.1.

$$\begin{array}{ccc} B_1 & \xrightarrow{\;A\;} & B_2 \\ \downarrow D_{1,n} & & \downarrow D_{2,n} \\ B_{1,n} & \xrightarrow[A_n]{} & B_{2,n} \end{array}$$

Fig. 32.1

We must read this diagram as follows: A stands for the equation to be solved $Au = f$, the so-called discretisation operators $D_{1,n}$ and $D_{2,n}$ transfer the problem to (usually) finite dimensional spaces, and there the approximating equation $A_n u_n = f_n$ is solved. We say that $u_n = A_n^{-1} f_n$ is obtained by the difference process, or approximation process. Since in what follows it is clear which norms are involved, we will not indicate this by any additional subscript. We next give conditions under which the approximation method gives a sequence u_n converging to the solution u of the equation $Au = f$.

Convergence theorem 32.1 *Let $u \in B_1$, $u_n \in B_{1,n}$, $f \in B_2$, $f_n \in B_{2,n}$ with $Au = f$, $A_n u_n = f_n$, for all $n \in \mathbb{N}$, $\| D_{2,n} f - f_n \| \to 0$ as $n \to \infty$. Suppose further that*

$$\| D_{2,n} Au - A_n D_{1,n} u \| \to 0 \quad \text{as } n \to \infty$$

and $\{A_n^{-1}\}$ is uniformly bounded. Then it follows that

$$\| D_{1,n} u - u_n \| \to 0 \quad \text{as } n \to \infty.$$

Proof. Let $\| A_n^{-1} \| \leqslant M < \infty$ for all $n \in \mathbb{N}$, then

$$\begin{aligned} \| D_{1,n} u - u_n \| &= \| A_n^{-1} A_n (D_{1,n} u - u_n) \| \\ &\leqslant M \| A_n (D_{1,n} u - u_n) \| = M \| A_n D_{1,n} u - f_n \| \\ &\leqslant M(\| A_n D_{1,n} u - D_{2,n} Au \| + \| D_{2,n} f - f_n \|) \to 0. \end{aligned}$$ ■

Definitions 32.1

1. *If in the scheme above $\| D_{2,n} Au - A_n D_{1,n} u \| \to 0$ holds, then we say that A and A_n are consistent at u.*
2. *If $\{A_n^{-1}\}$ is uniformly bounded, the approximation process is called stable.*
3. *Instead of $\| D_{1,n} u - u_n \| \to 0$ we also say that $(u_n)_{n \in \mathbb{N}}$ converges discretely to u, analogously for f_n, f.*
4. *If $f \in B_2$, $Au = f$ and if for each sequence $(f_n)_n$ converging discretely to f, it is true that $u_n = A_n^{-1} f_n$ converges discretely to u, we say that the method converges for f.*

Remarks 32.1

(i) With this notation we can also formulate the convergence theorem as: consistency and stability imply convergence.

(ii) If A and A_n are consistent, then this implies that the Fig. 32.1 above is commutative in the limit.

(iii) Two simple variations of the concepts of consistency and stability suggest themselves:

1. kth order consistency, that is, there exists some $c > 0$ with $\| D_{2,n} Au - A_n D_{1,n} u \| \leqslant c n^{-k}$, and
2. α-stability, that is, there exists some $M > 0$ with $\| A_n^{-1} \| \leqslant M n^{\alpha}$.

It is easily checked that with these concepts we can also state the theorem as: if $k > \alpha > 0$, $\beta = k - \alpha$, then it follows that kth-order consistency and α-stability imply βth-order convergence.

(iv) In the convergence theorem the choice of norms in B_1 and B_2 plays no role.

Example 32.1 We consider the heat equation

$$\frac{\partial u}{\partial t} - \frac{\partial^2 u}{\partial x^2} = f \quad \text{on } (0, \pi) \times (0, T), \tag{1}$$

with the initial condition

$$u(x, 0) = g(x), \quad x \in [0, \pi] \tag{2}$$

and the boundary conditions

$$u(0, t) = u(\pi, t) = 0, \quad t \in [0, T]. \tag{3}$$

We suppose that the solution u belongs to $C^4([0, \pi] \times [0, T])$, by §27 this can always be arranged under suitable assumptions on f and g. In order to set up a difference process for the numerical calculation of u we first cover the rectangle $[0, \pi] \times [0, T]$ with a lattice of width k in the t-direction and h in the x-direction. We denote by

$$G := \{(mh, jk) : m = 0, \ldots, \pi/h, j = 0, \ldots, T/k\} \backslash,$$

the set of lattice points of the rectangle $[0, \pi] \times [0, T]$, which do not lie on the edges $\{0\} \times [0, T]$ and $\{\pi\} \times [0, T]$ – this is because of the boundary condition (3). We next replace derivatives by difference quotients

$$\frac{\partial u}{\partial t} \sim \frac{w(x, t+k) - w(x, t)}{k}, \quad \frac{\partial^2 u}{\partial x^2} \sim \frac{w(x+h, t) - 2w(x, t) + w(x-h, t)}{h^2}$$

and then in conformity with (1) obtain the difference equation

$$\frac{w(x, t+k) - w(x, t)}{k} = \frac{w(x+h, t) - 2w(x, t) + w(x-h, t)}{h^2} + f(x, t),$$

which we consider at the lattice points $(x, t) \in G$. We solve for $w(x, t+k)$ and obtain

$$w(x, t+k) = (1 - 2\lambda)w(x, t) + \lambda w(x+h, t) + \lambda w(x-h, t) + kf(x, t), \tag{4}$$

where $\lambda := k/h^2$. Therefore we obtain the computing process, proceeding in the t-direction

$$\begin{aligned} &w(x, 0) = g(x) \\ &w(x, 1 \cdot k) = (1 - 2\lambda)w(x, 0) + \lambda w(x+h, 0) + \lambda w(x-h, 0) + kf(x, 0), \text{ etc.} \end{aligned} \tag{5}$$

Here for the lattice points (x, t) which lie on the forbidden edges $\{0\} \times [0, T]$ and $\{\pi\} \times [0, T]$, we put $w(x, t) = 0$, in accordance with (3).

In order to be able to apply the convergence Theorem 32.1 we write

$$B_1 = C_0([0, \pi) \times [0, T]);$$

it is the space of all continuous functions on $[0, \pi] \times [0, T]$ which satisfy (3) equipped with the sup norm; moreover

$$B_2 = C([0, \pi] \times [0, T]) \times C[0, \pi),$$

$$B_{1,n} = \mathbb{R}^d, d = \text{card } G = \text{number of lattice points in } G$$

$$B_{2,n} = \mathbb{R}^d \times \mathbb{R}^{d_1}, d_1 = \left[\frac{\pi}{h}\right] + 1 = \text{number of lattice points on } [0, \pi],$$

and we put the sup norm on both spaces $B_{1,n}, B_{2,n}$. On the Cartesian product $B_1 \times B_2$ we put the norm $\|\cdot\|_1 + \|\cdot\|_2$. As discretisation operators $D_{1,n}, D_{2,n}$ we take restriction to the lattice points in both cases. For the operators A and A_n we now have

$$A = \left(\frac{\partial}{\partial t} - \frac{\partial^2}{\partial x^2}, I\right),$$

where $Iz := u(x, 0)$ is the initial value operator. For $w \in B_{1,n}$ we define

$$L_{k,h}w := \frac{1}{k}[w(x, t+k) - \{(1 - 2\lambda)w(x, t) + \lambda w(x + h, t) + \lambda w(x - h, t)\}].$$

We can write (4) as

$$L_{k,h}w = f$$

and have

$$A_n := (L_{k,h}, I_h),$$

where $I_h w = w(x, 0)$, $x = mh$ is the initial value operator on the lattice points of $[0, \pi]$. By (5) we have shown the invertibility of A_n, therefore the operators A_n^{-1} exist.

We now show – under the assumption $(1 - 2\lambda) \geqslant 0$ – that the process (5) is stable, that is, that the sequence $\{A_n^{-1}\}$ is uniformly bounded. For this let $\|w\|^t := \max_{G_t} |w(x, \tau)$, where

$$G_t := \{(x, \tau) \in G: \quad 0 \leqslant \tau \leqslant t\}.$$

With

$$L_{k,h}(v)(x, t) = \frac{1}{k}(v(x, t+k) - [(1 - 2\lambda)v(x, t) + \lambda v(x + h, t) + \lambda v(x - h, t)]),$$

where v is an arbitrary function defined on the lattice, we have

$$|v(x,t+k)| \leqslant (1-2\lambda)|v(x,t)| + \lambda|v(x+h,t)| + \lambda|v(x-h,t)| + k|L_{k,h}(v)(x,t)|$$
$$\leqslant (1-2\lambda)\|v\|^t + 2\lambda\|v\|^t + k\|L_{k,h}(v)\|^t,$$

and therefore

$$\|v\|^{t+k} \leqslant \|v\|^t + k\|L_{k,h}(v)\|^t.$$

Here we have made essential use of $1-2\lambda \geqslant 0$.

Lemma 32.1 *Let g, f be arbitrary lattice functions and suppose that for all $t = nk$, $n = 0, 1, \ldots, T/k$ we have*

$$\|g\|^{t+k} \leqslant \|g\|^t + k\|f\|^t$$

Then it follows that

$$\|g\|^t \leqslant \int_0^t \|f\|^\tau \mathrm{d}\tau + \|g\|^0.$$

Proof. (Induction on n): $n=0$, clear. Passage from n to $(n+1)$:

$$\int_0^{t+k} \|f\|^\tau \mathrm{d}\tau + \|g\|^0 = \int_0^t \|f\|^\tau \mathrm{d}\tau + \int_t^{t+k} \|f\|^\tau \mathrm{d}\tau + \|g\|^0$$
$$\geqslant k\|f\|^t + \int_0^t \|f\|^\tau \mathrm{d}\tau + \|g\|^0$$
$$\geqslant k\|f\|^t + \|g\|^t \text{ (inductive hypothesis)}$$
$$\geqslant \|g\|^{t+k}$$ ∎

Thus $A_n = (L_{k,h}, I_h)$ we immediately have

$$\|w\|^T \leqslant T\|L_{k,h}(w)\|^T + \|w\|^0 \leqslant \max(T,1)[\|L_{k,h}w\|^T + \|w\|^0] = M\|A_n w\|, \tag{6}$$

or $\|A_n^{-1}\| \leqslant M$, which is the uniform boundedness of $\{A_n^{-1}\}$.

Next we show the consistency of the process, that is, if u is a solution of the differential equation then on the lattice G we have

$$\left| L_{k,h}(u) - \left(\frac{\partial u}{\partial t} - \frac{\partial^2 u}{\partial x^2}\right)\right| \to 0 \quad \text{as } k \to 0, h \to 0.$$

This follows easily from the Taylor expansion of the solution u

$$\frac{u(x,t+k) - u(x,t)}{k} = \frac{\partial u}{\partial t} + \frac{k}{2}\frac{\partial^2 \bar{u}}{\partial t^2},$$

first with respect to t, ($\bar{u}$ is the abbreviation for taking the function value at an intermediate point), then with respect to x

$$u(x \pm h, t) = u(x, t) \pm h\frac{\partial u}{\partial x} + \frac{h^2}{2}\frac{\partial^2 u}{\partial x^2} \pm \frac{h^3}{6}\frac{\partial^3 u}{\partial x^3} + \frac{h^4}{24}\frac{\partial^4 \bar{u}}{\partial x^4}$$

from which follows

$$\frac{u(x+h,t) - 2u(x,t) + u(x-h,t)}{h^2} = \frac{\partial^2 u}{\partial x^2} + \frac{h^2}{12}\frac{\partial^4 \bar{u}}{\partial x^4}.$$

Together this gives

$$\left| L_{k,h}(u) - \left(\frac{\partial u}{\partial t} - \frac{\partial^2 u}{\partial x^2}\right)\right| = \left|\frac{k}{2}\frac{\partial^2 \bar{u}}{\partial t^2} - \frac{h^2}{12}\frac{\partial^4 \bar{u}}{\partial x^4}\right| \leqslant N(k + h^2) \quad \text{(consistency estimate)},$$

where $N < \infty$ is the maximum on $[0, \pi] \times [0, T]$ of the derivatives of u. The convergence of the process now follows easily, either by the convergence theorem or directly:

$$\begin{aligned} \| u - w \|^T &\leqslant T \| L_{k,h}(u - w) \|^T + \| u - w \|^0 \quad \text{(by (6))} \\ &= T \| L_{k,h}(u - w) \|^T \end{aligned}$$

where we have

$$L_{k,h}(u - w) = L_{k,h}(u) - L_{k,h}(w) = L_{k,h}(u) - f = L_{k,h}(u) - \left(\frac{\partial u}{\partial t} - \frac{\partial^2 u}{\partial x^2}\right),$$

hence by the consistency estimate

$$\| u - w \|^T \leqslant T \cdot N(k + h^2) \rightarrow 0.$$

Example 32.2 We consider the wave equation

$$\frac{\partial^2 u}{\partial t^2} - \frac{\partial^2 u}{\partial x^2} = f \quad \text{on } (0, \pi) \times (0, T), \tag{7}$$

with the initial conditions

$$u(x, 0) = g_0(x), \quad \frac{\partial u(x, 0)}{\partial t} = g_1(x), \quad x \in [0, \pi] \tag{8}$$

and the boundary conditions

$$u(0, t) = u(\pi, t) = 0, \quad t \in [0, T]. \tag{9}$$

For simplicity we use a square lattice $k = h$

$$G = \left\{(mh, jh) | m = 0, \ldots, \frac{\pi}{h}, j = 0, \ldots, \frac{T}{h}\right\} \backslash,$$

again without the edges $\{0\} \times [0, T]$, $\{\pi\} \times [0, T]$.

We replace the derivatives by difference quotients

$$\frac{\partial^2 u}{\partial x^2} \sim \frac{v(x+h,t)+v(x-h,t)-2v(x,t)}{h^2},$$

$$\frac{\partial^2 u}{\partial t^2} \sim \frac{v(x,t+h)+v(x,t-h)-2v(x,t)}{h^2},$$

and by means of (7) construct the difference process

$$\frac{v(x,t+h)+v(x,t-h)-2v(x,t)}{h^2} = \frac{v(x+h,t)+v(x-h,t)-2v(x,t)}{h^2} + f(x,t). \tag{10}$$

Hence

$$v(x,t+h) = v(x+h,t)+v(x-h,t)-v(x,t-h)+h^2 f(x,t).$$

On a time slice this process calculates the values from the values on two earlier slices. Therefore for starting values we need the values on the two slices $t=0, h$. For this we also replace the derivative in the initial conditions by the difference quotient

$$\frac{\partial u(x,0)}{\partial t} \sim \frac{v(x,h)-v(x,0)}{h}$$

and obtain

$$v(x,0)=g_0(x), \quad v(x,h)=g_0(x)+hg_1(x). \tag{11}$$

We complete the computation (10), (11) in conformity with (a) by putting

$$v(0,t)=v(\pi,t)=0.$$

As in Example 32.1 we again introduce the difference operator

$$L_h(w)(x,t) := \frac{1}{h^2}(w(x,t+h)-w(x+h,t)-w(x-h,t)+w(x,t-h)),$$

where w is a function defined on the lattice satisfying the boundary conditions

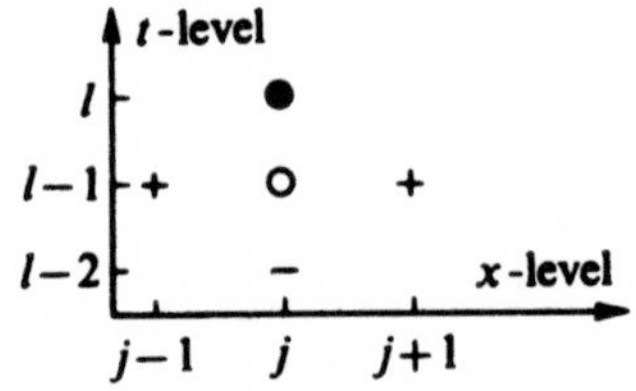

Fig. 32.2

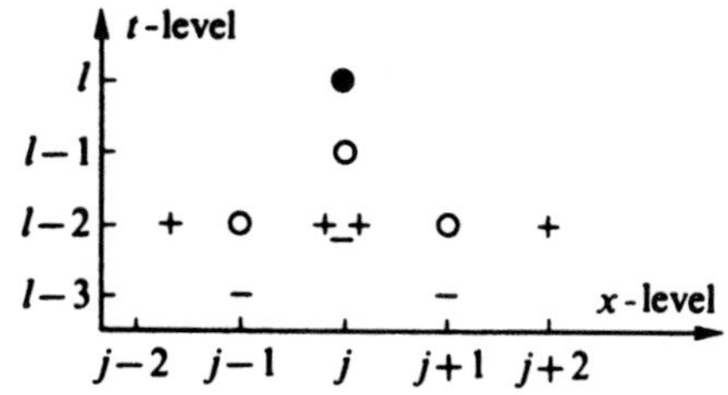

Fig. 32.3

$w(0, t) = w(\pi, t) = 0$ for all t. In addition we introduce two further abbreviations

$$w_{j,l} := w(jh, lh), \quad L_h w_{j,l} := L_h(w)(jh, lh).$$

With this notation we have

$$w_{j,l} = w_{j+1,l-1} - w_{j,l-2} + w_{j-1,l-1} + h^2 L_h w_{j,l-1}. \tag{12}$$

We illustrate this calculation by means of the following point scheme, see Fig. 32.2. Here ⊙ stands for the value to be calculated, $w_{j,l}$, + (respectively −) stands for $+w_{j+1,l-1}$, $+w_{j-1,l-1}$, $-w_{j,l-2}$ and ○ for $h^2 L_h w_{j,l-1}$. If we now calculate $w_{j+1,l-1}$ and $w_{j-1,l-1}$ by means of (12) and substitute these in the equation for $w_{j,l}$, we can indicate the result in our point scheme by means of two plus signs. We then obtain the scheme shown in Fig. 32.3.

A decisive observation is that two plus signs and one minus sign in the middle may be changed to a plus. If by means of (12) we once more substitute the w-values of the $(l-2)$ level, we obtain the scheme shown in Fig. 32.4.

We now see how the scheme reproduces itself downwards. We have only to control what happens when we come near the boundary. For this we consider the scheme shown in Fig. 32.5.

The plus on the edge disappears, since $w_{0,l} = 0$ for all l because of the boundary conditions. In the next elimination step therefore only one plus

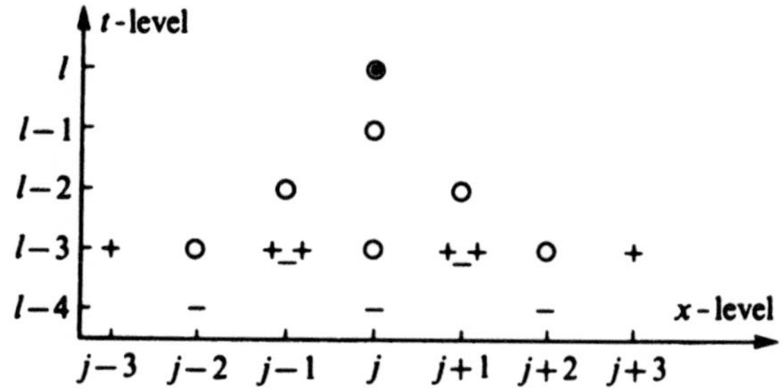

Fig. 32.4

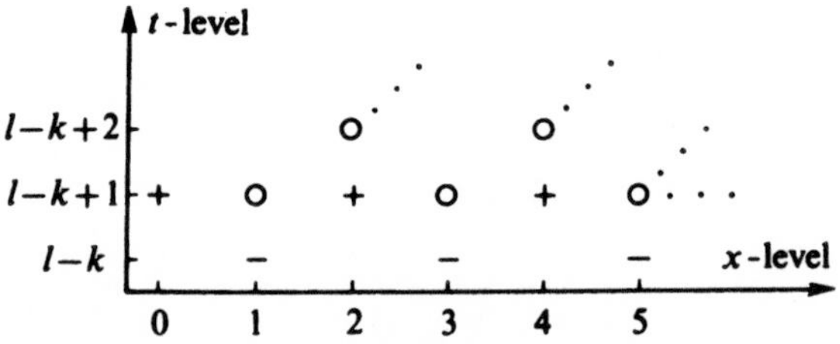

Fig. 32.5

arrives on the minus situated furthest to the left. In this way we obtain Fig. 32.6.

For we easily see how this reproduces itself further downwards. For present purposes it is only important to observe that the coefficients of the values of w (or $L_h w$) necessary for the calculation of $w_{j,l}$ are always equal to ± 1. Therefore

$$w_{j,l} = \underbrace{\sum_{k=0}^{\pi/h} \sigma_1(k) w_{k,0}}_{\text{Sum I}} + \underbrace{\sum_{k=0}^{\pi/h} \sigma_2(k) w_{k,1}}_{\text{Sum II}} + \underbrace{h^2 \sum_{i=0}^{T/h} \sum_{k=0}^{\pi/h} \sigma_3(i,k) L_h w_{i,k}}_{\text{Sum III}}, \tag{13}$$

where the σ_i $(i = 1, 2, 3)$ only take the values $\pm 1, 0, -1$. Sum I means that we sum on the line at level $t = 0$, sum II at $t = 1 \cdot h$, sum III on G. With the notation of Example 32.1 we have

$$B_1 = C_0([0, \pi] \times [0, T]), \quad B_2 = C([0, \pi] \times [0, T]) \times C[0, \pi] \times C[0, \pi],$$

$$B_{1,n} = \mathbb{R}^d, \quad B_{2,n} = \mathbb{R}^d \times \mathbb{R}^{d_1} \times \mathbb{R}^{d_1},$$

$$A = \left(\frac{\partial^2}{\partial t^2} - \frac{\partial^2}{\partial x^2}, I, I \circ \frac{\partial}{\partial t} \right), \quad A_n = (L_h, I_h^{t=0}, I_h^{t=h}),$$

where $I_h^{t=0} v = v(x, 0)$, $I_h^{t=h} v = v(x, h)$ are the initial value operators on the lines at level $t = 0$ and $t = h$. (10) and (11) show that we can uniquely solve

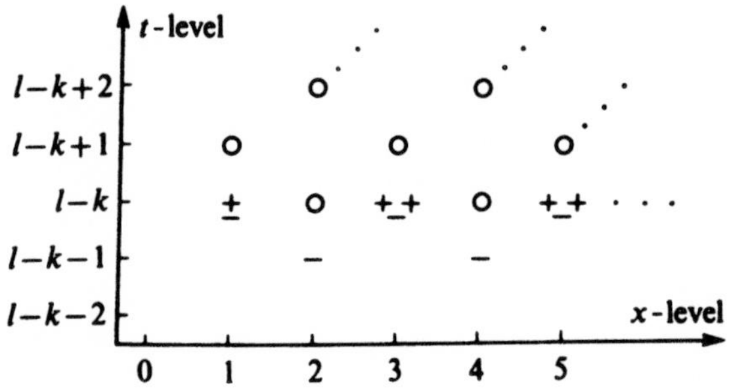

Fig. 32.6

the equations

$$L_h v = f, \quad I_h^{t=0} v = g_0, \quad I_h^{t=h} v = g_0 + hg, \tag{14}$$

or that A_n is invertible. For the solution $v = A_n^{-1}(f, g_i, g_1)$ of (14) it follows from (13) that

$$\begin{aligned}\max_G |v(x,t)| &\leq 2\left(\frac{\pi}{h}+1\right)\max|g_0| + \left(\frac{\pi}{h}+1\right)h\max|g_1| \\ &\quad + h^2\left(\frac{\pi}{h}+1\right)\left(\frac{T}{h}+1\right)\max|f| \\ &\leq (\text{constant})\cdot h^{-1},\end{aligned}$$

that is, we have h^{-1}-stability. The bad behaviour is caused by the summand $2\pi/h \max|g_0|$. All the same the process converges, since we have chosen the initial conditions $v(x,0) = u(x,0) = g_0$, that is, if $t = 0$ we have exact consistency $|v(x,0) - u(x,0)| = 0$, hence of each order. We demonstrate the convergence of the process directly; we could also proceed by showing consistency and then applying the modified convergence theorem, see Remark 32.1(iii).

Let $e(x,t) := u(x,t) - v(x,t)$, then for the difference operator L_h it follows by means of (13) that

$$|e(x,t)| \leq \sum_{\mathrm{I}} |e(x,0)| + \sum_{\mathrm{II}} |e(x,h)| + h^2 \sum_{\mathrm{III}} |L_h(e)(x,t)|.$$

Now $e(x,0) = 0$ for $x = 0, h, \ldots$, and because

$$u(x,h) = u(x,0) + h\frac{\partial u}{\partial t}(x,0) + \frac{h^2}{2}\frac{\partial^2 \bar{u}}{\partial t^2}$$

for $t = h$, it follows immediately that

$$e(x,h) = \frac{h^2}{2}\frac{\partial^2 \bar{u}}{\partial t^2},$$

and therefore

$$|e(x,h)| \leq h^2 M,$$

where M is again the maximum among all the derivatives of u appearing. As in Example 32.1 it follows (again assuming that the solution of the differential equation is four times continuously differentiable) that

$$\frac{1}{h^2}(u(x,t+h) + u(x,t-h) - 2u(x,t)) = \frac{\partial^2 u}{\partial t^2} + \frac{h^2}{12}\frac{\partial^4 \bar{u}}{\partial t^4},$$

$$\frac{1}{h^2}(u(x+h,t) + u(x-h,t) - 2u(x,t)) = \frac{\partial^2 u}{\partial x^2} + \frac{h^2}{12}\frac{\partial^4 \bar{u}}{\partial x^4},$$

and therefore

$$L_h(e) = L_h(u) - L_h(v) = L_h(u) - f$$
$$= L_h(u) - \left(\frac{\partial^2 u}{\partial t^2} - \frac{\partial^2 u}{\partial x^2}\right) = \frac{h^2}{12}\left(\frac{\partial^4 \bar{u}}{\partial t^4} - \frac{\partial^4 \bar{u}}{\partial x^4}\right).$$

so that

$$|L_h(e)| \leqslant h^2 M,$$

which is second-order consistency. Putting together we have

$$\max|e(x,t)| \leqslant h^2 M \sum_{\mathrm{I}} 1 + h^4 M \sum_{\mathrm{III}} 1 = h^2 M \frac{\pi}{h} + h^4 M \frac{\pi}{h}\frac{T}{h} = hM\pi(1 + hT),$$

and hence the process converges uniformly.

We return to the general considerations which led to the convergence Theorem 32.1. At this point we consider the concepts introduced more precisely, and we are particularly interested in the following three questions:

1. Are the hypotheses not only sufficient for convergence, but also necessary?

2. In fact we are not really interested in a sequence of elements from $B_{1,n}$ converging discretely to the solution, but rather in a sequence in B_1 converging to the solution. As a rule we can, however, embed the spaces $B_{1,n}$ in B_1. Which convergence assumptions must these embeddings fulfil in order that we can prove convergence in B_1 of the embedded sequence from the discrete convergence?

3. From the existence of a consistent, stable approximation to the operator A can we arrive at properties of this operator?

We begin with the necessity of stability.

Theorem 32.2 *If the process converges discretely for $f = 0$, see Definition 32.1.4. then the sequence $\{A_n^{-1}\}$ is uniformly bounded, that is, we have stability.*

Proof. Suppose that $\{A_n^{-1}\}$ is not uniformly bounded. Then there exists $\bar{f}_n \in B_{2,n}$ with $\|\bar{f}_n\| = 1$ and

$$\|A_n^{-1}\bar{f}_n\| = a(n) \to \infty \quad \text{as } n \to \infty.$$

We now put

$$f_n = \frac{1}{a(n)^{1/2}}\bar{f}_n, \quad f = 0, \quad u = 0, \quad u_n = A_n^{-1} f_n.$$

Then we have

$$\|f_n - D_{2,n} f\| = \|f_n\| \to 0,$$

hence the process converges discretely to u by assumption, that is,

$$\|u_n - D_{1,n}u\| = \|u_n\| \to 0,$$

However, on the other hand,

$$\|u_n\| = a(n)^{1/2} \to \infty$$

and so the assumption leads to a contradiction. ∎

Under an additional condition, namely the uniform boundedness of $\{A_n\}$ consistency also follows from convergence.

Theorem 32.3 *Let $u \in B_1$ and suppose that:*

(a) *$\{A_n\}$ is uniformly bounded,*
(b) *the process converges for $f = Au$. Then A and A_n are consistent at u.*

Proof. Let $u_n = A_n^{-1} D_{2,n} f$, then by hypothesis

$$\|D_{1,n}u - u_n\| \to 0.$$

Suppose further that $\|A_n\| \leqslant M < \infty$, then

$$\|A_n D_{1,n}u - D_{2,n}Au\| = \|A_n(D_{1,n}u - A_n^{-1}D_{2,n}Au)\| \leqslant M\|D_{1,n}u - u_n\| \to 0. \quad ∎$$

We turn next to the second question, namely how do we obtain convergence in B_1?

Theorem 32.4 *Let $i_n : B_{1,n} \to B_1$ be linear, then the following conditions are equivalent:*

(a) $\|u_n - D_{1,n}u\| \to 0$ *implies that* $\|i_n u_n - u\|_{B_1} \to 0$.
(b) $\{i_n\}$ *is uniformly bounded and for all* $u \in B_1$, $\|u - i_n D_{1,n}u\| \to 0$.

Proof. (a)⇒(b) Assume that $\{i_n\}$ is not uniformly bounded, then there exists $u_n \in B_{1,n}$ with

$$\|i_n u_n\| = a(n) \to \infty, \quad \|u_n\| = 1.$$

We put $v_n = u_n / a(n)^{1/2}$, $u = 0$, and it follows that

$$\|v_n - D_{1,n}0\| = \|v_n\| = \frac{1}{a(n)^{1/2}} \to 0$$

and therefore, by assumption, $a(n)^{1/2} = \|i_n v_n\| \to 0$. Therefore we obtain a contradiction. The remainder follows immediately by setting $u_n = D_{1,n}u$.

(b)⇒(a) This follows easily from the decomposition

$$u_n - i_n u_n = u - i_n D_{1,n}u + i_n(D_{1,n}u - u_n) \quad ∎$$

We next answer the third question.

Theorem 32.5 *Let the process be stable and suppose that the condition*

$$\| D_{1,n}u \| \to 0 \text{ implies that } u = 0,\ u \in B_1, \qquad (D_1')$$

is satisfied. Then for each $f \in B_2$ there exists at most one consistent solution of the equation $Au = f$.

Remark (D_1') is, for example, satisfied if for all $u \in B_1$ we have

$$\| D_{1,n}u \| \to \| u \|. \qquad (D_1)$$

Proof. Let $Au_1 = f = Au_2$ and let A, A_n be consistent at u_1 and u_2. With $u_n = A_n^{-1} D_{2,n} f$ the convergence theorem implies that

$$\| D_{1,n}(u_1 - u_2) \| = \| D_{1,n}u_1 - D_{1,n}u_2 \| \leqslant \| D_{1,n}u_1 - u_n \| + \| D_{1,n}u_2 - u_2 \| \to 0,$$

given (D_1'), therefore $u_1 - u_2 = 0$. ■

For later use we need the uniform boundedness theorem in a somewhat altered form and the Banach–Steinhaus theorem. We state these theorems here – for the proofs we refer, for example, to Akhieser & Glazman [1] and Dunford & Schwartz [1].

Uniform boundedness theorem *Let X be a Banach space and F a family of continuous functionals $\Phi_\alpha : X \to \mathbb{R}_+, \alpha \in F$, with the properties:*

(a) $\Phi_\alpha(x) \geqslant 0$,
(b) $\Phi_\alpha(x + y) \leqslant \Phi_\alpha(x) + \Phi_\alpha(y)$,
(c) $\Phi_\alpha(\lambda x) = |\lambda| \Phi_\alpha(x), \quad \lambda \in \mathbb{C}$ *($\lambda \geqslant 0$ is sufficient).*

Then the pointwise boundedness of F (that is, $\Phi_\alpha(x) \leqslant k_1(x) < \infty, \alpha \in F$) implies the uniform boundedness of F, that is, there exists a constant $k \geqslant 0$ with

$$\Phi_\alpha(x) \leqslant k \| x \| \quad \text{for all } x \in X,\ \alpha \in F.$$

Banach–Steinhaus theorem *Let X and Y be Banach spaces, $A_n : X \to Y$ linear and continuous. Then the following two conditions are equivalent:*

(a) *The A_n converge pointwise to a linear continuous map $A : X \to Y$.*
(b) *$\{A_n\}_{n \in \mathbb{N}}$ is uniformly bounded and there exists a dense subset D in X, such that if $x \in D$, then $\{A_n x'\}_{n \in \mathbb{N}}$ is a Cauchy sequence.*

We return to Fig. 32.1 and assume from now on that both the discretisation operators $D_{1,n}$ and $D_{2,n}$ and the approximation operators A_n are continuous.

Theorem 32.6 *Suppose that the condition (D_2) is satisfied for the discretisation* *operators $D_{2,n}$, that is,*

$$\lim_n \| D_{2n} f \| = \| f \| \quad \text{for all } f \in B_2. \tag{D_2}$$

Then we have

$$\| D_{2,n} f \| \leqslant k \| f \| \quad \textit{for all } f \in B_2,\ n \in \mathbb{N}. \tag{15}$$

Proof. We use the uniform boundedness theorem, with $\Phi_n(f) = \| D_{2,n} f \|$, then using ($D_2$) we obtain the pointwise boundedness of the family $F = \{ \| D_{2,n} \| \}$ and (15) is the uniform boundedness of F. ■

For the next theorem we consider the embedding operators $i_n: B_{1,n} \to B_1$ of Theorem 32.4.

Theorem 32.7 *Suppose that the discretisation operators $D_{1,n}: B_1 \to B_{1,n}$ are surjective, and suppose that the embedding operators $i_n: B_{1,n} \to B_1$ are right inverse to $D_{1,n}(D_{1,n} \circ i_n = I)$ and uniformly bounded. Then the pointwise boundedness* of $A_n \circ D_1$, *that is,*

$$\| A_n D_{1,n} \| \leqslant k_1(n) < \infty \tag{16}$$

implies the uniform boundedness of $\{A_n\}$, that is,

$$\| A_n \| \leqslant k < \infty, \quad n \in \mathbb{N}. \tag{17}$$

Proof. The uniform boundedness theorem applied to $\| A_n D_{1,n} u \|$ gives

$$\| A D_{1,n} u \| \leqslant k_2 \| u \|. \tag{18}$$

Suppose now that $\bar{u}$ is an arbitrary element from $B_{1,n}$. Because $D_{1,n}$ is surjective there exists $u \in B_1$ with $\bar{u} = D_{1,n} u$, and from (18) we obtain

$$\| A \bar{u} \| \leqslant k_2 \| u \|. \tag{19}$$

On the right-hand side of (19) we form the infimum over the class $\bar{u} \in B_1 / \ker D_{1,n} \simeq B_{1,n}$

$$\| A_n \bar{u} \| \leqslant k_2 \inf_{\bar{u} = D_{1,n} v} \| v \|.$$

In the class $\bar{u}$ we find the element $v = i_n \bar{u}$, since (right inverse!)

$$D_{1,n} v = D_{1,n} i_n \bar{u} = \bar{u},$$

and the uniform boundedness of $\{i_n\}$ leads to

$$\| A_n \bar{u} \| \leqslant k_2 \inf_{\bar{u} = D_{1,n} v} \| v \| \leqslant k_2 \| i_n \bar{u} \| \leqslant k_2 c \| \bar{u} \|,$$

which is (17) with $k = c k_2$.

Theorem 32.8 *Suppose that in addition to the assumptions for Theorem 32.7 condition (D_2) (respectively (15), see Theorem 32.6) is satisfied. Then the discrete convergence of A_n implies the uniform boundedness of $\{A_n\}$.*

Proof. Discrete convergence implies, see Definition 32.1, that for each $u \in B_1$ there exists some $f \in B_2$ with

$$\| A_n D_{1,n} u - D_{2,n} f \| \to 0 \quad \text{as } n \to \infty.$$

and by means of (15) we estimate

$$\begin{aligned} \| A_n D_{1,n} u \| &\leqslant \| A_n D_{1,n} u - D_{2,n} f \| + \| D_{2,n} f \| \\ &\leqslant A_n D_{1,n} u - D_{2,n} f \| + k \| f \| < \infty, \end{aligned}$$

which is (16); therefore Theorem 32.7 gives

$$\| A_n \| \leqslant k < \infty \quad (n \in \mathbb{N}).$$

■

Until now we have always started from the assumption that u is a solution of the equation $Au = f$. We can now reverse the situation and show that if the difference process given by u_n converges discretely to u, then u must be a solution of $Au = f$. Therefore we can use a difference process to show the existence of solutions for partial differential equations, see also Ladyženskaya & Uralceva [1], Ladyženskaya, Solonnikov & Uralceva [2], Ladyženskaya [3].

Theorem 32.9 *For the Fig.* 32.1 *we assume*:

(a) $D_{1,n}: B_1 \to B_{1,n}$ *is surjective*, $D_{2,n}$ *satisfies condition* (D_2), *the embedding operators* $i_n: B_{1,n} \to B_1$ *are uniformly bounded and right inverse to* $D_{1,n}$.
(b) A *is defined on the whole space* B_1.
(c) A *and* A_n *are consistent for all* $u \in B_1$.
(d) *If* $f \in B_2$ *the process given by* $u_n = A_n^{-1} D_{2,n} f$ *converges discretely to* u. *Then* u *is a solution of the equation* $Au = f$.

Proof. Consistency, that is, $\| D_{2,n} A_n - A_n D_{1,n} u \| \to 0$ implies the discrete convergence of $A_n D_{1,n} u$ to Au, from which by assumption (a) and Theorem 32.8 it follows that $\{A_n\}$ is uniformly bounded, that is, $\| A_n \| \leqslant k < \infty$, and we can estimate

$$\begin{aligned} \| D_{2,n}(Au - f) \| &\leqslant \| D_{2,n} Au - A_n D_{1,n} u \| + \| A_n D_{1,n} u - D_{2,n} f \| \\ &= \| D_{2,n} Au - A_n D_{1,n} u \| + \| A_n D_{1,n} u - A_n A_n^{-1} D_{2,n} f \| \\ &\leqslant \| D_{2,n} Au - A_n D_{1,n} u \| + k \| D_{1,n} u - u_n \|. \end{aligned}$$

Here the first summand tends to zero because of (c), and the second because of (d). Condition (D_2), applied again, allows us to conclude that

$$\| Au - f \| = \lim_n \| D_{2,n}(Au - f) \| = 0, \quad \text{that is, } Au = f. \qquad \blacksquare$$

It is possible to combine the assumptions in different ways and, by means of results already proved, obtain many other assertions. Here we state just one, in which we apply Theorem 32.4.

Theorem 32.10 *Let $D \subset B_2$ be a dense, linear subspace and assume:*

(a) *for all $f \in D$ there exists a consistent solution u of the equation $Au = f$,*
(b) *the process is stable,*
(c) *$D_{2,n}$ is uniformly bounded, that is, (15),*
(d) *there exist linear maps $i_n : B_{1,n} \to B_1$ with $\{i_n\}$ uniformly bounded, and $\| i_n D_{1,n} u - u \| \to 0$ for all $u \in B_1$.*

Then there exists a continuous linear solution operator $L: B_2 \to B_1$, so that for $f \in D$, $A \circ Lf = f$ holds.

Proof. The family $\{i_n A_n^{-1} D_{2,n}\}$ is uniformly bounded by our assumptions. If now $f \in D$, then there exists a consistent solution u, and the convergence Theorem 32.1 implies that $\| D_{1,n} u - A_n^{-1} D_{2,n} f \| \to 0$. Because of hypothesis (d) it follows from Theorem 32.4 that

$$\| u - i_n A_n^{-1} D_{2,n} f \| \to 0 \quad \text{for } f \in D, \tag{20}$$

therefore we can apply the Banach–Steinhaus theorem, and obtain the existence of a continuous linear map

$$L: B_2 \to B_1, \quad \left(Lf = \lim_n i_n A_n^{-1} D_{2,n} f \right),$$

for which because of (20) it is true that $u = Lf$ for $f \in D$, that is, $A \circ Lf = f$. $\blacksquare$

The statement of Theorem 32.10 implies that if $Au = f$ is uniquely solvable, then only for correctly posed problems does it make sense to look for consistent, stable approximation processes.

Let us look back once more to the convergence theorem. We have seen that the solutions for the (consistent, stable) calculation process $u_n = A_n^{-1} f_n$ converge to the solution u of the equation $Au = f$, if only the values f_n converge discretely to f. In computational practice this kind of convergence is very important since it is often not possible to work with the exact, discretised values $D_{2,n} f$, but only with neighbouring values f_n, for example when f itself is only known up to an approximation. The next theorem is concerned with the convenient behaviour of the process for 'small perturbations'. For this suppose that

$$Q_n : B_{1,n} \to B_{2,n} \text{ is linear and continuous with } \|Q_n\| \to 0,$$
$$A'_a := A_n + Q_n.$$

If we replaces A_n by A'_n we obtain a new approximation process.

Theorem 32.11 (perturbation theorem)

(a) *If A, A_n are consistent at u, and in addition $Q_n D_{1,n} u \to 0$, then A, A'_n are also consistent at u.*
(b) *If the family $\{A_n^{-1}\}$ is uniformly bounded, then for some $n_0 \in \mathbb{N}$ the family $\{A_n'^{-1} : n \geqslant n_0\}$ is also uniformly bounded.*

Proof. (a) We have

$$\|D_{2,n}Au - A'_n D_{1,n}u\| \leqslant \|D_{2,n}Au - A_n D_{1,n}u\| + \|Q_n D_{1,n}u\| \to 0.$$

(b) Suppose $\|A_n^{-1}\| \leqslant M$: we first show that $A_n + Q_n$ is invertible for $n \geqslant n_1$. For this we consider the operator $\mathrm{Id} + A_n^{-1} \circ Q_n = A_n^{-1} A'_n$.

Because $\|A_n^{-1} Q_n\| \leqslant M \|Q_n\|$ and $\|Q_n\| \to 0$, there exists some $n_1 \in \mathbb{N}$ with $\|A_n^{-1} Q_n\| < 1$ for $n \geqslant n_1$. Therefore by Lemma 12.1, if $n \geqslant n_1$ the operator $\mathrm{Id} + A_n^{-1} Q_n$ is invertible; hence for $n \geqslant n_1$, A'_n is also continuously invertible and in particular surjective. Moreover,

$$\|A'_n u_n\| = \|(A_n + Q_n)u_n\| \geqslant \|A_n u_n\| - \|Q_n u_n\| \geqslant \frac{1}{M}\|u_n\| - \|Q_n\| \, \|u_n\|$$
$$= \left(\frac{1}{M} - \|Q_n\|\right)\|u_n\| \geqslant \frac{1}{2M}\|u_n\| \quad \text{for } n \geqslant n_2.$$

For $n_0 = \max\{n_1, n_2\}$ it therefore follows that

$$\|A_n'^{-1}\| \leqslant 2M \quad \text{for } n \geqslant n_0,$$

which is the stability of $A_n'^{-1}$. ∎

Next we show how difference methods for linear partial differential equations fit into this theory. For this we suppose that $\Omega \subset \mathbb{R}^r$ is a region with sufficiently regular frontier $\partial\Omega$; here individual boundary pieces $\Gamma_i \subset \partial\Omega$, $i = 1, \ldots, s$ may be distinguished, with the possibilities that $\Gamma_i = \Gamma_j$ or $\bigcup_i \Gamma_i \neq \partial\Omega$.

Let F_0^1, F_0^2 be Banach spaces of functions $f : \Omega \to \mathbb{R}$,
F_i Banach spaces of functions $f_i : \Gamma_i \to \mathbb{R}$, $i = 1, \ldots, s$,

$L: F_0^1 \to F_0^2$ a linear differential operator, and

$l_i: F_0^1 \to F_i$ linear boundary operators, $i = 1, \ldots, s$.

Then the general boundary problem reads:

given $f \in F_0^2, f_i \in F_i, \quad i = 1, \ldots, s,$

we look for some $u \in F_0^1$ with $Lu = f$ and $l_i u = f_i, \quad i = 1, \ldots, s.$ (21)

We now equip $\bar{\Omega}$ with a rectangular lattice, whose edges $K(h)$ depend on a single variable $h \in \mathbb{R}^+$; they can, however, be of different lengths. We denote the set of lattice points by $\bar{\Omega}_h$. Suppose further that

$$S_h := \{x \in \bar{\Omega}_h : d(x, \partial\Omega) < h\}, \quad \Gamma_{i,h} \subset S_h, \quad i = 1, \ldots, s,$$
$$\Omega_h := \bar{\Omega}_h \backslash S_h.$$

We now want to make the problem (21) discrete. For this suppose:

$F_{0,h}^1, F_{0,h}^2$ are Banach spaces of functions $f: \bar{\Omega}_h \to \mathbb{R}$,

$F_{i,h}$ are Banach spaces of functions $f: \Gamma_{i,h} \to \mathbb{R}, \quad i = 1, \ldots, s,$

$D_{0,h}^i: F_0^i \to F_{0,h}^i$ are linear, $\quad i = 1, 2,$

$d_{i,h}: F_i \to F_{i,h}$ are linear, $\quad i = 1, \ldots, s.$

The precise definition of the discretisation operators and of the discrete function spaces depends on the nature of problem (21). If for example F is a space of continuous functions equipped with the sup-norm, then as discretisation we can as a rule take restriction to the vertices. We equip the space F_h with the max-norm and embed it in F, extending the vertex functions linearly. For 'kinky' regions difficulties on the frontier ($S_h \not\subset \partial\Omega$) may arise. If F is a Sobolev or L_2-space, we have to proceed otherwise. Examples will be given in the next subsections.

We now also make the operators L, l_i discrete, replacing derivatives by suitable difference operators, and consider the approximation operators:

$$L_h: F_{0,h}^1 \to F_{0,h}^2 \quad \text{linear,}$$
$$l_{i,h}: F_{0,h}^1 \to F_{i,h} \quad \text{linear,} \quad i = 1, \ldots, s.$$

With this the system of difference equations then reads: we look for some $u_h \in F_{0,h}^1$ with

$$L_h u_h = D_{0,h}^2 f \quad \text{and} \quad l_{i,h} u_h = d_{i,h} f_i, \quad i = 1, \ldots, s. \tag{22}$$

With the aid of the convergence theorem above we wish to show when the solutions of problem (22) converge to that of problem (21). For this we write

$$B_1 := F_0^1 \qquad B_2 := F_0^2 \times \mathop{\times}_1^s F_i \text{ with the norm } \|\cdot\|_{F_0^2} + \sum_1 \|\cdot\|_{F_i},$$

$$B_{1,n} := F_{0,h}^1, \qquad B_{2,n} := F_{0,h}^2 \times \mathop{\times}_1^s F_{i,h}, \quad h = \frac{1}{n},$$

$$\begin{aligned} A &: B_1 \to B_2, & u &\mapsto (Lu, l_1 u, \ldots, l_s u), \\ A_n &: B_{1,n} \to B_{2,n} & v &\mapsto (L_h v, l_{1,h} v, \ldots, l_{s,h} v), \\ D_{1,n} &: B_1 \to B_{1,n}, & u &\mapsto D_{0,h}^1 u, \\ D_{2,n} &: B_2 \to B_{2,n}, & u &\mapsto (D_{0,h}^2 u, d_{1,h} u, \ldots, d_{s,h} u). \end{aligned}$$

We can now carry over the theory above. The assumptions for the convergence of the process read:

(a) We have consistency for A and A_n, if

$$\| D_{0,h}^2(Lu) - L_h(D_{0,h}^1 u) \| \to 0,$$

and

$$\| d_{i,h}(l_i u) - l_{i,h}(D_{0,h}^1 u) \| \to 0, \quad i = 1, \ldots, s.$$

(b) The A_n^{-1} are uniformly bounded if there is some so that for all h

$$\| u_h \| \leqslant c\left(\| L_h u_h \| + \sum_1^s \| l_{i,h} u_h \| \right) = c \| A_n u_h \|,$$

or also, if there is some $c \geqslant 0$ so that for all h

$$\| u_h \| \leqslant c\left(\| D_{0,h}^2 f \| + \sum_1^s \| d_{i,h} f_i \| \right),$$

for all f, f_i, u_n which satisfy equations (22).

If these assumptions are satisfied, then by Theorem 32.1 $\| D_{0,h}^1 u - u_h \| \to 0$ as $h \to 0$ provided that u is a solution of (21) and u_h is a solution of (22). The other theorems may also be immediately carried over to the difference process (22) for the boundary value problem (21).

Exercises

32.1 Generalise the considerations of Example 32.1 (heat equation) to several space variables $x = (x_1, \ldots, x_r)$. Let k be the width in the t-direction and h in the x-direction, let $\lambda := k/h^2$. Show that one has stability if $1 - 2r\lambda \geqslant 0$.

32.2 Develop a difference process for the initial value problem $y' = y$, $y(0) = 1$, and prove the convergence of the process.

32.3 Solve the initial value problem $y' = cy$, $y(0) = a$ (a, c real constants) by means of a difference process, in which one approximates y' by $(1/h)[3y(x) - y(x+h) - 2y(x) - h)]$. Does the process converge, is the process consistent and of what order?

32.4 Develop a difference process for the initial value problem

$$y'' + y' - 2y = 0, \quad y(0) = a, \quad y'(0) = b.$$

§33 Difference processes for elliptic differential equations and for the wave equation

In this section we develop a method for the construction of difference processes for partial differential equations of elliptic type, and we generalise the process considered in Example 32.2 for the wave equation, to several space coordinates. We need some estimates, which we derive in the next subsection.

33.1 Some inequalities

The results which we set down at this point correspond to known results from the theory of partial differential equations, if we replace the sum by the integral sign and the difference by the differential operator.

Notation Let $\Omega \subset \mathbb{R}^r$ be open, connected and bounded, let e_i be the unit vector in the ith space coordinate. If $u:\Omega \to \mathbb{R}$, as usual write

$$\Delta_h^i u(x) := \frac{u(x + he_i) - u(x)}{h}, \quad i = 1, \ldots, r,$$

$$(\Delta^h u)^2 := \sum_1^r (\Delta_h^i u)^2.$$

Remark Because $u(x + he_i)$ is not defined for $x + he_i \notin \Omega$, we extend the function $u:\Omega \to \mathbb{R}$ to a function $\tilde{u}:\mathbb{R}^r \to \mathbb{R}$. The choice of the extension depends on the problem, mostly we can make do with the 0-extension.

Let $\bar{\Omega}_h$ be a square lattice in $\bar{\Omega}$ of width h.

$$S_h := \{x \in \bar{\Omega}_h : d(x, \partial\Omega) < h\}, \quad \Omega_h := \bar{\Omega}_h \backslash S_h.$$

If $x_1, \ldots, x_k \in \mathbb{R}^r$, $L[x_1, \ldots, x_k]$ denotes the subspace spanned by $x_1, \ldots, x_k$. We first introduce a formula for Δ_h^i, which corresponds to the rule for partial integration with respect to $\partial/\partial x_i$.

Theorem 33.1 *Let $u, v:\bar{\Omega}_h \to \mathbb{R}$ be such that $u|_{S_h} = 0 = v|_{S_h}$. Then*

$$\sum_{\Omega_h} u \cdot \Delta_h^i v = -\sum_{\Omega_h} \Delta_{-h}^i u \cdot v.$$

Proof. Since Ω is bounded, we can embed Ω in an r-dimensional cube W. We extend u, v to all of $\mathbb{R}^r$ by means of 0. It is then enough to show the formula for $\bar{W}_h$. The proof proceeds by induction on r.

$r=1$: then Ω is an interval (without loss of generality $\Omega=(0,1)$) and $\bar{\Omega}_h$ is the set of points

$$x=jh, \quad j=1,\dots,n=1/h.$$

Having assumed $v(0)=vn(h)=0=u(nh)=u(0)$, we have (of course here $i=l$)

$$\sum_{j=0}^{n-1} u\Delta_h^1 v=\frac{1}{h}\sum_{j=0}^{n-1}(u(jh)v(jh+h)-u(jh)v(jh))$$

$$=\frac{1}{h}\sum_{j=0}^{n-1}(u(jh-h)v(jh)-u(jh)v(jh))=-\sum_{j=0}^{n-1}\Delta_{-n}^1 uv.$$

$r>1$: we can consider $\bar{W}_h$ as the direct sum of an $(r-1)$-dimensional cube $\bar{W}_h^1$ and a one-dimensional cube $\bar{W}_h^2$. For these we write

$$\overline{W_h^1}:=\text{projection of } \bar{W}_h \text{ on } L[e_1,\dots,e_{i-1},e_{i+1},\dots,e_r]$$
$$\overline{W_h^2}:=\text{projection of } \bar{W}_h \text{ on } L[e_i].$$

It then follows that

$$\sum_{\bar{W}_h} u\Delta_h^i v=\sum_{\bar{W}_h^1}\sum_{\bar{W}_h^2} u\Delta_h^i v=\sum_{\bar{W}_h^1}\left(-\sum_{\bar{W}_h^2}\Delta_{-h}^i uv\right) \quad \text{(one-dimensional case)}$$
$$=\sum_{\bar{W}}-\Delta_{-h}^i uv.$$

■

Lemma 33.1 *Let* $u:\bar{\Omega}_h\to\mathbb{R}$, $u|_{S_h}=0$. *Then there exists some* $c>0$, c dependent only on Ω *with*:

$$h^r\sum_{\Omega_h} u(x)^2\leqslant ch^r\sum_{\Omega_h}(\Delta^h u)^2(x)$$

Proof. We may assume without loss of generality that

$$\bar{\Omega}_h=\bar{W}_h=\{x:x=(k_1h,\dots,k_rh), -n\leqslant k_i\leqslant n\},$$

and again carry out the proof by induction on r. $r=1$: since $u(-nh)=0$, we have

$$u(kh)=h\sum_{j=-n}^{k-1}\Delta_h^1 u(jh)$$

and therefore

$$u(kh)^2=h^2\left(\sum_{-n}^{k-1}\Delta_h^1 u(jh)\right)^2\leqslant h^2\left(\sum_{-n}^{k-1}1\right)\left(\sum_{-n}^{k-1}(\Delta_h^1 u(jh))^2\right)$$

$$\leqslant h^2 2n\sum_{-n}^{k-1}(\Delta_h^1 u(jh))^2\leqslant(2nh)h\sum_{-n}^{n-1}(\Delta_h^1 u(jh))^2$$

$$=(2nh)h\sum_{\bar{W}_h}(\Delta_h^1 u(jh))^2.$$

Since in the one-dimensional case $(\Delta^h u)^2 = (\Delta_h^1 u)^2$, it follows that

$$\sum_{\bar{W}_h} u(kh)^2 \leqslant \sum_{\bar{W}_h}\left(2nh\cdot h\sum_{\bar{W}_h}(\Delta^h u)^2\right) = 2nh\left(\sum_{\bar{W}_h} h\right)\sum_{\bar{W}_h}(\Delta^h u)^2$$
$$= (2nh)^2\sum_{\bar{W}_h}(\Delta^h u)^2.$$

Here $2nh$ depends only on the size of the region (length of the edge of the cube $\bar{W}_h$).

$r > 1$: let $\bar{W}_h^1$, $\bar{W}_h^2$ be defined as above, then

$$\sum_{\bar{W}_h} u^2 = \sum_{\bar{W}_h^1}\left(\sum_{\bar{W}_h^2} u^2\right) \leqslant \sum_{\bar{W}_h^1}\left((2nh)^2\sum_{\bar{W}_h^2}(\Delta_h^i u)^2\right) \leqslant \sum_{\bar{W}_h^1}(2nh)^2\sum_{\bar{W}_h^2}(\Delta^h u)^2$$
$$= (2nh)^2\sum_{\bar{W}_h}(\Delta^h u)^2.$$

Remark In the continuous case this is the Poincaré lemma.

Lemma 33.2 *Let* $u:\bar{\Omega}_h \to \mathbb{R}$, $u|_{S_h} = 0$, then

$$\max_{\Omega_h}|u| \leqslant c\left(h^r\sum_{\Omega_h}(\Delta_h^r\cdots\Delta_h^1 u)^2\right)^{1/2},$$

where c *again depends only on* Ω.†

Proof. Let $\bar{W}_h$ be as above, then we may argue analogously

$$|u(k_1h,\ldots,k_rh)|^2 \leqslant (2nh)h\sum_{j_1=-n}^{n}(\Delta_h^1 u(j_1h, k_2h,\ldots,k_rh))^2,$$

and further:

$$|\Delta_h^1 u(j_1h, k_2h,\ldots,k_rh)|^2 \leqslant (2nh)h\sum_{j_2=-n}^{n}(\Delta_h^2\Delta_h^1 u(j_1h, j_2h, k_3h,\ldots,k_rh))^2 \text{ etc.}$$

altogether

$$|u(k_1h,\ldots,k_rh)|^2 \leqslant (2nh)^r h^r\sum_{j_1=-n}^{n}\cdots\sum_{j_r=-n}^{n}(\Delta_h^r\cdots\Delta_h^1 u(j_1h,\ldots,j_rh))^2$$
$$= (2nh)^r h^r\sum_{\bar{W}_h}(\Delta_h^r\cdots\Delta_h^1 u)^2,$$

and the lemma follows with $c = (2nh)^{r/2}$. ■

As an immediate consequence we have:

† Again we put $u = 0$ on $C\bar{\Omega}_h$, so the differences Δ are defined everywhere.

Theorem 33.2 *Let* $u:\bar{\Omega}_h \to \mathbb{R}$, $u|_{S_h} = 0$, *then*

$$\max_{\Omega_h} |u| \leqslant c\left(h^r \sum_{\Omega_h} (u^2 + (\Delta_h^1 u)^2 + \cdots + (\Delta_h^r \cdots \Delta_h^1 u)^2) \right)^{1/2},$$

where c *only depends on* Ω.

Remark In the continuous case this theorem corresponds to the Sobolev lemma. This lemma actually leads to a sharper conjecture: the order of differentiation in the Sobolev lemma is $r/2$, while here the difference order is r.

33.2 Construction of a difference process for the Dirichlet problem

We consider the following problem: let $\Omega \subset \mathbb{R}^r$; be open, bounded and connected. Suppose given $f:\Omega \to \mathbb{R}$, we look for $u:\Omega \to \mathbb{R}$ with

$$Lu := \sum_{i,j=1}^{r} \frac{\partial}{\partial x_i}\left(a_{ij} \frac{\partial u}{\partial x_j} \right) + \sum_{i=1}^{r} \left(\frac{\partial}{\partial x_i}(a_i u) + b_i \frac{\partial u}{\partial x_i} \right) + au = f \quad \text{and } u|_{\partial\Omega} = 0.$$

Here suppose that the coefficient functions $a_{ij}, a_i, b_i, a \in C(\bar{\Omega})$ are real valued, and that the operator L satisfies the Ladyženskaya condition, that is, there exists some $d > 0$ with

$$\sum_{i,j} a_{ij} y_i y_j + \sum_i (a_i - b_i) y_i y_0 - a y_0^2 \geqslant d \sum_l y_i^2, \tag{L}$$

for all $y_0, y_1, \ldots, y_r \in \mathbb{R}$ and all $x \in \bar{\Omega}$, the coefficient functions may be dependent on x.

We may limit ourselves to the homogeneous boundary condition, for from

$$Lu = f, u|_{\partial\Omega} = g, \tag{1}$$

it follows that

$$L(u - g) = Lu - Lg = f - Lg := \bar{f}, \quad u - g|_{\partial\Omega} = 0.$$

We may therefore solve

$$Lv = \bar{f}, \quad v|_{\partial\Omega} = 0. \tag{1'}$$

If v is a solution of (1′), then $u = v + g$ is a solution of (1). Of course the assumption is that g is extendable from $\partial\Omega$ to Ω and that Lg exists, but we can guarantee this by the trace theorems of §8.

We now show heuristically how one arrives at the approximation operator L_h. Since we are only looking for weak solutions for our problem ($Lu = f, u|_{\partial\Omega} = 0$) – the strong ones we obtain by means of Weyl's lemma (Theorem 20.4), by means of suitable regularity assumptions – we have for all $v \in C_0^\infty(\Omega) = \mathscr{D}(\Omega)$

$$\int_\Omega \left[\sum_{i,j} a_{ij} \frac{\partial u}{\partial x_j} \frac{\partial v}{\partial x_i} + \sum_i \left(a_i u \frac{\partial v}{\partial x_i} - b_i \frac{\partial u}{\partial x_i} v \right) - auv \right] dx = -\int_\Omega f v \, dx, \quad u|_{\partial\Omega} = 0. \tag{2}$$

We now pass to the difference equation in (2) on the square lattice $\bar{\Omega}_h$

$$\sum_{\Omega_h} h^r\left[\sum_{ij} a_{ij}\Delta_h^j u\Delta_h^i v+\sum_i (a_i u\Delta_h^i v-b_i\Delta_h^i uv)-auv\right]=-h^r\sum_{\Omega_h} f\cdot v,$$
$$u|_{S_h}=0=v|_{S_h}.$$

Here we extend u, v and the coefficient functions $a_{ij}(x), a_i, b_i, a$ by 0 to the lattice $\mathbb{R}_h^r\backslash\Omega_h$. By Theorem 33.1 the last equation is equivalent to

$$h^r\sum_{\Omega_h}\left[-\sum_{i,j}\Delta_{-h}^i(a_{ij}\Delta_h^j u)v-\sum_i(\Delta_{-h}^i(a_i u)v-b_i(\Delta_h^i u)v)-auv\right]=-h^r\sum_{\Omega_h} fv.$$

With

$$L_h u:=\sum_{i,j}\Delta_{-h}^i(a_{ij}\Delta_h^j u)+\sum_i(\Delta_{-h}^i(a_i u)+b_i\Delta_h^i u)+au$$

– this is the approximation operator – we obtain

$$h^r\sum_{\Omega_h}(L_h u)v=h^r\sum_{\Omega_h} fv \quad \text{for all } v\in C_0^\infty(\Omega),\ u|_{S_h}=0. \tag{3}$$

This is equivalent to

$$L_h u(x)=f(x) \quad \text{for all } x\in\Omega_h,\ u|_{S_h}=0. \tag{4}$$

Proof. (4)⇒(3) is clear. (3)⇒(4): choose $v_x\in C_0^\infty(\Omega)$ with

$$v_x=\begin{cases}1 \text{ at } x\\ 0 \text{ at all other points of } \bar{\Omega}_h.\end{cases}$$

Then it follows that

$$\sum_{\Omega_h} L_h uv_x=L_h u(x),\quad \sum_{\Omega_h} fv_x=f(x).$$ ■

We call u_h a solution of (4). (4) is a system of linear equations with $|\Omega_h|$ equations and $|\Omega_h|$ unknowns. Unique solvability proves the following:

Theorem 33.3 *$L_h u_h=0$, $u_h|_{S_h}=0$ implies that $u_h=0$.*

Proof. Because $L_h u_h=0$ the Ladyženskaya condition implies that

$$\begin{aligned}0&=-\sum_{\Omega_h} L_h u_h u_h\\ &=-\sum_{\Omega_h}\left(\sum_{i,j}\Delta_{-h}^i(a_{ij}\Delta^h u_h)u_h+\sum_i(\Delta_{-h}^i(a_i u_h)u_h+b_i\Delta_h^i u_h u_h)+au_h u_h\right)\\ &=\sum_{\Omega_h}\left(\sum_{i,j} a_{ij}\Delta_h^j u_h\Delta_h^i u_h+\sum_i(a_i-b_i)\Delta_h^i u_h u_h-au_h u_h\right)\\ &\geqslant\sum_{\Omega_h} d\sum_1^r(\Delta_h^i u_h)^2.\end{aligned} \tag{L}$$

Therefore it follows that $\Delta_h^i u_h = 0$ and hence $u_h = \text{constant}$, since $\bar{\Omega}_h$ is connected. Because $u_h|_{S_h} = 0$ we therefore have $u_h = 0$. For small h connectedness of Ω implies connectedness of $\bar{\Omega}_h$. ■

We next apply the convergence theorem from §32. Since by Remark 32.1 (iv) the norms on B_1 and B_2 play no role, we may simply choose

$$B_2 := C(\bar{\Omega}), \quad B_1 := \{u \in C(\bar{\Omega}) : u|_{\partial\Omega} = 0\} = C_0(\bar{\Omega}).$$

We could also take the Sobolev spaces $\mathring{W}_2^1(\Omega)$ and $L_2(\Omega)$. For $B_{2,h}$ we take all lattice functions on $\bar{\Omega}_h$ and for $B_{1,h}$ all lattice functions u_h, which satisfy the boundary condition $u_h|_{S_h} = 0$. We choose the following norms.

$$\|f\|_{2,h}^2 = \sum_{\Omega_h} h^r f^2, \quad f \in B_{2,h},$$

$$\|u_h\|_{1,h}^2 = \sum_{\Omega_h} h^r (u_h^2 + (\Delta^h u_h)^2), \quad u_h \in B_{1,h}.$$

As discretisation operators $D_{1,h}, D_{2,h}$ we take the restriction operators. We now prove the stability and consistency of the process. As in the case $f = 0$ we conclude (Ladyženskaya condition (L))

$$-h^r \sum_{\Omega_h} f u_h = h^r \sum_{\Omega_h} \left[\sum_{i,j} a_{ij} \Delta_h^j u_h \cdot \Delta_h^i u_h + \sum_i (a_i u_h \Delta_h^i u_h - b_i \Delta_h^i u_h u_h) - a u_h u_h \right]$$

$$\leqslant d h^r \sum_{\Omega_h} \sum_1^r (\Delta_h^i u_h)^2 = d h^r \sum_{\Omega_h} (\Delta^h u_h)^2,$$

and hence

$$d h^r \sum_{\Omega_h} (\Delta^h u_h)^2 \leqslant h^r \sum_{\Omega_h} |f| \, |f| \, |u_h|.$$

If $e > 0$ because

$$(e^{1/2} u_h \pm e^{-1/2} f)^2 \geqslant 0,$$

it follows immediately that

$$|f| \, |u_h| \leqslant \tfrac{1}{2} e u_h^2 + \frac{1}{2e} f^2,$$

hence

$$d h^r \sum_{\Omega_h} (\Delta^h u_h)^2 \leqslant h^r \sum_{\Omega_h} \left(\tfrac{1}{2} e u_h^2 + \frac{1}{2e} f^2 \right)$$

$$\leqslant h^r \sum_{\Omega_h} \left(\frac{e}{2} c (\Delta^h u_h)^2 + \frac{1}{2e} f^2 \right) \quad \text{by Lemma 33.1.}$$

We now choose e so that $ce = d$, then we have

$$\frac{d}{2} h^r \sum (\Delta^h u_h)^2 \leqslant h^r \sum \frac{1}{2e} f^2$$

or

$$h^r \sum (\Delta^h u_h)^2 \leqslant \frac{1}{de} h^r \sum f^2.$$

Therefore again by Lemma 33.1 and the last estimate

$$\|u_h\|_{1,h}^2 = \sum h^r (u_h^2 + (\Delta^h u_h)^2) \leqslant (c+1) h^r \sum (\Delta^h u_h)^2 \leqslant \frac{c+1}{de} h^r \sum f^2$$

$$= \frac{c+1}{de} \|f\|_{2,h}^2.$$

This is stability in the chosen norms: $\|L_h^{-1}\| = \|A_n^{-1}\| \leqslant \sqrt{[(c+1)/de]}$

For the proof of consistency we assume that the solution u of the differential equation belongs to $C^3(\bar{\Omega})$, and that for the coefficients $a_{ij}, a_i \in C^2(\bar{\Omega})$, b_i, $a \in C(\bar{\Omega})$. We want to show that $\|L_h u - Lu\|_{2,h} \leqslant hM$, that is, consistency of the first order. Since Ω is bounded, for $g \in C(\bar{\Omega})$ we have

$$\sum_{\Omega_h} h^r g^2 \leqslant \max_{\Omega_h} (g^2) \sum_{\Omega_h} h^r \leqslant \max_{\Omega_h} (g^2) \cdot \mathrm{vol}(\Omega),$$

Therefore it suffices to show

$$\max_{\Omega_h} |L_h u - Lu| \leqslant hm,$$

(put $M_1 = M \cdot \mathrm{vol}(\Omega)$).

By Taylor expansion we have

$$\Delta_h^i u(x) = \frac{u(x+he_i) - u(x)}{h} = \frac{\partial u}{\partial x_i} + \frac{h}{2} \frac{\partial^2 \bar{u}}{\partial x_i^2},$$

$$\Delta_{-h}^i u(x) = \frac{u(x-he_i) - u(x)}{-h} = \frac{\partial u}{\partial x_i} - \frac{h}{2} \frac{\partial^2 \bar{u}}{\partial x_i^2},$$

$$\Delta_{-h}^i u(x) = \frac{\partial \bar{u}}{\partial x_i},$$

where the horizontal bar again denotes that the value of the function is taken at a point between $x - he_i$ and $x + he_i$. Hence it follows that

$$L_h u = \sum_{i,j} \left[\Delta_{-h}^i \left(a_{ij} \left(\frac{\partial u}{\partial x_j} + \frac{h}{2} \frac{\partial^2 \bar{u}}{\partial x_j^2} \right) \right) \right] + \sum_i \Delta_{-h}^i (a_i u) + \sum_i b_i \left(\frac{\partial u}{\partial x_i} + \frac{h}{2} \frac{\partial^2 \bar{u}}{\partial x_i^2} \right) + au$$

$$= \sum_{i,j} \left[\frac{\partial}{\partial x_i} \left(a_{ij} \frac{\partial u}{\partial x_j} \right) - \frac{h}{2} \frac{\partial^2}{\partial x_i^2} \overline{\left(a_{ij} \frac{\partial u}{\partial x_j} \right)} + \frac{h}{2} \frac{\partial}{\partial x_i} \overline{\left(a_{ij} \frac{\partial^2 \bar{u}}{\partial x_j^2} \right)} \right]$$

$$+ \sum_i \left[\frac{\partial}{\partial x_i} (a_i u) - \frac{h}{2} \frac{\partial^2}{\partial x_i^2} \overline{(a_i u)} \right] + \sum_i \left[b_i \frac{\partial u}{\partial x_i} + \frac{h}{2} b_i \frac{\partial^2 \bar{u}}{\partial x_i^2} \right] + au,$$

hence $\max_{\Omega_k}|L_h u - Lu| \leqslant hM$, with $M := \max$ among $a_{ij}, a_i, b_i u$ and its derivatives. We can now apply the convergence theorem from §32 and obtain first-order convergence for the process (4).

Exercises

33.1 Which additional conditions must one take in order to obtain the Ladyženskaya condition (L) from uniform strong ellipticity?

33.2 Let Ω be the rectangle with the vertices

$$A(-3,4),\quad B(3,4),\quad C(3,-4)\quad \text{and } D(-3,-4).$$

Set up a difference process for the solution of the Dirichlet problem

$$\frac{\partial^2 u}{\partial x^2} + \frac{\partial^2 u}{\partial y^2} = 0, \quad (x, y) \in \Omega,$$

$$u(x, y) = g, \quad (x, y) \in \partial\Omega \tag{D}$$

where for the boundary function g we put

$$\text{(a) } g = 1, \quad \text{(b) } g = y, \quad \text{(c) } g = x + y.$$

Compare the approximate solutions for $h = 1, \frac{1}{2}$ (width) with the exact solutions.

33.3 Let Ω be the open disc $x^2 + y^2 < 16$. Set up a difference process for the solution of the Dirichlet problem (D) where the boundary function g is chosen to be

$$\text{(a) } g = 0, \quad \text{(b) } g = 1, \quad \text{(c) } g = x.$$

Compare the approximate solutions for $h = 1, \frac{1}{2}$ with the exact solutions.

33.3 A difference process for the wave equation in several space variables

In Example 32.2 we studied a difference process for the wave equation in one variable. This process will now be generalised to several space variables. The proof of convergence becomes somewhat harder, and we can only achieve convergence in discrete Sobolev norms, as in the process for elliptic equations in §33.2.

Suppose that $\Omega \subset \mathbb{R}^r$ is a bounded region, $T \in \mathbb{R}^+$, and that we are given $f: \Omega \times [0, T] \to \mathbb{R}$, $g_0, g_1: \Omega \to \mathbb{R}$. We look for a function $u: \bar{\Omega} \times [0, T] \to \mathbb{R}$ with

$$\frac{\partial^2 u}{\partial t^2} = \sum_1^r \frac{\partial^2 u}{\partial x_i^2} + f,$$

$$u(\cdot, 0) = g_0, \quad \frac{\partial u(\cdot, 0)}{\partial t} = g_1, \quad u|_{\partial\Omega \times [0,T]} = 0. \tag{1}$$

We cover $\bar{\Omega}$ with a square lattice of width h and subdivide the interval $[0, T]$ with the division points $k\tau$ so that without loss of generality $N:= T/\tau \in \mathbb{N}$. $\bar{\Omega}_h, \Omega_h, S_h$ may be defined as in §33.1.

In order to make the following proof of stability somewhat more transparent we must handle double differences Δ^i_h, Δ^j_h – we introduce another notation for the difference operators Δ^i_h. For $u:\Omega \times [0, T] \to \mathbb{R}$ let

$$u_{x_i}(x, t) := \frac{u(x + he_i; t) - u(x, t)}{h},$$

$$u_{\bar{x}_i}(x, t) := \frac{u(x, t) - u(x - he_i, t)}{h},$$

$$u_t(x, t) := \frac{u(x, t + \tau) - u(x, t)}{\tau}$$

$$u_{\bar{t}}(x, t) := \frac{u(x, t) - u(x, t - \tau)}{\tau},$$

$$T^h_i u(x, t) := u(x + he_i, t), \quad T^\tau u(x, t) := u(x, t + \tau).$$

Further let

$Q := \Omega_h \times \{0, \ldots, N\tau\}$, $F := S_h \times \{0, \ldots, N\tau\}$, the surface of a cylinder.

For later use we collect some trivial identities

(a) $u_{x_i\bar{x}_i}(x, t) = u_{\bar{x}_i x_i} = \frac{1}{h^2}(u(x + he_i, t) - 2u(x, t) + u(x - he_i, t))$,

(b) $(uv)_{x_i} = u_{x_i} v + (T^h_i u) v_{x_i}$,

(c) $(uv)_{\bar{x}_i} = u_{x_i} v + T_i^{-h}(uv_{x_i})$,

(there are analogues of (a), (b), (c) for $u_t, u_{\bar{t}}, T^\tau$)

(d) $u_{t\bar{t}}(u_t + u_t) = (u_t)^2_{\bar{t}}$,

(e) $u_{x_i\bar{x}_i}(u_t + u_{\bar{t}}) = (u_{x_i}(u_t + u_{\bar{t}}))_{\bar{x}_i} - T_i^{-h}(u_{x_i}(u_t + u_{\bar{t}})_{x_i})$,

((d), (e) follow immediately from (b), (c)).

We consider the following difference process: we look for $v: Q \cup F \to \mathbb{R}$ with

$$L_{h,\tau} v := v_{t\bar{t}} - \sum_1^r v_{x_i\bar{x}_i} = f \quad \text{in } \Omega_h \times \{1, \ldots, (N-1)\tau\} \tag{2}$$

$v(x, 0) = g_0(x)$ for $x \in \Omega_h$, $v(x, \tau) = g_0(x) + \tau g_1(x)$ for $x \in \Omega_h$, $v(x, t) = $ in F.

In order to be able to apply the convergence Theorem 32.1, we employ the notation introduced at the end of §32. We label the portions of boundary

$\Gamma_1 = \Gamma_2, \Gamma_3$ of $\partial(\bar{\Omega} \times [0, T])$ by

$$\Gamma_1 := \Gamma_2 := \bar{\Omega} \times \{0\}, \Gamma_3 := \partial\Omega \times [0, T].$$

For the sake of simplicity (the norms on the spaces B_1, B_2 play no role in the convergence theorem) we again take

$$F_0^1 := C(\bar{\Omega} \times [0, T]) =: F_0^2$$
$$F_1 := F_2 := C(\bar{\Omega} \times \{0\}), \quad F_3 := C(\partial\Omega \times [0, T]).$$

For the differential and boundary operators we have:

$$L = \frac{\partial^2}{\partial t^2} - \sum_{i=1}^{r} \frac{\partial^2}{\partial x_1^2} : F_0^1 \to F_0^2,$$

$$l_1 = I|_{\Omega \times \{0\}} : F_0^1 \to F_1,$$

$$l_2 = \frac{\partial}{\partial t}\bigg|_{\Omega \times \{0\}} : F_0^1 \to F_2,$$

$$l_3 = I|_{\partial\Omega \times [0,T]} : F_0^1 \to F_3.$$

To make things discrete, take as lattice, respectively as boundary lattice,

$$\bar{Q}_{h,\tau} = \bar{\Omega}_h \times \{0, \dots, N\tau\}, \quad \Gamma_{1,h} = \Gamma_{2,h} = \Omega_h \times \{0\}, \quad \tau = 0,$$
$$\Gamma_{3,h} = S_h \times \{0, \dots, N\tau\}.$$

As the space $F_{0,h,\tau}^1$ we take all lattice functions on $\bar{Q}_{h,\tau}$ equipped with the norm

$$\| v \|_1^2 := \sum_{s=1}^{N} \sum_{\Omega_h} \tau h^r \left(v^2 + v_t^2 + \sum_1^r v_{x_i}^2 \right), \quad v \in F_{0,h,\tau}^1.$$

We equip $F_{0,h,\tau}^2(\bar{Q}_{h,\tau})$ with the norm

$$\| f \|_2^2 := \sum_{s=1}^{N} \sum_{\Omega_h} \tau h^r f^2, \quad f \in F_{0,h,\tau}^2(\bar{Q}_{h,\tau}),$$

$F_{1,h}(\Gamma_{1,h})$ with the norm

$$\| g_0 \|_3^2 := \sum_{\Omega_h} h^r \left(g^{20} + \sum_1^r (g_0)_{x_i}^2 \right), \quad g_0 \subset F_{1,h}(\Gamma_{1,h}),$$

$F_{2,h}(\Gamma_{2,h})$ with the norm

$$\| g_1 \|_4^2 : \sum_{\Omega_h} h^r g_1^2, \quad g_1 \in F_{2,h}(\Gamma_{2,h}),$$

$F_{3,h,\tau}(\Gamma_{3,h,\tau})$ with the norm

$$\| v \|_5^2 := \max_{\Gamma_{3,h,\tau}} |v|, \quad v \in F_{3,h,\tau}.$$

As discretisation operators we take the appropriate restrictions; explanation is needed only for $d_{3,h,\tau}: F_3(\partial\Omega \times [0, T]) \to F_{3,h,\tau}$, here for $d_{3,h,\tau}u$ we take one of the function values u lying nearest on $\partial\Omega \times [0, T]$.

Now for the definition of the approximation operators: they are defined by (2), that is,

$$L_{h,\tau}v = v_{t\bar{t}} - \sum_{1}^{r} v_{x_i x_i}: F^1_{0,h,\tau} \to F^2_{0,h,\tau},$$

$$l^1_h = I|_{\Gamma_{1,h}}: F^1_{0,h,\tau} \to F_{1,h}(\Gamma_{1,h}),$$
$$l^2_h = \Delta_\tau|_{\Gamma_{2,h}}: F^1_{0,h,\tau} \to F_{2,h}(\Gamma_{2,h}),$$
$$l^3_h = I|_{\Gamma_{3,h,\tau}}: F^1_{0,h,\tau} \to F_{3,h,\tau}(\Gamma_{3,h,\tau}),$$

therefore

$$A_n = (L_{h,\tau}, l^1_h, l^2_h, l^3_{h,\tau}).$$

It is easily checked that the system of equations (2) is uniquely solvable; indeed on each time level v is calculated from the values on the two preceding time levels, the values on the initial levels $t = 0$, 1τ are preassigned, that is, A_n is invertible.

The consistency of the process: we assume that u, the solution of (1), belongs to $C^3(\Omega \times [0, T])$. This can be assured by suitable assumptions on f, g_0, g_1 and $\partial\Omega$ (see §30). As in Example 32.2 we again have

$$\left| L_{h,\tau}u - \left(\frac{\partial^2 u}{\partial t^2} - \sum_{1}^{r} \frac{\partial^2 u}{\partial x_i^2} \right) \right| = \left| \frac{\tau}{6} \frac{\partial^3 \bar{u}}{\partial t^3} - \frac{h}{6} \sum_{1}^{r} \frac{\partial^3 \bar{u}}{\partial x_i^3} \right| \leqslant \max(\tau, h) \cdot M \to 0 \quad \text{as } \tau, h \to 0,$$

that is, we have consistency in the maximum norm. Because

$$\sum_{s=1}^{N} \sum_{\Omega_h} h^r \tau \cdot f^2 \leqslant \max_{\Omega_{h,\tau}} (f^2) \cdot \mathrm{vol}(\Omega \times [0, T]),$$

we also have consistency in our norm ($\Omega \times [0, T]$ is bounded!). Furthermore we see immediately that the boundary conditions for $t = 0$ are consistently approximated. For consistency on the surface of the cylinder $F := \Gamma_{3,h,\tau}$, we exploit the fact that we do not need to calculate with the exact restricted values, but that by the convergence Theorem 32.1 it is enough if the discrete values used converge to the restricted values. By our definition of $d_{3,h,\tau}u$ this is satisfied, since u is continuous and vanishes on $\partial\Omega \times [0, T]$.

In order to simplify the proof for the stability of the process, we incorporate the third boundary condition $u|_{\partial\Omega \times [0,T]} = 0$ into the definition of F^1_0, thus

$$F^1_0 := \{u \in C(\bar{\Omega} \in [0, T]) : u|_{\partial\Omega \times [0,T]} = 0\},$$

and similarly for the discretised space

$$F^1_{0,h,\tau}(\bar{Q}_{h,\tau}) = \{v \in R^{|Q|} : v|_{F=\Gamma_{3,h,\tau}} = 0\}.$$

We define the discretisation operators $D^1_{0,h,\tau}: F^1_0 \to F^1_{0,h,\tau}$ as restriction on $Q_{h,\tau}$ and as zero on $\Gamma_{3,h,\tau}$.

By the perturbation Theorem 32.11 the stability is not altered by this, because – as already mentioned – u is continuous and vanishes on $\partial\Omega \times [0, T]$. For stability by §32 it only remains to show that there exists $c > 0$, so that for all τ, h we have

$$\|v\|_1 \leqslant c(\|f\|_2 + \|g_0\|_3 + \|g_1\|_4).$$

We now require that

$$\lambda := \tau/h < \frac{1}{r}, \lambda = \text{const} \quad \text{(stability condition)} \tag{5}$$

then $a := 1 - \lambda r > 0$ and $1 - \lambda \geqslant a$.

For $(x, t) \in \Omega_h \times \{\tau, \dots, (N-1)\tau\}$ the solution v of (2) satisfies

$$L_{h,\tau}v - f = 0.$$

Since moreover for $(x, t) \in S_h \times \{\tau, \dots, (N-1)\tau\}$ we have

$$v_t + v_{\bar{t}} = \frac{1}{\tau}(v(x, t+\tau) - v(x, t-\tau)) = 0,$$

it follows that (if we extend v by 0 outside $Q \cup F$)

$$(v_t + v_{\bar{t}})(L_{h,\tau}v - f) = 0 \quad \text{for } (x, t) \in \{\tau, \dots, (N-1)\tau\}.$$

Therefore, using the identities (d), (e) we obtain

$$(v_{\bar{t}})^2_t - \sum_1^r (v_{x_i}(v_t + v_{\bar{t}}))_{\bar{x}_i} + \sum_1^r T_i^{-h}(v_{x_i}(v_t + v_{\bar{t}})_{x_i}) - f(v_t + v_{\bar{t}}) = 0.$$

For $p \leqslant N$ we now sum over $\bar{\Omega}_h \times \{\tau, \dots, (p-1)\tau\}$. The second summand disappears and we obtain

$$\tau \sum_{s=1}^{p-1} h^r \sum_{\Omega_h} (v_{\bar{t}})^2_t + \tau \sum_{s=1}^{p-1} h^r \sum_{\Omega_h} \sum_1^r T_i^{-h}(v_{x_i}(v_t + v_t)_{x_i}) - \tau \sum_{s=1}^{p-1} h^r \sum_{\Omega_h} f(v_t + v_{\bar{t}}) = 0. \tag{3}$$

Now we have for the first summand in (3)

$$\tau \sum_{s=1}^{p-1} h^r \sum_{\Omega_h} (v_{\bar{t}})^2_t = h^r \sum_{\Omega_h} (v_{\bar{t}})^2 |_\tau^{p\tau} = h^r \sum_{\Omega_h} (v_{\bar{t}})^2 |^{p\tau} + h^r \sum_{\Omega_h} (v_t)^2 |_\tau. \tag{4}$$

Here we employ $|_\tau^{p\tau}$ for: sum over $t = p\tau$ minus sum over $t = \tau$; $|^{p\tau}$ and $|_\tau$ are correspondingly defined. Furthermore we have

$$\begin{aligned}
&\tau\sum_{s=1}^{p-1} h^r \sum_{\Omega_h}\sum_{1}^{r} T_i^{-h}(v_{x_i}(v_t + v_{\bar t})_{x_i}) = \tau\sum_{s=1}^{p-1} h^r \sum_{\Omega_h}\sum_{1}^{r}(v_{x_i}(v_t + v_{\bar t})_{x_i})\\
&= \sum_{s=1}^{p-} h^r \sum_{\Omega_h}\sum_{1}^{r}(v_{x_i}(T^\tau v - T^{-\tau} v)_{x_i}) = \sum_{s=1}^{p-1} h^r \sum_{\Omega_h}\sum_{1}^{r}(v_{x_i}(T^\tau v_{x_i} - T^{-\tau} v_{x_i}))\\
&= h^r \sum_{\Omega_h}\sum_{1}^{r}(v_{x_i} T^\tau v_{x_i})|^{(p-1)\tau} - h^r \sum_{\Omega_h}\sum_{1}^{r}(v_{xi} T^{-\tau} v_{x_i})|^{\tau}\\
&= h^r \sum_{\Omega_h}\sum_{1}^{r}(v_{x_i})^2|_0^{p\tau} - h^r\sum_{\Omega_h}\tau\cdot\sum_{1}^{r}(v_{x_i}v_{x_i\bar t})|^{p\tau} - h^r\sum_{\Omega_h}\tau\cdot\sum_{1}^{r}(v_{x_i}v_{x_i\bar t})|^{0}.\\
&= h^r \sum_{\Omega_h}\sum_{1}^{r}(v_{x_i})^2|_0^{p\tau} - S_1 - S_2.
\end{aligned}$$

Here we have abbreviated the last two summands by S_1 and S_2. Substituting (4), (5) in (3) we obtain

$$h^r\sum_{\Omega_h}\left((v_{\bar t})^2 + \sum_{1}^{r}(v_{x_i})^2\right)\Bigg|^{p\tau} - h^r\sum_{\Omega_h}\left((v_t)^2 + \sum_{1}^{r}(v_{x_i})^2\right)\Bigg|^{0}$$
$$- S_1 - S_2 - \tau\sum_{s=1}^{p-1} h^r\sum_{\Omega_h} f(v_t + v_{\bar t}) = 0. \qquad (6)$$

We next estimate S_1 and S_2

$$\begin{aligned}
|S_1| &\leqslant h^r\sum_{\Omega_h}\frac{\tau}{h}\sum_{1}^{r}(|v_{x_i}|(|T_i^h v_t| + |v_t|))|^{p\tau}\\
&\leqslant \frac{\lambda}{2} h^r \sum_{\Omega_h}\sum_{1}^{r}(2(v_{x_i})^2 + (T_i^h v_t)^2 + (v_t)^2)|^{p\tau} = \lambda h^r\sum_{\Omega_h}\left(\sum_{1}^{r}(v_{x_i})^2 + r(v_t)^2\right)\Bigg|^{p\tau}\\
|S_2| &\leqslant h^r\sum_{\Omega_h}\frac{\tau}{h}\sum_{1}^{r}(\| v_{x_i}|(|T_i^h v_t| + |v_t|))|^0 \leqslant \lambda h^r\sum_{\Omega_h}\left(\sum_{1}^{r}(v_{x_i})^2 + r(v_t)^2\right)\Bigg|^{0}.
\end{aligned}$$

Furthermore we have for the third summand in (3), respectively for the last in (6),

$$\begin{aligned}
\left|\tau\sum_{s=1}^{p-1} h^r\sum_{\Omega_h} f(v_t + v_{\bar t})\right. &\leqslant \tau\sum_{s=1}^{p-1} h^r\sum_{\Omega_h}\left(f^2 + \frac{(v_t)^2 + (v_{\bar t})^2}{2}\right)\\
&\leqslant \tau\sum_{s=1}^{p} h^r\sum_{\Omega_h}(f^2 + (v_{\bar t})^2).
\end{aligned}$$

In the last estimates we have used several times that $|ab| < \frac{1}{2}(a^2 + b^2)$ for $a, b \in \mathbb{R}$. Thus (6) becomes

$$h^r \sum_{\Omega_h} \left((v_{\bar{t}})^2 + \sum_1^r (v_{x_i})^2 \right) \Bigg|^{p\tau}$$

$$= h^r \sum_{\Omega_h} \left((v_t)^2 + \sum_1^r (v_{x_i})^2 \right) \Bigg|^{0} + S_1 + S_2 + \tau \sum_{s=1}^{p-1} h^r \sum_{\Omega_h} f(v_t + v_{\bar{t}}),$$

then with the estimates above we have

$$h^r \sum_{\Omega_h} \left((v_{\bar{t}})^2 + \sum_1^r (v_{x_i})^2 \right) \Bigg|^{p\tau}$$

$$\leqslant h^r \sum_{\Omega_h} \left((v_t)^2 + \sum_1^r (v_{x_i})^2 + \lambda \sum_1^r (v_{x_i})^2 + r\lambda (v_t)^2 \right) \Bigg|^{0}$$

$$+ \lambda h^r \sum_{\Omega_h} \left(\sum_1^r (v_{x_i})^2 + \tau (v_{\bar{t}})^2 \right) \Bigg|^{p\tau} + \tau \sum_{s=1}^{p} h^r \sum_{\Omega_h} (f^2 + (v_{\bar{t}})^2),$$

and therefore

$$h^r \sum_{\Omega_h} \left((1 - \lambda r)(v_{\bar{t}})^2 + (1 - \lambda) \sum_1^r (v_{x_i})^2 \right) \Bigg|^{p\tau}$$

$$\leqslant h^r \sum_{\Omega_h} \left((1 + \lambda r)(v_t)^2 + (1 + \lambda) \sum_1^r (v_{x_i})^2 \right) \Bigg|^{0} + \tau \sum_{s=1}^{p} h^r \sum_{\Omega_h} (f^2 + (v_{\bar{t}})^2).$$

We rearrange the last inequality and obtain

$$h^r \sum_{\Omega_h} \left((v_{\bar{t}})^2 + \sum_1^r (v_{x_i})^2 \right) \Bigg|^{p\tau}$$

$$\leqslant h^r \sum_{\Omega_h} \left(\frac{1 + \lambda r}{a} (v_t)^2 + \frac{1 + \lambda}{a} \sum_1^r (v_{x_i})^2 \right) \Bigg|^{0} + \frac{\tau}{2} \sum_{s=1}^{0} h^r \sum_{\Omega_h} (f^2 + (v_{\bar{t}})^2)$$

$$\leqslant c_1 h^r \sum_{\Omega_h} \left((v_t)^2 + \sum_1^r (v_{x_i})^2 \right) \Bigg|^{0} + c_2 \tau \sum_{s=1}^{p} h^r \sum_{\Omega_h} \left((v_{\bar{t}})^2 + \sum_1^r (v_{x_i})^2 \right)$$

$$+ \frac{\tau}{a} \sum_{s=1}^{p} h^r \sum_{\Omega_h} f^2. \tag{7}$$

Here we have used the stability condition (S), $1 - \lambda r = a > 0$. Writing

$$y(p) = \tau \sum_{s=1}^{p} h^r \sum_{\Omega_h} \left((v_{\bar{t}})^2 + \sum_1^r (v_{x_i})^2 \right) = \| v \|_1^2,$$

$$F(p) = c_1 h^r \sum_{\Omega_h} \left((v_t)^2 + \sum_1^r (v_{x_i})^2 \right) \Bigg|^{0} + \frac{\tau}{a} \sum_{s=1}^{p} h^r \sum_{\Omega_h} f^2$$

(7) becomes the inequality

$$\frac{y(p)-y(p-1)}{\tau} \leqslant c_2 y(p) + F(p). \tag{8}$$

Lemma 33.3 *Let* $1 - c_2\tau \geqslant \frac{1}{2}$, *then for a solution of the system of inequalities* (8) *we have*

$$y(p) \leqslant e^{2c_2T}(y(1) + TF(p)). \tag{9}$$

Proof. We rewrite (8) and have

$$y(p) \leqslant \frac{1}{1-c_2\tau} y(p-1) + \frac{\tau}{1-c_2\tau} F(p).$$

We put $E = 1/(1 - c_2\tau)$ and obtain inductively that

$$y(p) \leqslant E^{p-1} y(1) + E\tau \sum_{s=2}^{p} E^{p-s} F(s).$$

However $p\tau \leqslant T$ and therefore

$$E^p = \left(1 + \frac{c_2\tau}{1-c_2\tau}\right)^p \leqslant e^{2c_2T},$$

hence

$$y(p) \leqslant E^{p-1} y(1) + E^{p-1}\tau \sum_{s=2}^{p} F(s) \leqslant e^{2c_2T}(y(1) + \tau p F(p))$$

$$\leqslant e^{2c_2T}(y(1) + TF(p)). \qquad \blacksquare$$

Furthermore by looking back to the initial conditions

$$\begin{aligned}
y(1) + TF(p) &\leqslant \tau h^r \sum_{\Omega_h} \left((v_{\bar{t}})^2 + \sum_1^r (v_{x_i})^2 \right)\Bigg|^{\tau} + c_3\tau \sum_{s=1}^{p} h^r \sum_{\Omega_h} f^2 \\
&\quad + c_3 h^r \sum_{\Omega_h} \left((v_t)^2 + \sum_1^r (v_{x_i})^2 \right)\Bigg|^{0} \\
&\leqslant c_4 h^r \left(\sum_{\Omega_h} (v_t)^2|^0 + \sum_{\Omega_h}\sum_1^r (v_{x_i})^2|^{\tau} + \sum_{\Omega_h}\sum_1^r (v_{x_i})^2|^0 + \tau \sum_{s=1}^{p}\sum_{\Omega_h} f^2 \right) \\
&= c_4 h^r \left(\sum_{\Omega_h} g_1^2 + \sum_{\Omega_h}\sum_1^r ((g_0 + \tau g_1)_{x_i})^2 + \sum_{\Omega_h}\sum_1^r ((g_0)_{x_i})^2 \right. \\
&\quad \left. + \tau \sum_{s=1}^{p}\sum_{\Omega_h} f^2 \right) \\
&= c_4 h^r \left(\sum_{\Omega_h} g_1^2 + \sum_{\Omega_h}\sum_1^r ((g_0)_{x_i} + \lambda(T_i^h g_1 - g_1))^2 \right.
\end{aligned}$$

$$+\sum_{\Omega_h}\sum_1^r((g_0)_{x_i})^2+\tau\sum_{s=1}^p\sum_{\Omega_h}f^2\Bigg)$$

$$\leqslant c_4'h^r\Bigg(\sum_{\Omega_h}g_1^2+\sum_{\Omega_h}\sum_1^r(g_0)_{\bar{x}_1}^2+\sum_{\Omega_h}\sum_1^r(((g_0)_{x_i})^2$$

$$+\lambda^2((T_i^h g_1)^2+g_1^2))+\tau\sum_{s=1}^p\sum_{\Omega_h}f^2\Bigg)$$

$$\leqslant c_5h^r\Bigg(\sum_{\Omega_h}g_1^2+\sum_{\Omega_h}\sum_1^r((g_0)_{x_ix_i})^2+\tau\sum_{s=1}^p\sum_{\Omega_h}f^2\Bigg).$$

If $p=N$ it therefore follows from (9) that

$$\|v\|_1^2=\tau\sum_{s=1}^N h^r\sum_{\Omega_h}\Bigg(v^2+(v_{\bar{t}})^2+\sum_1^r(v_{x_i})^2\Bigg)$$

$$\leqslant e^{2c_2T}c_5\Bigg(\|g_1\|_4^2+\|f\|_2^2+h^r\sum_{\Omega_h}\sum_1^r((g_0)_{x_i})^2\Bigg).$$

If we add the term $h^r\sum_{\Omega_h}g_0^2$ to the right-hand side, we can modify this last inequality to

$$\|v\|_1\leqslant c_6(\|f\|_2+\|g_0\|_3+\|g_1\|_4).$$

and this is the stability of the process in the chosen norms. The convergence theorem from §32 is applicable, and we obtain the convergence of the process in the discrete Sobolev norm.

§34 Evolution equations

In this section we consider the following problem; let B be a (B)-space, $D\subset B$ a linear subspace,

$$L(D,B):=\{L:D\to B:L\,\text{linear}\}.$$

$$T\in\mathbb{R},\quad 0<T<\infty,$$

$$A:[0,T]\to L(D,B)$$

$$f:[0,T]\to B\quad\text{continuous, } g\in B.$$

We look for u: $[0,T]\to B$ with

$$\frac{du}{dt}-A(t)u=f,\quad u(0)=g, \tag{1}$$

where

$$\frac{du}{dt}:=\lim_{h\to 0}\frac{u(t+h)-u(t)}{h}\quad\text{in } B.$$

Before we develop a difference process for problem (1) we first show how the examples from §32 present themselves in this context.

Example 32.1 There we consider the heat equation

$$\frac{\partial u}{\partial t} - \sum_{1}^{r} \frac{\partial^2 u}{\partial x^2} = f,$$

with the boundary conditions

$$u(\cdot, 0) = g, \quad u(x, t) = 0, \quad \text{for } x \in \partial\Omega.$$

We write

$$B = C_0(\Omega) := \{u \in C(\Omega) : u(x) = 0 \quad \text{for } x \in \partial\Omega\},$$

$$D = C_0^2(\Omega) := \{u \in C_0(\Omega) : u \text{ twice continuously differentiable}\}, g = g(\cdot),$$

$$A = \sum_{1}^{r} \frac{\partial^2}{\partial x_i^2} \text{(not dependent on } t\text{)},$$

and we immediately see how problem (1) with this notation represents the heat equation.

Example 32.2 There we considered the wave equation

$$\frac{\partial^2 u}{\partial t^2} - \frac{\partial^2 u}{\partial x^2} = f,$$

with the boundary conditions

$$u(\cdot, 0) = g_0, \quad \frac{\partial u}{\partial t}(\cdot, 0) = g_1, \quad u(x, t) = 0 \quad \text{for } x \in \partial\Omega.$$

We write

$$B = C_0(\Omega) \times C_0(\Omega), \quad D = C_0^2(\Omega) \times C_0(\Omega),$$

$$g = \begin{bmatrix} g_0 \\ g_1 \end{bmatrix}, \quad f = \begin{bmatrix} 0 \\ f(\cdot, \cdot) \end{bmatrix}, \quad A = \begin{bmatrix} 0 & 1 \\ \dfrac{\partial^2}{\partial x^2} & 0 \end{bmatrix}$$

again not dependent on t.

For a solution of (1) we obtain

$$u = \begin{bmatrix} u_1 \\ u_2 \end{bmatrix} \quad \text{with } \frac{du}{dt} - Au = f,$$

therefore

$$\left.\begin{aligned}\frac{du_1}{dt} - u_2 = 0 \\ \frac{du_2}{dt} - \frac{\partial^2 u_1}{\partial x^2} = f(\cdot,\cdot)\end{aligned}\right\} \quad \text{and} \quad \begin{cases} u_1(0) = g_0 \\ u_2(0) = g_1. \end{cases}$$

It is easily checked that u_1 is then a solution of the wave equation with the preassigned boundary conditions. Let us now return to our abstract problem (1). If we want to develop a difference process for problems of this kind, it is convenient to proceed in two stages, first we replace differentiation with respect to t by difference quotients, and then the operator A by a suitable difference operator. We will now make this precise.

34.1 The time-independent case

For simplicity of notation we first assume that A does not depend on t, that is, $A \in L(D, B)$. The modifications for the time-dependent case are not very hard; they are carried out below. We subdivide the interval $[0, T]$ at division points $n\tau$, where without loss of generality $N := T/\tau \in \mathbb{N}$, and write

$$t_n := n\tau \quad \text{for } n = 0, \ldots, N.$$

Suppose further that $g_\tau \in B$, $f_n \in B$ for $n = 0, \ldots, N$ and $A_\tau : B \to B$ is continuous and linear. We want to solve the problem by means of

$$u_\tau = (u_0, \ldots, u_N) \in B^{N+1}, \text{ such that}$$

$$\frac{u_{n+1} - u_n}{\tau} = A_\tau u_n + f_n, \quad n = 0, 1, \ldots, N-1, \; u_0 = g_\tau. \tag{2}$$

Definition 34.1 *The process* (2) *is called convergent, if for* g_τ, f_n *such that* $\max_n(\|g - g_\tau\|, \|f_n - f(t_n)\|) \to 0$ *it follows that the solution* u_τ *of* (2) *converges to the solution* u *of* (1), *that is,* $\max_n(\|u_n - u(t_n)\|) \to 0$.

Remarks 34.1

(i) It is easily checked that the linear extension of a convergent solution u_τ of (2) to all of $[0, T]$ converges in $C([0, T], B)$ to u.

(ii) The transition from A to A_τ involves making the space coordinates discrete.

Theorem 34.1 (2) *is explicitly solvable.*

Proof. We have

$$u_n = (I + \tau A_\tau)u_{n-1} + \tau f_{n-1} \quad (I : B \to B \text{ the identity map}).$$

We write $C(\tau) := I + \tau A_\tau$ and have

$$u_n = C^n(\tau)u_0 + \tau \sum_{j=0}^{n-1} C^{n-j}(\tau) f_j. \qquad \blacksquare$$

We now translate the problem (1) and equation (2) into the language of the theory developed in §32. For this suppose that B^{N+1}, $C([0,T],B)$ are given the max-norm and further that

$$D_\tau : C([0,T],B) \to B^{N+1}; \qquad u \mapsto (u(t_0),\ldots,u(t_N))$$
$$D'_\tau : B \times C([0,T],B) \to B^{N+1} \qquad (b,u) \mapsto (b, u(t_0),\ldots,u(t_{N-1}))$$
$$L: C([0,T],B) \to B \times C([0,T],B) \quad u \mapsto \left(u(0), \frac{\mathrm{d}u}{\mathrm{d}t} - Au \right)$$
$$L_\tau : B^{N+1} \to B^{N+1}$$
$$(u_0,\ldots,u_N) \mapsto \left(u_0, \frac{u_1 - u_0}{\tau} - A_\tau u_0, \ldots, \frac{u_N - u_{N-1}}{\tau} - A_\tau u_{N-1} \right).$$

Therefore we obtain the scheme appropriate to Fig. 32.1

$$\begin{array}{ccc} C([0,T],B) & \xrightarrow{L} & B \times C([0,T],B) \\ D_\tau \downarrow & & \downarrow D'_\tau \\ B^{N+1} & \xrightarrow{L_\tau} & B^{N+1} \end{array}$$

Fig. 34.1

The problem corresponding to (1) is

$$\text{find } u \in C([0,T],B) \quad \text{with} \quad Lu = (g,f),$$

which by using discretisations D_τ, D'_τ may be transformed into the problem corresponding to (2)

$$\text{find } u_\tau \in B^{N+1} \quad \text{with} \quad L_\tau u_\tau = D'_\tau(g,f).$$

Definition 34.2

1. (2) *is said to be consistent with* (1) *at* $u \in C([0,T],B)$ *if*

$$\max_n \left\| \left(\frac{u(t_n) - u(t_{n-1})}{\tau} - A_\tau u(t_{n-1}) \right) - \left(\frac{\mathrm{d}u}{\mathrm{d}t} - Au \right)(t_{n-1}) \right\| \to 0$$

or

$$\max_n \left\| \frac{1}{\tau}(u(t_n) - C(\tau)u(t_{n-1})) - \left(\frac{\mathrm{d}u}{\mathrm{d}t} - Au \right)(t_{n-1}) \right\| \to 0 \quad \text{as } \tau \to 0.$$

2. (2) *is said to be stable, if there exists some* $C \geqslant 0$, *so that for all* $g \in B$, $f \in C([0, T], B)$ *we have*

$$\|u_\tau\| \leqslant C(\|g\| + \max_n \|f(t_n)\|),$$

where u_τ *is a solution of* (2) *for* g, f.

Remark 34.2 One sees that the notions defined here: convergence of the process (2), stability of (2), consistency of (2) and (1) at u, are equivalent to the notions defined in §32: discrete convergence, stability of the process, consistency of L_τ and L at u. Use Fig. 34.1.

Theorem 34.2 *Let u be a solution of* (1) *corresponding to* g, f, *let* (2) *and* (1) *be consistent at u and let* (2) *be stable. Then the process* (2) *converges.*

The proof follows immediately from Remark 34.2 and the convergence Theorem 32.1. ■

We wish to reformulate this theorem in order to obtain a form more usable in practice.

Theorem 34.3 (2) *is stable if and only if there exists an* $M < \infty$, *so that for all* τ, n *with* $0 \leqslant n\tau \leqslant T$ *we have* $\|C^n(\tau)\| \leqslant M$.

Proof. Sufficiency is clear; we have an explicit description of the solution, see Theorem 34.1.

Necessity: Put $f = 0$, take $u_0 \in B$ to be arbitrary and let $u_\tau = (u_0, \ldots, u_N)$ be the solution of (2) for f, u_0. Since (2) is stable it follows that

$$\|u_n\| \leqslant C\|u_0\| \quad \text{for } n = 0, \ldots, N$$

because $u_n = C^n(\tau)u_0$ therefore

$$\|C^n(\tau)u_0\| \leqslant C\|u_0\|.$$

Since this holds for all $u_0 \in B$ the family $\{C^n(\tau)\}$ is uniformly bounded. ■

Definition 34.3 *From now on we abbreviate the stability criterion introduced in the theorem above as* $\{C^n(\tau)\}$ *is uniformly bounded.*

Thus the convergence theorem for the process (2) reads:

Theorem 34.4 *Let u be a solution of* (1) *corresponding to* f, g; *let* (2) *and* (1) *be consistent at u and let* $\{C^n(\tau)\}$ *be uniformly bounded. Then the process* (2) *converges.*

Before we carry over Theorem 32.10 for the correct posing of the problem, we wish to introduce the notion of the generalised solution. For this suppose that $D \subset B \times C([0,T],B)$ is a linear subspace. If for all $(g,f) \in D$ there exists a uniquely determined solution u of (1), we can define on D a solution operator E with $E(g,f) = u$. If D is dense in $B \times C([0,T],B)$ and E is continuous, we can extend E to the whole space as $\tilde{E}$. We call $\tilde{u} = \tilde{E}(g,f)$ a generalised solution, and $\tilde{E}$ the generalised solution operator.

We have:

Theorem 34.5 *Let (2) be stable, $D \subset B \times C([0,T],B)$ dense, and for all $(g,f) \in D$ let there exist a consistent solution u of (1) corresponding to (g,f). Then the process converges for all $(g,f) \in B \times C([0,T],B)$ to a generalised solution, and the generalised solution operator is linear and continuous.*

Proof. If we define $i_\tau : B^{N+1} \to C([0,T],B)$ as the linear extension on $[0,T]$, we immediately see that $\|i_\tau\| = 1$ for all τ and that

$$\|i_\tau D_\tau u - u\| \to 0 \quad \text{for all } u \in C([0,T],B).$$

Moreover $\|D'_\tau\| = 1$ for all τ. Therefore the theorem follows from Theorem 32.10 for the correct posing of the problem. ∎

Remarks 34.3 (*i*) We can also formulate the last theorem only for the homogeneous equation ($f = 0$). For this we consider the scheme

$$\begin{array}{ccc} C([0,T],B) & \xrightarrow{L} & B \times \{0\} \\ {\scriptstyle D_\tau}\downarrow & & \downarrow{\scriptstyle D'_\tau} \\ B^{N+1} & \xrightarrow{L} & B \times \{0\}^N. \end{array}$$

The deciding fact is that $B \times \{0\}$ is again a (B)-space. If $f \neq 0$ we cannot simply take $B \times \{f\}$, but must at least consider $B \times \{cf : c \in \mathbb{R}\}$. As above, here we can of course also apply Theorem 32.10 for the correct posing of the problem and obtain:

Theorem 34.6 (Lax) *Let $f = 0$, (2) be stable, $D \subset B$ dense and for all $g \in D$ suppose there exists a consistent solution u of (1) corresponding to g. Then for all $g \in B$ the process converges to a generalised solution and the generalised solution operator is continuous.*

Remark 34.3 (*ii*) The other theorems from §32 may also be carried over for problems (1), (2). We leave the formulation of the results to the reader. In §34.3 we will once more come back to the perturbation Theorem 32.11.

We can give some simple sufficient conditions for stability:

1. Stability follows from $\| C(\tau) \| \leqslant 1$, since $\| C^n(\tau) \| \leqslant \| C(\tau) \|^n \leqslant 1$.
2. Stability also follows from the condition: there exists some $a \in \mathbb{R}^+$ with $\| C(\tau) \| \leqslant 1 + \tau a$, since

$$\| C^n(\tau) \| \leqslant \| C(\tau) \|^n \leqslant \left(1 + \frac{Ta}{N} \right)^N \leqslant e^{Ta} \quad \left(N = \frac{T}{\tau} \right).$$

3. Stability follows from $\| A_\tau \| \leqslant a$, since

$$\| C(\tau) \| = \| I + \tau A_\tau \| \leqslant 1 + \tau a.$$

Example 34.1 Once more we consider the equation $\partial u/\partial t = \partial^2 u/\partial x^2$ in $\Omega = (0, 1)$ for various boundary conditions.

(a) Boundary conditions: $u(x, 0) = g(x)$, $u(1, t) = 0$ and $\partial u/\partial x(0, t) + au(0, t) = 0$. We borrow the difference process from Example 32.1

$$v(x, t + k) = (1 - 2\lambda)v(x, t) + \lambda v(x + h, t) + \lambda v(x - h, t),$$

with $\lambda = k/h^2 \leqslant \frac{1}{2}$ and the boundary values (made discrete)

$$v(x, 0) = g(x), \quad v(1, t) = 0 \quad \text{and} \quad v(0, t) = \frac{v(h, t)}{1 - ha}.$$

It follows immediately that

$$\| C(k) \| \leqslant \max \left\{ 1, \frac{1}{1 - ha} \right\},$$

and hence

$$a \leqslant 0 \subsetneq \| C(k) \| \leqslant 1 \subsetneq \text{stability};$$

$$a > 0 \subsetneq \| C(k) \| \leqslant \frac{1}{1 - ha} \leqslant 1 + b'h \leqslant 1 + bk^{1/2},$$

with suitable constants b, b'. The sufficient criteria 1–3 are therefore not satisfied if $a > 0$.

(b) Boundary conditions: $u(x, 0) = g(x)$, $\partial u/\partial x(0, t) = 0$ and $\partial u/\partial x(1, t) = 0$. The difference process then reads

$$v(x, t + k) = (1 - 2\lambda)v(x, t) + \lambda v(x + h, t) + \lambda v(x - h, t)$$
$$v(x, 0) = g(x), \quad v(h, t) - v(0, t) = 0 \quad \text{and} \quad v(1 - h, t) - v(1, t) = 0.$$

For $\lambda = k/h^2 \leqslant \frac{1}{2}$ it follows immediately that $\| C(k) \| \leqslant 1$ and hence the stability of the process.

34.2 The time-dependent case

If the operator A is dependent on t, we must obviously take care of this in setting up the difference process for (1). We obtain the slightly altered difference process – in comparison to (2) –

$$\frac{u_{n+1}-u_n}{\tau}=A_\tau(t_n)u_n+f_n,\quad n=0,\ldots,N-1,\ u_0=g_\tau. \tag{2'}$$

Here we also immediately see that the system of equations is explicitly solvable. Indeed we have

$$u_n=\left(\prod_{k=0}^{n+1}C(\tau,t_k)\right)u_0+\tau\left(\sum_{k=0}^{n-1}\left(\prod_{j=k+1}^{n-1}C(\tau,t_j)\right)f_k\right),$$

where $C(\tau,t_k)=I+\tau A_\tau(t_k)$.

The scheme of Banach spaces from §34.1 (Fig. 34.1) may be carried over immediately. The only changes arise in the definition of L, L_τ. There we have to replace A (respectively A_τ) by $A(t)$ (respectively $A_\tau(t_{n-1})$). The definitions of convergence and stability remain the same, in the definition of consistency we replace $C(\tau)$ by $C(\tau,t_{n-1})$ and A by $A(t_{n-1})$. However, the stability of the process may not be so easily characterised. But at least we still have:

Theorem 34.7 *The condition*

$$\left\|\prod_{j=k}^{i}C(\tau,t_j)\right\|\leqslant M<\infty\quad \textit{for all } \tau, 0\leqslant k\leqslant i\leqslant N-1,$$

is sufficient for the stability of (2′), *where* M *is independent of* τ,k,i. *If* $k=0$ *the condition is necessary.*

Proof. Suppose that the condition above is satisfied. Then: If $u_\tau=(u_0,\ldots,u_N)\in B^{N+1}$ with $L_\tau(u_0,\ldots,u_N)=(u_0,f_0,\ldots,f_{N-1})$, then from the explicit representation of the solution above we can deduce the inequalities

$$\|u_n\|\leqslant M\|u_0\|+\tau M\sum_{k=0}^{n-1}\|f_k\|\leqslant M\|u_0\|+\tau MN\max_k(\|f_k\|)$$
$$\leqslant M(1+T)\|(u_0,f_0,\ldots,f_{N-1})\|$$

and so $\{L_\tau^{-1}\}$ is uniformly bounded, that is, (2′) is stable. Suppose now that (2′) is stable, that is, $\|L_\tau^{-1}\|\leqslant K$, then

$$\|L_\tau^{-1}(u_0,0,\ldots,0)\|\leqslant K\|u_0\|\quad \text{for all } u_0\in B.$$

From the explicit representation of the solution we see that

$$L_\tau^{-1}(u_0,0,\ldots,0)=\left(u_0,C(\tau,t_0)u_0,\ldots,\left(\prod_{j=0}^{N-1}C(\tau,t_j)\right)u_0\right).$$

Therefore

$$\max_{i=1,\dots,N}\left\|\left(\prod_{j=0}^{i-1} C(\tau,t_j)\right)u_0\right\| \leqslant K\|u_0\| \quad \text{for all } u_0\in B,$$

and hence the condition follows with $k=0$. ∎

The stability condition may be sharpened by weakening the notion of convergence on the image spaces of the operators L_τ. If for $(u_0, f_0,\dots,f_{N-1})\in \operatorname{im} L_\tau$ we define the norm by

$$\|(u_0, f_0,\dots,f_{N-1})\|_1 = \|u_0\| + \sum_{k=0}^{N-1}\frac{\|f_k\|}{N},$$

then we have:

Theorem 34.8 *$\{L_\tau^{-1}\}$ is uniformly bounded if and only if*

$$\left\|\prod_{j=k}^{i} C(\tau,t_j)\right\| \leqslant M<\infty \quad \textit{for all } \tau, 0\leqslant k\leqslant i\leqslant N-1$$

where M is independent of τ, k, i.

Proof. Let the condition above be satisfied, then if $u_\tau=(u_0,\dots,u_N)\in B^{N+1}$ with $L_\tau(u_0,\dots,u_N)=(u_0,f_0,\dots,f_{N-1})$, it follows from the explicit representation of the solution (see above) that

$$\begin{aligned}\|u_n\| &\leqslant M\|u_0\| + \tau M\sum_{k=0}^{n-1}\|f_k\| \leqslant M\|u_0\| + \tau M\sum_{k=0}^{N-1}\|f_k\|\\ &= M\|u_0\| + TM\sum_{k=0}^{N-1}\frac{\|f_k\|}{N} \leqslant M(1+T)\|(u_0,f_0,\dots,f_{N-1})\|_1,\end{aligned}$$

thus $\{L_\tau^{-1}\}$ is uniformly bounded.

Suppose now that $\|L_\tau^{-1}\|\leqslant K$, then for all $u_\tau=(u_0,f_0,\dots,f_{N-1})\in B^{N+1}$

$$\|L_\tau^{-1}(u_0,f_0,\dots,f_{N-1})\| \leqslant K\|(u_0,f_0,\dots,f_{N-1})\|.$$

As above, the condition then follows for $k=0$.

Now write $u_\tau=(0,\dots,0,f_l,\dots,0)$, then from the explicit representation of the solution it follows that

$$\max_{l+1\leqslant i\leqslant N}\tau\left\|\left(\prod_{j=l+1}^{i-1} C(\tau,t_j)\right)f_l\right\| \leqslant K\|u_\tau\|_1 = K\frac{\|f_l\|}{N}$$

and therefore

$$\left\|\left(\prod_{j=l+1}^{i} C(\tau,t_j)\right)f_l\right\| \leqslant \frac{K}{T}\|f_l\| \quad \text{for } l\leqslant i\leqslant N-1.$$

Since $f_l\in B$ is arbitrary the assertion follows.

As in §34.1 we check immediately that:

1. $\|C(\tau,t_j)\| \leqslant 1$,
2. $\|C(\tau,t_j)\| \leqslant 1+\tau a$,
3. $\|A_\tau(t_j)\| \leqslant a$,

are sufficient criteria for the stability of the process (2′). The other theorems from §34.1 and §32 may again be easily carried over.

34.3 Stability behaviour of the perturbed process

From now on we again simplify the written expressions by assuming that A does not depend on t. The appropriate modifications for the time-dependent case should give the reader no difficulty.

Theorem 34.9 *Let* $u_{n+1} = C(\tau)u_n, u_0 = g$ *and* $\|C^n(\tau)\| \leqslant M$ (*that is,* the process is stable), *further let* $Q(\tau)$ *be linear, continuous and* $\|Q(\tau)\| \leqslant H < \infty$ *for* $0 < \tau \leqslant \tau_0$. *Then the process* $u'_{n+1} = (C(\tau)+\tau Q(\tau))u'_n$, $u'_0 = g$ *is also stable.*

Proof. $u'_n = (C(\tau)+\tau Q(\tau))^n u'_0$. Since $C(\tau)$, $Q(\tau)$ do not necessarily commute, we must write out the product at length. For each $j=0,\dots,n$ we obtain $\binom{n}{j}$ summands, each of which contains the product of j copies of $\tau Q(\tau)$ and $n-j$ of $C(\tau)$. Such a summand has the form

$$S = \underbrace{C(\tau)C(\tau)}_{1}\tau Q(\tau)\underbrace{C(\tau)\dots}_{2}\tau Q(\tau)\underbrace{\dots C(\tau)}_{j+1},$$

where at most $j+1$ groups of $C(\tau)$ exist, since only j $\tau Q(\tau)$ are available. Since by assumption $\|C^n(\tau)\| \leqslant M$ for all $n=0,\dots,N$, we can estimate the norm of each group of $C(\tau)$ in terms of M, and hence obtain

$$\|S\| \leqslant M^{j+1}\tau^j H^j$$

and therefore

$$\|(C(\tau)+\tau Q(\tau))^n\| \leqslant \sum_{j=0}^{n}\binom{n}{j} M^{j+1}\tau^j H^j$$

$$= M(1+\tau MH)^n \leqslant M\left(1+\frac{T}{N}MH\right)^N \leqslant Me^{MHT} \qquad \blacksquare$$

Remark If we already know that $\|C(\tau)\| \leqslant 1+\tau K$, $K<\infty$, the last theorem is of course trivial, since then

$$\|C(\tau)+\tau Q(\tau)\| \leqslant \|C(\tau)\| + \tau\|Q(\tau)\| \leqslant 1+\tau(K+H),$$

and by one of the simple stability criteria introduced in §34.1 we conclude that stability also holds for the perturbed process.

Theorem 34.10 *With the same assumptions and notation as before we have*

$$\|u_n - u_n'\| \leqslant THM^2 e^{HMT} \|u_0\|.$$

Proof.

$$\|u_n - u_n'\| \leqslant \|C^n(\tau) - (C(\tau) + \tau Q(\tau))^n\| \ \|u_0\| \leqslant \sum_{j=1}^{n} \binom{n}{j} M^{j+1} H^j \tau^j \|u_0\|$$

$$= M \sum_{j=1}^{n} \binom{n}{j} M^j H^j \tau^j \|u_0\| \leqslant M(e^{HMT} - 1)\|u_0\|$$

$$\leqslant MHMTe^{a}\|u_0\|, 0 \leqslant a \leqslant HMT \quad \text{(mean value theorem)}$$

$$\leqslant THM^2 e^{HMT}\|u_0\|.$$ ■

Remark The last two theorems are completely distinct from the perturbation theorem presented in §32. There we were concerned with small perturbations, while here the perturbations appearing can be very large. The operator L_τ was indeed defined in terms of single components by

$$\frac{1}{\tau}(u_n - C(\tau)u_{n-1}).$$

Replacing $C(\tau)$ by $C(\tau) + \tau Q(\tau)$ we obtain

$$\frac{1}{\tau}(u_n - C(\tau)u_{n-1}) + Q(\tau)u_{n-1}.$$

Here we have a linear perturbation of L_τ which, however, in no way needs to tend to zero.

Example 34.2 We consider the differential equation

$$\frac{\partial u}{\partial t} - a(x,t)\frac{\partial^2 u}{\partial x^2} - b(x,t)\frac{\partial u}{\partial x} - c(x,t)u = f(x,t),$$

with the boundary conditions

$$u(x,0) = g(x) \quad \text{and} \quad u(0,t) = 0 = u(1,t),$$

where $x \in \Omega = (0,1)$, $t \in [0,T]$ and $0 \leqslant a \leqslant a(x,t) \leqslant A$, $|b(x,t)| \leqslant B$, $|c(x,t)| \leqslant C$. We cover $\Omega \times [0,T]$ with a rectangular lattice of width k in the x-direction and h in the t-direction, and put

$$u_j^n := u(jk, nh), N := 1/k.$$

Let the difference process be defined by

$$\frac{1}{h}(u_j^{n+1}-u_j^n)-\frac{a_j^n}{k^2}(u_{j+1}^n-2u_j^n+u_{j-1}^n)-\frac{b_j^n}{2k}(u_{j+1}^n-u_{j-1}^n)-c_j^n u_j^n=f_j^n,$$

$$u_j^0=g_j,\quad u_0^j=0=u_N^j.$$

As usual we check the consistency of the process by means of the Taylor expansion of the solution of the differential equation. We check stability under the assumptions

$$Bk\leqslant 2a,\quad 2Ah\leqslant k^2.$$

We have

$$u_j^{n+1}=\left(1-\frac{2h}{k^2}a_j^n\right)u_j^n+hc_j^n u_j^n+hf_j^n+\left(\frac{h}{k^2}a_j^n-\frac{h}{2k}b_j^n\right)u_{j+1}^n$$
$$+\left(\frac{h}{k^2}a_j^n+\frac{h}{2k}b_j^n\right)u_{j-1}^n;$$

because $A\leqslant k^2/2a$ we have

$$1-\frac{2h}{k^2}a_j^n\geqslant 0$$

and

$$\frac{h}{k^2}a_j^n\pm\frac{h}{2k}b_j^n\geqslant 0\quad\text{because } B\leqslant\frac{2a}{k},$$

therefore

$$|u_j^{n+1}|\leqslant\max_j|u_j^n|(1+hC)+h|f_j^n|,$$

which is the stability of the process.

34.4 Several step processes

For practical purposes first-order convergence is not always good enough. In difference processes it is possible to improve the convergence properties by raising the order of consistency (see Remark 32.1 (iii) associated with the convergence theorem). Here we start from the easily proved fact that for C^{m+k}-functions we can give difference quotients, which approximate $f^{(m)}$ to order k. For example we show that $f'(x)$ is approximated by

$$\frac{1}{12h}(8f(x+h)-8f(x-h)+f(x-2h)-f(x+2h))$$

to order 4 if $f\in C^5$. If we replace the time derivative in the evolution equation

$du/dt - Au = f$ by such a (better) difference quotient we obtain a process of the form:

$$u_{n+q} = C_{q-1}(\tau)u_{n+q-1} + \cdots + C_0(\tau)u_n + \tau f_n, \quad n = 0, \ldots, N - q, \tag{3}$$

where we again assume that A does not depend on t. The modifications in the time-dependent case proceed analogously to §34.2 and are left to the reader. We also wish to subordinate these processes to the Banach space scheme developed above in order to apply the theorems from §32. Here we must know that in such a q-step process we need the values on the first q time levels in order to start the process. We assume for the sake of simplicity that we already know the solution on the first q levels. This assumption involves no limitation of the practical applicability of the process under consideration, since for all the processes discussed by us, and for all the processes which are acceptable in practice, it is permissible to calculate using approximate values, if these only converge to the exact discretised values. We may even take the starting value u_0 on all q levels, since the solution u is continuous, and hence $u_i \to u_0$ for $\tau \to 0$, $i = 1, \ldots, q-1$ (note: q is a fixed number, independent of τ). As a rule, however, we obtain these values by one of the single-step processes with narrowed step width, in order not to nullify the effect of the increased consistency by badly discretised initial values. The reader should once more look at the proof of the convergence theorem in §32: the error was estimated precisely in terms of consistency and discretisation errors.

For the q-step process (3) the Banach space scheme has the form

$$\begin{array}{ccc} C([0,T],B) & \xrightarrow{L} & B \times C([0,T],B) \\ D_\tau \Big\downarrow & & \Big\downarrow D'_\tau \\ B^{N+1} & \xrightarrow{L_\tau} & B^{N+1} \end{array}$$

where L and D_τ are defined as in §34.1, while for D'_τ and L_τ certain modifications need to be made.

$$\begin{aligned} D'_\tau : B \times C([0,T],B) &\to B^{N+1} \\ (u_0, f) &\mapsto (u_0, u_1, \ldots, u_{q-1}, f_0, \ldots, f_{N-q}) \\ L_\tau : B^{N+1} &\to B^{N+1} \\ (u_0, \ldots, u_N) &\mapsto (u_0, \ldots, u_{q-1}, (1/\tau)(u_q - C_{q-1}u_{q-1} -, \ldots, - C_0 u_0)) \\ &\qquad \ldots, (1/\tau)(u_N - C_{q-1}u_{N-1} -, \ldots, - C_0 u_{N-q}). \end{aligned}$$

We will now demonstrate a stability criterion for the operator L_τ by means of a small trick; the criterion has essentially the same form as that developed for one-step processes. We introduce additional variables in equation (3) to obtain a one-step process over B^q. For this let

$$C = \begin{bmatrix} C_{q-1} & \cdots & C_0 \\ I & & \\ & \vdots & 0 \\ 0 & & I \; 0 \end{bmatrix},$$

$\bar{u}_n = (u_{n+q-1}, \ldots, u_n)^T, \bar{f}_n = (f_n, 0, \ldots, 0)^T$, ($T$ stands for transposed vector).

With this notation equation 3 becomes

$$\bar{u}_{n+1} = C\bar{u}_n + \tau \bar{f}_n, \quad n = 0, \ldots, N-q. \tag{4}$$

As in §5.1 we now obtain the explicit formula for the solution

$$\bar{u}_n = C^n u_0 + \tau \sum_{j=0}^{n-1} C^{n-j} \bar{f}_j.$$

We equip B^q exactly as B^{N+1} with the max-norm and in what follows distinguish between

$$\|\cdot\|\text{-norm on } B, \quad \|\cdot\|_q\text{-norm on } B^q, \quad \|\cdot\|_{N+1}\text{-norm on } B^{N+1}.$$

Theorem 34.11 (3) *is stable if and only if there exists* $M < \infty$, *so that for all* τ, n *with* $0 \leqslant n\tau \leqslant T$, *we have*

$$\|C^n(\tau)\|_q \leqslant M.$$

Proof. Sufficiency: It follows from the explicit solution formula that

$$\|\bar{u}_n\|_q \leqslant M\left(\|\bar{u}_0\|_q + \tau \sum_{j=0}^{n-1} \|\bar{f}_j\|_q \right)$$
$$\leqslant \left(\|\bar{u}_0\|_q + \tau N \max_j \|\bar{f}_j\|_q \right) \leqslant M(1+T)\left(\|\bar{u}_0\|_q + \max_j \|\bar{f}_j\|_q \right),$$

and therefore because

$$\|f_j\| = \|\bar{f}_j\|_q$$
$$\|u_n\| \leqslant M(1+T)\|(u_0, \ldots, u_{q-1}, f_0, \ldots, f_{N-q})\|_{N+1}.$$

This is the uniform boundedness of the operators $\{L_\tau^{-1}\}$, hence the stability of (3). Necessity: Let $\|L_\tau^{-1}\| \leqslant K < \infty$, and $u_0, \ldots, u_{q-1} \in B$ be arbitrary. It follows that

$$\|L_\tau^{-1}(u_0, \ldots, u_{q-1}, 0, \ldots, 0)\|_{N+1} \leqslant K\|(u_0, \ldots, u_{q-1}, 0, \ldots, 0)\|_{N+1}$$
$$= K\|(u_0, \ldots, u_{q-1})\|_q$$

and thus

$$\| C^n(\tau)\bar{u}_0 \| q \leqslant K \| \bar{u}_0 \| q.$$

Since $u_0, \ldots, u_{q-1}$ and therefore $\bar{u}_0$ were arbitrary, the assertion follows. ■

Hence the problem of checking the stability of difference processes for evolution equations, both for one-step and several-step processes, is reduced to estimating powers of operators between Banach spaces.

References

Adams, R.A. [1]. *Sobolev spaces.* New York: Academic Press 1975

Agmon, S. [1]. *Lectures on elliptic boundary value problems.* Princeton: Van Nostrand-Reinhold 1965

Agmon, S. [2]. The coerciveness problem for integro-differential forms. *J. Anal. Math.* **6** (1958) 182–223

Agmon, S., Douglis, A. & Nirenberg, L. [1]. Estimates near the boundary for solutions of elliptic partial differential equations satisfying general boundary conditions. *Comm. Pure Appl. Math.* I: **12** (1959) 623–727; II: **17** (1964) 35–92

Agranovič, M.S. [1]. Elliptic singular integro-differential operators. *Usp. Mat. Nauk* **20**, 5 (1965) 1–122

Akhiezer, N.I. & Glazman, I.M. [1]. *Theory of linear operators in Hilbert space.* London: Pitman 1981

Aronszajn, N. [1]. On coercive integro-differential quadratic forms, University of Kansas 1954. Technical Report No 14, pp. 94–106

Aubin, J.P. [1]. *Approximation of elliptic boundary value problems.* New York: Wiley 1972

Berezanskii, J.M. [1]. *Expansions in eigenfunctions of selfadjoint operators.* Providence: Am. Math. Society 1962

Bers, L., John, F.F. & Schechter, M. [1]. *Partial differential equations.* New York: Wiley 1964

Bourbaki, N. [1]. *Espaces vectoriels topologiques.* Paris: Herman 1955

Bourbaki, N. [2]. *Intégration.* Paris: Herman 1956

Carrol, R.W. [1]. *Abstract methods in partial differential equations.* New York: Harper & Row 1969

Calderon, A.P. & Zygmund, A. [1]. On the existence of certain singular integrals. *Acta Math.* **88** (1952) 85–134

Douady, A. [1]. Un espaces de Banach dont le groupe linéaire n'est pas connexe. *Indag. Math.* **27** (1965) 787–9

Douglis, A. & Nirenberg, L. [1]. Interior estimates for elliptic systems of partial differential equations. *Comm. Pure App. Math.* **8** (1955) 503–38

Dunford, N. & Schwartz, J.T. [1]. *Linear operators I, II.* New York: Wiley 1958, 1963

Edwards, R.E. [1]. *Functional analysis, theory and applications.* New York: Holt, Rinehart & Winston 1965

Fichera, G. [1]. *Linear elliptic differential systems and eigenvalue problems.* Lecture Notes in Mathematics Vol. 8. New York: Springer 1965

Floret, K. [1]. *p*-integrale Abbildungen und ihre Anwendung auf Distributionsrüume. Diplomarbeit, Heidelberg 1967

Friedman, A. [1]. *Partial differential equations of parabolic type*. Englewood Cliffs: Prentice Hall 1964
Gagliardo, E. [1]. Properietà di alcune classi di funzioni in piu variabili. *Ricerche Mat.* **7** (1958) 102–37
Gårding, L. [1]. Dirichlet's problem for linear elliptic partial differential equations. *Math. Scand* **1** (1953) 55–72
Gelfand, I.M. & Silov, G.M. [1]. *Generalized functions I, II*. New York: Academic Press 1964, 1968
Gilbarg, D. & Trudinger, N.S. [1]. *Elliptic partial differential equations of second order*. New York: Springer 1977
Gohberg, I.T. & Kreĭn, M.G. [1]. Foundations of defect numbers, radical numbers and indices of linear operators. *Usp. Mat. Nauk* **12**, 2 (1957) 43–118
Greub, W.H. [1]. *Linear algebra*, 3rd edn. New York: Springer 1967
Halmos, P. [1]. *Measure theory*. New York: Van Nostrand 1950
Hellwig, G. [1]. *Partial differential equations*. Stuttgart: Teubner 1977
Hestenes, M.R. [1]. Extension of the range of a differentiable function. *Duke Math. J.* **8** (1941) 183–92
Heuser, H. [1]. *Funktionalanalysis*. Stuttgart: Teubner 1975
Holmann, H. & Rummler, H. [1]. *Alternierende Differentialformen*. Mannheim: Bibliographisches Institut 1972
Hörmander, L. [1]. *Linear partial differential operators*. New York: Springer 1976
Janssen, R. [1]. Differenzenverfahren. Funktionalanalytische Beschreibung und Stabilitätsdefinitionen. Diplomarbeit Kiel 1979
Kato, T. [1]. *Perturbation theory for linear operators*. Corrected printing of 2nd edn. New York: Springer 1980
Köthe, G. [1]. *Topological vectorspaces I*. New York: Springer 1969
Krasnoselskiĭ, M.A., Perov, A.I., Provolockii, A.J. & Zabreĭko, P.P. [1]. *Plane vector fields*. New York: Academic Press 1966
Kreĭn, S.G. [1]. *Linear differential equations in Banach space*. Providence: Am. Math. Society 1971
Kreiss, H.O. [1]. Über die Stabilitätsbedingungen für Differenzengleichungen die partielle Differentialgleichungen approximieren. *BIT* **2** (1962) 153–81
Ladyženskaya, O.A. & Uralceva, N.N. [1]. *Linear and quasilinear elliptic equations*. New York: Academic Press 1968
Ladyženskaya, O.A., Solonnikov, V.A. & Uralceva, N.N. [2]. *Linear and quasilinear equations of parabolic type*. Providence: Am. Math. Society 1968
Ladyženskaya, O.A. [3]. *The mixed problem for hyperbolic equations*. Moscow: GOS 1953
Lawruk, B.R. [1]. The index of the operator for a boundary value problem of a system of elliptic, linear, differential equations of second order. *Dokl. Acad. Nauk SSSR* **111**, 2 (1956) 287–90
Lawruk, B.R. [2]. The dependence of the index of the operator for a boundary value problem of a system of linear, elliptic differential equations of second order on the highest coefficients. *Dokl. Acad. Nauk SSSR* **121**, 6 (1958) 970–2
Lawruk, B.R. [3]. On the unique solvability of a general boundary value problem for homogeneous linear systems of second order differential equations of elliptic type with constant coefficients in a half space, *Ann. Polon. Math.* **14** (1963) 85–95
Lewy, H. [1]. An example of a smooth linear partial differential equation without solution. *Ann. Math.* (2) **66** (1957) 155–8
Lions, J.L. [1]. *Optimal control of systems governed by partial differential equations*. New York: Springer 1971
Lions, J.L. [2]. *Lectures on elliptic partial differential equations*. Bombay: Tata 1957

Lions, J.L. [3]. *Quelques méthodes de résolution des problèmes aux limites non linéaires.* Paris: Dunod 1969

Lions, J.L. & Magenes, E. [1]. *Non-homogeneous boundary value problems and applications,* Vol. I–III. New York: Springer 1972

Lopatinskii, Ya. B. [1]. On a method of reducing boundary problems for a system of differential equations of elliptic type to regular integral equations. *Ukraïn. Mat. Ž.* **5** (1953) 123–51

Mäulen, J. [1]. Bemerkungen zur Theorie harmonischer Funktionen und Vektorfelder. Dissertation Stuttgart 1975

Meyers, N. Serrin, J. [1]. *H = W. Proc. Nat. Acad. Sci USA.* **51** (1964) 1055–6

Michlin, S.G. [1]. On a class of singular integral equations. *Dokl. Acad. Nauk* **24**, 4 (1939) 315–17

Miranda, C. [1]. *Partial differential equations of elliptic type,* 2nd edn. New York: Springer 1970

Morrey, C.B. [1]. *Multiple integrals in the calculus of variations.* New York: Springer 1966

Naïmark, M.A. [1]. *Linear differential operators.* Providence: Am. Math. Society 1968

Natanson, I.P. [1]. *Theory of functions of a real variable.* Providence: Am. Math. Society 1965

Nečas, J. [1]. *Les méthodes directes en théorie des équations elliptiques.* Prague: Academia 1967

Nirenberg, L. [1]. Remarks on strongly elliptic partial differential equations. *Comm. Pure Appl. Math.* **8** (1955) 648–74

Palais, R.S. [1]. *Seminar on the Atiyah–Singer index theorem.* Princeton: University Press 1965

Richtmyer, R.D. & Morton, K.W. [1]. *Difference methods for initial value problems,* 2nd edn. New York: Wiley 1967

Rudin, W. [1]. *Principles of mathematical analysis,* 3rd edn. New York: McGraw-Hill 1976

Rudin, W. [2]. *Real and complex analysis.* New York: McGraw-Hill 1966

Rudin, W. [3]. *Functional analysis.* New York: McGraw-Hill 1973

Saks, S. [1]. *Theory of the integral,* 2nd edn. Warsaw: Monografie Matematyczne 1937

Šapiro, Z. Ja. [1]. On general boundary value problems of elliptic type, *Isv. AN, ser. Matem.* **17** (1953) 539–62

Schechter, M. [1]. *Principles of functional analysis.* New York: Academic Press 1971

Schechter, M. [2]. *Spectra of partial differential operators.* Amsterdam: North-Holland 1971

Schechter, M. [3]. *Modern methods in partial differential equations.* New York: McGraw-Hill 1977

Schechter, M. [4]. Solution of the Dirichlet problem for systems not necessarily strongly elliptic. *Comm. Pure Appl. Math.* **12** (1959) 241–7

Schwartz, L. [1]. *Théorie des distributions.* Paris: Heman 1966

Slobodeckiĭ, L.N. [1]. Generalized Sobolev spaces and their application to boundary problems for partial differential equations. *Usp. zap. Leningr. gos. ped. inta, A.I. Hercena* **197** (1958) 54–112, English translation

Sobolev, S.L. [1]. *Applications of functional analysis in mathematical physics.* Providence: Am. Math. Society 1963

Taylor, A.E. [1]. *Introduction to functional analysis.* New York: Wiley 1958

Triebel, H. [1]. *Höhere Analysis.* Berlin: Deutscher Verlag der Wissenschaften 1972

Triebel, H. [2]. *Interpolation theory, function spaces, differential operators.* Amsterdam: North-Holland 1978

Vekua, I.N. [1]. On the theory of singular integral equations. *Soobšenija AN Gruz. SSR* III **9** (1942) 869–76

Vekua, I.N. [2]. *Systems of elliptic differential equations of first order and boundary value problems*

Volevič, L.R. [1]. Solvability of boundary value problems for general elliptic systems. *Mat. Sbor. T.* **68** (110) 3 (1965) 373–416

Volevič, L.R. [2]. Elliptic operators on compact manifolds and the theory of harmonic integrals, Lectures at the Caciveli Summer school, 1964, **1**, 147–98

Volevič, L.R. & Panejach, B.P. [1]. Some distribution spaces and imbedding theorems. *Usp. Mat. Nauk* **20** (1965), 3–74

Warner, F.W. [1]. *Foundations of differentiable manifolds and Lie groups.* London: Scott, Foresman 1971

Weidmann, J. [1]. *Linear operators in Hilbert spaces.* New York: Springer 1980

Wloka, J. [1], *Funktionalanalysis und Anwendungen.* Berlin de Gruyter 1971

Wloka, J. [2]. Partielle Differentialgleichungen. Vorlesungsausarbeitung von H.C. Zapp, Kiel 1975

Wloka, J. [3]. Differenzenverfahren zur Lösung partieller Differentialgleichungen. Vorlesungsausarbeitung von R. Janssen, Kiel 1979

Yosida, K. [1]. *Functional analysis*, 5th edn. New York: Springer 1978

Function and distribution spaces

$A^l(\Omega)$ 75
$B^1(S, H)$ 384
$C^l(\Omega)$ 2
$C^l(\bar{\Omega})$ 2
$C^{l,\lambda}(\Omega)$ 2
$C_0^0(\Omega)$ 10
$\tilde{C}^{k,\kappa}(\Omega)$ 75
$\mathring{C}^k(\bar{\Omega})$ 109
$\mathring{C}^k(\mathbb{R}^r)$ 110
$\mathscr{D}(\Omega) = C_0^\infty(\Omega)$ 6
$\mathscr{D}'(\Omega)$ 10
$\mathscr{D}'_F(\Omega)$ 10
$\mathscr{D}'(\Omega, H)$ 390
$\mathscr{E}(\Omega) = \mathscr{E}^\infty(\Omega) = C^\infty(\Omega)$ 2
$\mathscr{E}^l(\Omega)$ 2
$\mathscr{E}'(\Omega)$ 19
H^l 91
$H^l(\Omega)$ 93
H^{-l} 270
$L_p(\Omega)$ 3
$L_1^{\mathrm{loc}}(\Omega)$ 3
L_2^l 91
$L_p(S, H)$ 383
$\mathscr{S}$ 26
$\mathscr{S}'$ 26
$W_2^l(\Omega)$ 61
$\mathring{W}_2^l(\Omega)$ 66
$W_2^l(M)$ 87
$W_2^l(\partial\Omega)$ 89
$W_2^l(0) = \mathbb{C}$ 154
$W_2^{-l}(\Omega)$ 270
$W^{2m}(B) = W^{2m}(\{b_j\}_1^m)$ 222
$W^{2m}(B)^*$ 224
$W_2^{1,\delta}(\Omega)$ 370
$\mathring{W}_2^{1,\delta}(\Omega)$ 371
$W_2^1(0, T) = W(0, T)$ 392
$W_2^k((0, T); V)$ 404
$\mathscr{Z} := \mathscr{F}\mathscr{D}$ 91

Index

acceptence of initial conditions 435
adjoint differential operator, formal 139
admissible coordinates 55
admissible transformation 54
antidual map 165
antidual space 165
antisymmetric form 344
approximation process 463
a priori estimate 181
assumptions for the regularity theorem 20.4 (strongly elliptic case) 314–15
assumptions for the solution theorem 21.1 (strongly elliptic case) 336–8
automorphism group 176

Bessel differential equation 213
Bessel function 213
Bochner integral 384
boundary value problem
  adjoint 223
  general elliptic 187
  mixed for the Laplace operator 354
  natural 415
  self adjoint 223
  stable 415
  third 159–61
  with skew derivative 159

Cauchy–Riemann operator 141
characteristic roots 150
$C^{k,\kappa}$-atlas 58
$C^{k,\kappa}$-diffeomorphism 47
$\tilde{C}^{k,\kappa}$-diffeomorphism 77
closed range theorem 170
compact map 366
compatibility conditions
  hyperbolic 443
  parabolic 406

condition
  Agmon's 281; for Δ^2 303
  Ladyženskaya 484
  Lopatinskiĭ–Šapiro (L.Š.) 150
cone 36
cone property 37
  uniform 37
cone vertex 36
congruent 37
consistency 463
  *k*th order 463
Convergence
  almost uniform 379
  in $\mathscr{D}(\Omega)$ 11
  in measure 380
  of a difference process (discrete) 463
  of sequences 11
convolution 20
  integral 9
  of distributions 20
correction operators
  $Z_h^\tau, \tilde{Z}_h^\tau$ 317
  $Z_h^{s,\tau}, \tilde{Z}_h^{s,\tau}$ 319
countability axiom, second 58
covering
  locally finite 4
  open 4
covering condition (= Lopatinskiĭ–Šapiro condition) 150

defect numbers 168
difference process 463
  for evolution equations 498
  for general boundary value problems 478
differentiable manifold 58
differentiation of a distribution 14
Dirac delta distribution 11

direct method 343
Dirichlet problem 159
for strongly elliptic differential operators 305
general 211
Dirichlet system 214
discretisation operators 462
distribution 10
of finite order 10
tempered 26

eigenspace 166
eigenvalue 166
for elliptic operators 209
eigenvector 166
elementary solutions 35–36
elliptic 140
properly 144
strongly 140
strongly uniform 146
uniform 146
extension operators, $F^{\Omega}_{\Omega'}$ 94
exterior space problem 373

finite function 6
finitely valued function 376
Fourier function 25
Fourier inversion formula 28
Fourier–Laplace transform 32
Fourier method for hyperbolic equations 452–3
Fourier method for parabolic equations 432–3
Fourier norm 93
Fourier transformation of a distribution 30
Fredholm operator 168
Friedrich's inequality 278
(F)-space 18
fundamental functions 6

Gårding inequality 273
abstract 373
for differential operators 293
Gauss–Stokes formula 257
Gelfand triple 262
globally smoothable 206
Green formula
first 219
function, classical 359–61
second 231
Green solution operator 236–9
Gronwall's integral inequality 436

Hadamard estimate 403
Heat equation 426

index of a map 168
integral inequalities 436

(k, κ)-regular 48
(k, κ)-smooth 48
Kolmogorov compactness criterion 4

λ-Hölder continuous 2
Leibniz product rule 15
lemmas
Ehrling (abstract) 114; (for W-spaces) 115
Gochberg–Kreĭn 183
Gronwall 436
Poincaré (for $W_2^{1,\delta}$) 371
Sobolev 107
Weyl: (abstract) 182; for elliptic differential operators 188; for strongly elliptic differential operators 324
Lewy's operator 142
Lipschitz continuous 2
locally smooth distribution 12

manifold, differentiable 58
$C^{k,\kappa}$ 59
measurable function
strong 376
weak 376
mixed boundary value problem for the Laplace operator 354
mixed initial value
Dirichlet problem 423
Neumann problem 425
multiplication of a distribution 15

negative norm 267
$N^{k,\kappa}$-property 38
Neumann problem 159–60
normal coordinates 53
normal boundary value operator 214
normal transformation 53

order
of a distribution 10
of a differential operator 139
Orthogonal space 169

paracompact 4
parallelepiped 42
Parseval equation 31
partition of unity 7
on manifolds 75
subordinate 7
periodic distribution 34
Poincaré inequality
first 116
second 117
Poincaré lemma for $W_2^{1,\delta}$ 371
polar coordinates 372

principal part of a differential operator 140
principal part polynomial 140
'Principe du recollement de morceaux' 12
pullback operator 80
pyramid 43

Radon measure 10
reduction
  left 174
  right 175
refinement of a covering 4
regularisation 9
regulariser
  left 170
  local 206
  right 170
  smoothing 182
regular value of an operator 166
representation of functionals
  on W_2^l 268
  on $\mathring{W}_2^l$ 268
  on sesquilinear forms 271
restriction of a distribution 12
restriction operator, $R_{\Omega'}^{\Omega}$ 94

scalar or scheme 262
scale property of W- and H-spaces 106
Schauder operator 367
Schauder scheme 180
segment property 36
self-adjoint differential operator (formally) 147
separably valued function 376
sesquilinear form 271
several step process 507
smoothable map 182
smoothing operator 182
  global 207
smoothing regulariser 182
spectral value 166
  problem for elliptic differential operators 209
  residual 209
spectrum, Sp(A) 166
stability, α- 463
  of a difference process 463
standard cone 37
star-shaped 60
step process, q- 507
submanifold 59
subspace adjoint to $W^{2m}(B)$ 223
  determined by boundary values 223
support 6
  of a distribution 14

theorems
  Agmon 295
  Aronszajn 301
  Banach–Steinhaus 474
  Brouwer 361
  Convergence of difference processes 463
  Egorov 379
  extension for $\mathring{W}_2^l(\Omega)$
  extension (Calderon–Zygmund) 95
  extension (Hestenes) 100
  Gårding 291
  Lax 501
  Lax–Milgram 271
  main theorem for elliptic boundary value problems 188
  perturbation 478, 505
  Pettis 377
  regularity for hyperbolic differential equations 450; (abstract) 442
  regularity for parabolic differential equations 419; (abstract) 406
  Riesz–Schauder spectral 166
  Schauder fixed point 367
  solutions for hyperbolic differential equations 449; (abstract) 437
  Solutions for parabolic differential equations 416–17; (abstract) 397
  Solutions for strongly elliptic differential equations 339
  transformation 80
trace operator 121
  image 129
  kernel 130
  theorem 126; inverse 129
translation operator 20
transmission condition 357
transmission problem 356
transverse boundary form 312

variational method 343
V-coercive
  abstract 274
  differential operators 279
V-elliptic
  abstract 273
  differential operators 279
vibrating string 461

wave equation 454 *et seq.*
weak equation 274
  solvability 274
weakly sequentially compact 133
weak solvability 274

CPSIA information can be obtained at www.ICGtesting.com
Printed in the USA
BVOW03s0008250314

348675BV00001B/19/P